Mark Crowder

PROPERTIES AND CHARACTERIZATION OF AMORPHOUS CARBON FILMS

PROPERTIES AND CHARACTERIZATION OF AMORPHOUS CARBON FILMS

Edited by

John J. Pouch and Samuel A. Alterovitz

NASA Lewis Research Center
Cleveland, Ohio 44135
USA

TRANS TECH PUBLICATIONS

ISBN 0-87849-604-1

Volumes 52 & 53 of
Materials Science Forum
ISSN 0255-5476

Distributed in the Americas by

Trans Tech Publications
Old Post Road
Brookfield VT 05036
USA

and worldwide by
Trans Tech Publications
Segantinistr. 216
CH-8049 Zürich
Switzerland

TABLE OF CONTENTS

SECTION I. GENERAL REVIEW

SECTION II. PREPARATION METHODS

SECTION III. CHARACTERIZATION

SECTION IV. APPLICATIONS

Introduction

Amorphous carbon (also known as diamondlike carbon) has been available for almost 20 years. During the 1980's, the interest in these films increased. In the late 1980's, the Materials Research Society (both the American and the European MRS) and other societies held sessions devoted to this area. In many cases, proceedings of these conferences were published. However, papers submitted to these proceedings were limited in length and scope. On the other hand, several review articles were published. The papers did not cover, in detail, all the aspects of this field. It was evident that a gap existed. There was no single book describing, in detail, the most important work done in the last several years in the field. In order to fill this gap, Dr. Fred Wohlbier (Trans Tech Publications) invited us to edit this book.

There are two main purposes to this book: first, to give a scientist starting to work in the field a way to know, in detail, the state-of-the-art; second, to give researchers in the field a more complete description of work done in laboratories all over the world without looking for the original publications. In addition, the book can serve as a reference book for people not working in the area.

The book is divided into four sections: general reviews, preparation methods, characterization and applications. Each section contains several invited chapters. In many cases, a chapter could be classified under more than one section. Most of the chapters summarize the work that has been done over several years. These chapters contain preparation, properties and, in many cases, applications. Therefore, in these cases, the classification of the chapters may look somewhat arbitrary.

The first section contains general review chapters starting with Dr. Aisenberg, a pioneer in the field. Large effort in this area is described by Prof. Koidl and his group, while Drs. Tsai and Green review a large number of preparation methods, properties and applications. Dr. Robertson gives a general overview and describes his own work in the theoretical area. We also invited a chapter in the area of diamond films, which is a new area and is closely related to amorphous carbon films. This chapter is given by the Penn State group, which is one of the largest groups in the U.S.A.

The section covering preparation methods includes chapters on film preparation using the plasma technique (Catherine), ion beam techniques (Hirvonen, Mirtich, Lifshitz) and other methods (Renschler). The effects (Berg) encountered in the film preparation are also described. It should be mentioned that preparation methods are also given in detail in the first section (Aisenberg, Koidl, Green) and in some chapters in the subsequent two sections.

Each of the first four chapters in section III describes the application of a single characterization technique to amorphous carbon films. The four techniques are optical (Smith), real time ellipsometry (Collins), Raman spectroscopy (Yoshikawa) and nuclear magnetic resonance (Petrich). The other chapters in this section describe the effects of radiation on amorphous carbon (Savvides, Kalish and Ingram) and a combination of characterization methods (Fitzgerald, Thompson and Gonzalez-Hernandez).

The last section describes some of the applications of amorphous carbon films. Each of the first two chapters (by Profs. Enke and Woollam) describes several applications, e.g. optical, electrical and mechanical. Tribological and mechanical properties of metal-containing (Klages) and normal carbon films (Miyoshi and LaFontaine), sensors (Olcaytug) and photoconduction (Itoh) are described in the remainder of this section.

The book does not cover all the areas dealing with amorphous carbon films. There is a limit to both the scope of this book and the ability to collect up to date material. However, we believe that we covered the most important aspects of preparation, characterization and applications.

The invited chapters are of high scientific quality and up to date results are presented. We believe that the book will be used extensively by researchers already in the field, new comers, and researchers that need a general reference to this new and developing area.

We would like to thank all authors for their contributions, without which this book could not be published.

John J. Pouch
Samuel A. Alterovitz
Cleveland, Ohio, U.S.A.

November 1989

Materials Science Forum Vols. 52 & 53 (1989) pp. 1-40

ION BEAM AND ION-ASSISTED DEPOSITION OF DIAMOND-LIKE CARBON FILMS

S. Aisenberg (a) and F.M. Kimock (b)
(a) Data Associates, 36 Bradford Road, Natick, MA 01760, USA
(b) Air Products and Chemicals, Inc., 7201 Hamilton Boulevard
Allentown, PA 18195, USA

ABSTRACT

This chapter will describe and analyze the differences between diamond-like carbon (DLC) and hydrogenated DLC (H-DLC) and the role of energetic ions in forming the diamond-like materials. It will show that neither hydrogen or high temperature substrates are needed to deposit diamond-like carbon. The extension to ion-assisted deposition will be described. Some applications based upon the unique properties of the resulting materials will be discussed.

1. INTRODUCTION

There is considerable recent interest in the various forms of hard carbon that have unique properties similar to carbon in the true diamond crystal form. Associated with these properties are a number of potentially useful and unique applications.

Many methods have been described for producing thin films of this hard carbon, and many different names for the resulting materials have been created. Included are diamond-like carbon, diamondlike carbon, DLC, a:carbon, a:C, i:carbon, i:C, H-carbon, a-C:H, and even diamond. Another name is IBD DLC (Ion Beam Deposited Diamond-Like Carbon). The different names often refer to different modifications of the deposition process; in some instances different names have been used by different researchers for the same process. Unfortunately the same names have also been used to describe different forms of carbon.

Understanding the basic differences between some of these forms of carbon is important for both development and application of these unique materials.

In this chapter we will describe the initial low pressure ion beam deposition work used for deliberately producing useful thin films of diamond-like carbon (DLC). We will elucidate similarities and differences between this process and some of the subsequent methods used to produce hard carbon films. We will point out that transparent diamond-like carbon films can be made without the deliberate use of hydrogen and without the use of heated substrates. The discussion will focus on the factors most important in producing these special materials.

2. BACKGROUND

2.1 Early Studies of Ion Beam Deposition

In 1968-1969, Aisenberg and his associates began work on ion beam deposition of thin films for electronic device fabrication under sponsorship by the former NASA Electronics Research Center [1,2].

Initially, Si films were deposited by extracting a beam of ions from a $SiCl_4$/Ar plasma. Because contamination of the Si films with Cl atoms was a concern, the ion beam equipment was modified to generate a pure Si^+ ion beam by sputtering Si atoms from Si electrodes within the ion source Ar plasma.

Using this method, epitaxial Si thin films were deposited on Si(111) substrates, at slightly above room temperature. The deposition rate was about 10 Angstroms per second. The single crystal nature of the Si films was verified by reflection electron diffraction (1° incident angle, 100 keV electron beam) performed on five samples. Analysis showed that all deposits were single crystal with (111) planes; beam scanning showed that the films were uniform and without polycrystalline regions [2].

Both n-type and p-type Si (111) films were deposited. Production of n-type films required higher ion energy (130 V). Thermal emf measurements were used to determine whether the deposited films were n-type or p-type. A surprising result was the observation of unusually high thermal emf's of 10-40 V for some p-type films, for temperature differences of about 54-61°C [2]. It is suggested that these materials could be of use in some sensor devices, and in energy conversion.

The doped Si films were used to demonstrate the feasibility of making thin film field effect transistors (FET) on low temperature substrates by ion beam deposition. Other films of potential use in electronic devices were also deposited by the ion beam method, including silicon dioxide, silicon nitride, titanium, molybdenum, aluminum, and aluminum oxide [1-10].

Fabrication of semiconductor devices required preparation of films with low or controlled amounts of impurities. In terms of contamination control, the operation of the high vacuum plasma ion beam deposition process had advantages over previously developed thin film techniques. First, the level of gaseous contaminants in the vacuum system was reduced by pumping the system down to 1×10^{-8} Torr before beginning the process. In addition, maintenance of the deposition chamber pressure in the 1×10^{-4} to 1×10^{-6} Torr range, along with the use of an energetic particle beam to sputter chemisorbed and physisorbed gases helped keep the substrate clean during deposition. Finally, sputtering of n-type and p-type doped Si electrodes into a pure Ar plasma within the ion source produced a deposition beam with a minimum amount of contamination.

2.2 Ion Beam Deposited Diamond-Like Carbon (DLC)

One of the more interesting results of this work occurred when a beam of carbon ions (generated by sputtering carbon electrodes into an Ar plasma) was used to produce insulating films of carbon. An insulating film was required in order to demonstrate that a complete thin film field effect transistor could be produced using ion beam technology. The use of insulating elemental carbon avoided problems of stoichiometry measurement and control which were encountered in the deposition of silicon dioxide and silicon nitride films.

It was predicted that non-equilibrium deposition of carbon ions would provide the carbon atom surface mobility that could lead to special carbon films. The objective of the initial work was to prove that the proposed technique would indeed produce insulating carbon films. There were no efforts to obtain high deposition rates.

Early proof of the existence of the transparent carbon films was easy since bright optical interference fringes were observed for films deposited onto Si substrates. Counting fringes proved to be a useful way to determine film thickness. The transparency of the films indicated the absence of free electrons and predicted that the films would also be electrically insulating. Verification of high film resistivities was subsequently made by a simple study of current-voltage characteristics. It was also found that the thin, transparent carbon films remained intact after dissolving the Si substrate in acid.

There was concern about an accurate and descriptive name for this new form of carbon material. It was determined that these carbon films had many of the properties of natural diamond but that the films were predominantly amorphous and not single crystal. Professional caution prevented use of the name "diamond" but it was felt that "diamond-like" would identify the properties as being similar to diamond without being presumptuous. Thus, the name plus the concept of "diamond-like carbon" (DLC) material were introduced [2,5,6].

In 1976, Spencer et al. at Bell Laboratories used essentially the same low pressure carbon and argon ion beam technique and demonstrated that these diamond-like carbon films could be duplicated. They provided confirming data about the optical, dielectric, and chemical resistance properties of DLC, along with crystal structure information by x-ray and electron diffraction [11,12].

2.3 Hydrogenated Diamond-Like Carbon (H-DLC)

It was realized early on that the deposition rate of diamond-like carbon from the ion beam system was limited by the availability of carbon atoms within the ion source plasma. While plasma sputtering of carbon electrodes produced pure films, it generally resulted in slow deposition rates (20-40 Angstroms/min) because of the very low sputter yield of carbon [13].

As part of the program to increase the deposition rate, hydrocarbon gases such as methane or acetylene were added to the argon gas provided to the plasma source. In some cases, the film deposition rate was increased by up to a factor of 30 by this approach. (See Table I, Section 4.4.)

The properties of films formed from plasma pyrolysis of the hydrocarbon gas within the ion source were found to be similar to the properties of films generated from solid carbon. Subsequent analysis of films deposited from hydrocarbon/argon ion beams showed that these films contained significant levels of hydrogen, and thus were a different material than the original diamond-like carbon.

Since the original work of Aisenberg and colleagues, many investigators have reported deposition of hard carbon films exhibiting some properties similar to those of diamond. In a landmark work in 1978, Holland and Ohja described the formation of hard carbon films from a hydrocarbon plasma using a capacitive RF discharge technique at pressures of about 0.1 Torr [14]. Chemical analysis of similar films produced by this technique has revealed that up to 50 atomic percent of hydrogen can be incorporated into the structure [15].

Many variations on this hydrocarbon plasma approach have followed, and are the subject of an excellent review [16]. Because of the unique properties and capabilities of ion beams, interest in forming this hard hydrogenated carbon material by direct ion beam deposition has been re-kindled [17-25].

Thus, we believe that after many years of research it has been clearly shown that the diamond-like materials can be broadly grouped into two categories, namely diamond-like carbon (DLC) which is substantially hydrogen-free, and hydrogenated diamond-like carbon (H-DLC) which contains significant amounts of hydrogen, in the range of approximately 20-50 atomic percent [15].

The vast majority of the more recent work on hard carbon films has involved the deliberate introduction of hydrogen. This has led many researchers to speculate that it is necessary to incorporate hydrogen into carbon films to obtain the diamond-like properties. The initial DLC work [1-12] which deliberately excluded hydrogen and other impurities and some of the more recent work in direct carbon ion beam [26-41] and ion-assisted deposition of carbon [42-52] should result in a reexamination of the assumption that hydrogen incorporation is necessary. Hydrogen addition may be sufficient, but is not necessary to stablize sp^3 bonding in diamond-like carbon films. We also point out that the presence of large quantities of hydrogen in carbon is not necessarily a limitation, provided that the resulting material has the properties desired for specific applications. However, the name used for this material should be selected to properly describe its composition and differentiate it from the original DLC.

3. COMPOSITION AND STRUCTURE OF DIAMOND-LIKE CARBON MATERIALS

3.1 Composition and Structure of Diamond

In understanding the synthetic diamond-like carbon materials, it is useful to compare their composition and structure to natural diamond. Crystalline elemental carbon occurs naturally in two major allotropes, diamond and graphite. Apart from their chemical identity (elemental carbon), diamond and graphite differ in almost every other regard. Diamond, for example, is the hardest known material, and is an electrical insulator. Pure diamond is usually nearly colorless, and has a high refractive index (2.42) which accounts for the brilliance of cut diamond gemstones. Graphite is a soft, black material with a greasy feel. Depending on its orientation, it can be a good electrical conductor.

The difference in properties between diamond and graphite are directly traceable to differences in the bonding structures. The diamond crystal lattice is extremely dense (3.5 g/cm^3) and strongly bonded. Each carbon atom is bonded by covalent bonds to four other carbon atoms in a tetrahedral geometry ("sp^3" bonding). Diamond has the highest atom number density of all known materials. Graphite is composed of a planar structure of weakly bonded sheets of hexagonal rings. Each ring is built of carbon atoms with three nearest neighbors arranged in an equilateral triangle ("sp^2" bonding). Graphite is much less dense (2.6 g/cm^3) than diamond.

At low pressures or temperatures diamond is thermodynamically unstable with respect to graphite. Fortunately, the slow kinetics of this process allow diamond and graphite to coexist at room temperature and atmospheric pressure.

Other amorphous forms of carbon, such as glassy carbon, are also known. These other forms of carbon, usually exhibit properties intermediate between those of graphite and diamond.

Regarding the chemical composition of diamond, a comprehensive reference on diamond by Field states that for natural diamond "only nitrogen and boron are known with certainly to be incorporated in the diamond lattice [53]." The various types of natural diamonds have very small amounts of gaseous impurities, some in the form of inclusions. For example, most natural diamonds are Type Ia, and and can contain "substantial amounts of nitrogen, on the order of 0.1% [53]." Almost all synthetic diamonds are Type Ib and contain nitrogen at concentrations up to 500 ppm. Type IIa have insufficient nitrogen to be easily detected. Type IIb diamonds have such a low concentration of nitrogen that some of the boron acceptors are not compensated and the diamond has p-type semiconductor properties [53]. Synthetic diamonds made at high temperature and pressure with metal catalysts may contain "nickel or iron inclusions which occupy as much as 10% of the crystals [53]."

With respect to elements detected in natural diamond, the measured impurities that are reported range up to 2.8% nitrogen [53]. No values for hydrogen content are given. This suggests that the concentration of hydrogen is small enough to be not significant, and probably less than that of nitrogen.

3.2 Chemical Composition of DLC and H-DLC

Similar to true crystalline diamond, DLC films are composed almost completely of carbon, and are substantially hydrogen-free. A few atomic percent of argon is usually trapped within DLC films as a result of bombardment by energetic Ar^+ ions. Because of the tight packing density of these carbon films, it is not possible for trapped Ar atoms to diffuse out of the structure under normal conditions. Perhaps the best evidence that the hydrogen-free material can be prepared has been provide by Rabalais et al. [37,38] who used a mass-selected C^+ ion beam to deposit DLC films in an ultrahigh vacuum environment at a pressure $< 1 \times 10^{-9}$ Torr. Under these conditions, even the possible reaction of residual hydrogen-containing gases (H_2, H_2O) with the growing film is negligible.

On the other hand, H-DLC's are known to contain about 20-50 atomic percent of hydrogen in their structure [15]. Film properties such as optical transparency, hardness, and electrical resistivity are all influenced by the hydrogen content. As found in DLC films, Ar has been observed to remain trapped within H-DLC films for periods of several years [54].

3.3 Structure of DLC and H-DLC

The diamond-like materials have some structural similarities. Both DLC and H-DLC are metastable materials that are generated under the influence of energetic particle bombardment. Each material normally exists in predominantly amorphous form. Finally, all diamond-like carbons exhibit significant extensive sp^3 bonding, which is thought to be responsible for the diamond-like properties.

However, in other ways the DLC's and H-DLC's are structurally very different. Diamond-like carbon may be composed of different sp^3-bonded structures including microcrystalline or polycrystalline diamond, small clusters, and amorphous phases. The relative size of the crystallites, and the fractions of crystalline and amorphous phases are highly dependent on deposition parameters. X-ray diffraction line broadening measurements on direct ion beam deposited DLC have indicated the presence of diamond crystallites in the 50-100 Angstrom range. Isolated crystallites up to 5 microns in diameter have been observed [11,12]. Electron diffraction studies of crystalline regions within these DLC films have shown a lattice constant close to that of cubic diamond [6,11,12]. The deposition of highly oriented microcrystalline diamond-like films having discernable registry with a Si(111) substrate has also recently been observed [55]. Finally, diamond crystallites up to a few microns in size have been produced by ion-assisted deposition of sputtered [42-48] and evaporated carbon [52].

Hydrogen is responsible for stabilizing the sp^3 bonding in H-DLC. As a result of the extensive hydrogen incorporation, H-DLC's are nearly completely amorphous although microcrystallites have been observed on occasion [56].

It may be instructive to compare the H-DLC films with polymers since they each contain substantial amounts of hydrogen. Polymers usually consist of rings or chains containing various amounts of carbon, hydrogen, oxygen, and nitrogen. Polymers may have crystalline structure and can exhibit some properties of crystals (such as crystalline x-ray patterns). When crystalline, they are denser than when disordered, and have a higher refractive index [57].

Similar to many polymers, H-DLC films are composed of carbon and hydrogen atoms. However, these films have a structure unlike any known polymers. The H-DLC materials exhibit the highest atom number densities of any of the hydrocarbons. Angus and Jansen have shown that the H-DLC's can be described as a mechanically constrained random covalent network [54]. Their range of compositions (H/C ratio) provides the limiting values of atom number density and cross-linking that can be attained in a hydrocarbon framework. It is this combination of high atom number density, mechanical constraint, and sp^3 bonding that gives H-DLC its diamond-like properties.

4. ION BEAM DEPOSITION TECHNOLOGY

4.1 Properties of Ion Beams

The use of ion beams in film deposition technology has expanded greatly over the past 15 years. The driving forces behind the use of ion beams lie in the enhanced control available with ion beams compared to other plasma sources. For example, an ion beam can be produced with a narrow energy distribution and specified direction. Important parameters such as beam energy and ion current density can be controlled almost independently over a wide range of process conditions. This is in sharp contrast to most plasma techniques in which bombardment conditions are controlled by a variety of parameters including plasma power, gas pressure, gas composition, flow rate, and system geometry.

In addition, the separation of the ion-generating plasma volume from the substrate minimizes interaction between high energy plasma electrons and the substrate. Thus, high energy particle impact events occur only as a result of impingement of well-defined beam ions. Finally, with high efficiency ion sources [58] or differential pumping schemes [1-12,26], ion-substrate interactions can be made to occur at pressures of 1×10^{-4} to 1×10^{-5} Torr or less. Operation in this high vacuum regime results in maximum control over the fluxes of energetic and/or depositing species.

For a more complete discussion, the reader is referred to an excellent review on the technology and applications of ion beams in film deposition [59].

In order to take full advantage of the control offered by the ion beam deposition technique, it is important to maintain the ion beam energy, beam current, and chemical identity of the ion species while transporting the beam to the substrate or target. In this regard, it is most important to minimize the pressure in the region of beam transport. Loss of beam "control" as evidenced by a broadened energy and spatial distribution of the beam ions is usually a result of interaction of the beam ions with the chamber background gas. The most important mechanisms operating here are resonant (and to some extent non-resonant) charge exchange and energy transfer collisions [59].

Briefly, the mean free path for resonant charge exchange for Ar^+ ions increases from about 1 x 10^{-2} Torr-cm at 100 eV to about 1.5 x 10^{-2} Torr-cm at 1000 eV energy. (Mean free paths for non-resonant charge exchange are much larger.) This means that the mean free path for 100% charge exchange of Ar^+ ions at 1 x 10^{-4} Torr is about 100 cm. Put another way, for a chamber pressure of about 1 x 10^{-4} Torr, 90% of the Ar^+ beam ions can be expected to maintain their charge for a distance of 10 cm downstream of the ion source. The energy loss mean free path for Ar^+ is about 0.1 Torr-cm at 100 eV, about ten times greater than the charge exchange mean free path. Thus charge exchange is likely before much energy loss takes place. In practice, this means that it is possible to deliver beams of useful current and fairly narrow energy spread through a vacuum chamber across reasonable distances. Nevertheless, it is important to understand and account for both loss mechanisms when conducting deposition studies.

4.2 Physics of Direct Ion Beam Deposition

It is the impact of ions with energies in the few eV to keV range onto the surface of a growing film which leads to the production of metastable materials with unique properties. The formation of diamond-like materials is but one example of this [43]. Primary ion impact energies used to generate metastable materials are normally in the range of 30-1000 eV.

Many theoretical studies have shown that impact of a particle having energy in the range of a few hundred eV produces lattice displacements which are confined to about the top 10 Angstroms of the solid [60]. Using this as a guide, it is straightforward to show that the effective power density pumped into this surface layer during ion beam deposition is the order of 1 million W/cm^3 (for a 100 eV beam at 1 mA/cm^2). For another point of comparison, it is useful to consider that only 1 eV of energy corresponds to an effective temperature of about 11,606 K. Excess energy which is not removed from the substrate by sputtered atoms is then drained from the surface region by the cold underlying substrate.

Thus, an ion beam is able to deliver to a surface (at much lower bulk temperatures than conventional thermal processes) high energy which can produce chemical reaction, diffusion, atom surface mobility, crystallization, and films with unique and advantageous properties.

In direct ion beam deposition, an ion beam of controlled composition, energy, and flux is directed onto a substrate. The impacting ions in this technique are used to supply both the deposition atoms and the energy required for improved film formation.

The energy of an impinging ion beam can cause many processes which are useful for film deposition. The degree to which each of these processes occurs is largely dependent upon the ion energy, as discussed below.

A. At energies of tens of eV, an ion beam can cause preferential sputtering of weakly bound or physisorbed gases or impurities. Because of this effect, a low energy ion beam can be used to clean a surface of these species without appreciably damaging the underlying substrate. This also means that during the deposition process, these unwanted impurity atoms can be continuously removed. In this way a surface environment characteristic of much lower pressures is effectively produced.

B. In the range of about 50 eV, physical sputtering of most materials begins [13]. This process can be used to clean strongly bound contaminants such as surface oxides prior to deposition to improve adhesion of a subsequently deposited film.

C. Direct ion beam deposition at energies in excess of about 40 eV can induce formation of interfacial structures which develop prior to deposition of the desired film. This usually improves adhesion. Formation of a surface carbide layer in the case of DLC deposition is one example of this [37,38].

D. Impacting ions in the range of 50-1000 eV produce lattice motion in the top layers of a solid. This motion occurs as a result of isolated impact events, and is therefore described as non-equilibrium and not a steady state process. In many cases, the existence of non-equilibrium conditions, even for short times, gives surface atoms mobility that would otherwise not be available at low substrate temperatures. It has been reported that the ion flux at the surface can enhance the diffusion coefficient of surface atoms by five orders of magnitude [61]. This means that there is a unique opportunity for lateral motion of surface atoms before they are bound into advantageous stationary sites.

Note that if there are impurities or gases adsorbed onto the surface, they will absorb some of the non-equilibrium surface energy and will shorten the time each energetic atom is mobile. This is one of the reasons that it is desirable to deposit films at low pressures.

E. Because the sputter yield is related to the bond strength and structure of a solid material, undesirable film structures can be sputtered and removed preferentially, leaving the more tightly bound atoms on the surface. This process is somewhat analogous to fractional distillation of liquids. This is believed to be a mechanism operating in the removal of "graphitic domains" in DLC deposition [11,12].

F. Ions with moderate energy can undergo shallow implantation into a growing film. For example, simple range calculations show that a 100-eV carbon ion will be stopped within 1-10 Angstroms of a carbon surface [54]. This may be an important mechanism in film growth.

There are suggestions that "thermal spikes" exist in the vicinity of the impacting ion presumably because of the high energy and small dimensions of each ion, as discussed by Weissmantel [62].

Calculations based upon ion current densities show that the chances of overlapping spikes within reasonable times are very small. If the deposition rate is about 1 monolayer per second, (about 1 to 3 Angstroms per second) and if all the growth rate is due to arrival of solid ions, the average time lag for another ion arriving within a radius of 3 Angstroms of a prior ion is about 0.3 seconds. Even if the energetic ion arrival rate is 10 times the growth rate of solid atoms, the average time between neighboring energy impact events is about 0.03 seconds. In this time, the original energy deposited by the prior ion certainly has dissipated. Thus any energy deposition spikes must be considered as isolated events.

In addition, the possibility of thermal spikes being an important mechanism for resulting in DLC and similar special materials must explain the effect of angle of ion incidence on the film properties. In a related process involving ion impact, the observed angular dependence of sputtering yield suggests that thermal processes are not important in sputtering [63].

Thus, we believe that while the possibility of thermal spikes cannot be ruled out, the thermal spike mechanism need not be invoked to explain the deposition of special metastable materials such as hard carbons. The concepts of enhanced mobility, preferential sputtering, and shallow implantation are probably sufficient.

G. Ions at energies of hundreds of eV can begin to cause significant atomic displacements below the surface which may damage the structure of the deposited film.

H. Ions having energies above 1 keV can undergo implantation deep into the film structure.

Since it is normally desirable to minimize structural damage of the deposited film, ion beams less than 400 eV are normally used for direct deposition.

Figure 1 illustrates the effect of the energetic ion flux on surface atoms.

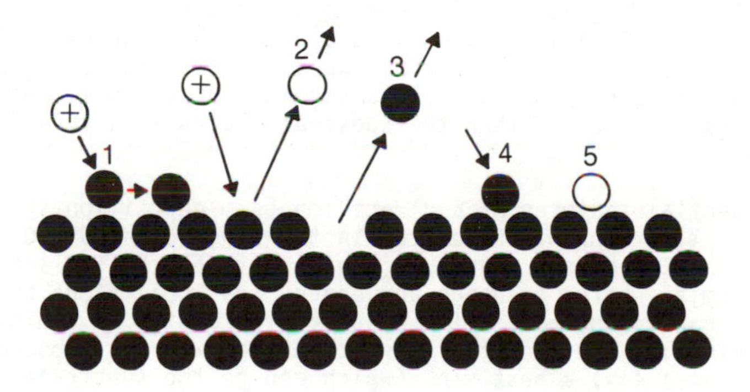

FIGURE 1. Effect of energetic ion flux on surface atoms. (1) Impinging ion providing energy for lateral motion of surface atom into advantageous site; (2) sputtering of impurity gas atom from surface; (3) sputtering of loosely bonded solid atom from surface; (4) solid atom (or ion) depositing on surface; (5) gas atom contaminating surface.

Ion beam deposition also has other unique features which are worthy of note. It appears that the charges carried by the particles and the resulting electrical charge on the substrate may have a role in the nucleation and growth of the film. It is reported that an electric field applied in the plane of the substrate will improve the continuity of evaporated metal films

[64]. Charges introduced on an insulating substrate or an insulating film being grown will have an electric field component in the plane of the substrate due to non-uniform charge distribution.

There is an important factor to be considered in connection with deposition onto insulating substrates. For insulating materials, surface potential is difficult to control by changing the potential of a metal substrate holder. Even if the substrate is a conductor, the growth of an insulating film will soon render the top of the substrate insulating and it will be difficult to control the potential. For this reason, one mode of ion beam deposition has been described in which an alternating potential is applied to the substrate in addition to a dc bias [2,9,10]. In this method, ions and electrons from the ion source plasma bombard the substrate on alternating cycles, resulting in no net charge buildup. Another strategy is to space-charge neutralize the ion beam downstream of the source by injecting electrons into the beam from a thermionic filament or hollow cathode [58].

One interesting effect can explain the unusually smooth nature (reported from electron microscopic examination) of the ion beam deposited films. As the energy of the non-equilibrium surface atoms decays down to substrate energy these surface atoms have a much higher average energy than the substrate. This high average surface energy corresponds to high free surface energy which is equivalent to high surface tension. A system tends to go to the lowest energy state. The total surface energy is equal to the product to the surface tension and the surface area, and according to the principal of virtual work, the drive to minimum total energy corresponds to a drive to minimum area. The driving force is proportional to the surface tension and in the case of non-equilibrium atoms the surface tension is unusually large and produces a strong drive to minimum surface area. The surface area includes surface irregularities and pinholes. This stronger drive to minimum area will help produce surfaces smoother than the substrate, with less irregularities and less pinholes.

Thus the essential features of a direct ion beam deposition system are (i) a source of deposition ions having energy in the range of about 40 to several hundred eV; (ii) a high vacuum transport region between the ion source and substrate to minimize charge exchange and energy loss collisions between the ion beam and the background gas; (iii) a substrate located in a high vacuum environment to minimize the amount of surface energy which is absorbed by surface contaminants; (iv) a heat sink (which can be the substrate itself) to drain excess energy from the substrate surface.

In the simplest set-up, the substrate is located directly within the beam transport region, and is bombarded by ions which are non-selectively extracted from the ion source plasma [1-12,17-27]. The beam ions are then a combination of inert gas, deposition atoms, clusters, molecular fragments, etc. The exact composition of the beam is dependent on the discharge characteristics within the ion source.

Greater control is afforded by the use of mass-selected ion beam deposition, in which the ion beam is prefiltered by a mass separator [29-41]. This technique permits precise control over the beam species, and allows for deposition in ultrahigh vacuum conditions which minimize substrate contamination [35,37,38,55]. While extremely pure films are produced, the deposition rate is usually sacrificed so that this approach may not be practical for many industrial applications.

It is clear that one of the most important parameters is the energy deposited per arriving atom (ion) [65]. The impinging ion is probably charge neutralized within a few Angstroms of the surface but arrives with its original kinetic energy which it shares with the neutral atoms on the surface. The energy sharing takes place by means of a series of binary collisions in which the daughter atoms each have considerable kinetic energy before this energy is lost to the substrate. The rate of energy loss due to collisions of surface atoms is slow because the daughter atoms also have non-equilibrium energy and the efficiency of energy transfer is proportional to the energy difference. This results in the production of daughter swarms with the high mobility needed to produce metastable structures.

In the following sections, we will discuss the concept of energy per condensing atom [51] as this parameter can be readily measured in both direct ion beam deposition and ion-assisted deposition studies. In mass-filtered ion beam deposition, the energy per condensing atom is simply the primary ion beam energy multiplied by the sticking coefficient. If the ion beam is composed of more than one ion type (i.e. inert gas ions plus film deposition ions), the energy per condensing atom is greater since extra energy is delivered to the surface by ions which are not incorporated into the film.

For the sake of completeness, we note that energy can be imparted to a surface from a number of other sources, including (i) high substrate temperature, (ii) exothermic surface chemical reactions, (iii) interactions with electronically excited atoms or molecules produced in a plasma discharge, (iv) impact by energetic electrons, and (v) impact by sputtered atoms having energy up to a few eV. None of these methods is capable of providing surface atoms with as much kinetic energy as is supplied by the momentum transfer of energetic ion bombardment.

4.3 Physics of Ion-Assisted Deposition

The extension of the physics of ion beam deposition to the technique of ion plating and the related concept of non-equilibrium ion-assisted deposition were described by Aisenberg and Chabot in 1973 [8]. The concept of ion-assisted deposition of thin films was also described by Aisenberg in two U.S. patents issued in 1975 and 1976 [9,10]. Included in these documents were discussions of the physical processes involved in producing the advantageous and unique properties of diamond-like carbon and other materials.

The physics of ion plating is very similar to that of ion beam deposition since the kinetic energy and electrical charge supplied by ions are the important factors involved in producing special film structures. The usual concept of ion plating is that the film material is carried to the surface by ions. It was shown [8] that the degree of ionization of most plasmas is small, and much less than 1%. Also, the measured ion current density is not sufficient to explain the observed film deposition rate if it is assumed that the film material is supplied only by impacting ions. A more accurate description of ion plating is that the film material is also deposited by neutral atoms or groups of atoms, and the improved film properties are produced by the ion current component attracted to the negatively biased substrate. Note that negative ions can also be used to deposit material on positively biased substrates. Disadvantages of this approach are that the efficiency of negative ion production is normally low, and a positively biased substrate will be bombarded by a large flux of plasma electrons, resulting in undesired substrate heating.

Similar to ion plating, ion-assisted deposition can produce films at a much faster rate and over a larger area compared to the original ion beam method, because there is no need to produce ions of the solid deposition material. In this case, the non-equilibrium ion energy is provided by a gaseous ion source operated in parallel with the source of atoms or clusters of the material to be deposited [8-10]. This configuration still allows the deposition chamber to be maintained at high vacuum, permitting retention of ion energy and minimizing substrate contamination. An illustration of the ion-assisted deposition method is presented in figure 2.

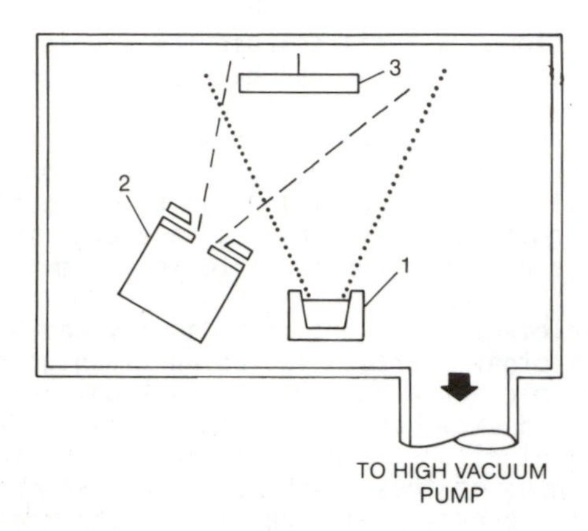

FIGURE 2. Ion-assisted deposition using larger supply of solid material from one source, plus non-equilibrium substrate atom energy supplied by energetic ion source. (1) Source of vaporized solid material; (2) ion beam source; (3) substrate located in high vacuum region.

Many variations of the ion-assisted deposition technique are possible. For example, the deposition flux can be composed of laser evaporated material [48], thermal atoms or clusters [52], sputtered atoms [42-51], or even a low energy ion beam [24,25]. In each case, a high energy ion beam is used to deliver an energy assist to fine-tune the properties of the deposited film.

We also point out that in practice, magnetron sputtering and ion beam sputtering are ion-assisted processes. In these methods, additional surface energy is provided by the fraction of high energy primary ions which strike the substrate after scattering from the sputtering target. In addition to the kinetic energy of the sputtered ions, the energy assist from scattered primary ions may be responsible for the improved properties of sputtered films compared to evaporated films [66,67].

In ion-assisted deposition, the energy per condensing atom is the ratio of the ion flux to the condensing atom flux, multiplied by the ion beam energy.

4.4 Direct Ion Beam Deposition of DLC and H-DLC

Diamond-like carbon films were fabricated by direct ion beam deposition by Aisenberg and associates as early as 1969 [2]. The method used a beam of carbon and argon ions which were extracted from a remote plasma into a low pressure (1×10^{-4} to 1×10^{-6} Torr) deposition chamber. The cathode and anode electrodes were made of high purity carbon and were water-cooled.

The original equipment [1,2,6,9,10] for producing diamond-like carbon is illustrated schematically in figure 3.

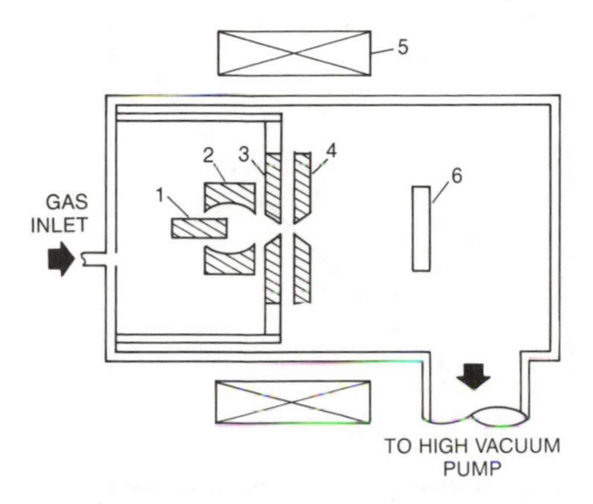

FIGURE 3. Carbon ion beam deposition system with Penning discharge as a source of ions, separated from the substrate in the high vacuum deposition chamber by a differential pumping opening. (1) Carbon cathode; (2) carbon anode; (3) differential pumping and extraction aperture; (4) plasma-ion extraction electrode; (5) electromagnet; (6) substrate located in high vacuum region.

The source of the carbon ions for the deposition was provided by sputtering of a high purity carbon cathode by argon ions in a Penning discharge. These sputtered carbon atoms were ionized in the argon plasma.

An axial magnetic field (provided by an external water-cooled electromagnet) provided a Penning type discharge which was used to operate the plasma at lower gas pressure. It was shown by Penning and Moubis that magnetic fields could be used to increase the plasma density [68]. The magnetic field deflects electrons and causes them to take longer paths on the way to the anode and thus produce more ions and electrons before being collected. Typical pressure within the Penning discharge source was about 1×10^{-2} to 5×10^{-2} Torr.

The plasma ions were extracted into the lower pressure deposition chamber through an extraction opening (about 1.5 mm diameter) in the structure separating the plasma chamber and the deposition chamber. The extraction opening also served as a differential pumping aperture. With a 1.5 mm diameter aperture, the pressure in the substrate deposition chamber was maintained at less than 1 x 10^{-4} Torr, by a 6" oil diffusion pump. The relationship between the gas pressure in the deposition chamber and the pressure within the ion source chamber is presented in figure 4.

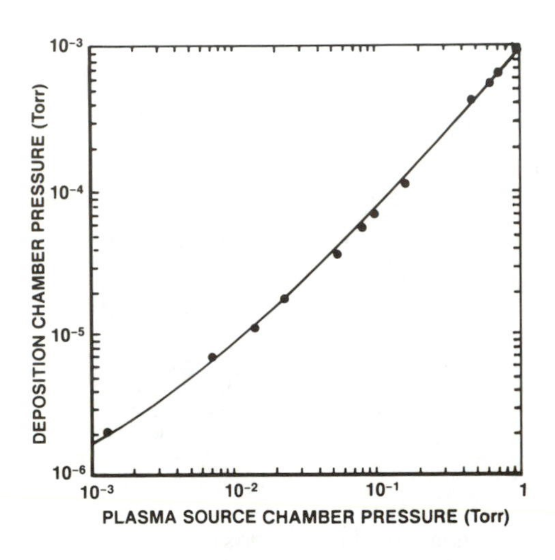

FIGURE 4. Pressure in deposition chamber as a function of plasma ion source chamber pressure [2].

Larger ion extraction openings resulted in higher ion beam current, larger deposition area, and faster deposition, although the deposition chamber pressure was also increased.

Ions were drawn from the ion source plasma into the main deposition chamber by an arc struck between the anode and extraction electrode. The substrate was biased at a negative potential to attract positive ions and to provide the ion kinetic energy. Figure 5 shows the measured relationship between the ion current to the substrate and the extraction arc current. Higher extraction arc currents allowed more of the ion source plasma to be drawn into the deposition chamber, which resulted in a larger ion current at the substrate.

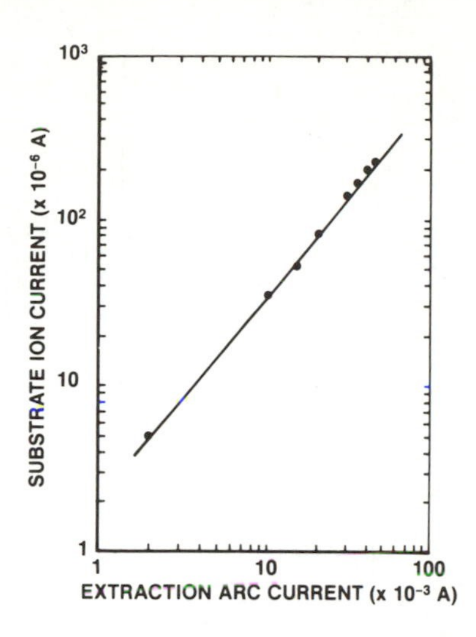

FIGURE 5. Ion current to the substrate as a function of extraction arc current [2].

The axial magnetic field also aided extraction of the carbon and argon ions from the plasma source into the deposition chamber, and helped direct the ions to the substrate. The diameter of the carbon ion beam impinging onto the biased substrate was modified by changing the magnetic field strength by control of the electromagnet current.

The differential pumping opening and the extraction electrode were lined with bushings of high purity carbon in order to prevent the introduction of unwanted elements into the ion beam. As a result, the diamond-like carbon films were free of impurities with the exception of trace amounts of argon.

Similar ion sources with planar carbon cathodes were subsequently used by Spencer, et al. [11,12] and Moravec, et al. [26,69] to deposit DLC films.

In order to demonstrate that the deposition rate of diamond-like carbon could be increased in spite of the limitations of the sputtering rate of the carbon cathode, carbon vapor was provided by introduction of hydrocarbon gas into the Ar plasma discharge. This was done in spite of the concern about reducing the purity of the plasma and resulting films by hydrogen addition.

Figure 6 illustrates a typical result achieved by operating the carbon source on a gas feed of 20 volume % CH_4, 80 volume % Ar [70]. The deposition still occurred at a pressure less than 1 x 10^{-4} Torr.

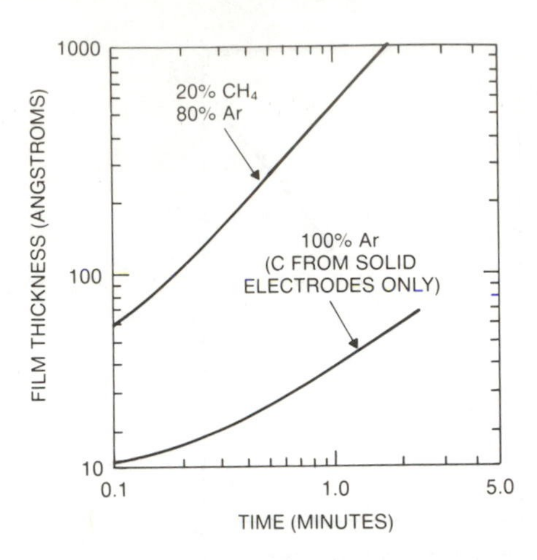

FIGURE 6. Increase of carbon film deposition rate by addition of hydrocarbon gas to argon provided to ion beam source [70].

Table I shows some of the early deposition rates obtained in 1969 [1,2,70] and 1971 by Aisenberg and associates for several materials using the ion beam source illustrated in figure 3. The larger sputter yields of Ti and Si relative to C are responsible for the faster deposition rates.

TABLE I. Typical Deposition Rates.

Film Material	Cathode	Plasma Gas	Deposition Rate
DLC	C	Ar	~ 20 A/min
H-DLC	C	CH_4 + Ar	> 600 A/min
TiC	Ti	CH_4 + Ar	~ 120 A/min
Si	Si	Ar	~ 600 A/min

Note that a deposition rate of 1 Angstrom/sec corresponds to 60 Angstroms/min, 3600 Angstroms/hour, or 0.36 microns/hour.

Since this original demonstration, many other groups have prepared H-DLC films by direct deposition from hydrocarbon ion beams [17-19] and hydrocarbon/argon ion beams [20-25]. Gridless [17,18] and Kaufman-type ion beam sources [19-25] have been used. In all cases, magnetic confinement of a filament supported discharge is used to reduce the operating pressure, and increase the degree of ionization within the source, while maintaining relatively large (>10 cm^2) deposition area. Deposition rates up to 10 Angstroms per second have been achieved [17].

Even more interesting, it has been reported that low hydrogen content [17,18] and even crystalline films [71] can be deposited from hydrocarbon ion beams when the substrate temperature is elevated to about 400°C. The combination of elevated substrate temperature and high negative substrate bias (>200 V) is probably responsible for the removal of hydrogen atoms. The faster deposition rate and absence of hydrogen are attractive, but the high substrate temperature limits some of the applications of this method.

Because it offers the maximum control over ion beam parameters and deposition environment, mass-selected carbon ion beam deposition has been studied by a number of researchers [29-41,55]. In most systems, C^+ ions are selectively filtered (by a magnetic sector) from a mixture of ions generated by electron bombardment of small molecules such as CO. Deposition of DLC from negatively charged carbon ions has also been demonstrated [39-41]. The filtered ion beam is passed through a differentially pumped transport region before delivery to the substrate. With sufficient differential pumping, the substrate can be maintained at pressures as low as 1×10^{-9} Torr. Rabalais, et al. used this method and characterized the growth of monolayers of DLC films under ultrahigh vacuum conditions by a variety of surface analysis techniques [37,38]. While the mass-selected ion beam deposition has produced the highest quality DLC films, the cost of the deposition system, the small size (<3 cm^2) of the ion beam, and limited current which can be delivered at low ion energy will probably limit its use to special applications.

In comparing and understanding the production of different forms of diamond-like carbon by different deposition techniques, one factor appears to be the most important. This is the energy of the impinging ions. If the ion energy is high enough, then a carbon film with some diamond-like properties will be produced.

Under ultrahigh vacuum deposition conditions, and on a room temperature substrate, a normally incident mass filtered C^+ ion beam of energy between 30 and 180 eV will produce a carbon film with some diamond-like structure [37]. The lower energy limit can be rationalized based on the displacement energy for graphitic carbon, which is about 25 eV. The carbon atom displacement energy for diamond is about 80 eV. Thus collisions which are capable of delivering energy in excess of 80 eV to the lattice atoms can disrupt the diamond structure. Assuming that about half of the impinging ion energy is transferred upon its first collision with the lattice, then a practical upper limit on ion energy would be about 160 eV. Qualitatively, this explains why films made at C^+ energies in excess of 180 eV begin to display an amorphous or graphitic structure [37]. For mass-filtered, normal incidence C^+ ion beam deposition of DLC, the optimum energy appears to be about 100-120 eV [35,55]. It is worth noting that because of the low self-sputter yield of carbon, hard carbon films with some degree of diamond structure have been deposited by C^+ ion beams at energies as high as 600 eV [29,30,36].

This optimum energy is also consistent with the earlier DLC studies which did not use a mass-filtered beam. In the work of Aisenberg, et al. [1-10], Spencer, et al. [11,12] and Moravec et al. [26,69] ion beams composed of a mixture of Ar^+ and C^+ ions at energies between 30 eV and 100 eV were found to produce diamond-like carbon. In these systems, extra kinetic energy was provided to the growing film by Ar^+ bombardment. Assuming that the Ar^+ flux was several times greater than the C^+ flux, the energy delivered per condensing carbon atom is close to that found to be optimum by the mass-selected approach.

To our knowledge, mass-filtered ion beam deposition of H-DLC films has not yet been reported. The ion flux extracted from a hydrocarbon plasma is a complicated mixture of hydrogen, hydrocarbon, and fragment ions. When argon is added to the hydrocarbon plasma, Ar^+ ions are extracted also. In this case it is difficult to make a direct comparison with the C^+ ion beam studies. However, it is interesting to note that beam energies in the range of 100 eV can produce H-DLC [20-25].

4.5 Ion-Assisted Deposition of DLC

In comparison to direct ion beam deposition, DLC films can be deposited over larger areas and at potentially faster rates by ion-assisted deposition. In this case, a vaporization source can be optimized to deliver a high flux of carbon atoms, while a second inert gas ion beam source delivers a large flux of ions to the substrate [8-10]. The resulting films can be of good quality and substantially hydrogen-free if the atom and ion sources are operated under clean high vacuum conditions. Ion beam sputtering [41-47], magnetron sputtering [49-51], laser evaporation [48], and electron beam evaporation [52] have all been used as atomization sources for ion-assisted deposition of hard carbon films. It is believed that ion bombardment of the vapor-deposited carbon increases the diamond-like component of the film by giving surface atoms extra kinetic energy to move into sp^3-bonded sites, and by preferential sputtering of amorphous or sp^2-bonded graphitic domains.

Weissmantel et al. [42-44] reported that when ion beam sputtered carbon films were co-bombarded with 600-1000 eV Ar^+ ions, the films became highly transparent in the visible region. The films had properties similar to those reported by Aisenberg et al. [1-10] and by Spencer et al. [11,12]. Crystalline regions were found in the films, but were difficult to reproduce. Particle fluxes were in the range of 0.9 to 2×10^{15} Ar^+/cm^2s and 1.7×10^{14} C atoms/cm^2s, thus the ion energy delivered per condensing C atom was very high, in the range of 500-1000 eV/atom. Pellicore et al. [45] also reported improvements in visible transparency by a similar dual ion beam sputtering technique.

In a related method, Savvides [49-51] used a magnetic field to extract Ar^+ ions from the discharge of an unbalanced magnetron sputtering source. In this way, the substrate was provided with carbon atoms (sputtered from the cathode) and Ar^+ ions. A small percentage of the total ion flux may have been carbon ions as well. It was demonstrated that films became more transparent, electrically insulating, and dense as the ratio of Ar^+ ions to condensing carbon atoms increased. The highest ratio of Ar^+ ion energy per condensing carbon atom investigated was estimated to be 75 eV/atom, and at this condition, the diamond-like properties were not yet maximized. We point out that due to the relatively high pressure (7×10^{-3} Torr) used in this study, the effective ion energy delivered to the condensing C atoms was probably much lower. Unfortunately, in this method, the fluxes of Ar^+ ions and C atoms are not independently variable, thus it is assumed a higher energy per condensing C atom could not be achieved.

Finally, Ogata et al. [52] prepared diamond-like carbon by bombarding electron beam evaporated carbon films (at an incident angle of about 45 degrees) with a variety of inert gas ions. They investigated a wide range of ion energies and inert gas ion and carbon atom arrival rates. Under certain conditions, large clusters of diamond crystals (as identified by Raman spectroscopy) a few micrometers in size were observed. Diamond crystals were formed under the condition where the average ion energy per condensing carbon atom was about 45-75 eV. Considering the difference in the system configurations and geometry, this energy range is not too different from that which is optimum for production of DLC by direct ion beam deposition.

5. PROPERTIES OF DLC AND H-DLC

The following is a summary of some of the observed properties of diamond-like carbon (DLC) films as deposited using the direct ion beam deposition process [1-10]. Confirmation of the reported properties was provided by Spencer, et al. [11,12] and by others in subsequent years. Some properties of ion beam deposited H-DLC are also included in this discussion.

Similar to the RF plasma process, ion beams of different hydrocarbons have been found to produce H-DLC [20,22]. Film properties are nearly independent of the type of precursor hydrocarbon, but depend strongly on ion energy and elemental composition of the hydrocarbon gas [72]. Since hydrocarbon ions of energy on the order of 100 eV disintegrate upon surface impact [72], the insensitivity of H-DLC film properties to hydrocarbon type is not surprising. Both ion beam energy and elemental composition influence the hydrogen content of H-DLC, which is believed to control many properties.

Since no external heating of the substrate is required with ion beam deposition of DLC and H-DLC, coating of temperature sensitive parts is possible. Typically, the substrate temperature rises to only about 30°C during deposition due to the energy carried by ion bombardment. DLC and H-DLC have been deposited on glass, quartz, silicon, NaCl, KCl, CaF_2, germanium, plastics, stainless steel and other metals, mylar, paper, and a wide variety of other substrates that can be held under high vacuum.

One important question related to the deposition of thin films is the strength of the bond to the substrate, since this effects the usefulness of the film. Diamond-like films display the highest adhesion to carbide-forming substrates such as Si, Ge, Mo, and W. In general, DLC films have good adhesion to most substrates when the substrates are properly cleaned by ion bombardment (to remove surface impurities) prior to deposition.

Some typical data for the adhesion of 1000-Angstroms thick DLC films to various substrates are presented in Table II.

TABLE II. Adhesion of DLC films to various substrates.

Substrate	Film Adhesion (p.s.i.)
Silicon (111) and (100)	>8,000 (1)
Polished Stainless Steel	>5,000
Fused Silica	>2,500 (2)
CR-39 optical plastic	>2,000 (2)

(1) Maximum measurable adhesion with available method.
(2) Adhesion of film exceeded cohesive strength of substrate.

The adhesion measurements were made using a standard Sebastian perpendicular pull-tester. For reference, the adhesion of Scotch Tape according to ASTM Standard Test #D-3359 (for optical coatings) is about 50 pounds per square inch (p.s.i.). The adhesion of ion beam DLC to a wide variety of metals is about 5,000 p.s.i. (24,400 kg/m^2). The adhesion of ion beam DLC and H-DLC is quite similar for most substrates.

Tests had been made of the adhesion of DLC to mylar after flexing. It was shown that after automatically flexing a DLC-coated mylar strip 100,000 times through an angle of 90 degrees around a 0.635-cm radius rod, the DLC film on the mylar had not lost adhesive strength.

The thickness of the DLC films is normally made to be about 1000 Angstroms, but has been made up to 10 microns. For thicker films, the films may separate from the substrate, probably due to internal compressive stress. Thicker films may delaminate under thermal cycling, probably due to the difference in thermal expansion between DLC and most substrates. It was observed in some thermal cycling tests that pieces of a silicon substrate remained attached to the thick DLC film after delamination. This illustrates that in some cases the carbon-silicon interfacial bond was stronger than the bonding within the bulk Si.

Observation of free standing DLC films about 1000 Angstroms thick (obtained by dissolving the substrates) showed that they are flexible. Also there was no apparent curl in these free standing films, indicating a minimum of built-in stress.

Ion beam DLC and H-DLC films are unusually smooth; scanning electron microscope images show that the films are conformal and often smoother than the substrate. The films are also unusually free of pinholes. This is probably due in part to the ion energy, and large value of free surface energy (surface tension) associated with this non-equilibrium process of ion beam deposition. On porous substrates, the film surface mirrors the substrate porosity.

Low porosity and barrier properties of DLC were demonstrated by showing a resistance to sodium ion migration. This was done by measuring the change of surface states of silicon substrates after a solution of NaCl was placed on the surface of a diode made with an insulating film of DLC [2,6]. The capacitance-voltage characteristics did not change significantly even after the diode was biased and heated, showing that there was no significant diffusion of sodium through the DLC to the silicon surface [2]. Figure 7 shows the small change in C-V characteristics that resulted when diamond-like carbon was used as a barrier to sodium ion migration. For the case of a silicon oxide film treated under the same conditions, the expected large shift of capacitance-voltage characteristics was observed [2,6]. The dense DLC and H-DLC films are also a barrier to the transmission of moisture and gas [54].

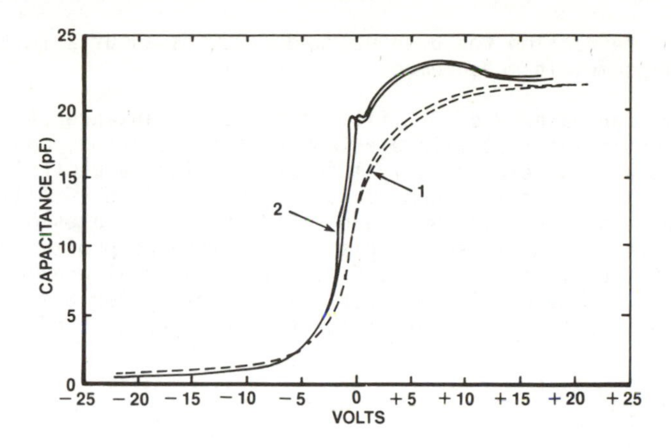

FIGURE 7. Measured capacitance as a function of applied voltage for a diode made with insulating DLC film on silicon illustrating the effectiveness of DLC as a barrier to Na^+ ion migration. (1) C-V characteristic before test; (2) after forward bias at 300°C for one minute [2].

Pin-on-disk measurements (steel on DLC) yield a friction coefficient of about 0.1 for tests done in room air. The friction coefficient is dependent upon the moisture content of the air environment.

The films are very hard as shown by scratch tests. Moh's hardness values of 9 for DLC are typical [45]. Vicker's microhardness values of 2,000-5,000 kg/mm^2 have been measured for DLC and H-DLC films using light loaded indenters. However, accurate determination of thin film hardness by indentation measurement requires that the indentation be less than 20% of the film thickness so that the substrate properties are not a factor. Most measurements have not met this criteria, so the published Vicker's hardness values are often of little meaning.

The density of ion beam DLC is typically about 2.4 g/cm^3 from float-sink tests. Ion beam H-DLC films typically have a density about 1.8 g/cm^3 [21,24].

Both DLC and H-DLC films have been observed to be chemically inert to all known solvents including hydrofluoric, nitric, sulfuric, and acetic acid, bases, acetone, trichloroethane, chloroform, and ethyl acetate.

The DLC films produced by low pressure ion beam deposition are optically transparent and films up to several thousand Angstroms in thickness can be water clear or pale yellow [6,11]. IR absorption of DLC films in the 1-10 micron wavelength range is very low (see Section 6.1). Measurements show low IR absorption for H-DLC films except at wavelengths where C-H vibrations are excited. Films produced at higher deposition chamber pressure (e.g. greater than 10^{-3} Torr) frequently exhibit reduced transparency.

The index of refraction for both DLC and H-DLC is about 2 in the visible range, from ellipsometric measurements.

The optical transparency of DLC films suggests an absence of free electrons that could interact with electromagnetic radiation, and implies that the films should also be electrically insulating. Indeed DLC films with ac and dc resistivities up to 10^{12} ohm-cm have been deposited [6,11]. The films are often less insulating when deposited at faster rates or at higher pressures. Figure 8 presents an example of a resistance profile of a DLC film deposited on a Si substrate [2]. The film covered a total area of about $1 cm^2$, but most of the deposition was confined to the center 6 mm. The resistivity, measured at the center of the film (1 micron thick) was greater than 10^{11} ohm-cm.

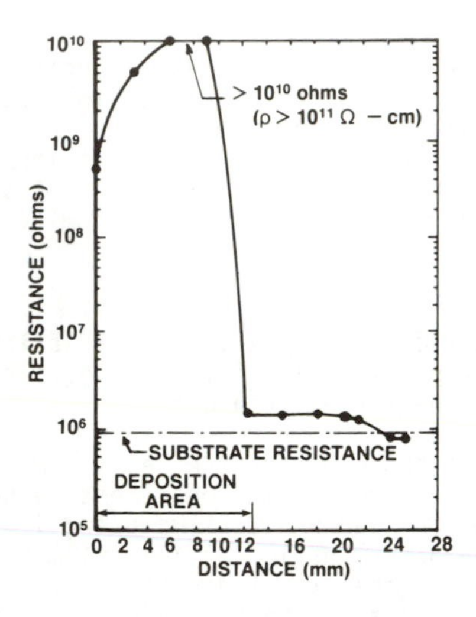

FIGURE 8. Measured resistance profile of DLC film on a silicon substrate [2].

The resistivity of H-DLC films is variable from about 10^7 ohm-cm to 10^{16} ohm-cm, and increases with increasing hydrogen content.

DLC films have a high dielectric breakdown voltage of about 1×10^6 V/cm. Capacitance measurements demonstrated a high dielectric constant with an average of about 8 [6].

Studies of DLC used as the insulating film for silicon diodes showed a resistance to radiation. No significant shift of current-voltage characteristics or increase in leakage current was observed, even after irradiation with 1 megarad from a cobalt-60 source [2,6].

In biological studies the surface potential of DLC was shown to be about 0.1 volt negative, as determined by the Kelvin vibrating electrode method.

6. APPLICATIONS

The applications of DLC and H-DLC arise to a large extent because of the combinations of the unique properties possessed by these materials. The major useful properties of the ion beam deposited diamond-like films include the following: (i) transparency, (ii) electrical insulation, (iii) high breakdown voltage, (iv) high thermal conductivity, (v) chemical resistance to most acids, bases, and solvents, (vi) barrier properties, (vii) hardness, (viii) good adhesion to many substrates, (ix) remarkable smoothness, (x) wear resistance, (xi) low coefficient of friction, (xii) deposition on substrates at slightly above room temperature.

Most of the applications can be grouped under the categories of optical, electrical, thermal, chemical, and mechanical. Some of the applications benefit from combinations of properties, such as electrical and thermal, or optical and mechanical.

6.1 Optical

An initial application for the transparent DLC films was as a protective coating for plastic lenses. The hardness of the coating, good adhesion to plastics high transmission in the visible (water-clear to slightly yellow in color), and the low deposition temperature made this an attractive application. DLC coatings on CR-39 plastic lenses were scratch tested by rubbing with SiC grit. Haze buildup on the lens was measured as a function of the number of abrasive strokes. The number of strokes for the appearance of visible haze was extended by about a factor of ten, indicating that the DLC coating could indeed improve scratch resistance of these lenses [8].

Diamond-like coatings of controlled thickness have been used to improve the transmission of Ge infrared optics [45,73]. Because the index of refraction of DLC and H-DLC (about 2) is intermediate between that of Ge and air, a diamond-like coating is capable of improving overall transmission by reducing interface reflection losses. DLC and H-DLC coatings can be used to reduce reflection losses of Si solar cells. The effect of enhanced infrared transmission for a Si wafer coated with a thin film of DLC is illustrated in figure 9. The high hardness of the diamond-like films also imparts abrasion protection to these infrared optical materials.

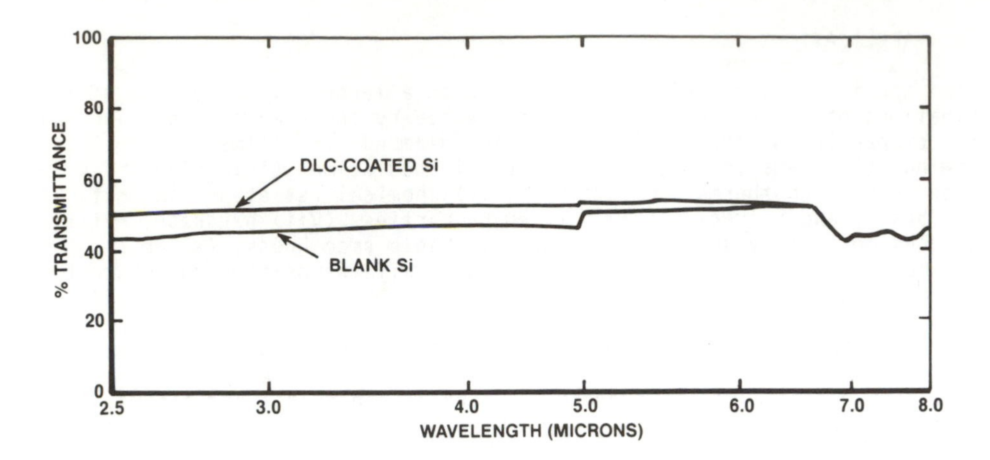

FIGURE 9. Infrared transmission spectra of Si wafer coated with 600 Angstroms of DLC, and uncoated Si wafer, illustrating increased transmission as a result of refractive index matching.

The use of ion beam DLC as a protective coating for optical elements in high power lasers was suggested for a number of reasons. Included were adhesive strength of the coating to optical materials such as glass and CaF_2, low optical absorption of DLC throughout the visible and infrared spectral regions, expected abrasion protection, and the capability to form a hermetic coating with chemical resistance to acids, bases, and solvents [74].

Low optical loss in the near infrared region was verified by measuring the adsorption of DLC deposited on CaF_2. Figure 10 presents an infrared transmission spectrum for a 2,500-Ångstrom thick DLC film deposited on a CaF_2 disk [74]. Two curves, one for the coated disk and one for the uncoated CaF_2 disk are superimposed. The actual difference in transmission was found to be less than the width of the recorder trace, or less than 1%. The absence of any C-H absorption bands in the near infrared region (3.1 - 3.7 microns wavelength) for this DLC film should be noted.

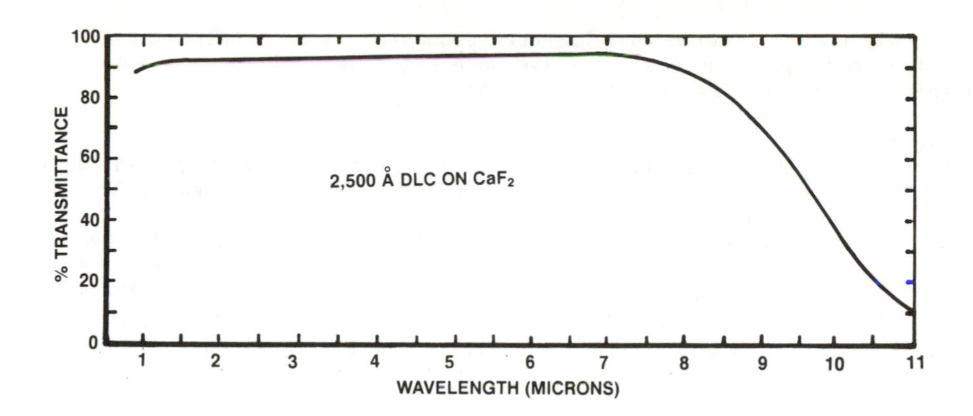

FIGURE 10. Infrared transmission spectra of a 2,500-Angstroms thick DLC coating on a CaF_2 window. Traces for the coated and uncoated window are superimposed.

Plasma formation at the surface of mirrors and windows inside the laser cavity also causes problems in high power lasers. Small defects on the surface of laser optics can be the initiation sites for the breakdown plasma. The breakdown plasma then degrades these optical surfaces and reduces the power threshold for subsequent use. In some cases, DLC films can be smoother than the substrate surface. Thus, it is believed that a DLC coating on the intracavity optics can increase their breakdown power threshold by providing a smoother surface and reducing protrusions that would otherwise enhance breakdown.

Finally, because DLC films are resistant to chemical attack, they are candidates as protective coatings for laser optics which come in contact with corrosive gases. Note, however, that the various forms of hard carbon are susceptible to attack by reactive ions and atoms of oxidizers such as fluorine and oxygen.

Another proposed optical application for DLC is as a transparent protective coating for optical disks [75]. The DLC coating is a hermetic barrier and will protect the recording medium from degradation and extend the archive lifetime.

Optical sensors immersed in corrosive fluids or environments can also benefit from the protection provided by thin optically transparent diamond-like films applied over the exposed surfaces.

There have been suggestions that transparent DLC on softer polished gem stones would give them the surface hardness of diamond, while retaining the appearance of the gem. This approach should be effective against mild abrasion, but will not be effective against scratches that break through the film to reach the softer substrate and gouge the protective film away.

Another related application that has been suggested is as transparent protective coatings on decorative silverware and jewelry to prevent the tarnishing and reduce the need for cleaning.

6.2 Electrical

If the electrical properties (intrinsic high resistivity, wide band gap) and thermal properties of true diamond material can be approached by DLC, then there will be many advantages to using DLC in device applications. For example, DLC films with high resistivity, $>10^{11}$ ohm-cm, have already been deposited (see figure 8). Ion beam deposited DLC films can exhibit a thermal conductivity greater than 700 W/m-K [76]. This combination of high thermal conductivity and high electrical resistivity makes DLC attractive as a dielectric layer for high temperature devices. To our knowledge, DLC films with a band gap greater than about 2 eV have not yet been demonstrated.

As an example of the use of ion beam deposited DLC in electronic device applications, thin film silicon field effect transistors (FET) with insulating DLC were fabricated. In the initial work, the problem of producing ohmic contacts was avoided by using a coplanar FET design in which the action occurs by carrier modulation effects. This design did not necessarily offer the advantage of very short carrier distances.

One FET type had a regular coplanar structure with the active silicon thin film deposited on the bottom. The other FET type had an inverted coplanar structure with the active silicon layer deposited as the top layer. The demonstration devices that were fabricated showed that active FET units resulted and that the inverted FET's had a higher transconductance, as was expected [2,6]. The details of the geometry of the field effect transistors are illustrated in figure 11.

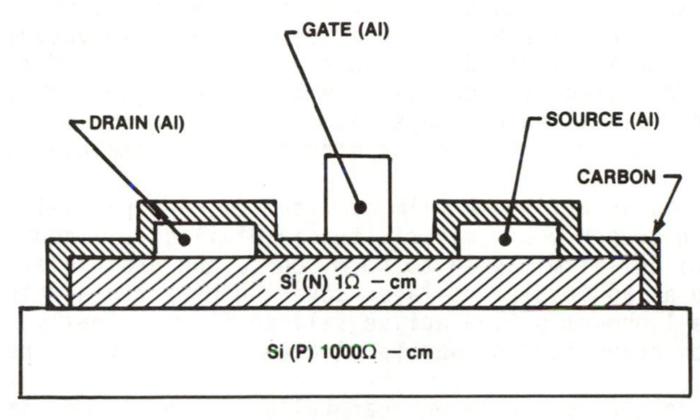

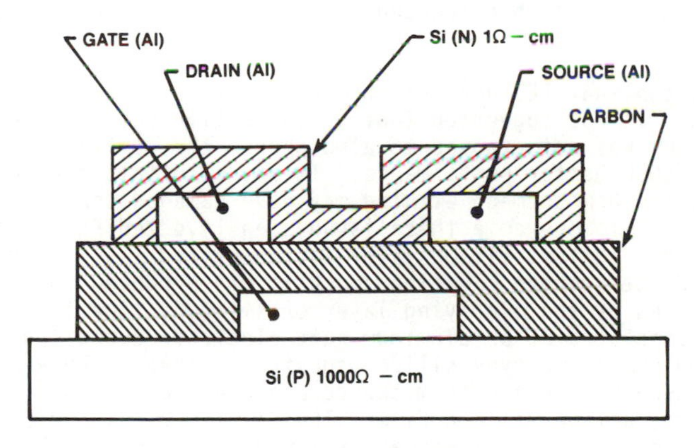

FIGURE 11. Geometry of active thin film field effect transistors fabricated using ion beam deposited n-type silicon and ion beam deposited DLC as the intermediate insulator film [2].

The films were deposited with a special ion beam source similar to the one described for DLC deposition [2,6,9,10], with the addition of multiple cathodes rather than one carbon cathode. These cathodes could be electrically switched to provide sputtering of different atoms into the ion source plasma. For FET fabrication, cathodes of carbon and 1 ohm-cm, n-type doped Si were used sequentially to deposit the different layers. Aluminum metal electrodes were deposited by standard thermal evaporation (in the same vacuum chamber) since forming these metal electrodes was not a critical part of the device fabrication. The deposited silicon films were n-type as determined by thermal emf values compared with known n-type silicon. The geometry of the various film layers was defined by stencil masks etched in thin stainless steel.

The regular coplanar FET had a film of 1 ohm-cm, n-type silicon deposited on a 1000 ohm-cm, p-type higher resistivity (insulating) silicon substrate, followed by an aluminum film masked to form source and drain electrodes. DLC was then deposited as an insulating layer covering over the aluminum electrodes and the 1 ohm-cm n-type active silicon film. Finally, a gate electrode made from evaporated Al was fabricated at the top of the structure.

The actual deposition sequence was carefully planned to permit stencil masks to be changed between film layers. After completion of a layer, the system was let up to atmosphere and the stencil was rotated to the new position. The vacuum system was then closed and pumped down to about 10^{-8} Torr. In order to remove the system contamination caused by each venting, the ion source was operated behind a shutter for several minutes to sputter clean the electrodes. The shutter was then opened, and the prior FET film layer was cleaned by a few minutes of ion bombardment by ions of about 200 to 400 eV energy. The ion energy was then reduced to about 40 to 100 eV to permit film deposition.

The resulting coplanar FET devices showed active but low transconductances. It was suggested that if the active face of the 1 ohm-cm, n-type silicon layer was not exposed to atmosphere, the active film would be cleaner and would show better performance. It was also felt that cleaning the active surface by ion bombardment etching may have damaged the silicon and reduced performance. To overcome these issues, an inverted FET geometry was employed in which the active silicon layer was deposited last. Thus, the active surface did not require cleaning since it was produced under high vacuum and was buried by the overlying layer of silicon. The inverted coplanar devices consisted of an aluminum gate electrode deposited on an (insulating) 1,000 ohm-cm, p-type silicon substrate, covered by an insulating film of DLC, followed by the aluminum source and drain electrodes, and finally the active 1 ohm-cm, n-type silicon film. The Si film was about 5,000 Angstroms thick, and the DLC film was about 1,000 Angstroms thick. The inverted FET devices had measured transconductances of about 75 micro-mho [2,6], which could probably be improved upon with further research. This early program was designed only to demonstrate feasibility.

It is expected that DLC will play a special role in applications where the power density is high. An example of this will be found in the next generation of high density integrated circuits. These devices will permit shorter signal transit distances and reduce communication time between transistors and memory cells, resulting in higher computation rates. Because of the ability of the resulting devices to operate at higher temperatures, cooling requirements can be reduced and the operating lifetime can be much greater. This will be important in the new supercomputers where the high power density currently makes sophisticated cooling systems a requirement.

Another important application can be for semiconductor circuits used in space environments. Here, the high temperature capability will permit lighter cooling systems to be used. Also, the radiation resistance of DLC devices will permit a reduction in the weight of radiation shields. Both of these features can permit the weight and cost of satellites to be reduced, and increase operating life. This radiation resistance property will also be important for a variety of military applications, and nuclear power plant electronics.

It is possible that the hydrogenated form of diamond-like carbon may not be as useful as a gate dielectric for application in microelectronics. Among the reasons for this are low reported band gap (1 eV), low resistivity (10^7 ohm-cm), high fixed insulator charge density (10^{11} - 10^{13}/cm^2) and high interface state density (10^{12}-10^{14}/eV) [23]. Also, unlike DLC, H-DLC is thermally decomposed at temperatures greater than about 400°C, which can be a disadvantage for standard processing.

6.3 Thermal

The power dissipation of a solid state device such as a microwave oscillator or power transistor can be significantly increased if it is mounted on a substrate (such as copper) that permits thermal energy to be efficiently transferred away from the device. Heat is ultimately removed from the system via conduction to a cooling fluid or by radiation loss.

If the device must be electrically isolated from the cooling substrate, it is desirable to use an insulating layer with high breakdown voltage (so that it can be made thin) and high thermal conductivity. If the insulating layer becomes too thick, and has insufficient thermal conductivity, a large temperature drop between the device and the substrate is produced. Thus, heat removal is impeded and the power capability of the device is limited. Phonon impedance mismatch between the insulating layer and underlying substrate may also limit heat transfer. For example, these problems are currently encountered in power devices which commonly use BeO as the insulating layer.

Diamond is a logical choice as the insulating layer for these applications. The thermal conductivity of Type IIa diamonds at room temperature is the highest of all known materials, about 2000 W/m-K, nearly five times that of copper [53,77]. Room temperature conductivity values for ion beam DLC greater than 700 W/m-K have been measured [76]. The thermal conductivity of BeO is about 250 W/m-K at room temperature.

Because DLC films (i) can exhibit high thermal conductivity, (ii) can be made thin while providing the necessary electrical isolation, and (iii) do not require intermediate bonding layers (which may cause phonon impedance mismatch) to adhere to most substrates, they should find use as insulating layers in many high power or high density device applications.

Thick chips of natural diamonds have been used as "heat spreaders" for laser diodes. In this application, the diamond chip not only removes heat in a direction perpendicular to the plane of the device, but also transfers heat laterally. (Electrical insulation is normally not critical.) If the heat spreads in a region of high thermal conductivity such as diamond, then the heat flux and the temperature drop between the device and the heat sink substrate are much reduced. It has been shown that in order to produce significant lateral heat flow (spreading) in the diamond, the thickness of the diamond layer must be several times greater than the dimensions of the heat producing device [78]. Provided that DLC films can be made thick enough, they may find use as heat spreaders. However, because of the thickness required (usually >100 micrometers) this application will probably be better fulfilled by polycrystalline diamond films prepared by low pressure CVD techniques [79-81].

6.4 Chemical

The chemical resistance to most acids, bases, and solvent permits the use of DLC and H-DLC as protective coatings for various applications. Only two key application areas (on which experimental work has already been performed) are discussed here, although there may be many others.

6.4.1 In-Vivo

The inert nature of DLC suggested its use as a blood compatible coating for use in the artificial heart program [82,83]. Additionally, use of diamond-like films as tissue compatible coatings for other artificial organs or biological implants such as hip joints and plastic replacement arteries or veins used in bypass operations has been suggested. When using foreign materials in organ implants, it is necessary to insure that the material does not harm the tissue or blood it contacts. Also, inert materials are needed to prevent rejection reactions in which the body fluids attack the implanted material. Fortunately, when implanted in tissue, encapsulation can occur to isolate the foreign body.

The case of interaction with blood is a more difficult problem than reactions with tissue, since the formation of clots can be accelerated by the surface of foreign bodies. These blood clots can break loose and lodge downstream in vital organs such as the brain, lungs, or kidneys. This can be fatal as found in actual cases in which artificial hearts were implanted.

Thus, the roles of in-vivo protective coatings are (i) to be biologically inert to blood contact, (ii) to prevent harmful components (e.g. plasticizers, monomers, fillers) of the artificial organs from leeching into the blood, and (iii) to prevent blood components from leeching into the plastic to degrade the performance and lifetime. Because the chemical composition is similar to organic material, and since the dense lattice provides an excellent diffusion barrier, DLC was considered a nearly ideal coating. It was also expected that the smooth surface and inert nature of DLC would hinder the formation of blood clots. Unfortunately, it was found that when plastic substrates had surface structural defects or inclusions larger than 1000 Angstroms, a DLC coating of about 1000 Angstroms thickness could not cover these defects, and blood clots would eventually form. This result suggested that thicker protective coatings and/or much smoother test substrates were required. At the time of the tests, plastic substrates without defects could not always be supplied [82,83]. If artificial hearts made of plastics are to be practical in the long term, then plastic surfaces without clot nucleating defects are needed, even if the plastic material and the diamond-like coating are intrinsically blood compatible.

6.4.2 Fiberoptics

DLC coatings have been shown to be an effective hermetic seal (moisture barrier), capable of extending the life of optical fibers.

Glass fibers are initially very strong when drawn. However, their strength is greatly reduced when the surface is touched due to formation of surface microcracks. To overcome this problem, polymer coatings are applied to the glass fibers during the drawing process to protect the fiber surface against mechanical damage. Failure due to a stress corrosion mechanism is also important. When subjected to stresses of stretching or bending, microcracks grow until eventually the fiber fractures. The presence of water molecules in the cracks enhances the breaking of silicon-oxygen bonds, further reducing the fiber strength. Since all known polymers transmit water to some degree, even the polymer coatings used to prevent mechanical abrasion will not protect the glass fiber surface from water vapor. It was suggested that DLC films applied to the surface of the glass fibers during the drawing process would be a satisfactory barrier against water vapor because of the high atomic density of the DLC lattice.

A coating system was designed and built to permit in-line deposition of DLC or H-DLC onto an optical fiber after it was drawn, but prior to application of the protective polymer coating [84-87]. Quartz fibers were coated with DLC as they moved down the axis of a cylindrical deposition chamber. The fiber passed from the atmospheric pressure drawing stage through a high vacuum coating stage via a small entrance aperture. After traversing the DLC coating stage, the fiber was passed through an exit aperture, and returned to atmospheric pressure. At no time did the fiber touch any solid material which could damage the fiber surface. The necessary deposition chamber vacuum was maintained by differential pumping stages at the entrance and exit ports. In the deposition chamber, coaxial carbon electrodes served as the source of carbon for the DLC coating. The electrodes were sputtered in a magnetically enhanced Ar discharge operated at a pressure of about 2-5 x 10^{-3} Torr. The thin fiber did not reach high temperatures.

With this system, DLC coatings up to a thickness of about 300 Angstroms were applied. The films could not be made thicker because of the limited discharge power and cooling capability in the coating section, and because the fiber could not be drawn too slowly. This 300-Angstrom DLC coating improved the fiber strength as determined by measurements of the breaking strength as a function of strain rate (dynamic fatigue). This data permitted calculation of parameters that predicted an improvement in strength under constant strain (static fatigue) relative to the case where only polymer coatings were used [84,85]. The benefit of the DLC coating is illustrated in Table III, which presents the percent loss of strength (relative to freshly drawn fiber) versus time. The static stress required to produce failure as a function at time is also given.

TABLE III. Calculated percent loss of strength for coated quartz fibers, and static stress (k.s.i.) required to produce failure in Suprasil-2 optical fibers as a function of time. DLC coating thickness = 300 Angstroms, indium coating thickness = 1,000 Angstroms [84,85].

Coating material	Time to Failure 1 second	1 hour	1 day	1 year	10 years
polymer	0% (570)	33% (380)	46% (310)	60% (230)	63% (210)
DLC + polymer	0% (570)	21% (450)	30% (400)	44% (320)	46% (305)
Indium + polymer	20% (460)	21% (450)	25% (430)	26% (420)	28% (410)

The insulating protective coating of DLC has advantages over metallic coatings when interferences due to EMP (Electromagnetic Pulses) occur. (A conducting coating can act as an antenna for electrical pickup.) For uses where EMP is not a problem, metal protective films have the advantages that they are good moisture barriers, and can be deposited faster and/or thicker because of their larger sputtering yields compared to carbon. Coatings of 1000 Angstroms of indium were also applied using the differentially pumped ion-plasma coating system. The data in Table III show that for the indium coatings there was even greater improvement in strength under static fatigue than for the thinner DLC coating [84,85]. The improved stability achieved with In is probably related both to the increased thickness, and ductility of this material.

6.5 Mechanical

Because of the hardness, smoothness, and low friction coefficient of DLC, it was anticipated that DLC films on the edges of cutting and razor blades would improve performance. Experimental results [8] showed that DLC on industrial cutting blades reduced the cutting force required to slit paper by more than a factor of four, presumably by reducing friction between the paper and the blade. Second, the coated blades could cut about 100 times longer than the uncoated blades before the same cutting force was reached. It is anticipated that DLC and H-DLC films on other cutting tools (e.g. knives, scalpels, microtomes) can improve their cutting ability, and extend the period of sharpness under use.

Diamond-like coatings on certain machine tools used under high force or load may also provide some degree of protection. However, if high temperatures occur at the cutting edge, the diamond-like films will be rapidly degraded by oxidation. Also, H-DLC films begin to decompose by losing hydrogen at about 400°C. The diamond-like coatings are probably not appropriate for application on tool bits for machining iron or steel, because chemical reaction between the carbon film and iron will occur. Machining non-ferrous materials and ceramics is a possibility.

Another important application of hard carbon coatings will be for wear protection of magnetic recording disks and thin film media. H-DLC coatings have already demonstrated the ability to protect Winchester disks from head crashes [88-90]. The diamond-like coating also protects the disk against chemical reaction with lubricating fluids. For more information, the reader is referred to an excellent review article on the field [91]. There will be many other wear-related applications (e.g. ball bearings, bearing races, valves, seals, etc.) which can benefit from the application of a smooth, low friction diamond-like coating.

The rigidity of H-DLC films has resulted in a novel application associated with loudspeakers. Because of the stiffness of the H-DLC film relative to the added mass, the response of coated titanium tweeters in the high frequency range (20-35 kHz) is significantly improved compared to that of the uncoated tweeter [92].

7. SUMMARY

The essential features of the high vacuum ion beam deposition process include: (i) high kinetic energy (40 to 400 eV) of the incident ions; (ii) low gas coverage on the substrate surface due to the high vacuum in the deposition chamber; (iii) reduced frequency of collisions between beam ions and gas molecules on the way to the substrate, and therefore retention of the high ion energy; (iv) use of ion sputtering to clean the substrate prior to deposition; (v) electrical surface charge, (vi) high average energy or mobility of surface atoms resulting in bonding in advantageous sites; (vii) high surface tension or free surface energy resulting in a drive to minimum surface area, producing very smooth films with low pinhole density.

Ion-assisted deposition is a method capable of increasing the deposition rate and area while retaining the film purity attained by the original direct ion beam deposition process. The basic concept of ion-assisted deposition, originally proposed for DLC, is applicable to many other materials.

The energy of the bombarding ions is the most important factor governing the formation of the diamond-like carbon materials. If the average carbon surface energy is high enough, then carbon with some diamond properties will be produced. Ion impact energies per condensing carbon atom in the range of about 50 to 150 eV are sufficient to form the diamond-like carbon materials.

DLC films were first deposited from a beam of carbon and argon ions generated by sputtering carbon electrodes into a pure argon plasma. Addition of hydrocarbon gas into the argon plasma increases the deposition rate significantly, but also introduces hydrogen into the resulting film. This H-DLC, however, has many properties similar to DLC although it is not the same material. In both DLC and H-DLC, many of the properties of natural diamond are retained even though the films are mostly amorphous and not crystalline.

It has been demonstrated that DLC and H-DLC can be deposited on substrates slightly above room temperature. This fact, along with the unique set of optical, electronic, and mechanical properties of the diamond-like films will result in many applications for these materials.

8. REFERENCES

1. S. Aisenberg, and V. Rohatgi, "Study of the Deposition of Single Crystal Silicon, Silicon Dioxide and Silicon Nitride on Cold-Substrate Silicon," Interim Report, prepared for NASA Electronics Research Center under Contract #NAS 12-541 (April, 1968).

2. S. Aisenberg and R. Chabot, "Study of the Deposition of Single Crystal Silicon, Silicon Dioxide, and Silicon Nitride on Cold-Substrate Silicon," Final Report prepared for NASA Electronics Research Center under Contract #NAS 12-541 (October, 1969).

3. S. Aisenberg, "The Deposition of Single Crystal Silicon Films on Cold Single Crystal Silicon Substrates," Presented at the 1968 Government Microcircuit Applications Conference (GOMAC), Gaithersburg, MD (October, 1968).

4. S. Aisenberg and R. Chabot, "Ion Beam Deposition of Thin Films of Insulating Carbon," Presented at the 1970 Government Microcircuit Applications Conference (GOMAC), New Jersey (October 6-8, 1970).

5. S. Aisenberg and R. Chabot, "Ion Beam Deposition of Thin Films of Diamond-like Carbon," Proceedings of the 17th National Vacuum Symposium, Washington, D.C. (October 20-23, 1970).

6. S. Aisenberg and R. Chabot, J. Appl. Phys. 42, 2953 (1971).

7. S. Aisenberg and R. Chabot, J. Vac. Sci. Technol. 8, 1 (January/February, 1971).

8. S. Aisenberg and R. Chabot, J. Vac. Sci. Technol. 10, 104 (1973).

9. Sol Aisenberg, "Apparatus for Film Deposition," U.S. Patent No. 3,904,505 (September 9, 1975).

10. Sol Aisenberg, "Film Deposition," U.S. Patent No. 3,961,103 (June 1, 1976).

11. E. G. Spencer, P. H. Schmidt, D. C. Joy, and F. J. Sansalone, Appl. Phys. Lett. 29, 118, 1976.

12. D. C. Joy, E. G. Spencer, P. H. Schmidt, F. J. Sansalone, Proc. 24th Ann. Electron Microscopy Soc. America Mtg. p. 646 (1976).

13. G. Carter and J. S. Colligon, Ion Bombardment of Solids, (American Elsevier, New York, 1968).

14. L. Holland and S. M. Ojha, Thin Solid Films, 38, L17 (1976).

15. J. C. Angus, J. E. Stultz, P. J. Shiller, J. R. MacDonald, M. J. Mirtich, and S. Domitz, Thin Solid Films, 118, 311 (1984).

16. J. C. Angus, P. Koidl, and S. Domitz, in Plasma-Deposited Thin Films, J. Mort and F. Jansen, editors, (CRC Press, Boca Raton, Florida, 1986)

17. T. Mori, and Y. Namba, J. Vac. Sci. Technol. A1, 23 (1983).

18. Y. Namba, and T. Mori, J. Vac. Soc. Technol. A3, 319 (1985).

19. D. C. Ingram, and A. W. McCormick, Nucl. Instrum. Meth. Phys. Research, B34, 68 (1988).

20. D. B. Kerwin, I. L. Spain, R. S. Robinson, B. Daudin, M. Dubus, and J. Fontenille, Thin Solid Films, 148, 311 (1987).

21. J. C. Angus, M. J. Mirtich, E. G. Wintucky, in Materials Formation by Ion Implantation, S. T. Picraux, and W. T. Choyke, editors, (Elsevier, Amsterdam, 1982), p. 433.

22. D. Nir, and M. Mirtich, J. Vac. Sci. Technol. A4, 560 (1986).

23. V. J. Kapoor, M. J. Mirtich, and B. A. Banks, J. Vac. Sci. Technol. A4, 1013 (1986).

24. M. J. Mirtich, D.M. Swec, and J.C. Angus, Thin Solid Films, 131, 245 (1985).

25. M. J. Mirtich, J. S. Sovey, and B. A. Banks, U.S. Patent No. 4,490,229 (December 25, 1984).

26. T. J. Moravec and T. W. Orent, J. Vac. Sci. Technol. 18, 226 (1981).

27. A. I. Maslov, A. I., G. K. Dmitriev, and Y. D. Chistyakov, Pribory i Tekhnika Eksperimenta, 3, 146 (1985).

28. S. S. Wagal, E. M. Juengerman, and C. B. Collins, Appl. Phys. Lett. 53, 187 (1988).

29. J. H. Freeman, W. Temple, and G. A. Gard, Vacuum, 34, 305 (1984).

30. J. H. Freeman, W. Temple, D. Beanland, and G. A. Gard, Nucl. Instrum. Meth. 135, 1 (1976).

31. J. Anttila, J. Koskinen, M. Bister, and J. P. Hirvonen, Thin Solid Films, 136, 129 (1987).

32. J. Anttila, J. Koskinen, R. Lappalainen, and J. P. Hirvonen, Appl. Phys. Lett. 50, 132 (1987).

33. J. Anttila, J. Koskinen, J. Raisanen, and J. P. Hirvonen, Nucl. Instrum. Meth. Phys. Res. B9, 352 (1985).

34. J. Koskinen, J. P. Hirvonen, and A. Anttila, Appl. Phys. Lett. 47, 941 (1985).

35. E. F. Chaikovskii, V. M. Puzikov, and A. V. Semenov, Sov. Phys. Crystallogr. 26, 122 (1981).

36. T. Miyazawa, S. Misawa, S. Yoshida, and S. Gonda, J. Appl. Phys. 55, 188 (1984).

37. J. W. Rabalais and S. Kasi, Science, 239, 623 (1988).

38. S. R. Kasi, H. Kang, and J. W. Rabalais, J. Chem. Phys. 88, 5914 (1988).

39. J. Ishikawa, Y. Takeiri, and T. Takagi, Rev. Sci. Instrum. 57, 6 (1966).

40. J. Ishikawa, Y. Takeiri, K. Ogawa, and T. Takagi, J. Appl. Phys. 61, 17 (1987).

41. J. Ishikawa, K. Ogawa, K. Miyata, H. Tsuji, and T. Takagi, Nucl. Instrum. Meth. Phys. Research, B21, 205 (1987).

42. C. Weissmantel, G. Reisse, H. J. Erler, F. Henny, K. Bewilogua, U. Ebersbach, and C. Schurer, Thin Solid Films, 63, 315 (1979).

43. G. Gautherin and C. Weissmantel, Thin Solid Films, 50, 135 (1978).

44. C. Weissmantel, K. Bewilogua, K. Brewer, D. Dietrich, V. Ebersbach, H. J. Erler, B. Rau, and G. Reisse, Thin Solid Films, 96, 31 (1982).

45. S. F. Pellicore, C. M. Peterson, and T. P. Henson, J. Vac. Sci. Technol. A4, 2350 (1986).

46. M. Kitabatake and K. Wasa, J. Appl. Phys. 58, 1693 (1985).

47. M. Kitabatake and K. Wasa, J. Vac. Sci. Technol. A6, 1793 (1988).

48. S. Fujimori, T. Kasai, and T. Inamura, Thin Solid Films, 92, 71 (1982).

49. N. Savvides and B. Window, J. Vac. Sci. Technol. A3, 2386 (1985).

50. N. Savvides, J. Appl. Phys. 58, 518 (1985).

51. N. Savvides, J. Appl. Phys. 59, 4133 (1986).

52. K. Ogata, Y. Andoh, and E. Kamijo, Nucl. Instrum. Phys. Res. B33, 685 (1988).

53. J. E. Field, editor, Properties of Diamond, (Academic Press, New York, 1979).

54. J. C. Angus and F. Jansen, J. Vac. Sci. Technol. A6, 1778 (1988).

55. J. L. Robertson, S. C. Moss, Y. Lifshitz, S. R. Kasi, J. W. Rabalais, G. D. Lempert, and E. Rapoport, Science, 243, 1047 (1989).

56. K. Kobayashi, N. Mutsukura, and Y. Machi, J. Appl. Phys. 59, 910 (1986).

57. O. J. Sweeting, in The Science and Technology of Polymer Films, Vol. I, O. J. Sweeting, editor, (Interscience Publishing, NY, 1968,) p. 35.

58. H. R. Kaufman, J. Vac. Sci. Technol. 15, 272 (1978).

59. J. M. E. Harper, J. J. Cuomo, and H. R. Kaufman, J. Vac. Sci. Technol. 21, 737 (1982).

60. N. Winograd, Prog. Solid St. Chem. 13, 285 (1982).

61. A. H. Eltoukhy and J. S. Greene, Appl. Phys. Lett. 33, 343 (1978).

62. C. Weissmantel, in Thin Films From Free Atoms and Particles, K. J. Klabunde, editor, (Academic Press, NY, 1985), p. 158.

63. G. K. Wehner and Gerald S. Anderson, in Handbook of Thin Film Technology, L. I. Maissel and Reinhard Glang, editors, (McGraw-Hill Book Company, NY, 1970), pp. 3-1 to 3-38.

64. K. Chopra, Appl. Phys. Lett. 7, 140 (1965).

65. T. Takagi, Thin Solid Films, 92, 1 (1982).

66. B. A. Banks, S. K. Rutledge, J. Vac. Sci. Technol. 21, 807 (1982).

67. C. Weissmantel, J. Vac. Sci. Technol. 18, 179 (1981).

68. F. M. Penning, J. H. A. Moubis, Koniki. Ned. Akad. Wetenschap. Proc. 43, 41 (1940).

69. T. J. Moravec, Rome Air Force Development Center Report No. RADC-TR-85-65 (1985).

70. Partly derived from brochure IBD-101 dated 12/17/71, from the Space Sciences Division of Whittaker Corporation. This is a product description brochure describing an Ion Beam Deposition system (IBD-101) and typical performance, based upon the initial work by Aisenberg and Chabot of the Space Sciences Division).

71. Y. Namba, J. Wei, T. Mohri, E. A. Heidarpour, J. Vac. Sci. Technol. A7, 36 (1988).

72. J. Wagner, F. Pehl, P. Koidl, Appl. Phys. Lett. 48, 106 (1986).

73. P. Koidl, A. Bubenzer, B. Dischler, Proc. Soc. Photo-Opt. Instrum. Eng., 381, 186 (1983).

74. M. L. Stein and S. Aisenberg, in Laser Induced Damage in Optical Materials, NBS Special Publication 638, (U.S. Government Printing Office, Washington, D.C., 1981), p. 482.

75. S. Aisenberg and M. Stein, "Novel Materials for Improved Optical Disk Lifetimes," Proceedings of the 15th International Technical Symposium of the Society of Photo and Electro-Optic Engineers, (SPIE), August, 1981.

76. M. Yoshida, K. Ogawa, H. Tsuji, J. Ishikawa, T. Takagi, Proc. 11th Symp. on ISIAT87, Tokyo (1987).

77. G. A. Slack, J. Phys. Chem. Solids 34, 321 (1973).

78. J. Doting, J. Molenaar, Semi-Therm, Proc. IEEE, 113 (1988).

79. R. C. Devries, Annu. Rev. Mater. Sci., 17, 161 (1987).

80. J. C. Angus, C. L. Hayman, Science, 241, 913 (1988).

81. K. E. Spear, J. Am. Ceram. Soc., 72, 171 (1989).

82. S. Aisenberg and M. Stein, "Diamond-like Carbon Films-Factors Leading to Improved Biocompatibility," Extended Abstracts of the 13th Biennial Conference on Carbon (July 18-22, 1977), Irvine, CA., p. 87.

83. S. Aisenberg and M. Stein, "Ion Beam Deposited Carbon Films and Factors Important to Improved Biocompatibility," Proceedings: AAMI 13th Annual Meeting, p. 6 (1978).

84. M. Stein, S. Aisenberg, and J. M. Stevens, in Advances in Ceramics, edited by Bendow and Mitra (American Ceramic Society, Columbus, Ohio, 1981), Vol. 2, p. 124.

85. S. Aisenberg, Radiation curing VI Conference Proceedings (Society of Manufacturing Engineers, Dearborn, Michigan, 1982), Chap. 12, pp. 14-32.

86. S. Aisenberg and M. L. Stein, "Process for Coating Optical Fibers," U.S. Patent No. 4,402,993 (September 6, 1983).

87. S. Aisenberg and M. L. Stein, " Apparatus for Coating Optical Fibers," U. S. Patent No. 4,530,750 (July 23, 1985).

88. B. M. Meyerson, R. V. Joshi, R. Rosenberg, V. V. Patel, "Silicon/Carbon Protection of Metallic Magnetic Structures," U.S. Patent No. 4,647,494 (March 3, 1987).

89. H. E. Aine, "Thin Film Recording and Method of Making," U.S. Patent No. Re.32,464 (July 28, 1987).

90. S. Hoshino, K. Fujii, N. Shohata, H. Yamaguchi, Y. Tsukamoto, M. Yanagisawa, J. Appl. Phys, 65, 1918 (1989).

91. H. Tsai, D. B. Bogy, J. Vac. Sci. Technol. A5, 3287 (1987).

92. N. Fujimori, Jpn. Revs. in New Diamond, p. 36 (1988).

9. ADDITIONAL REFERENCES

The following references related to diamond-like carbon are not in the open literature but are taken from contract reports and presentations. They can contain technical details not available otherwise. Many of these references are not widely known and may be of interest if copies can be obtained from the sponsoring agencies.

A1. S. Aisenberg, "The Ion Beam Deposition of Single Crystal Silicon Films on Cold Single Crystal Silicon Substrates," Proceedings of the 15th National Vacuum Symposium, Pittsburgh, PA (October, 1968).

A2. S. Aisenberg and R. Chabot, "An Ion Beam Deposition Technique for the Formation of Thin Films of Insulating Carbon," Presented at the Second Annual Symposium of the New England Section of the Surface Division of the American Vacuum Society, Waltham, MA (May, 1970).

A3. S. Aisenberg and R. Chabot, "Ion Beam Deposition of Thin Films of Diamond-like Carbon," Invited paper presented to the New England Chapter, Thin Film Division, American Vacuum Society (December, 1970).

A4. S. Aisenberg and R. Chabot, "Investigation of the Properties of Thin Insulating Films Depositied with an Ion Beam System," Final Report #AFCRL-TR-73-0176 for Air Force Cambridge Research Laboratories under Contract #F19628-72-C-0291 (January, 1973).

A5. S. Aisenberg, "Ion Beam Deposited Carbon Films," Invited Seminar presented at Pennsylvania State University, Department of Material Sciences (May, 1971).

A6. S. Aisenberg and R. Chabot, "Deposition of Carbon Films with Diamond-Properties," Proceedings of the 10th Biennial Conference on Carbon. Lehigh University (July, 1971).

A7. S. Aisenberg and R. Chabot, "Versatile Coating Systems for Ion Beam Deposition, Ion Plating, Sputter Deposition and Sputter Etching," Proceedings of the 19th National Symposium of the American Vacuum Society (October, 1972).

A8. S. Aisenberg and R. W. Chabot, "Physics of Ion Plating and Ion Beam Deposition," Proceedings of the 19th National Symposium of the American Vacuum Society (October, 1972).

A9. S. Aisenberg and R. W. Chabot, "Ion Beam Deposited Carbon Coatings for Bio-Compatible Materials," Final Report under Contract NIH-73-2919 for Division of Blood Diseases and Resources, National Heart and Lung Institute, NIH (December, 1973).

A10. S. Aisenberg and R. W. Chabot, "Ion Beam Deposited Carbon Coatings for Biocompatible Materials," Comprehensive Report prepared for National Heart and Lung Institute, NIH on Contract #NIH-N01-HB-3-2919, Report No. SSD-P-71111CR (November, 1974).

A11. R. W. Chabot and S. Aisenberg, "Blood Compatibility of Ion Beam Deposited Carbon Coatings," Proceedings of the 28th Annual Conference on Engineering in Medicine and Biology (1975).

A12. S. Aisenberg and M. Stein, "Continued Studies of Ion Beam Deposited Carbon as a Blood Compatible Material," Annual Report, SSD-P-836A for Contract NIH-N01-HB-3-2919 National Institutes of Health (1977).

A13. M. Stein and S. Aisenberg, "Evaluation of Ion Deposited Carbon Films," Extended Abstracts of the 14th Carbon Conference (1979).

A14. S. Aisenberg and M. Stein, "The Use of Ion Beam Deposited Diamond-like Carbon for Improved Optical Elements for High Powered Lasers," Proceedings of the 12th Annual Symposium on Optical Materials for High Power Lasers (1980).

A15. S. Aisenberg and M. Stein, "The Moisture Protection of Strong Optical Fibers," Interim Report, RADC-TR-80-252, Rome Air Development Center (1980).

A16. S. Aisenberg, M. Stein, J. Stevens, and B. Bendow, "Ion Deposited Hermetic Coatings for Optical Fibers," Presented at the Fiber Optic Sensor Systems Workshop (May 1981).

A17. M. Stein, S. Aisenberg, and J. Stevens, "Ion-Plasma Deposition of Carbon-Indium Hermetic Coatings for Optical Fibers," Proceedings of the Conference on Lasers and Electro-Optics (June 1981).

A18. M. Stein, S. Aisenberg, B. Bendow, and BDM Corporation: "Studies of Diamond-like Carbon Coatings for Protection of Optical Components," Proceedings of the 13th Annual Symposium on Optical Materials for High Power Lasers (November 1981).

A19. S. Aisenberg and M. Stein, "The Moisture Protection of Optical Fibers," Final Report, Contract #F19628-78-C-0180, RADC/ESM, Hanscom AFB (December 1983).

A20. S. Aisenberg, "Technology of Diamond-like Carbon and its Applications," Invited lecture presented at a meeting of the American Association for Crystal Growth, Mid Atlantic Section, NJ (October 22, 1987).

Materials Science Forum Vols. 52 & 53 (1989) pp. 41-70

PLASMA DEPOSITION, PROPERTIES AND STRUCTURE OF AMORPHOUS HYDROGENATED CARBON FILMS

P. Koidl, Ch. Wild, B. Dischler, J. Wagner and M. Ramsteiner
Fraunhofer-Institut für Angewandte Festkörperphysik
Eckerstrasse 4, D-7800 Freiburg, FRG

ABSTRACT

Plasma deposition and properties of a-C:H are reviewed. The role of process gas, plasma chemistry and plasma surface interaction is discussed. It is shown that the deposition energy of the hydrocarbons is the most critical deposition parameter in determining the film properties. Hard a-C:H films with precursor-independent properties are obtained if the deposition energy is large enough to allow efficient fragmentation of the energetic hydrocarbons at the film surface. Otherwise, soft polymerlike films are obtained at low deposition energies. As a key for the understanding of the a-C:H properties, hydrogen incorporation, carbon bonding and network structure - as studied by IR and Raman spectroscopy - are discussed.

I. INTRODUCTION

Amorphous hydrogenated carbon films (a-C:H) have attracted much interest. The properties of these non-crystalline hydrocarbons may cover a wide range. They are intermediate between the properties of diamond, graphite and hydrocarbon polymers. Under appropriate deposition conditions, very hard, chemically inert and optically transparent films can be prepared. It is this combination of diamondlike characteristics that has stimulated much of the work on a-C:H. Deposition and properties of a-C:H have been reviewed previously [1-4].

Here we will discuss plasma deposition of a-C:H with emphasis on the energy dependent non-equilibrium processes at the growth surface. It will be shown how the film properties can be varied by changing the deposition energy. In addition, chemical composition, bonding and network structure will be discussed as the basis for the understanding of this unusual class of carbonaceous material.

II. PLASMA DEPOSITION OF a-C:H

II.1 PLASMA DEPOSITION TECHNIQUE

For the deposition of "diamondlike" a-C:H films several techniques are appropriate [1]. Common preparation techniques include deposition from hydrocarbon plasmas, direct ion-beam deposition and carbon sputtering within a hydrogen atmosphere. A common feature of these techniques is the exposure of the growing film to a flux of energetic (several 100 eV) ions. Due to this ion bombardment the growth of a-C:H occurs under nonequilibrium conditions - a prerequisite for the formation of a-C:H films with diamondlike properties.

In the following we will concentrate on the deposition from rf excited hydrocarbon plasmas onto negatively self-biased substrates - a widely used technique which was pioneered by Holland and coworkers [5]. Rf self biasing is preferrable over dc biasing since insulating a-C:H films have to be deposited, sometimes on insulating substrates. The film properties can be varied as a function of the deposition parameters [7]. In addition, the homogeneous coating of large substrates is possible with deposition rates up to about 50 Å/s.

A schematic picture of the apparatus used for plasma deposition of a-C:H is shown in figure 1 [7]. It is basically identical to a rf-diode sputtering system. The substrate is placed on a 4 in. diam. electrode (cathode) which is capacitively connected to a 13.56 MHz rf-generator. The large counter electrode is formed by the grounded walls of the plasma chamber.

Due to the different mobilities of ions and electrons in the plasma, a high potential drop is formed across the cathode dark space. The time averaged value of this sheath potential is given by the sum of negative self bias voltage of the cathode and the plasma potential $U_S = U_B + U_P$. In an asymmetric system with small capacitively coupled cathode and large grounded anode the plasma potential reduces to a small fraction of the rf-amplitude. Thus, the bias voltage is a good approximate measure of the sheath potential [7]. It can be measured by means of an inductively coupled voltmeter. Figure 2 shows the

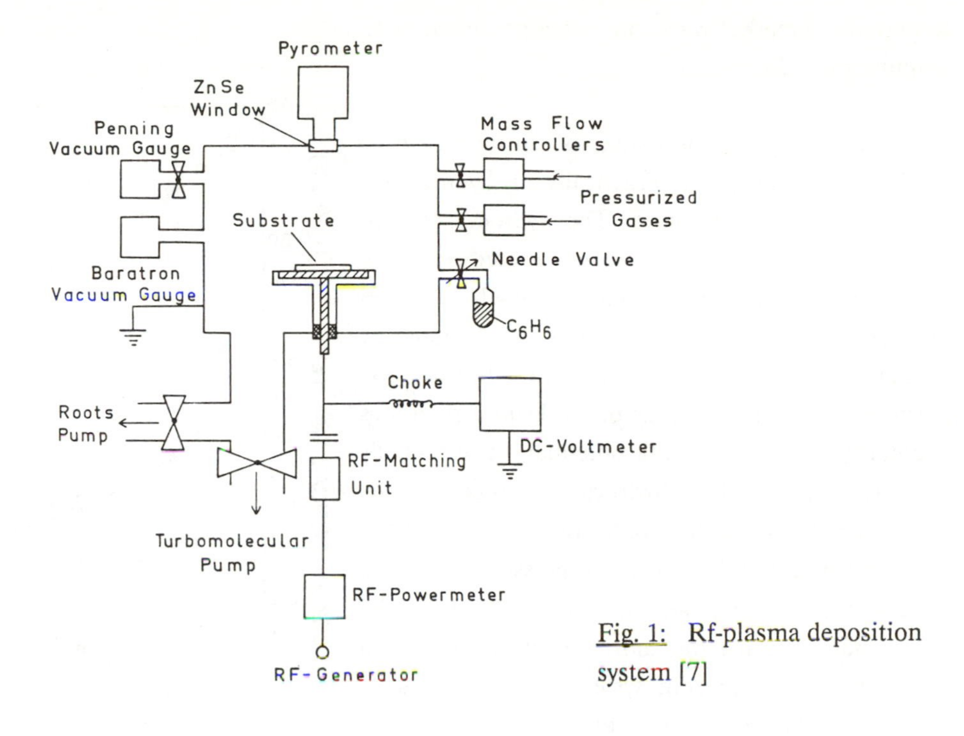

Fig. 1: Rf-plasma deposition system [7]

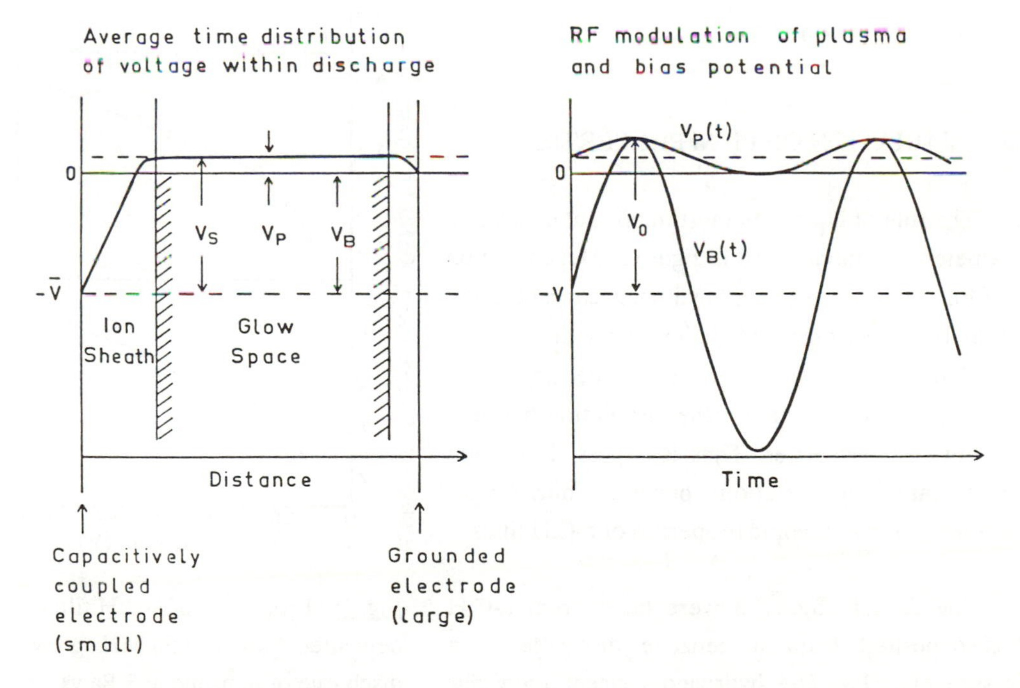

Fig. 2: Spatial and temporal potential distribution in a capacitively coupled rf-discharge system with asymmetric electrode area [7]

spatial and temporal variation of the potentials in an asymmetric rf discharge.

The rf glow discharge is sustained in a hydrocarbon atmosphere. Practically any hydrocarbon gas or vapour can be used. The hydrocarbon ions generated in the glow space are accelerated across the ion sheath towards the cathode. Growth of a-C:H proceeds at the cathode-mounted substrate by condensation of these energetic hydrocarbon particles. The impact energy of the hydrocarbon ions is determined by the sheath potential, i.e. essentially the bias voltage, and the hydrocarbon pressure which determines the mean free path. In the course of the present work, the hydrocarbon pressure was kept close to the optimum value of about 3 Pa. Thus, the bias voltage became the most important deposition parameter. In the present system U_B can be varied between approximately 50 V and more than 1000 V by variation of the rf power.

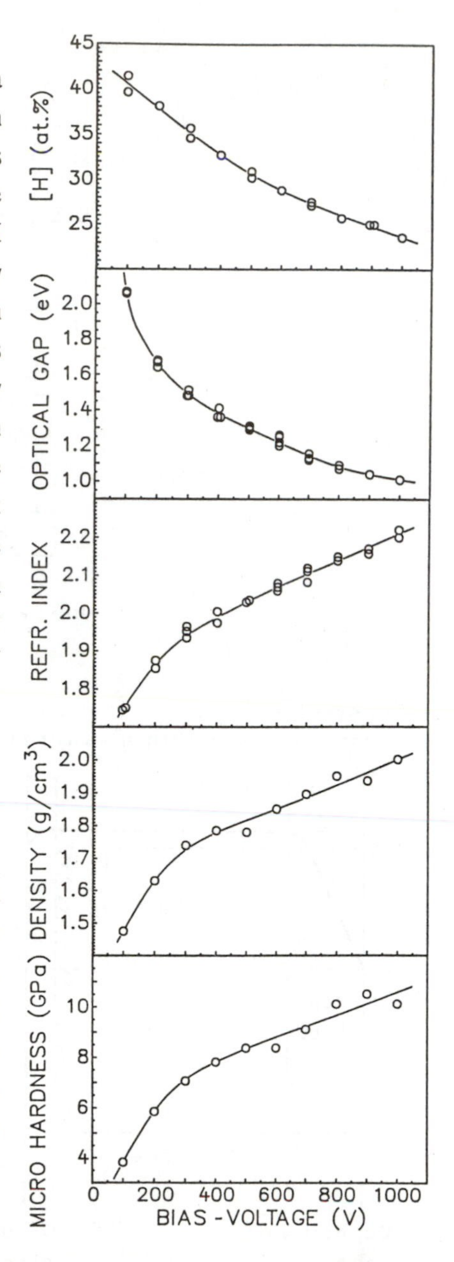

Fig. 3: Properties of a-C:H films deposited from a 13.56 MHz glow discharge of benzene at 3 Pa vs. substrate bias voltage

II.2 VARIATION OF FILM PROPERTIES

The role of U_B as the most important deposition parameter is demonstrated in figure 3. Depending on U_B film properties such as hydrogen concentration, optical bandgap, refractive index, mass density and micro hardness can be varied over a wide range. This variability allows to analyze the correlation between different film properties. Thus it is possible to gain insight into the relation between microscopic structure and macroscopic properties of a-C:H films.

The data in figure 3 were taken from a-C:H films deposited from a benzene discharge at a pressure of 3 Pa. The hydrogen concentration was determined by nuclear reaction analysis using the

$^{1}H(^{15}N,\alpha\gamma)^{12}C$ reaction [6]. Values between 24 and 41.5 at.% were obtained corresponding to H/C-ratios between 31.5 and 71 %. The concentration of hydrogen as a bond terminating atom plays a central role for the properties of a-C:H since it has direct influence on the degree of cross-linking in the amorphous network. Some of the variations in figure 3 can be explained by the change in hydrogen concentration. The increase of mass density with increasing U_B, e.g., is mainly due to the replacement of hydrogen atoms by carbon keeping the number density of atoms constant [1]. The change in mass density is also reflected in the variation of the refractive index. The optical bandgap E_{opt} in figure 3 (i.e. the "Tauc"-gap [1]) decreases from more than 2 eV to about 1 eV as a function of U_B. According to Robertson [2] E_{opt} is determined by the size of π bonded clusters of fused aromatic rings embedded in the amorphous network. A decrease of E_{opt} should be due to an increase of cluster sizes. Since hydrogen atoms can only be attached to the outer carbon atoms of the π bonded clusters, it appears reasonable, that the cluster size decreases with increasing H/C-ratio. The film hardness has been determined using a nano indenter by evaluating the dependence of indentation depth vs. load [8]. The hardness appears also to be correlated with the hydrogen content and thus with the degree of three dimensional cross-linking of the a-C:H network.

II.3 PROCESS CHARACTERIZATION

II.3.1 EXPERIMENTAL

In order to get insight into the elementary processes and mechanisms leading to the formation of a-C:H films or having influence on their properties, we investigated the plasma deposition of a-C:H by a combination of techniques including mass spectroscopy, optical emission spectroscopy, probe measurement, and reflection interferometry [9-12].

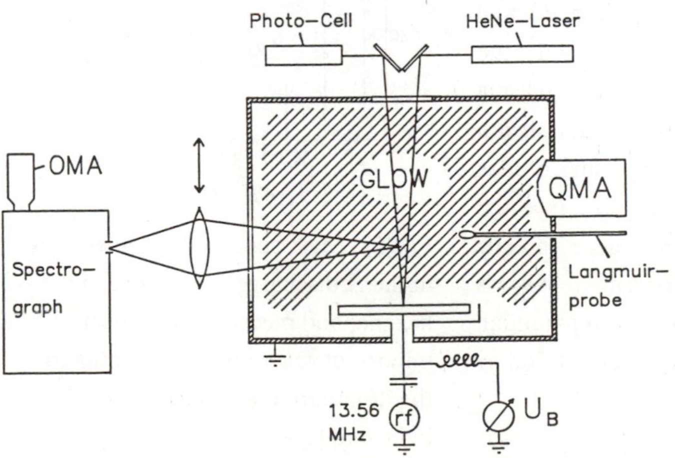

Fig. 4: Experimental set up for characterization of the plasma deposition process including quadrupole mass analyzer (QMA), optical emission spectrometer with spectrograph and optical multichannel analyzer (OMA), Langmuir probe and reflection interferometer for film growth monitoring

Figure 4 shows the experimental set up used for process characterization. A differentially pumped quadrupole mass analyzer (QMA) was attached to the reactor chamber. It was operated without electron beam ionization to measure directly the mass distribution of positive ions in the plasma. Optical emission spectroscopy was performed using a 0,5 m monochromator and an optical multichannel detection system. Spatially resolved spectra were taken by scanning the focussing lens in front of the quartz window parallel to the axis of the plasma column [9-11]. Plasma parameters were determined by Langmuir probe measurements. Errors due to the condensation of an insulating layer on the probe were avoided by electrically heating the probe to about 900°C [11]. The thickness and optical properties of the growing film were monitored by measuring in situ the reflection of a HeNe-laser beam.

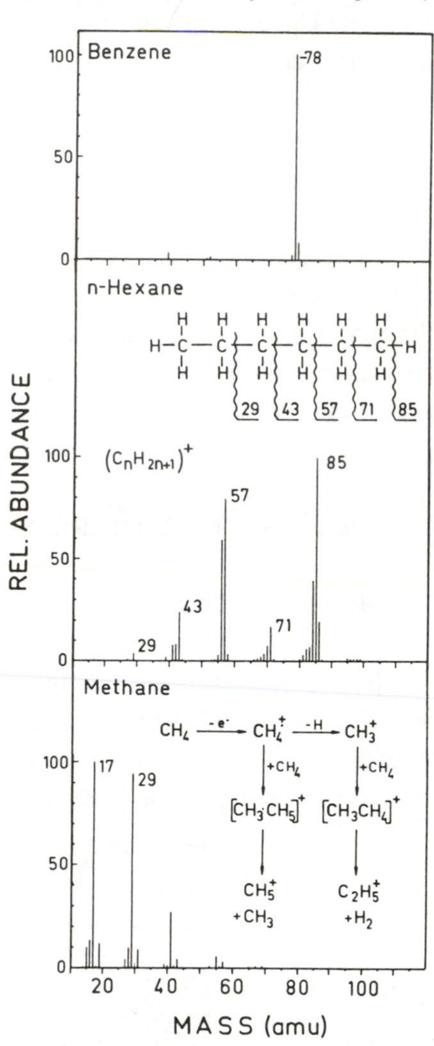

Fig. 5: Mass spectra of positive ions in the rf discharge in benzene, n-hexane, and methane. The most important reaction paths leading to the dominant mass peaks are indicated [11]

II.3.2 PLASMA CHEMISTRY

By evaluating the measured current voltage characteristic of the heated Langmuir probe the plasma density, electron temperature, and the plasma potential were obtained [11]. Typical values for a benzene discharge (bias voltage 500 V, pressure 3 Pa) were $N_e = 7.5 \times 10^9$ cm^{-3} and $kT_e = 2.2$ eV. Plasma potentials between 20 and 60 V were determined depending on the contamination of the stainless steel plasma chamber with an insulating film. The sheath potential $U_S = U_B + U_P$ is given by the sum of bias voltage and plasma potential (figure 2). Since the bias voltage is much simpler and more directly determined than the plasma potential, U_B is usually taken as a measure for the sheath potential. Except for very low bias voltages, this is justified by the above numbers.

Typical mass spectra of positively charged ions in the rf discharge are shown in figure 5 for three different hydrocarbon process gases: benzene, n-hexane and methane. The spectra were recorded for identical deposition parameters. Benzene is rather stable and $C_6H_6^+$ is found as the dominant ion. All the other species are at least one order of magnitude less abundant. N-Hexane tends to break up into fragments due to the cleavage of C-C σ-bonds resulting in a relatively high concentration of $C_3H_x^+$, $C_4H_x^+$ and $C_5H_x^+$ besides $C_6H_x^+$. The ion mass spectrum of methane in contrast shows a high relative abundance of polymers. These are created by ion-molecule reactions in the glow space. Possible reaction paths are indicated in figure 5. Only little change is found in the relative intensities of the various ion species when the plasma parameters U_B and P are varied. Thus the ion mass distribution in the glow discharge depends essentially on the type of hydrocarbon gas.

II.3.3 PLASMA-SURFACE INTERACTION

To discriminate between processes occurring in the glow space and those taking place at the growth surface, spatially resolved optical emission spectroscopy was used [9-11]. Typical emission spectra of a rf glow discharge in benzene and methane are shown in figure 6a and 6b, respectively. The spectra were recorded with the glow region focused on the entrance slit of the spectrometer. Common to both spectra is the emission from atomic and molecular hydrogen and from CH radicals. The spectrum of the benzene discharge shows in

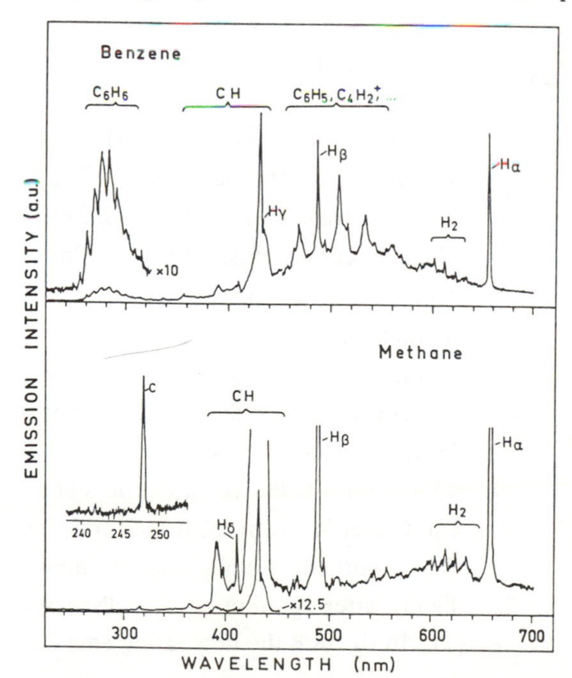

Fig. 6: Optical emission spectrum of rf glow discharge in benzene (a) and methane (b). The insert in (b) shows the emission of atomic carbon in the vicinity of the cathode [10].

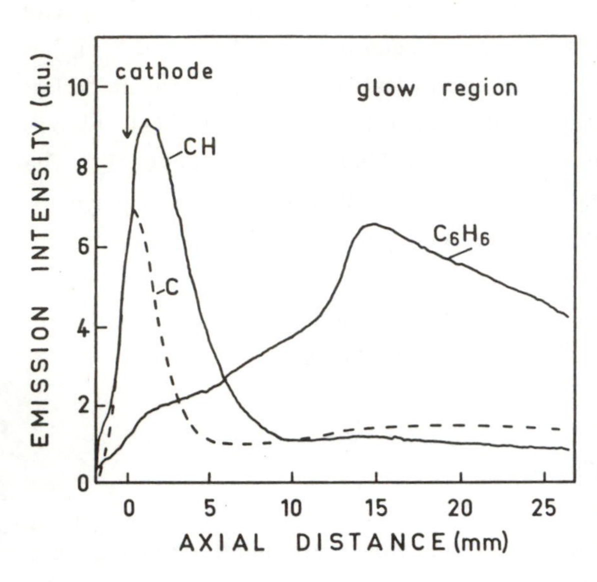

Fig. 7: Spatial variation of CH, C, and C_6H_6 optical emission intensity in a benzene discharge

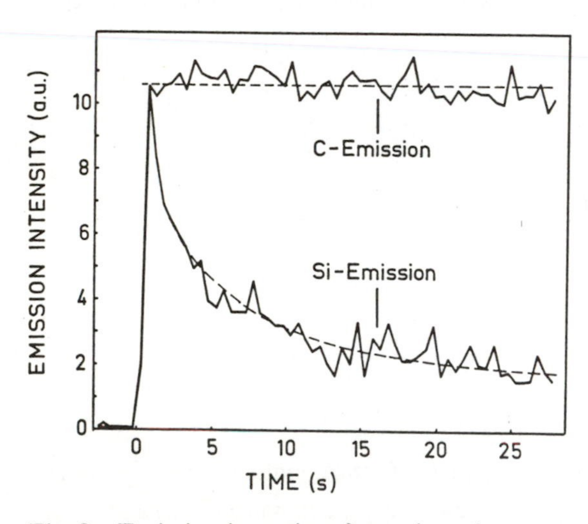

Fig. 8: Emission intensity of atomic carbon and silicon (sputtered off the substrate) during ignition of the hydrocarbon discharge vs. time [11]

addition emission from benzene (C_6H_6) and from benzene fragments such as C_6H_5 and $C_4H_2^+$ [9]. In the vicinity of the cathode (substrate) also emission from atomic carbon is observed (figure 6b). The intensities of the C, CH and C_6H_6 emission vary strongly along the plasma column. This spatial variation is plotted in figure 7. The C_6H_6 emission band shows increasing intensity with increasing distance from the cathode with a maximum just beyond the sheath. The CH and C emission intensities, in contrast, exhibit a pronounced maximum at the cathode (substrate) and fall off quite rapidly with increasing distance. The maximum in C and CH emission at the cathode is independent of the hydrocarbon process gas - e.g. C_6H_6, C_6H_{12}, C_6H_{14}. Even for methane, which shows strong CH emission in the glow region, a second maximum is found at the cathode.

Two processes might be responsible for the occurrence of these low mass fragments, namely, sputtering of the already deposited a-C:H film or impact-induced fragmentation of the energetic hydrocarbon particles impinging on the cathode. To clarify the origin of these fragments, the temporal dependence of the emission signals was studied for a-C:H deposition on silicon. The C and Si (from substrate) emission was recorded simultaneously immediately after ignition of the discharge [11]. In figure 8 the decreasing emission

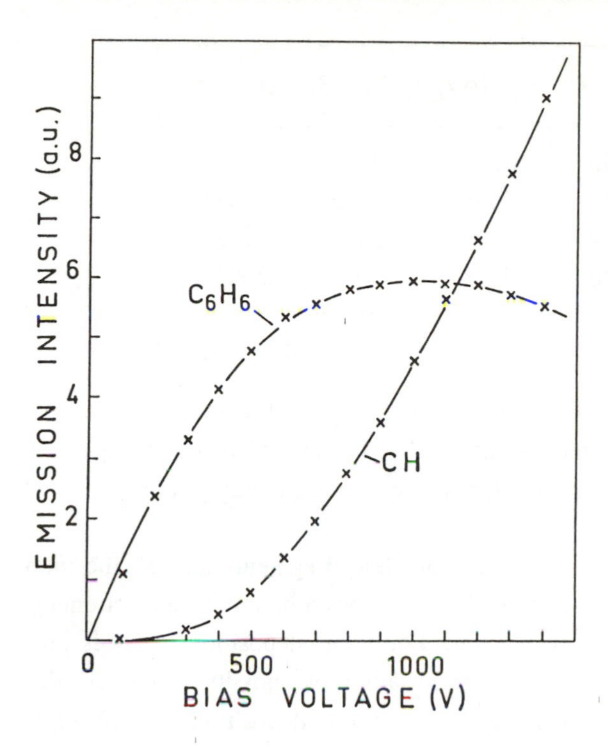

Fig. 9: Intensity of optical emission from C_6H_6 (in glow space) and CH (above cathode) vs. bias voltage [9]

from Si atoms sputtered off the substrate reflects the growth of an a-C:H layer on the Si substrate. The immediate and constant C signal clearly shows that impact-induced dissociation, and not sputtering, is responsible for the observed carbon atoms above the cathode. The observed CH radicals at the cathode were also shown to be dissociation products [10]. The dissociation probabiliy at the cathode increases superlinearly with the impact energy. This is shown in figure 9, which illustrates the different excitation mechansism for the CH and C_6H_6 emission [9]. Backscattering of excited dissociation products is a well known feature in the deposition of energetic particles [13].

II.3.4 ROLE OF PROCESS GAS

The properties of various a-C:H films, deposited from different process gases at a common bias voltage of 400 V are collected in table 1. Surprisingly, the film properties are independent of the type of process gas. Especially remarkable are the constant values of hydrogen content and carbon bonding ratio, for films deposited e.g. from benzene (H/C=1, sp^2-hybridization) or methane (H/C=4, sp^3-hybridization).

This structural "lost memory" effect in hard a-C:H is most naturally explained by the efficient fragmentation of the energetic hydrocarbons at the growth surface. The small fragments, such as C, C_2 or CH may be regarded as the actual precursors for a-C:H growth. Plasma chemistry in the glow space which leads to strongly gas-dependent ion mass distributions is not important for the properties of hard a-C:H.

Process Gas	n	E_{opt}(eV)	[H] (at.%)	sp^3:sp^2
Benzene	2.00	1.39	31.9	71:29
Cyclo-Hexane	2.02	1.44	31.0	72:28
n-Hexane	2.04	1.41	31.7	72:28
Methane	2.08	1.36	30.3	72:28

$U_B = 400$ V, $P = 3$ Pa

Table 1: Dependence of a-C:H film properties on type of hydrocarbon process gas. n denotes the refractive index, E_{opt} the optical bandgap, [H] the hydrogen content, and sp^3/sp^2 the sp^3/sp^2 carbon bonding ratio in the films

For complete fragmentation of the film forming hydrocarbons a minimum impact energy is required. For lower deposition energies only partial dissociation is possible. Under this condition, i.e. in the deposition of soft polymerlike films, larger units of the process gas molecules can be incorporated in the film. The properties of polymerlike films therefore generally do depend on the type of process gas. This is illustrated in figure 10 where the infrared absorption due to CH vibrations is shown for a-C:H films deposited at 400 and 100 V bias from benzene and n-hexane. It will be shown later that the form of the CH absorption band reflects the bonding situation in the hydrocarbon network. While the CH bands of the high bias films are identical, large differences are seen in the 100 V bias films deposited from n-hexane or benzene.

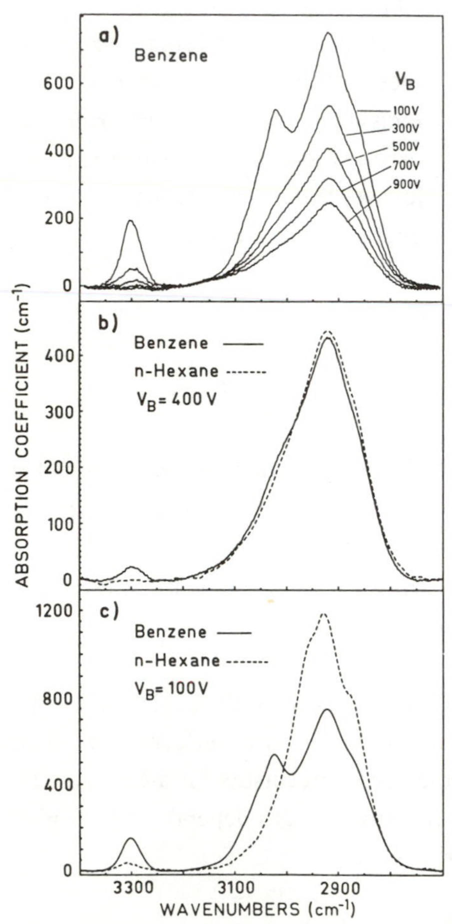

Fig. 10: CH stretch vibrational absorption spectra of various a-C:H films deposited from
(a) benzene at bias voltages as indicated,
(b) benzene and n-hexane at 400 V bias,
(c) benzene and n-hexane at 100 V bias.
The hydrocarbon pressure was 3 Pa in all cases [12]

While the properties of hard a-C:H are independent of the type of process gas, the deposition rate depends significantly on the process gas. Generally an increase in deposition rate has been observed with increasing hydrocarbon molecule mass [12, 14]. This has been shown to be a combined effect of the increase in carbon transport per ion and the decreasing ionization energy of the hydrocarbon with increasing mass [12].

Especially high deposition rates of up to ≈ 50 Å/s have been obtained with benzene. This advantage and the simple plasma composition with dominating $C_6H_6^+$ abundance [11] are the reasons why many studies on the a-C:H deposition / property relations have been made with benzene.

The growth of a-C:H normally proceeds with a constant deposition rate. However, delayed deposition has been reported on substrates that are easily sputtered by carbon [15] or those which form carbides by dissolving carbon [16]. It is therefore desirable to monitor the film growth during deposition. This is easily achieved by measuring the reflection e.g. of a HeNe laser off the growing film [11]. By analyzing the time dependent reflectivity film thickness and in addition the refractive index and optical gap are obtainable in an on-line process [11].

II.4 DEPOSITION ENERGY

In an rf discharge system the impact energy of the film forming hydrocarbon ions is determined by the bias voltage and the inelastic collisions in the ion sheath. Macroscopically the average impact energy has been controlled by the bias voltage and pressure. However, there are other parameters like rf-frequency, reactor geometry, and type of process gas that influence the impact energy. Wild and Koidl [17] have shown that the ion transport through an rf-modulated sheath depends on two parameters: the ratio of the sheath thickness over the mean free path $\alpha = d/\lambda$ and a plasma scaling parameter $\eta = eU_B/m\omega^2d^2$, where m is the ion mass, ω the rf-frequency and d the thickness of the electrode ion sheath.

In order to determine the impact energies of the film forming hydrocarbons, we measured the energy flux onto the cathode by analyzing the temperature rise of the cathode during deposition. For quantitative results the thermal properties of the cathode, i.e. thermal capacity and heat conduction to the environment, were determined by electrically heating the cathode using a heating resistor. The temperature was measured by an optical pyrometer. Figure 11 displays the deposited power during a-C:H growth in a benzene discharge (P = 2.25 Pa) as a function of U_B^2. Since both the average impact energy *and* the deposition rate are proportional to U_B, a linear dependence of the deposited power on U_B^2 is expected. This is indeed what we find in figure 11. Assuming a sticking coefficient of one, an average impact energy of $E=0.39eU_B$ can be derived from the slope in figure 11. For lower sticking coefficients the above factor of 0.39 has to be reduced accordingly. Thus for the investigated conditions the average energy of the impinging hydrocarbons is considerably smaller than U_B. Their energy is reduced by inelastic collisions during transit through the sheath. Thereby charge exchange is known to be the most important interaction process especially in cases, where ions are accelerated through a background of neutrals of the same species (symmetric charge exchange).

In addition to the calorimetric measurements the energy of ions was measured directly by placing an ion energy analyzer behind the cathode into a differentially pumped vacuum chamber, connected via a 200 μm diam. orifice with the plasma chamber. Experimental details are given in reference 17. Figure 12 shows as an example the ion energy distribution at the cathode of a benzene discharge ($U_B=500$ V, $P=2.25$ Pa). It can be seen, that the distribution of ion energies covers the whole range from 0 to eU_B. The average ion energy was found to be $0.38eU_B$, which is in good agreement with the calorimetric results and shows that the sticking coefficient must be close to unity. The peaks in the ion energy distribution were shown to be due to the interaction of the ions with the rf modulated electric field combined with the creation of thermal ions in the cathode sheath by charge exchange.

From the data presented in figure 3 and 10a it can be inferred, that the transition towards "polymerlike" a-C:H (soft, low mass density, low refractive index, high E_{opt}) takes place at bias voltages between 100 and 200 V. This corresponds to average impact energies in the range 40 - 80 eV, as expected by considering the molecular bonding energies.

In conclusion it was found that the impact energy of the film forming hydrocarbons is the most important deposition parameter. By a variation of impact energies the properties of a-C:H films can be varied over a wide range. Ion bombardment affects primarily the hydrogen content. Increasing impact energy reduces the hydrogen incorporation - presumably by preferential re-sputtering. As a result, the degree of three-dimensional cross-linking in the hydrocarbon network increases leading to harder and denser a-C:H films. For high deposition energies ($\gtrsim$ 100 eV) efficient fragmentation of the hydrocarbon molecules occurs upon impact. Due to this fragmentation hard, strongly cross-linked a-C:H films with precursor independent properties are obtained. In contrast, at low impact energies ($\lesssim$ 50 eV) insufficient fragmentation of the film forming hydrocarbons leads to hydrogen-rich, polymerlike a-C:H films. The structure and bonding of these polymerlike films are significantly influenced by the type of precursor gas used for deposition.

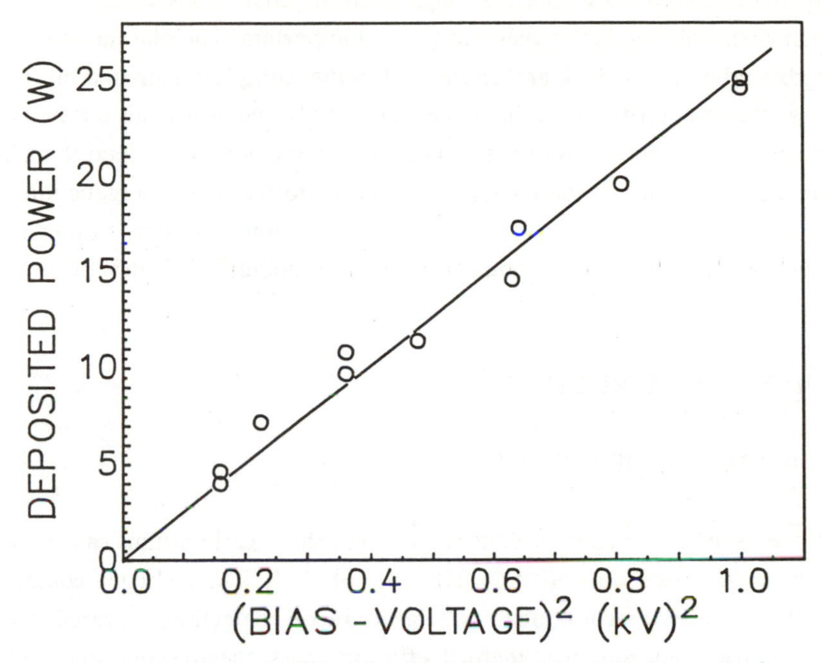

Fig. 11: Calorimetric determination of power deposited to cathode vs. square of bias voltage (benzene discharge, P=2.25 Pa)

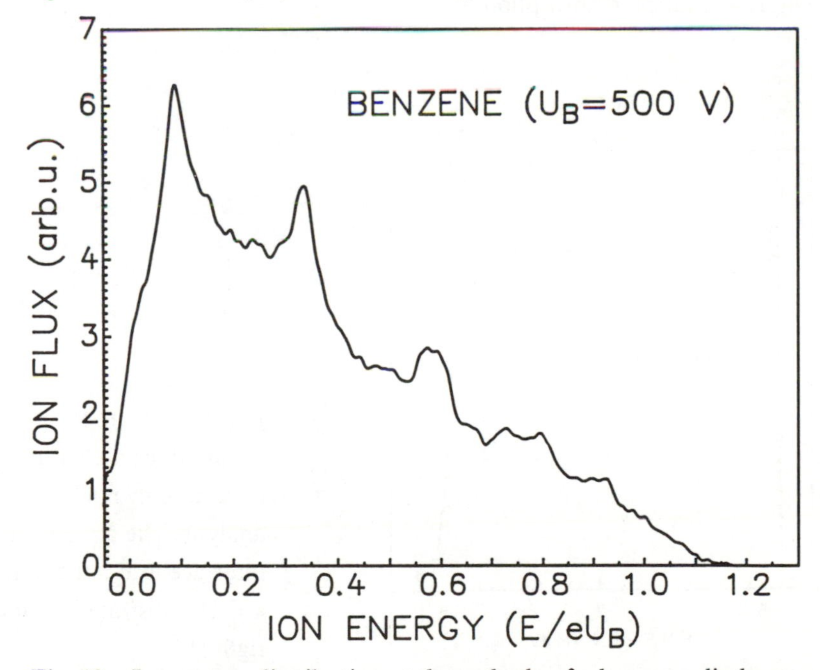

Fig. 12: Ion energy distribution at the cathode of a benzene discharge (U_B=500 V, P=2.25 Pa)

For the plasma deposition of hard, strongly cross-linked a-C:H films the type of process gas or plasma chemical reactions turned out to be unimportant. The plasma simply acts as a source of hydrocarbon ions, which are accelerated in the cathode sheath and hit the surface with an energy that is determined by the bias voltage and by inelastic collisions in the sheath. In this picture the rf plasma deposition is equivalent to direct ion-beam deposition. The only difference is the distribution of ion energies. In contrast to the monoenergetic ions from an ion source, the ions in an rf plasma show a broad distribution with some distinct structures due to the interaction of the hydrocarbon ions with the rf modulated electric field.

III. BONDING AND STRUCTURE

III.1 HYDROGEN INCORPORATION

It has been shown in numerous studies that amorphous carbon films can contain large amounts of hydrogen, exceeding 50 at % (H/C = 1.0) [1, 4]. The hydrogen content can be determined by nuclear reaction analysis, elastic recoil detection, infrared absorption spectroscopy, NMR spin counting, thermal effusion mass spectroscopy and combustion analysis [1, 4]. We will describe the two most widely used techniques, namely nuclear reaction analysis and infrared absorption.

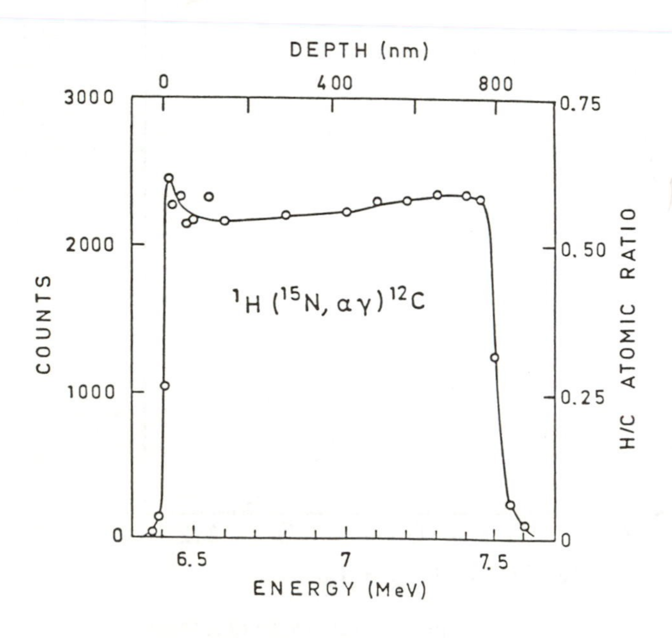

Fig. 13: Hydrogen profile in a 0.8 μm thick a-C:H film measured by nuclear reaction analysis. The air/a-C:H interface is at left and the a-C:H/substrate interface at right [18]

In the nuclear reaction experiment ^{15}N ions hit the sample and the reaction $^{1}H(^{15}N,\alpha\gamma)^{12}C$ which exhibits a resonance in the cross section at 6.4 MeV allows to determine the hydrogen content by counting the emitted 4.4 MeV γ radiation [18]. The sharpness of the resonance allows high resolution depth profiling. The depth scale is obtained from the known energy loss of the ^{15}N ions per unit pathlength. Figure 13 shows the hydrogen profile of a typical hard a-C:H film prepared by plasma deposition from benzene [18]. Besides a surface absorption peak, the hydrogen concentration is constant over the film thickness of 800 nm. The measured hydrogen concentration of 35 at.% (H/C = 0.54) is very [illegible] absorption of the C-H stretch [illegible] the IR-method is a powerful tool to study relative concentrations, an absolute calibration is very difficult. Most other methods measure the total hydrogen content, whereas the IR-method measures only the chemically bonded hydrogen. It has been reported, that about 30 to 50 % of hydrogen is not directly bonded to carbon and thus not infrared active [1]. This "weakly" bonded hydrogen may be chemisorbed at internal surfaces [22] or exist in internal clusters [23].

With infrared spectroscopy not only the concentration of bonded hydrogen can be studied but also the details of bonding. From the fine structure of the CH vibrational bands (see below) it has been shown that hydrogen in hard a-C:H is mainly incorporated as monohydride [21] with a typical $CH:CH_2:CH_3$ ratio of 60:40:0. In contrast, the larger hydrogen content in soft, polymerlike films results in a higher relative concentration of di-hydride. Typical data as obtained by IR spectroscopy are given in table 2.

property	a-C:H	
	(hard)	(polymerlike)
$sp^3 : sp^2 : sp^1$	68 : 30 : 2	53 : 45 : 2
$CH_3 : CH_2 : CH$	0 : 40 : 60	25 : 60 : 15
optical gap (eV)	1.1	3.0
refractive index	2.1	1.65
density (g/cm^3)	1.9	1.3
[H] total (at.%)	27	50

Table 2: Bonding and hydrogen incorporation in hard and polymerlike a-C:H. Other relevant properties are included for comparison

III.2 BONDING IN a-C:H

It is evident that the type of carbon bonding is instrumental for the physical and chemical properties of the a-C:H films, especially the optical and mechanical properties [1, 4]. Carbon can form bonds with three different hybridisations, namely tetrahedral sp^3 (like in diamond), trigonal sp^2 (like in graphite) or linear sp^1 bonds (like in acetylene), all of which have been found in a-C:H. Quantitative data on carbon bonding in a-C:H have first been obtained by infrared spectroscopy [19, 24]. The importance of sp^3 and sp^2 bonding in a-C:H has been confirmed by electron energy loss spectroscopy (EELS) [25, 26] and later by nuclear magnetic resonance [27-29].

In figure 14 we show the IR absorption of an a-C:H film with bands from C-H stretching vibrations in the 3300 to 2850 cm^{-1} region and a superposition of C-C stretching and C-H bending modes in the 1620 to 700 cm^{-1} region. The second derivative spectrum in figure 14 reveals unresolved structure under the broad bands. An expanded view of the C-H stretch region is given in figure 15, also showing the deconvolution into individual bands. In the as grown sample we find 68 % sp^3 sites, 30 % sp^2 sites and 2 % sp^1 sites [19]. A complete list with the frequencies for C-H and C-D stretching and bending as well as C-C stretching vibrations was given previously [20, 21]. In table 2 the data for carbon and hydrogen bonding are compared for hard and polymerlike films.

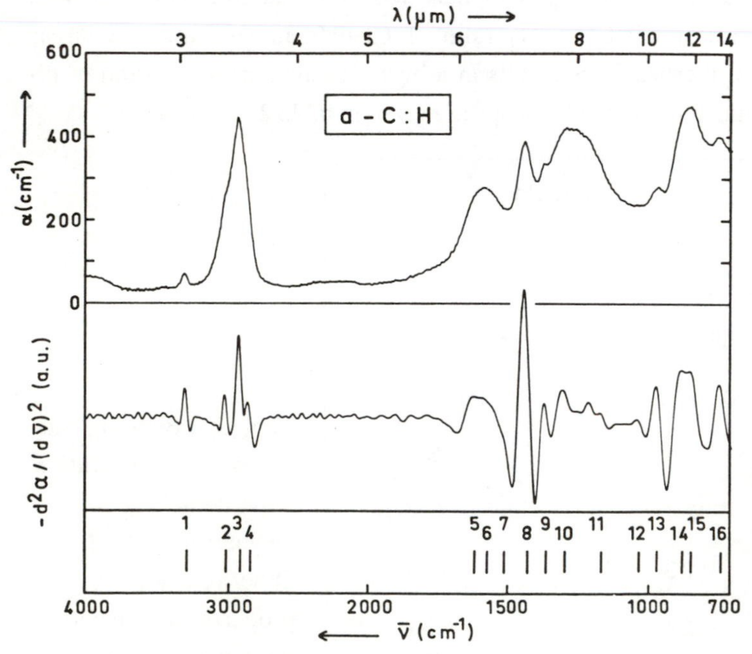

Fig. 14: Infrared absorption from an a-C:H film (above) and its second derivative (below) with numbering for individual lines (bottom) [24]

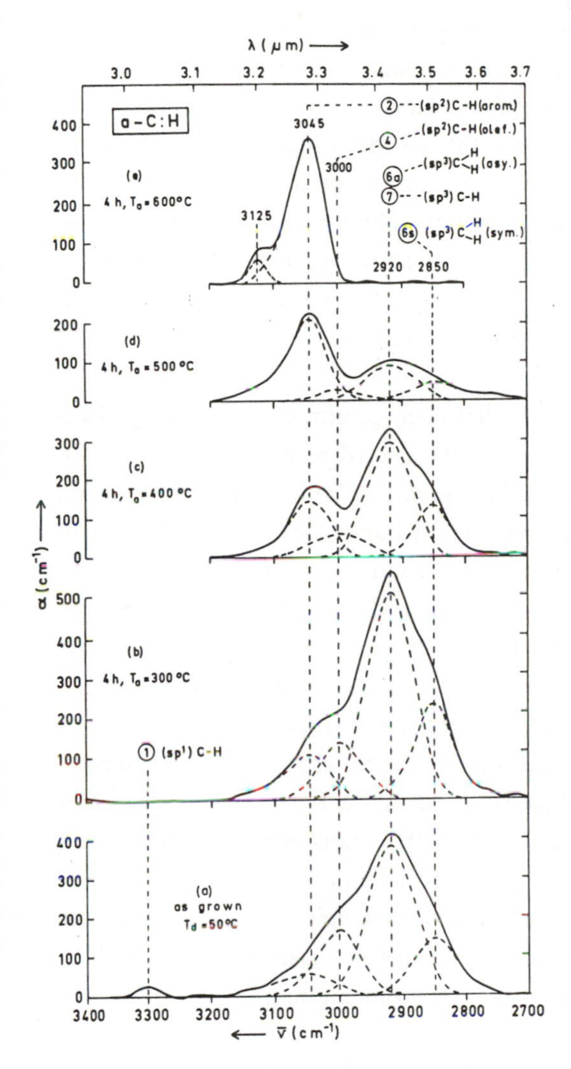

Fig. 15: C-H stretching absorption from an a-C:H film, showing the changes in bonding induced by thermal treatment. Spectrum (a) is for the as grown film, deposited at a substrate temperature of 50°C. Spectra in (b) to (e) were obtained after 4h anneal in vacuum at temperatures of 300°, 400°, 500° and 600°C, respectively [19]

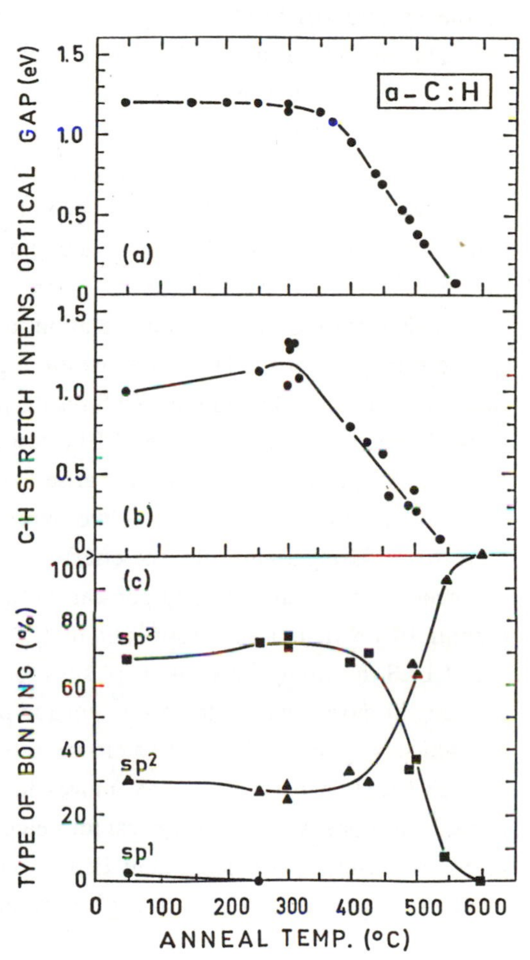

Fig. 16: Optical gap (a), integrated C-H stretch intensity (b) and percentage of sp^3, sp^2 and sp^1 carbon sites (c) as a function of anneal temperature [21]

In figure 15 and figure 16 we show results of thermally induced changes in hard a-C:H films, observed after a 4 hour vacuum anneal at temperatures up to 600°C. The heat treatment transforms sp^3 sites and olefinic sp^2 sites into aromatic sp^2 sites [19]. This is shown in the spectra (b) to (e) of figure 15 and plotted in figure 16(c). The a-C:H films are thermally stable up to 300°C in agreement with the thermal desorption spectra described below. Above this temperature hydrogen is effused (figure 16(b)) and a drastic change in the bonding occurs (figure 16(c)). Recent models predict [2], that the formation of polyaromatic clusters causes a decrease of the optical gap (figure 16(a)). The slight increase of C-H stretch intensity at 300°C (figure 16(b)) has been observed for several samples and may be due to a transformation of chemisorbed hydrogen into chemically bonded hydrogen.

This picture of the bonding situation in a-C:H has been confirmed by applying EELS on the same type of a-C:H films only differing in thickness. The IR results [19] are shown in figure 15 and figure 16 and the EELS results [25, 26] in figure 17. The dielectric loss function ε_2 is obtained from the measured EELS loss spectra via a Kramers-Kronig analysis [25]. In the ε_2 curves, two peaks below 8 eV arise from π - π^* interband transitions, while the broad peak above 8 eV arises from σ - σ^* interband transitions and collective excitation of all σ and π electrons. Using a sum rule, the effective number of π and σ electrons is obtained. Taking the average π electron density per carbon as a measure of the fraction of sp^2 hybridized carbon, EELS yields a sp^3 to sp^2 ratio of 2/3 to 1/3, in agreement with the optical data. Also with EELS the strong increase of sp^2 sites for anneal temperatures above 300°C is observed. Looking in more detail at the π - π^* portion of the curves in figure 17 reveals a peak near 6 eV, which is typical for localized π electrons (e.g. the $^1A_{1g}$ - $^1E_{1u}$ transition of benzene occurs at 7 eV) [25]. A peak near 4 eV moves to 3 eV and becomes dominant after the 540°C anneal. An ε_2 peak at 3 eV is typical for delocalized π electrons in polyaromatic molecules or graphitic clusters. An independent EELS confirmation for this delocalization was obtained by measuring the dispersion of the π excitations for variable momemtum transfer [25].

A very direct way to determine the sp^3:sp^2 carbon bonding ratio is ^{13}C nuclear magnetic resonance spectroscopy (NMR) [27-29]. However, in contrast to IR-spectroscopy, NMR is not applied as a routine technique, since large a-C:H sample masses ($\gtrsim$ 0.2 g) are needed. For a-C:H films prepared in different ways sp^3:sp^2 ratios between 80 : 20 [27] and 35 : 65 [29] were determined by NMR.

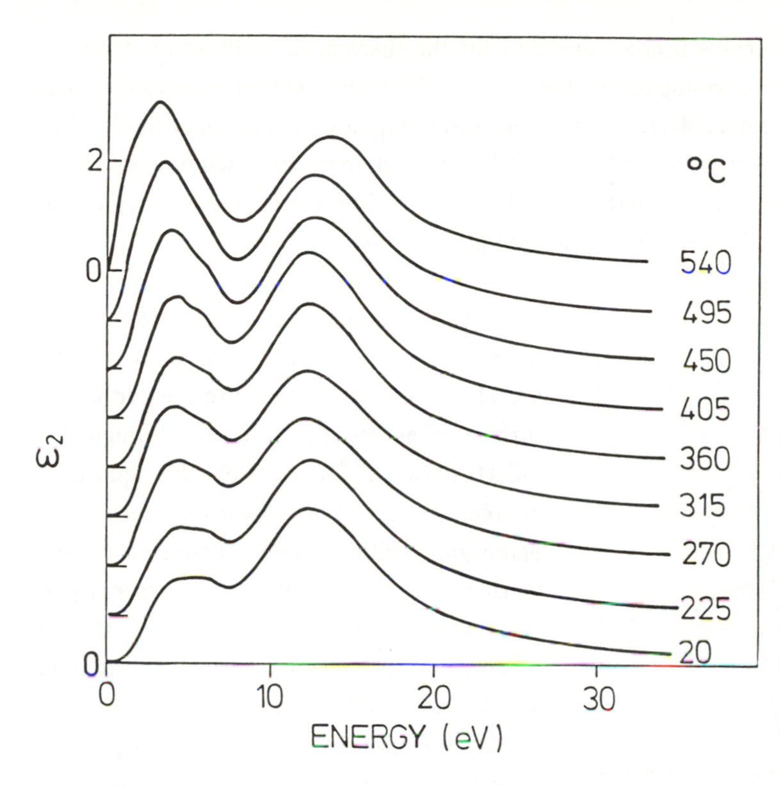

Fig. 17: Imaginary part of the dielectric function (ε_2 spectra) from EELS analysis of a-C:H films, as deposited and annealed at the temperatures as indicated [25]

III.3 NETWORK CROSS-LINKING AND THERMAL DECOMPOSITION

So far we discussed hydrogen incorporation and the local bonding situation. It has been shown that the film properties depend strongly on the deposition energy during growth. With increasing deposition energy, density and hardness increase; in addition, the content of hydrogen as a bond terminating atom decreases indicating a stronger cross-linking of the a-C:H network. In the following we will describe thermal desorption spectroscopy, which has proven useful in the study of the network structure and the thermal decomposition behaviour of a-C:H [30, 31]. Thermally induced gas evolution has been studied from a-C:H films deposited on silicon substrates as described in II.1. The benzene pressure was kept constant at 3 Pa and the bias voltage was varied between 50 V and 900 V.

For the effusion measurements, the samples were placed in an evacuated quartz tube which was heated up to 700°C with a slope of 20°C/min. The residual gas in the system was analyzed with a quadrupole mass spectrometer.

Figure 18 shows effusion spectra for a-C:H films deposited on Si with bias voltages between 50 and 900 V. The film volume was constant in all samples. It is seen that the threshold temperature for gas effusion - and thus for the thermal decomposition process - increases with increasing bias voltage from 300°C (U_B = 50 V) to 600°C (U_B = 900 V). The mass of the desorbed molecules decreases with increasing U_B. a-C:H films deposited at low bias ($\lesssim$ 300 V) release hydrogen, methane and higher hydrocarbons; medium bias films ($U_B \approx$ 400 V) desorb H_2 and CH_4, whereas from hard a-C:H films deposited at $U_B \gtrsim$ 500 V, only H_2 is observed on the linear intensity scale in figure 18.

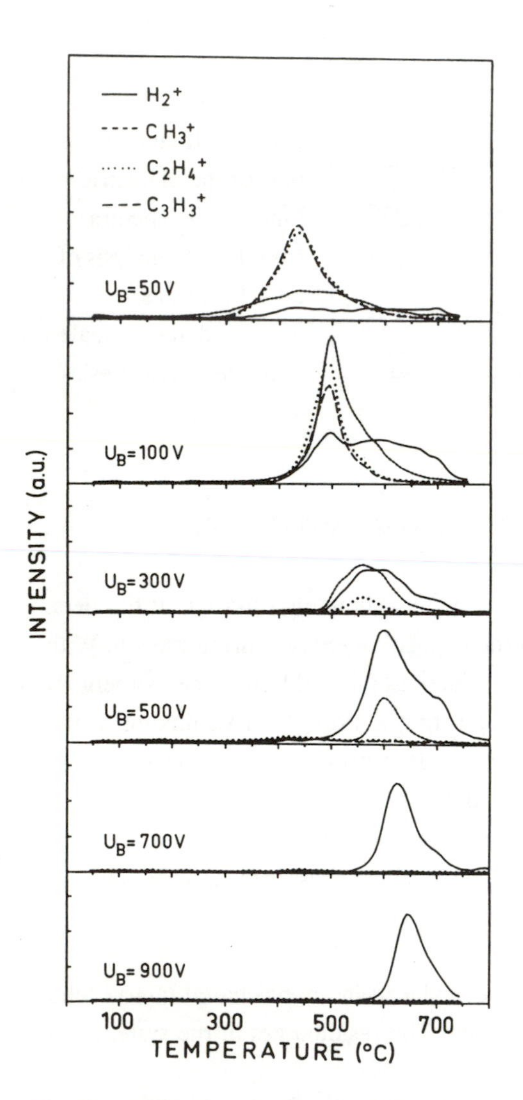

Fig. 18: Effusion spectra of a-C:H films deposited at different bias voltages [30]

The question arises wether the released molecules are formed in the bulk of the a-C:H film or wether they are created at the surface after atomic diffusion through the a-C:H network. To answer this question, double layers of hydrogenated and deuterated films were prepared using benzene or deutero-benzene as process gas, respectively. Figure 19 shows the effusion spectra of those double layers deposited at 400 V and 800 V bias. The individual sub-layers have an equal thickness of 0.4 μm. An atomic diffusion of hydrogen followed by surface recombination would imply a H_2:HD:D_2 ratio of 1:2:1. The missing HD intensity observed with the U_B = 400 V film clearly shows that molecules are formed in each individual layer followed by molecular transport to the surface. The slightly increased HD intensity observed from the 800 V film may be due to ion mixing at the a-C:H/a-C:D interface. Molecular migration through the a-C:H network has also been demonstrated for methane [31].

Hard a-C:H films deposited at high bias voltage release mainly hydrogen but practically no hydrocarbons. This might be due either to an inhibited detrapping process of hydrocarbons or to inhibited diffusion. To clarify this question, effusion from double layers deposited at different bias voltages was studied. Two silicon substrates were covered with 0.5 μm of a-C:H at 400 V. After removing one of the samples from the deposition chamber, a second a-C:H layer with 0.2 μm thickness was deposited at U_B = 600 V onto the remaining sample. Figure 20 shows the effusion spectra of hydrogen and methane, as monitored by the CH_3^+ signal, from both samples. The 400 V single layer releases hydrogen and methane, as expected from figure 18. The sample with the hard top layer, however, only desorbs hydrogen. This demonstrates that hydrocarbon diffusion is inhibited in high bias films by the reduced pore size of the strongly crosslinked hydrocarbon network.

The present data are consistently explained by a a-C:H network structure with U_B-dependent packing density. As the deposition energy increases, less hydrogen is bonded in the a-C:H structure. Consequently, the degree of cross-linking increases and the pore size in the amorphous material decreases. It is the limited pore size which prevents hydrocarbon diffusion in high bias films while the films remain permeable for hydrogen.

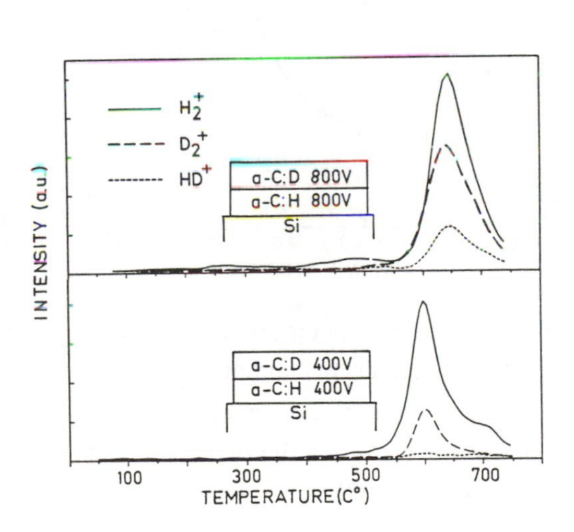

Fig. 19: Effusion spectra of a-C:D/a-C:H double layers, each 0.4 μm thick [30]

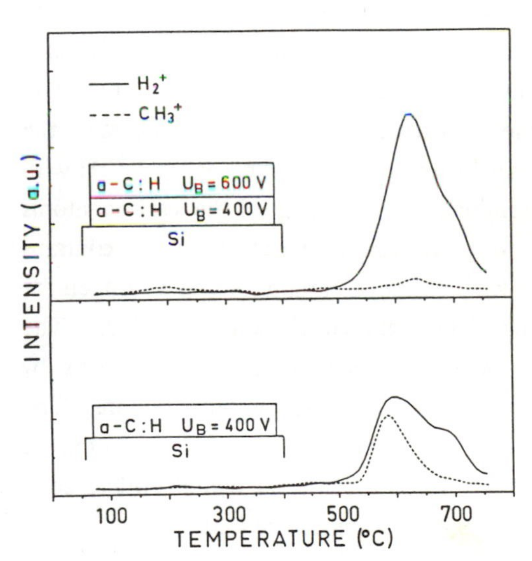

Fig. 20: Effusion spectra of (a) 0.47 μm a-C:H layer deposited at U_B=400 V; (b) equivalent sample with 0.2 μm a-C:H top layer deposited at U_B=600 V [30]

IV. STRUCTURAL PROPERTIES STUDIED BY RAMAN SPECTROSCOPY

IV.1 THERMALLY INDUCED CHANGES

It is well documented that Raman spectroscopy is a useful technique to monitor the thermally induced changes in the a-C:H structure, especially the build-up of graphitic medium range order during annealing [32]. As an example, figure 21 displays a series of Raman spectra of as deposited hard a-C:H and of films annealed at temperatures up to 600 °C. Upon annealing the peak at 1520 cm^{-1}, which is the dominant feature in as deposited a-C:H, transforms into a relatively sharp peak at 1600 cm^{-1} (G peak). An additional broader peak appears at ≈ 1350 cm^{-1} (D peak). This thermally induced change in the Raman spectrum reflects the transformation of hard and dense a-C:H with dominant sp^3 bonding of the carbon atoms into more graphitic carbon which contains only sp^2 bonded carbon and no hydrogen for the highest anneal temperature [20, 33]. The G peak has been assigned to scattering by graphitic optic zone center phonons. The feature labelled D has been attributed to scattering by disorder activated optical zone edge phonons of graphite [33].

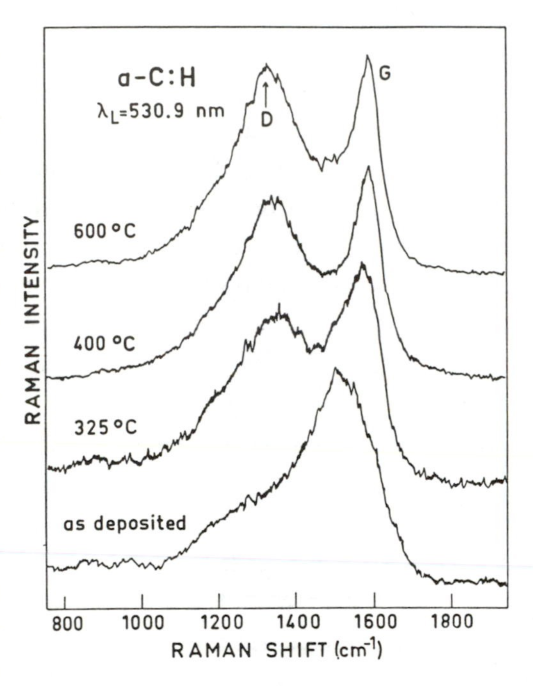

Fig. 21: Raman spectra of as-deposited a-C:H (bottom curve) and of a-C:H annealed for 4 h in vacuum at different temperatures given in the figure (middle and top curve) [32]

IV.2 INTERFACIAL BONDING

Among the many aspects of bonding in a-C:H, the bonding situation at the a-C:H/substrate interface is of special practical importance. It is well known that a-C:H films show strong adhesion to carbide forming substrates, like silicon. But it is only recently that quantitative data on interfacial bonding became available. Photoelectron spectroscopy [28, 34-36], Auger electron spectroscopy [28, 34-36], and secondary ion mass spectroscopy [34, 36] have been used to study the a-C:H/substrate interface. From this work evidence has been obtained for the formation of an interfacial carbide layer for a-C:H deposited on crystalline silicon [34-36] and germanium [35, 36]. Further, Raman spectroscopy has been used to investigate the a-C:H on Si and a-C:H on Ge interface. This was done by studying a series of a-C:H films deposited on these materials with decreasing film thicknesses down to about 10 Å [32, 37].

In figure 22a series of Raman spectra of a-C:H films deposited on crystalline silicon is displayed. The film thicknesses range from 550 Å (figure 22a) to 11 Å (figure 22d). The spectrum of the thickest film (figure 22a) shows the typical Raman spectrum of a-C:H with the main peak at ≈ 1560 cm^{-1}. In addition, scattering by the silicon substrate is observed at 520 cm^{-1} (O_Γ phonon) and ≈ 900 cm^{-1} (2TO phonons). The spectra displayed for the thinner films (figure 22b-d) were obtained by subtracting the Raman spectrum of bare silicon from that of the coated substrate to suppress the overlapping third-order Raman spectrum of the silicon substrate. With decreasing layer thickness the main a-C:H peak shifts to lower frequencies. This low frequency shift, which amounts to 70 cm^{-1} for the thinnest layer of ≈ 11 Å, was found to be independent of the exciting photon energy [38].

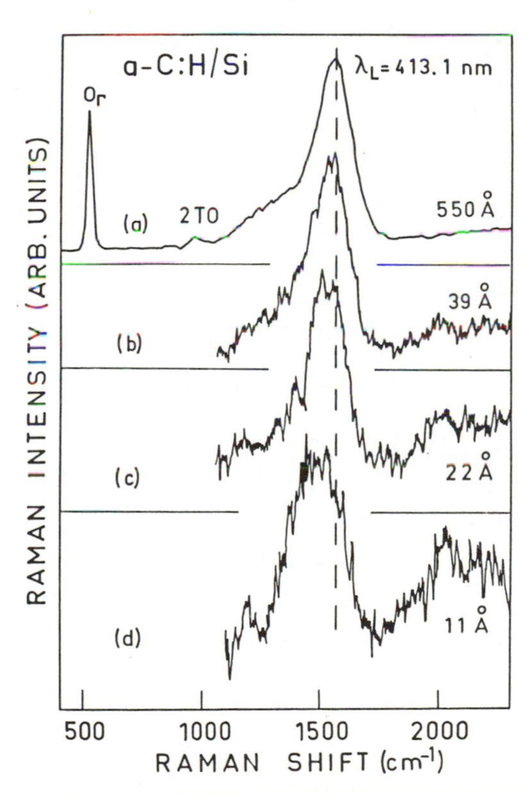

Fig. 22: Raman spectra of hard a-C:H deposited on crystalline silicon at various thicknesses given in the figure. The dashed line marks the frequency of the main Raman peak for thick (≈ 1 μm) films [37]

The low-frequency shift of the main a-C:H Raman peak relative to its position for thick (≈ 1 μm) films is plotted versus layer thickness in figure 23a for a-C:H on crystalline silicon and germanium. It is evident that this shift is substrate dependent. The maximum shift observed for a-C:H on Ge amounts to only ≈ 1/3 of that for a-C:H on Si. This indicates that this low-frequency shift reflects the properties of the a-C:H/substrate interface. As discussed above, photoelectron spectroscopy gave evidence for a carbide layer at the interface [34-36]. Therefore the vibrational properties of thick (≈ 1 μm) a-$Si_{1-x}C_x$:H and a-$Ge_{1-x}C_x$:H films have been studied by Raman spectroscopy [37]. In these films a frequency down-shift of the C-C vibrational band was observed with decreasing carbon content. As plotted in figure 23b this down-shift is linear in film composition and significantly larger for a-$Si_{1-x}C_x$:H than for a-$Ge_{1-x}C_x$:H. The magnitude of this shift is comparable to the frequency down-shifts found for thin a-C:H films on Si and Ge, respectively (see figure 23). From this comparison it was concluded that also Raman spectroscopy provides evidence for the formation of an amorphous carbide layer at the a-C:H substrate interface [37]. The average Si and Ge concentration is up to ≈ 40 % within the first ≈ 10 Å for a-C:H on Si and a-C:H on Ge, respectively. The thickness of the interfacial carbide layer can be estimated to ≈ 20 Å based on these Raman measurements. It compares well with the width of ≈ 40 Å estimated from photoelectron spectroscopy [34] and the value of 20 Å deduced from high resolution transmission electron microscopy [39].

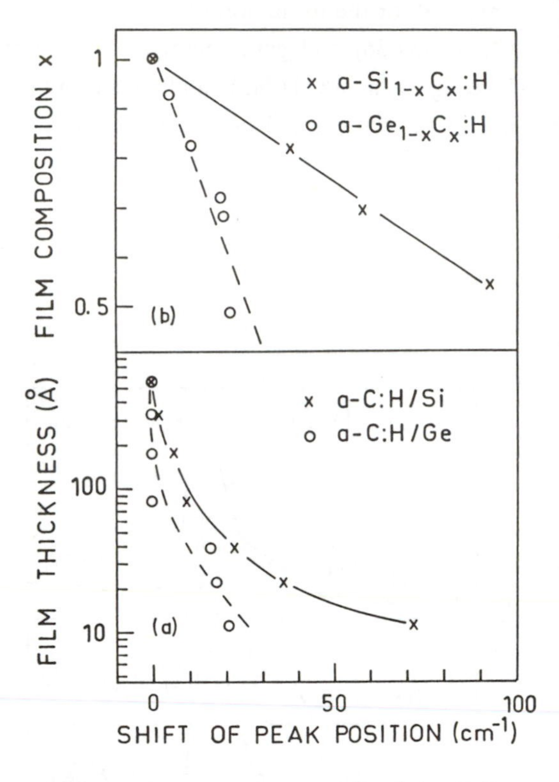

Fig. 23: Frequency down-shift of the C-C vibrational band relative to its position for thick (≈ 1 μm) a-C:H films (a) as a function of ilm thickness and (b) as a function of alloy composition. The full and the dashed curves are drawan to guide the eye [37]

IV.3 MEDIUM RANGE ORDER

As discussed by Bredas and Street [40, 41] and Robertson [2], sp^2 bonded carbon atoms tend to segregate in small graphitic clusters in a-C:H. Based on this finding it became possible to model the optical band gap of hard and dense a-C:H [2, 42], which is much lower than one would expect from the large fraction (≈ 2/3) [19, 25] of "diamond-like" bonded sp^3 carbon. According to Robertson [2, 42] in a-C:H with an optical band gap of 1.3 eV these graphitic clusters may contain up to about 20 fused sixfold rings.

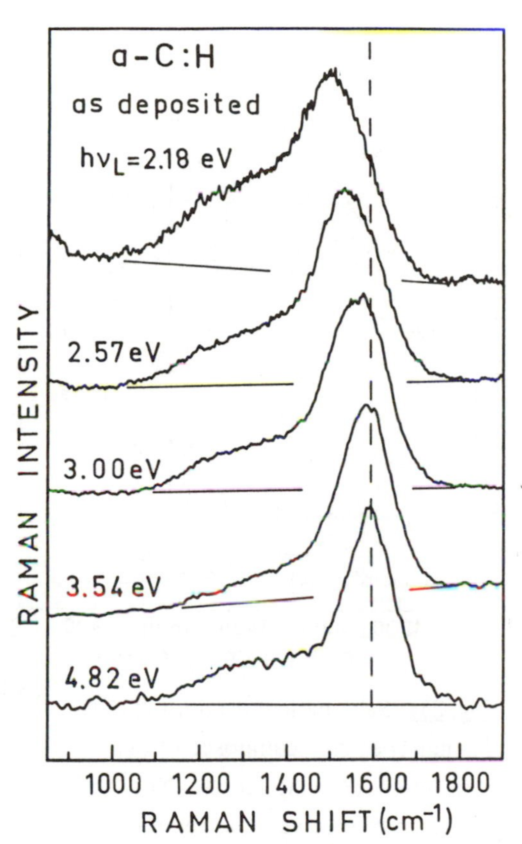

Fig. 24: Raman spectra of plasma-deposited hard a-C:H excited at different incident photon energies given in the figure. The drawn lines indicate the underlying luminescence background, the dashed line marks the peak frequency observed for excitation at 4.82 eV [44]

Recently, resonant Raman scattering on such hard and dense a-C:H films gave an additional experimental verification of the concept discussed above [32, 43-46]. Figure 24 displays a series of Raman spectra of hard and dense a-C:H prepared by plasma deposition. The photon energies used to excite these spectra range from 2.18 to 4.82 eV. The spectra consist of a main peak at ≈ 1500 - 1600 cm^{-1} and a low-frequency shoulder at ≈ 1300 cm^{-1}. The main Raman peak shows a well defined high-frequency shift from 1500 to 1600 cm^{-1} with increasing photon energies up to 3.5 eV. Simultaniously the relative intensity of the 1300 cm^{-1} shoulder decreases. As shown in reference 43 these changes are due to a resonant enhancement of the Raman scattering efficiency of the high frequency portion (≈ 1600 cm^{-1}) of the spectrum with increasing photon energy. Increasing the incident photon energy further to 4.82 eV, the main Raman peak remains fixed at 1600 cm^{-1} and the relative intensity of the low-frequency shoulder increases again.

This high-frequency shift of the main peak in the a-C:H Raman spectrum was found to occur only in films consisting of both sp^3 and sp^2 bonded carbon. It was not observed in material which was converted into purely sp^2 bonded graphite-like carbon by thermal treatement (see Section IV.1) [32, 43, 44].

Because it is difficult to interpret the changes in the Raman spectrum of a-C:H upon variation of the exciting photon energy within a homogeneous structural model, a heterogeneous structure has been suggested involving two phases [32, 43, 44]. These two phases differ in their electronic and vibronic properties. The individual vibrational modes are essentially localized within the corresponding phase. As a consequence, the experimentally observed Raman spectrum is a superposition of two distinct vibronic Raman spectra which are overlapping as displayed schematically in figure 25a. As shown for a wide range of materials including e.g. amorphous silicon, the variation of the Raman scattering efficiency as a function of incident photon energy is closely related to the electronic band structure of the material [47]. With the two phases having different electronic structure also the photon energy dependence - or dispersion - of the individual scattering efficiencies are expected to be different. Therefore a variation of the photon energy used to excite the Raman spectrum may cause a change of the relative contributions of the tow phases to the Raman spectrum. Such a change in the relative intensity, in turn, can lead to an apparant peak shift in the measured Raman spectrum as shown schematically in figure 25b.

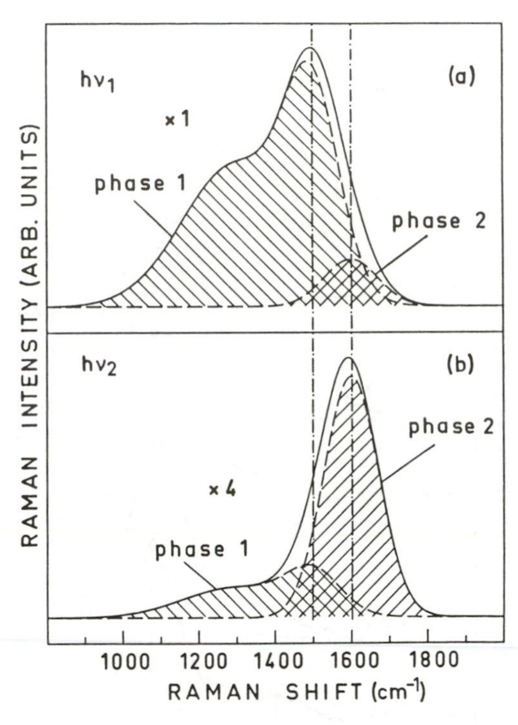

Fig. 25: Schematic presentation of a Raman spectrum composed of two different underlying spectra. Note the apparent change in peak frequency from (a) to (b) when the relative intensities of the two components are changed

To account for the observed photon energy dependent shift in the Raman spectrum of a-C:H, one has to assume that the scattering efficiency of phase 2, which contributes to the high frequency portion of the Raman spectrum, is strongly resonant for photon energies increasing from 2.2 to 3.5 eV. The Raman scattering efficiency of phase 1, in contrast, which contributes to the low-frequency portion of the Raman spectrum, has to be much less resonant in the above range of photon energies.

Based on the heterogeneous structural model introduced by Robertson [2, 42] it is straight forward to identify the two phases mentioned above with graphitic clusters of sp^2 bonded carbon and with the interconnecting network of sp^3 bonded carbon, respectively [43]. The comparison of the Raman spectra of hard and dense a-C:H (figure 24) which contains both sp^2 and sp^3 bonded carbon with the spectra of sp^2 bonded graphitic carbon (figure 21, top curve) strongly suggests to identify phase 2 with the sp^2 bonded carbon clusters and phase 1 with the interconnecting sp^3 carbon network.

Consequently the graphite-like phase 2 contributes dominantly to the high frequency portion of the a-C:H Raman spectrum (G-peak), whereas the low frequency shoulder arises from phase 1 plus some possible contribution from the graphitic disorder mode (D-peak) from the larger clusters of phase 2. With this assignment also the above postulated requirements for the dispersion of the Raman scattering efficiencies are fulfilled. For sp^2 bonded carbon π - π^* transitions occur in the energy range 3.5 - 6.5 ev [25]. The Raman scattering efficiency can be taken proportional to $|\partial\varepsilon(\omega)/\partial\omega|^2$ where $\varepsilon(\omega)$ denotes the energy dependent dielectric function [47]. Therefore one expects a resonant enhancement of scattering by the graphitic clusters (phase 2) at the onset of these π - π^* transitions ($\approx$ 2 - 3.5 eV) where $|\partial\varepsilon(\omega)/\partial\omega|^2$ shows a maximum. This is fully consistent with the observed enhancement of the high frequency portion in the Raman spectrum of a-C:H for photon energies of $\approx$ 3.5 eV [43]. The σ - σ^* transitions occuring in both sp^2 and sp^3 bonded carbon lie at higher energies $\gtrsim$ 6 eV [25]. Therefore one expects only for the highest photon energy used (4.8 eV) an enhancement of scattering by sp^3 bonded carbon (phase 1), which contributes to the low-frequency portion of the Raman spectrum. This is again consistent with the experimental finding, that the relative intensity of the low frequency shoulder in the Raman spectrum of a-C:H increases with increase of the photon energy from 3.5 to 4.8 eV (see figure 24).

The resonant Raman experiments discussed above [32, 43, 44] have been confirmed recently by two independent groups [45, 46]. These authors, however, used a much smaller range of photon energies (1.9 to 2.7 eV) and discussed their data exclusively on the basis of clusters of sp^2 bonded carbon. It should be noted that the present Raman data were obtained from the same a-C:H films for which dominant sp^3 bonding had definitely been established by IR spectroscopy [19] and EELS [25]. This together with the complex resonance behaviour over the full energy range (up to 4.8 eV) strongly support our explanation that both sp^2 carbon clusters and the sp^3 bonded interconnecting carbon network contribute to the Raman spectra.

REFERENCES

1. Angus, J.C., Koidl, P., and Domitz, S.: chapter IV in "Plasma Deposited Thin Films", Mort, J., and Jansen, F., Eds. (CRC Press, Boca Raton, 1986)
2. Robertson J.: Adv. Phys., 1986, 35, 317
3. Tsai, H., and Bogy, D.B.: J. Vac. Sci. Technol., 1987, A5, 3287
4. "Amorphous Hydrogenated Carbon Films", Koidl,.P., and Oelhafen, P., Eds., Proc. European Mat. Res. Soc., Vol 17 (Les Editions de Physique, Paris, 1987)
5. Holland, L., and Ojha, S.M.: Thin Solid Films, 1976, 38, L17
6. Baumann, H.: private communication
7. Bubenzer, A., Dischler, B., Brandt, G., Koidl, P.: J. Appl. Phys. 1983, 54, 4590
8. Pethica, J.R., Koidl, P., Gobrecht, J., and Schüler, C.: J. Vac. Sci. Technol., 1985, A3, 239
9. Wagner, J., Wild, C., Pohl, F., and Koidl, P.: Appl. Phys. Lett., 1986, 48, 106
10. Wagner, J., Wild, C., Bubenzer, A., and Koidl, P.: Mat. Res. Soc. Symp. Proc., 1986, vol 68, 205
11. Wild, C., Wagner, J., and Koidl, P.: J. Vac. Sci. Technol., 1987, A5, 2227
12. Wild, C., Koidl, P., and Wagner, J.: in Ref. 4, p. 127
13. Selwyn, G.S., and Kay, E.: Plasma Chem. Plasma Processes, 1985, 5, 183
14. Anderson, L.P., Berg, S., Norström, H., Olaison, R., and Towta, S.: Thin Solid Films, 1979, 58, 155
15. Anderson, L.P., and Berg, S.: Vacuum, 1978, 28, 449
16. Tschersich, K.G., and Littmark, U.: in Ref. 4, p. 161
17. Wild, C., and Koidl P.: Appl. Phys. Lett., 1989, 54, 505
18. Baumann, H., Rupp, T., Bethge, K., Koidl, P., and Wild, C.: Ref. 4, p. 343
19. Dischler, B., Bubenzer, A., and Koidl, P.: Solid State Commun., 1983, 48, 105
20. Dischler, B., Sah, R.E., Koidl, P., Fluhr, W., and Wokaun, A.: Proc. 7th Int. Symp. on Plasma Chemistry, Timmermans, T.J., Ed. (Eindhoven 1985) p. 45
21. Dischler, B.: Ref. 4, p. 189
22. Nyaiesh, A.R., and Nowak, W.B.: J. Vac. Sci. Technol., 1983, A1, 308
23. Reimer, J.R., Vaughan, R.W., Knights, J.C., and Lujan, R.A.: J. Vac. Sci. Technol., 1981, 19, 53
24. Dischler, B., Bubenzer, A., and Koidl, P.: Appl. Phys. Lett., 1983, 42, 636
25. Fink, J., Müller-Heinzerling, Th., Pflüger, J., Scheerer, B., Dischler, B., Koidl, P., Bubenzer, A., and Sah, R.E.: Phys. Rev. B, 1984, 30, 4713
26. Fink, J., Müller-Heinzerling, T., Pflüger, J., Bubenzer, A., Koidl, P., and Grecelins, G.: Solid State Commun., 1988, 47, 687
27. Kaplan, S., Jansen, F., and Machonkin, J.: Appl. Phys. Lett., 1985, 47, 750

28. Grill, A., Meyerson, B.S., Patel, V.V., Reimer, J.A., and Petrich, M.A.: J. Appl. Phys., 1987, 61, 2874
29. Jansen F., Machonkin, M., Kaplan, S., and Hark, S.: J. Vac. Sci. Technol., 1985, A3, 605
30. Wild, C., and Koidl, P.: Appl. Phys. Lett., 1987, 51, 1506
31. Wild, C., and Koidl, P.: in Ref. 4, p. 207
32. Wagner, J., Ramsteiner, M., Wild, Ch., and Koidl, P.: Ref. 4, p. 63
33. Dillon, R.O., Wollam, J.A., and Katkanant, V.: Phys. Rev. B., 1984, 29, 3482
34. Sander, P., Kaiser, U., Altebockwinkel, M., Wiedmann, L., Benninghoven, A., Sah, R.E., and Koidl, P.: J. Vac. Sci. Technol., 1987, A5, 1470
35. Ugolini, D., Oelhalfen, P., and Wittmer, M.: Ref. 4, p. 287
36. Sander, P., Wiedmann, L., Benninghoven, A., and Sah, R.E.: Ref. 4, 305
37. Ramsteiner, M., Wagner, J., Wild, Ch., and Koidl, P.: Solid State Commun., 1988, 67, 15
38. Ramsteiner, M., Wagner, J., Wild, Ch., and Koidl, P.: J. Appl. Phys., 1987, 62, 729
39. Oelhafen, P. and Wittmer, M.: unpublished results
40. Bredas, J.L., and Street, G.B.: J. Phys. C, Solid State Phys., 1985, 18, L651
41. Bredas, J.L., and Street, G.B.: in Ref. 4, p. 237
42. Robertson, J., and O'Reilly, E.P.: Phys. Rev. B, 1987, 35, 2946
43. Ramsteiner, M. and Wagner, J.: Appl. Phys. Lett., 1987, 51, 1355
44. Wagner, J., Ramsteiner, M., Wild, Ch., and Koidl, P.: Phys. Rev. B, 1989, to appear
45. Yoshikawa, M., Katagiri, G., Ishida, H., Ishitani, A., and Akamatsu, T.: Appl. Phys. Lett., 1988, 52, 1639; J. Appl. Phys., 1988, 64, 6464; Solid State Commun., 1988, 66, 1177
46. Tamor, M.A., Haire, J.A., Wu, C.H., and Hass, K.C.: Appl. Phys. Lett., 1989, 54, 123
47. See, e.g., Cardona, M. in "Light Scattering in Solids II", edited by M. Cardona and G. Güntherodt (Springer, Berlin, 1982), p. 19

Materials Science Forum Vols. 52 & 53 (1989) pp. 71-102

STRUCTURE AND PHYSICAL PROPERTIES OF AMORPHOUS HYDROGENATED CARBON (a-C:H) FILMS

H. Tsai
HMT Technology Corp.
1220 Page Ave., Freemont, CA 94538, USA

ABSTRACT

The electronic structure and bonding of amorphous carbon and the effect of hydrogenation were discussed first so that a base for understanding of properties of a-C:H is formed. Then the review discusses the following properties of a-C:H films: density, electrical conductivity, residual stresses, adhesion to the substrate, hardness and tribological properties.

1. STRUCTURE AND BONDING OF a-C:H FILMS

Before discussing a-C:H films, it is necessary to have a brief review of unhydrogenated amorphous carbon films, because it is fundamental to the understanding of structure and bonding characteristics of a-C:H films.

It is known that evaporated a-C is amorphous on a 0.5nm scale and contains largely trigonally bonded sp^2 sites and perhaps some tetrahedrally bonded sp^3 configuration. Sputtered and ion-beam a-C is generally similar to evaporated a-C in its structure, except the particle bombardment during deposition tends to increase the number of sp^3 bonding in the films and make the films harder [1][2][3]. The effect of bombardment on electrical resistivity can, however, be dramatical. Resistivity $\geq 10^{-11}$ Ωcm has reportedly been obtained in sputtered a-C films with and without simultaneous ion beam bombardment [4]. Unfortunately, no unambiguous data by direct measurement of sp^3 sites in a-C are available yet, because unlike the a-C:H films, ^{13}C nuclear magnetic resonance (NMR) could not detect sp^3 sites in unhydrogenated amorphous carbon [5].

At present, estimation of sp^3 concentration in a-C have been made only from indirect measurement. For DC magnetron sputtered carbon films, 5% sp^3 bonding was derived from Raman spectroscopic data [6], whereas a range of sp^2/sp^3 ratio from 1 to 1/3 was reported in another study as obtained from optical measurements [7]. Savvides et al. noticed in their study [3] that a gradual sp^2-to-sp^3 transition in the bonding configuration was observed mainly due to the degree of ion bombardment, which is, in turn, related to the ion-to-condensing-atom arrival rate ratio Ar^+/C, and the energy of the impinging ions. For a-C films by condensation of the plasma stream from a vacuum arc, the fraction of sp^3 bond was found to be 85%, as determined from the ratio of the π^* peak integral of the film to that of graphite in the EELS spectra [8]. All these discrepancies may largely be due to the nonequilibrium condition at film growing surface in a plasma or ion beam process. However, in another extreme case in which the evaporated and sputtered a-C films are deposited on substrates at liquid nitrogen temperature, no evidence of sp^3 bonding was found in the films by using Raman and electron energy loss spectroscopy (EELS) [9]. The differences between all these observations demonstrate the process dependence of sp^2/sp^3 ratio and the uncertainty of current determination of sp^2/sp^3 concentration in unhydrogenated a-C. Consequently, the understanding about the structure of carbon films and the origin of diamond-like property have been greatly hampered. At present, it is hard to make a proper judgement on various structural models of a-C films proposed in the period of last 30 years.

Kakinoki et al proposed a 50% sp^2 + 50% sp^3 model for evaporated a-C [10], but most of the recent works suggest a much lower proportion of sp^3 bonding. One of the continuous-random-network models by Beeman et al. [11], which contains 91% sp^2 sites, was found to fit the experimental data better than those with other sp^2 concentrations. The existence of sp^3 bonds in a-C, was confirmed by most indirect, experimental evidence. Further, from the energetic standpoint, a structure, consisting of predominating sp^2 regions with occasional sp^3 atoms allowing changes in orientation of the planes, seems more reasonable than a pure sp^2 structure requiring either a large number of dangling bonds or bending/warping of sp^2 planes [2].

Another issue in the structural modelling of a-C is the medium-range order. Robertson and O'Reilly [2] believed that in order to relieve the strain, the sp^2 sites are arranged in compact clusters (~15nm in diameter) rather than having distributed the bond-angle disorder throughout the structure so that the π bonding energy is maximized [2]. However, it is still uncertain whether these sp^2 clusters are interlinked by sp^3 or surrounded by dangling bonds.

Due to all these uncertainties about the a-C structure, it is not surprise to have controversy about the role of hydrogen in a-C:H films. However, it is quite certain that the incorporated hydrogen tends to satisfy the dangling sp^3 and sp^2 bonds in the amorphous carbon. This should reflect in a decrease of unpaired electron states of the dangling bonds.

A decrease in spin density of hydrogenated carbon as measured by electron spin resonance (ESR) is an evidence for the passivation of dangling bonds by hydrogen [12]. But it is not clear to what extent hydrogen passifies the dangling bonds and the extent it saturates π (graphite) bonds. Upon hydrogenation, the band gap of a-C:H remains small (e.g., 1.2 eV[12]) relative to that of sp^3 bonded carbon (5.5 eV). This is quite distinct from silicon, where hydrogenation increases the band gap of amorphous Si from 1.2 to 2.0 eV, which is a significant fraction of that of sp^3 bonded crystalline Si (3.1 eV). The difference in behavior is obviously due to the allotropy of carbon or the presence of unsaturated sp^2 bonding states in amorphous carbon.

Another distinction from the behavior of silicon, as disclosed by optical absorption spectra [12], is that the effect of hydrogen in carbon is different at low and high concentration. At first stage of hydrogenation, the incorporated hydrogen primarily passifies the unpaired electrons in nonbonding orbitals, i.e., dangling bonds, and removes these states from the gap. As a result, the band gap is widened while the density of states in the bands remains almost constant. If additional hydrogen is available, further hydrogenation reduces the unsaturated graphitic states in the bands while the band gap remains nearly unchanged [5][12]. When a-C:H films are prepared by the plasma decomposition of hydrocarbons, more hydrogen, up to 60%, can be incorporated into the carbon films. This results in further widening of the band gap due to the additional reduction in the graphitic states at the band edge as shown by NMR measurement [5].

Whereas the C-C bonding in the films can be characterized by using Raman spectroscopy, electron energy loss spectroscopy (EELS) and extended X-ray absorption fine structure (EXAFS)[13], the C-H bonding in a-C:H films had been studied by using IR absorption in the CH stretch region [14-16]. One of these studies [16] showed that high-energy ion bombardment induces a decrease in hydrogen and an increase in sp^3 content, and hydrogen is consequently linked mainly to carbon as sp^3 CH groups:

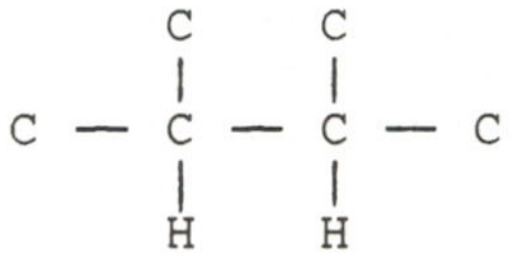

```
          C     C
          |     |
     C -- C  -- C -- C
          |     |
          H     H
```

Under low ion bombardment, large numbers of C-H bonds are in sp^3CH_3, sp^3CH_2, and sp^2CH_2 configurations:

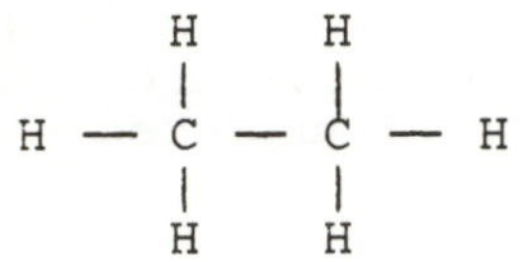

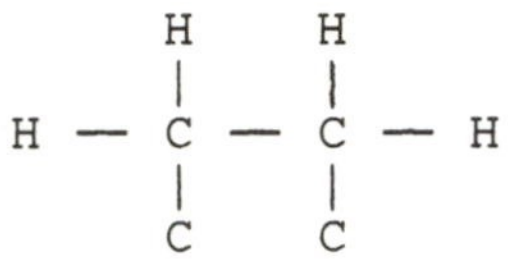

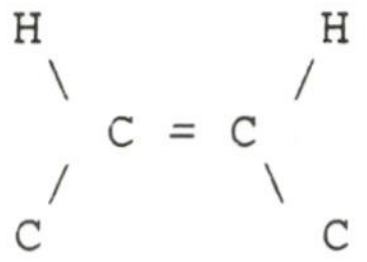

```
     H    H              H    H              H         H
     |    |              |    |               \       /
H -- C -- C -- H    H -- C -- C -- H           C = C
     |    |              |    |               /       \
     H    H              C    C              C         C
```

This observation confirms a common belief that high ion bombardment causes weaker bonds such as single bonded hydrogen to be removed and strongly bonded sp^3 and sp^2 carbon atoms will survive in the growing film (bond strength of C-C is 607 KJ/mol vs 337.2 KJ/mol for C-H [17]).

Another observation about ion bombardment effect was made by Kaplan, Jansen and Machonkin [5]. They suggested that since double bond hydrogenation is an exothermic process, formation of sp^2 bonding is favored over sp^3 bonding under higher ion energy condition. This conjecture was confirmed by a later study using ellipsometric and optical measurements [18].

Most studies agree that the hydrogenation of carbon film is proceeded by bonding hydrogen with both sp^3 and sp^2 carbon, but preferentially with sp^3. The ratio of two bondings depends on the deposition condition. For instance, the ^{13}C-NMR spectrum of an ion beam sputtered a-C:H film (35 at. %H) shows that the sp^2/sp^3 ratio, with and without attached H, is about 1.5 [12], whereas for plasma deposited a-C:H films (39 at. %H), this ratio is 0.5 [19].

Graphitic sp^2 bonds in a-C:H films are identified by synchrotron-radiation [20] and EXAFS spectra of EELS [19] as unsaturated π-bonding at near-edge of the band. The persistent presence of sp^2 bonding in a-C:H is responsible to its optical and electronic properties distinctive from those of hydrogenated amorphous silicon (a-Si:H).

Since the bonded hydrogen tends to attach to carbon predominantly in sp^3 configuration [15], the sp^3/sp^2 ratio increases with hydrogen content. Consequently, the a:C-H films become more transparent, less dense and softer as hydrogen content increases. A reversal in property change may occur when hydrogen is removed by annealing the films.

As mentioned before, ESR measurement revealed a reduction in number of dangling bonds due to hydrogenation. In another study [21], the ESR signal from as-deposited a-C:H films was interpreted as arising from carbon-rich aromatic ring structure. One study on a-C:H films by electron-diffraction also provided the evidence of polymer component present in the films [22]. Moreover, the a-C:H films prepared by a plasma activated CVD process have been shown by spectroscopic analysis to exhibit a polymer-like structure with strong cross-linking between the polymer chains [23]. Although x-ray analysis shows no evidence of crystalline region in the films, the polymer structure demonstrates high stability against heat treatment and chemical attack. This may be due to the strong cross-linking between polymers which seems to increase with increasing DC bias of the plasma. At higher DC bias, the structure of a-C:H films appeared to be more or less independent to the hydrocarbon used in the process.

The presence of hydrocarbon groups was also observed in a-C:H films by DC magnetron plasma deposition with minimal ion bombardment [24]. The authors suggest that as minimum hydrocarbon dissociation occurs in the gas phase and at the substrate surface, polymer films were formed accordingly. Smith [25] attempted to model a-C:H by an effective medium approximation (EMA) based on the assumption that as-deposited a-C:H film contains amorphous polymeric, sp^3 and sp^2 components. The high electrical resistivity and optical transparency can be understood as the result of dominance of polymeric and sp^3 components. On the other hand, the growth of graphitic sp^2 component upon annealing causes a transition to a softer, more conducting and less transparent film.

It is quite clear now that the properties of a-C:H films are determined primarily by the hydrogen content. The hydrogen content in a-C:H films can be measured by gravimetric or combustion method [5][26][27], proton NMR spin counting [5], Rutherford backscattering (RBS) analysis [28], proton recoil spectra [29] and elastic recoil detection analysis [30].

The hydrogen in a-C:H may be present in either bound or unbound state. Support for the presence of unbound hydrogen was given by the change in H/C ratio on the film surface [31]. As measured by nuclear reaction analysis, hydrogen was found to leave the sample in the time interval between deposition and analysis. Consequently, the authors concluded tentatively that much of the hydrogen in their a:C-H films is not chemically bonded to the carbon [31]. The amount of unbound or chemisorbed hydrogen in a-

C:H films had been determined by differential scanning calorimetry (DSC)[32]. Trapped molecular hydrogen may be chemisorbed on the microscopic internal surfaces which are considered to exist in the film.

In summary, whereas a-C films are believed to comprise predominantly of graphitic bonding, the basic structure of a-C:H film is most likely to be a random network of sp^3/sp^2 C-C and C-H bonded atoms and voids. For example, in one study by using electron diffraction radial distribution function (RDF) and measurement on a microdensitometer trace, the crystallite size of short-range sp^3 bonded diamond structure was estimated to be less than 0.6 nm [26]. Farther neighbors of carbon atom are regarded as highly disordered. No graphite-like regions are observed on a-C:H film by using scanning tunneling microscopy [33]. The roles of hydrogen in a-C:H are twofold: (1) passivates dangling bonds; (2) forms C-H bonds and thus modifies the configuration and bond state of carbon network. Accordingly, the density, optical and electrical properties, hardness and stress of carbon film are changed by hydrogenation. Most of these properties (except optical, for which readers are referred to a review by Angus et al. [34]) will be covered in the following review. In order to have certain reference for comparison, some of the important properties and structural/bonding feature for diamond, graphite and amorphous carbon are listed in Table I. Table II gives a summary of physical properties of a-C:H films published in the decade of 1976-1986.

2. FILM DENSITY

Generally, the density of a film may be considerably smaller than the bulk material due to voids between islands or vacancies at the grain boundaries [49]. For amorphous solids, the situation of density is even more complicated, because microscopic as well as macroscopic voids may exist throughout the whole volume. Transmission electron microscopic (TEM) micrographs of sputtered amorphous carbon overcoat films [50] show a microstructure of darker islands surrounded by low density regions, which had been identified as a network of voids [51]. Moreover, an ESR result [21] provides indirectly the evidence of porous structure of a-C:H film. The observed destruction of the ESR signal by exposing film to air is apparently due to the interaction between the film and the oxygen molecules. Since oxygen must be in intimate contact with the unpaired electron and thus the film must be quite porous according to this experimental result.

There is experimental evidence revealing that the density of carbon film changes with deposition condition. For a-C films by DC magnetron sputtering [3], the highest density ($2.1-2.2 g/cm^3$) obtained at the lowest sputtering power is comparable to that of single-crystal graphite (Table I). Increasing the sputtering power causes a sharp decrease in film density to about $1.9 g/cm^3$

and thereafter decreasing gradually to about $1.6 g/cm^3$ [3]. Similarly, the densities of RF sputtered carbon films were also observed to decrease with increasing RF power and so deposition rate due probably to the entrapment of gases [52]. A simultaneous argon ion bombardment with sputter deposition can increase the density of carbon film to $2.2 g/cm^3$ from $2.1 g/cm^3$ which was obtained by sputtering alone [4]. This enhancement of density is likely due to a reduction of trapped gas and an increase in sp^3 fraction of carbon bonding. It is worth noting that since the films are amorphous, the observed densities of carbon films, which although are much less than that of diamond (Table I), are extremely high when compared with density of crystalline graphite. This may indicate the existence of diamond sp^3 bonding in the films.

For a-C:H films the presence of hydrogen further complicates the situation and the density depends not only on the hydrogen content, but also on whether the hydrogen is bonded to carbon or simply "buried" in the film. As a result the reported densities of a-C:H films vary over a wide range: 1.02-2.67 g/cm^3 as shown in Table II. Generally, the density decreases with increasing hydrogenation, as shown by a study on a series of ion beam sputtered carbon films [12]. The film density as measured by sink-float method starts at $2.25 g/cm^3$ for unhydrogenated and gradually decreases to 1.6 g/cm^3 for the maximally hydrogenated film.

The film density can be measured not only by using gravimetric techniques, i.e. direct measurement of volume and mass [24], and sink-float method using halogenated hydrocarbon liquids [25], but also by using RBS method [29][53]. This nuclear physical measurement gives the areal density or the number of carbon atoms per square centimeter. Film density can then be determined by combining with measurement of carbon film thickness by either ellipsometry [29] or a profilometer [53].

3. ELECTRICAL CONDUCTIVITY

Generally, the temperature dependence of the conductivity, σ in amorphous semiconductor can be expressed as:

$$\sigma = \sigma_o \exp [-(E_A/k_B T^n)], \qquad (1)$$

where E_A denotes activation energy and k_B is Boltzman constant. For n=1, the conduction is mainly by thermal activation. If mobility of electron $\sigma_o > 10^3\ \Omega^{-1}cm^{-1}$, the conduction is by activation to extended states beyond the conduction band and valence band mobility edge, E_c and E_v respectively, while

$\sigma_o < 10^3 \Omega^{-1}$ cm^{-1} means that conduction occurs by hopping between localized states on near-neighbor sites. If n<1, this suggests that conduction is by a variable range hopping in localized states around the Fermi level E_F [1]. This occurs when $N(E_F)$ is low so

that as the temperature falls, an electron must hop (tunnel) to more distant sites to find a level within $\sim k_BT$ of its energy.

Eq(1) at n=1/4 is the classic power law suggested by Mott [54] for variable range hopping.

Apparently, electrical conduction of amorphous semiconductors is largely determined by localized electron energy states, which fall within an energy range called the mobility gap. Amorphous semiconductors such as a-Ge and a-Si were found to obey Mott relation (Eq. (1) at n=1/4) when the temperature is low enough. However, for evaporated amorphous carbon the hopping conduction mechanism seems to occur only in a narrow range of low temperatures: below 8^oK [35] and 5^oK when through the film [55]. In the case of a-C:H films, the situation appears to be more complicated because the electrical conduction is not simply by activation and can vary over 11 orders of magnitude just by changing the deposition temperature [56].

For those a-C films sputtered at 95^oK, the electrical resistivity is found to be fitted well between 20 to 300^oK by Mott relation, whereas those a-C films sputtered at room temperature or above showed even larger deviation from Mott relation than that of a-C films evaporated at room temperature [57]. This indicates that the better fit to the hopping conduction as a characteristic of amorphous semiconductor or equivalently to greater degree of amorphousness obtained by sputtering at 95^oK should not be attributed to difference in preparation (evaporation or sputtering), but instead to the different temperatures of substrate T_d. A strong dependence of conductivity on T_d was also

observed in several cases of a-C:H films [27][56].

It is generally accepted that all a-C films consist of a

mixture of sp^2 graphite bonds and sp^3 diamond bonds. Hauser [36] suggested that graphite bonds act as the localized hopping states. Thus, the disparity in electrical conduction behavior between films sputtered at room temperature and 95^oK may simply be due to a greater concentration of graphite bonds in room-temperature deposited films than those deposited at 95^oK. He also pointed out that this view is supported by the annealing experiments [58].

Upon annealing of evaporated a-C films at 450^oK, the

resistivity drops from that at 77^oK to $10^{-2}\,\Omega$cm by seven orders of magnitude while no crystallization was detected by electron diffraction [35]. Amorphous carbon films after annealling or being deposited at room temperature and above tend to have imperfect amorphous structure, i.e., with microscopic graphite regions, or simply having more graphite bonds than a perfect amorphous carbon should have. The resultant increase in the number of localized states leads to a large increase in conductivity, a small increase in optical absorption and no change in electron diffraction, since the microcrystals of graphite are too small (<1nm [35]) to detect. Therefore, electrical resistivity appears to be the most sensitive indicator for the degree of amorphousness of carbon films.

Further evidence on the correlation between hopping conductivity and graphite bonds in amorphous carbon has been given by the conductivity measurement of C-implanted amorphous diamond, which demonstrated that the same state can be achieved either by adding diamond bonds to graphite by evaporation or sputtering, or by generating graphite bonds in diamond by implantation [58]. Annealing the implanted layer for 8 hr. in vacuum at 300°C yields resistivity (300°K) of $\sim 3\,\Omega$cm, or in a decrease of one order of magnitude, while no crystallization was observed. Annealing 2 hr. at 765°C yields resistivity (300°K) of $1.8 \times 10^{-3}\,\Omega$cm which is almost temperature independent and behaves like graphite. However, this annealed layer is very hard like diamond [58]. The authors suggested that this dual behavior may be due to the overlapping of localized states which causes metallic conductivity and whereas many diamond bonds still remain in the layer resulting in a high hardness.

The anomalous increase in resistivity with decrease of thickness for evaporated a-C films thinner than 50 nm can also be understood in terms of a similar reasoning. Films thinner than 50 nm may have higher degree of amorphousness as the thickness is reduced. Consequently, the resistivity of thinner films approaches $50\,\Omega$cm, i.e., 500 times higher than that of the thicker ones, which is $0.1\,\Omega$cm when measured at room temperature [57].

During hydrogenation of amorphous carbon films, as mentioned in Section 1, the first task for incorporated hydrogen is to passify the dangling bonds in the gap. As a result, the a-C:H film can be treated as a wide band-gap amorphous semiconductor, using the theory developed primarily for amorphous silicon or a-Si:H.

Generally, a-C:H films have higher resistivity than unhydrogenated a-C films by up to 12 orders of magnitude [44], likely due to stabilization of sp^3 bonds and passivation of dangling bonds. The formation of C-H bonds results in a reduction in the density of states in the gap, which in turn, leads to an increase in resistivity.

The conduction mechanism in a-C:H films is more complex and thus unclear in many aspects. The observed, non-linearity of plots of $\log\sigma$ vs $1/T^{1/4}$ for as-deposited a-C:H films had ruled out that the variable-range hopping conduction between localized states near the Fermi level be the predominant mechanism of conduction in these films [27]. On the other hand the simple conduction via extended states in either the valence or conduction band was also precluded because of the curvature of the $\log\sigma$ vs

1/T plots and the low preexponential value ($\sigma_o=10^{-4}-10^{-2}\Omega^{-1}cm^{-1}$ [56]) in Eq. (1).

The deviation from both the Mott relation and the simple conduction by thermal activation suggest that the possible mechanism for the electrical conduction in these semiconducting a-C:H films appears to involve a thermal excitation of electrons into a broad range of localized energy states at one of the band edges. Therefore, conduction is likely to occur via thermally activated hopping [1]. Such mechanism would also be consistent with the low values of σ_o: about 100-10,000 times lower than those for extended state conduction, since the mobility of electrons within the mobility gap is very low [56].

Apparently, electrical conduction in both a-C and a-C:H films is more or less a phenomenon associated with the localized states in the mobility gap. Since the electronic structure is associated with the microstructure of the films, which, in turn, is determined by the deposition condition. It is thus not surprising that electrical resistivity of a-C:H films have strong dependence on deposition parameters, such as RF power and substrate temperature T_d in the glow-discharge decomposition of hydrocarbons [45] [27], and bias voltages in either ion plating [59], or RF decomposition of hydrocarbons [60]. Further, the studies on the effects of T_d on the conductivity were often combined with the attempts to understand the conduction mechanism of a-C:H films.

The increase in conductivity with increasing T_d may be ascribed to an increase in the density of localized states [27]. At lower deposition temperatures and thus with less localized states in the gap, the conduction in the a-C:H films even at elevated temperatures is most likely to occur via thermally activated hopping in a region where the density of localized states is fairly high [27]. However, with increasing density of localized states at higher T_d, the electrical properties of a-C:H films became more like a-C films. The conduction mechanisms other than thermally activated hopping may come into effect under certain condition, e.g., Anderson [44] showed that two a-C:H films by decomposition of acetylene in plasma at T_d = 500 and 550^oK to obey the relation of extended state conduction mechanism when measured above about 500^oK. The observed value of σ_o (between 1 and $10^2\Omega^{-1}cm^{-1}$) is also consistent with extended state conduction occurs in this temperature range. The activation energy of Eq.(1) in this case is 1.46 eV for the T_d = 500^oK specimen. Since the optical gap was measured to be 2.2 eV, the Fermi level appears to lie near the middle of the gap [44]. Same conclusion was reached also by a later study on a-C:H films [56].

Meyerson and Smith [56] suggested that fundamental changes in the bonding and band structure occur as T_d increases above 250°C, when optical gap was observed to drop rapidly. They ascribed such change to a decreasing incorporation of hydrogen in a-C:H film at higher T_d. With less hydrogen present, the number of sp^2 grows at the expense of sp^3 bond. As a result, both the electrical resistivity and the optical gap decreases and they suggested that at higher deposition temperature, e.g., $T_d > 425$°C, the a-C:H films would turn to be like a-C films and behave more graphitic as far as their electrical and optical properties are concerned. They also attributed the disparity in 3 to 5 orders of magnitude between their σ(RT) values and Anderson's to the difference in hydrogen content in the films.

However, their conclusion on T_d effect by correlating with incorporation of hydrogen was not supported by the hydrogen content, C_H measurement of a-C:H films in a more recent work by Jones and Stewart [27]. According to the latter authors, the hydrogen content of their a-C:H films lies between 30% and 45%, and shows no systematic variation with T_d and the gas used. Jones and Stewart, therefore, suggested that the increase in the electrical conductivity produced by increasing T_d or by doping is neither due to the hydrogen content, or to a shift in the Fermi level with respect to the valence states, but rather to a modification of the localized state density which would be associated with a change in the structure of the films. The authors speculated that the hydrogen may be bonded in different configuration for different T_d samples, similar to the amorphous silicon. Thus, change in T_d can alter the relative proportion of hydrogen in these configurations and thus the electrical conductivity without changing the value of C_H.

Carbon films with very low hydrocarbon content prepared by decomposition of acetylene in a RF plasma show a strong dependence of the film resistivity on the RF power: decrease of ρ in 2 orders of magnitude as RF power varies from 30 to 70 W [45]. An increase in electrical conductivity with increasing T_d was also observed in this study, confirming the results obtained in the previous works [27, 44, 56].

The electrical conductivity of carbon films produced by RF plasma decomposition of methane was found to depend strongly on the negative self-bias voltage of the electrode, V_e [60]. The electrical conductivity at room temperature σ(RT) varies from 10^{-11} to $10^{-4}\ \Omega^{-1}cm^{-1}$ as V_e changes from -20 to -620 V. The greatest change of (RT) in 3 orders of magnitude occurs between V_e = -100 and -150 V. Both activation energy for electrical conduction, E_A and optical gap E_{opt} decrease with decreasing V_e. However, $E_{opt} - E_A$ remains roughly constant when V_e changes. This implies that the Fermi level remains fixed relative to the edge of valence bond [60]. This is the same conclusion obtained from the dependence of E_A and E_{opt} on T_d in a previous study [27].

4. RESIDUAL STRESSES AND ADHESION

Residual stresses and adhesion to substrate determine the stability of the overcoat/substrate composite and thus the durability of the component, since the composite may fail by cracking, delamination, and buckling before the damage by wear occurs.

Residual stresses in the films are developed during the deposition process because either of differential thermal expansion or of intrinsic stresses arising from lattice mismatch (crystal mismatch is not applicable for the amorphous films as a-C:H), impurities incorporation at the interface, structural reordering, etc. The total stress in the films, which may be measured by substrate deflection techniques or x-ray lattice measurements, is a function of film thickness. Consequently, thin stressed films may adhere while thick film may fail because of buckling.

Since a porous structure will not sustain a high stress, the deposit structure is important to stress generation. Any mechanism which impedes atomic rearrangement will cause the development of high stresses. Thus, impurity incorporation may play a major role in stress retention. The magnitude of intrinsic stress may be affected by a number of processing parameters, such as film deposition rate, angle of incident, presence of residual gases, and deposition temperature. Ion bombardment during deposition appears to create a high compressive stress similar to the peening effect on the surface. By controlling the amount of ion bombardment during deposition, it is possible to sputter films with zero intrinsic stress [61].

The conventional method for measuring the film residual stresses is the beam-bending technique. The stress in a beam curved by a film on one surface can be determined by a modified Stoney equation [62],

$$\sigma_s t = E_B T^2 6R(1-\nu_B) \qquad (2)$$

where σ_s - stress in the film, t - film thickness, E_B-elastic modulus of substrate, ν_B - Poisson's ratio of substrate, T -

substrate thickness, and R - radius of curvature of substrate, which, in turn, can be determined by

$$R = (x_1-x_3)^2/4(y_1 + y_3 - 2y_2) \qquad (3)$$

where (x_1-x_3) is the distance along the reference plane between the extreme points of measurement and y_1, y_2, and y_3 are the corresponding vertical distances from the reference plane. The curvature may also be measured with the bending cantilever plate technique as described in a study on the stresses in Si_3N_4 films on silicon [63] and those on carbon films [64]. Changes in the curvature of the sample were determined by clamping one end of the sample and detecting the movement of the free end with a capacitance probe.

Providing there is no intrinsic stress, a film with a smaller coefficient of thermal expansion (CTE), α than that of the substrate will be under compression when cooled down to room temperature from the deposition temperature. A film with $E\alpha$ product that is closer to that of the substrate is thus preferred. Generally, the stress originating from thermal mismatch is predictable and easy to analyze. However, the magnitude and sign of residual stress in the film frequently varies with the deposition process. In other words, the overall stress in the film is not solely caused by thermal mismatch; the sign of stress might be contrary to the expectation. As an example, normal SiO_2 films on Si are under compression because of a less CTE than that of silicon. However, the CVD-deposited SiO_2 films formed by the reaction of SiH_4 and O_2 are found to be in tension [65].

Films under compression are stronger than films under tension and therefore, more desirable in terms of structural strength. Tensile stress in the coating is most undesirable since it may initiate cracks. When the coatings are under compression, cracks may not be generated but wrinkles or buckling may occur. The origin of the tensile stresses has been postulated by various models, but few models are applicable to the interpretation of the intrinsic compressive stress, frequently seen in the carbon films.

The compressive stress is essentially related to the volume expansion of the coating during the deposition process. High compressive stresses have been observed in a-C films prepared by electron beam evaporation [64]. Compressive stress has also been a prevailing phenomenon in the a-C:H films prepared by ion beam plating [66] or plasma decomposition [67][68]. This might be caused by tilted sp^3 bonds and/or by hydrogen or residual gas inclusion [66]. In the case of a-C:H films prepared by a DC plasma in ethylene, as reported by Enke [67], the compressive stress in the films was so high that it was impossible to produce good films thicker than 1 μm. This intrinsic compressive stress was believed to be caused by the large amount of incorporated hydrogen in the films [67]. It varies with the deposition conditions, among which self-bias voltage and hydrocarbon pressure are the most important parameters. The two-dimensional maps show that the highest stresses are observed in the a-C:H films with diamondlike properties, with a decrease towards the graphitelike region and the polymerlike region. This may be due to less incorporation of hydrogen in the graphitelike region, and to the softness of polymeric films [67].

In a study of a-C:H films prepared by RF plasma decomposition of methane [68], an empirical linear relationship was obtained between the compressive stress in the films, σ_i and $P^{1/2}/V_B$, where P is the hydrocarbon gas pressure between 1 and 16 mTorr (0.13-2.13Pa), and V_B, the negative DC self-bias voltage, is between 400 and 1600 V. The quantity $V_B/P^{1/2}$ has been characterized as the average particle energy, $\bar{E}$ impinging on the substrate during deposition [47]. When this energy increases, the degree of cracking of hydrocarbons in the growing films also increases, causing the hydrogen content in the films to decrease. The observed relationship between σ_i and $P^{1/2}/V_B$ confirms that σ_i is related to the hydrogenation of the films so that films produced at high $\bar{E}$ should have low σ_i and vice versa. The compressive stresses observed in this study are in the range of 0.6 to 2.2 GPa [68], which is in the same order of magnitude with Enke's measurement [67].

It has been revealed that the compressive internal stresses in carbon films may cause delamination and thus generate buckling patterns such as wavy wrinkles with well-defined shapes and widths. These had been observed in carbon films grown by electron beam evaporation onto glass [64], by ion beam plating using benzene on glass or silicon [66] and by ion beam deposition [69]. When uniform films were carefully prepared, wrinkles appeared at thickness greater than 300 nm. When the film thickness was not uniform, wrinkles appeared at thickness of about 50 nm. The size of the wrinkles tend to increase with increasing film thickness [64]. However, the generation of wrinkles in carbon films seems to be not only associated with film thickness, but also with the gas atmosphere around the film. As long as the films were kept in a vacuum, no wrinkles occurred even for a 500 nm thick film. However, exposure of the film to a gas (which gas is not important) causes generation of wrinkles even in the films which are thinner than 100 nm. It has been suggested that gas atoms diffuse into the interface between the coating and the substrate, reducing the adherence and thus initiating the wrinkles [64, 70]. Similarly, it was observed with an optical microscope that the evolution of the buckling and stress relief process in an a-C:H film usually started some time after exposure to air and was accelerated greatly by humidity. In most cases, buckling began at the film edge or at defects and propagated by spreading from a few centers and not by generating new buckling centers [69]. The fact that these a-C:H films have good adherence with the substrate, and the buckling only occurs under certain conditions indicates the existence of high compressive stress in the films.

The buckling phenomena of carbon films have been used to evaluate such mechanical parameters of the films as the internal stress, the strain, Young's modulus, and adhesion energy, based on the information of wrinkle geometry [64], or on the formation and stability of wrinkles [71]. In the latter case, internal compressive stresses of 3 to 6 GPa are estimated to exist in the a-C:H films of 200 nm thickness, with adhesion energies estimated in the range of 4 to 7 Nm^{-1} [71]. Both are more than an order of magnitude higher than those measured previously by Matuda et al [64] and by Kinbara et al [70].

Diamondlike carbon films with exceptionally low stress (<10 MPa) have been produced by a hybrid process using a bias sputtering plus plasma decomposition of butane [72]. The density of the film is high (2.8 g/cm^3). The films have distinguished dielectric and optical properties (n=2.02-2.4 in the infrared, and E_o=3eV), and are highly resistant to chemical attack. The low hydrogen content (<1 at. %) in the films as measured by secondary ion mass spectrometry (SIMS) was suggested to account for the low stress in the films [72].

In contrast to the previous findings, a reduction of compressive stress with an increase in hydrogen content in a-C:H films was observed in a recent study [73]. Kerwin et al showed that an order of magnitude decrease in compressive stress occurs when a hydrogen content of the film increased up to about 20 at.% for both sputtered and ion-beam deposited films. The hydrogen content of films measured by nuclear reaction analysis was in the same order of magnitude as those measured by Enke [67].

Recently, the internal stress in a-C:H films was also studied by Nir [74] who found that internal stresses increase with increasing bombarding energy of argon ions. His a-C:H films were also prepared by either ion beam methods or by sputter deposition with or without the addition of hydrogen. Nir claimed that the contribution of hydrogen, either in the gaseous phase or inside the films as measured by nuclear reaction analysis, appeared to be small and inconsistent. However, a strong dependence of internal stress on bombarding energy of ion was observed. The author thus suggested that the compressive stresses in the films were attributed mainly to bombardment of energetic argon ions rather than hydrogen content [74].

Adhesion can be defined as the capacity of a thin film to remain completely attached to the substrate when the couple is subjected to a mechanical shear or tensile stress. Thus adhesion characterizes the rupture strength of the interface between the coating and the substrate. This is an important property because any protective coating is only as good as its adhesion to the substrate. Among the various techniques for testing the adhesion, the scratch test appears to be the one that has led consistently to meaningful results [75, 76]. The scratch test consists of introducing stresses at the interface by a moving indenter (such as a diamond tip) under an increasing load. The minimum load to damage the coating (either by adhesive or by cohesive failure) is called the critical load L_C and is determined by optical or

electron microscopy and/or by acoustic emission (AE). There is good correlation between the onset of the AE signal and the microscopical observation of the first occurrence of damage [77]. The scatter in L_C measured on the same specimen was reported

generally within 10% [76].

The major application of scratch tests was for thin, hard and well-adherent metallurgical coatings, such as carbides, nitrides, or oxides deposited by chemical vapor deposition (CVD) and physical vapor deposition (PVD), as sputtering, ion plating, and ion implantation techniques [76, 78]. The CVD TiC coatings on steel studied by the use of scratch tests have a thickness range of 3 to 22 μm [78].

In the case of CVD, the film and substrate materials form a transition zone by interdiffusion, whereas a pseudodiffusion zone beneath the substrate surface can be produced by an ion beam plating due to ion bombardment. Therefore, very good adhesion can generally be achieved by CVD and ion plating techniques [76].

Since sputtered species have velocities nearly an order of magnitude higher than those of species produced by thermal evaporation, good film-to-substrate adhesion is usually obtained by sputtering. The advantage of sputtering is its low deposition temperature, which makes it a more flexible process for many applications [79].

Layers of a-C:H formed by ion plating are generally found to be strongly adherent when the film thickness is below about 1 μm [66]. By using ion plating from benzene, the a-C:H films deposited on pure iron, grey cast iron, 16MnCr5 steel, and HG10 hard alloy in thickness of 1 to 10 μm have proved to be well adherent and smooth, although with increasing thickness the coatings show an increasing tendency to crack and burst under loading [80]. This is not unusual for the films, such as a-C:H, with high compressive stresses, e.g. if stress level was above 1.5 GPa, delamination would occur quickly (within minutes or hours) for some samples, whereas samples with high hydrogen content, prepared by ion beam deposition, showed low compressive stress and excellent adherence [73].

5. HARDNESS AND ITS MEASUREMENT

For a-C:H films, particularly when used as a protective overcoat in tribological applications, hardness is one of the most important properties of this material. In general, high intrinsic hardness is associated with high cohesive energy, short bond length, and a high degree of covalent bonding. The latter effect can be demonstrated with the three compounds TiC, TiN and TiO having the same crystal structure and similar lattice parameters. The decrease in hardness in going from TiC to TiN to TiO corresponds to the decrease of covalent contribution to the bonding as shown by a decrease in overlapping between p and d states. Combining the first two factors, cohesive energy and bond length, the hardness can be correlated with cohesive energy per molar volume [81]. However, the hardness of a material is not only determined by the strength of interatomic forces but also by the operative deformation mechanism. Therefore, the hardness value can vary with the microstructure and other structural constituents (void, impurities, defects, texture, etc.)[81].

Most hardness measurements are done by indentation techniques. For thin films, the intrinsic hardness can correctly be measured if both the indentation-size effect and the influence of the substrate have been properly considered. Although there is an effect of surface preparation on the microhardness measurement, it is always true that the measured hardness increases as the size of the indentation decreases. In other words, as the indentation depth increases, the hardness decreases until it reaches a constant value. It has been suggested that this size effect is primarily due to extreme localized work hardening, which causes an increase of yield stress within a very small volume since all glide planes are active and intersecting [82]. To obtain a meaningful result of bulk material hardness, a minimum indentation load is required to guarantee a plastic deformation so that the hardness is independent of the applied load. It is, however, difficult to meet this requirement when measuring hardness of thin films. Therefore, when small loads are normally used in thin film samples, one has to be aware that the extremely high hardness value might be erroneous due to the elastic properties of the film.

To eliminate the substrate-induced artifacts in hardness measurements, the ratio of indentation depth to film thickness should be limited below a critical value of 0.07 to 0.2. The most unfavorable case is that of a hard coating on a softer substrate [83].

To measure the hardness of thin films, the use of very small indentation sizes has put severe requirements on the resolution of optical measurements, which often introduces large errors in diagonal measurements. Therefore, it has been recognized that depth-sensing instruments are needed instead, in order to provide the necessary resolution and repeatability in the measurement of thin-film hardness. For this purpose, the ultramicrohardness testers or Nanoindentor [84] were developed. In the past decade, the resolution in depth measurement has been enhanced from 10 nm (interferometer-capacitor) [85] in 1978 to 5 nm by an inductive displacement transducer (1984) [86] or by a capacitance bridge (1982) [87], and further down to 0.2-0.3 nm with a force resolution of about 0.5 μN by a capacitance gauge [84]. By using this latest instrument [84], elastic displacements during the penetration of the diamond indenter are determined from the data obtained by unloading. Young's modulus can then be calculated. In addition, the elastic contribution to the total displacement can be removed in order to calculate the hardness. This kind of mechanical characterization of thin films by a Nanoindenter or ultramicro-indentation has also been done recently on implanted amorphous alumina and a-C:H film by researchers at Oak Ridge

National Laboratory [88] and at Philips Laboratory [89], respectively. Moreover, hardness was shown to be dependent on strain rate and repeatable hardness values can usually be obtained only when the indentation depth is greater than ~20 nm [90].

There are other methods than the indentation tests for measuring the hardness values. Some of them might be relevant to the hardness evaluation of a-C:H films and will be discussed here.

(1) Microscratch Test

The earliest attempt to measure microhardness by a microscratch tester is called Bierbaum Microcharacter [91], in which a diamond indenter is drawn across a polished metal surface under a load of 3 g. The width λ of the resulting groove is measured with a microscope in μm. The microhardness number, which is roughly equal to Vickers hardness [92] is then

$$B_i = 10^4/\lambda^2. \quad (4)$$

For scratch test on coatings, the adhesion effect should be taken into account.

(2) Abrasion Tests

A small, high-speed abrasive wheel is pressed against the test surface under a fixed load for a standard time. The size of the wear scar is the measure of hardness of the sample [92]. The complication of the abrasion method for hardness measurement of coatings is that adhesion is important to the interpretation of the results [93]. It was found that by use of a group of nine specimens tested on a commercial lapping machine simultaneously, one is able to measure the hardness of brittle specimens as well as abrasive powders [94]. Since the wear is measured by weighing, the use of materials with a higher density gives greater accuracy. Thus porous materials should be avoided because they lead to irregular results. In another study of hardness of nitrogen-implanted steels, the surface hardness was obtained from a verified relationship between wear rate and hardness [95].

For hard abrasives the volume of wear is found to be nearly proportional to the load and independent of the area [96]. For non-hardened materials the relative wear resistance, which is equal to the ratio of the volume of wear on the reference material to the volume of wear on the test material, is found to be proportional to the hardness [97]. This thus provides a measure for evaluating hardness, which characterizes the resistance to plastic deformation of the material.

(3) Simplified Model of Film Deformation

When the films are too thin for the direct measurement of hardness, the use of a simple model of film deformation during indentation allowed Jonsson and Hogmark [83] to estimate the characteristic hardness of the thin film from the conventional microhardness tests of both coated and uncoated substrates and a knowledge of the film thickness. The model was verified for chromium films with thickness greater than 200 nm on four different substrate materials.

6. TRIBOLOGICAL PROPERTIES

It is well known that in both abrasive and adhesive wear situations, the wear rate generally is inversely proportional to the yield pressure or hardness [98]. Therefore, the hardness is of prime importance for hard coatings used in wear protection. However, the hardness of a thin film, such as an overcoat on a magnetic media disk for data storage, is often too thin to be measured directly. In some cases, even if it were measurable, it may not be representative of the property of interest. In certain cases, the wear rate can be independent of hardness, and Archard's equation, which correlates wear rate W with the applied pressure P and the hardness of the material H,

$$W = k_o\ (P/H) \qquad (5)$$

is not followed [99]. Among the others, fracture might play a part in the wear of brittle solids [100], and fracture toughness should thus be included in the wear rate equation [101]. An example of the complication related to the transition from ductile to brittle wear occurs in erosive wear, i.e., it is the reversal in the dependence of erosive wear resistance on hardness of steels at high particle energy [99]. Furthermore, in sliding wear applications, mechanical wear can sometimes be reduced by choosing a coating of high ductility, because wear particles do not form even as extreme deformation proceeds [102]. Thus, soft coatings are very effective in preventing delamination wear. A mathematical model of subsurface crack propagation in delamination wear is used to explain the phenomenon that increased hardness sometimes is related to increased wear [103].

In a recent tribological study of submicron thin films for magnetic recording application, a series of wear tests were carried out by a Al_2O_3-TiC slider over a spinning disk at a speed of 100 cm/sec and 150 mN load. A variety of thin films: carbon, SiO_2, Co-30 at % Ni, and Cr were deposited by RF magnetron sputtering on 130 mm disks [104]. Two kinds of wear mode were observed. Abrasive grooves with significant plastic deformation are observed in the severe wear regime, whereas in mild wear regime several micro-cavities were formed exhibiting brittle fracture characteristics. In none of the cases was a correlation between the wear resistance and the microhardness of the material found. Therefore, the microhardness measurement was considered to be inadequate for the purpose of evaluation of macroscopic properties, such as wear resistance, of the material. However, the load-indentation depth curve was found useful to detect the presence of microdefects and blistering within the thin film and at the interface, respectively. The defect density measured in this way appeared to be closely related to the wear resistance. The specific wear rate appeared to increase with increasing compressive stress in the a-C film. Highest compressive stress occurred in the carbon film when sputtered on glass. Consequently, highest wear rate and film separation from glass substrate were observed apparently due to the high stress in the carbon film [104].

Since wear rate is not necessarily related to the film hardness in an explicit manner, the wear properties of a film are best evaluated directly by measuring the rate of removal of the film material, while subjecting the film surface to a controlled abrasion. This method has been adopted in studying the wear resistance of a variety of the amorphous films, including hydrogenated and unhydrogenated carbon films [105]. It was noted that the abrasive wear resistance of different unhydrogenated films can be correlated with the atomic bond strength of the material, i.e., it decreases in the order of carbon, SiNx, silicon and germanium. For hydrogenated films, the wear rate was found to vary systematically with the amount of hydrogen incorporated in the films during the deposition process. The presence of hydrogen bond diminishes the degree of crosslinking of the network because hydrogen has only one bond and therefore acts as a network terminator. The incorporation of hydrogen into the films thus induces a substantial increase in wear rate of the film. When the hydrogen concentration in amorphous films is increased, the decrease in the wear resistance is most pronounced in the carbon films which have the highest wear resistance among the unhydrogenated films [105].

The friction coefficient, μ of a-C:H is usually low. The μ values at 1% relative humidity (RH) from 0.01 to 0.23 were measured between a steel ball and a-C:H film prepared by RF plasma decomposition of ethylene [67]. The μ value decreases with decreasing RH and exhibits extremely low values of 0.005-0.02 under vacuum. This resembles diamond except that μ of diamond remains below 0.07 even for an RH of nearly 100%, whereas the friction coefficient of a-C:H sliding on steel was found to be strongly dependent on the RH in the test atmosphere [67]. According to another study, at low RH (<1%), very low μ values (0.01-0.02) were obtained. Above 1% RH the μ increases rapidly and reaches about 0.2 at RH$\sim$100% [106]. This friction behavior is the opposite of that in graphitic films which have higher friction and pronounced wear at low humidity. The friction between a-C:H and diamond or between two a-C:H films is less sensitive to humidity than the friction between a-C:H and steel and was measured to be less than 0.05 in both dry and humid atmosphere [106]. The friction is a function of the shear strength of the elastic contact area [77]. Thus the low values of μ of a-C:H and diamond are not easily understood, since no layered structure exists in these materials.

Because of the combination of high hardness and low friction coefficient, the a-C:H films prepared by plasma decomposition of C_2H_4 show negligible wear in pin-on-disk tests even after more than $4x10^6$ rotations in some cases [106]. The wear life of the film was found virtually to be independent of the rotation speed. However, as the load increases, the life decreases sharply, since the film was destroyed quite fast by friction or breaking away. Wear life was also found to be adversely affected by the surface roughness of the coated substrate. To counteract this effect, the film needed to be appreciably thicker than 0.3 μm [106]. By examining the wear track during the life tests, the wear of a-C:H films was found to be caused mainly by being locally rubbed away. This is entirely different from the behavior of a MoS_2 film. The wear tests between a steel pin and a steel substrate, both with and without a-C:H coatings, showed that the wear rate for those with coatings was about two orders of magnitude lower than those without coatings. Therefore, on a polished substrate, the a-C:H films offer good protection against wear under moderately high loads and temperatures [106].

In a more recent study of polymeric layers of a-C:H prepared by plasma-activated CVD, similar results to the previous works were obtained on friction tests of a steel ball sliding on an a-C:H-coated disk [89]. The films with H/C ratio of 0.3-0.5 exhibited an amorphous structure with strong cross-linkage between the polymer [107]. The microstructure of polymeric films depends strongly on the type of hydrocarbon and the RF power used in the deposition process. However, all type of polymeric films, showed essentially the same frictional behavior [89]. It is quite unusual that extremely low friction ($\mu \sim 0.02$) was achieved by a-C:H vs iron under ultrahigh vacuum ($\leq 10^{-8}$ Torr), since for most materials in ultrahigh vacuum (UHV) a very high friction has normally been reported if the sliding surfaces were very clean. A significant increase of friction coefficients in UHV and dry N_2 (up to 0.7 and 0.5 respectively) occurred after a-C:H films were annealed at 550°C. This is certainly related to the hydrogen content in the film, since nearly all hydrogen was removed at temperatures of around 600°C and a pure carbon film was left. Moreover, material transfer during sliding was observed when tested in dry O_2 and dry N_2, which at least explains the high friction ($\mu = 0.6$) observed in dry O_2 [89].

Another tribological study done in Philips Laboratory was reported recently [108]. The metal-containing a-C:H films were deposited by a reactive RF sputtering process using an Ar/C_2H_2 or Ar/C_2H_4 plasma. In order to improve the adhesion to the steel substrate, a thin metal layer of about 0.1 µm thickness was first deposited in a pure Ar atmosphere. Then hydrocarbon gas was slowly introduced so that a metal gradient was formed over the thickness. The H/C ratio was around 0.2-0.5, whereas the ratio of metal to carbon,

$$x_{Me} = \frac{c_{Me}}{c_{ME} + c_{carbon}} \qquad (6)$$

was measured by electron probe microanalysis (EPMA). The friction coefficient μ between the disk coated with these a-C:H films and uncoated steel ball generally increases with increasing metal concentration at low humidity. For instance, a small concentration of iron ($x_{Fe} = 0.041$) leads already to a considerable increase in μ under UHV (from <0.02 to 0.14). However, at higher RH (>1%) the μ values became lower than that of metal-free a-C:H films, because of the drastical increase in μ of metal-free a-C:H with RH [108].

All the test results demonstrated that tribological behaviors are closely related to the testing environment, such as humidity, atmosphere, temperature, material transfer, airborne contaminants, debris buildup, etc. Furthermore, neither wear resistance nor friction coefficient are intrinsic material properties, the wear problem of a-C:H films is thus related to both sides of the wear interface. For example, according to Khrushchev, little or no abrasive wear occurs when the hardness ratio H_a/H_M of the wear pair is in the range of 0.72 to 1.12 (where H_a denotes the hardness of the abrading species and H_m--abraded material) [98]. Therefore, the tribological properties of a-C:H films could not be treated properly if its counterpart in the wear pair has not been taken into consideration.

References

1. Robertson, J.: Advances in Physics, 1986, 35, 317.

2. Robertson, J, and O'Reilly, E.P.: Physical Review, 1987, 35B, 2946.

3. Savvides, N. and Window, B.: J. Vac. Sci. Technol. 1985, A3, 2386; Savvides, N.: J. Appl. Phys. 1986, 59, 4133.

4. Banks, B.A. and Rutledge, S.K.: J. Vac. Sci. Technol. 1982, 21, 807.

5. Kaplan, S., Jansen F. and Machonkin, M.: Appl. Phys. Lett, 1985, 47, 750.

6. Tsai, Hsiao-chu, Bogy, D.B., Kundmann, M.K., Veirs, D.K., Hilton, M.R. and Mayer, S.T.: J. Vac. Sci. Technol. 1988, A6, 2307.

7. Savvides, N.: J. Appl. Phys.: 1985, 58, 518.

8. Berger, S.D. and McKenzie, D.R.: Philos, Mag. letters, 1988, 57, 285.

9. Wada, N., Gaczi, P.J. and Solin, S.A.: J. Non-cryst. Solids, 1980, 35/36, 543.

10. Kakinoki, J., Katada, K., Hanawa, T. and Ino, T.: Acta Cryst. 1960, 13, 171.

11. Beeman, D.; Silverman, J., Lynds, R. and Anderson, M. R.: Phys. Rev. 1984, 30B, 870.

12. Jansen, F., Machonkin, M., Kaplan, S. and Hark, S.: J. Vac. Sci. Technol. 1985, A3, 605.

13. Tsai, Hsiao-chu and Bogy, D.B.: J. Vac. Sci. Technol. 1987, A5, 3287.

14. Dischler, B., Bubenzer, A. and Koidl, P.: Appl. Phys. Lett. 1983, 42, 636; and Solid state Commun. 1983, 48, 105.

15. Nadler, M.P., Donovan, T.M. and Green, A.K.: 1984, 116, 241.

16. Couderc, P. and Cahterine, Y.: Thin Solid Films, 1987, 146, 93.

17. Weast, R.C.: Handbook of Chemistry and Physics, 1983, CRC Press, 63rd ed. p.F-180.

18. Alterovitz, S.A., Warner, J.D., Liu, D.C. and Pouch, J.J.: J. Electrochem. Soc. 1986, 133, 2339.

19. Fink, J., Miller-Heinzerling, T., Pfluger, J., Bubenzer, A., Koidl, P. and Crecelius, G.: Solid State Commun. 1983, 47, 687.

20. Wesner, D., Krummackrer, S., Carr, R., Sham, T.K., Strongin, M., Eberhardt, W., Weng, S.L., Williams, G., Howells, M., Kampas, F., Heald, S. and Smith, F.W.: Phys. Rev. 1983, 28B, 2152.

21. Miller, D.J., and McKenzie, D.R.: Thin Solid Films, 1983, 108, 257.

22. McKenzie, D.R., Botten, L.C. and McPhedran, R.C.: Phys. Rev. Lett. 1983, 51, 280.

23. Memming. R.: Thin Solid Films, 1986, 143, 279.

24. McKenzie, D.R., McPhedran, R.C., Savvide, N. and Botten, L.C.: Philos. Mag. B, 1983, 48, 341.

25. Smith, F.W.: J. Appl. Phys. 1984, 55, 764.

26. Craig, S. and Harding, G.L.: Thin Solid Films, 1982, 97, 345.

27. Jones, D.I. and Stewart, A.D.: Philos. Mag. B, 1982, 46, 423.

28. Ojha, S.M., Norstrom, H. and McCulluch, D.: Thin Solid Films, 1979, 60, 213.

29. Ingram, D.C., Woollam, J.A. and Bu-Abbud, G.: Thin Solid Films, 1986, **137**, 225.

30. El-Hossary, F.M., Fabian, D.J. and Sofield, C.J.: Thin Solid Films, 1988, 157, 29.

31. Angus, J.C., Stultz, J.E., Shiller, P.L., MacDonald, J.R., Mirtich, M.J. and Domitz, S.: Thin Solid Films, 1984, 118, 311.

32. Nyaiesh, A.R. and Nowak, W.B.: J. Vac. Sci. Technol., 1983, A1, 308.

33. Elings, V. and Wudl, F.: J. Vac. Sci. Technol., 1988, A6, 412.

34. Angus, J.C., Koild, P. and Domitz, S.: Carbon Thin Films in "Plasma Deposited Thin Films" (J. Mort and F. Jansen, ed.), 1986, CRC Press, Boca Raton, Florida, p.89.

35. Morgan, M.: Thin Solid Films, 1971, 7, 313.

36. Hauser, J.J.: J. Non-Cryst. Solids, 1977, 23, 21.

37. Andersson, L.P., Berg. S., Norstrom, H., Olaison, R. and Towta, S.: Thin Solid Films, 1979, 63, 155.

38. Ojha, S.M., Norstrom, H. and McCullich, D.: Thin Solid Films, 1979, 60, 213.

39. Woollam, J.A., Natarajan, V., Lamb, J., Azim Khan, A., Bu-Abbud, G., Mathine, D., Rubin, D., Dillon, R.O., Banks, B., Pouch, J., Gulino, H., Domitz, S., Liu, D.C. and Ingram, D.: Thin Solid Films, 1984, 119, 121.

40. Mirtich, M.J., Swec, D.M. and Angus, J.C.: Thin Solid Films, 1985, 131, 245.

41. Natarajan, V., Lamb, J.D., Woollam, J.A., Liu, D.C. and Gulino, D.A.: J. Vac. Sci. Technol., 1985, A3, 681.

42. Koeppe, P.V., Kapoor, V.J., Mirtich, M.J., Banks, B.A. and Gulino, D.A.: J. Vac, Sci. Technol., 1985, A3, 2327.

43. Whitmell, D.S. and Williamson, R.: Thin Solid Films, 1976, 35, 255.

44. Anderson, D.A.: Philos. Mag. 1977, 35, 17.

45. Pirker, K., Schallauer, R., Fallmann, W., Olcalyug, F., Urban, G., Jachimowicz, A., Kohl, F. and Prohaska, O.: Thin Solid Films, 1986, 138, 121.

46. Bewilogua, K., Dietrich, D., Pagel, L., Schurer, C. and Weissmantel, C.: Surf. Sci. 1979, 86, 308.

47. Bubenzer, A., Dischler, B., Brandt, G. and Koidl, P.: J. Appl. Phys. 1983, 54, 4590.

48. Dischler, B., Bubenzer, A. and Koidl, P.: Appl. Phys. Lett. 1983, 42, 636; also; Solid State Commun. 1983, 48, 105.

49. Thun, R.E. in "Physics of Thin Films" (G. Hass ed.), Vol. 1, Academic Press, New York, 1963, 187.

50. Agarwal, S.: "Structure and Morphology of RF Sputtered Carbon Overlayer Films" (unpublished).

51. Staudinger, A. and Nakahara, S.: Thin Solid Films, 1977, 45, 125.

52. Marinkovic, Z. and Roy, R.: Carbon, 1976, 14, 329.

53. Antilla, A., Koskinen, J., Bister, M. and Hirvonen, J.: Thin Solid Films, 1986, 136, 129.

54. Mott, N.F. and Davis, E.A.: "Electronic Processes in Non-Crysatlline Materials", 1971, Clarendon, Oxford, p. 197-219.

55. Adkins, C.J., Freake, S.M. and Hamilton, E.M.: Philos, Mag. 1970, 22, 183.

56. Myerson, B. and Smith, F.W.: J. Non-Cryst. Solids, 1980, 35 & 36, 435.

57. Hauser, J.J.: Solid State Commun. 1975, 17, 1577.

58. Hauser, J.J. and Patel, J.R.: Solid State Commun. 1976, 18, 789.

59. Weissmantel, C., Bewilogua, K., Dietrich, D., Erler, H.-J., Hinneberg, H.-J, Klose, S., Nowick, W. and Reisse, G.: Thin Solid Films, 1980, 72, 19.

60. Has, Z., Mitura, S., Clapa, M. and Szmidt, J.: Thin Solid Films, 1986, 136, 161; also Staryga, E., Lipinski, A., Mitura, S. and Has. Z.: Thin Solid Films, 1986, 145, 17.

61. Mattox, D.M.: in "Deposition Technologies for Films and Coating", (R.F. Bunshah ed.), Noyes, Park Ridge, N.J., 1982, 63.

62. Jaccodine, R.J. and Schlegel, W.A.: J. Appl. Phys. 1966, 37, 2429.

63. EerNisse, E.P.: J. Appl. Phys. 1977, 48, 3337.

64. Matuba, N., Baba, S. and Kinbara, A.: Thin Solid Films, 1981, 81, 301.

65. Pliskin, W.A.: J. Vac. Sci. Technol. 1977, 14, 1064.

66. Weissmantel, C., Schurer, C., Frohlich, F., Grau, P. and Lehmann, H.: Thin Solid Films, 1979, 61, L5.

67. Enke, K.: Thin Solid Films, 1981, 80, 227.

68. Prince, E.T., and Romach, M.M.: J. Vac. Sci. Technol. 1985, A3, 694.

69. Nir, D.: Thin Solid Films, 1984, 112, 41.

70. Kinbara, A., Baba, S., Matuba, N. and Takamisana, K.: Thin Solid Films, 1981, 84, 205.

71. Gille, G. and Rau. B.: Thin Solid Films, 1984, 120, 109.

72. Zelez, J.: J. Vac. Sci. Technol. 1983, A1, 305.

73. Kerwin, D.B., Spain, I.L., Robinson, R.S., Daudin, B., Dubus, M. and Fontenille, J.: Thin Solid Films, 1987, 148, 311.

74. Nir, D.: Thin Solid Films, 1987, 146, 27.

75. Benjamin, P. and Weaver, C.: Proc. R. Soc. London, 1960, A254, 163.

76. Hintermann, H.E.: Wear, 1984, 100, 381.

77. Miyoshi, K., Burkley, D.H., Pouch, J.J., Alterovitz, S.A. and Sliney, H.E.: Surface and Coating Technology, 1987, 33, 221; also Inst. Mech. Eng. 1987, C222, 621.

78. Steinmann, P.A. and Hintermann, H.E.: J. Vac. Sci. Technol. 1985, A3, 2394.

79. Ramalingam, S.: Thin Solid Films, 1984, 118, 335.

80. Weissmantel, C., Bewilogua, K., Breuer, K., Dietrich, D., Ebersbach, U., Erler, H.-J., Rau, B., and Reisse, G.: Thin Solid Films, 1982, 96, 31.

81. Sundgren, J.-E. and Hentzell, H.T.G.: J. Vac. Sci. Technol. 1985, A3, 2259.

82. Pethica, J.B., Hutchings, R. and Oliver, W.C.: Philos. Mag. 1983, A48, 593.

83. Jonsson, B. and Hogmark, S.: Thin Solid Films, 1984, 114, 257.

84. Doerner, M.F. and Nix, W.D.: J. mat. Res. 1986, 1, 601.

85. Nishibori, M. and Kinosita, K.: Thin Solid Films, 1978, 48, 325.

86. Wierenga, P.E. and Franken, J.J.: J. Appl. Phys. 1984, 55, 4244.

87. Newey, D., Wilkins, M.A. and Pollock, H.M.: J. Phys. 1982, E15, 119.

88. Oliver, W.C. and McHargue, C.J.: Thin Solid Films, 1988, 161, 117.

89. Memming, R., Tolle, H.J. and Wierenga, P.E.: Thin Solid Films, 1986, 143, 31.

90. Doerner, M.F., Gardner, D.S. and Nix, W.D.: J. Met. Res., 1986, 1, 811.

91. Bierbaum, C.H.: Trans. Am. Soc. Steel Treat., 1930, 18, 1009.

92. McClintok, F.A. and Argon, A.S.: Mechanical Behavior of Materials, 1966, Addison-Wesley, Reading, MA, p.451-453.

93. Maissel, L.I. and Glang, R.: Handbook of Thin Film Technology, 1970, McGraw-Hill, New York, p.12

94. Rabinowicz, E.: Lubr. Eng. 1977, 33, 378.

95. Bolster, R.N. and Singer, I.L.: Appl. Phys. Lett. 1980, 36, 208.

96. Richardson, R.C.D.: Proc. Inst. Mech. Engrs. London, 1967, 182, 410.

97. Khruschov, M.M. and Babichev, M.A.: Russ. Eng. J., 1964, 44(6), 43.

98. Peterson, M.B. and Ramalingam, S. in "Fundamentals of Friction and Wear of Materials", (A.A. Rigney ed.), American Society of Metals, Metal Park, OH, 1980, 331.

99. Hornbogen, E.: Wear, 1975, 33, 251.

100. Moore, M.A.: in ref. 98, p.73.

101. Evans, A.G. and Wilshaw, T.R.: Acta Metall., 1976, 24, 939.

102. Kramer, B.M.: Thin Solid Films, 1983, 108, 117.

103. Fleming, J.R. and Suh, N.P.: Wear, 1977, 44, 39.

104. Tsukamoto, Y., Yamaguchi, H. and Yanagisawa, M.: IEEE Trans. Magnetics, 1988, 24, 2644.

105. Jansen, F. and Machonkin, M.A.: Thin Solid Films, 1986, 140, 227.

106. Dimigen, H. and Hubach, H.: Philips Tech. Rev., 1983/84, 41, 186.

107. Memming, R.: Thin Solid Films, 1986, 143, 279.

108. Dimigen, H., Hubsch, H. and Memming, R.: Appl. Phys. Letter. 1987, 50, 1056.

Table I. Structural/bonding characteristics and physical properties of diamond, graphite and amorphous carbon (a-C).

	Structure/bonding	Hardness kg mm^{-2}	Density g·cm^{-3}	Optical gap E_o,eV	Resistivity Ωcm	Reference
Diamond	isotropic sp^3	10^4	3.52	5.5	10^{18}	[1]
Graphite	hexagonal layers of sp^2, van der Waals along c-axis	soft	2.3	0	$4x10^{-5}$ ($\perp$c)	[1]
Evaporated a-C	random network of sp^2 cluster linked by sp^3, upper limit of sp^3 $\sim$ 12%.	20-50	$\sim$2.0	0.4-0.7	0.1	[1, 35]
Sputtered and ion beam a-C	similar to evaporated a-C, sp^2/sp^3 ratio varies in a wide range.	1350 - 2400	1.7-2.8	0.4-3.0	0.2-10^{11}	[3, 4, 7, 36]

Table II. Physical properties of hydrogenated carbon (a-C:H) films.

Source of Carbon	H/C Ratio	Density (g/cm^3)	Optical energy gap, E_o(eV)	Electrical resistivity (Ω cm)	Hardness HV or HK (kg/mm^2)
Carbon target sputtered by ion beam [12]	0-0.54	2.25-1.6	<1.2	10^5 - 10^{10}	2000g to scratch
A variety of hydrocarbon in a RF plasma [37]		1.9-2.0		10^9	3400HV
CH_4 or C_4H_{10} in a RF plasma [38]	0.29-0.42	2.0-2.67		10^{12}	
CH_4 in a RF plasma [39]			1.3-2.7	up to 10^{13}	
Dual beam using CH_4 +Ar [40]	1.0	1.8	0.34	3.35 x 10^6	
CH_4 in a DC plasma [5]	0.45-0.89	1.46-1.7	1.7		700-1250 (g to scratch)
CH_4 in a RF plasma [5]	1.38-1.56	1.02-1.17	4-4.1		160-250 (g to scratch)
CH_4 or C_2H_4 in a RF plasma [41]			2.7	>10^{13}	4-7 Mohs
Ion beam using 25% CH_4 + Ar source [42]	0.08		0.9-1.1	8.1 x 10^6	
C_2H_4 in a DC plasma [43]				>2 x 10^5	>2800HV

C_2H_2 in a RF plasma [44]		1.7	1.5-2.6	10^{12}	
C_2H_2 in a DC plasma [26]	0.61	1.27	1.55	$>10^{7}$	
C_2H_2 in a DC plasma [24]		1.2-1.3	1.8	$>10^{8}$	
C_2H_2 in a DC plasma [25]		1.35	2.2		
C_2H_2 in a RF plasma [45]	very low (H)		0.65-1.1	10^{2}-10^{5}	
DC ionization of C_6H_6 at ion energy of [46] (A) 250 eV (B) 800 eV			(A)>2	(A) $>10^{10}$	(A) ~5000HV (B) ~3000HV
C_6H_6 in a RF plasma [47][48]	0.5	1.5-1.8	0.8-1.8	10^{12}	1250-1650HK

Materials Science Forum Vols. 52 & 53 (1989) pp. 103-124

THE MICROSTRUCTURE OF CARBON THIN FILMS

D.C. Green, D.R. McKenzie and P.B. Lukins
School of Physics, University of Sydney, NSW, 2006, Australia

ABSTRACT

Amorphous carbon shows a wide variety in its physical behaviour from being electrically conducting to highly insulating, from light absorbing to transparent and from soft to very hard. This range of behaviour can be understood on the basis of the structure of the amorphous network. Techniques for determining the structure of carbon films are discussed. Forms of amorphous carbon and poorly crystalline carbon are then considered, using the results of the techniques best suited to that form of carbon. The observed structures range from poorly crystalline graphite through structures containing odd-membered rings of atoms to structures closely related to that of amorphous silicon, containing a very high fraction of tetrahedrally bonded atoms.

INTRODUCTION

Carbon is unique amongst the Group IV elements in its ability to bond in a variety of electronic configurations. Within the simple hybridization model, these configurations are denoted sp, sp^2 and sp^3 and are illustrated schematically [1] in figure 1.

In common with the remainder of the Group IV elements (Si, Ge etc) carbon forms the "diamond" structure that is, the zincblende structure with like atoms [2] as shown in figure 2. In this structure, the carbon atoms are sp^3 hybridized and have four bonds tetrahedrally oriented.

A related structure also showing sp^3 bonding is the so called hexagonal diamond or lonsdaelite. This form is found in meteorites and has been synthesized in the laboratory [3]. Hexagonal diamond has the wurtzite structure [4] with like atoms and is distinguished by the existence of a six-fold symmetry axis along the c-direction compared with a three-fold symmetry axis in cubic zincblende. There is no measurable difference between the first interatomic distance in the two forms.

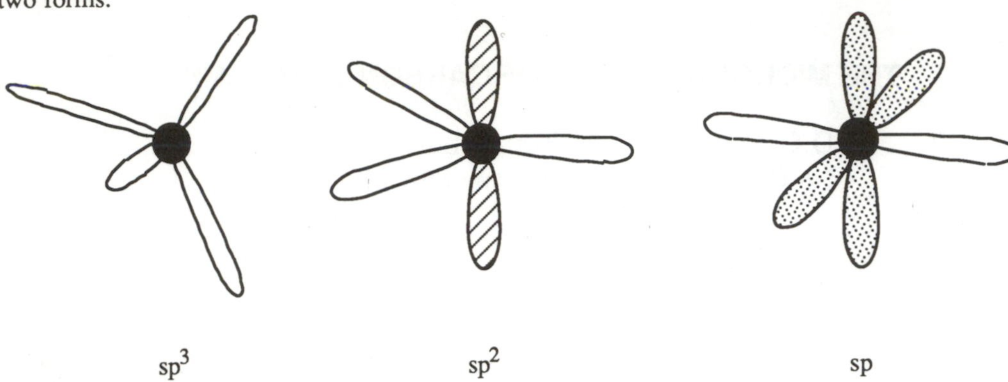

FIGURE 1 Schematic diagram of the hybridizations of s and p atomic orbitals permitted in carbon. The σ orbitals are shown un-shaded and the π orbitals are shown shaded.

Unlike the other Group IV elements, elemental carbon is capable of bonding with sp^2 hybridization of atomic orbitals in which three bonds are coplanar and directed to the vertices of a triangle. Structures derived from atoms in this hybridization may contain sheets of atoms having hexagonal symmetry. One particular stacking arrangement of these sheets gives rise to the Bernal graphite structure where the bonding between the sheets is of Van der Waals type and the stacking sequence is ABAB as shown in figure 2(b). Considerable variations on this structure can occur such as distortions of the sheets (bending, buckling etc), misalignment of the stacking sequence as well as conventional defect structures such as point defects and slip planes.

FIGURE 2 Two of the crystallographic forms of carbon. (a) cubic diamond and (b) graphite.

Carbon also forms triple-bonded compounds such as acetylene C_2H_2 and "carbyne" $(C{\equiv}C)_n$ where the carbon atoms have sp hybridization but in the solid state the presence of elements other than carbon is generally required to stabilize the structure.

Non-crystalline and poorly crystalline carbons may show mixtures of sp^2 and sp^3 hybridizations. Those forms showing a predominance of sp^2 may be classified according to the degree of perfection of the graphite like sheets (the spatial extent and stacking pattern). For those forms showing a predominance of sp^3 hybridization, the local environment of a carbon atom is tetrahedral although there are observed departures from the diamond structure, even locally, caused by distortions of the bond angle and the freedom of rotation about the axes of the bonds. In this respect these forms resemble the structures of amorphous silicon and amorphous germanium.

PREPARATION

Thermal Evaporation

Coatings of amorphous carbon are produced by evaporation from hot surfaces of carbon. This is commonly achieved by passing a high current between graphite electrodes in contact in vacuum. The ends of the electrodes are heated sufficiently to cause evaporation of carbon. The same result is achieved in a more controllable way by electron beam heating of a carbon target. Under ideal conditions these coatings contain little hydrogen or oxygen contamination. These materials have an electrical resistivity considerably higher than that of graphite in the a-b plane and are properly classed as semiconductors [5].

Sputtering

Carbon has a relatively low sputtering yield (0.12 with argon at 500eV [6]), but useful deposition rates can be achieved by using a magnetron usually with argon as the discharge gas. The energy of the sputtered carbon vapour at the substrate depends on the pressure of the background gas at which the magnetron is operated. Spectroscopic studies of emission from sputtering discharges [7] have shown that at all but the lowest attainable pressures the vapour is essentially thermal in energy and as a result the material is essentially similar to the amorphous carbon produced by thermal evaporation. Hydrogen is readily incorporated into the film if it is present in the deposition gas [8].

Glow Discharge CVD of Hydrocarbons

When hydrocarbon gases are decomposed in a glow discharge plasma operated at frequencies in the range from DC to GHz, a hydrogenated form of amorphous carbon is produced. The properties of these materials are quite dependent on preparation conditions and they range in hardness from soft "polymeric" materials to quite hard coatings which have variously been termed "diamond-like carbon", i-carbon and a-C:H. The term "diamond-like carbon" when applied to this material does not imply any structural relationship to diamond or the presence of sp^3 bonding other than that which involves hydrogen. Under conditions of high substrate temperature and high hydrogen dilution of the gases in the glow discharge, diamond crystals may be produced, resulting in a polycrystalline coating of a substrate [9].

Ion Beam Deposition of Carbon

If a carbon-containing gas, usually a hydrocarbon, is introduced into a Kauffmann-type ion gun a beam of ions and ionized molecular fragments is produced which may be used to grow an amorphous film of carbon containing hydrogen. The important parameters determining the end product material are the source gas chemistry and the energy of the ions leaving the source [10]. At low energies of impact the films show substantial retention of chemical groups characteristic of the source gas. At higher energies, in most cases over 500eV or so, the product is independent of the impact energy and source gas and resembles in hardness and other properties a-C:H prepared by glow discharge CVD.

Vacuum Arc Deposition

Under appropriate conditions, an arc may be sustained in vacuum on a planar cathode well seperated from the anode [11]. The cathode spot of such a vacuum arc is a very hot plasma, is very highly ionised and acts as a source of ions and electrons. The cathode spot of a vacuum arc on a graphite cathode is then a source of singly and doubly ionized (≈60eV per charge state) carbon atoms. Such a plasma, when filtered of neutral carbon and graphite macroscopic particles by means of a curved magnetic filter, may be used to form an amorphous carbon film of exceptional properties [12]. It appears that under some conditions crystalline diamond is formed [13], and under others a form of amorphous carbon containing a very high fraction of sp^3 bonding is produced [14]. This form of amorphous carbon is termed amorphous diamond or a-D. A cubic form of crystalline carbon has also been reported with this technique which has a density exceeding that of cubic diamond and is termed C_8 [15]. The essential difference then between vacuum arc and ion beam depositions is that the former gives rise to a stream of ionised atoms which are used to grow the film while the latter gives rise to a stream of ionised molecular fragments for the same purpose.

Pyrolysis

Complex hydrocarbon polymer resins can be thermally modified to form an interesting form of carbon called glassy carbon. During the heat treatment, which is usually done in an inert atmosphere, the polymers are successively dehydrated, reduced to hydrogenenated carbon chain structures and finally dehydrogenated and annealed leaving a tangled ribbon-like network of sp^2 bonded carbon sheets [16]. These ribbons are entangled and cross linked with a small fraction of sp^3 bonded carbon. The material is extremely inert and very hard.

TABLE 1 DENSITY VALUES FOR FORMS OF CARBON

CARBON FORM	PREPARATION	DENSITY (kg m^{-3})	REFERENCE
Cubic C_8	arc	4100	[15]
Diamond	natural	3520	[2]
a-D	vacuum arc	3000	[17]
Graphite	natural	2260	[16]
Glassy Carbon	pyrolysis	800-1600	[16]

METHODS OF INVESTIGATING STRUCTURE

Methods of investigating the structure of amorphous or poorly crystalline carbon may be classified according to whether they probe atomic structure directly or whether they probe the electronic structure. These two features of structure are, of course, intimately linked and information about one may be used to make conclusions about the other.

The question of structure may be posed as one of the local bonding arrangements of atoms, particularly the distributions of atoms among the sp^2 and sp^3 hybridizations. Therefore methods which probe this distribution will be of great use and will be discussed.

Evaluation of Dielectric Function

The complex dielectric function $\hat{\varepsilon} = \varepsilon_1 + i\,\varepsilon_2$ is a measure of the response of a material to an imposed electric field. This function may be measured by optical techniques (either photometric or ellipsometric) or by electron energy loss techniques. In principle, both optical [18] and electron optical [19,20] methods may be used to determine $\hat{\varepsilon}$ over a wide energy range. However, it is often more convenient to carry out measurements in the vacuum ultraviolet using the electron energy loss techniques.

The optical dielectric function in the energy range (1 - 10eV) is dictated principally by the electronic band structure. A measurement of $\hat{\varepsilon}$ may be used to infer the relative numbers of atoms in sp^2 and sp^3 hybridizations using the sum rule for the number of electrons per unit volume

$$n_{eff} = \frac{WM_0}{\rho} \frac{\varepsilon_0 m}{2\pi^3 h^2} \int_0^{E_m} E\,\varepsilon_2(E)\,dE$$

where W is the atomic weight, ρ is the density (kg m^{-3}), M_0 is the atomic mass (kg), m is the electron mass (kg) and E is the energy in eV. The method relies on having an energy separation of the absorptions due to electronic transitions between the π bonding orbitals and the π^* anti-bonding orbitals and the absorption due to transitions involving σ and σ^* orbitals (see figure 3).

When applied to graphite using the dielectric function determined from measurements with the electric vector parallel to the sheets of atoms (i.e. in the a-b plane) the expected result of one π electron per atom is obtained. In the case of graphite the method works well because of the clear energy separation between the π and σ absorption regions. When applied to a-C:H [18], an obvious problem arises because of the energy overlap of the absorption regions, so that the calculated number of π electrons per atom is a function of the upper limit of the integral, E_{max}. The interpretation of these results also requires caution because of the polarisation sensitivity of the absorptions corresponding to transitions between π and π^* bands. The method has been recently criticised [21] on the basis of an error in the analysis of Taft and Philipp [22]. Using the corrected analysis [23] the absorption due to the π electrons is not saturated until well over 7eV, hence it is not possible to "factor out" σ-σ* transitions effectively. When these problems are taken into account, the method loses quantitative accuracy and should be regarded as an approximate technique only.

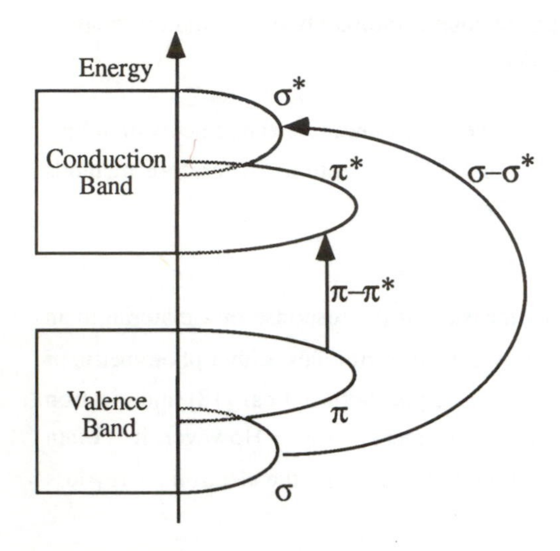

FIGURE 3 A schematic of the density of states of carbon showing the relation between the π and σ bonding orbitals and the π* and σ*anti-bonding orbitals. Also shown are the π–π* and σ-σ* transitions referred to in the text.

Electron Energy Loss Near Edge Structure

This method of probing the electron energy levels provides a fingerprint of the presence of π orbitals which give rise to a characteristic sharp pre-ionization feature on the K core loss edge as seen in figure 4. This method cannot be criticised for the same reasons as the method based on the dielectric function, as the contribution from the σ* states to the leading edge of the K core loss is likely to be small in all forms of carbon [1]. The method may be made quantitative by making use of fully graphitised carbon (randomly orientated) as a standard.

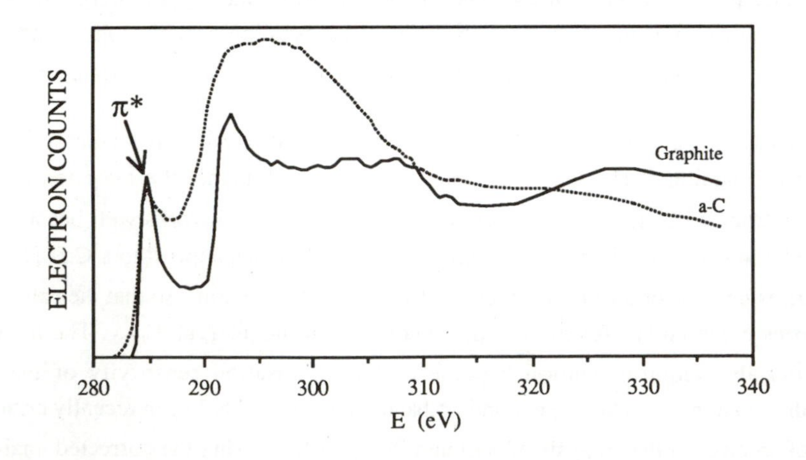

FIGURE 4 Electron energy loss sprectrum [14] for an evaporated amorphous carbon specimen in the vicinity of the carbon K-edge. Shown also, is the same region for a graphite standard.

Nuclear Magnetic Resonance

Nuclear magnetic resonance (NMR) spectroscopy is a powerful method in the study of bonding type and short range structure. Most recent solid-state ^{13}C NMR of amorphous carbons has been recorded using the proton-enhanced cross-polarization method with high power decoupling [24] and, in some cases, with delayed decoupling. Furthermore, magic-angle spinning [25] has been used to minimize spectral broadening due to dipolar interactions and chemical shift anisotropy.

The observed shifts for graphites (≈150ppm relative to TMS) [26,27] and diamonds (≈39ppm) [27,28] are consistent with the accepted models of graphite as an extended, layered aromatic structure and diamond as a tetrahedrally bonded network structure. Henrichs et al. [28] also observed very long spin-lattice relaxation times for diamonds and interpreted their data for powdered diamonds in terms of the presence of paramagnetic centres.

Vibrational Spectroscopy

Infra-red and Raman spectroscopies provide information on vibrational modes and, conversely, on the structure through a comparison with vibrational frequencies of known structures. Amorphous carbons show IR absorption bands in the range 700-1800 cm^{-1} [18,29-33]. In the case of a-C:H, the intense bands near 3000 cm^{-1} due to CH stretching modes are particularly useful in assessing the extent and character of carbon-hydrogen bonding. Both conventional dispersive and Fourier transform spectrometers have been used. The Raman spectrum of diamond has a single peak at 1332 cm^{-1} whereas, for graphite, a characteristic feature is the intralayer vibration at 1580 cm^{-1} [34,35].

Direct Imaging

Modern electron microscopes, under appropriate conditions, are capable of resolving individual atoms in a thin film specimen. This technique is useful in revealing the presence of sp^2 bonded layers if the stacking of these layers occurs to any extent. The characteristic interlayer distance of ≈3.4 Å is seen easily as fringes when the layers are oriented all but normal to the beam [36]. The interpretation of images becomes controversial if the extent of the layers is small, however. The image then reveals only a "speckle" pattern characteristic of amorphous materials. The spacing of the maxima and minima of intensity in such a pattern is characteristic of the contrast transfer function of the microscope and does not necessarily contain information about the interatomic spacings present in the specimen.

Electron Diffraction

For amorphous or poorly crystalline materials a useful approach in unravelling structure is the evaluation of the reduced density function (RDF). The RDF, G(r), is obtained by Fourier sine transformation of the reduced elastic scattering intensity,

$$\varphi(s) = \frac{(I(s)-Nf^2)}{Nf^2}\, s$$

where $s=2\sin\theta/\lambda$ and Nf^2 is the scaled atomic scattering intensity.

Convenient means of determining $\varphi(s)$ are neutron, x-rays, and electron diffraction. The specimen volume required decreases in the order neutron, x-ray and electron diffraction. In the case of electron diffraction, care must be taken to eliminate the considerable intensity of inelastic electrons due to plasmon scattering processes. This can conveniently be done using a commercial electron energy loss spectrometer as described recently [37].

Peaks in the RDF will, in principle, correspond to interatomic distances in the amorphous material since,

$$G(r) = 4\pi r\,(\rho(r) - \rho_o) = 8\pi \int_0^{s_{max}} \varphi(s)\, e^{-\beta s^2} \sin(2\pi sr)\, ds$$

where $\rho(r)$ is the local density, ρ_o is the average density, β is an artificial temperature factor and s_{max} is determined by the maximum collection angle of the microscope. The G(r) is asymptotic to zero at large r $(\rho(r) \approx \rho_o)$ in an amorphous specimen since there is no long range order.

Caution is sometimes required for the interpretation of experimental G(r). Oscillations below 1.2Å may be ignored since they and other spurious periodic features are caused by non-ideal conditions in the experimental $\varphi(s)$. With care in interpretation, valuable structural information can be obtained from the G(r) curve obtained from experimental intensity data.

Interpretation and Modelling

To assist with the interpretation of an experimental G(r), it is useful, especially for poorly crystalline specimens, to calculate $\varphi(s)$ and thence G(r) for small crystallites randomly oriented to the incident beam. This is conveniently done using the Debye formula for the scattered intensity,

$$I(s) = f^2 \sum_{j=1}^{N} \sum_{k=1}^{N} \sin(2\pi s r_{jk})$$

where r_{jk} is the seperation of atoms j and k and the summations are done for the atoms in a single crystallite. This is clearly an approximation to the polycrystalline state however it has been shown to be a valuable interpretative tool for other materials [38, 39]. The advantage of this approach, over direct evaluation of the RDF, is that the calculated intensity function can be processed as if it were experimental data.

Although there have been many attempts to model the network structure of amorphous silicon and amorphous germanium [40], there have been relatively few attempts to model the structure of amorphous carbon. Beeman et al. [41] have presented 4 "mixtures" of sp^2 and sp^3 bonded carbon atoms and compared them with then available experimental scattering and atomic distribution data [42,43]. Their models were of a constrained, random nature and the four values of the fraction of four-coordinated atoms were 0, 9.1%, 51.4% and 100%. The atomic positions within the structure were "relaxed" computationally to an equilibrium configuration using a valence force field model. Prior to this work, attempts to interpret the observed diffraction of x-rays by glassy carbon were made by Franklin [44], and Ergun [45] who incorporated strain into a random layer lattice model [46]. A domain model for glassy carbon was proposed [47] in which microcrystallites of graphite were envisaged to be connected by a proportion of sp^3 bonded carbons. The latter is known to predict sp^3 fractions which are too high [48,49].

THE STRUCTURE OF SOME AMORPHOUS CARBON FORMS

Glassy Carbon

Glassy carbon when viewed under the electron microscope at high resolution reveals a tangled network of fringes of spacing ≈3.35Å, the interlayer spacing in graphite. These fringes show that glassy carbon is essentially sp^2 bonded. The extent of the fringes increases with increasing treatment temperature [16], showing a tendency toward fully developed crystalline graphite. However , it is impossible to fully graphitise glassy carbon even at temperatures as high as 3000°C because of the high strain energy of the tangled sp^2 network. Electron imaging cannot provide an accurate estimate of the average dimensions of the ribbons in the "a,b plane" because any particular ribbon will pass in and out of the focal plane of the microscope.

Electron diffraction provides valuable information about the nature of the sp^2 bonded ribbons. In figure 5(a), we compare the G(r) functions obtained by Mildner and Carpenter [49] using neutron diffraction with those of Green [50] using energy filtered electron diffraction data [51]. The latter can be seen in figure 5(b) to exhibit intralayer order extending to at least 15 Å with the peaks corresponding to interlayer distances being poorly coordinated. This is consistent with the turbostratic model for glassy carbon. It should be noted that the strength of the diffraction peak corresponding to this first interlayer distance- the (002) reflection- showed some variation in intensity over specimen. The intralayer distances giving rise to the peaks in G(r) are shown in figure 6(a) and are tabulated together with their infinite crystal coordination numbers in Table 2. Also listed in the table are the average coordination numbers expected for atoms in an infinite length ribbon of width corresponding to only one hexagonal ring as shown in figure 6(b).

The peak in the experimental G(r) of figure 5(a) corresponding to the smallest interlayer distance (3.35Å) is usually weak, showing that this distance has a low coordination number. The G(r) curves obtained with electron diffraction can be used to estimate the dimensions of the sp^2 bonded ribbons in the a,b plane, since the coordination number of the peak can be obtained both, experimentally by appropriate integration of G(r) [37], and analytically as a function of the postulated ribbon dimensions [50]. On the basis of this comparison, the data of figure 5(b) is consistent with ribbons of dimensions 20Å by 100Å. Raman spectroscopy has been used on similar samples [36] to estimate the size of the correlated regions and the value of 30Å was found.

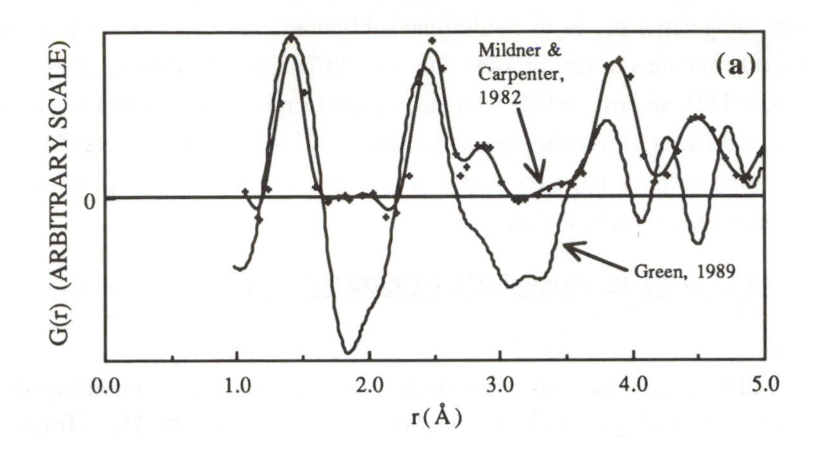

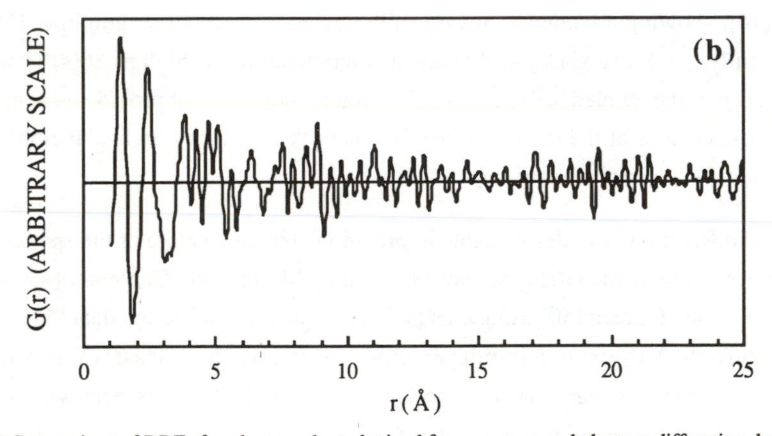

FIGURE 5 (a) Comparison of RDFs for glassy carbon obtained from neutron and electron diffraction data.
(b) RDF obtained from the same electron diffraction data shown on larger scale showing substantial intralayer order still present at 15Å.

TABLE 2 INTRALAYER INTERATOMIC DISTANCES AND COORDINATION NUMBERS

NUMBER	DISTANCE (Å)	INFINITE CRYSTAL	NARROW RIBBON
1	1.418	3	2.5
2	2.456	6	4.0
3	2.836	3	1.5
4	3.752	6	3.0
5	4.254	6	2.0

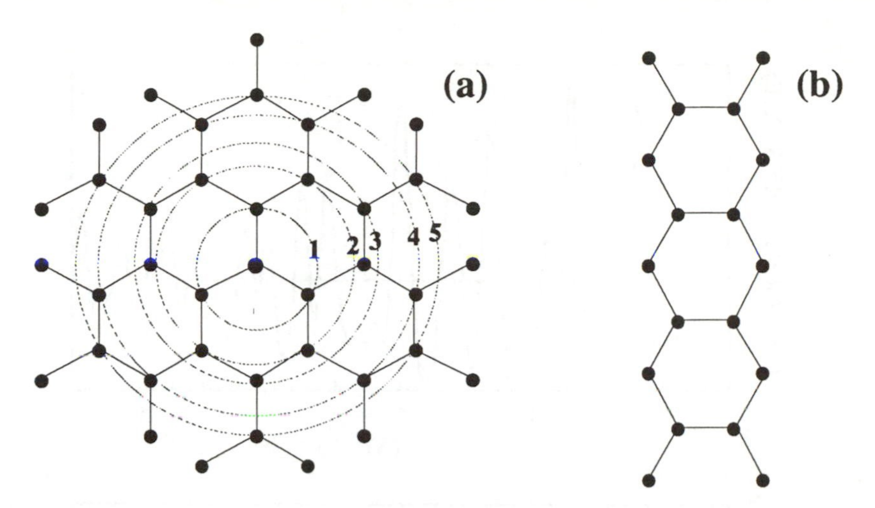

FIGURE 6 (a) A schematic diagram showing the first five intralayer interatomic distances exhibited in glassy carbon. (b) A section of infinite, narrow ribbon discussed in the text.

Electron micrographs show what appears to be considerable bending of the ribbons about axes parallel to a,b plane of the graphite-like ribbbons. However, by considering the scattering from randomly oriented graphite crystallites consisting of 2 layers, the effect of even slight curvature on the G(r) of such a structure is quite marked, as shown in figure 7, where the assemblies of atoms and their G(r) are shown. The low coordination of interlayer distances and the persistence of intralayer distances to long range would therefore be attributable to a poor registration of the layers in the direction normal to the plane due to, for example, slip planes, stacking faults etc and not due to curvature. That is, the sp^2 bonded ribbons are flat over regions of size ≈15Å. These flat regions may lie between regions, of smaller extent, where strong curvature is present. The ribbons are understood to cross-link at either weak or strong confluences [16]. Weak confluences can be visualized as places where a ribbon peels into two thinner ones (breaking the Van der Waals bonds between the layers), while a strong confluence is a place where the ribbon is severely disrupted. It would be expected that sp^3 bonded atoms would be most abundant in the vicinity of strong confluences.

The structure of glassy carbon has been modified with a view to improving its wear characteristics by ion bombardment [52]. Raman spectroscopy revealed evidence for increased disorder after bombardment while EELS data was consistent with graphitization having occurred [36]. This apparent contradiction was resolved by noting the relatively strong Raman scattering for the graphite disorder mode and viewing the ion-bombardment as causing local defects within the graphitic ribbons or sheets but leaving the overall layered structure essentially unaltered. The more highly resolved Raman spectra for glassy carbon and microcrystalline graphite [53] testify to the greater degree of order in these materials than in a-C.

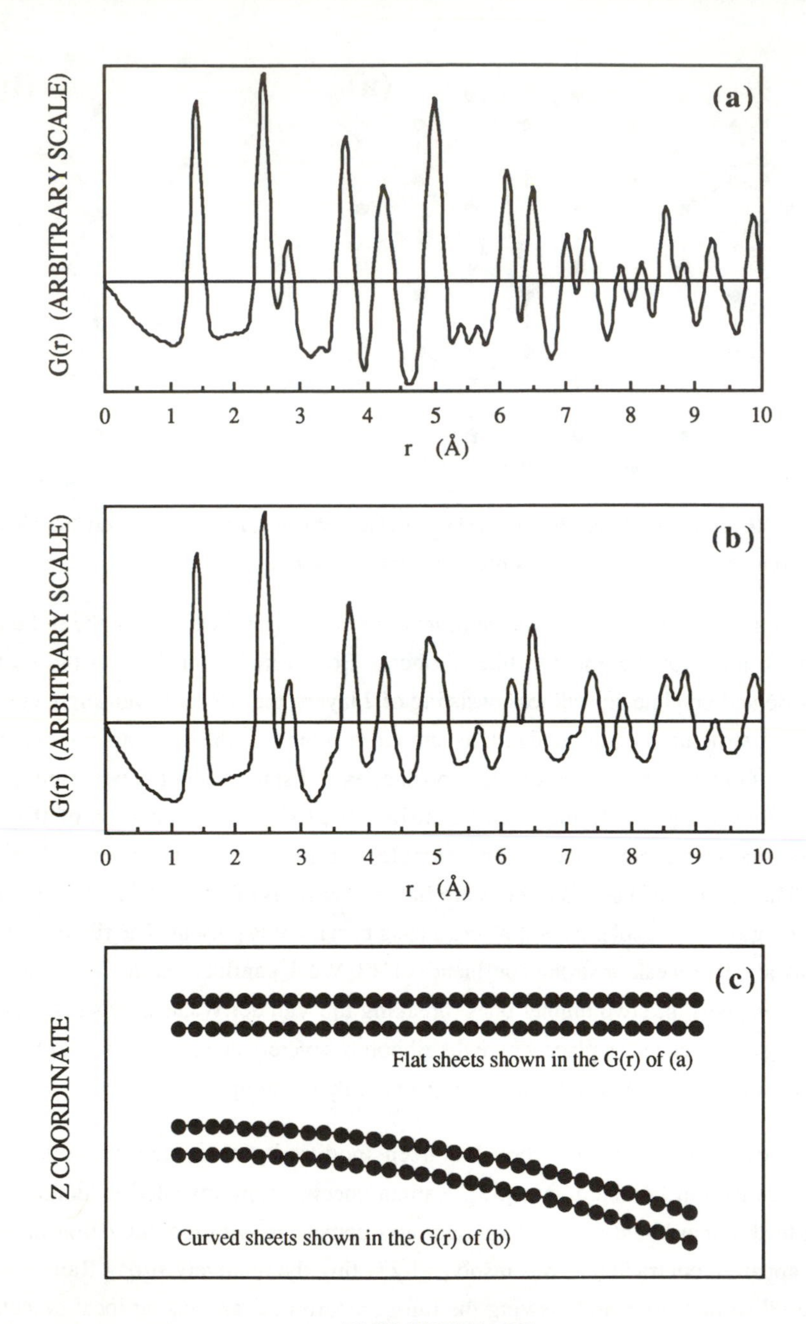

FIGURE 7 The influence of curvature on the RDF calculated for a double layer model for glassy carbon. The G(r) shown are for double layers of 792 atoms set in the graphite configuration with 33 x 6 unit cells. The coordinates used for calculating (a) were curved as shown in the schematic (c) and used to calculate (b). The X and Z directions correspond to a plane parallel to the a-c plane and the atom positions are to scale.

Evaporated Amorphous Carbon

The G(r) of thermally evaporated a-C, prepared without hydrogen, shows many features in common with glassy carbon, but with important differences. The first and second principal peak positions in the G(r) of several authors are listed in Table 3 together with crystalline graphite and diamond data and the 356 atom random network model of Beeman et al. [41]

TABLE 3 FIRST AND SECOND INTERATOMIC DISTANCES FOR EVAPORATED AMORPHOUS CARBONS

PREPARATION	DISTANCE 1 (Å)	DISTANCE 2 (Å)
E-beam evaporation, [54]	1.47±0.02	2.48±0.03
Arc evaporation, [42]	1.51	2.53
Arc evaporation, [43]	1.43	2.51
<AVERAGE>	< 1.47 >	< 2.51 >
Graphite	1.42	2.44
340 atom (9% sp^3) model, [41]	1.42	2.43
356 atom (51% sp^3) model, [41]	1.51	2.52
Diamond	1.55	2.52

The distances are intermediate between those for graphite and diamond, leading Kakinoki to propose a model for evaporated a-C in which domains of sp^2 and sp^3 bonding are intermingled, in approximately equal numbers. The value of 1.5Å is, however, now regarded with some caution [1]. In spite of this, either of the models containing mixtures of 3-fold and 4-fold coordinated atoms gives rise to first and second interatomic distances which are comparable with the corresponding experimental results particularly in the case of equal proportions. The slightly poorer fit of the 9% sp^3 fraction may point to this proportion being towards the low end of plausible values. On the other hand, evidence from another source suggests that the estimate of 50% sp^3 bonding in a-C may be too high. An examination of the carbon K-edge spectrum of a-C [14] gives an estimate of the fraction of sp^2 bonded carbon as ≈92% on the basis of a comparison of the integrated area under the π^* feature of randomly oriented fully crystalline graphite with that of a-C.

These RDFs may be assessed against those of simple planar network models which are based entirely on distorted sp^2 bonding with the incorporation of five and seven membered rings in addition to the more abundant six membered rings. The RDFs obtained for a these ring systems are shown in figure 8 together with experimentally determined G(r) for evaporated a-C as well as as-prepared and annealed amorphous hydrogenated carbon (see below). A model which mixes 5 and 7 membered rings with 6 membered rings has features which qualitatively agree with the features in the experimental G(r). In this regard the second peak is broadened thereby smearing out the third distance. Furthermore, the fifth distance is greatly weakened. In conclusion, the structure of evaporated amorphous carbon is predominantly sp^2 bonded carbon.

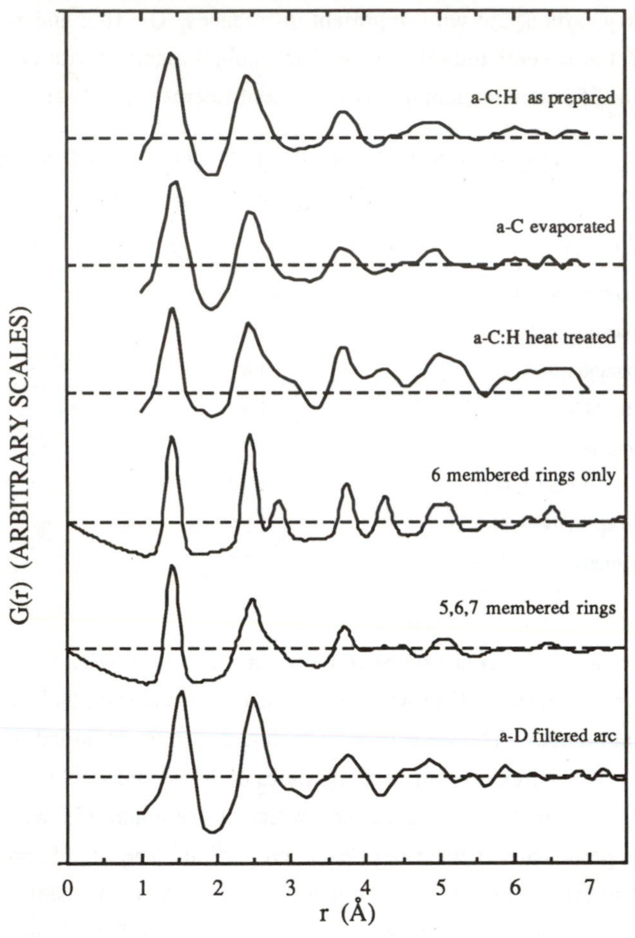

FIGURE 8 The reduced density functions for amorphous hydrogenated carbon, amorphous carbon, amorphous hydrogenated carbon annealed, a planar network comprised of 7 six membered rings, a planar five, six and seven membered rings network of small extent with a distribution 12%,86%,2%, and amorphous diamond.

Amorphous Hydrogenated Carbon

There are many similarities observed in the structure of a-C:H and a-C. The observed G(r) for a-C:H, prepared by glow discharge of methane, bears strong structural similarity to that for a-C, as shown in figure 8.

The similarities between the two materials goes further. Scattering by plasmon processes (collective oscillations of the electron clouds against the atom cores) gives rise to a peak in the electron energy loss spectrum at a characteristic energy - the plasmon energy. The value of the plasmon energy is a key indicator of the electron density which in turn reflects on the atomic density. As Table 4 shows, the plasmon energies of a-C and a-C:H are the same within experimental uncertainty.

TABLE 4 PLASMON ENERGIES FOR CARBON FORMS

CARBON FORM	PLASMON ENERGY (eV)
Graphite, [55]	27.0
CVD a-C:H, [56]	23.0±0.8
Evaporated a-C, [14]	23.6
Arc a-D, [14]	32±0.8
Diamond, [57]	33

As already mentioned, energy loss near edge structure (ELNES) is a strong indicator of the presence of sp^2 bonded material. The ELNES spectra [14,58] for a-C and a-C:H show comparable π^* transition features.

There is a marked difference between the two materials, however, when the low angle region of the electron diffraction patterns are examined. Figure 9(a) shows diffraction results for a-C and a-C:H together with those for amorphous diamond, while figure 9(b) shows the corresponding regions for two graphite models of different extent in the c-direction. Although complicated by the "straight-through" beam of the electron microscope, there is a subtle distinction [54] between the nature of the shoulders of the a-C and the as-prepared a-C:H curves below $s=0.3Å^{-1}$. This feature, corresponding to the 002 ($s\approx0.3Å^{-1}$) reflection in graphite is more pronounced for a-C than for a-C:H indicating that the former is more ordered in the c-direction. Upon annealing the feature strengthens slightly to exhibit a more highly resolved band as does, the maximum at $s\approx0.45Å^{-1}$ indicating a partial ordering within the layers and, to a lesser extent, some increased correlation in the c-direction.

There are significant differences in the electronic structure of a-C and a-C:H. These differences are seen in the optical properties of the two materials. a-C:H is much less absorbing in the visible part of the spectrum than is a-C. This is understood in terms of the removal of states in the band gap by the hydrogen. Electronically, the hydrogen is localising the electrons to smaller conjugated hydrocarbon structures. The effect of this process of electron localisation can easily be seen if ultraviolet and visible spectra are measured for a series of aromatic hydrocarbons [59].

The principal conclusion then of electron diffraction and energy loss studies is that locally, a-C:H and evaporated a-C are similar. However, in a-C, the stacking of sp^2 bonded structures would appear to extend a greater distance in the direction normal to the sheets.

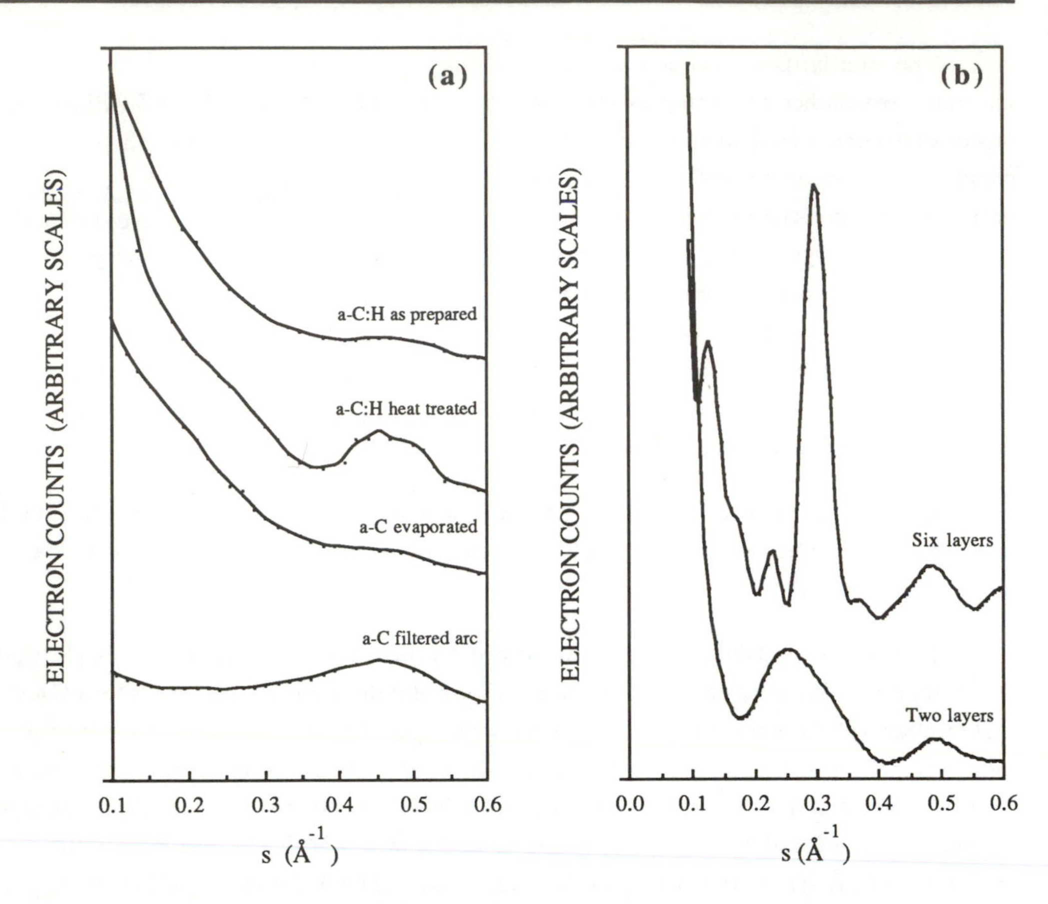

FIGURE 9 Small angle regions of the electron diffraction patterns of (a) a-C and a-C:H and (b) two graphitic models containing 100 atoms in two layers and 300 atoms in six layers.

Solid-state ^{13}C NMR spectra of hydrogenated amorphous carbons [29,60-62] generally contain two broad bands centred at ≈40ppm and ≈130ppm corresponding to aliphatic or sp^3 carbons and aromatic or sp^2 carbons respectively. These bands are typically ≈50-100ppm wide reflecting the complex distribution of bonding arrangements in these materials. In contrast, the line widths for amorphous polymers [63, 64] can be of the order of only a few ppm. Dilks et. al [60] have obtained the full and delayed decoupling ^{13}C spectra for hydrogenated amorphous carbons formed from RF plasmas in ethane, ethylene and acetylene. They were able to identify five spectral sub-bands and, using standard shift compilations, assign these to particular carbon types. Kaplan et al. [61] confirmed that a-C:H films prepared by ion beam sputtering and both DC and RF glow discharges in methane exhibit quite different NMR spectra and sp^2/sp^3 ratios. Decreasing optical band gap was found to correlate well with increasing relative proportion of sp^2 carbons. For films prepared by RF glow discharge in methane, Yamamoto et al. [62] found that the proportion of sp^2 carbon increases with increasing RF power. Hence, the existence of a maximum in the relation between film hardness and RF power can be understood in terms of the balance between hydrogen content and carbon type. Carbon-rich a-SiC:H films possess substantial unsaturated carbon character [65]. As the carbon content of these films decreases, the saturated carbon band tends to lower chemical shift indicating an increase in Si-C bonding at the expense of unsaturated carbon bonding. Naturally occurring carbonaceous materials such as fossil fuels have been extensively studied [66] by ^{13}C NMR and it has been shown that quantitatively useful information can be gained even for high carbon content materials.

The Raman spectra of a-C and a-C:H both contain a broad band at $\approx$1550 cm^{-1} assigned to graphitic structures and a shoulder at $\approx$1350 cm^{-1} attributed not to C-C single bonds but to a graphitic disorder mode [67,68]. The apparent absence of a peak near 1332 cm^{-1} is not inconsistent with the presence of singly-bonded carbon because of the latter's relatively lower Raman scattering efficiency [67].

Amorphous Carbons from Arc-Plasma Beams

As discussed above, carbon deposits formed from fully ionised plasmas obtained from the cathode spot of vacuum arcs under appropriate conditions demonstrate unusual properties. The electron energy loss near edge spectrum enables an estimate of the fraction of sp^3 bonded carbon present in a specimen of a-D. A typical result is 85% sp^3 bonded carbon [14].

The small angle scattering of electrons shows that this material is fundamentally different to amorphous carbons prepared by other techniques. Figure 9 shows the electron diffraction pattern of this material in the small angle region. There is very little diffracted intensity in this region and hence no tendency to form the layered structure characteristic of amorphous carbons containing significant amounts of sp^2 carbon. In this respect the diffraction pattern resembles those of tetrahedrally bonded amorphous semiconductors such as amorphous silicon and amorphous germanium.

In other respects, too, this analogy is carried through. The G(r) for filtered vacuum arc amorphous carbon (see figure 8) shows peaks in the positions expected for an "amorphous diamond" or a-D structure. The mean angle between bonds to nearest neighbour atoms may be evaluated trigonometrically by assuming small variation in the first interatomic distance. The observed breadth of the first interatomic distance peak is due almost entirely to thermal vibration once instrumental broadening effects have been taken into account [37]. Further, the variation in the bond angle can be implied from the width of the second G(r) peak and this is presented for a variety of carbon forms together with some results for a-Ge in Table 5. The amorphous tetrahedrally-bonded networks of amorphous germanium and amorphous diamond should differ in the degree to which the tetrahedral angle deviates from its average value in the structure. This should occur because of the greater resistance of the tetrahedral angle to bending in diamond than the other sp^3 bonded materials. Indeed, the bond-bending force constants are in the ratio 4.9 : 0.44 : 0.37 for carbon, silicon and germanium, respectively [69].

TABLE 5 BOND ANGLES AND THEIR VARIATIONS IN AMORPHOUS MATERIALS

FORM	BOND ANGLE (a)	HALF ANGLE FOR HALF MAXIMUM
glassy carbon, [50]	119°	12°
a-C:H heat treated, [54]	120°	17°
a-C:H as prepared, [54]	118°	16°
a-C evaporated, [54]	115°	15°
a-D, [54]	110°	8°
a-Ge evaporated, [37]	109°	11°
2052 atom a-Ge model, [70]	109.5°	10.6° (b)

(a) Estimated uncertainty of ± 4% in experimental values
(b) RMS value

Influence of Thermal Annealing

The heat treatment of a-C:H has been studied using several techniques. The G(r) of a-C:H which has been heated to 500°C in vacuum shows changes consistent with the elimination of odd-membered rings and hence a G(r) which resembles to some extent that for glassy carbon. This process may be viewed as a 2-D crystallization of the amorphous structure. The third inter-atomic distance of a perfect graphite layer shown in figure 8 appears as a shoulder on the peak corresponding to the second inter-atomic distance, and the fourth and fifth inter-atomic distances show as separate well-formed peaks. At the same time, there is an increase in the interlayer ordering as estimated from the shape of the small angle scattering shown in figure 9.

Infrared, Raman and electron spin resonance spectroscopies have been successfully used in the characterization of amorphous carbons and in the effects of heat treatment on these materials. The IR spectrum of evaporated a-C [30] contains three broad bands, two associated with C-C modes and a third due to aromatic C-C stretching. Table 6 shows the assignments for the IR spectra [18,31] of a-C:H films prepared by DC glow discharge in acetylene and RF plasma deposition from benzene. By comparing the integrated intensities of the deconvoluted peaks corresponding to the CH stretching modes near $3000m^{-1}$, Dischler et al. [32] have estimated the relative proportions of sp^3, sp^2 and sp carbons in as-prepared and annealed films. It was found [18,31,32] that as-prepared films contained mainly aliphatic with some olefinic and aromatic carbons and traces of acetylenic groups (e.g. $sp^3:sp^2:sp \approx 68:30:2$ [32]). Although hydrogen bonds mainly to sp^3 carbons in these films, the relative proportions of mono- and di-hydrogenated carbon groups is a little uncertain since both the sp^3 CH stretch and the sp^3 CH_2 asymmetric stretch occur at about $2920cm^{-1}$. As the annealing temperature increases, weakly bound components and ≡ CH groups are quickly removed followed by olefinic groups which are converted to aromatics and then finally aliphatics are also converted to aromatics. By ≈600°C the films are essentially graphitic, presumably with some aliphatic linking groups.

TABLE 6 INFRARED ABSORPTION BANDS OF a-C:H

WAVENUMBERS (cm^{-1})	ASSIGNMENT
3300	≡CH CHstretch
3050	aromatic CH stretch
2960	=CH CH stretch
2920	⋗CH CH stretch
2920	>CH_2 asymmetric CH stretch
2860	>CH_2 symmetric CH stretch
1625	C=C stretch
1570	aromatic C-C stretch
1460	CH_2 scissor
1450	CH_3 asymmetric bend
1430	aromatic C-C stretch
1375	CH_3 symmetric bend
1300	C-C stretch
880	aromatic CH bend
800	aromatic CH bend

The Raman spectra of both a-C and a-C:H show changes consistent with the above observations. As a-C and a-C:H are annealed, the graphitic peak at ≈1550cm^{-1} narrows and shifts to higher frequency [67,68] indicating a reduction in bond-angle disorder accompanying the 2-D crystallization. Furthermore, the decrease in the size of the disorder peak near 1350cm^{-1} at temperatures exceeding ≈900°C has been associated with the formation of a fully developed microcrystalline graphitic phase. Electron spin resonance [71] shows that the size of the regions over which electrons are delocalized increases with the time and temperature of heat treatment in a manner consistent with the growth of regions of graphitic order. Calculations by Beeman et al. [41] of the phonon density of states for diamond, graphite and several random network models are consistent with the above results.

CONCLUSIONS

Carbon is intriguing in its ability to form a great diversity of structures and forms. Methods of preparing amorphous and poorly crystalline forms of carbon, and techniques for elucidating their structure, have been presented. The majority of carbons examined exhibit structures consistent with regions of substantial graphite-like bonding with some tetrahedrally coordinated carbon atoms interspersed. However, forms of amorphous carbon with almost entirely diamond-like local order do exist, particularly in specimens prepared from highly ionized plasma beams.

ACKNOWLEDGEMENTS

The authors acknowledge the financial assistance of His Royal Highness Prince Nawaf bin Abdul Aziz of the Kingdom of Saudi Arabia through the Science Foundation for Physics at the University of Sydney, Australian Research Grants Scheme for the electron microscope, and the Commonwealth Postgraduate Research Awards. Dr B. Pailthorpe and J. Tansell-Green are thanked for their valuable assistance with the figures.

REFERENCES

1 Robertson, J.W.: Advances in Physics, 1986, 35, 317
2 Kittel, C.: "Introduction to Solid State Physics", 5th Edition, 1976, J. Wiley and Sons
3 Hanneman, R.E, Strong, H.M. and Bundy, F.P.: Science, 1967, 155, 955
4 Burns, G.: "Solid State Physics", 1985, Academic Press
5 Hauser, J. J.: Sol. St. Comm., 1975, 17, 1577
6 Vossen, J. L. and Cuomo, J.J .: "Thin Film Processes", 1978, Edited by J.L. Vossen and W. Kern, Academic Press
7 Ball, L.T., Falconer, I.S., McKenzie, D.R. and Smelt, J.M.: J. Appl. Phys., 1986, 59, 720
8 McKenzie, D.R., McPhedran, R.C., Botten, L.C., Savvides, N. and Netterfield, R.P.: Applied Optics, 1982, 21, 3615
9 Saito, Y., Matsuda, S. and Nogita, S.: J. Mat. Sc. Lett., 1986, 5, 565
10 Ugolini, D., Eitle, J., Oelhafen, P. and Wittmer, M.: Appl. Phys. A, 1989, 48, 549
11 Aksenov, I.I., Belous, V.A., Padalka, V.G. and Khoroshikh, V.M.: Sov. J. Plasma Phys, 1978, 4, 425
12 Aksenov, I.I., Vakula, S.I., Padalka, V.G., Strel'nitskii, V.E. and Khoroshikh, V.M.: Sov. Phys. Tech. Phys, 1980, 25, 1164
13 Spencer, E.G., Schmidt, P.H., Joy, D.C. and Sansalone, F.J.: Appl. Phys. Lett., 1976, 29, 118
14 Berger, S.D., McKenzie, D.R. and Martin, P.J.: Phil. Mag. Lett., 1988, 57, 285
15 Matyushenko, N.N., Strel'nitskii, V.E. and Gusev, V.A.: J.E.T.P. Lett, 1979, 30, 199
16 Jenkins, G. and Kawamura, K.: "Polymeric Carbons - Carbon Fibre, Glass and Char", 1976, Cambridge University Press
17 Swift, P.D.: Ph.D. Thesis, University of Sydney, Australia (in preparation) density by float/sink method, 10% void fraction assumed
18 McKenzie, D.R., McPhedran, R.C., Savvides, N. and Botten, L.C.: Phil. Mag. B, 1983, 48, 341
19 Fink, J., Muller-Heinzerling, Th., Pfluger, J., Scheerer, B., Dischler, B., Koidl, P., Bubenzer, A. and Sah, R.E.: Phys. Rev. B, 1984, 30, 4713
20 Fink, J., Muller-Heinzerling, Th., Pfluger, J., Bubenzer, A., Koidl, P. and Crecelius, G.: Sol. St. Comm., 1983, 47, 687
21 Gao, C., Wang, Y.Y., Ritter, A.L. and Dennison, J.R.: Phys. Rev. Lett., 1989, 62, 945
22 Taft, E.A. and Philipp, H.R.: Phys. Rev., 1965, 138, A197
23 Sonnenschein, R., Hanfland, M. and Syassen, K.: Phys. Rev. B, 1988, 38, 3152
24 Pines, A., Gibby, M.G. and Waugh, J.S.: J. Chem. Phys., 1973, 59, 569
25 Andrew, E.R.: Prog. Nucl. Magn. Reson. Spectrosc., 1971, 8, 1
26 Carver, G.P.: Phys. Rev. B, 1970, 2, 2284
27 Retcofsky, H.L. and Friedel, R.A.: J. Phys. Chem., 1973, 77, 68
28 Henrichs, P.M., Cofield, M.L., Young, R.H. and Hewitt, J.M.: J. Mag. Reson., 1984, 58, 85
29 Lukins, P.B., McKenzie, D.R. and Vassallo, A.M.: Unpublished data
30 Knoll, J. and Geiger, J.: Phys. Rev. B, 1984, 29, 5651

31 Dischler, B., Bubenzer, A. and Koidl, P.: Appl. Phys. Lett., 1983, 42, 636
32 Dischler, B., Bubenzer, A. and Koidl, P.: Solid State Commun., 1983, 48, 105
33 Nadler, M.P., Donovan, T.M. and Green, A.K.: Thin Solid Films, 1984, 116, 241
34 Tuinstra, F. and Koenig, J.L.: J. Chem. Phys., 1970, 53, 1126
35 Lespade, P., Al-Jishi, R. and Dresselhaus, M.S.: Carbon, 1982, 20, 427
36 Prawer, S. and Rossouw, C.J.: J. Appl. Phys., 1988, 63, 4435
37 Cockayne, D.J.H. and McKenzie, D.R.: Acta Cryst., 1988, A44, 870
38 Anstis, G., Lake, M.R. and Liu, Z.Q.: Ultramicroscopy, 1988, 26, 65
39 Lake, M.R.: Ph.D. Thesis, University of Sydney, Australia, 1989
40 Polk, D.E.: J. Non-Cryst. Sol., 1971, 5, 365
41 Beeman, D., Silverman, J., Lynds, R. and Anderson, M.R.: Phys. Rev. B, 1984, 30, 870
42 Kakinoki, J., Katada, K. Hanawa, T. and Ino, T.: Acta Cryst., 1960, 13, 171
43 Boiko, B.T., Palantik, L.S., and Deveryanchenki, A.S.: Sov. Phys. Dokl., 1968, 13, 237
44 Franklin, R.E.: Acta Cryst., 1950, 3, 107
45 Ergun, S.: Acta Cryst., 1973, A29, 605
46 Warren, B.E.: Phys. Rev., 1941, 9, 693
47 Stenhouse, B.J. and Grout, P.J.: J. Non-Cryst. Sol., 1978, 27, 247
48 Summerfield, G.C., Mildner, D.F.R. and Carpenter, J.M.: J. Non-Cryst. Sol., 1983, 57, 289
49 Mildner, D.F.R. and Carpenter, J.M.: J. Non-Cryst. Sol., 1982, 47, 391
50 Green, D.C.: Ph.D. Thesis, University of Sydney, Australia (in preparation)
51 Specimen prepared by S. Prawer, diffraction data collected by D. Dwarte
52 Farrelly, M. and Pollock, J.T.A.: Mater. Forum, 1987, 10, 198
53 Solin, S.A. and Nemanich, R.J.: Phys. Rev. B, 1979, 20, 392
54 McKenzie, D.R., Martin, P.J., White, S.B., Liu, Z., Sainty, W.G., Cockayne, D.J.H. and Dwarte, D.M.: Euro-MRS Meeting Strassbourg June 1987, Les Editions de Phys., 17, 203
55 Edgerton, R.F.: "Electron Energy Loss Spectroscopy in the Electron Microscope", 1986, Plenum Press, New York
56 Martin, P.J., Filipczuk, S.W., Netterfield, R.P., Field, J.S., Whitnall, D.F. and McKenzie, D.R.: J. Mat. Sc. Lett., 1988, 7, 410
57 Daniels, J., Festenberg, C.V., Raether, H., and Zepenfeld, K.: Springer Tracts in Mod. Phys., 1970, 54, 77
58 McKenzie, D.R., McPhedran, R.C., Savvides, N. and Cockayne, D.J.H.: Thin Solid Films, 1983, 108, 247
59 Clar, E.: "Polycyclic Hydrocarbons", 1964, Academic Press
60 Dilks, A., Kaplan, S. and Van Laeken, A.: J. Polym. Sci. - Polym. Chem. Ed., 1981, 19, 2987
61 Kaplan, S., Jansen, F. and Machonkin, M.: Appl. Phys. Lett., 1985, 47, 750
62 Yamamoto, K., Ichikawa, Y. Nakayama, T. and Tawada, Y.: Jpn. J. Appl. Phys., 1988, 27, 1415
63 Schaefer, J., Stejskal, E.O. and Buchdahl, R.: Macromolecules, 1977, 10, 384
64 Earl, W.L. and Vander Hart, D.L.: Macromolecules, 1979, 12, 762
65 Yamamoto, K., Ichikawa, Y., Fukada, N., Nakayama, T. and Tawada, Y.: Thin Solid Films, 1989, 173, 253
66 Gerstein, B., Dubois Murphy, P. and Ryan, L.: in "Coal Structure", Edited by R. Meyers, Academic Press, New York, 1982
67 Wada, N., Gaczi, P.J. and Solin, S.A.: J. Non-Cryst. Solids, 1980, 35-36, 543
68 Dillon, R.O. and Woollam, J.A.: Phys. Rev. B, 1984, 29, 3482
69 Tomassini, N., Amore Bonapista, A., Lapiccirella, A., Lodge, K.W., and Altmann, S.: J. Non-Cryst. Sol., 1987, 93, 241
70 Popescu, M.: J. Non-Cryst. Sol., 1985, 75, 477
71 Miller, D.J. and McKenzie, D.R.: Thin Solid Films, 1983, 108, 257

Materials Science Forum Vols. 52 & 53 (1989) pp. 125-150

ELECTRONIC STRUCTURE AND BONDING OF a-C:H

J. Robertson

Central Electricity Research Laboratories
Kelvin Avenue, Leatherhead, Surrey, KT22 7SE, UK

ABSTRACT

The electronic properties of hydrogenated amorphous carbon (a-C:H) are reviewed, with an emphasis on the close relationship between electronic strucure and atomic structure. It is shown how a-C:H must contain three-fold coordinated sp^2 sites in addition to four-fold coordinated sp^3 sites, that the sp^2 sites tend to be segregated in small graphitic clusters and that the size of these clusters determines the optical band gap. Electronic structure calculations on model structures are used to interpret optical, luminescence, photoemission, electron energy loss, X-ray absorption, electron spin resonance and electrical conductivity data.

INTRODUCTION

Carbon can form a number of crystalline and non-crystalline solids, each with very distinctive and useful properties, some of which are listed in Table 1. Hydrogenated amorphous carbon (a-C:H) is of interest because it forms hard, chemically inert, optically transparent and electronically dopable thin films. Its properties have been reviewed previously [1-4]. These properties are discussed here in terms of its underlying chemical bonding and electronic structure, by using the results of calculations and by drawing comparisons to other forms of non-crystalline carbon.

The two crystalline forms of carbon are diamond and graphite. Diamond is a hard, wide band gap insulator, in which each atom is four-fold coordinated and sp^3 bonded. Each of the four valence electrons of a carbon atom is assigned to a tetrahedrally-directed sp^3 hybrid orbital, which makes a strong σ band to an adjacent atom. Graphite is an anisotropic metal, in which each atom is sp^2 bonded and three-fold coordinated in a hexagonal layer structure. Three of each atom's valence electrons are assigned to trigonally directed sp^2 orbitals, which form σ bonds, and the fourth electron lies in a pπ orbital normal to the σ bonding plane. This orbital forms weak, delocalized π bonds with its neighbours.

TABLE 1. PROPERTIES OF VARIOUS FORMS OF CARBON.

	Density gm/cm^3	Hardness kg/mm^2	% sp^3	at %H	Gap eV
Diamond	3.515	10^4	100		5.5
Graphite	2.267		0		-0.04
Glassy C	1.3-1.55	800-1200	∿0		0.01
a-C, sputtered	∿2	20-50	<5		0.4-0.7
a-C:H, hard	1.6-2.0	1500-5000	40-70	10-40	0.8-1.7
a-C:H, soft	1.2-1.6	Low	50-70	35-60	1.7-4
Polyethylene	1.0	Low	100	67	6

The various forms of non-crystalline carbon can be considered to be intermediate between graphite, diamond and polyethylene, $(CH_2)_n$. Carbon fibres are graphitic units organised with a fibrous structure [5,6]. Glassy carbon is also based on an imperfect graphitic layer structure [6]. The correlation length of structural order of atoms along the layers, known as L_a, is of order 40Å [7], indicating considerable medium range order and so it cannot really be classed as a true amorphous material [1]. True amorphous carbon can be prepared by the electron beam evaporation or sputtering of graphite [8-10]. Its structure has been studied by electron and X-ray diffraction [9-11] and has been modelled by Beeman et al. [12] in terms of continuous random networks containing various proportions of sp^2 and sp^3 sites. The modelling suggests that a-C contains an upper limit of 5% of sp^3 sites. The Raman spectrum of a-C shows two features, a broad main peak at 1550 cm^{-1} and the development of a subsidiary peak at ∿1350 cm^{-1} [13], close to the positions observed in microcrystalline graphite, and identified with the existence of a three-coordinated lattice and finite graphitic islands, respectively [14]. These studies suggest that a-C consists mainly of disordered graphitic layer clusters, 15-20Å in diameter, with little correlation between each cluster. The two recent structural simulations using first principles [15] and realistic empirical inter-atomic potentials [16] generally support this model.

A-C:H can be prepared by various techniques such as plasma deposition or ion beam deposition from hydrocarbon gases or by sputtering of graphite in a hydrogen-containing atmosphere. The material formed depends strongly on the deposition conditions. The key deposition parameters are the bias voltage and the temperature of the substrate. A high substrate bias voltage favours dehydrogenation and a high temperature above about 200°C favours graphitization. Koidl et al. [2,17-21] distinguish two types of a-C:H, hard and soft. A negative substrate bias of over about 400 V produces hard a-C:H, which is a hard network solid with mixed sp^3, sp^2 bonding, a band gap under 1.7 eV, a density over 1.7 gm cm^{-3}, and considerable chemically bonded hydrogen, largely as ≡CH groups. On the other hand, a low bias of under about 400 V produces soft a-C:H which is essentially a low density solid polymeric hydrocarbon. It contains both sp^3 and sp^2 groups and more hydrogen, largely as $=CH_2$ groups [20]

The sp^3:sp^2 content of a-C:H is most directly measured by magic angle spinning C^{13} nuclear magnetic resonance (NMR) [22], for which proton decoupling can also locate the positions of the hydrogens [23]. In practice, most analysis has been carried out using infra-red spectroscopy on the C-H bond stretching modes [18-20].

It is also possible to produce an unhydrogenated form of a-C which is almost 100% sp^3 bonded, by deposition of a filtered ion beam from a cathode spot source [24].

Finally, microcrystalline films of diamond can be prepared by chemical varpour deposition or microwave discharge in a methane-hydrogen mixture [25,26]. A large excess of atomic hydrogen present in the discharge is believed to selectively etch away any sp^2 sites which may deposit [27].

THE π ELECTRONIC STRUCTURE

Table 1 gives an optical gap of 0.5 eV for a-C [7] and 1.2-1.7 eV for hard a-C:H, while that of graphite is -0.04 eV, due to a slight band overlap. It should first be pointed out that the existence of any band gap, particular in a-C, is in many ways remarkable. Disorder generally acts to reduce or eliminate band gaps, as it does in, say, a-Si. Thus, one might not expect a gap in a-C if it is considered to be disordered graphite.

Figure 1 shows the calculated electronic density of states (DOS) for a single layer of graphite, diamond and various models of a-C. These were calculated using the tight-binding method using the interaction parameters of Table 2, which were found by fitting to the band structures of graphite and diamond [1,28]. A single layer of graphite in Fig. 1(a) is seen to be a zero band gap semiconductor and its DOS at the Fermi level (E_F=0) is seen to be due solely to the π electrons, shown by the dashed line. In diamond, Fig. 1(b), all the electrons are in σ bonds and the band gap is 5.5 eV. Figures 1 (c-f) shows the DOS of the continuous random network models of a-C of Beeman et al. [12], containing various proportions of sp^2 and sp^3 sites. The networks which contain any sp^2 are seen to be metallic, with E_F lying in a high density of π states. This is contrary to experiments and shows that the models cannot fully represent the structure of a-C. It suggests that merely introducing sp^3 sites does not automatically create a gap and this requires the sp^2 sites to have specific spatial correlations.

The band gap depends entirely on the π states. The key feature of the π states is that they are half-filled (Fig. 1), so that creating a gap in their spectrum at E_F tends to lower the total energy of the occupied states and stabilise the resulting structure. These effects can be analysed in the Huckel model. This is a simplified tight-binding Hamiltonian which retains only the π orbitals and their nearest neighbour interaction, $\beta = V(p_z\pi)$. Generally, intervening sp^3 sites will block any π interaction passing through that site. This model therefore maps the original network of a-C or a-C:H into a series of isolated π-bonded clusters. Finding the stable structure is now reduced to maximizing the total π electron binding energy per site (E_{tot}) of each cluster.

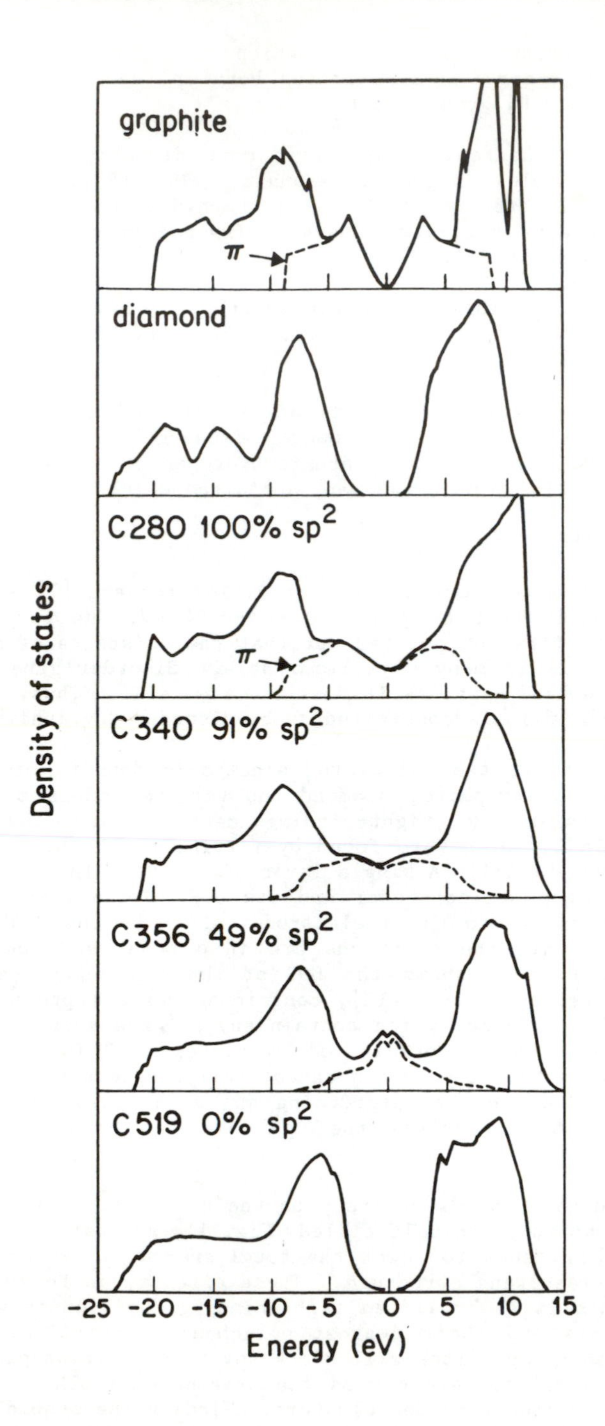

Figure 1. Calculated DOS of various crystalline and random carbon lattices

TABLE 2. TIGHT-BINDING PARAMETERS FOR C-C AND C-H BONDS, IN eV. ORBITAL ENERGIES ARE REFERRED TO E_F, 4.7 eV BELOW THE VACUUM LEVEL

	E(s)	E(p)	E(s*)
C	-5.35	0.45	
H	-2.3		

	V(ss)	V(sp)	V(pσ)	V(pπ)	$V(p_z\pi)^+$	$V_2(p_z\pi)^\phi$	V(s*p)
C-C	-4.55	5.2	5.45	-1.6	-2.9	0.15	4.5
C-H	-7.5	8.9					

$^+$ between two π states

ϕ second neighbour interaction between two π states

Fig. 2(a) and (c) show the π electron DOS calculated for the single graphite layer and the C280 network model (100% sp^2), the latter calculated ignoring the dihedral angle dependence of the π interaction. The similarity of these spectra to the π electron DOS in Fig. 1(a) and (c), respectively, justifies the use of the Huckel model.

We have investigated the binding energy and band gap of a wide range of possible structures (Table 3) and we conclude that the structure of π-bonded regions conform to a number of rules, similar to those applying to organic molecules:

(a) Clusters tend to be planar. π orbitals on adjacent sites are then parallel to one another, maximizing their interaction.

(b) There should be an even number of sites in each cluster. Otherwise, a state exists at E_F with low binding energy, which tends to reduce E_{tot} and eliminate the gap.

(c) Clusters tend to grow. An adjacent pair of π orbitals (c.f. ethylene) has $E_{tot}/\beta=1$, while if joined into a chain of N sites they have $E_{tot}/\beta \to 4/N$ as $N \to \infty$.

(d) Six-fold rings are strongly favoured, whether isolated or in clusters. An isolated six-fold ring (c.f. benzene) has $E_{tot}/\beta = 4/3$ which is 1/3 per site greater than that of three isolated π bonds. Thus, aromatic groups are favoured over olefinic groups. Also, 4,5,7 and 8-fold rings are all unfavoured. The isolated rings give a state near E_F. This behaviour persists in fused ring structures, as illustrated in Figure 2(b), as the local density of states of a 'quadrupole' of two five-fold and two seven-fold rings set in an otherwise infinite graphite layer is seen to have a sizeable DOS at E_F. We note, however, that 5- and 7-fold rings can occur, but only as an adjacent 'dipole' pair, as this azulene configuration remains aromatic, with states avoiding E_F and with an E_{tot} only slightly less than the ideal configuration, napthalene (Table 3).

TABLE 3. π ELECTRON-BINDING ENERGY PER SITE, E_{tot}

group	name	E_{tot}/β
	polyacetylene	1
$(\equiv-)_n$	carbynes	2
	benzene	1.333
	napthalene	1.368
	azulene	1.336
	quinoid	1.240
	polybenzoid	1.403
	graphite	1.616

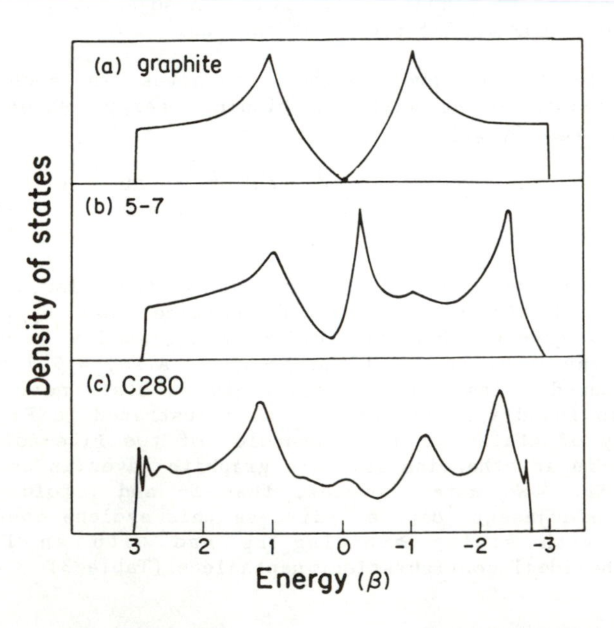

Figure 2. π electron DOS of (a) single graphite layer, (b) 5-7 ring quadrupole in a graphite layer and (c) C280 random network model

(e) Compact clusters are favoured, such as graphitic islands. These clusters minimize the number of edge sites so that the total energy per site tends towards the value of E_{tot}/β = 1.616 of an infinite graphite layer.

(f) Quinoid rings tend to dissociate into separate aromatic and olefinic fragments. A quinoid ring formed by linking sp^2 sites to the 1 and 4 positions of a benzene ring is marginally less stable than a separate benzene ring and a π bond.

A recent first-principles simulations of the atomic and electronic structure of a-C using the Car-Parrinello method [15] generally supports these conclusions. A small sample of 54 atoms in a periodic lattice with a density set at 2 gm/cm^3 and given a particular quenching sequence was found to consist of 85% sp^2 sites and 15% sp^3 sites. The sp^2 sites were found to be clustered into buckled, largely graphitic layers, which also contained several adjoining 5- and 7-fold ring configurations, as in azulene. The sp^3 sites were also found to be slightly clustered.

In summary, it is energetically favourable to have graphite-like islands. The range of island sizes is determined by the interplay between the energetics of the π bonds and the strain energy of the underlying σ-bonded network.

Furthermore, control of the band gap is qualitatively different than in σ-bonded systems like a-Si. There, the gap is determined by the short-range order, that is bond lengths, bond angles etc., and as this is set by the chemical bonding requirements of the atom, the gap in all forms of a-Si is similar to that of the respective crystal, c-Si. In a π-bonded system such as amorphous carbon, the gap is determined by medium range order (MRO), that is the degree of clustering. This can vary even when the proportion of sp^2 and sp^3 sites remains constant.

THE OPTICAL GAP

Figure 3 shows the calculated band gap of compact (upper line) and linear (lower line) aromatic clusters for increasing cluster size. The gap of compact clusters, the most stable type, decreases slowly and irregularly with increasing number M of rings in the cluster, falling off roughly as

$$E_g = 2\beta M^{-1/2} \qquad (1)$$

for large M. We set β = -2.9 eV from fitting the π bands of graphite.

As the optical absorption spectrum of amorphous semiconductors tails into the gap region, the optical gap is defined either as the energy at which the optical absorption coefficient $\alpha = 10^4$ cm^{-1} (E_{04}) or as the Tauc gap, E_T, found by fitting α to the equation $\alpha E = B(E-E_T)^2$ for $\alpha \simeq 10^4$-10^5 cm^{-1}. Figure 4 shows the optical absorption spectrum of a-C and a-C:H, around their absorption edges. Using equation 1, the E_{04} gap of a-C, 0.5 eV [8], corresponds to aromatic clusters of ~120 rings, or L_a = 18Å. This is consistent with Wada's estimate of 15-20Å graphitic islands, deduced from the Raman spectrum [13].

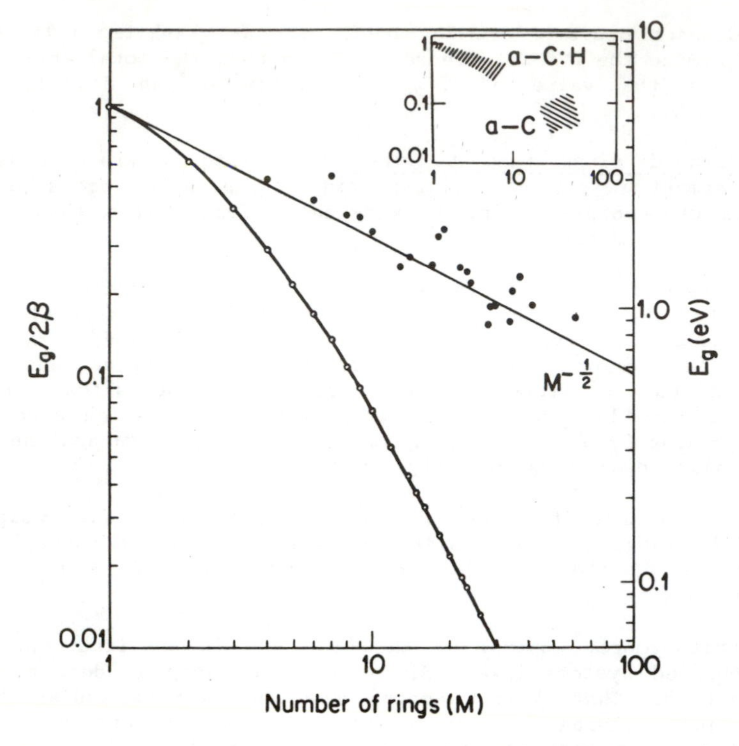

Figure 3. Band gap of compact (upper curve) and linear (lower curve) aromatic clusters

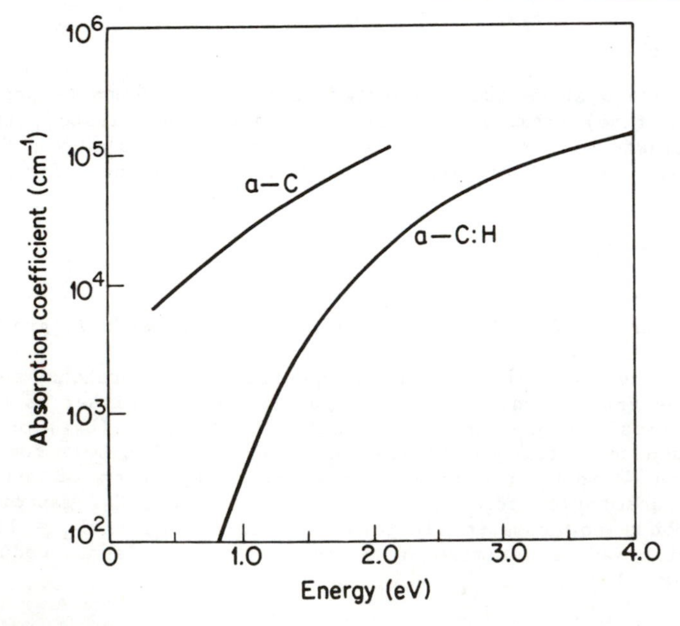

Figure 4. Optical absorption spectra of a-C (Hauser [8]) and soft a-C:H (Smith [29])

The optical gap of a-C:H is strongly dependent on deposition conditions, as already noted, but it does not exceed about 4 eV, Table 1 [17-22, 27-32]. Figure 1 shows that the gap of σ states in carbon always exceed 5 eV. Figure 5 shows the calculated local DOS at hydrogen sites for various configurations of bonded hydrogen, using the tight-binding parameters given in Table 2. It is clear that C-H states also lie at least 2.5 eV away from E_F. Therefore, the optical gap must again be determined by π-bonded domains, and even soft a-C:H <u>must</u> contain sp^2 sites. A typical gap of 1.3 eV for hard a-C:H is much less than the π gap of ethylene or benzene, for both of which $E_g > 4$ eV. Therefore, we again require considerable clustering of sp^2 sites, as either aromatic islands or long olefinic chains. We consider olefinic chains to be unlikely as their binding energy is much less than that of equivalent aromatic units. Hence, we conclude with Bredas and Street [33] that the sp^2 sites in a-C:H form aromatic clusters. The gap of 1.3 eV corresponds to compact clusters of $M \simeq 20$. This is found in a sample in which $\sim$30% of carbon sites are sp^2 bonded [19,34], indicating the very high degree of clustering present.

The calculations show that the gap depends on the sp^2 cluster size and hence on the sp^2:sp^3 ratio and not explicitly on the hydrogen concentration. This is indeed found experimentally [22]. Thus, the role of hydrogen in an optical context is indirect, to decrease the sp^2 site concentration and thereby reduce the sp^2 cluster size.

<u>GAP STATES</u>

The electronic states of an amorphous semiconductor fall into two classes, extended states and localized states. These lie at different energies and are separated by an energy called the mobility edge [35]. The extended states are propagating, band-like states found in regions of high density of states. Localized states are states created in the gap by the disorder. The separation of the valence and conduction band mobility edges defines the mobility gap or pseudogap (Figure 6) and is analogous to the crystalline band gap.

It is also convenient to further classify localized states as either tail states or deep states. Tail states lie adjacent to the mobility edges. Deep states lie around midgap and are generally associated with specific "defect" bonding configurations.

The nature of the tail and defect states in amorphous carbons can be seen from our π cluster model [36,37]. Clusters either produce states only well away from E_F or produce some states near E_F. The local gap of the first type of cluster decreases with increasing cluster size, so tail states are created by larger than average clusters. All clusters with an odd number of sites and some with an even number of sites produce a state or states near midgap (at E_F in the Huckel model). These clusters are defined as the defects.

The width of the optical absorption edges in Figure 4 indicates that a considerable range of cluster sizes exists in both a-C and a-C:H. This implies that there are strong spatial fluctuations in the local band edge energies in these materials and it is of interest to describe the effects of this on the position of the mobility edges in these materials.

Most amorphous semiconductors are σ-bonded, and the fluctuations have length scales of order of the bond length. The positions of the

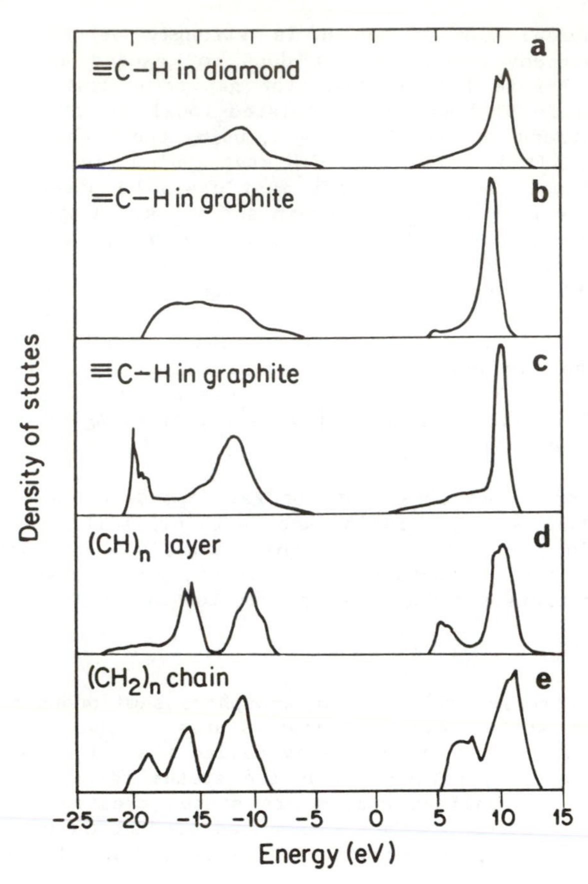

Figure 5. Calculated DOS at hydrogen sites for various configurations

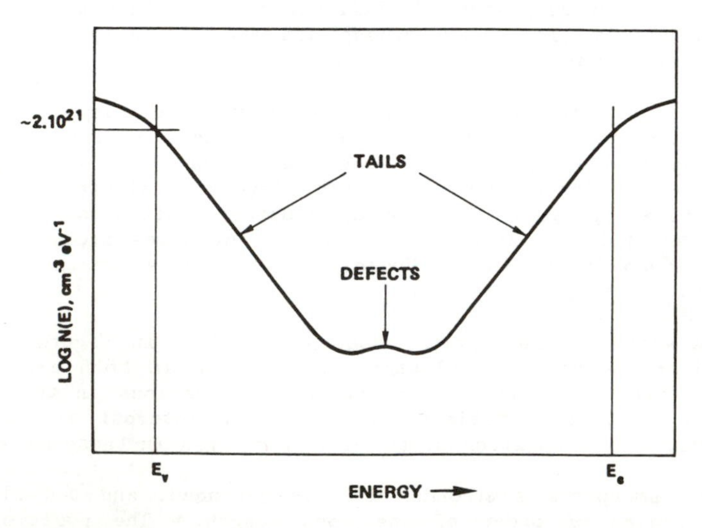

Figure 6. Schematic of the gap state spectrum in an amorphous semiconductor

mobility edges are then set by considerations of quantum mechanical tunnelling [36]. In practice they lie at energies at which band edge tailing causes g, the ratio of the density of states to the free electron density of states [37], is given by

$$g \approx 1/3 \tag{2}$$

On the other hand, clusters in a-C can be much wider, and classical percolation arguments might be more appropriate.

Figure 7 shows a schematic band model of a-C and a-C:H. Consider a-C first. Its cluster boundaries are defined by its few sp^3 sites and by the discontinuities in the π bonding plane, and are therefore narrow. Thus, the boundaries do not act as strong tunnel barriers and the principal interest is in the fluctuations due to cluster size. The classical position of a mobility edge in such a mixture of clusters is at an energy sufficiently deep into the band that it intersects enough clusters to form a percolation path across the sample. This occurs at a probability of

$$p_c \simeq 1/3 \tag{3}$$

of passing between adjacent clusters, independent of their size or shape [38,39]. The similarity of the criteria g and p_c in equations (2) and (3) suggests that the mobility edge positions in a non-barrier system like a-C are esentially independent of fluctuation scale length. Furthermore, the E_{04} optical gap also tends to separate energies at which $g \simeq 1/3$. Thus, we expect the mobility gap E_μ to equal the optical gap E_{04}, in a-C, as it does in a-Si:H [40]. Thus,

$$E_\mu \simeq E_{04} \sim 0.5 \text{ eV} \tag{4}$$

for sputtered a-C.

Turning to a-C:H, this contains a much larger fraction of sp^3 sites. A full valence electron calculation found that the wavefunction of an isolated π cluster in an sp^3 matrix decays rapidly outside it, with a decay length $\lambda \simeq 1$Å, for energies within 1 eV of E_F. This is usually much less than the intercluster spacing, so the sp^3 matrix will clearly act as a considerable tunnel barrier for π states. The matrix will tend to localize the π states and confine them to their clusters. The mobility edges are then forced deeper into the bands than given by equation 2. Unfortunately, it is not possible to give a more quantitative analysis of their position except to say for a-C:H

$$E_\mu > E_{04} \tag{5}$$

The situation is reminiscent of quantum confinement in AlGaAs superlattices. The confinement is confirmed by the presence of a long-lived luminescence up to 2.6 eV in hard a-C:H well above its $E_{04} \simeq 1.3$ eV [41,42]. Confinement is even stronger in soft a-C:H, because of the large amount of

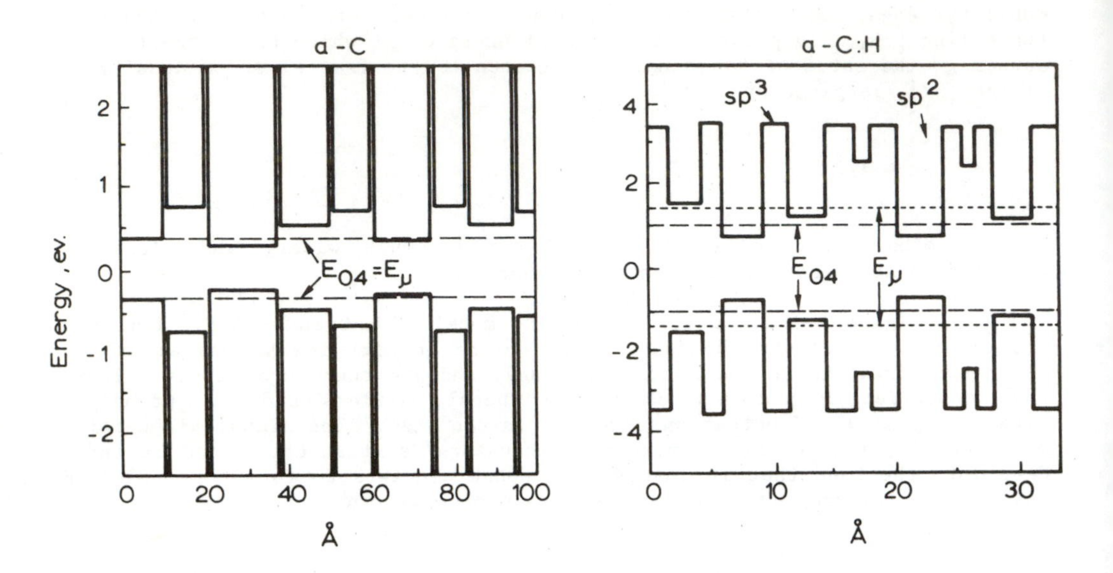

Figure 7. Schematic band diagram of (a) a-C and (b) a-C:H

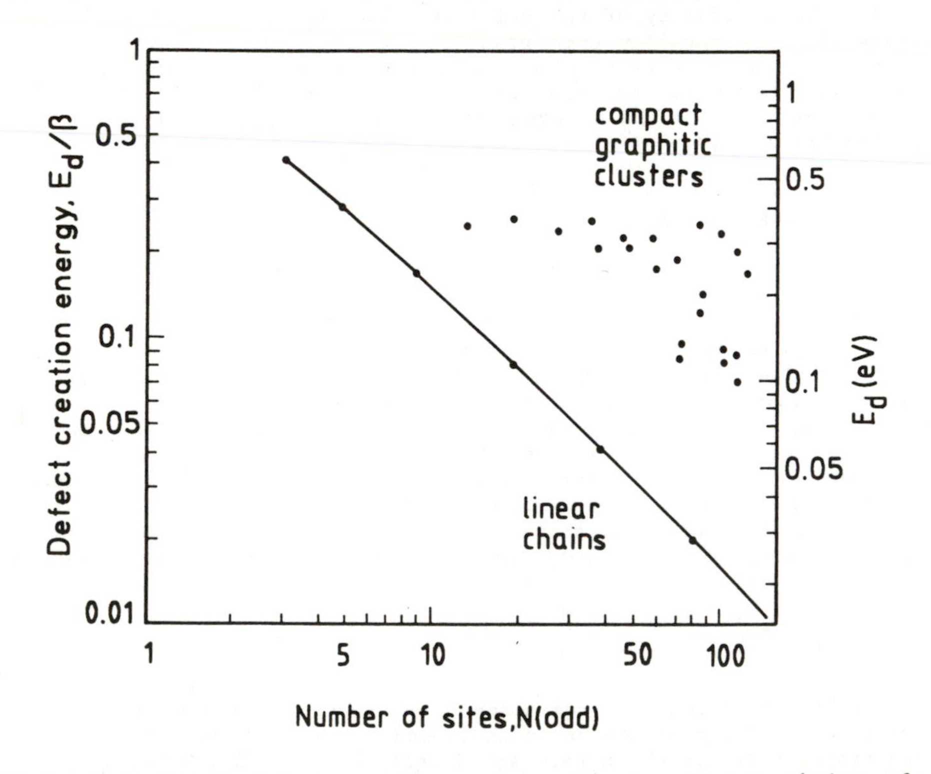

Figure 8. Calculated defect creation energy E_d for linear chains of π states and for compact graphitic clusters, taking β = -1.4 eV

low density polymeric material, and indeed the luminescence efficiency is now increased by $\sim 10^4$ above that in hard a-C:H [41].

DEFECT STATES

Defects in σ-bonded amorphous semiconductors tend to be under- and over-coordinated sites, and specifically dangling bonds (i.e. trivalent sites) in a-Si:H. The nature of defects in π bonded amorphous semiconductors can be different. The existence of an optical gap in a-C and a-C:H indicates that our rules regarding medium range order are well obeyed. Any configuration defect which breaks these rules can therefore be considered to be a defect. In the Huckel model, the energy spectrum of the π states is symmetric about $E_F = 0$, so any cluster with an odd number of sites has at least one state at E_F. These states broaden into a defect band across the gap when more distant interactions are included.

It is possible to define a defect creation nergy E_d for such defects,

$$E_d = N(\bar{E}_N - E_N) \tag{6}$$

where N is the number of sites in the cluster, $\bar{E}_N$ is the total π energy per site E_{tot} of the cluster and E_N is the value of E_{tot} for an equivalent even membered cluster. Fig. 8 shows the variation of E_d with N for linear olefinic chains and compact aromatic clusters, calcuated in the Huckel model. We must use $\beta = -1.4$ eV for total energies, different to that used for one-electron energies, because of over-simplifications in the Huckel model. Fig. 8 illustrates two important differences between π and σ defects. Firstly, E_d is not single-valued for π defects but decreases slowly with increasing cluster size. Secondly, the typical value of E_d for aromatic defects, 0.4 eV, is significantly less than that of the equivalent σ defect for carbon, the dangling bond, whose energy might be expected to be roughly half of the C-C bond energy, or 1.8 eV. This lower E_d suggests that π defects will dominate in amorphous carbons.

The half-filled defect levels are paramagnetic and can be detected by electron spin resonance (ESR), which finds 10^{19}-10^{20} spins cm^{-3} in a-C and 10^{16}-10^{19} cm^{-3} in a-C:H [13,43-46]. The spin density is lowest in soft a-C:H. High temperature annealling does not reduce the spin density in a-C[13], unlike in a-Si, while it tends to increase the spin density of a-C:H due to dehydrogenation and ultimately due to graphitization.

ELECTRICAL CONDUCTIVITY

The conductivity of sputtered and evaporated a-C tends to occur by hopping of electrons between localized states near E_F. At low temperatures, a variable range hopping occurs, giving rise to a $T^{1/4}$ law

$$\sigma = \sigma_o \exp(-(T_n/T)^n) \tag{7}$$

with n = 1/4. For thin films in which the hopping range exceeds the film thickness, the resistance increases and σ obeys a $T^{1/3}$ law with n = 1/3. The constants T_4 and T_3 are given by

$$T_4 = 16/(k\lambda^3 N(E_F)) \text{ and } T_3 = 8/(kd\lambda^2 N(E_F))$$

where k is Boltzmann's constant, λ is the wavefunction decay length, $N(E_F)$ is the density of states at the Fermi level and d is the film thickness of the thin films [35]. Conductivity measurements on both thick and thin films permitted Hauser [8] to determine both $N(E_F)$ and

$$\lambda \simeq 12Å \tag{8}$$

for a-C. This value of λ is just less than the mean cluster size in a-C (15-20Å). $N(E_F)$ was found to be 10^{18}-10^{19} ev^{-1} cm^{-3}, so $N(E) \sim 2.10^{18}$-2.10^{19} cm^{-3}, similar to that found by ESR.

Recently, it has been noted that σ does not fit the $T^{1/4}$ law over a wide temperature range but rather

$$\sigma = \sigma_o (T/T_o)^n$$

with n = 15-17, suggesting that each electron hop involves not one but many phonons [47].

The conductivity of a-C:H has been measured by a number of workers [31,32,48-50]. Anderson [31] found that the conductivity fitted

$$\sigma = \sigma_o \exp(-E_\sigma/kT) \tag{9}$$

above 500K, and gave an activation energy of E_σ = 1.1 eV, which was half the optical gap, 2.2 eV. This is consistent with E_F lying in midgap and the optical gap equalling the mobility gap in this soft a-C:H, contrary to the earlier discussion. At lower temperatures, σ also fits equation (9) but with a lower value of E_σ and σ_σ. Meyerson and Smith [48] found $\sigma_o \simeq 10^{-2}$-10^{-4} Ω^{-1} cm^{-1}, typical of hopping from E_F up to localized band tail states. It is likely that the conductivity data for a-C:H is consistent with a multiphonon trapping mechanism, as in a-C [47], but this has not been tested.

A potentially important property of a-C:H is that it can be electronically doped. Its conductivity can be varied by many orders of magnitude by including B_2H_6, NH_3 or PH_3 in the deposition gas stream [32,49]. The doping is by the substitutional mechanism, as thermopower measurements confirm that the group III element, boron, introduces holes and gives p-type doping while group V elements, N and P, introduce electrons and give n-type doping.

It is likely that doping occurs by the dopant substituting for a carbon at a sp^3 site. This would lead to the band model shown in Figure 9. In undoped a-C:H, E_F lies in the minimum of the DOS at midgap. N-type doping introduces partially occupied donor states in the antibonding σ^* band at $\sim$3 eV. The donor electrons fall down to E_F, raising it above midgap into the conduction band tail. The efficiency of doping is quite low, 1% of

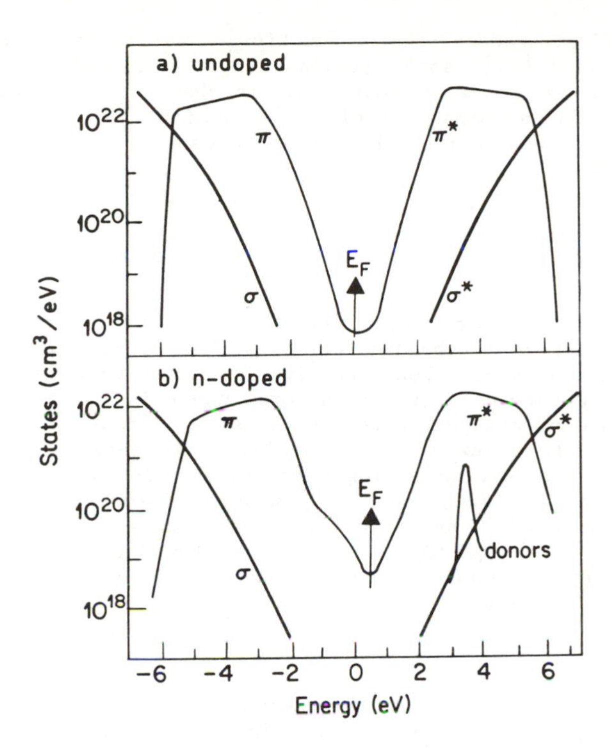

Figure 9. Schematic of the origins of states around the gap for (a) undoped and (b) n-type a-C:H

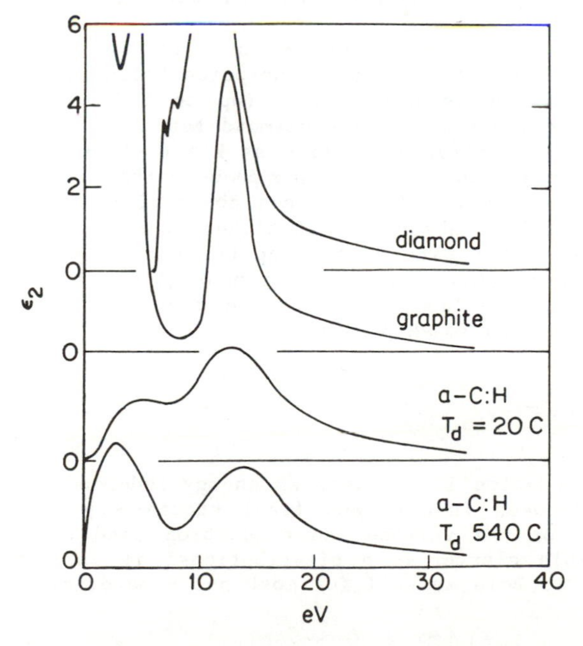

Figure 10. Dielectric functions of graphite, diamond and a-C:H [59]

dopant is needed to reduce the conductivity activation energy, E_σ in equation 9, to 0.3 eV [49], so E_F remains stuck in localized states in the mobility gap. It is probable that this is due to the autocompensation mechanism, known to accompany doping in a-Si:H [51,52], whereby the introduction of a substitutional dopant creates a defect, e.g. for phosphorus

$$P_3^0 \rightleftarrows P_4^+ + D^- \qquad (10)$$

Here, the subscripts represent coordination and the superscripts charge. A phosphorus atom at a normally-bonded, trivalent, non-doping site is P_3^0, and a tetravalent phosphorus is P_4. Reaction (10) indicates that as the phosphorus donor level is well above E_F, the phosphorus donor is ionized, P_4^+, and that the donor electron does not just enter the existing density of midgap states, but it tends to create a new defect state for it to enter. The equilibrium represents processes occurring during deposition. In this way doped films possess a higher density of defects than undoped films, a process which is well studied in a-Si:H [53], and is probably true for a-C:H. The law of mass action can be applied to equation (10), and the low doping efficiency arises because the equilibrium causes most phosphorus atoms to enter trivalent rather than tetravalent sites.

OPTICAL PROPERTIES

The optical properties in the region of the optical gap are of most importance for applications of a-C:H, in particular its low optical absorption coefficient and the ability to vary the refractive index to provide antireflective coatings on a variety of window materials. Those features have been analysed in considerable detail as a function of deposition parameters for films produced in Koidl's group [2,17-21].

One can identify three regions in the optical absorption coefficient, α, in the optical gap. Absorption below 0.5 eV is dominated by lattice vibrations. α is higher here (up to 500 cm^{-1}) than in diamond (<10 cm^{-1}) [20] for two reasons; the diamond lattice has a high symmetry and so its vibrations are infrared inactive at $k = 0$ while disorder permits C-C vibrations to be infra-red active in amorphous carbon [12], and secondly the 40% hydrogen content of a-C:H introduces absorption at 0.36 eV due to C-H bond stretching models. Absorption at higher energies is due to electronic transitions. The second part of the gap up to $\alpha \simeq 10^4$ cm^{-1} is the Urbach absorption tail in which α increases exponentially with photon energy, and the third region is the Tauc tail in which α increases more slowly and is found to follow the relationship [35]

$$\alpha E = B(E-E_T)^2 \qquad (11)$$

The validity of equation 11 requires an energy-independent momentum optical matrix element between initial and final states $\langle i|\partial/\partial x|f\rangle$ and parabolic energy-dependence of the valence and conduction band DOS [40]. Given the variety of possible cluster size distributions, it is remarkable that the Tauc law appears to hold so well for most a-C:H materials [17,29-32,54-56].

The value of the optical gap E_T was related earlier to the average size of the π clusters in the material.

In the wide band optical spectra, the excitations of π and σ electrons show up as two largely separate contributions. This may be used as a means of estimating the relative concentration of sp^2 and sp^3 sites. Figure 10 compares the imaginary part of the dielectric function, ϵ_2, for diamond [57], graphite for E⊥C [58] and a-C:H deposited at 25°C and at 540°C [59]; the later curve being used to represent dehydrogenated a-C.

Each curve consists of two peaks, at ~4 eV and 13 eV. Absorption due to σ electrons is very weak below 8 eV, both in graphite and diamond [58, 59]. A similar behaviour in amorphous carbon will allow absorption in the low energy peak, 0-8 eV, to be assigned solely to π excitations and that in the high energy peak, 8-30 eV, to σ electrons. Quantitatively, the number of electrons contributing to a peak can be found from the sum rule

$$n_{eff} = (m/2\pi^2 N_A e^2 \hbar^2) \int_{E_1}^{E_2} E\epsilon_2(E)dE \qquad (12)$$

where N_A is number of carbon atoms per unit volume, and E_1, E_2 are the cut-off energies. This approach was used by Fink et al. [34,59] to calculate $n_{eff}(\pi)$ and $n_{eff}(\sigma)$ and thus the proportion of sp^2 sites in a number of amorphous carbons. They found almost 100% sp^2 for a-C [59] and roughly 32% sp^2 for 25°C deposited hard a-C:H [34]. Savvides [54,55] used a riskier version of this to estimate the proportion of sp^2 sites in his ion beam deposited samples. Using data for 0-8 eV, he estimated $n_{eff}(\pi)$ and then used the sample density to find the normalisation factor, $n(\pi) + n(\sigma)$. Nevertheless, he was able to detect increasing sp^3 content with increasing ion bombardment.

PHOTOEMISSION

Photoemission spectroscopy gives the valence band density of states weighted by the appropriate cross sections. In a-C:H, the cross section of hydrogen states is much lower than that of carbon states at the relevant energies, so photoemission gives the local DOS at carbon. Varying the photon energy varies the relative cross section of carbon 2s and 2p electrons, so that ultraviolet photoemission (UPS) at 20 eV or 40 eV emphasises the p states, X-ray photoemission at 1486 eV emphasises the s states, while photoemission around 100 eV from a synchrotron source gives a similar weighting to s and p states.

Figure 11 compares the 120 eV photoemission spectra of Wesner et al. [63] for graphite [61], diamond [62], a-C and a-C:H. Their a-C:H sample was deposited below 100°C while the 'a-C' curve corresponds to a a-C:H sample annealled at 500°C. The spectra can be compared with the calculated spectra in Figure 1 of graphite, diamond and the random network models of a-C. The upper DOS peak at 0-12 eV is due to p states and the lower peak at 12-22 eV is due to s states. The upper peak contains both the σ and π p states. The π states are apparent as a shoulder at 2 eV in the graphite spectrum and as weak shoulder in the a-C spectrum. This shoulder is not visible in the a-C:H

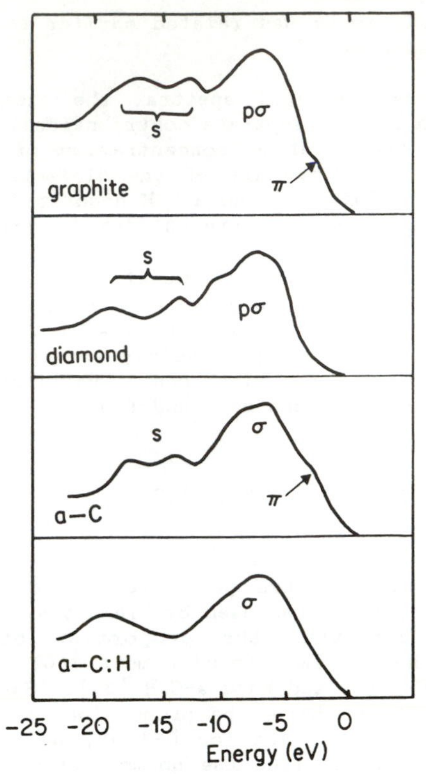

Figure 11. Photoemission spectra of graphite, diamond, a-C and a-C:H [63]

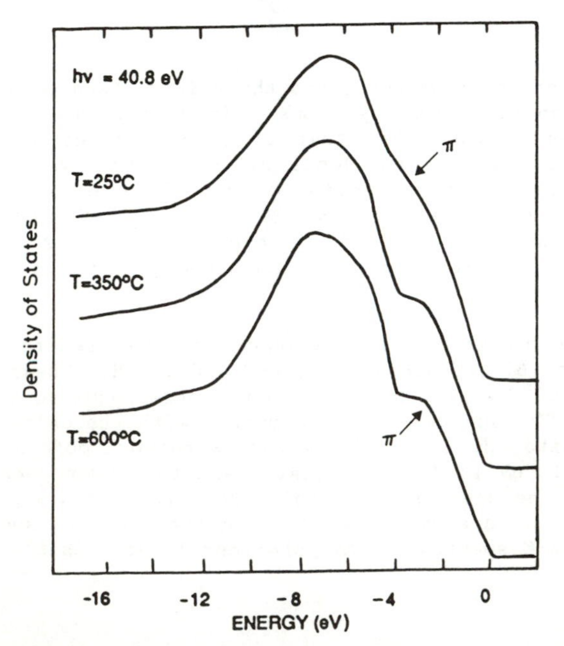

Figure 12. Photoemission spectra of a-C:H annealled at various temperatures [65]

spectrum of Wesner et al. [63], or that of Oelhafen et al. [64], but it can be seen to appear with annealing in the spectra of Oelhafen and Ugolini [65], shown in Figure 12. This feature confirms that the valence band maximum of a-C:H is π-like.

The shape of the s peak gives information on the network topology of the random network. The s peak of a network containing only 6-fold rings consists of two peaks, as seen for diamond and graphite in Fig. 1(a) and (b). However, introducing isolated 5- and 7-fold rings, as in the C519 or Polk model [12,66] in Fig. 1(e) washes out the dip between the two peaks [67]. Thus, it is interesting that the s-band shows two peaks for a-C and one peak for a-C:H. The a-C result is consistent with its dominant π bonding which strongly favours 6-fold rings. On the other hand, a single peak is found for a-C:H because its sp^3 bonded regions allow odd-membered rings.

The photoemission cross-section of hydrogen-centred states is too low for the local DOS at hydrogen to be visible in Fig. 11. However, the a-C:H spectra may contain some hydrogen related features. The calculated local DOS of sp^3 carbon sites varies with the number of hydrogen neighbours, Figure 13. The s band peak tends to move to higher binding energies. A similar movement is seen in the photoemission spectrum in Figure 11; the two s peaks of a-C not only merge into one, the upper peak also tends to vanish.

X-RAY ABSORBTION SPECTRA

Excitations of carbon 1s core electrons into the unoccupied conduction band states creates absorption in the X-ray region above 285 eV which can be used as a sensitive probe for the presence of sp^2 bonding. The first 10 eV of this spectrum is frequently called the X-ray near edge absorption spectra (XANES) and the higher energy oscillations are called the extended X-ray absorbtion fine structure (EXAFS).

The XANES spectra of Fink et al. [34] for graphite, diamond, a-C and a-C:H are compared in Figure 14. The first peak at ∿285 eV is due to transitions from the 1s orbital to unoccupied 2π* states, and the second feature starting at 289 eV is due to transitions to unoccupied σ* states. The presence of a 285 eV peak in a-C and a-C:H confirms the presence of π states and sp^2 sites in these materials. In fact, the measurement of XANES using an energy loss analyser attachment to an electron microscope provides perhaps the most readily available and unequivocal test of the presence of sp^2 sites in carbon samples. Thus, the final spectrum in Figure 14 is that of an unhydrogenated a-C prepared by ion-beam filtered deposition from a cathode spot source, by Berger et al. [24]. It is apparent that this type of a-C contains very few sp^2 sites and that hydrogenation is not the only way to remove sp^2 sites.

The XANES spectra can only be used with great care to extract the actual sp^2: sp^3 ratio in a sample. This is because the XANES spectrum is proportional to the conduction band DOS, but distorted by the interaction between the core hole and the excited electron [68,69]. This interaction produces an excitonic enhancement and shift of the s→π* peak and modifies the onset of the s→σ* transitions.

The EXAFS spectra of a-C and a-C:H have been measured by two groups [34,70]. Fink et al. [34] interpreted the spectra of a-C as consistent with a graphitic island model but with a large bond length disorder. Comelli et

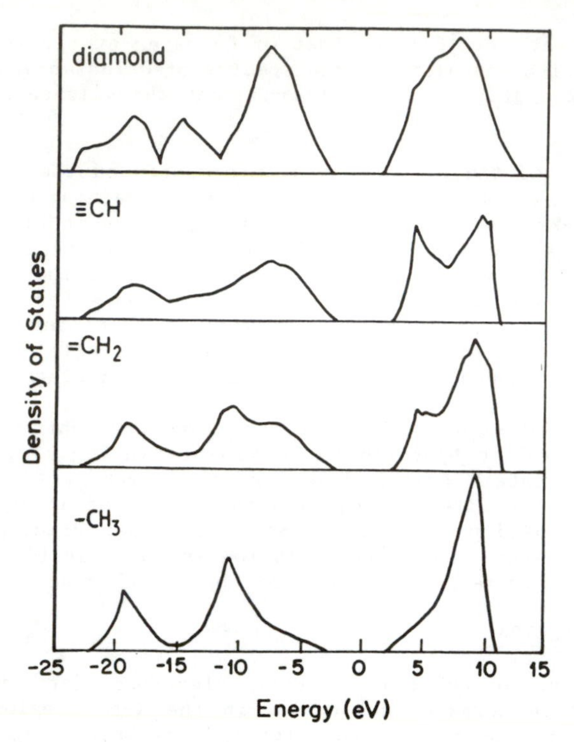

Figure 13. Calculated DOS at a carbon sp^3 site with various numbers of hydrogen neighbours

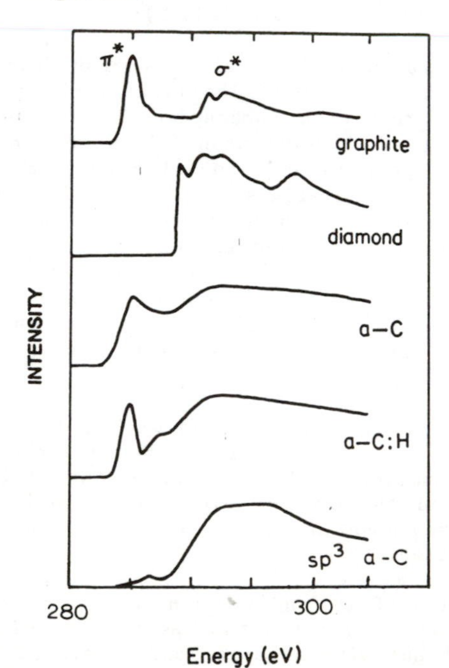

Figure 14. Carbon 1s absorption spectra (XANES) of graphite, diamond, a-C, and a-C:H [34], and of a-C prepared by filtered ion-beam [24]

al. [70] found a slightly different spectrum for a-C (Ar bombarded a-C:H) and gave a very different interpretation. The EXAFS intensity is proportional to the first neighbour coordination number. Normalizing their spectra to that of graphite, Comelli et al. [70] proposed that a-C had a mean carbon-carbon coordination of 2, a bond length of 1.44 A, intermediate between that of diamond (1.54A) and graphite (1.42A) and much less bond length disorder. While the source of the problem cannot be identified, it is difficult to accept so low a coordination number for a-C, as it is in conflict with all other data on this material.

ELECTRON ENERGY LOSS SPECTRA

The electron energy loss (EELS) function is defined as

$$-\text{Im}\ (1/\epsilon) = \epsilon_2/(\epsilon_1{}^2+\epsilon_2{}^2) \tag{12}$$

where Im is the "imaginary part of", ϵ_1 is the real part of the dielectric function and ϵ_2 is the imaginary part of the dielectric function. The loss function can be calculated from the optical spectra or measured directly. In general, the loss can refer to the valence electrons or the core electrons. The core loss spectrum was discussed earlier as X-ray absorption. The valence electron loss spectra of graphite [58], diamond [57], a-C and a-C:H [59] are compared in Fig. 15.

The loss function can show both one-electron features (band-to-band transitions) and many electron features (plasma oscillations). Plasma oscillations produce a peak in the loss function if the particular plasmon energy E_p occurs where ϵ_1 is small. Thus, the loss function of carbon in general consists of two peaks, a lower peak due to π plasmon oscillations and an upper peak due to all the valence electrons, $\sigma + \pi$. E_p is given by

$$E_p = ((4\pi e^2 h^2/m) N_a n_{eff})^{1/2} \tag{13}$$

Such E_p values are compared with the measured values in Table 4.

TABLE 4. PLASMON ENERGIES IN CARBON, IN eV

	π		$\sigma + \pi$	
	Exp	Theory	Exp	Theory
Graphite	7.2	12.5	25.2	25.1
Diamond		-	30-32	31
a-C	6		25	
a-C:H, T_d = 25°C	7		20.8-24	
T_d = 600°C	6		21	

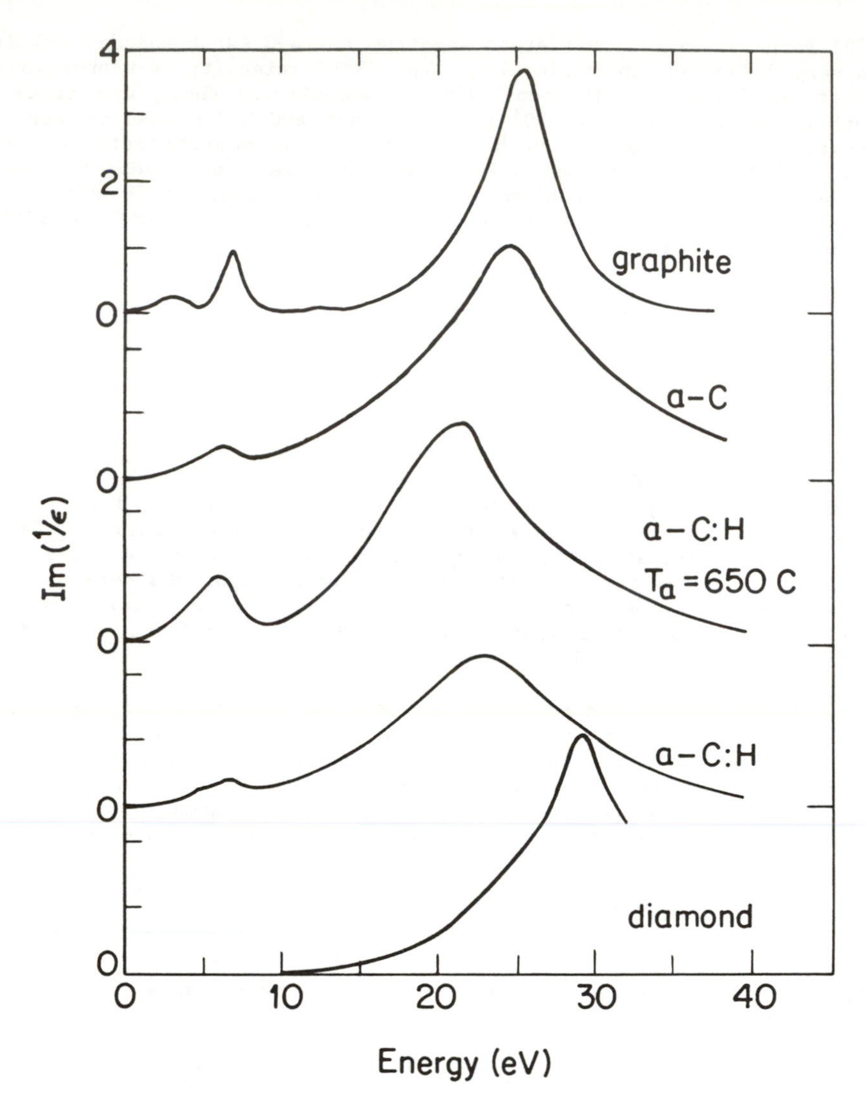

Figure 15. Energy loss spectra of the valence electrons of graphite [58], diamond [60], a-C and a-C:H [59]

The comparison is good in graphite and diamond for the $\sigma + \pi$ plasmon. However, the π plasmon in graphite occurs at a lower energy than expected because its expected position of 12.5 eV coincides with the main $\sigma \to \sigma^*$ one-electron peak in ϵ_2, which has the effect of moving it to lower energies. The $\sigma + \pi$ plasmon energies of a-C and a-C:H given in Table 4 are consistent with their atomic densities, showing an internal consistency in these optical spectra. The π plasmon energies are again depressed in energy. Thus, the π valence plasmon in amorphous carbons should only be used as an indication of the presence of sp^2 and not of their concentrations.

REFERENCES

1) Robertson, J.: Adv Phys, 1983, 32, 361

2) Angus, J.C., Koidl, P. and Domitz, S.: In "Plasma Deposited Thin Films", 89, Ed. J. Mort and F. Jansen (CRC Press, Boca Raton USA, 1986)

3) Sundgren, J.E. and Hentzell, H.T.G.: J. Vac Sci Technol A, 1986, 4, 2259

4) Tsai, H. and Bogy, D.B.: J. Vac Sci Technol A, 1987, 5, 3287

5) Dresselhaus, M.S., Dresselhaus, G., Sugihara, K., Spain, I.L. and Goldberg, H.A.: "Graphite Fibers and Filaments", (Springer, Berlin, 1988)

6) Jenkins, G.M. and Kawamura, K.: "Polymeric Carbons" (Cambridge UP, Cambridge, 1976)

7) Mildner, D.F.R. and Carpenter, J.M.: J. Non Cryst Solids, 1982, 47, 391

8) Hauser, J.: J. Non-cryst Solids, 1977, 23, 21

9) Boiko, B.T., Palantik, L.S. and Deveryanchenki, A.S.: Sov Phys Dokl, 1968, 13, 237

10) Kakinoki, J., Katada, K. and Hanawa, T.: Acta Cryst, 1960, 13, 171

11) Cervinka, L., Dousek, F.P. and Jausta, J.: Philos. Mag. B., 1985, 51, 603

12) Beeman, D., Silverman, J., Lynds, R. and Anderson, M.R.: Phys Rev, 1984, B, 30, 870

13) Wada, N., Gaczi, P.J. and Solin, S.A.: J. Non Cryst Solids, 1980, 35, 543

14) Nemanich, R.J. and Solin, S.A.: Phys Rev B, 1979, 20, 392

15) Galli, G., Martin, R.M., Car, R. and Parrinello, M.: Phys Rev Lett, 1989, 62, 555

16) Tersoff, J.: Phys Rev Lett, 1988, 61, 2879

17) Bubenzer, A., Dischler, B., Brandt, G. and Koidl, J.: J. App Phys, 1983, 54, 4590

18) Dischler, B., Bubenzer, A. and Koidl, P.: App Phys Let, 1983, 42, 636

19) Dischler, B., Bubenzer, A. and Koidl, P.: Solid State Commun, 1983, 48, 105

20) Dischler, B.: In "Hydrogenated Amorphous Carbon", Ed. P. Koidl and P. Oelhafen, Proc European Materials Research Society, Vol. 17, 1987, 189

21) Koidl, P. and Wild, Ch.: Proc Mat Res Soc, 1988, EA-15, 41

22) Kaplan, S., Jansen, F. and Machonkin, M., App Phys Let, 1985, 47, 750

23) Grill, A., Meyerson, B.S., Patel, V.V., Reimer, J.A. and Petrich, M.A.: J. Appl Phys, 1987, 61, 2874

24) Berger, S.D., McKenzie, D.R. and Martin, P.J.: Philos Mag Lett, 1988, 57, 285

25) DeVries, R.C.: Ann Rev Mat Sci, 1987, 17, 161

26) Papers in Proc Mat Res Soc, 1988, EA-15, 41

27) Hsu, W.L.: J. Vac Sci Technol, 1988, A6, 1803

28) Robertson, J. and O'Reilly, E.P.: Phys Rev B, 1987, 35, 2946

29) Smith, F.W.: J. App Phys, 1984, 55, 764

30) McKenzie, D.R., McPhedran, R.C., Savvides, N. and Botten, L.C.: Phil Mag B, 1983b, 48, 341

31) Anderson, D.A.: Phil Mag, 1977, 35, 17

32) Jones, D.I. and Stewart, A.D.: Philos Mag B, 1982, 46, 423

33) Bredas, J.L. and Street, G.B.: J. Phys C, 1985, 181, L651

34) Fink, J., Muller-Heinzerling, T., Pfluger, J., Bubenzer, A., Koidl, P. and Crecelius, G.: Solid State Commun, 1983, 47, 887

35) Mott, N.F. and Davis, E.A.: "Electronic Processes in Non-Crystalline Materials", (Oxford University Press, 2nd ed., 1979)

36) Robertson, J.: Philos Mag Lett, 1988, 57, 143

37) Robertson, J. and O'Reilly, E.P.: p 257 in Ref. 20

38) Kirkpatrick, S.: Rev Mod Phys, 1973, 45, 574

39) Scher, H. and Zallen, R.: J. Chem Phys, 1968, 53, 3758

40) Jackson, W.B., Kelso, S.M., Tsai, C.C., Allen, J.W. and Oh, S.J.: Phys Rev B, 1985, 31, 5187

41) Wagner, J. and Lautenschlager, P.: J. App Phys, 1986, 59, 2044

42) Ramsteiner, M. and Wagner, J.: App Phys Lett, 1987, 51, 1355

43) Gambino, R.J. and Thompson, J.A.: Solid State Commun, 1970, 34, 15

44) Miller, D.J. and McKenzie, D.R.: Thin Solids Films, 1983, 108, 257

45) Orzesko, S., Bala, W., Fabisiak, K. and Rozploch, F.: Phys Stat Solidi, 1984 81a, 579

46) Jansen, F., Machonkin, M., Kaplan, S. and Hark, S.: J. Vac Sci Technol, A, 1985, 3, 605

47) Shimakawa, K and Miyake, K.: Phys Rev Lett, 1988, 61, 994

48) Meyerson, B. and Smith, F.W.: J. Non Cryst Solids, 1980, 35, 435

49) Meyerson, B. and Smith, F.W.: Solid State Commun, 1980, 34, 531

50) Meyerson, B. and Smith, F.W.: Solid State Commun, 1982, 41, 23

51) Street, R.A.: Phys Rev Let, 1982, 49, 1187

52) Robertson, J.: Phys Rev B, 1986, 33, 4399

53) Stutzmann, M., Biegelsen, D.K. and Street, R.A.: Phys Rev B, 1987, 35, 5666

54) Savvides, N.: J. App Phys, 1985, 58, 518

55) Savvides, N.: J. App Phys, 1986, 59, 4133

56) Savvides, N.: In Ref. 20

57) Roberts, R.A. and Walker, W.C.: Phys Rev, 1967, 161, 730

58) Taft, E.A. and Phillip, H.R.: Phys Rev, 1965, 138, A197

59) Fink, J., Muller-Heinzerling, T., Pfluger, J., Scheerer, B., Dischler, B., Koidl, P., Bubenzer, and Sah, R.E.: Phys Rev B, 1984, 30, 4713

60) Phillip, H.R. and Taft, E.A.: Phys Rev, 1964, 136, A1445

61) Bianconi, A., Hagstrom, S.B.M. and Bachrach, R.Z.: Phys Rev B, 1977, 16, 5543

62) Pate, B.B., Spicer, W.E., Ohta, T. and Lindau, I.: J. Vac Sci Technol, 1980, 17, 1087

63) Wesner, D., Krummacher, S., Car, R., Sham, T.K., Strongin, M., Eberhardt, W., Weng, S.L., Williams, G., Howells, M., Kampas, F., Heald, S. and Smith, F.W.: Rev B, 1983, 28, 2152

64) Oelhafen, P., Freehouf, J.L., Harper, J.M. and Cuomo, J.J.: Thin Solid Films, 1984, 120, 231

65) Oelhafen, P. and Ugolini, D.: In Ref. 20

66) Polk, D.E.: J. Non-Cryst Solids, 5, 365

67) Thorpe, M.F., Weaire, D. and Alben, R.: Phys Rev B, 1973, 7, 3777

68) Mele, E.J., Risko, J.J.: Phys Rev Let, 1979, 43, 68

69) Morar, J.F., Himpsel, F.J., Hollinger, G., Hughes, G. and Jordan, J.L.: Phys Rev Lett, 1985, 54, 1960

70) Comelli, G., Stohr, J., Robinson, C.J. and Jark, W.: Phys Rev B, 1988, 38, 7511

Materials Science Forum Vols. 52 & 53 (1989) pp. 151-174

THE CHEMICAL VAPOR DEPOSITION OF DIAMOND

W.A. Yarbrough (a), A. Inspektor (b) and R. Messier (c)

(a) Materials Research Laboratory, The Pennsylvania State University
University Park, PA 16802, USA
(b) Nuclear Research Center Negev, P.O. Box 9001, 84190 Beer Sheva, Israel
(c) Also, Department of Engineering Science and Mechanics

ABSTRACT

The synthesis of well crystallized diamond has been achieved by numerous techniques including various glow discharge plasma techniques (microwave, RF and DC) as well as thermally activated or "hot wire" techniques. These coatings can be differentiated from the "diamondlike"cabon and hydrocarbon films by Raman spectroscopy and x-ray diffraction, demonstrating long range order. Many important issues remain to be resolved before a reliable asessment can made of the impact these developments will have. Reviewed in this work are ideas and experiments relevant to the basic issues of nucleation, methods and mechanisms of growth, characterization and possible technological applications.

INTRODUCTION

Ever since the work of W. G. Eversole [1] in the early 1960's, there have been sporadic reports that diamond can be synthesized from gas phase hydrocarbon precursors. Early in this decade, Japanese scientists [2-4] confirmed earlier Soviet [5,6] and pioneering American experiments [7-9] which showed that stable diamond growth was possible provided certain conditions were established and maintained during the growth process. Although kinetically stable growth of non-crystalline or poorly crystallized forms of carbon, boron nitride, boron carbide, and other compositions have been known for some time, the growth of well crystallized metastable forms is rare. The stable growth of well crystallized diamond, now accomplished by many techniques and in many laboratories, may herald a new age of metastable materials synthesis for specialized products, or may represent a peculiarity of carbon and carbon chemistry. This question remains an important and open issue, but even if the correct appraisal is the latter, the development of diamond and diamond composite coatings, has had and will continue to have a profound effect on several strategically important industries in the U.S. and abroad.

DIAMOND AND DIAMOND-LIKE CARBON

Although this discussion is concerned only with the chemical vapor deposition of diamond, it is important to observe that many of the physical vapor deposition techniques which have been used to prepare diamond-like carbon films, have also been used to prepare films which clearly contain nanocrystalline diamond in addition to various disordered carbon structures. If one considers only those reports of films containing essentially no hydrogen (~1 at. % or less), there are several which show clear evidence for the presence of diamond. However, most such films show evidence for the presence not only of diamond but of non-crystalline carbon, as well as chaoite, lonsdaleite, and other phases which have yet to be identified [10-15]. Hence, differentiation between what one refers to as diamond and what is better termed diamond-like carbon remains an occasionally contentious issue in the materials community.

Two means of characterization can be used to differentiate between the different carbon coatings. One of these is classical x-ray diffraction. A diamond coating for the purpose of the present discussion will be sufficiently well crystallized (i.e. have crystallites of the order of ~10 nm or larger and present in sufficient quantity) to show an X-ray diffraction (XRD) pattern characteristic for polycrystalline diamond. The x-ray diffraction lines may be weak, greatly broadened or show asymmetry resulting from strain, defects, or small crystallite size, but they should, nevertheless, be present. In addition, preferred crystallographic orientation has been frequently reported in such films and consequently the relative intensities of the as-deposited film may not correlate with the reference pattern. Although electron diffraction (particularly selected area diffraction from transmission electron microscopy) has been used to demonstrate the presence of diamond, its sensitivity is sufficiently great to introduce the possibility that the diamond seen represents only a small fraction of the carbon present. In addition, significant errors and uncertainties can arise as a result of the extremely small volume of material being sampled and the strong possibility of texture or preferred orientation in vapor phase deposited thin films [16,17]. The second technique, which has proved quite useful and powerful in the characterization of carbon coatings, is Raman spectroscopy. A diamond film or coating for the present purpose will show the characteristic Raman shift for diamond at 1332 cm^{-1} [18]. Other peaks may be present, and the diamond shift may be broadened by defects, and/or small crystallite size. In addition this peak may be shifted by as much as 4 to 8 cm^{-1} (either higher or lower) depending upon preparation conditions and substrate, but again, it should, nevertheless, be present. Those carbon coatings and layers which do not exhibit the characteristic x-ray diffraction pattern of diamond, or its characteristic Raman shift, but which do have many of the physical properties of diamond including hardness, chemical resistance, IR transparency, etc., are those which have become known as diamond-like carbon [19-21].

PREPARATION METHODS

Plasma Assisted Deposition

One of the very first methods found to produce well crystallized diamond used a 2.45 GHz microwave power source to produce a glow discharge plasma in mixtures of 1-3% methane in hydrogen and at gas pressures of 1 to 8 KPa (8 to 60 Torr) [3]. A substrate, placed within the glow discharge region, is heated to temperatures of 700 to 1000°C, depending on power dissipation in the plasma and the nature of the substrate. Using these conditions well crystallized diamond can be nucleated and grown on many different substrates at commonly used power levels from less than 500 W to ~1 kW. Production of the glow discharge is normally achieved through the use of a waveguide to carry microwave energy to a resonant cavity in or through which a gas tight tube, or housing, of nonconductive material (typically fused silica) is installed.

Two such configurations are diagrammed in Figures I and II. The successful production of diamond using microwave plasma assisted deposition rapidly led to the development of additional plasma glow discharge techniques which cover the frequency spectrum from 2.45 GHz to DC, with much patent activity [22-26]. One of the difficulties experienced in the use of many of the plasma techniques has been the distortion introduced, particularly at microwave frequencies, by the presence of electrically conducting surfaces in the plasma. Hence, uniform deposition on curved or convoluted metals or carbides can be a major applications engineering challenge.

High Growth Rate Plasma Methods

The pressure range over which diamond can be grown has been extended to atmospheric with high power continuous discharge or thermal plasma techniques [27,28]. At atmospheric pressure the temperatures of the various species electrons, ions and neutral atoms or molecules-become nearly the same with gas temperatures, reaching as high as 5,000 to 8,000°C [29]. At these temperatures and energy levels, the stable species in the gas phase tend to be the various simple hydrocarbon radicals, various ions and atomic hydrogen. Diamond growth rates approximately two orders of magnitude greater than those seen using low pressure, have been reported by several

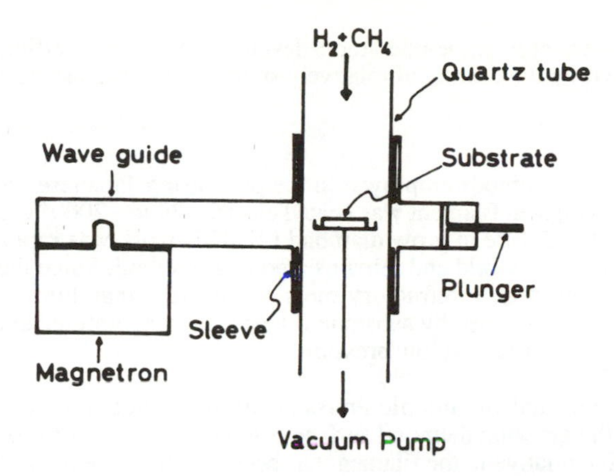

Figure I. Tubular Microwave Deposition System. Microwave energy is carried to a resonant cavity through a rectangular waveguide. Passing through the resonant cavity at 90° to the wave propagation direction is an evacuated fused silica tube which contains the sample and gas mixture for deposition. (from Ref. 3)

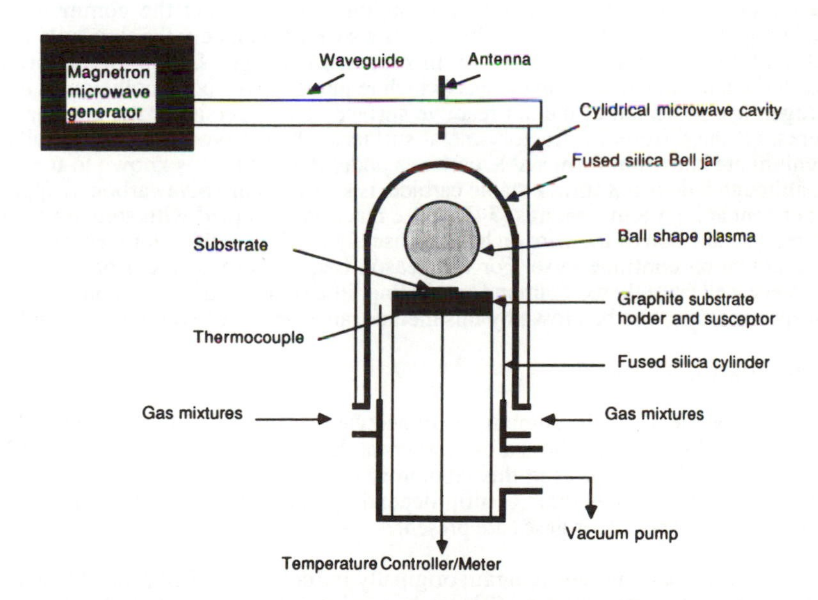

Figure II. Bell Jar Microwave Deposition System. Microwave energy is focused into a resonant cavity which contains an evacuated fused silica bell jar. The substrate to be coated is placed inside the bell jar on its own support, and deposition gases are passed through the chamber. (from Ref. 109)

laboratories. However the very high gas temperatures developed necessitate efficient water cooling of the substrate and to date only the coating of relatively simple substrates has been reported.

Hot wire methods

Among the first of the methods employed in the pioneering Japanese work was a simple thermal method in which a tungsten filament was heated electrically to ~2000° C and used in a low pressure, dilute CH_4/H_2 gas mixture to grow diamond [2]. This work was rapidly reproduced in numerous laboratories around the world and remains a popular method. Since the original work it has been found that all of the highly refractory metals, tungsten, tantalum, molybdenum and rhenium, will produce diamond on a nearby substrate if heated to temperatures in excess of ~2000° C in dilute methane hydrogen mixtures at low pressure.

It has been suggested that thermionic emission from the hot refractory surface and/or electron bombardment of the growing diamond surface is important to the process and electrically biasing the substrate positive relative to the filament has been reported to enhance both growth rate and nucleation density [30,31]. However efforts to reproduce the reported effects of biasing have not been generally successful [32], and the issue of whether thermionic emission or the simple presence of the hot refractory metal (or metal carbide) is more important remains undecided. In more recently reported work, the presence of a DC plasma is reported to exist in very similar experiments which was believed to enhance the concentration of atomic hydrogen in the gas mixture [33]. Clearly some of the conflict in the reported results may arise from differences in whether or not a plasma is created, and if it is, where it is located.

Relatively simple thermodynamic calculations do show that the commonly observed "threshold" temperature of ~2000° C for the wire is also approximately the threshold temperature for significant hydrogen dissociation in this pressure range. Given the relatively slow recombination of atomic hydrogen in the gas (which requires a three body collision) it is likely that once hydrogen is dissociated at the hot reactive surface, it will persist in the gas phase at lower temperatures. Of the different refractory metal surfaces which have been successfully used, all except rhenium are known to form stable carbides under the conditions known to form diamond. Rhenium, although it does not form a stable carbide, is known to dissolve carbon to approximately 1 weight per cent at high temperatures [34]. These reactions, coupled with sometimes significant secondary recrystallization of the wire on heating, usually result in significant mechanical distortion and embrittlement on continued use. For this reason frequent replacement of the wire is often necessary. Very well crystallized diamond, exhibiting little or no evidence of non-diamond carbon by Raman spectroscopy, can be grown by this method, an example of which is seen in Figure III.

Combustion Growth of Diamond

Diamond synthesis by the combustion of acetylene was originally reported by Hirose and Kondo [35] early in 1988 and confirmed by workers at the Naval Research Laboratory [36,37] and at General Electric [38] as well as in this laboratory [39]. Oxyacetylene flames are expected to produce gas temperatures of ~3000° C [40], depending upon the proportions of acetylene and oxygen, and whether or not other gases are present.

SOLGASMIX, a computer program originally introduced by Eriksson [41] and based on work by White, Johnson, and Dantzig [42], can be used to calculate the equilibrium composition of a multicomponent system provided thermodynamic data exist for the species likely to be present. This method uses the formation reactions to establish stoichiometric relationships between each of the possible species (including the elements) and the Gibbs free energy equation ($\Delta G = \Delta H + T\Delta S$) to determine the total free energy of the system as a function of composition. This function is minimized, at the temperature and pressure of interest, by the Lagrangian method of undetermined multipliers using the mass balance constraints established by the formation reactions and the user specified initial composition of the system. This particular method has been applied to problems in chemical vapor deposition in efforts to determine the parameter space in which various solid as well as gas phase species can be expected [43,44]. Table 1 shows a comparison of the expected equilibrium yields obtained using this method (at pressures of 0.0658 atm or 50 Torr and 1.00 atm

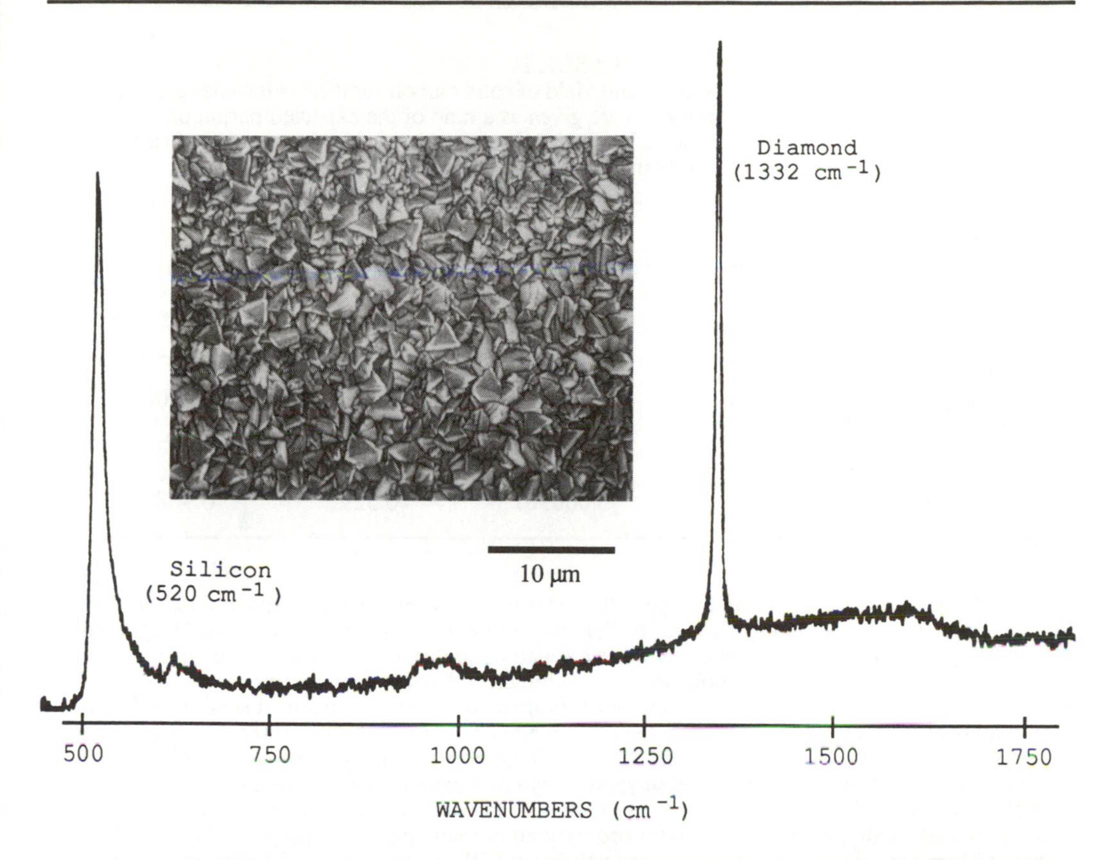

Figure III. Raman spectrum and scanning electron photomicrograph of a well crystallized diamond film on silicon by thermal CVD from tantalum wire. A 1 vol. % mixture of CH_4 in H_2 used with a wire/substrate separation of ~5 mm. Filament temperature measured to be 2050°C by optical pyrometry without an emissivity correction.

or 760 Torr) of various gas phase species and solid carbon at an assumed substrate surface and boundary layer temperature of 1273° K (1000° C). It has been noted that the yield of solid carbon from flames which produce well crystallized diamond is very low as a fraction of the acetylene consumed [38,39]. Reference to Table 1 clearly suggests the explanation for this and suggests that in many of the experiments in which oxygenated species are used or oxygen is added, the vapor phase carbon supersaturation, relative to the solid, is reduced. Inasmuch as the microwave plasma assisted CVD of diamond from mixtures of CO and H_2 is also known [45,46], the chemistry of diamond deposition by combustion processes need not be radically different. At the temperature of acetylene-oxygen combustion much of the hydrogen liberated will persist as the monoatomic species; hence, the flame deposition of diamond is not necessarily inconsistent with many current ideas on the mechanism of diamond growth from gases. The use of combustion to grow diamond does represent a potentially important development from both technological and scientific points of view. Its technological importance is related not only to the relatively low initial capital investment required, but also to the observation that diamond can be grown in the ambient atmosphere using relatively portable equipment. In addition, it offers a new experimental field in which the mechanism(s) and kinetics of diamond formation by vapor phase deposition can be studied.

TABLE 1

Expected (equilibrium) gas phase species and yield of solid carbon for different initial gas phase compositions at 1273K. Gas phase activities are given as a ratio of the expected partial pressure for that species to the total system pressure. The yield of solid carbon is given per mole of carbon precursor in the initial composition.

Species	Equimolar C_2H_2-O_2		1% CH_4 in H_2	
	1 atm.	0.0658 atm.	1 atm.	0.0658 atm.
H_2	0.331	0.333	0.991	1.00
CH_4	0.00103	0.000687	0.00923	0.000619
CO	0.662	0.666	-	-
CO_2	0.00322	0.000214	-	-
H_2O	0.00265	0.000176	-	-
C (solid)	0.0144	0.000967	0.529	0.968

The demonstration experiment typically consists of a water cooled sample holder, or stage, an oxyacetylene torch, and mass flow controllers for regulation of the gas flows. Small fragments, 0.5 to 1 cm in diameter, of single crystal silicon wafers, and other materials have been used as substrates and most of these are polished with submicron diamond powder prior to deposition; however, the nucleation of diamond without this treatment has also been observed. Substrate temperatures are similar to those used in the other methods, typically 800 to 1000° C. Both cutting (multiple nozzle) and brazing or welding torches (single nozzle) can be used with similar results. The ratio of O_2 to C_2H_2 in a premixed oxygen/acetylene flame has been varied over the range of ~0.85 to ~1.1 with the most consistent results being obtained at ratios close to unity. Total gas flow rates vary with nozzle size and with ordinary equipment are in the range of 1.8 to 3.0 slm (standard liter per minute), giving linear gas velocities of the order of 50 to 200 m/sec. The torch flame consists (at an O_2 to C_2H_2 ratio slightly less than unity) of three distinguishable zones. The first (closest to the nozzle) is colorless and transparent and is where the premixed gases emerging from the nozzle are heated to a high enough temperature to ignite and burn, producing a sharply conical and stable flame front, 1 to 5 mm from the torch tip depending on flow rate and nozzle size. The second zone is a bright white or incandescent combustion zone in which the primary combustion of acetylene with premixed oxygen occurs. The extent of this second zone depends very sensitively on the O_2/C_2H_2 ratio used and somewhat less sensitively on the total flow rate and nozzle size. Diamond deposition can be observed when the substrate is placed in this region of the flame. The third zone (farthest from the nozzle and by far the largest volume of the flame) is blue and transparent. Deposition is not observed in this portion of the flame. The second zone of a premixed oxygen deficient oxyacetylene flame is where the principal reaction products H, H_2 and CO are formed. The third zone is essentially a diffusion flame in which the products of initial combustion burn with atmospheric oxygen [47].

THE NUCLEATION OF DIAMOND ON NON-DIAMOND SURFACES

Nucleation is of prime importance and the nucleation density affects the resulting growth morphology and its evolution. Compared with the cauliflower-like top surfaces and columnar cross-sectional morphologies of amorphous and nanocrystalline thin films (prepared typically at low temperatures and thus low adatom mobilities), well crystallized diamond films show a large grain size polycrystalline structure with long range order in its atomic structure. Formation of the amorphous to nanocrystalline columnar structure has been described by a fractal-like model based on random aggregation of the adatoms, leading to clusters, and on the ensuing random competition of intermediate clusters (forming an evolving, with film thickness, hierarchy of larger clusters with the smaller clusters as their substructure) for cone growth [48,49]. For such materials the boundary regions between the nanocrystallites can be a significant percentage of the total volume and, thus, have a significant effect on resulting film properties. This is consistent with the observation that the amount of non-diamond carbon appears to increase as the crystallite size decreases. Correspondingly, in order to explain the long range order in micrometer scale polycrystalline diamond films, first, formation and nature of nucleation sites, and, then, the growth process, leading to non-random insertion of carbon species, together with the stabilization of sp^3 hybridization, should be considered. The range of crystallite size and morphologies is related to the initial nucleation density as determined by the nature of the substrate and by surface preparation methods. Renucleation on the growing surface and twinning can alter crystal evolution. The resulting columnar crystals evolve in a competitive fashion leading to a columnar morphology in which the individual columns are not made up of smaller sub-columns as for the nanocrystalline type morphologies, but are single crystal diamond. Thus, since the percentage of intercrystalline material decreases with increasing film thickness, film quality should increase with film thickness. This later result has been reported recently by Y. Sato and his coworkers at NIRIM [50].

The wide range of substrate materials used for diamond deposition are, generally, classified into: 1) diamond crystals; 2) carbides and carbide forming materials (such as Si, Mo, SiC, and WC); and, 3) substrates which do not form carbides (Cu and Au).

Angus and Hayman[58] discussed the role of various diamond faces in directing carbon atom addition to specific sites in homoepitaxial growth on diamond crystals. They suggest that: "the rate of addition of single-carbon-atom species to the principal surfaces are in the order {100}>{110}>{111}. The rapid addition of single carbon atoms to {100} planes will lead to faceted crystals with {111} faces. Rapid addition of atoms to {111} planes will lead to faceted crystals bounded by {100} cube faces".

Diamond formation on Si represents the second case. The formation of ß-SiC buffer layer, with a partial matching between the diamond lattice and the ß-SiC lattice, has been cited as possibly necessary for diamond nucleation. However, the surface energy of diamond is strongly affected by chemisorption of various gases, notably hydrogen, and various possible reconstructions [52-56]. Badzian suggests that other lattice matching rules may explain diamond deposition on other carbide forming substrates as well [57].

Spitsyn has argued [51] that heterogeneous nucleation on a partially covered substrate is affected by the ratio between adsorption energy of the crystallized material on the substrate (E_{as}) and adsorption energy on the growing film (E_{ad}) [51]. When $E_{ad} > E_{as}$ three dimensional nuclei and non-uniform coverage is expected; two dimensional nucleation is preferred if $E_{ad} \approx E_{as}$; and if $E_{ad} < E_{as}$ strong epitaxial interaction may lead to a different structure in the first few atomic layers of the film. The values of E_{ad} and E_{as}, in turn, are determined by intrinsic properties of the involved materials and by experimental conditions such as substrate temperature, species present on the surface, surface states, dangling bonds and chemical processes on the surface.

Various surface treatments have been reported to increase nucleation density and affect diamond quality. The most common technique used to enhance diamond nucleation on many different materials is to polish or to abrade the surface with a fine diamond powder. Although the typical procedure calls for removing residual diamond detritus by various cleaning methods prior to deposition, its complete removal can only rarely be assured. Hence, two major schools of thought have been advanced to explain the greatly enhanced nucleation rate which is often observed. The first maintains simply that polishing with diamond leaves diamond particles, and/or other carbonaceous residues, in and on the surface. The subsequent "nucleation" is simply the result of diamond growing on diamond. The second argues, particularly in the case of silicon, that the mechanical and crystallographic damage done by the polishing enhances nucleation by creating high energy damage sites on the surface. A variation on this latter idea is the suggestion that such polishing, just prior to deposition, removes surface contaminants and/or strongly adherent films which, if not removed may inhibit nucleation. Figure IV shows diamond growing along a scratch (produced by diamond) in the surface of a silicon wafer. The correlation in many experiments between the scratches left by polishing with diamond powder, and the subsequent nucleation and

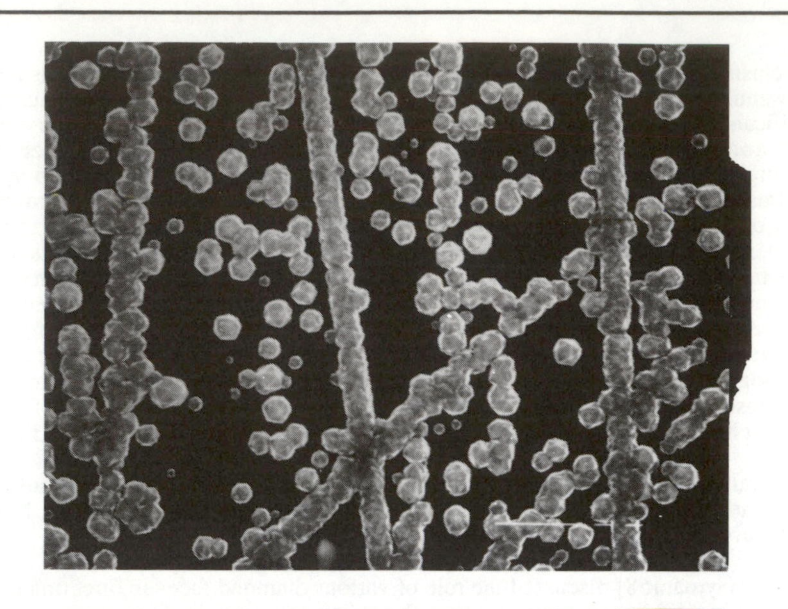

Figure IV. Diamond Nucleation on diamond polished silicon. Note that along scratches left in the silicon surface from diamond polishing, the nucleation density of diamond is extremely high. Elsewhere on the surface a much lower density of nucleation is seen.

growth of diamond crystals is unmistakable. It should be noted that these ideas, although emphasizing quite different effects, are not mutually exclusive and may well depend on the details of a given experiment.

Polishing with other materials besides diamond does not have nearly the same efficacy as diamond [59,60]. For example, in the hot wire deposition of diamond a series of silicon wafers were polished using various abrasives and then cleaned using an ultrasonic bath and a highly alkaline detergent. The abrasives used included submicron cubic boron nitride (Borazon™, General Electric Co., Worthington, Ohio), α- and β- silicon carbides, tantalum carbide, tungsten carbide, aluminum oxide, silicon nitride, and others). The treated samples were placed in a tantalum hot wire deposition system and attempts made to grow diamond. In the case of the diamond polished sample nucleation and growth was observed within 5 minutes using previously conditioned tantalum wire. With all of the other abrasive materials, no evidence of any nucleation or deposition could be observed even after 30 minutes. In an independent set of experiments, conducted using the "bell jar" microwave system seen in Figure II, the effects of seeding silicon surfaces with various powders was studied [60]. Without seeding or polishing the surface, diamond nucleation could be observed only after 8 to 10 hours in the reactor. With the various powders tested only diamond produced evidence of growth immediately, with all other powders requiring greater or less periods of time in the reactor before a weight increase and evidence of growth could be obtained. One of the more provocative experiments, however, was to polish a 3 to 5 mm wide path across a 2.54 cm diameter silicon wafer with diamond with a small cotton swab, followed by 15 minutes in an ultrasonic bath with a strongly alkaline detergent. The wafer was then rinsed, using an additional 15 minutes in an ultrasonic bath with deionized water. Defining the path polished across the wafer as the east-west direction, a second path was polished across the wafer in the north-south direction using submicron Borazon™ powder as was done with the diamond. This wafer was then placed in the same tantalum filament reactor used for the earlier polishing experiments and diamond grown on the wafer. High density nucleation was obtained only on the regions of the wafer which had been polished with diamond, and which had not been over polished with boron nitride. This is shown in Figure V.

Moreover, when the region of intersection between the two polishing steps was examined by scanning electron microscopy it could be seen that although some diamond nucleation took place in this area, it is correlated with the scratches running east and west, the diamond polishing direction, and not with the scratches running north and south. These experiments clearly lend substantial support to the thesis that it is residual diamond, or some other carbonaceous byproducts of polishing with diamond, which enhances nucleation. Whichever explanation one prefers for the effect, it nevertheless remains true that many surfaces which one might wish to coat cannot be polished or damaged in any way and still retain their value, e.g. in semiconductor electronics. Consequently a current research topic in many laboratories addresses the means by which nucleation can be enhanced without this step.

Of all the methods used, clearly the one most often used at the present time is plasma assisted chemical vapor deposition, and most typically, microwave plasma assisted chemical vapor deposition. Of the approximately 60 papers presented at the First International Conference on the New Diamond Science and Technology in Tokyo, Japan, in October 1988, which dealt with some aspect of CVD diamond, fully 26 used some kind of plasma chemical vapor deposition. Consequently an understanding of plasmas and the processes which occur within them represents an important step towards a fuller understanding of the process.

Figure V. Diamond deposition on silicon polished with diamond and cubic boron nitride. Note that the highest density of nucleation is in the areas polished with diamond and not post polished with cubic boron nitride. Little or no diamond nucleation occurs in the unpolished regions of the wafer.

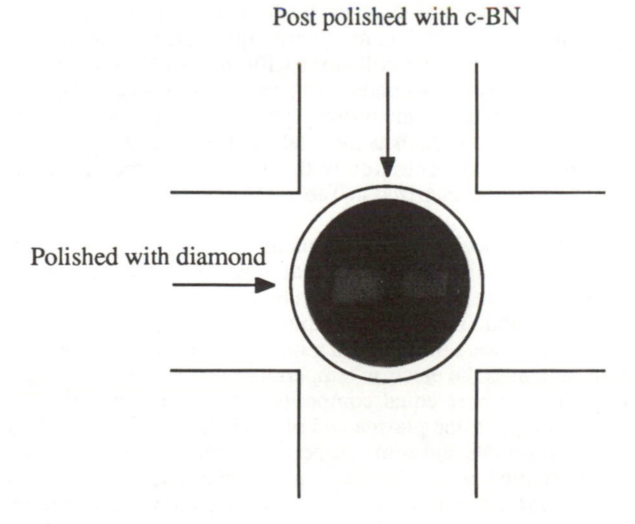

THE PLASMA STATE

Low pressure, metastable, diamond films are formed in continuous competition with other carbon phases. Starting with homogeneous reactions in the gas phase, and continuing through heterogeneous processes at the surface of the growing film, the competition can be thought to conclude only after the sp^3 diamond structure is stabilized and covered with another layer. There are many elementary processes involved in diamond formation and these can be classified in three groups: 1) dissociation of the hydrocarbon source and the formation of related intermediate species; 2) transport to the substrate with reaction at the growth surface; and 3) carbon addition and stabilization as diamond on the surface of the growing film. Knowledge of the reaction pathways and the ability to control the rate determining step are crucial for a consistent reproduction of coating characteristics. The following discussion addresses the major aspects of plasma chemistry in the bulk and at the surface of the growing film.

PLASMA CHEMICAL REACTION CHAIN

"Plasma" is a state of matter in which a significant number of atoms or molecules are electrically charged or ionized. It is an electrically quasi-neutral, reactive assembly of positive and negative ions, free electrons, radicals, atoms, and molecules which collide with each other and with the wall [61-63]. The plasma is formed when an electromagnetic field is applied across a gas. Because of the mass difference between electrons and other charged species and because of the relatively low electron-neutral collision rate in a low pressure gas, the energy of the applied electric field is transferred mainly to free electrons. A low pressure plasma state is characterized by a high electron temperature and a low gas temperature [61-66].

In a sequence of inelastic collisions, the accelerated electrons initiate a sequence of plasma-chemical reactions which consist of two main steps: "primary" and "secondary" reactions [61,67-68]. Primary reactions are electron-neutral encounters in which atoms and molecules of the reactant gases and of the carrier gas undergo excitation, ionization and dissociation (fragmentation) to form various radical and ion-radical species. Electron impact ionization is a major source of charged species in the plasma. Also of particular importance are photoionization processes and excitation collisions in which excited species in a metastable state with a relatively long lifetime are formed [69]. The energy of the "metastable" is relatively high (e.g. the metastable state of Ar is 11.56 eV). Consequently, Penning ionization collisions with metastable species are an additional source of charged particles in the plasma. Secondary processes are reactions between the primary products and with fresh gases. The nature of the primary product is determined by the nature of the reactants and by experimental parameters such as the total gas pressure, net power input, residence time in the plasma, and the reactant concentration in the gas feed. Some of the more significant primary and secondary reactions are summarized in Table 2 [61].

Two general approaches to the mechanism of secondary reactions in the plasma are found in the literature: ion-molecule and radical-molecule reactions. Some degree of control of the ion/radical balance in the plasma can be achieved by controlling the substrate's bias, the input power, and the feed gas composition. Thus, for example, the addition of hydrogen increases the concentration of free radicals and lowers the electron temperature in the plasma [61]. The addition of argon, on the other hand, increases the electron temperature and enhances ionization processes. By defining a plasma deposition where equal competition between ion and radical precursors exists, control of the electron energy in the plasma and of the balance of ions and radicals is a key factor in determining the deposition rate and some properties of the film. Detailed discussion of the ion/radical balance in high frequency discharges is given elsewhere [61]. Ion-molecule and radical-molecule reactions in the plasma are discussed by Bell [70], Havens et. al. [71] and others[64]. Physical processes for the production of ions, atoms and radicals in plasmas are reviewed by McDaniel [72] and Kaufman [73].

TABLE 2.
Major Processes in the Plasma[61]

		rate constants [$cm^3 molec.^{-1} s^1$]
<u>Primary reactions</u>		
impact ionization	$A + e \rightarrow A^+ + 2e$	10^{-10} - 10^{-11}
excitation	$A + e \rightarrow A^* + e$	10^{-6} - 10^{-10}
dissociation	$AB + e \rightarrow A + B + e$	10^{-8} - 10^{-11}
dissociative attachment	$AB + e \rightarrow A^- + B$	10^{-11} - 10^{-12}
dissociative recombination	$AB^+ + e \rightarrow A + B$	10^{-7}
dissociative ionization	$AB + e \rightarrow A^+ + B + 2e$	
<u>Secondary reactions</u>		
ion-neutral collision	$A^+ + BC \rightarrow AB^+ + C$	
radical-neutral collision	$A\cdot + BC \rightarrow AB\cdot + C$	
charge transfer	$A + B^+ \rightarrow A^+ + B$	
associative detachment	$A^- + B \rightarrow AB + e$	
Penning ionization	$A + B^* \rightarrow A^+ + B + e$	
electron transfer	$A + B \rightarrow A^+ + B^-$	

Formation rates and properties of the deposit and of gas born solid precipitates are determined by the secondary processes in the plasma bulk and at the plasma surface boundary. In the plasma bulk, the secondary reactions are homogeneous in nature. At the plasma surface boundary both homogeneous and heterogeneous reactions, with adsorbed molecules and fragments, prevail. Because of the difference between the mobility of electrons and ions in the plasma any surface in contact with an electrodeless discharge assumes a negative potential with respect to the plasma potential. The potential drop, across a Debye sheath at the plasma-surface boundary, accelerates positive ions towards the surface [61,63,64]. Consequently, a driving force for plasma surface boundary processes is the impact of plasma species on the surface. Phenomenologically, these processes can be summarized as follows [63,74]: The accelerated ions together with excited neutral atoms and molecules from the plasma bulk, strike the surface, inducing structural changes in the surface (such as formation of altered layer and sputtering) and emission of secondary electrons back to the plasma. The emitted electrons are accelerated into the plasma, thereby enhancing excitation and ionization processes in the near surface plasma region. The result is the formation of a plasma layer with a high concentration of active species. The boundary of this plasma-surface layer is the region where the energy of the emitted secondary electrons decay, through inelastic collisions with gas constituents, to the steady state energy of the electrons in the plasma bulk.

PLASMA PROCESSES IN A DIAMOND FORMING SYSTEM

Diamond films are formed in hydrocarbon/hydrogen plasmas when a high supersaturation of hydrogen atoms (i.e. well above the thermal equilibrium value in hydrocarbon/hydrogen mixtures) as well as an elevated surface temperature, typically 700-1000 ^oC, are present. Study of the plasma composition by optical emission spectroscopy and by mass spectrometry show that the plasma bulk includes CH radicals (431.4 nm), C_2 (Swan band, 563.6, 516.5 and 473.7nm), H atoms (Balmer series: H_α:656.2nm, H_β:486.1nm, and H_γ: 434.0nm), and excited H_2 molecules. Compared with the bombardment induced stabilization of DLC coatings, the formation of diamond films is accomplished by controlling chemistry in the gas phase and at the surface of the growing film. The nature of the starting monomer has very little effect on plasma characteristics and on the properties of the resulting film. Sato *et al.* [71] investigated morphological changes of microwave CVD diamond as a function of the starting monomer [CH_4, C_2H_2, C_2H_6, i- and n-C_4H_{10}] and their concentration in the feed. SEM observation of the deposited particles, together with x-ray diffraction and with Raman shift measurements show "similar growth features when compared as a function of C/H ratio". Spitsyn [51] discussed the equilibrium partial pressure of C_nH_m compounds over the metastable diamond (phase a) and that over graphite (phase b). He concluded that "the number of polyatomic molecules over phase b is much smaller than over phase a", and consequently, "carbon containing compounds with more than 4-5 carbon atoms are unsuitable for the defect free growth of diamond" [51]. These observation suggest that regardless of the starting material, similar 1 to 2 carbon atom containing species were produced and apparently became the building blocks of the diamond films. Accordingly, the role of hydrogen is, first, to inhibit the polymerization of the hydrocarbon source in the plasma bulk. This is achieved by diluting the hydrocarbon (to about 1% of the starting gas feed) and by lowering the electron energy of the plasma.

A direct electronic mechanism (DEM), direct excitation to the lowest repulsive state ($b^3\Sigma^+_u$) is often considered as the major mechanism for plasma enhanced dissociation of hydrogen [73].

$$(1)\quad e + H_2 \rightarrow e + H_2\,(b^3\Sigma^+_u) \rightarrow e + 2H$$

The threshold of the DEM process is 8.85 eV, with a peak in the dissociation cross-section near 16-17 eV (at 4.5 x 10^{-17} cm^2) and the dissociation rate constant (Kd) is 2 x 10^{-10} cc^3 molecules^{-1} sec^{-1} at E_k= 2 eV, and 11x10^{-10} cc^3 molecules^{-1} sec^{-1} at E_k= 3 eV [82]. Capitelli and Molinari [76] discussed a joint vibroelectronic (JVE) dissociation of the hydrogen molecule as a "ladder-climbing mechanism across the vibrational manifold" up to and across v'=14 which is the last bound vibrational level of the ground electronic state of hydrogen. They suggested [76] that the JVE mechanism is dominant in a low pressure (<5 mbar) hydrogen plasma and that an increased contribution of rotational excitation exists in the moderate pressure (5-55 mbar) range.

Additional important sources of atomic hydrogen in a plasma are dissociative recombination on the wall:

$$(2)\quad H_2^+ + e \rightarrow 2H$$

and, ion-molecule reactions,

$$(3)\quad H_2^+ + H_2 \rightarrow H_3^+ + H$$

the latter being an exothermic process with large cross-section (of the order of 10^{-15} cm^2) and low threshold energy [73]. Data on plasma dissociation of hydrogen at pressures up to a few mbar were analyzed by Bell [77]. Processes in the moderate pressure range (5-55 mbar) were treated by Capezzuto *et al.* [78]. For details of the kinetics of the dissociation process in non-equilibrium plasmas see Capitelli and Molinari [76].

Rupture of the C-H bond to form R and H radicals is the first and rate determining step of the dissociation process:

$$(4)\quad RH + e \xrightarrow{K_1} R + H + e$$

Capezzuto *et al.*[79], and Capitelli and Molinari [76] investigated plasma cracking of CH_4, C_2H_6, C_2H_4, and n-C_4H_{10} in hydrocarbon-hydrogen plasmas and showed that the decomposition process follows close to zero order kinetics with approximately the same rate constants.

When hydrogen is present in the plasma an additional two step path is open for the dissociation process:

$$(5)\quad H_2 \xrightarrow{K_2} H + H$$

$$(6)\quad RH + H \xrightarrow{K_3} R + H_2$$

Combination of these equations gives the expression for the formation rate of hydrogen atoms, the disappearance rate of the monomer, and the formation rate of R radicals in the plasma bulk:[76]

$$(7)\quad \frac{d(H)}{dt} = K_1(RH) + 2K_2(H_2) - K_3(RH)(H)$$

and

$$(8)\quad -\frac{d(RH)}{dt} = \frac{d(R)}{dt} = K_1(RH) + K_3(RH)(H) = (RH)[K_1 + K_3(H)]$$

Indeed, optical emission studies of CH_4/H_2 and of C_2H_5OH/H_2 gas feed systems show that the intensity of the C_2 band head increases as the percentage of the hydrocarbon source in the system is increased [80] whereas the intensity of both H_α and H_β lines decreases [80,81].

In summary, the plasma bulk reactions in a diamond forming plasma are controlled by two competing processes: first, the dissociation of the starting monomer into various intermediate fragments and by-products, such as CH_x (x=1,2,3) and C_2, and second, the depletion of atomic hydrogen in the plasma. Initially, the concentration of the intermediate species and the formation rate of the deposit increase with increasing monomer concentration in the feed. As more monomer is introduced to the plasma the concentration of atomic hydrogen decreases. This in turn, 1) reduces the monomer's decomposition rate, and, 2) increases the amount of non-diamond carbon component in the film. Raman spectra of films prepared using different concentrations of CH_4 are shown in Figure VI. The non-diamond carbon content of the film is represented by the Raman scattering modes around 1580 cm^{-1} [12]. It is easily seen that the intensity of 1580 cm^{-1} band increases with increasing hydrocarbon concentration in the feed.

Several plasma processes can contribute to an increase of the hydrocarbon dissociation rate and of the atomic hydrogen concentration in the plasma bulk. Since the early work of Wood, Delaplace and Finch, (see Coffin [82]) it is evident that oxygen and water vapor enhance the production of hydrogen atoms. Shaw [83] reported an eight-to-tenfold increase of the atomic hydrogen yield in wet (saturated with 2% water vapor) hydrogen, and a substantial increase of the minimum mw-power required to sustain a "dry" hydrogen discharge. Variation of the dissociation rate of hydrogen with water vapor concentration in a hydrogen glow discharge at a pressure of 0.258 torr and a current of 15 mA are presented by Zaitsev et. al. [64]. The maximum dissociation rate was found at water concentration of ~0.15%.

As the addition of CO_2 or O_2 to the gas feed does not change significantly the rate of C-H dissociation [85,86] it is evident that the major feature of these systems is the formation of new active intermediates in the plasma. For example reaction of hydrogen with excited (metastable) oxygen atoms ($O'(^1D)$) yield OH and H radicals or OH and R radicals when the decomposition of RH is involved whereas the decomposition of CO_2 in CO_2/H_2 creates new species such as CO, O, OH and H_2O. Similar products, although due to somewhat different reactions are present in other C/H/O systems. Each of these open new reaction paths in the plasma-bulk and at the plasma-surface boundary to increase the dissociation rate of the starting monomer and to accelerate the preferential etching of graphitic components.

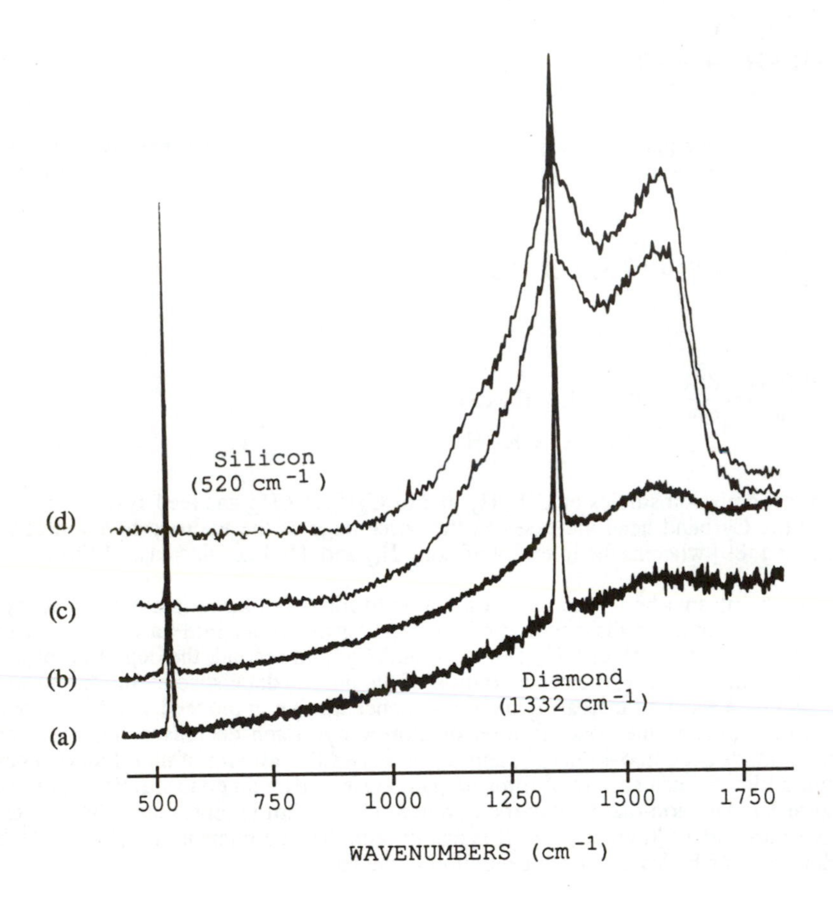

Figure VI. Raman spectra of several CVD diamond films at increasing methane concentrations in hydrogen of (a) 0.5%, (b) 1.0%, (c) 2.0% and (d) 3.0%. As the methane concentration is increased, with all other operating parameters held as constant as possible, evidence is seen for an increasing amount of non-diamond in the coatings obtained.

A similar effect of increasing the concentration of H atoms is observed in rare gas-hydrogen discharges (e.g. in Ar/H_2 mixture) where process (9) take place:

(9) $Ar^+ + H_2 \rightarrow ArH^+ + H$

This reaction, together with energy transfer from the highly excited Ar metastable state (11.56 eV):

(10) $Ar^* + H_2 \rightarrow Ar + H_2^* \rightarrow Ar + 2H$

explain the enhanced production of H atoms when Ar gas is admixed to a H_2 discharge. The accelerated formation of H atoms can be extended to other noble gases as well.

ON THE QUESTION OF DIAMOND VS. GRAPHITE

Professor John Angus of Case-Western Reserve University, one of the pioneers in the vapor phase synthesis of diamond [7,8], observed recently that he was told that 15 years ago he had been thought foolish for pursuing this goal and now that it has been achieved, he was considered a prophet. The commentor went on to observe that he hoped that for Professor Angus's sake the process was not a cyclical one [87]. Much of the skepticism faced by the early workers in the field stems from the well recognized thermodynamic instability of diamond relative to graphite in its many forms and numerous ideas and mechanisms have been suggested to explain its formation in the various experiments. Many of these ideas center on the role of atomic hydrogen in the process and particularly on its presence in the gas phase at partial pressures and temperatures where it is not a stable species, i.e. at "superequilibrium" partial pressures. Which of the many roles attributed to atomic hydrogen is considered to be of primary importance appears to depend on whether the theorist is emphasizing kinetic or thermodynamic considerations. Among these various roles are principally the preferential gasification of graphite and/or other reactive carbons, and the stabilization of sp^3 vs. sp^2 bonding at the growth surface. Although the precise nature of the reaction mechanisms responsible for the growth of diamond are not known, much speculation has centered on various carbon radical species [24], notably $CH_3\cdot$, and on an acetylene reaction mechanism originally proposed by Frenklach and Spear [25]. Both experimental work [26] and simple thermodynamic calculation show that these are the major carbon species present in at least some of these experiments [27].

Once diamond is nucleated, the formation of graphite and/or other non-diamond phases on the growth surface must be inhibited. Graphitic contaminants can be obtained from two sources: 1) graphite and polymers formed in the plasma and/or directly at the growing surface; and 2) graphitization of previously deposited diamond. The activation energy for graphitization of the diamond octahedral {111} surface is 253±18 Kcal/mole, and of dodecahedral {110} surface is 174±12 Kcal/mole, in vacuum at 1500° C [88,89]. Hence at a typical deposition temperature of 1000° C (1273K) the thermal graphitization of previously deposited diamond will occur only very slowly if at all.

Purity control of the film is then achieved by: 1) kinetic factors such as the inhibition of graphite formation in the plasma bulk, 2) surface conditioning (e.g. passivation of surface dangling C bonds by hydrogen atoms which is a major process to stabilize the structure of the growing diamond film], 3) controlling the plasma to surface potential to control the ion bombardment of the growing film (actually, ion bombardment which is essential in DLC deposition may induce long range structural changes and, by local heating, graphitization of the diamond film) and, 4) selective etching of the graphitic phase. The etching rate of graphite in a hydrogen plasma is orders of magnitudes faster than diamond (0.13 mg/cm^2.hr vs. 0.006 $mg/cm^2 \cdot hr$) [90]. The difference between the respective etching rates reflects the difference between the chemical reactivities of graphite and diamond towards gasification by hydrogen. It should be noted, however, that the reactivity of various graphites with atomic hydrogen can vary over 3 to 4 orders of magnitude depending on the nature of the graphite and its previous history [91]. Indeed one means of rendering graphite relatively unreactive towards gasification by hydrogen is by exposure to atomic hydrogen at low temperatures [92]. Although the kinetic arguments appear to rationalize the preferential growth of diamond in most experiments, in their simplest formulation they do not explain why other relatively unreactive forms of carbon, e.g. Lonsdaleite or some DLC materials are not also observed as reaction products, particularly at substrate temperatures below ~600° C.

Another approach to the problem emphasizes the special role played by surfaces in almost all CVD processes. Solid-gas equilibria are, of course, mediated by surfaces, and a rigorous thermodynamic description of the system must include the surface as one of the elements in the system. In many systems, the composition of the solid surface is taken to be the same as the bulk phase, and consequently the assumption can be made that the surface contributes only "excess" thermodynamic quantities, which are of importance only in those cases where surface area to volume ratios are large (e.g., nucleation phenomena, colloidal systems, etc.). The vapor phase deposition of carbon from hydrocarbon/hydrogen mixtures can be argued to represent a special case where the solid surface is no longer compositionally the same as the bulk and consequently must be treated as a separate phase in the system. From this point of view the key to diamond formation is maintaining the diamond structure surface as the thermodynamically stable surface during the growth process, a role often assigned to atomic hydrogen. Under the conditions used in most experiments the diamond surface would be hydrogenated. From this point of view cyclohexane in the "chair" conformation would be the model of a "pure" hydrogenated (111) diamond surface just as adamantane is sometimes pictured as the smallest possible diamond crystal with its surfaces hydrogenated [58]. For an illustration see Figure VII. The (111) diamond and (0001) graphite surfaces are pictured (plan view) in Figure VIII. Using group additivity rules for the estimation of thermodynamic properties [93-95] it is possible to estimate the enthalpies of formation for these surfaces by describing each in a way similar to how three dimensional crystals are described. First a unit cell of each surface is defined such that integral translation along each axis in the xy plane is sufficient to develop the surface to any desired extent. A "mole" of surface is then simply defined as Avogadro's number of such unit cells. By this formalism and the use of group additivity principles, the enthalpy of formation for each surface can be generated. Table 3 shows the results obtained by one of us (Yarbrough) indicating that the hydrogentated principal surfaces of diamond have lower enthalpies of formation than the hydrogenated principal surfaces of graphite.

Using the values estimated for the (0001) graphite and (111) diamond surfaces and assigning the entropy of the bulk solid to each respective surface, a calculation can be made indicating which surface is likely to dominate in equilibrium with hydrogen and/or atomic hydrogen at various temperatures and pressures. The assignment of bulk entropies to each respective surface obviously introduces error. In the case of the graphite (0001) surface the error is expected to be quite small as no new vibrational modes or electronic states are expected. This is not true in the case of the diamond (111) surface. The assumed entropy (that of the bulk solid) may be significantly lower than the true surface entropy as additional vibrational modes are likely both as a consequence of the interruption of the lattice symmetry and as a consequence of the presence of hydrogen at the surface. Consequently the entropy assumed for the (111) diamond surface represents a conservative estimate. Using these values it becomes possible to estimate the relative amounts of hydrogenated diamond (111) and graphite (0001) expected at equilibrium in the presence of atomic hydrogen at various partial pressures and temperatures through the equilibrium:

$$(11) \quad C_2\,[\text{graphite}, (0001)] + H \rightleftarrows C_2H\,[\text{diamond}, (111)]$$

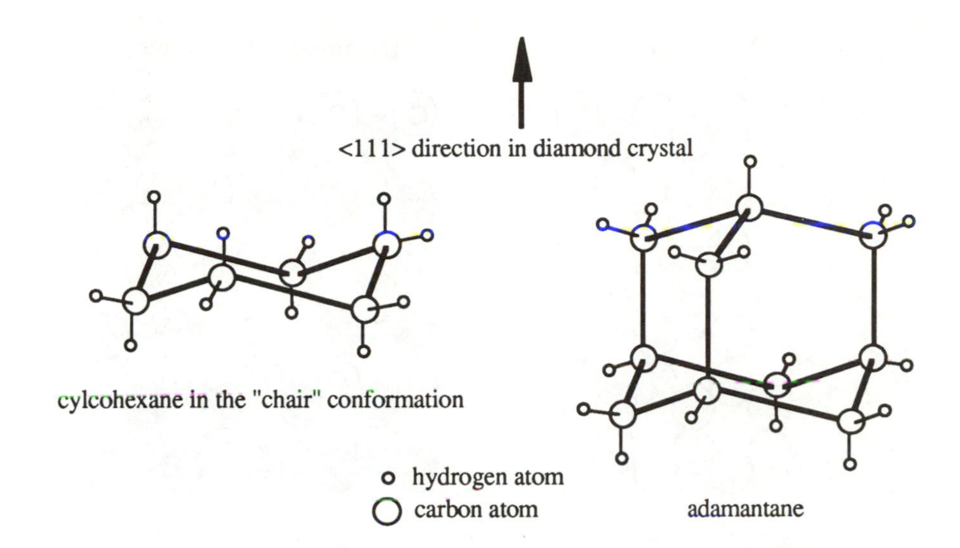

Figure VII. Two compounds often used to illustrate stereochemical relationships in diamond are cyclohexane and adamantane whose structures are shown above. Adamantane is sometimes pictured as the smallest possible crystal of diamond with its external surfaces hydrogenated. In this sense one might also think of cyclohexane in the chair conformation as representative of the (111) diamond surface.

TABLE 3.
Estimated Surface Enthalpies of Diamond and Graphite
(25° C, ergs cm^{-2})

Hydrogenated Diamond		"Clean" Diamond [from ref.96]	Hydrogenated Graphite	
(111)	-177	+5400	(1120)	+1990
(100)	-11(-543)[b]	+9400	(1010)	+282
(110)	-294	-	(0001)	0.0000[a]
(100)	-115 (2x2 reconstructed)			

a. Reference State.
b. Corrected for non-bonding steric (H) repulsion at 25° C using value of +9.90 kcal/mole for cyclooctane. Value in parenthesis is enthalpy without correction for repulsion. For conformational enthalpies see ref. 97.

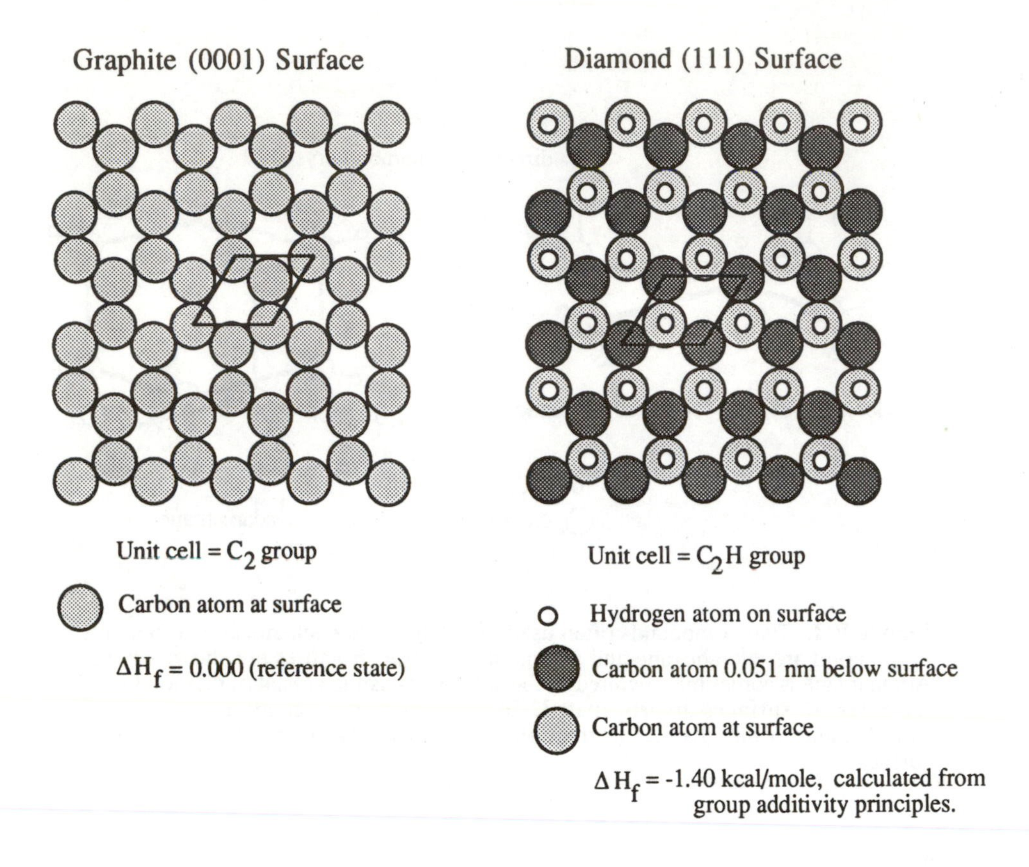

Figure VIII. A comparison of the (0001) graphite and (111) diamond surfaces (plan view) illustrating the similarity in structure. Rhombic unit cells are outlined for each surface. The unit cell or mesh for the (0001) graphite surface consists of two sp^2 carbon atoms and is taken as the reference state by analogy with graphite as the reference state for elemental carbon. The unit cell for the diamond surface consists of two sp^3 hybridized carbon atoms and a bound hydrogen atom.

This kind of estimate assumes, of course, that solid carbon is a stable phase in the system and ignores all vapor-solid equilibria except that between the solid surfaces and atomic hydrogen. Alternatively, a similar expression may be written for an equilibrium involving molecular hydrogen:

$$(12)\quad 2C_2\,[\text{graphite, (0001)}] + H_2 \leftrightarrows 2C_2H\,[\text{diamond, (111)}]$$

In the former case it is found that reasonable agreement exists between theory and experiment with substrate temperatures in excess of ~1100° C resulting in essentially all basal plane graphite and at ~1000° C or less the (111) diamond surface is indeed the stable surface at reasonable partial pressures of atomic hydrogen. In the latter case, represented by expression (12), graphite (0001) is by far the dominant surface at almost all temperatures and pressures.

Enhanced etching of non-diamond phases is believed to occur in C/H/O ternary systems [98,99]. It has also been suggested that the presence of oxygenated species in the gas phase might suppress the formation of some of the reactive species which are thought to lead to graphitic solids, specifically C_2H_2 [98]. In addition, it might be argued that oxygen addition to the gas phase will simultaneously reduce the supersaturation of carbon in the vapor phase, relative to the solid, and enhance the abstraction of hydrogen from the growth surface (through enhancement of both the steady state concentration of atomic hydrogen and the introduction of hydroxyl radicals as an additional reactive species). Raman spectra of films from gas mixtures of $CH_4+H_2+CO_2$ and $CH_4+H_2+O_2$ do show a decrease of the Raman peak at ~1580 cm^{-1} as the O/C ratio in the feed gas increases [80], suggesting that the quality of diamond films improves with the addition of oxygen. Therefore, it may be possible to obtain both better quality diamond films and high deposition rate by varying the concentration of the oxygen in the gas mixture [80,81].

Although the kinetic arguments appear to rationalize the preferential growth of diamond compared to graphite, in their simplest formulation they do not explain why well crystallized diamond should grow preferentially to other relatively unreactive forms of carbon, e.g., the diamond-like carbons or Lonsdaleite, particularly at low substrate temperatures. In addition, the reactivity of "graphite" towards gasification by atomic hydrogen can vary over as wide a range as three to four orders of magnitude, indeed with some unreactive forms being produced by low temperature exposure to atomic hydrogen. Similarly, the stabilization of diamond surfaces by atomic hydrogen appears to rationalize much of the data satisfactorily except that there are now experiments reported, at low substrate temperatures [32,101,102] and in combustion flames at low pressure and oxygen/acetylene ratios greater than unity [39], where no form of solid carbon is thought to be thermodynamically stable and it would appear that some appeal to kinetics must be made to explain the deposition. Hence, much work is left to be done before we can say with confidence that we understand the growth of diamond from vapor phase precursors. Furthermore, the issue of whether the vapor phase growth of diamond involves generalizable principles or is simply a peculiarity of carbon chemistry remains open and debatable.

PROPERTIES AND APPLICATIONS

Diamond and the diamond-like carbons and hydrocarbons have many properties which make them not only scientifically interesting but also commercially valuable. Concentrating on the properties of diamond, which are not shared by the diamond-like carbons, one can readily see why there is an enormous interest expressed and evidenced worldwide in CVD diamond. Paramount of these in the minds of many is its possible use as a high temperature semiconductor material. Its outstanding electrical properties include a high hole mobility (1600 $cm^2V^{-1}sec^{-1}$), electron saturation velocity (2.7 V/sec), high breakdown strength (10^7 V/cm), low dielectric constant (5.6) and high band gap (5.45 eV). Although the band gap is indirect, the possibility of obtaining it as a direct gap semiconductor through epitaxially induced strain [103], if realized, would make possible a wide range of applications in electrooptics. Applications under current development include transistors for high temperature and high frequency operation. Already Schottky diodes have been reported[104] using polycrystalline diamond films, with properties similar to single crystal natural diamond.[105] Demonstration of a MESFET transistor is the logical next step. Consequently much effort is focused on understanding the nature, distribution and types of defects found in CVD Diamond, and their causes [106-108]. One goal is a material with a resistivity similar to the best natural diamond. To date the best polycrystalline CVD diamond films have values in the range of 10^{10} to 10^{12} ohm-cm. Another goal is to dope it both n-type and p-type. At present only boron doped p-type diamond has been successfully prepared with conductivities of around $10^3(\Omega\text{-cm})^{-1}$. Clearly a large amount of work remains to be done for CVD diamond to achieve its potential as a semiconductor material. Also, for the best performance and widest range of applications it will be necessary to grow diamond heteroepitaxially and at high levels of crystal perfection.

Diamond has been suggested for those applications in which IR and/or optical transparency, mechanical damage resistance, and environmental inertness are of importance. With the recent advances in the growth of diamond by CVD techniques, two different approaches to the preparation of diamond suitable for IR applications have been discussed. These are first as a coating on another material which is suitable as an IR window material but which lacks the desired environmental or damage resistance; and second, by itself where the entire window or lens might consist of CVD diamond. A central difficulty with the first approach is that many of the more attractive IR window materials, e.g. the sulfides and selenides, would themselves be reactive at the conditions normally used for the preparation of CVD diamond. These conditions include substrate temperatures in excess of 600° C and the presence of atomic hydrogen. Although to date atomic hydrogen (or a similarly reactive radical species) appears be necessary to stabilize diamond in preference to graphite, it may be possible to grow satisfactory diamond films at lower temperatures. One of us (Yarbrough) has reported the synthesis of nanocrystalline diamond at temperatures less than 600° C [100] and there have been reports of diamond synthesis at temperatures of as low as 350° C [109]. Whether reduced substrate temperatures will make the coating of potentially reactive substrates feasible remains to be seen. Free standing diamond films would, until very recently, have been thought unlikely given the low growth rates (0.1 to 2.0 mm/hr) common to the low pressure plasma and "hot filament" techniques. However in recent months high growth rate methods (~100 μm/hr) have been reported which make the growth of diamond layers as thick as 1 mm practical. Also lower growth rate methods (5-20μm/hr) are reasonable so long as the growth technique is both stable (over the days required for growth) and inexpensive (equipment capital costs). Whether or not these materials can be made suitable will depend on the relative importance of the various mechanisms of IR absorption in CVD diamond. Such thick, free standing coatings could be used also in applications such as cutting and wear surfaces and thermally reactive substrates.

IR absorption in diamond is a sensitive function of several parameters which include effects due to the type and distribution of defects in polycrystalline material, crystallite size, morphology, impurities, and grain boundary effects. CVD diamond typically contains some non-diamond carbon which may be present in different forms including graphitic material as well as disordered diamond-like material. In addition, variously prepared films can contain impurities such as silicon, oxygen, metals, and hydrogen depending upon the preparation method and exact conditions. Diamond-like carbon, prepared by either PVD methods or by plasma decomposition of hydrocarbons, has been investigated as an IR transmitting protective coating. It has been shown that this material is also transmitting in the IR [110, 111]. However, to date it is readily prepared only as a thin film, and consequently its ability to provide protection against impact damage is limited. If much of the non-diamond carbon present in CVD diamond is truly diamond-like then its presence is not likely to be detrimental and may even be beneficial, however if non-diamond carbon is present as graphite or "glassy" carbon then IR transmittance is likely to suffer.

In addition to electronic and IR related applications as discussed above, diamond films are being considered for applications in which their extreme properties of hardness, thermal conductivity, acoustic velocity, chemical inertness, and biocompatibility are used either singly or in combination. Thus, diamond films have the potential of either replacing current coatings or creating new coatings applications.

The extent to and rate at which diamond coatings will have a technological impact will depend on advances in our general scientific understanding of the diamond deposition process and specific knowledge of preparation-characterization-property relations. For instance, for cutting tools, bearings, and other wear, friction, and hardness applications that take advantage of diamond's mechanical properties, the reproducible fabrication of highly adherent coatings is not yet achieved. For optical application the current primary limitation is the high optical scatter due to the rough film top surface due to the large, μm-sized, faceted crystallites. Control of crystallite morphology size and evolution is essential. And for electronic applications large area heteropitaxial diamond, doped both n- and p-type, are essential for widespread applications. Furthermore, low defect densities and high purities will be required: even better than that currently achieved in homoepitaxial diamond. Thus, the scientific challenge is to understand the nucleation and growth of diamond in sufficient detail so that a realistic assessment can be made of those applications which are achievable and those which may never be achieved.

ACKNOWLEDGEMENTS

Many of the results reported and the ideas discussed in this chapter were obtained through collaboration with numerous researchers in the diamond and related materials program at the Pennsylvania State University and the authors owe much to each and all of their colleagues. This work was supported by the Office of Naval Research with funding through the Strategic Defense Initiative's Office of Science and Technology, as has been true of much of the research in CVD diamond synthesis in the United States. In addition, this work has been supported by the Diamond and Related Materials Research Consortium of the Pennsylvania State University which is a non-profit consortium of private companies interested in sponsoring research in CVD diamond, cubic boron nitride, and related materials. The authors would also like to specifically thank the editors John Pouch and Samuel Alterovitz for their patience, and the support and help of Ms. Vicki Zimmerman without whose help the preparation of the manuscript would have been impossible.

REFERENCES

1. Eversole, W. G.: U. S. Patent no. 3,030,188, 1961.
2. Matsumoto, S., Sato, Y., Kamo, M. and Setaka, N.: *Jpn. J. Appl. Phys.*, 1982, 21, L183.
3. Kamo, M., Sato, Y., Matsumoto, S. and Setaka, N.: *J. Crys. Growth*, 1983, 62, 642.
4. Matsumoto, S.: *J. Mat. Sci. Letts.*, 1985, 4, 600.
5. Derjaguin, B. V. and Fedoseev, D. V.: *Nauka.*, 1977, Moscow (in Russian).
6. Spitsyn, B.V., Bouilov, L. L., and Derjaguin, B. V.: *J. Cryst. Growth*, 1981, 52, 219.
7 Angus, John C., Will, H. A. and Stanko, W. S.: *J. Appl. Phys.*, 1968, 39, 2915.
8. Angus, J. C.: U. S. Patent 3,630,677, 1971.
9. Knox, B. E. and Vedam, K.: *Coating Science and Technology*, Final Report, Contract no. AFCRL-TR-75-0152, 1975.
10. Sokolowski, M., Sokolowska, A., Gokieli, B., Michalski, A., Rusek, A., and Romanowski, Z.: *J. Cryst. Growth*, 1979, 47, 421.
11. Vora, H., and Moravec, T. J.: *J. Appl. Phys.*, 1981, 52 (10), 6151.
12. Namba, Y. and Mori, T.: *J. Vac. Sci. Technol.*, 1985, A3 (2), 319.
13. Mori, T. and Namba, Y.: *J. Vac. Sci. Technol.*, 1983, A1 (1), 23.
14. Kitabatake, M. and Wasa, K.: *J. Appl. Phys.*, 1985, 58 (4), 1693.
15. Kobayashi, K., Mutsukura, N. and Machi, Y.: *Thin Solid Films*, 1988, 158, 233.
16. Kitahama, K.: *Appl. Phys. Lett.*, 1988, 53 (19), 1812.
17. Rice, S. B. and Treacy, M. M. J., "Specimen Preparation for Transmission Elcetron Microscopy of Materials," *Mat. Res. Soc. Symp. Proc.*, 1988, 115, 15.
18. Solin, S. A. and Ramdas, A. K.: *Phys. Rev. B.*, 1970, 1(4), 1687.
19. Angus, J. C., Koidl, P. and Domitz, S.: Ch. 4 in *Plasma Deposited Thin Films*, (Mort, J. and Jansen, F., eds.) CRC Press Inc., Boca Raton, Fl., 1986.
20. Robertson, J.: *Advances in Physics*, 1986, 35(4), 317.
21. Tsai, H. and Bogy, L. B.: *J. Vac. Sci. Technol.*, 1987, A5(6), 3287.
22. Matsumoto, S., Hino, M., Moriyoshi, Y., Nagashima, T. and Tsutsumi, M.: U. S. Patent no. 4,767,608, 1988.
23. Sumitomo Electric Co., Jap. Patent no. JP 58:126972, 1983.
24. Sumitomo Electric Co., Jap. Patent no. JP 60:127299, 1985.
25. Sumitomo Electric Co., Jap. Patent no. JP 59:232991, 1985.
26. NEC. Corp., Jap. Patent no. JP 60:127300, 1985.
27. Kurihara, D., Sasaki, K., Kawarada, M. and Koshino, N.: *Appl. Phys. Lett.*, 1988, 52 (6), 439.
28. Mitsuda, Y., Yoshida, T. and Akashi, K.: *Rev. Sci. Instrum.*, 1989, 60 (2), 249.
29. Von Engel, A., *Electric Plasmas: Their Nature and Uses*, Taylor and Francis Ltd., New York and London, 1983, 162.
30. Sawabe, A. and Inuzuka, T.: *Appl. Phys. Lett.*, 1985, 46, 146.

31. Sawabe, A. and Inuzuka, T.: *Thin Solid Films*, 1986, 137, 89.
32. Feldman, A., Farabaugh, E. N. and Sun, Y. N.: *SPIE Proceedings on Laser Induced Damage in Optical Materials*, SPIE, Bellingham, WA, 1987.
33. Fugimori, N., Ikegaya, A., Imai, T., Fukushima, K. and Ohta, N.: Paper no. 114, 175th Meeting of the Electrochemical Society, Los Angeles, CA, May 7-12, 1989.
34. Hansen, M. and Anderko, K.: *Constitution of Binary Alloys*, McGraw-Hill, New York, 1958, 378.
35. Hirose, Y. and Kondo, N.: Japan Applied Physics Society Meeting, March 29, 1988.
36. Hanssen, L. M., Carrington, W. A., Butler, J. E. and Snail, K. A., *Mater. Lett.*, 1988, 7, 289.
37. Carrington, W. A., Hanssen, L. M., Snail, K. A., Oakes, D. B. and Butler, J. E.: *Metal. Trans. A* (in press).
38. Kosky, P.G.: Private Communication, Feb. 16, 1989 (Publication Pending).
39. Yarbrough, W. A., Stewart, M. A. and Cooper, J. A. Jr.: Paper No. B1-7, International Conference on Metallurgical Coatings, San Diego, CA, April 17-21, 1989. Also *Surface and Coatings Technology*, in press.
40. Steffensen, R. J., Agnew, J. T. and Olsen, R. A.: "Tables for Adiabatic Gas Temperature and Equilibrium Composition of Six Hydrocarbons," Engineering Extension Series No. 122, Purdue University, Lafayette, Indiana, 1966.
41. Eriksson, G.: *Chemica Scripta*, 1975, 8, 100.
42. White, W. B., Johnson, W. M. and Dantzig, G. B.: *J. Chem. Phys.*, 1958, 28 (5), 751.
43. Wang, M. S. and Spear, K. E.: *Proceedings of the Ninth International Conference on Chemical Vapor Deposition*, The Electrochemical Society, 1984, 98.
44. Kingon, A. I. and Davis, R. F.,: *Proceedings of the Conference on Emergent Process Methods for High-Technology Ceramics*, Plenum Press, New York, 1984, 317.
45. Ito, K., Ito, T. and Hosoya, I.: *Chemistry Letters*, The Chemical Society of Japan, 1988, 589.
46. Toshima, H., Kotaki, T., Yaguchi, Y. Amada, Y. and Matsumoto, O.: Paper No. P2-10, First International Conference on the New Diamond Science and Technology, Tokyo, Oct. 24-26, 1988.
47. Pecsok, R. L., Shields, L. D., Cairns, T. and McMilliam, I. G.: *Modern Methods of Chemical Analysis*, 2nd ed., 1976, John Wiley and Sons, New York, 243.
48. Messier, R. and Yehoda, J. E.: *J. Appl. Phys.*, 1985, 58, 3739.
49. Messier, R.: *J. Vac. Sci. Technol.*, 1986, A4, 490.
50. Sato, Y., Hata, C. and Kamo, M.: Paper No. 1-15, First International Conference on the New Diamond Science and Technology, Tokyo, Oct. 24-26, 1988.
51. Spitsyn, B. V.: *Growth of Crystals, Vol. 13*, E. I. Givargizov (ed.), Consultants Bureau, New York, pp. 58-66.
52. Pate, B. B.: *Surface Science*, 1986, 165, 83.
53. Pepper, S. V.: *J. Vac. Sci. Technol.*, 1982, 20 (3), 643.
54. Pepper, S. V.: *J. Vac. Sci. Technol.*, 1982, 20 (2), 213.
55. Takai, T., Halicioglu, T. and Tiller, W. A.: *Surface Science*, 1985, 164, 341.
56. Belton, D. N., Harris, S. J., Schweig, S. J., Weiver, A. M. and Perry, T. A.: *Appl. Phys. Lett.*, 1989, 54, 416.
57. Badzian, A. R. and Badzian, T.: *Thin Solid Films*, (in press).
58. Angus, J. C. and Hayman, C. C.: *Science*, 1988, 241, 913.
59. Yarbrough, W. A., Kumar, A. and Roy, R.: Paper No. N3.8, Fall Meeting, Materials Research Society, Boston, MA., 30 Nov-5 Dec, 1987.
60. Bachmann, P. K., Drawl, W., Knight, D., Weimer, R. and Messier, R. F.: *Diamond and Diamond-Like Materials Synthesis: Extended Abstracts*, Materials Research Society, Pittsburgh, PA, April, 1988, 99.
61. Inspektor, A.: *Surf. and Coat. Techn.*, 1987, 33, 31.
62. Bell, A. T.: *Techniques and Applications of Plasma Chemistry*, (Mollahan, J. R. and Bell, A. T., eds.), Wiley-Interscience, New York, 1979.
63. Thornton, J. T.: *Thin Solid Films*, 1983, 107, 3.
64. Venugopalan, M. and Avni, R.: Ch. 4 in *Thin Films from Free Atoms and Particles*, (Klabunde, J., ed.) Academic Press, New York, 1986.
65. Chapman, B.: *Glow Discharge Processes*, Wiley, New York, 1980.

66. Reinberg, A. R.: *Ann. Rev. Mater. Sci.*, 1979, 9, 341.
67. Smith, D. and Adams, N. G.: *Pure Appl. Chem.*, 1984, 56, 175.
68. Weakliem, M. A.: *Semiconductors and Semimetals*, Vol. 21A, Academic Press, New York, 1984.
69. Kondratev, V. N.: Ch. 8 in *Chemical Kinetics of Gas Reactions*, Pergamon, New York, 1964.
70. Bell, A. T.: *Topics in Current Chemistry*, Vol. III, (Veprek, S. and Venngopalan, M., eds.) Springer, Berlin, 1980, 43.
71. Havens, M. R., Biolsi, M. E. and Mayhan, K. G.: *J. Vac. Sci. Technol. A*, 1976, 13, 575.
72. McDaniel, E. W.: *Collision Phenomena in Ionized Gases*, Wiley, New York, 1964.
73. Kaufman, F.: *Adv. Chem. Ser.*, 1969, 80, 29.
74. Khait, Y. L.: *Plasma Processing and Synthesis of Materials*, (Szekeley, J. and Apelian, D., eds.) North-Holland, Amsterdam, 1984, 263.
75. Sato, Y., Kamo, M. and Setaka, N.: Paper No. S7-01, *Proceedings of the Eigth Int. Symp. on Plasma Chemistry*, Vol. 1, Tokyo, 1987, 2446.
76. M. Capitelli and M. Molinari,: *Topics in Current Chemistry*, (Veprek, S. and Venngopalan, M., eds.) Springer, Berlin, 1986, 58.
77. Bell, A. T.:*Ind. Eng. Chem. Fund.*, 1972, 11, 209.
78. Capezzuto, P., Cramarossa, F., d'Agostino, R. and Molinari, E.: *J. Phys. Chem.*, 1975, 79, 1487.
79. Capezzuto, P., Cramarossa, F., d'Agostino, R. and Molinari, E.: *Beitr. Plasma Phys.*, 1977, 17, 205.
80. Inspektor, A., Liou, Y., McKenna, T. and Messier, R.: *Surface and Coatings Technology*, (in press).
81. Mucha, J. A., Flamm, D. L. and Ibbotson,J.: *J. Appl. Phys.*, in press.
82. Coffin, F. E.: *J. Chem. Physics*, 1958, 30, 593.
83. Shaw, T. M.: *J. Chem. Phys.*, 1959, 31, 1142.
84. Zaitsev, V. V.: *Russ. J. Phys. Chem.*, 1977, 51, 316.
85. Capezzuto, P., Cramarossa, F., d'Agostino, R. and Molinari, E.: *Rev. Phys. Appl.*, 1977, 12, 1205.
86. Capezzuto, P., Cramarossa, F., d'Agostino, R. and Molinari, E.: *Combustion and Flame*, 1978, 33, 251.
87. Angus, J. C.: Personal Communication, 14 September 1988.
88. Davies, G. and Evans, T.: *Proc. R. Soc. Ser. A*, 1972, 328, 413.
89. Evans, T. and James, R.: *Proc. R. Soc. Ser. A*, 1964, 277, 260.
90. Setaka, N,: *J. Mater. Res.*, 1989, 4, 664.
91. Hsu, W. L., *J. Vac. Sci. Tchnol. A*, 1988, 6 (3), Part II, 213.
92. Auciello, O., *Nucl. Instr. Meth. Phys. Res.*, B13, 1986, 561-566.
93. Benson, S. W., and Buss, J. H., *J. Chem. Phys.*, 1958, 29 (3),546.
94. Benson, S. W., Cruikshank, F. R., Golden, D. M., Haugen, G. R., O'Neal, H. E., Rogers, A. S., Shaw, R. and Walsh, R., *Chem. Rev.*, 1969, 69, 279.
95. Hine, J., *Structural Effects on Equilibria in Organic Chemistry*, Wiley-Interscience, New York, 1975.
96. Harkins, W. D., *J. Chem. Phys.*, 1942, 10, 268.
97. Kitiagorodsky, A. I., *Molecular Crystals and Molecules*, Academic Press, New York, 1973, 398-408.
98. Kawato, T. and Kondo, K.: *Jap. J. Appl. Physics*, 1987, 26 (9), 1429.
99. Saito, Y., Sato, K., Tanada, H., Fujita, K. and Matuda, S.: *J. Mater. Sci.*, 1988, 23, 842.
100. Yarbrough, W. A. and Roy, Rustum, in *Diamond and Diamond-Like Materials Synthesis: Extended Abstracts*, Materials Research Society, Pittsburgh, PA, April, 1988, 33-38.
101. Pinneo, J. M., Yokota, S. and Ravi, K. V., Paper No. T-4, *Final Program and Abstracts*, Third Annual SDI/OST Diamond Symposium, Crystal City, Va., July, 1988.
102. Liou, Y., Inspektor, A., Weimer, R. and Messier, R., *Appl. Phys. Lett.*, (in press).
103. Pickett, W. E., in *Diamond and Diamond-Like Materials Synthesis: Extended Abstracts*, Materials Research Society, Pittsburgh, PA, April, 1988, 69-72.
104. Gildenblat, G. Sh., Grot, S. A., Wronski, C. R., Badzian, A. R., Badzian, T. and Messier, R., *Appl. Phys. Lett.* (in press).
105. Glover, G. H., *Solid State Electronics*, 16, 973 (1973).

106. Badzian, A. R., *Advances in X-Ray Analysis, Vol. 31*, C. S. Barrett, J. V. Gilfrich, R. Jenkins, J. C. Russ, J. W. Richardson, and P. K. Predecki (eds.), Plenum Press, New York, 1987, 113-128.
107. Matsumoto, S. and Matsui, Y. *J. Mat. Sci.*, 1983, 18, 1785.
108. Williams, B. E., Glass, J. T., Davis, R. F., Kobashi, K. and Kawate, Y., in *Diamond and Diamond-Like Materials Synthesis: Extended Abstracts*, Materials Research Society, Pittsburgh, PA, April, 1988,59-62.
109. Liou, Y., Inspektor, A., Weimer, R. and Messier, R. *Applied Physics Letters,* 1989, 55 (7), 631.
110. Holland, L. and Ojha, S. M, *Thin Solid Films*, 1978, 48, L21-L23.
111. Pellicori, S. F., Peterson, C. M. and Henson, T. P., *J. Vac. Sci. and Tech.*, 1986, A4(5), 2350-2355.

Materials Science Forum Vols. 52 & 53 (1989) pp. 175-196

DEPOSITION OF AMORPHOUS HYDROGENATED CARBON FILMS IN LOW AND HIGH FREQUENCY DISCHARGES

Y. Catherine

Lab. des Plasmas et des Couches Minces - U.A. CNRS 838
IPCM, 2 Rue de la Houssinière, F-44072 Nantes Cedex 03, France
and
IRESTE, La Chantrerie, CP 3003, F-44087 Nantes Cedex 03, France

ABSTRACT

The deposition of amorphous hydrogenated carbon films from methane, benzene and their mixture with hydrogen and helium has been studied as a function of glow discharge frequency. Discharge processes were investigated using grid probes and optical emission spectroscopy. The deposited films were characterized using FTIR, Raman and XPS spectroscopy. Also investigated were the film optical constants, density and internal volume stress.

Low and high frequency discharges result in quite different excitation processes and ion fluxes. The low frequency plasma enhances processes with high energy thresholds. In these discharges ions can follow the variations of the RF field and reach high kinetic energy values.

As regards the high frequency regime it appears to be very efficient for dissociation and ionization processes with threshold energies in the ten eV range. Similar structures and hydrogen concentrations were obtained for film deposited in both type of discharge. It appears that the properties of films deposited under sufficient ion bombardment energy are independent of the discharge type. However plasmon loss and density measurements revealed that for the highest ion fluxes and energies used in this work the heating of the layer might become important and induce graphitization and formation of voids by hydrogen effusion. No direct relation between internal stress and hydrogen concentration was found but this compressive stress is sensitive to the energy released by the ions impinging on the substrate.

1. INTRODUCTION

Plasma deposition of carbon films (a-C:H) is a particularly interesting field owing to the fact that these carbons are typically denser, harder, more transparent, and resistant to chemical attack than any other carbonaceous films [1,2,3]. These semi-transparent hard carbon coatings can be obtained by several techniques such as ion beams [4,5,6], cathodic sputtering [7,8,9] and plasma chemical vapor deposition [2,3,10,11,12,13,14]. Many different discharge types have been explored for their applicability in the field of plasma surface modification and of plasma deposition in particular. By far the most widespread deposition process for carbon films comprises the use of a low pressure R.F. or D.C. glow discharge in hydrocarbons or their mixture with inert gases. The subject of this paper is the description of the quality of carbon coatings determined by parameters like refractive index, density, optical gap and internal stress, with regards to the plasma parameters, and how it can be influenced by controlling the intensity and nature of the particle flux impinging on the substrate.

For this purpose amorphous carbon films have been deposited using low (25-100 kHz) and high (13.56 MHz) frequency discharges in CH_4, C_6H_6 and their mixture with He, Ar or H_2. The key variables of the low and high frequency plasmas were deduced from electrical and optical measurements and particle flux from current-voltage characteristics of grid probes. The influence of growth conditions on film characteristics were studied by electron spectroscopy for chemical analysis (XPS), UV-visible – near IR spectrophotometry and X-ray spectroscopy.

2. ELECTRICAL CHARACTERISTICS OF HIGH AND LOW FREQUENCY DISCHARGES IN HYDROCARBONS.

2.1. THE REACTOR SYSTEM AND GAS HANDLING

A diode type plasma reactor operating at frequencies of 25-125 kHz or 13.56 MHz was utilized for carbon film depositions (figure 1). The diameter of each circular stainless steel electrode was 12.7 cm and the electrode spacing 6.5 cm ; the upper electrode being grounded.

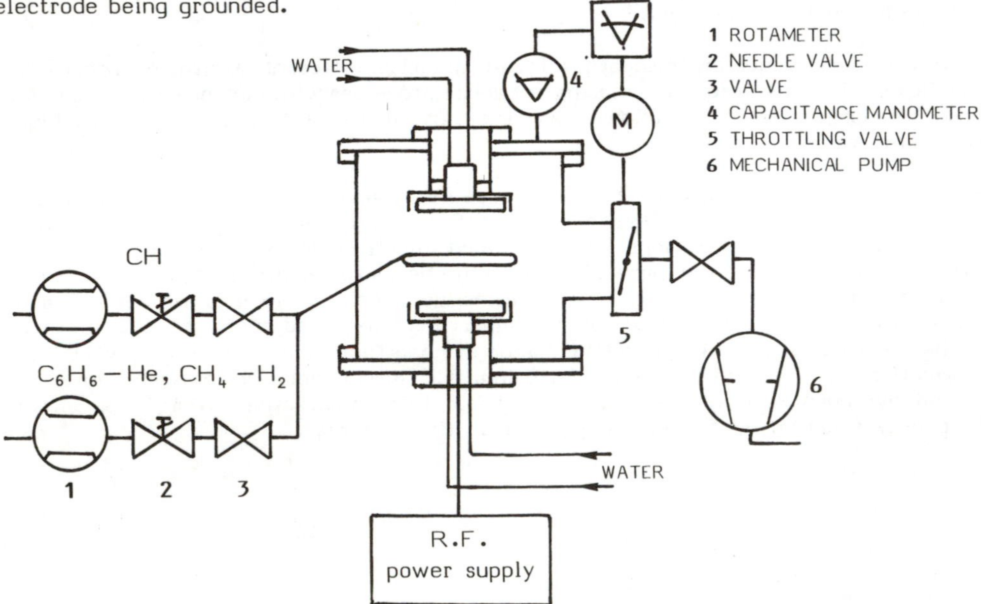

Figure 1 : Outline of the plasma reactor.

The electrical power was supplied to the lower electrode via an L-type capacitive matching network for the high frequency discharge (13.56 MHz) and via an inductive transformer for the lower frequencies. Both the forward and the reflected power were measured.

The gas feed rates were controled by needle valves and rotameters. The entire system was pumped down with a rotary pump (ultimate vacuum 0.13 Pa ($\sim 10^{-3}$ torr). The pressure was measured with a capacitance manometer which drived a throttling valve. This system enables keeping a stable pressure in the vacuum chamber even in long deposition runs.

2.2. EXPERIMENTAL DETAILS

The general deposition conditions are summarized in table 1.

TABLE 1 : Experimental deposition conditions

Substrate	Si, glass, silica
Discharge frequency	25 - 125 kHz ; 13.56 MHz
Discharge power	0 - 100 V (0. - 0.8 W/cm²)
Total pressure	7 - 70 Pa (50 - 500 mtorr)
Deposition temperature	60°C
Gases	CH_4, CH_4(5%) - He, C_6H_6 C_6H_6 (7.6%) – He, CH_4 – H_2

2.3. THE 13.56 MHz DISCHARGE AND THE SELF-BIAS VOLTAGE

The behaviour of ions and electrons in an RF discharge depends on the plasma frequency (equation 1) :

$$f_{p_{i,e}} = \frac{1}{2\pi}\left(\frac{n_{i,e}\,e}{m_{i,e}\,\varepsilon_o}\right)^{1/2} \qquad (1)$$

where $n_{i,e}$ and $m_{i,e}$ are respectively the ion and the electron density and mass. Assuming that in hydrocarbon plasmas $n_i \simeq n_e \simeq 10^{10}$ cm^{-3} the calculated values of f_{pi} are found to be 5.4 MHz for CH_4 ions and 2.37 MHz for C_6H_6 ions, while in both cases f_{pe} is about 895 MHz. Thus in the 13.56MHz discharge, the R.F. field frequency is sufficiently high that the ions require many R.F. cycles to cross the sheath while electrons are able to follow this R.F. field. Because the powered electrode is capacitively coupled, the steady-state D.C. current must be zero. If the powered electrode area is smaller than the grounded part of the system it develops a negative self bias voltage V_B relative to ground [15]. Such a situation is shown in figure 2. The net result is ion bombardment of the substrate which is placed on the powered cathode during film growth. For a non collisional sheath the mean ion energy is thus given by equation 2 :

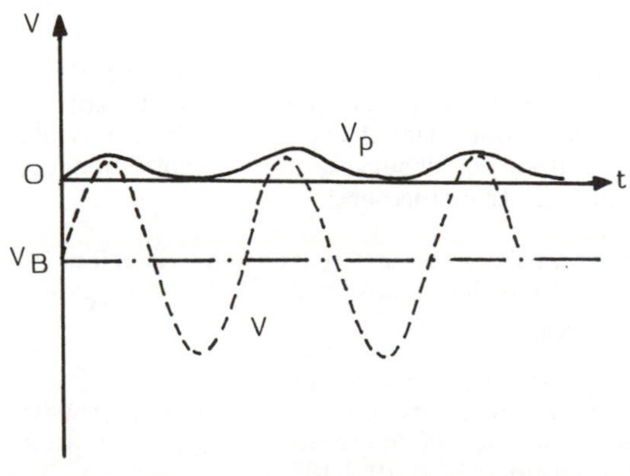

Figure 2 : R.F. modulation of plasma and self bias voltage V_B.

$$E_i = e(\overline{V}_p - V_B) \simeq e|V_B| \quad (2)$$

Values of the self bias voltage V_B for CH_4, C_6H_6 and mixtures with He are reported in figure 3. It is found that V_B is a linear function of the square root of the ratio of the ratio of the R.F. power over the discharge pressure :

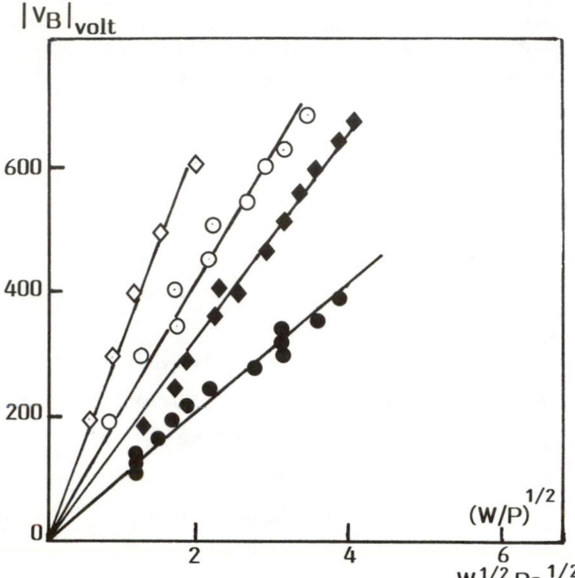

Figure 3 : Plot of the self bias potential against the parameter $(W/P)^{1/2}$.

● CH_4 , ◆ C_6H_6 , O CH_4 – He, ◇ C_6H_6 – He

(After CATHERINE and COUDERC [11]. Reprinted by permission of the publisher Elsevier Sequoia S.A.).

$$V_B \sim (W/p)^{1/2} \quad (3)$$

Bubenzer et al. [3] demonstrated that in their benzene discharge the ion energy depends on both bias potential and pressure with the proportionality :

$$E_i \sim V_B P^{-1/2} \quad (4)$$

Combining equations (3) and (4) leads to the relation (5)

$$E_i \sim \frac{W}{P}^{1/2} \quad (5)$$

This result is similar to the relation between ion energy and discharge parameters (power, pressure) deduced by Zarowin [16] who modelized the discharge by an equivalent electric network with capacitive sheaths connected in series with a resistive plasma body. Thus the key variables which control the energy of the ions impinging on the substrates are the self bias voltage (or the power) and the discharge pressure. Obviously if the geometry of the discharge changes , and more particularly the ratio of the powered electrode over the grounded area, there will be also a modification of the bombarding energy of ions.

2.4. THE LOW FREQUENCY DISCHARGE

In the low frequency discharge the transit time for ions in the sheath is much shorter than the period of the R.F. field and the ion flux will reflect the temporal dependence $V_P(t)$ of the plasma potential [17, 18]. This behaviour is illustrated in figure 4. No D.C. offset voltage is observed in this case because the coupling of the power is obtained through a transformer.

In these discharge the mean current density is an increasing function of the R.F. power (figure 5) while the R.F. voltage increases only slightly as examplified by the $\overline{V}(\overline{J})$ characteristics (figure 6).

Similar variations have been observed in an argon discharge at 380 kHz [18]. In this type of discharge the secondary electrons accelerated in the cathodic sheath are thought to take an important part in sustaining the discharge itself [19]. A relation between voltage and current density has been obtained by Gill [18] :

$$V = \frac{V_i}{\gamma}\left(1 + k\frac{\lambda^2}{eD^2}L\bar{J}\right) \qquad (6)$$

Where k is the volume recombination rate, L the electrode gap, λ the mean free path, D the diffusion coefficient, V_i the ionisation potential of the gas and γ the secondary emission coefficient of the material of the cathode under ion bombardment. This relation is linear and agrees with the results of figure 6. Furthermore the intercept with the $\bar{V}$ axis allows an estimate of γ. From this extrapolation we found $\gamma = 0.053$ for CH_4 and $\gamma = 0.025$ for C_6H_6 –He in the case of a carbon coated electrode. Measurements made in Ar always give linear $\bar{V}(\bar{J})$ characteristics and values of γ for stainless steel electrodes of :

$\gamma = 0.075$ (this work),

$\gamma = 0.04$ (Gill[18]).

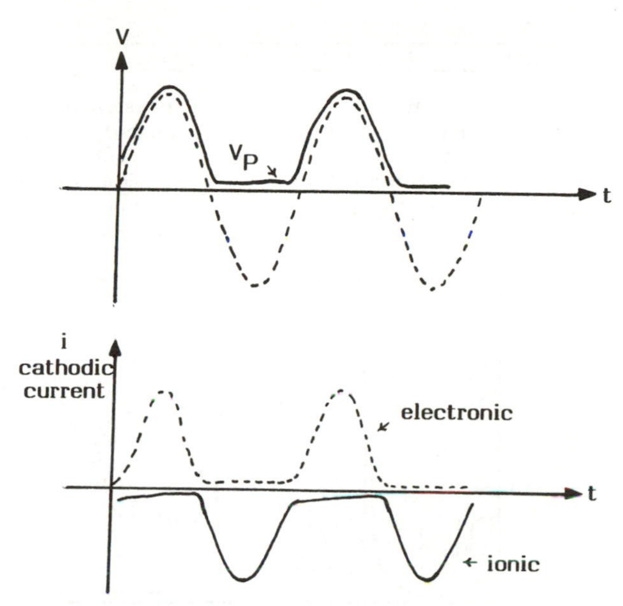

Figure 4 : Temporal dependence of the R.F. voltage, plasma potential V_p and R.F. current at the cathode for a 50 kHz discharge.

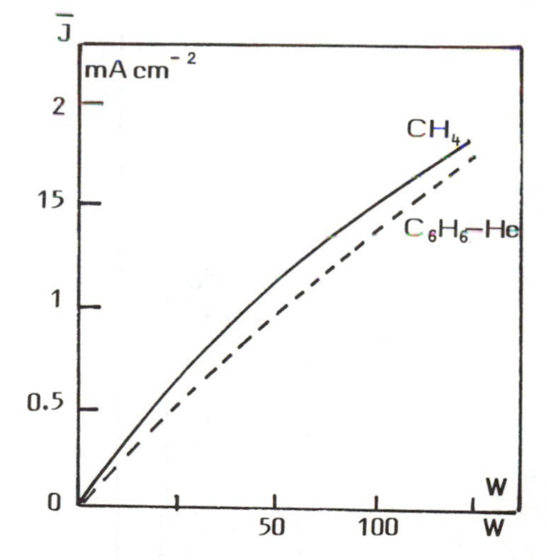

Figure 5 : R.F. power dependence of the R.F. current density.

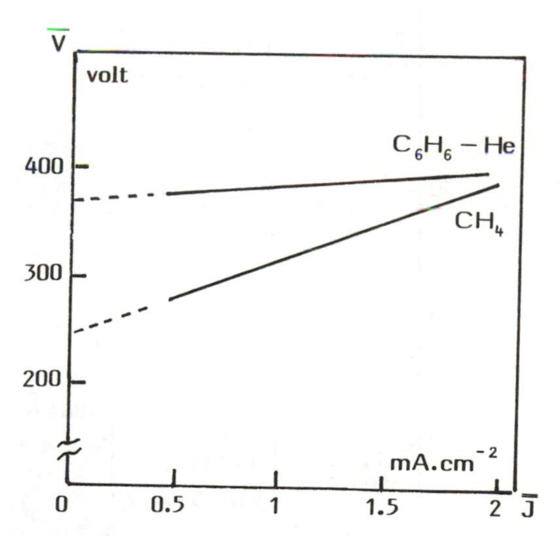

Figure 6 : Current-voltage characteristic of a 50 kHz discharge in hydrocarbon gases.

(After CATHERINE and COUDERC[11]).(Reprinted by permission of the publisher Elsevier Sequoia S.A.).

Accordingly it appears that the current density and the pressure are the most significative parameters of the low frequency discharge.

3. OPTICAL PROPERTIES OF THE LOW AND HIGH FREQUENCY DISCHARGES IN HYDROCARBONS.

Empirically it is known that the physicochemical properties of the formed layers depend strongly on the plasma conditions, especially on the nature and intensity of the applied electric high frequency field and the state variables in the plasma. In order to get detailed information, a plasma diagnostic using optical emission spectroscopy was performed on low and high frequency discharges in CH_4 . Figure 7 shows an emission spectrum of a CH_4 plasma for a – C : H deposition.

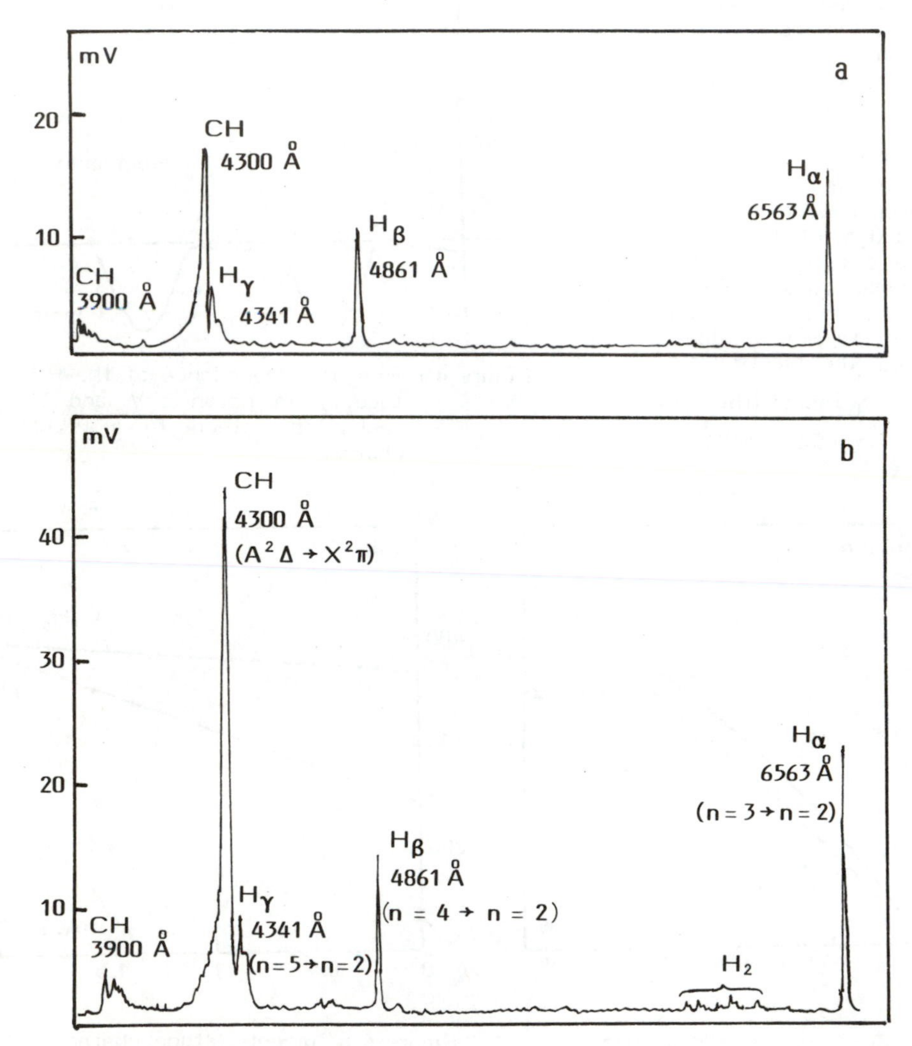

Figure 7 : Typical emission spectra of CH_4 plasmas :
a/ 25 kHz discharge ; 50 W ; 80 mtorr (10.4 Pa)
b/ 13.56 MHz discharge ; 20 W ; 70 mtorr (9.3 Pa).

Similar spectra are observed for both low and high frequency discharges. They agree quite well with previously published spectra [20, 21, 22]. Electronic transitions observed in the range 380 - 680 nm in the emission spectra of low and high frequency discharges in CH_4 are tabulated in table 2. The excitation energies are those given by Donahue et al. [23] and Aarts et al. [24].

TABLE 2 : Emissive species in CH_4 glow discharge.

Emission species	Wavelength	Excitation process	Excitation Energy
$CH(B^2\Sigma \longrightarrow X^2\pi)$	390 nm system		
$CH(A^2\Delta \longrightarrow X^2\pi)$	430 nm system	$CH_4 + e \rightarrow CH^* + H + H_2$	13.4 eV
$H(3 \rightarrow 2)$	656.3 nm	$H_2 + e \rightarrow H + H^* + e$ $H + e \rightarrow H^* + e$ $CH_4 + e \rightarrow CH_3 + H^* + e$	16.6 eV 12.1 eV 21.9 eV
$H(4 \rightarrow 2)$	486.1 nm	$H_2 + e \rightarrow H + H^* + e$ $H + e \rightarrow H^* + e$ $CH_4 + e \rightarrow CH_3 + H^* + e$	17.25 eV 12.75 eV 22.0 eV
$H(5 \rightarrow 2)$		$H_2 + e \rightarrow H + H^* + e$ $H + e \rightarrow H^* + e$ $CH_4 + e \rightarrow CH_3 + H^* + e$	17.9 eV 13.06 eV 22.6 eV

The H_2 emission bands were also observed over the examined wavelengths and are more intense in the 13.56 MHz discharge than in the 25 kHz discharge.

Emission intensities of CH^*(431.4 nm), H_β (486.1 nm) and H_2 (463.4 nm) were measured in both low and high frequency modes. In these experiments 5 % Ar was added to the CH_4 gas phase and it was found that such a few percent of Argon did not modify the whole emission spectrum. The emission intensity of the argon line studied (750.4 nm, 13.5 eV of excitation energy) may be related to the argon concentration by the relation [25] :

$$I_{Ar} = C_{Ar} k_{Ar} n_e [Ar] \quad (7)$$

where C_{Ar} is a constant which depends only on the spectroscopic arrangement, k_{Ar} is the rate constant for excitation of an argon atom to the excited state, n_e is the electron concentration in the discharge, and [Ar] is the argon concentration. Since I_{Ar} is proportional to the electron density in the discharge a comparison of the evolutions of I_{CH^*}, $I_{H\beta}$ and I_{H_2} as a function of I_{Ar} will give informations about the excitation processes in the low frequency and high frequency discharges.

Figure 8 shows a log-log plot of I_{CH^*} (431.4 nm), $I_{H\beta}$ (486.1 nm) and I_{H_2} (463.4 nm) as a function of I_{Ar} for increasing discharge powers in both cases.

For the 25 kHz discharge the intensities of the H_β and H_2 (463.4 nm) lines scale linearly with the argon line intensity. This evolution indicates that the excited hydrogen atoms and molecules are created from CH_4 though a single electron impact. The following reactions may thus be postulated [26] :

$$e + CH_4 \longrightarrow CH_4^* \longrightarrow CH_3 + H^* + e \quad (8)$$

with a rate coefficient $k \sim 8\,10^{-11}$ cm^3/s and :

$$e + CH_4 \longrightarrow CH_4^* \longrightarrow CH_2 + H_2^* + e \quad (9)$$

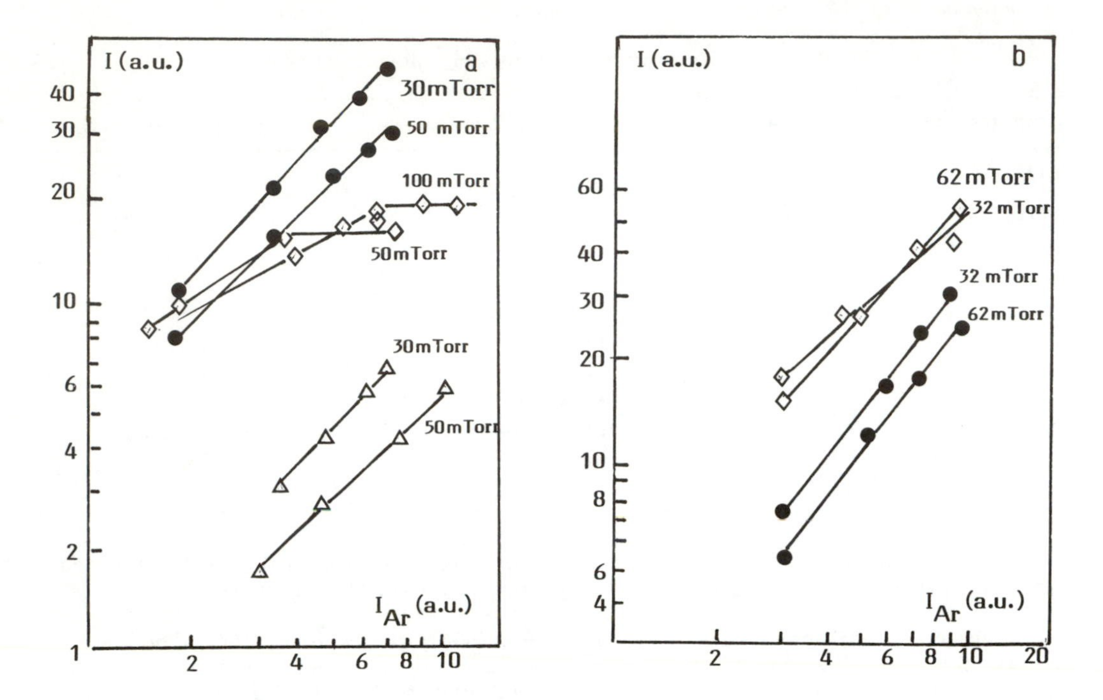

Figure 8 : Log-log plot of emission intensities I_{CH^*} (◊), $I_{H\beta}$ (●), I_{H_2} (Δ) as a function of I_{Ar} for increasing powers : **a** : 25 kHz ; **b** : 13.56 MHz.

The log-log plot of I_{CH^*} versus I_{Ar} and the low frequency discharge power does not give a straight line. This may be the result of several reaction pathways with different threshold for the excitative dissociation of CH_4 [25].

On the other hand in the 13.56 MHz discharge the log-log plot of I_{CH^*} and $I_{H\beta}$ versus I_{Ar} for various discharge powers give straight lines but while the intensity of the CH line I_{CH^*} is proportional to I_{Ar}, the intensity of the Hβ line is roughly proportional to $I_{Ar}^{1.4}$. We may thus suppose that excited CH^* is still the result of the reaction :

$$CH_4 + e \longrightarrow CH_4^* \longrightarrow CH^* + H + H_2 + e \qquad (10)$$

but that H_β is created by reaction (8) and also by the two following reactions :

$$H_2 + e \longrightarrow H + H^* + e \qquad (11)$$

$$H + e \longrightarrow H^* + e \qquad (12)$$

which need two electron impacts upon the CH_4 molecule, i.e : the creation of H atoms or H_2 molecules from CH_4 followed by the excitation reaction.

As a conclusion it is possible to say that the low frequency discharge favors processes with higher energy threshold than the high frequency discharge.

4. ION FLUX AT SUBSTRATE LOCATION IN HYDROCARBON DISCHARGES

Comparison of ion beam and R.F. plasma deposition indicates that films produced by both method are predominantly amorphous and that their electrical and structural properties depend on the accelerating voltage or the self-bias voltage [27, 28, 29]. Therefore the knowledge of the ion flux and energy at the substrate surface is of major importance for the control of the growth of the films and their structural properties. The charged particle flux arriving at the powered electrode was measured by grid probes designed to give meaningful results in thin film deposition applications [30]. Typical values of ion fluxes at the active electrode of low and high frequency discharges are shown in figure 9. It is clear from figure 9 that the 13.56 MHz discharge results in a higher ionization of the gas phase than the low frequency discharge. Typical ion fluxes are in the range 10^{13} - $3 10^{14}$ cm^{-2} s^{-1}. However the ion energy is greater at low frequencies with direct coupling, than for high frequencies with capacitive coupling [17, 31].

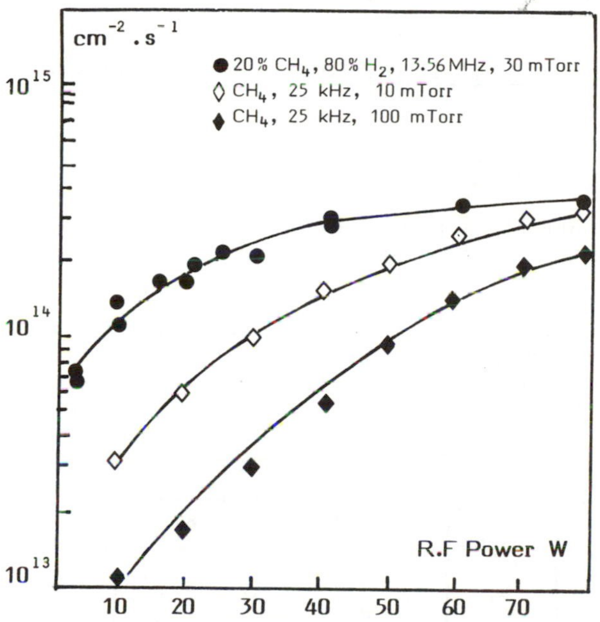

Figure 9 : Ion flux at the powered electrode of various discharges (from reference [30]). Reprinted by permission of the publisher "Les Editions de Physique".

It is interesting to know the contribution of ions to the growth of the layers. This can be done by measuring the mass deposition rates which are represented by equation 13 :

$$R = \frac{\Delta m}{S \Delta t} \qquad (13)$$

where Δm is the deposited film mass, S the covered surface and Δt the duration of the deposition run. The total flux of deposited carbon may thus be estimated by relation 14 :

$$\Phi_o = \frac{6.02\ 10^{23}\ R}{M} \qquad (14)$$

Taking M = 12 g for carbon atoms an estimate of the ratio Φ_i / Φ_o of the ion flux over the total flux of particles sticking on the substrate can be obtained. Results are reported in table 3.

TABLE 3 : Ratios of ion flux to mass flux Φ_i / Φ_o as a function of the power injected in a 13.56 MHz plasma.

Gas	24 % CH_4 – 76 % H_2					
Power (watt)	10	20	30	40	60	80
Power density W/cm²	0.09	0.18	0.26	0.35	0.53	0.70
$-V_B$ (Volts)	200	320	380	450	540	600
Φ_i / Φ_o	0.14	0.15	0.15	0.17	0.18	0.17

The ratio Φ_i / Φ_0 has a value between ~ 0.14 - 0.18 as the power density increases. It shows that ions are not the main species involved in mass transport from the gas phase to the substrate. However the energies carried to the surface of the substrate by ions are typically in the range 10-100 eV (and even more) for high frequency plasmas and in the range 100-600 eV for low frequency plasmas. These energies are much larger than particular energies which characterize a solid i,e :

- surface atom binding energy ~ 3 - 10 eV
- sputtering threshold energy ~ 10 - 40 eV.

Thus the bombardment of surfaces by ions in the range of several hundred electron-volts, with fluxes that produce noticeable sputtering rate will induce modifications in coating composition and structure.

5. STRUCTURE AND PROPERTIES OF PLASMA DEPOSITED a-C:H FILMS

5.1. BONDING AND STRUCTURE

5.1.1. FTIR spectroscopy

As stated by Angus and his coworkers [32] "the key question in understanding the unusual properties of carbon films is the bonding and structure of the amorphous network".

FTIR spectroscopy provides information on the presence of CH, CH_2 and CH_3 groups. Furthermore the bonding type of these hydrogenated carbon atoms can be inferred [33]. Absorption spectra in the wavenumber range 4000-2800 cm^{-1} are shown in figure 10 for both low and high frequency plasmas.

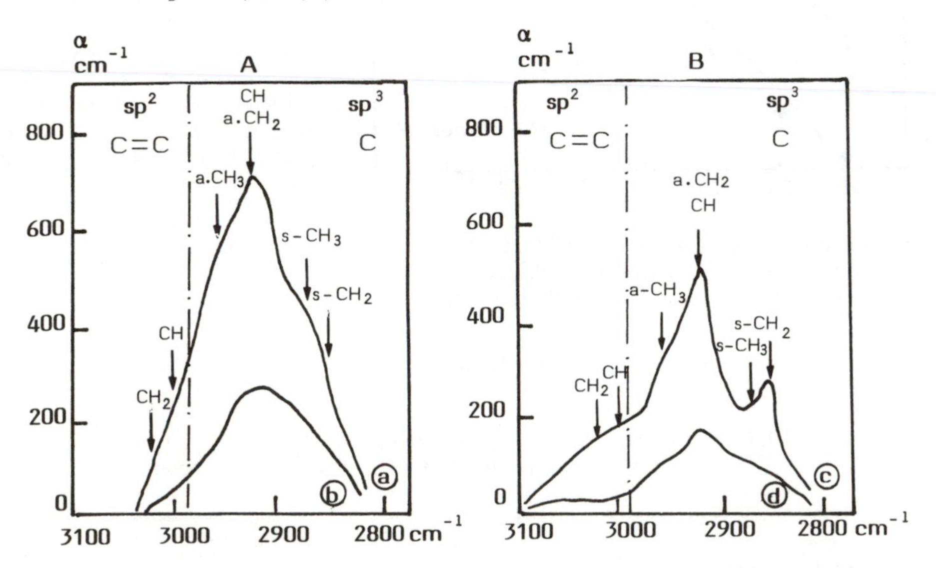

Figure 10 : Assignment of stretching vibrations of C — H bonds for films grown (A) in CH_4 gas and (B) in C_6H_6 — He gas at a frequency of 13.56 MHz and at various powers and pressures: curve a, 40 W, 53.3 Pa (0.4 Torr); curbe b, 50 W, 5.3 Pa (0.04 Torr) ; curve c, 50 W, 26.6 Pa (0.2 Torr) ; curve d, 100 W, 8 Pa (0.06 Torr) (from reference 36).

Reprinted with permission of the publisher, Elsevier Sequioa S.A..

The contributions of the various individual C − H vibrations to the total absorption band can be deduced from attributions previously made by several authors [33,34]. The total absorption band of each spectrum is the enveloppe of individual contributions of $\geq C-H$, $C{<}^{H}_{H}$, $-C{\leq}^{H}_{H}\!H$ and $C=C{<}^{H}_{H}$ groups. Making use of the analysis given by Nadler et al. [34] and Couderc et al. [36] the following points may be set forth :

- Spectra **a** and **c** correspond to films formed under low energy ion bombardment (high pressure and low power), and under these conditions the absorption due to C − H bonds of sp^2 carbon is enhanced as compared with the corresponding absorptions of spectra **b** and **d** of films obtained under higher ion energy bombardment.

- The I.R. absorption of the film obtained from C_6H_6 − He at 13.56 MHz under low ion energy flux (curve **c**) exhibits rather well defined peaks (2925 and 2850 cm^{-1}). This is characteristic of a polymeric structure [35].

The deposited films contained typically between 5 − 33 % of hydrogen which compares with the values given by several authors as can be seen in table 4.

TABLE 4 : Hydrogen content of a-C:H films

Reference	Deposition Technique	Method of analysis	[H]
Ingram et al.[37]	RF plasma CH_4	RBS	45 %
Jones et al. [38]	RF plasma C_2H_6, CH_4, C_2H_4 C_2H_2	Thermal Treatment	30-45%
Warner et al. [39]	L.F. plasma (30 kHz) CH_4	Nuclear reaction IR	35-42%
Dischler et al. [34]	RF plasma C_6H_6	IR	13-38 %
Ojha et al. [2]	RF plasma CH_4, C_4H_{10}	RBS	23-30%
This work	RF and LF plasmas $C_6H_6-H_2$, CH_4	FTIR	5-33%

The preferential sputtering of hydrogen and hydrocarbon structures is thought to be one of the processes involved in the formation of a-C:H films [32], since the C − H bond energy (3.5 eV) is significantly less than the C − C bond energies (7.41 eV for diamond) and low mass atoms are more efficiently sputtered than high mass atoms. In addition carbon has a very low sputtering yield compared to other materials [32]. Such a preferential sputtering process is illustrated by the decrease of the C − H absorption band integrated intensities A_{CH} as a function of an increasing ion bombardment energy : typically A_{CH} is roughly divided by a factor of 5 as the reduced parameter $V_B P^{-1/2}$ (self bias voltage/(pressure)$^{1/2}$) changes from 10 up to 200 V $Pa^{-1/2}$ [36]. Similar results can be found in other published works [37].

However IR spectroscopy cannot give information about unbound hydrogen ; non negligible amounts of such hydrogen have been found in a-C:H films deposited with similar techniques [38, 39]. Annealing hydrocarbon films under vacuum leads to the evolution of hydrogen and causes structural changes [33,36,40]. Figures 11**a** and **b** represent the variations of C – H absorption bands (2800-3100 cm^{-1}) for films deposited in a 50 kHz discharge at two deposition pressures and which suffered 1 hour annealing treatments at two different temperatures. It can be seen that for the high pressure film (0.5 torr, 66.6 Pa), hydrogen is nearly completely released from the layer at an annealing temperature of 600°C, whereas for the low pressure film (0.1 torr, 13.3 Pa) some hydrogen is still present in the structure at an annealing temperature of 800°C.

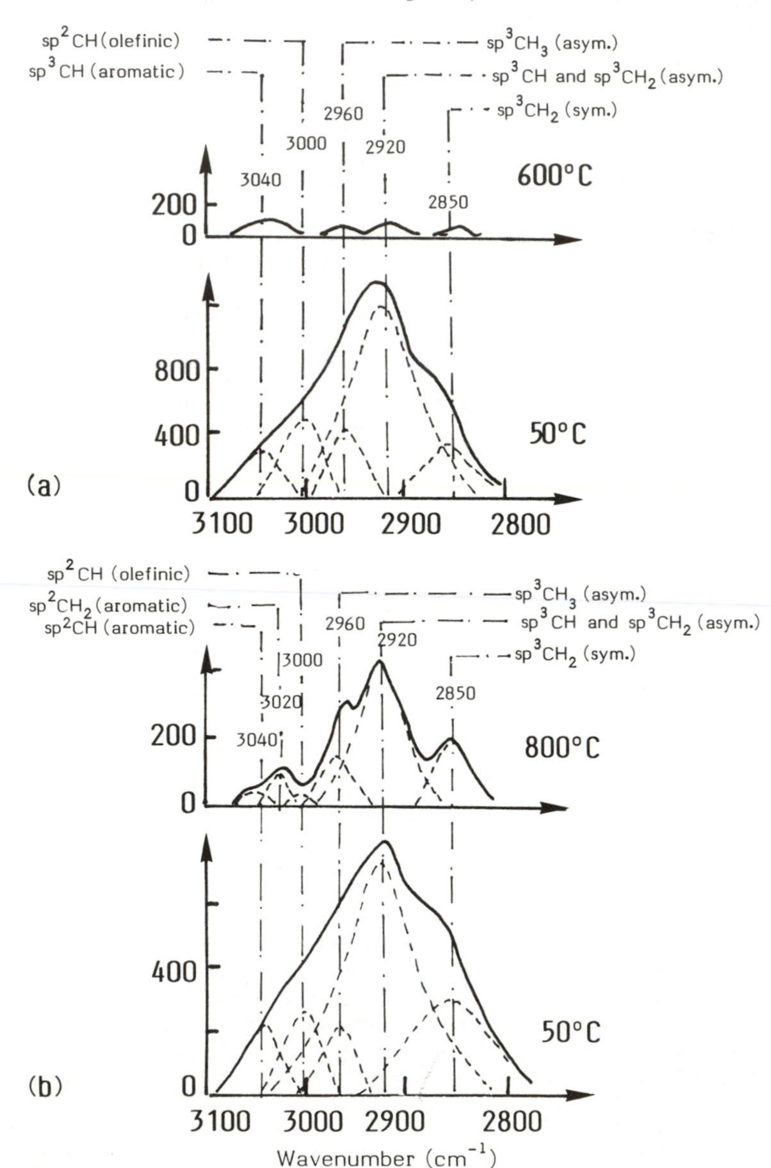

Figure 11 : FTIR spectra of annealed a-C:H films grown in CH_4 (frequency : 50 kHz, power : 50 W) at two pressures (**a**) 66.6 Pa (0.5 torr) ; (**b**) 13.3 Pa (0.1 torr) (from reference 36)

Four point probe resistivity measurements were made on the annealed films and are reported in table 5.

TABLE 5 : Resistivity of annealed a-C:H films (from reference 37).

Pressure Pa (torr)	Temperature (°C)	Resistivity Ω.cm
13.3 (0.1)	600	3.13
13.3 (0.1)	800	0.15
66.6 (0.5)	400	0.67
66.6 (0.5)	800	$1.06\ 10^{-3}$

These results show that temperature induced graphitization is less pronounced for the film deposited under high ion energy bombardment (Low pressure) than under low ion energy flux (high pressure). A resistivity value of $1.06\ 10^{-3}$ Ωcm for the high pressure film annealed at 800°C indicates a large graphitization of the structure (graphite resistivity $\sim 10^{-3}$ Ωcm) with dominant sp^2 bonding, while a resistivity of 0.15 Ωcm for the low pressure film gives evidence of non negligible amounts of sp^3 bonding and dispersed hydrogen in the structure [37].

5.1.2. Raman spectroscopy

Raman spectroscopy has been applied for structural investigation. The Raman spectra of films deposited in a CH_4 plasma (50 kHz) are reproduced in figure 12. Similar spectra are obtained at 13.56MHz. A general agreement is obtained with recent results [41,42,43]. However quantitative date on sp^3 vs. sp^2 bonding is difficult to obtain since the Raman diffusion cross section for diamond ($9.1\ 10^{-7} cm^{-1} sr^{-1}$) is low compared with that of graphite ($5\ 10^{-5} cm^{-1} sr^{-1}$). Furthermore fluctuations of bond lengths and angles of sp^3 and sp^2 moieties have similar effects on the Raman spectrum [43]. These spectra are composed of a D line at $\sim 1300\ cm^{-1}$ and a G line at $\sim 1550\ cm^{-1}$ [41]. The results for our films are given in Table 6. The G line position varies between 1350 – 1550 cm^{-1} while the D line position changes from 1270 up to 1305 cm^{-1} depending upon deposition conditions.

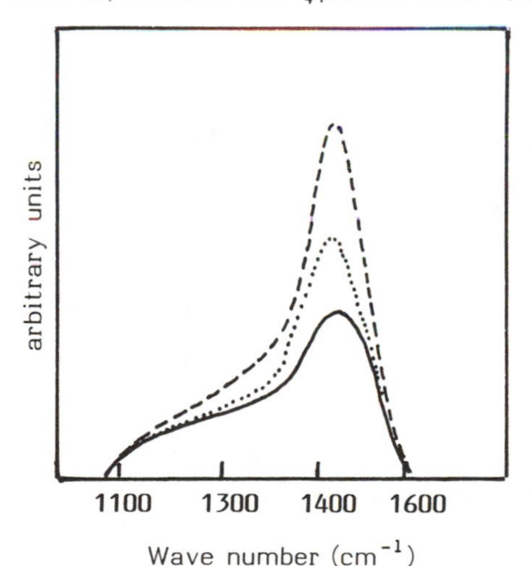

Figure 12 : Raman spectra of a-C:H films (---- 25 W, 26.6 Pa ; ..... 25 W, 13.3 Pa ; —— 100 W, 13.3 Pa)

(From ref. 36) with permission of the Publisher : Elsevier - Sequoia.

According to the model of Beeman et al. [42] we suppose that the films contain bond-angle disorder. The line positions thus yield information concerning bond angle disorder and also coordination. The low average values for the D lines may require the presence of fourfold-coordinated bonds as well as disorder.

TABLE 6 : Raman characteristics of a-C:H films deposited in CH_4

Deposition conditions		D line Position (cm^{-1})	D line FWHM (cm^{-1})	G line Position (cm^{-1})	G line FWHM (cm^{-1})	$\frac{I_D}{I_G}$
50 kHz	100 W ; 0.1 torr	1304	150	1547	92.5	0.60
	25 W ; 0.1 torr	1291	162	1535	87.5	0.45
	25 W ; 0.2 torr	1275	120	1535	85	0.31
13.56 MHz	50 W ; 0.2 torr	1272	1372	1532	98.6	0.34
	25 W ; 0.08 torr	1288	160.5	1535	72.2	0.56

5.1.3. XPS spectroscopy

In the last years photoelectron spectroscopy has been developped to a powerful tool to study the electronic structure of condensed matter. A number of photoelectron spectroscopy studies has been performed on various form of carbon [44, 45]. Recently electron spectroscopy measurements on hydrogenated amorphous carbon films have been published [46, 47, 48, 49]. Figure 13 shows the XPS spectrum, measured with MgKα excitation, of the carbon-1s line and its associated characteristic energy losses for a plasma deposited (50 kHz) a-C:H film.

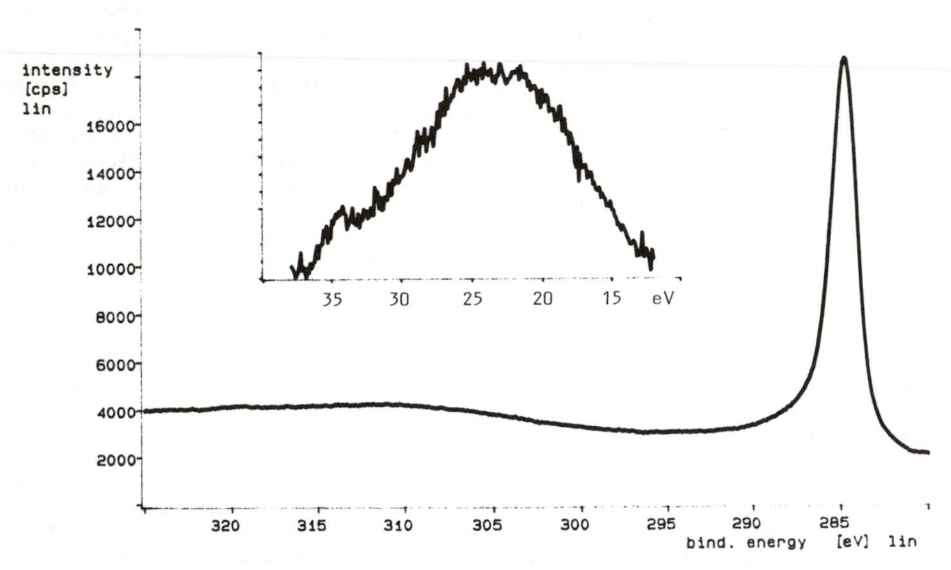

Figure 13 : XPS spectrum of a carbon film deposited in CH_4 with a 100 W, 50 kHz discharge and a pressure of 10 mtorr (1,33 Pa). The plasmon energy loss is given relative to the C_{1s} peak energy.

The binding energy of the carbon 1s core level of various a-C:H films always falls in the 284.6-285 eV range. Charging effects are probably responsible of the small shift observed. The analysis of the percent diamond likeness by the method described by Warner and coworkers [47] based on measurements of the asymmetry in the C_{1s} XPS line shape due to delocalized π bonds must be used with care since the C_{1s} line peaks at ~ 290.3 eV when carbon is linked to oxygen (C – O bond).

Plasmon energy loss spectroscopy has been used to characterize carbon films [46, 50, 51]. In the present work none of the deposited a-C:H films, either in the low or high frequency discharge show the π plasmon loss at 6.3 eV which is observed in highly oriented pyrolytic graphite [44, 50, 51].

Positions relative to the C_{1s} carbon peak of the main maximum $\hbar\omega_p$ of the loss spectrum for layer deposited in various plasma conditions are given in table 7.

TABLE 7 : Energy of the main maximum of plasmon loss spectra of amorphous hydrogenated carbon films.

FILM N°	GAS	PRESSURE mtorr (Pa)	FREQUENCY	POWER DENSITY W/cm²	DEPOSITION ELECTRODE	MAXIMUM $\hbar\omega_p$ (eV)	CARBON ATOMIC DENSITY cm^{-3}
C 1	5% CH_4-H_2	50 (6.5)	13.56 MHz	0.27	RF	23.5	$1\,10^{23}$
C 2	5% CH_4-H_2	50 (6.5)	13.56 MHz	0.35	Ground	20.4	$7.5\,10^{22}$
C 3	17% CH_4-H_2	80 (10.6)	13.56 MHz	0.27	RF	23.7	$1.02\,10^{23}$
C 4	CH_4	10 (1.33)	50 kHz	0.44	Ground	23	$9.6\,10^{22}$
C 5	CH_4	10 (1.33)	50 kHz	0.44	RF	23.3	$9.9\,10^{22}$
C 6	6% C_6H_6–He	600 (79.8)	50 kHz	0.44	RF	21.7	$8.5\,10^{22}$
C 7	C_6H_6	40 (5.3)	13.56 MHz	0.9	RF	18.8	$6.4\,10^{22}$
C 8	CH_4	200 (26.6)	13.56 MHz	0.18	RF	21.9	$8.7\,10^{22}$
C 9	CH_4	50 (6.5)	13.56 MHz	0.27	RF	19.6	$7\,10^{22}$

For most of the films the major loss peak has a binding energy 21-24 eV higher than the C – 1s core level line, but several volts below the graphite plasmon loss peak at 26 eV. We never observed the diamond peak at 31 eV. The small peak which can be seen in the plasmon loss spectrum of figure 13 comes from the Ar – 2s line due to a light cleaning of the surface for oxygen removal before acquiring the spectrum.

The plasmon energy may be expressed by :

$$\hbar\omega_p = \hbar\,(4\pi e^2\, n/m^*)^{1/2} \qquad (15)$$

with n characterizing the density of electrons participating in the oscillation [52]. It is known that in materials having simple band structures and if $\hbar\omega_p$ is large compared with interband energies, the free electron mass m can be used as the electron effective mass m* in relation 15 [50]. An estimate of the atomic density of the surface of the carbon layer may thus be obtained from the valence electron density n and the observed plasmon energy since the valence electrons per carbon atom are four irrespective of the nature of bonding. The results obtained for the films of table 7 are listed in the last column of this table. For films deposited on the activated electrode at mean power densities and low pressures, the atomic density is $n_a \sim 10^{23}$ atom / cm^3, while deposition in other conditions (ground electrode at 13.56 MHz, high pressure or very high powers) results in lower atomic densities: $n_a \sim 6.8\ 10^{22}\ cm^{-3}$. Deposition on the ground electrode of an RF discharge or at high pressures is known to give highly hydrogenated deposits with low densities. The effect of increasing hydrogen content, which by itself would enhance the electron density is over compensated by the decrease of the film density.

For a very high RF power density (C8 film) the plasmon peak is located at even lower energy values. This is probably the result of an increased deposition temperature due to the high flux and energy of ions bombarding the growing film. It can be compared with film (C9) which suffered a post deposition annealing treatment. Similar trends for as deposited and annealed films were observed by several authors [52,53]. Figure 14 confirms these assessments by clearly showing that a-C:H films having the greatest density $\rho \sim 1.95\ g/cm^3$ are those deposited for self-bias voltages in the range 350-450 V, that is for medium range ion flux and energy. Another peculiarity which can be deduced from table 7 is that in the 50 kHz discharge both the ground and active electrode give films with similar atomic densities showing thereby the symetrical role of both electrodes at these low frequencies.

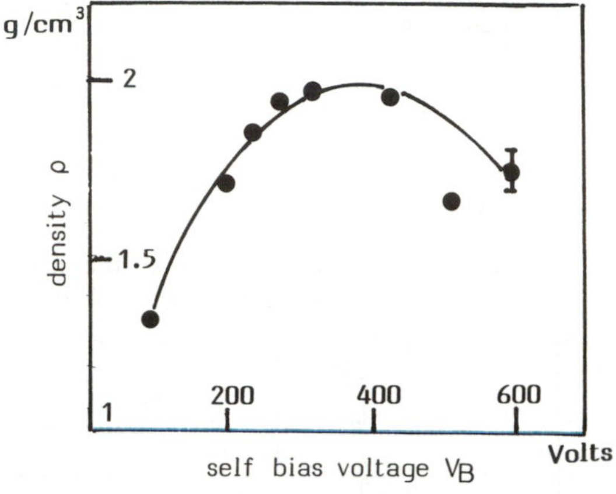

Figure 14: Evolution of a-C:H film density as a function of the self bias voltage V_B of a 13.56 MHz discharge in 20% CH_4 – 80% He.

5.2. OPTICAL PROPERTIES

5.2.1. The absorption edge

Among the various interesting properties of amorphous hydrogenated carbon films are their relatively good transparency in the IR and the possibility of varying the refractive index and short wavelength absorption.

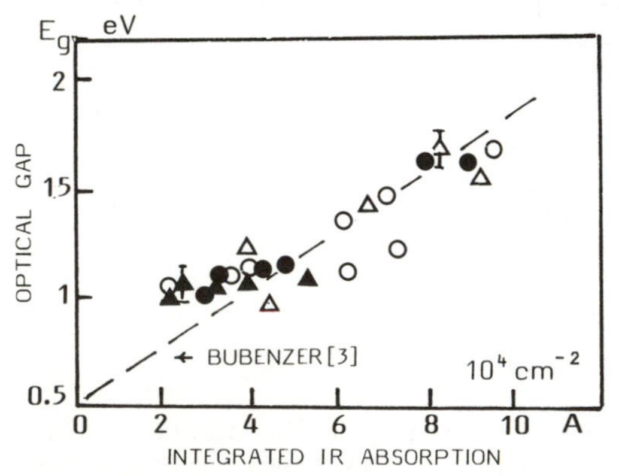

Figure 15: Relation between the optical E_g and the hydrogen content of the films.
50 kHz (full symbols), 13.56 MHz (open symbols) : ○ ● CH_4, △ ▲ C_6H_6 – He.
From reference 36 with permission of the publisher : Elsevier Sequoia.

The electronic absorption edge has been measured by several authors [3, 47, 54]. The absorption of the a-C:H films deposited in this study, except for the exponential absorption tail, is well described by the relation :

$$(\alpha h\nu)^{1/2} = B(E\ \ - h\nu) \qquad (16)$$

where $h\nu$ is the photon energy, α the absorption coefficient and B a constant. The optical gap is taken from the extrapolation of this curve towards the abscissa. The edge depends strongly on the deposition parameters. A decrease from 1.8 eV down to 0.8 eV is observed in various gases (CH_4, CH_4–He, C_6H_6) when the ion energy at substrate is increased [36,55].

The strong variations of E_g probably reflect differences in sp^2 and sp^3 bonding ratio but also hydrogen incorporation in the film. Large hydrogen concentrations appear to favor larger optical gaps. We have observed (figure 15) that the optical gap increases nearly linearly with the hydrogen content of the film irrespective to the nature of the discharge (50 kHz or 13.56 MHz) and of the gas phase (C_6H_6 – He, CH_4). Dischler et al [55] also reported a linear increase of E_g with the bounded hydrogen concentration for films deposited in C_6H_6. Another parameter that appears to influence the optical gap is the deposition temperature [56,57]. E_g has also been found to decrease with post deposition annealing (figure 16). For highly hydrogenated films (high pressure deposition) E_g decrease steeply of about 55 % of its initial value as the annealing temperature is raised from room temperature up to 400°C. On the other hand for the lightly hydrogenated film (low pressure deposition) the relative variation of the optical gap in the same temperature range is only 35 %. This is in agreement with a larger dispersion of hydrogen in the matrix for the low pressure film [36].

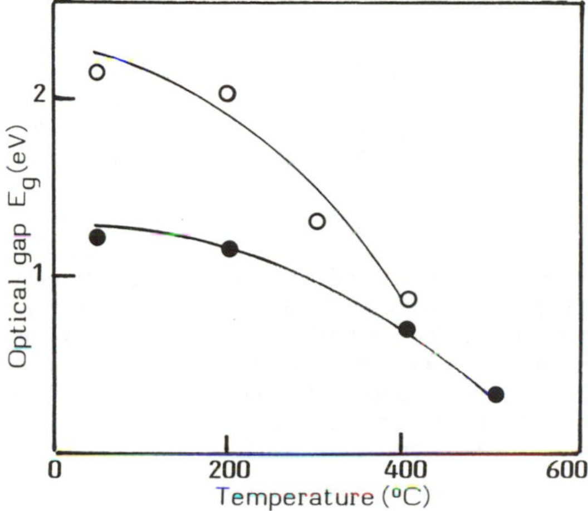

Figure 16 : Evolution of E_g with post annealing treatment : (○) 0.5 torr, (●) 0.1 torr (13.56MHz).

5.2.2. Optical constants

The optical properties of an isotropic absorbing medium are conveniently expressed in terms of the complex dielectric function

$$\varepsilon = \varepsilon_1 + i\,\varepsilon_2 \qquad (17)$$

The dielectric constants ε_1 and ε_2 are real functions of photon frequency ν [58]. These dielectric constants are related to the refractive constants n and k according to:

$$\varepsilon_1 = n^2 - k^2 \quad ; \quad \varepsilon_2 = 2nk \qquad (18)$$

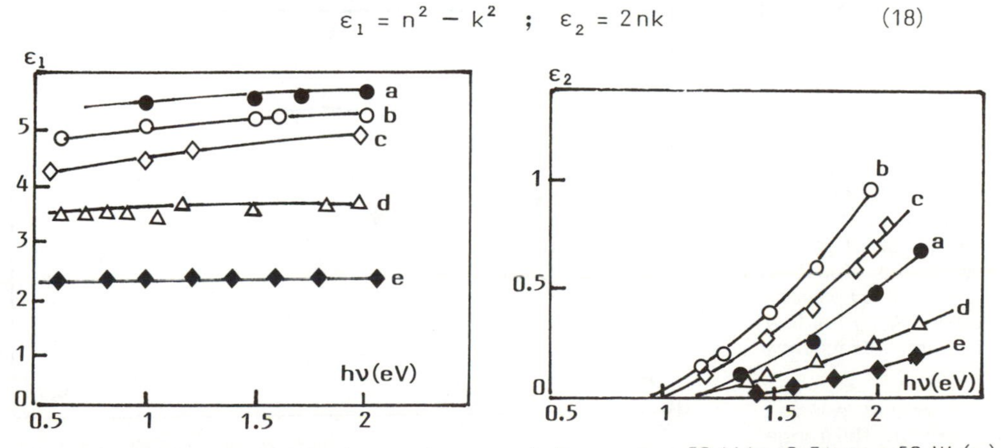

Figure 17 : Dielectric constant ε_1 and ε_2 vs photon energy; 50 kHz, 0-3 torr : 50 W (**a**), 75 W (**b**) ; 13.56 MHz : 0.05 torr, 50 W (**c**), 0.2 torr, 25 W (**d**); inductive 5 MHz 0.2 torr, 100 W (**e**).

The films examined in this work were deposited on glass substrates and the dielectric constants ε_1 and ε_2 were determined in the photon energy range 0.5-2.5 eV (wavelength region 0.4 - 2.5 µm). Figure 17 shows plots of the dielectric constants versus photon energy. While the investigations were made on a small photon energy range the results indicate that the spectral behaviour of the films shows little ressemblance to that of graphite or glassy carbon [59]. All films have similar spectral variations. However films grown using the low frequency discharge or in the low pressure range of the 13.56 MHz discharge have higher values of ε_1 and ε_2.

This is consistent with the fact that films which suffer increasing ion-bombardment will loose part of their hydrogen. It is known that deshydrogenation of carbon films results in a larger refractive index [58]. This view is also supported by values obtained with the inductive coupling. This coupling associates the lowest refractive index with the highest hydrogen content [36].

5.3. MICROSTRUCTURE

The majority of our hydrocarbon films were found to be amorphous. Sometimes crystalline inclusions in the amorphous matrix were observed. In most cases, the crystalline phases were sparselly distributed in the matrix yielding diffraction patterns that consisted of weak rings or spots superimposed on a strong central background. We identified these inclusions as graphite or chaoïtes crystallites. In agreement with Vora and Moravec [28] no conclusive evidence of significant formation of cubic diamond in films deposited in either type of discharge was obtained in this work.

5.4. INTRINSIC STRESS IN a-C:H FILMS

Useful applications of amorphous hydrogenated carbon films are sometimes complicated by delamination and spontaneous buckling, partly due to the large compressive stress in the film [36, 59, 60].

It is important to understand the origin of the compressive stresses in carbon films and find ways to reduce them. The internal stress of films deposited onto long thin glass strips, 75 µm in thickness, were measured by the curvature method [61]. The internal stress of the deposited film is sufficiently high to cause the curvature of the strip. The sagitta δ at the center of the strip was measured and the stress was obtained from :

$$S = \frac{4ED^2\delta}{3L^2(1-\nu)T} \qquad (18)$$

where E is the Young's modulus for glass, D the thickness of the strip, L the length of the substrate, ν the Poisson's coefficient and T the film thickness [61]. Figure 18 shows the variations of the internal stress of carbon films as a function of the ratio: power/pressure (W/P) for low and high frequency plasmas.

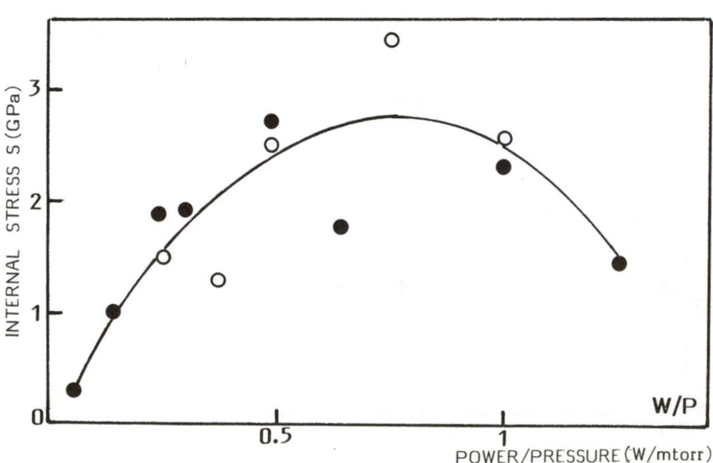

Figure 18 : Internal stress of a-C:H films deposited from CH_4 : ○ 13.56 MHz ; ● 50 kHz.

Internal stress values are in the range 0.2-3 GPa and of the same order of magnitude than other values given in the literature [60,62]. In agreement with the conclusions of Nir [62] we do not find any clear dependence of the stress with the hydrogen content of the films. The compressive stress can be mainly attributed to the energy released by the ions bombarding the growing layer. As the bombarding energy increases (i.e. W/P increases) the internal compressive stress increases (up to ~ 3 GPa for W/P ~ 0.8 W/mtorr). However a further increase of the ion energy and flux results in a lowering of this stress probably because an excess ion energy heats up the layer and changes its structure to a more graphitic one. Furthermore, contrary to other plasma deposited films such as silicon nitride [63] it was not possible to find a region within parameter space where films with tensile stress could be deposited.

6. CONCLUSION

The influence of the frequency of the discharge, process gas, ion flux on plasma deposited a-C:H films has been investigated. It has been found that hard cross linked films could be deposited in the high frequency as well as in the low frequency type of discharge. For high impact energies of the film forming particles the film properties are independant of the process gas and the type of discharge used (even if plasma processes are quite different as it has been shown by emission spectroscopy).

The structure of a-C:H has been evaluated by FTIR, Raman spectroscopy, ESCA. These measurements have shown that hard films are correlated with dispersed hydrogen in the carbon matrix and an increase of the amount of sp^3 relative to sp^2 carbon atoms. However too large ion fluxes lead to more graphitic structures due to some over heating of the layer during deposition. Optical properties such as absorption edge, refractive index and I.R. absorption have also been determined. They also correlate with the amount of hydrogen and the carbon structure giving films with low densities and refractive indices if discharge conditions induce a low energy ion bombardment of the substrate. The deposited films can be highly stressed (compressive) and it was not possible to find deposition conditions for which films have tensile stress.

ACKNOWLEDGEMENTS

The author is greatly indebted to Drs. P. Couderc and A. Pastol for their contribution to part of the experimental work. We also acknowledge the CNRS-SPI GRECO 57 for financial support.

REFERENCES

1) Weissmantel, C. : Thin Solid Film, 1979, 58, 101.
2) Ohja, S.M., Norström, H., Mc Culluch, D. : Thin Solid Films, 1979, 60, 213.
3) Bubenzer, A., Dischler, B., Brandt, G., Koidl, P. : J. Appl. Phys., 1983, 54, 459.
4) Weissmantel, C. : J. Vac. Sci. Technol., 1981, 18, 179.
5) Mirtich, M.J., Swec, D.H., Angus, J.C. : Thin Solid Films 1985, 13, 245.
6) Miyazawa, T., Misawa, S., Yoshida, S., Gonda, S. : J. Appl. Phys., 1984, 55, 188.
7) Craig, S., Harding, G.L. : Thin Solid Films, 1982, 97, 345.
8) Savvides, N., Window, B. : J. Vac. Sci. Technol., 1985, A 3, 2386.
9) Wyon, C., Gillet, R., Lombard, L. : Thin Solid Films 1984, 122, 203.
10) Berg, S., Andersson, L.P. : Thin Solid Films 1979, 58, 117.
11) Catherine, Y., Couderc, P. : Thin Solid Films, 1986, 144, 265.
12) Pouch, J.J., Warner, J.D., Liu, D.C., Alterowitz, S.A. : Thin Solid Films, 1988, 157, 97.
13) Mutsukura, N., Katoh, Y., Machi, Y. : J. Appl. Phys., 1986, 60, 3364.
14) Kroesen, G.M.W. : Ph. D. Thesis, University of Eindhoven, The Netherlands, 1986.
15) Vossen, J.L. : J. Electrochem. Soc., 1979, 126, 319.
16) Zarowin, C.B. : J. Vac. Sci. Technol., 1984, A2, 1537.
17) Köhler, K., Horne, D.E., Coburn, J.W. : J. Appl. Phys., 1985, 58, 3950.

18) Gill, M.D. : Vacuum, 1984, 34, 357.
19) Turban, G. : "Décharges luminescentes excitées en radiofréquence" in "Interaction Plasmas froids-Matériaux", Journées d'Etudes, "Oléron 81" édité par le GRECO 57 du CNRS et Les Editions de Physique, 1987, 79.
20) Kampas, F.J., Corderman, R.R. : J. Non-Crystall. Solids, 1983, 59-60, 683.
21) Andersson, L.P., Berg, S., Norström, M., Olaison, R., Towta, S. : Thin Solid Films, 1979, 63, 155.
22) Kobayashi, K., Mutsukura, N., Machi, Y. : J. Appl. Phys., 1986, 59, 910.
23) Donahue, D.E., Schiavone, J.A., Freund, R.S. : J. Chem. Phys., 1977, 67, 769.
24) Aarts, J.M.F., Beenakker, C.I.M., De Heer, F.S. : Physica, 1971, 53, 32.
25) Kampas, F.J. : J. Appl. Phys., 1983, 54, 2276.
26) Hashimoto, K., Sakamoto, T., Matuda, N., Baba, S., Kinbara, A. : Proc. Int. Ion Engineering Congress - ISIAT 83 and IPAT 83, Kyoto, 1983, p. 1125.
27) Weissmantel, C. : Thin Solid Films, 1982, 92, 55.
28) Vora, H., Moravec, T.J. : J. Appl. Phys., 1981, 52, 6151.
29) Catherine, Y. : "Croissance de couches minces sous flux d'ions" in "Interaction Plasmas froids-Matériaux", Journées d'Etudes, Oléron 87, Editeur GRECO 57 Les Editions de Physique, 1987, p. 319.
30) Catherine, Y., Pastol, A. : E-MRS Meeting, Strasbourg, France, Symposia Proceedings, vol. XVII, "Amorphous Hydrogenated Carbon Films" edited by P. Koidl and P. Oelhafen, Les Editions de Physique, 1987, p. 145.
31) Briaud, P., Turban, G., Grolleau, B. : Mat. Res. Symp. Proc., 1986, 68, 109.
32) Angus, J.C., Koidl, P., Domitz, S. : "Hydrogenated Amorphous Carbon Films" in Plasma Deposited Thin Films, edited by J. Mort and F. Jansen, CRC Press, 1986, p. 89.
33) Dischler, B., Bubenzer, A., Koidl, P. : Solid State Commun., 1983, 48, 105.
34) Nadler, M.P., Donovan, T.M., Green, A.K. : Thin Solid Films, 1984, 116, 241.
35) Bellamy, L.J. : The Infrared Spectra of Complex Molecules, Methuen, London, 1964.
36) Couderc, P., Catherine, Y. : Thin Solid Films, 1987, 146, 93.
37) Holland, L., Ojha, S.M. : Thin Solid Films, 1978, 48, L 21, L 23.
38) Nyaiesh, A.R., Nowak, W.B. : J. Vac. Sci. Technol., 1983, A 1, 308.
39) Tait, N.R.S., Tolfree, D.W.L., John, P., Odeh, I.M., Thomas, M.J.K., Tricker, M.J., Wilson, J.I.B., England, S.B.A., Newton, D. : Nucl. Instrum. Methods, 1980, 176, 433.
40) Craig, S., Harding, G.L. : Thin Solid Films, 1982, 97, 345.
41) Dillon, R.O., Woollam, J.A., Katkanant, V. : Phys. Rev., 1984, B 29, 3482.
42) Beeman, D., Silveman, J., Lynds, R., Anderson, M.R. : Phys. Rev., 1984, B 30, 870.
43) Wagner, J., Ramsteiner, M., Wild, C., Koidl, P. : E-MRS Meeting Symposia Proceedings, Vol. XVII, "Amorphous hydrogenated carbon films", edited by P. Koidl and P. Oelhafen, Les Editions de Physique, Strasbourg, 1987, p. 219.
44) Mc Feely, F.R., Kowalczyk, S.P., Ley, L., Cavell, R.G., Pollak, R.A., Shirley, D.A. : Phys. Rev., 1974, B 9, 5268.
45) Leder, L.B., Suddeth, J.A. : J. Appl. Phys., 1960, 31, 1492.
46) Moravec, T.J., Orent, T.W. : J. Vac. Sci. Technol., 1981, 18, 226.
47) Warner, J.D., Pouch, J.J., Alterovitz, S.A., Liu, D.C., Landford, W.A. : J. Vac. Sci. Technol., 1985, A 3, 900.
48) Sawabe, A., Inuzuka, T. : Thin Solid Films, 1986, 137, 89.
49) Ugolini, D., Oelhafen, P., Wittmer, H. : "Amorphous Hydrogenated Carbon Films", edited by P. Koidl and P. Oelhafen, Les Editions de Physique, Strasbourg, 1987, p. 287.
50) Miyazawa, T., Misawa, S., Yoshida, S., Gonda, S. : J. Appl. Phys., 1984, 55, 188.
51) Oelhafen, P., Freeouf, J.L., Harper, J.M.E., Cuomo, J.J. : Thin Solid Films, 1984, 120, 231.
52) Fink, J., Müller-Heinzerling, Th., Pflüger, J., Bubenzer, A., Koidl, P., Crecelius, G. : Solid State Commun., 1983, 47, 687.
53) Weissmantel, C., Bewilogua, K., Schürer, C., Brever, K., Zscheile, H. : Thin Solid Films, 1979, 61, L 1 - L 4.
54) Savvides, N. : J. Appl. Phys., 1985, 58, 518.

55) Dischler, B., Bubenzer, A., Koidl, P. : Appl. Phys. Lett., 1983, 42, 636.
56) Smith, F.W. : J. Appl. Phys., 1984, 55, 764.
57) Catherine, Y., Couderc, P., Zamouche, A. : 4ème Colloque Intern. sur les plasmas et la pulvérisation cathodique, CIP 82, Nice, 1982, suppl. n° 212, Le Vide Les Couches Minces, p. 125.
58) Mc Kenzie, D.R., Mc Phedran, R.C., Savvides, N., Botten, L.C. : Phil. Mag., 1983, B 48, 341.
59) Enke, K. : Thin Solid Films, 1981, 80, 227.
60) Gillet, G., Rau, B. : Thin Solid Films, 1984, 120, 109.
61) Brenner, A., Senderof, S. : J. Res. Natl. Bur. Stand., 1949, 42, 105.
62) Nir, D. : Thin Solid Films, 1987, 146, 27.
63) Claassen, W.A.P. : Plasma Chemistry and Plasma Processing, 1987, 7, 109.

Materials Science Forum Vols. 52 & 53 (1989) pp. 197-216
Copyright Trans Tech Publications, Switzerland

PREPARATION AND PROPERTIES OF HIGH DENSITY, HYDROGEN FREE HARD CARBON FILMS WITH DIRECT ION BEAM OR ARC DISCHARGE DEPOSITION

J-P. Hirvonen, J. Koskinen, R. Lappalainen and A. Anttila
University of Helsinki, Department of Physics
SF-00170 Helsinki, Finland

ABSTRACT

In recent years, the beneficial effects of energetic ions on the growth process of thin films have been very well established [1,2]. In fact novel materials such as cubic boron nitride and diamondlike carbon, otherwise obtained only at a high temperature and pressure, can be produced by increasing the energy of ions deposited [3-6].

Direct ion beam deposition offers the best way to study the deposition parameters under very controlled circumstances: ion energy, angle of incidence, ion current density, and control of impurities at an isotopic level can be determined in a unique way.

Ion beam deposited diamondlike coatings (*i*-C) have been fabricated using direct deposition of mass-separated C^+- and $C_nH_m^+$-ions in an energy range of E_i=10 eV - 2 keV. The films deposited using hydrocarbon ions incorporate a high hydrogen concentration, and have properties comparable to those of a-C:H. However, using a pure carbon beam hydrogen free, high density films with improved mechanical properties are produced.

The disadvantage of the slow growth rate of the direct ion beam deposition can be avoided using an arc discharge deposition of carbon. The structure and properties of these films are close to those produced with the direct deposition.

1. DIAMONDLIKE COATINGS

1.1 GENERAL

Carbon has three main allotropes: graphite, amorphous carbon, and diamond. Graphite has a hexagonal lattice structure with sp^2-bonded two dimensional planes interconnected by weak forces. Amorphous carbon is a transitional form of carbon which has no long range order and the local structure varies depending on the growth procedure. The structure is said to be consisted of sp^1-bonded aromatic chains [7]. New allotropic forms are produced by partial crystallization during heating of the amorphous carbon to 1300 K [7]. Diamond is a metastable phase of carbon at the room temperature and pressure. Diamond has a cubic crystal lattice (fcc with 000, 1/4 1/4 1/4 basis, a=0.3467 nm (298 K), or sometimes hexagonal [8]) with a four fold sp^3 covalent bond structure. In addition to this carbon has been observed to exist in other exotic allotropes. An example of this being a 60-atom complex macro molecule [9].

The stable phase of carbon at 300 K and 10^5 Pa is graphite. The diamond phase is stable at 300 K in pressures greater than 1.6×10^9 Pa, but the conversion of graphite to diamond structure demands pressures and temperatures of much higher magnitude [10]. Hydraulic compression shells and various catalysts have been used to produce synthetic diamonds at an industrial level since the 1960s.

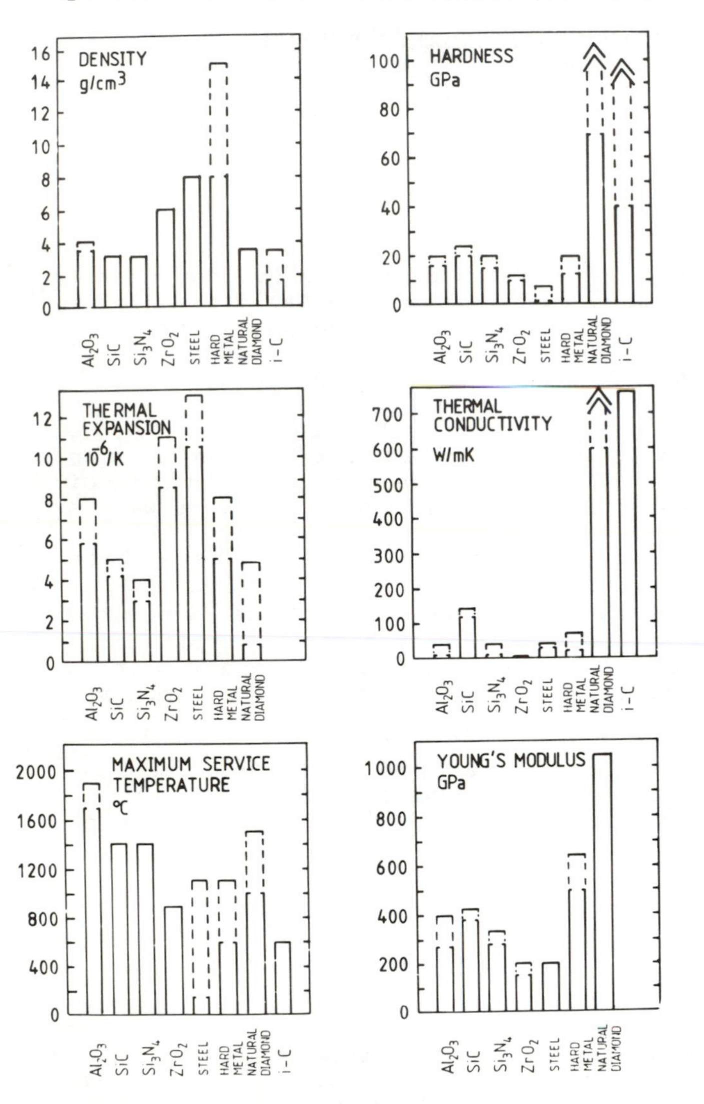

The past few years has seen a considerable amount of interest in diamond thin films grown by chemical vapour deposition (CVD) methods [11]. The main driving force for this is the search for a procedure to grow single crystalline films for semiconductor applications. At present epitaxial growth has succeeded on natural diamond substrates [12].

1.2 PROPERTIES OF NATURAL DIAMOND

Due to the firm sp^3-bonded crystal structure natural diamond has several physical properties which represent extreme values in the entire family of elemental solids a list

Fig. 1. Selected physical properties of diamond as compared to ceramics, steels, and hard metal [16], and i-C [6,23,24,29].

of some of them is as follows [8]: Its density is 3.515 g/cm³ or 1.762×10^{23} at/cm³. In an impurity free diamond lattice with a low defect concentration the thermal conductivity varies between 2000 - 4000 W/Km (T+293 K) due to the lattice conduction. Diamond is the hardest known material as determined by scratching with any other material. The thermal expansion coefficient is low $(0.8\pm0.1)\times10^{-7}$ (293 K). Diamond is chemically inert and it is not attacked by acids at room temperature. The electrical resistance is greater than 10^{20} Ωcm (except in the very rare nitrogen deficient p-type IIb diamond

10^5 - 10^7 Ωcm) and the optical band gap is 5.4 eV.

Diamond is optically transparent throughout the whole optical spectrum excluding absorption between 4 - 6 μm wave lengths in the IR region. Fig. 1 illustrates selected properties of natural diamond as compared to conventional ceramics, steel and *i*-C.

1.3 SURFACE GROWTH UNDER ION BOMBARDMENT

Ion beam deposition is a vacuum process where a layer is grown from the incident energetic ions. In the growing layer the kinetic energy of the ions may induce irreversible physical and chemical processes with high activation energy [13,15]. The growth of a layer under ion bombardment is a process where a number of different phenomena take place: penetration of the energetic ion, collision cascade of the target atoms, erosion of the surface, defect production, recombination and transport etc. At present no theoretical models exist that could yield the structure of the growing surface using the deposition parameters data. Binary collision and molecular dynamic computer simulations generally performed in metals have given insights into the atom transport and defect concentration in the collision cascade and also to the temperature and density fluctuations in the lattice [13-15]. The collision cascade can be subdivided into two consecutive periods: ballistic and thermal spike phases [13]. During the collision cascade some of the surface atoms gain enough energy to be ejected from the surface. The self-sputtering coefficient of carbon is less than 1 with all ion energies, E_i, which means that a carbon layer can be collected with any ion energy [17,18]. The surface sputtering is inversely proportional to the binding energy U_o of the surface atom. Because of the different binding energies of atoms e.g. in metal alloys an enrichment of the sputtered target in the less volatile component can be observed [19]. Accordingly, preferential sputtering of loosely bound carbon atoms may occur in favour of the four fold sp^3-bonding during the growth of a carbon layer on ion bombardment.

In the thermal spike phase the energy of the cascade atoms is equipartitioned and such parameters as temperature and pressure can be approximated. Applying the model of Seitz et al. [20] and using the thermal conductivity and specific heat of natural diamond crude approximations of the temperature in the cascade volume can be obtained. An $E_i \leq 1$ keV ion produces a thermal spike with a temperature about 10^3 K in the cascade [6,20]. The moving atoms in the cascade can produce a shock wave creating a pressure of the order of 10^9 Pa and a 50% increase of the local density [14,21]. In the case of diamond its high thermal conductivity results in the heat being rapidly dissipated from the cascade to the surrounding lattice. This prevents thermal transport and rearrangement of the atoms in the cascade area to much shorter times than in metals for example. Consequently, a highly disordered structure results unless the rearrangement is enhanced by procedures such as raising the substrate temperature.

The overlapping of the cascades is highly unlikely as the quenching rate of the spike is high and the immediate ion bombardment activated processes have been completed in less than 10^{-11} s. During this time the nearest cascades produced by a 1 mA/cm^2 ion flux density are about 10 μm apart from each other and thus the cascades are independent of each other. Therefore the formation of the material is determined by the local collision cascades characterized by the energy of the bombarding ion E_i.

1.4 GENERAL STRUCTURE OF *i*-C

The carbon layers produced under ion bombardment are generally referred as diamondlike films [3]. This is due to similarities in properties, such as mechanical, optical, electrical and chemical as compared to natural diamond. However, one has to keep in mind that many of the unique properties of natural diamond are due to its crystalline lattice structure.

One of the most striking properties of *i*-C is the highly disordered structure. The reported diffraction patterns generally consist of diffuse rings that arise from the amorphous structure or small crystals [6,22]. According to electron microscopy and diffraction studies the structure may consist of small crystals less than 10 - 100 nm in diameter in an amorphous matrix.

The nature of the bonds can be detected using infrared, x-ray photoelectron, nuclear magnetic resonance etc. spectroscopies. The ratio of sp^3/sp^2 has been 3/2 - 2/1 and sp^1 normally being absent [23,31,32]. The density of the films varies in a broad range 1.7 - 3.5 g/cm^3 [23,24,29,30].

Heat treated films normally convert to pure sp^2 amorphous structure at temperatures above 500 - 600 °C [31]. Small 100 nm diamond crystals have been observed to nucleate in electron beam annealing experiments [6].

Most of the *i*-C studies have been performed using hydrocarbon plasmas or beams and the film material is frequently referred to as hydrogenated amorphous carbon (a-C:H). Indeed the films produced contain hydrogen between 0 - 60 at.% [33]. According to the IR absorption spectra it can be concluded that most of the hydrogen is bonded to carbon terminating to the dangling bonds of the four fold bonding of the carbon atoms. The a-C:H may have a polymeric structure especially when benzene plasmas are used [6].

In conclusion the structure and properties of *i*-C vary in a broad range depending on the deposition parameters. Instead of having only well defined graphite, diamond, or amorphous structure, *i*-C seems to exist in a large number of different metastable phases.

1.5 DIFFERENT METHODS OF PRODUCING *i*-C

The range of different methods to produce hard carbon layers cover almost the entire arsenal of various thin film deposition methods. It can be subdivided into two groups, high temperature CVD and low temperature ion assisted deposition processes. The former methods are based on the chemical reactions between the gas phase and growing surface. The latter methods are performed at temperatures generally between 300 - 700 K. In ion beam methods the stabilization of sp^3-bonds is due to the energetic bombardment processes and these methods are examined in more detail below. The ion beam methods can be subdivided into methods where i) solid carbon is used as a source material, ii) hydrocarbon gas is the source of carbon atoms or iii) into a mass analyzing method where practically any carbon compound can be used to produce carbon ion beams:

i) carbon beam method [3,34], sputtering dual beam method [23,36], dc- and rf-plasma sputtering [36], carbon evaporation combined with ion beam [37], co-axial (pulsed) plasma method [38]

ii) rf- and dc- (magnetron) plasma deposition by cracking hydrocarbon [6,39,46], hydrocarbon ion beam deposition [6,41]

iii) mass-analyzing method [24-30,35].

In the methods of the group i) carbon is collected onto the surface as neutral atoms or ions. In the carbon beam deposition the carbon ions are sputtered from a graphite cathode using an inert gas sustained glow discharge. The C^+-ions are extracted with an acceleration voltage. In the sputtering dual-beam method the film is deposited with carbon atoms which are either sputtered from a graphite target or evaporated from graphite e.g. with a laser beam. To enhance the dynamic processes on the growing surface an Ar^+-ion beam is directed onto the surface. Additional H_2 gas can be fed to the deposition atmosphere or ion source to produce a-C:H. In these methods inert gas ions (Ar^+) bombard the growing surface and a few at.% of the inert gas may be incorporated in the layer.

A graphite cathode may be eroded with a plasma arc and a pure carbon plasma can thus be obtained. This can be utilized in a pulsed plasma method where the low energy C^+-ions (E_i= 50 - 100 eV) in the high density plasma hit the substrate surface [38].

In the methods of the group ii) the substrate is bombarded with various ionized hydrocarbon molecules, neutrals and inert ions (Ar^+) with ion energies typically below 1 keV. Most of the research and even commercial production of the *i*-C layers has been done with rf-plasma methods. The deposition conditions are generally characterized by the gas pressure and compound, rf-power density and bias voltage. Plasma deposition equipment is relatively simple and efficient but the most serious drawback of the method is the lack of independent control of the deposition parameters. In the ion beam method (ion plating) the hydrocarbon ions are produced in a separate ion source and the ion energy and the angle of incidence are independently controlled. In these methods hydrogen is always incorporated in the films.

The only way to control all the deposition parameters independently is to use pure carbon ion beams. On the laboratory scale this can be done using a conventional isotope separator by retarding the C^+ ions to the desired energy. Layers have been grown with density close to the natural diamond using monoenergetic C^+-ions [30]. The main drawback of this method is the rather low ion current.

2. PREPARATION AND PROPERTIES OF CARBON COATINGS DEPOSITED USING MASS-ANALALYZED ION BEAMS

2.1 GENERAL

Mass-separated $^{12}C^+$-ions were deposited on single crystal silicon and various metal substrates including hard metal (WC-Co cemented carbide). For comparison, some films with $C_2H_2^+$ -beam were made. The schematic picture of the deposition system is shown in Fig. 2. The carbon ion beam was extracted from a dual plasmatron or low voltage arc ion source with CO_2 as a feed gas. The ion beam was first accelerated up to the enrgy of 40 keV, after which it was mass-analyzed using the magnet of 90 degrees. Prior to hitting the substrate the energy of ions was decreased with the retarding arrangements shown in Fig. 3 [40]. The beam was swept between two

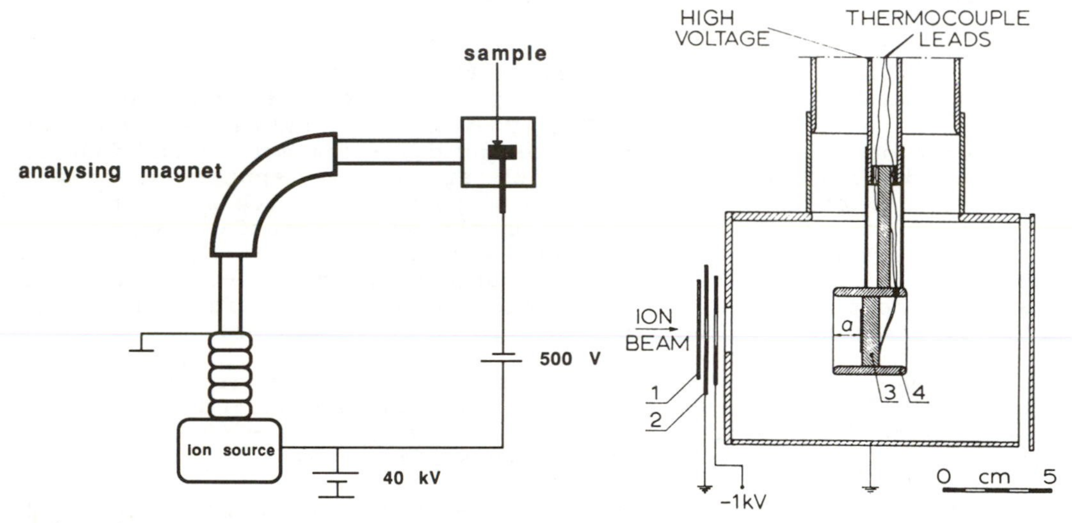

Fig. 2. A schematic picture of of the direct deposition.

Fig. 3. The ion retarding system [40].

stabilizing pins (1) before entering through the round hole aperture (2). The retardation voltage was connected to the high voltage cylinder (4). The final deposition energy varied from 10 ev to 2 keV. Most of the films used in tribological measurements were produced using an energy of 500 eV or higher that resulted in the good adhesion to the substrate. Due to the low intensity of the ion beam the typical deposition rate was only 0.07 Å/s. The film thicknesses varied from 0.1 µm to several micrometers. The hardness and tribological measurements were performed usually on the thickest samples. The depositions were carried out at room temperature. Typical deposition on the silicon substrate is shown in Fig. 4.

The purity of the films was determined with the Rutherford backscattering spectroscopy (RBS) on the selfsupporting film. In Fig. 5 the RBS spectra from evaporated and ion beam deposited carbon films are shown. As can be seen the film produced with the mass analyzed ion beam contains few impurities and only the ^{12}C isotope.

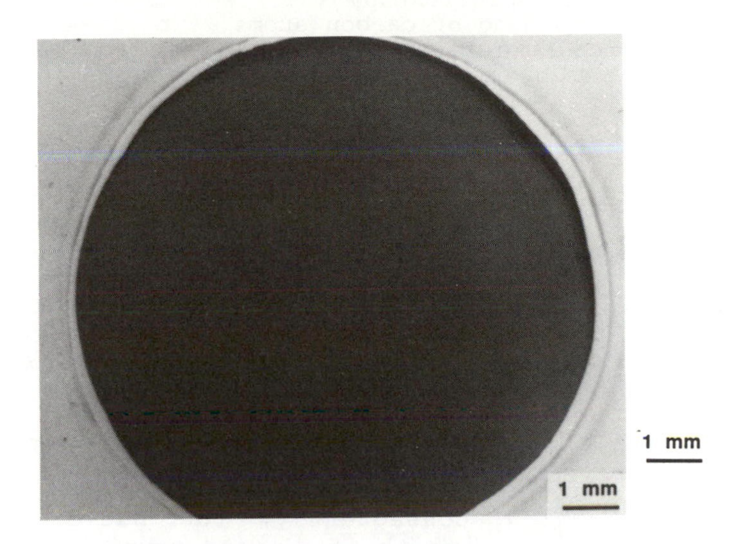

Fig. 4. A typical deposition on a silicon substrate.

The hydrogen concentration was determined with the forward recoil spectroscopy (FRES). As expected the film deposited with hydrocarbon ion beam had a high (about 20 at. %) hydrogen concentration, Fig. 6, whereas the hydrogen concentration of the film deposited with the pure carbon beam was below the detection limit of the technique (about 0.1 at. %).

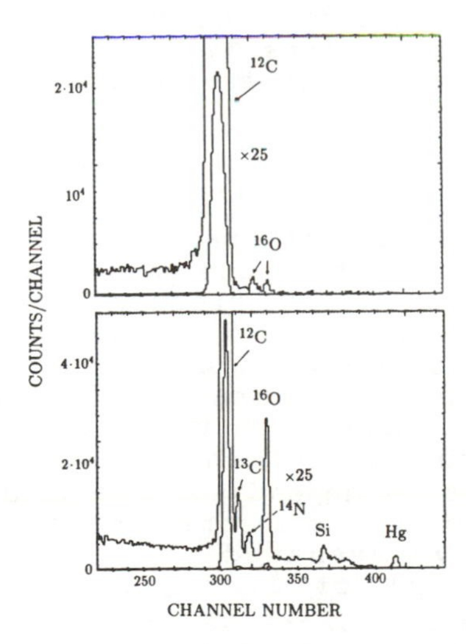

Fig. 5. RBS spectra from a self supporting i-C (up) and evaporated carbon film (down).

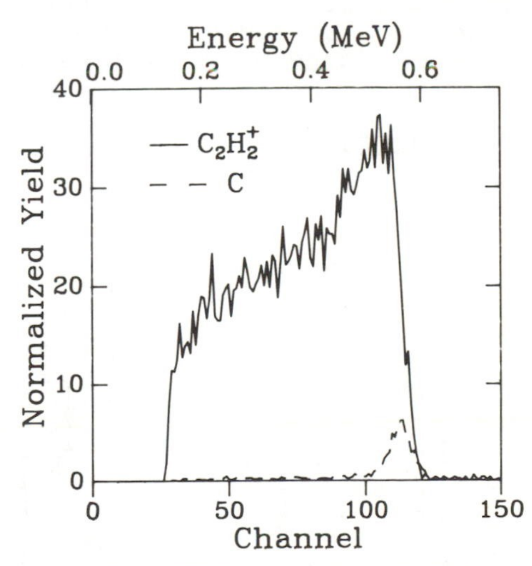

Fig. 6. FRES spectra of i-C films deposited using C_2H_2- *and* C_+- *beams.*

The microstructure of the materials were studied using electron and x-ray diffraction. These studies indicated an amorphous structure [29]. The Auger electron spectroscopy (AES) was used to probe the nature of the chemical bonding of carbon atoms. The line shape was typical to other diamondlike films fabricated with ion beams [41] indicating the contribution of sp^3 bondings.

2.2 DC RESISTANCE

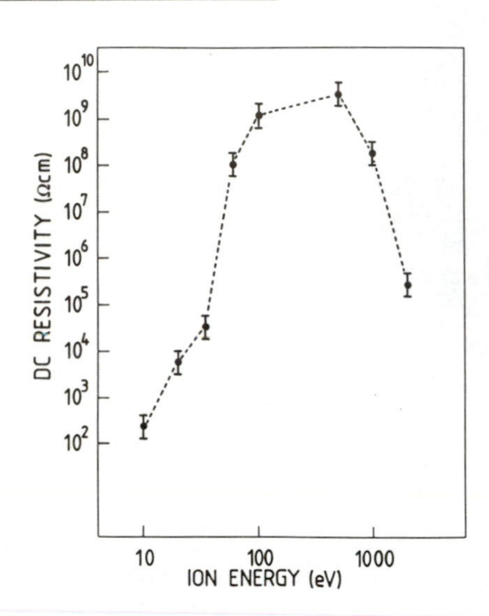

The dc resistance of the films was measured using an Au-pin and the WC-Co substrate as electrodes. The measurements were done using an electric field strength of 10 V/μm in the film. The highest resistance, 3×10^9 Ωcm, was obtained from a film deposited at 500 eV (Fig.7). Similar results have been reported using negative C^--ion beam deposition [24,30]. The rapidly cooling thermal spikes induce an amorphous structure. Consequently, the nucleation and growth of diamond crystals in C^+-ion bombardment is easily disturbed and terminated. This was demonstrated by

Fig. 7. DC resistance of a film as a function of the deposition energy.

Freeman et al. [42] as C^+-ions were deposited onto a natural diamond substrate at 700°C. Generally epitaxial diamond growth was observed. In addition some amorphous inclusions were observed to grow initiated by impurities or defects of the substrate diamond.

2.3 DENSITY

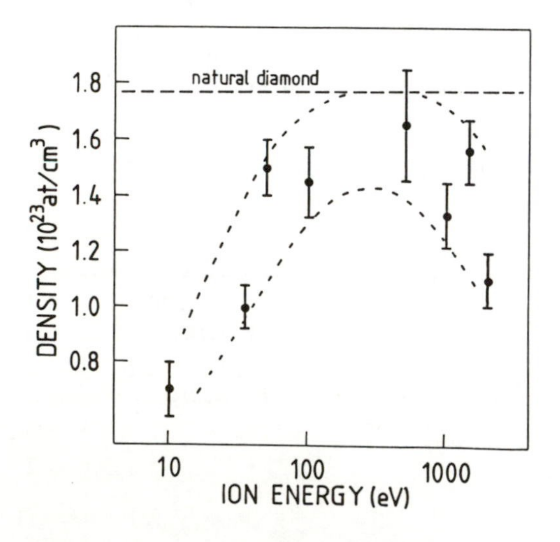

The density measurements have revealed that the density of the *i*-C layer is strongly dependent on the atomic composition (hydrogen content) and deposition parameters. When a dual beam sputter deposition with added H_2 gas is used the density depends on the hydrogen content of the film. The unhydrogenated film had a density of 2.25 g/cm^3 and the density decreased to 1.6 g/cm^3 with 35 at.% of hydrogen [23].

Fig. 8. Density of an i-C film as a function of the deposition energy.

However, using direct C^+ -ion beam deposition densities between 3.3 and 3.5 g/cm^3 were obtained [24,26,29,30]. The density of the films was measured using RBS combined with Dektak profilometer thickness measurements. Fig.8 illustrates the density as a function of ion energy. The most dense films were obtained using E_i=100 - 500 eV. The error bars arise mainly from the variations of the coating thickness which brings uncertainty to the comparison of the RBS and Dektak measurements. Nevertheless it can be noted that a density ≥ 1.6×10^{23} at/cm^3 can be reached with E_i=500 eV. This value is somewhat less than the density of a natural diamond, 1.76×10^{23} at/cm^3. The density measurements correspond closely to the values published recently by Ishikawa et al. [30]. In addition a strong dependence of the density on the ion energy has been observed. When E_i= 50 - 1000 eV it is possible to obtain densities close to natural diamond [26,29]. Molecular dynamic calculations and experiments show that ion bombardment can increase the density of the growing film with a knock-on of the surface atoms [43]. If the carbon impinging of the surface is not in the form of energetic ions a less dense or porous structure is produced with hydrogen terminating some of the sp^3-bonds [23].

To summarize: The maximal density of the *i*-C film is achieved only when a large enough proportion of incoming carbon is in the form of ions with a sufficiently high kinetic energy.

2.4 HARDNESS

The maximum values of the Vickers hardness obtained for the *i*-C layers are about the same as for the natural diamond [25].

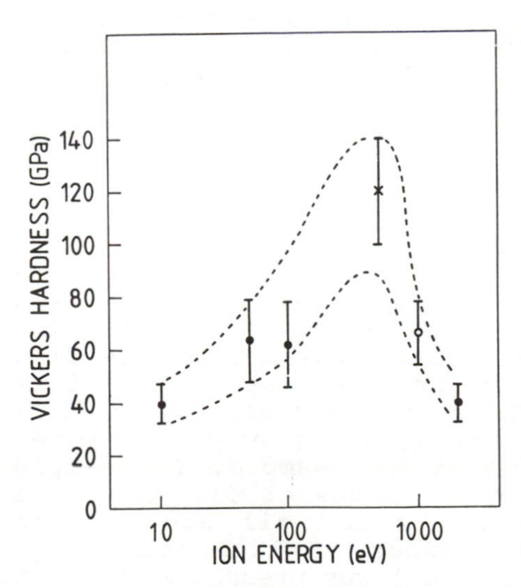

The Vickers hardness of the films was measured using a microhardness tester with an optical microscope. The results are shown in Fig. 9. The load used varied between 20 - 200 g and was chosen so that the indentation depth was less than half of the film thickness. Each hardness value is the average of 10 measurements. The error bars arise from the statistical uncertainty of optical measurement of the indentation area. The hardest *i*-C film was HV(200g)=120±20 GPa with E_i=500 eV. This value is of the order of that for natural diamond [8].

Fig. 9. Vickers hardness as a function of the deposition energy.

The temperature stability of the hardness of the film deposited at E_i=500 eV on WC-Co was studied by annealing the sample in vacuum (5 mPa) followed by a hardness measurement. The 1 hour annealings were performed in 100 °C steps starting from 100 °C. No change in the hardness was observed after the annealing up to 600 °C. After the 700 °C annealing the film became softer and was finally peeled off during the 800 °C annealing. The hardness was observed to be strongly dependent on E_i [29]. This is qualitatively in agreement with the data obtained using C_6H_6 in the ion beam deposition method [6]. For hydrogen containing layers hardness values between 20 - 60 GPa are reported which match the value of 32 GPa obtained using a mass-separated $C_nH_m^+$-ion beams [28]. Fig. 10 shows hardness measurements for pure *i*-C as compared to a hydrogen containing film obtained using a continuous hardness test. On inspection of the depth of the indentation and the form of the curves during unloading a clear difference in the hardness and elastic behaviour of the layers can be observed [27]. There is a strong elastic recovery in the pure *i*-C layer. This recovery cannot be excluded from the hardness values obtained by optical inspection of the indentation area. This may account for a 30 % exaggeration in the hardness values [6]. Similar elastic properties have been obtained in the hydrogen containing *i*-C films [6].

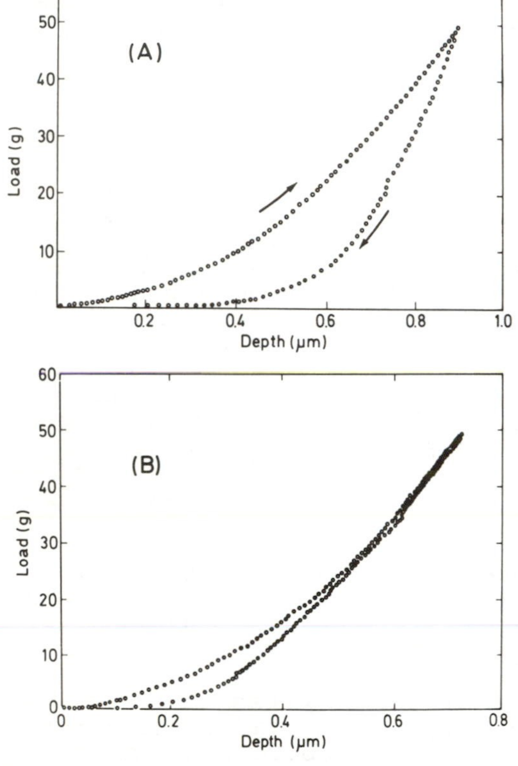

Fig. 10. Hardness measurements on the i-C films deposited using C_2H_2 *- (A) and* C^+ *-ions (B), respectively.*

2.5 FRICTION AND WEAR

2.5.1 THREE BODY ABRASIVE WEAR

Abrasive wear measurements and scratch tests were performed to find the deposition parameters yielding the most wear resistant *i-C* film [29]. The sample was worn using an 11 mm diameter 1 mm wide Cu-wheel rotating on the sample. The sample was kept in a vessel filled with a slurry of castor oil and silicon carbide powder of 1200 mesh (Fig. 11). The wheel was pressed with 0.1 N force and was rotating at a constant speed of 40 rpm. The electrical resistance between the substrate and

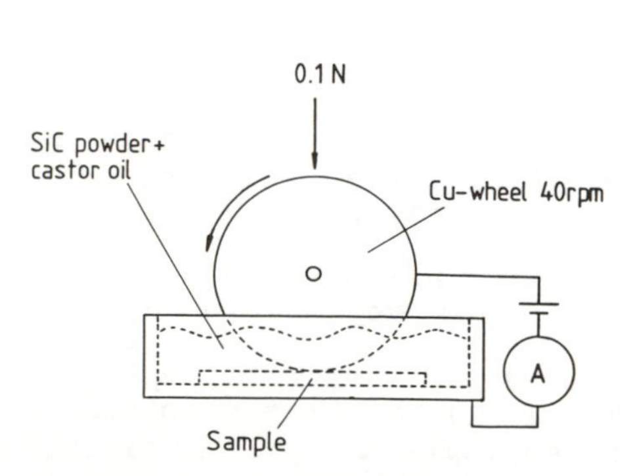

Fig 11. An abrasive wear measurement.

the Cu wheel was monitored, and a test was terminated automatically when a hole into the *i*-C film was produced. The wearing of the film is a sequence of two processes: a) abrasive wear of the *i*-C film with a constant rate (about 120 min for a 1 μm deep track), b) cracking off of the rest of the *i*-C layer when only a 0.1 - 0.6 μm thick layer is left. The wear tracks were measured with a Dektak profilometer. To deduce the abrasive wear rate the wear volume was considered to be the track produced by the first process. The wear rate is the wear volume divided by the time. The wear rate for each sample is the average of three separate wear experiments and the results are shown in Fig. 12. The *i*-C coatings deposited with $E_i \leq$ 50 eV cracked immediately when the wear test was initiated. The most wear resistive coatings were deposited with E_i=100 - 500 eV.

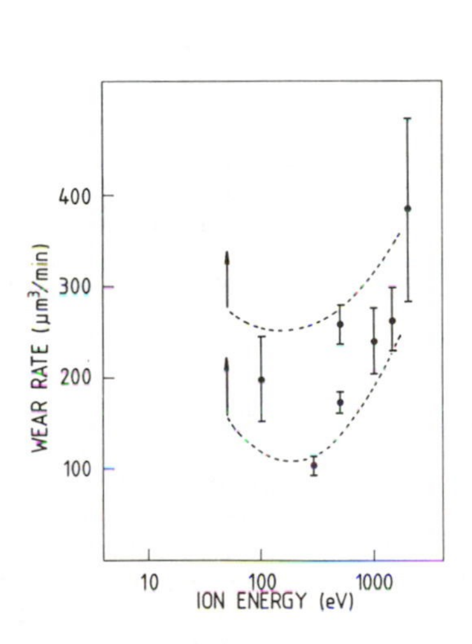

Fig. 12. Wear rate of i-C films as a function of the deposition energy.

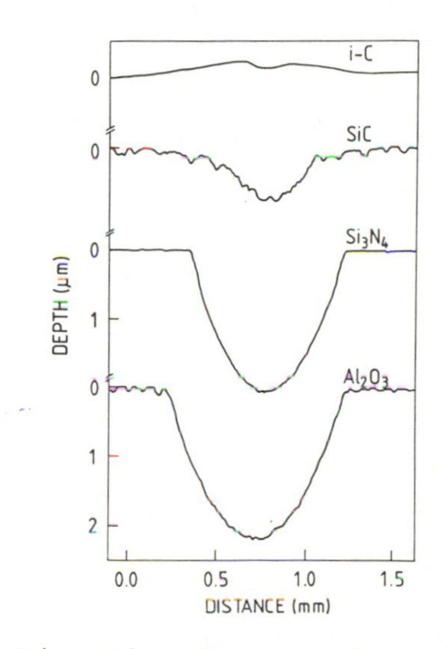

Fig. 13. Wear tracks on ceramic samples.

Table 1. The wear rates compared to the i-C film.

	Hardness (GPa)	Compressive strength (MPa)	Wear rate compared to *i*-C
Al_2O_3	16-20	2100	290±20
Si_3N_4	15-20	4000	230±30
SiC	24-27	2100-2300	40±10
WC-Co	14-16		60±10
natural diamond	98-103	8900	0

For comparison the same wear experiment was performed using three different construction ceramics, Al_2O_3, SiC, and Si_3N_4; WC-Co cemented carbide, and natural diamond. Some of the wear tracks are shown in Fig. 13. The wear rate of the ceramic samples compared to an E_i=1 keV deposited *i*-C coating (wear rate 260±40 μm^3/min) together with the mechanical properties provided by the manufacturers are shown in Table I. No sign of wear was observed on the natural diamond (001) surface. The ion energy E_i= 100 - 500 eV corresponds to the hardest and densest coating having also the highest electical resistance [29]. These films are also the most wear resistant ones. The low coefficient of friction evidently contributes to the high wear resistance of the *i*-C layers because friction is an important parameter in determining the manner in which stress is transferred from the wearing particles to the surface [44].

2.5.2 DRY SLIDING AGAINST STEEL AND Si_3N_4

One of the most interesting and important properties of materials from the tribological point of view is of course friction. Of all allotropes of carbon graphite has been shown to have very low friction in environments involving some humidy. This is attributed to the low shear strength of the layered structure sliding occurring between two graphite surfaces. Thus the final value of the friction coefficient is independent of the counterpart.

The friction of crystalline diamond, on the contrary, is dependent on the lattice direction and also on the humidity of the environment [45]. At present, no unambiguous knowledge about the sliding behavior of diamondlike films is available. The situation is even more confused by the wide variety of deposition systems and properties of films. Some experimental data have, however, been published [46,47]. Contrary to graphite the friction of a diamondlike film has been shown to increase with increasing humidy of the environment [46], the lowest value being as small as 0.01.

We have studied unlubricated sliding properties of hydrogen free, ion beam deposited diamondlike films. The friction and wear was examined using a pin-on-disc test. As a pin hardened chromium steel or Si_3N_4 balls 6 mm in diameter were employed. The normal force was 155 g giving a Hertzian surface pressure of 1120 and 951 MPa on a uncoated substrate for the Si_3N_4 pin and steel pin, respectively. In all cases the wear and friction measurements were extended to the 10 000 revolutions corresponding to a sliding distance of 157 m. The sliding speed was 13 cm/s. During the measurements the friction force was continuously monitored with the calibrated load cell. For comparison, the wear and friction of the same pins on the uncoated substrate, WC-Co cemented carbide, were tested.

The morphology of wear tracks and pins were examined with optical and scanning electron microscopy (SEM). Possible material transfer between the sliding surfaces was investigated with the energy dispersive spectroscopy (EDS).

Shown in Fig. 14 is the friction coefficients of hardened steel on coated and uncoated WC-Co cemented carbide as a function of the number of revolutions in the pin-on-disc test. On uncoated WC-Co

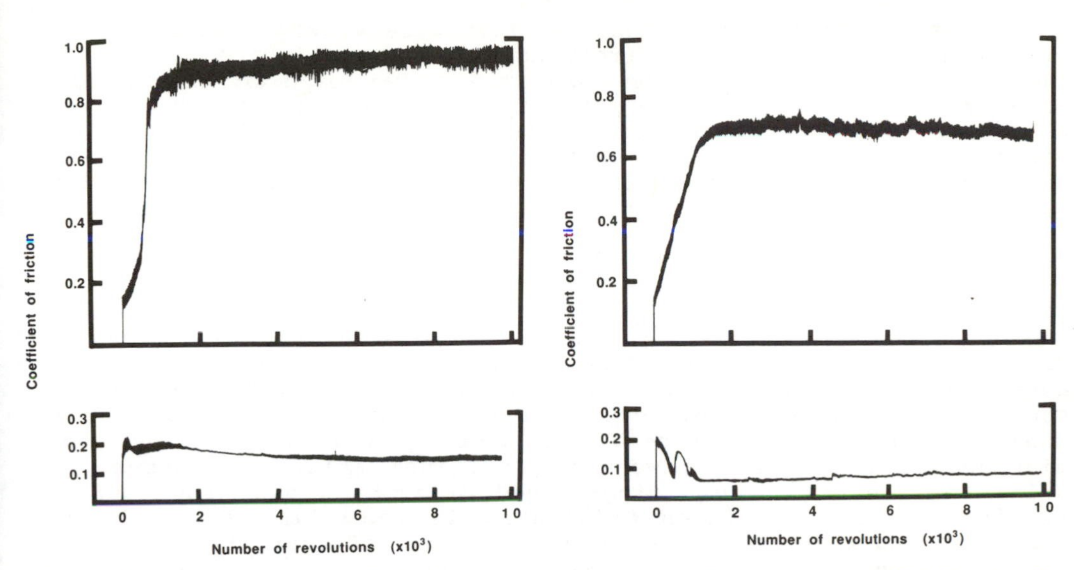

Fig. 14. Friction coefficient steel on WC-Co (up) and i-C film (down).

Fig. 15. Friction coefficient of Si_3N_4 *on WC-Co (up) and i- C film (down).*

the friction coefficient had a relatively low initial value, 0.15, that increased rapidly to the steady state value about 0.9. This high value of friction was caused by the wear of the steel pin and the subsequent sliding of the pin against the depositions of the same material on the wear track as was verified later. On the diamondlike film the initial friction coefficient was a little higher, about 0.2. This decreased to the steady state value of 0.14 and remained unchanged to the very end of the test. This low friction was also an indication of the minimal surface damage of the wear track.

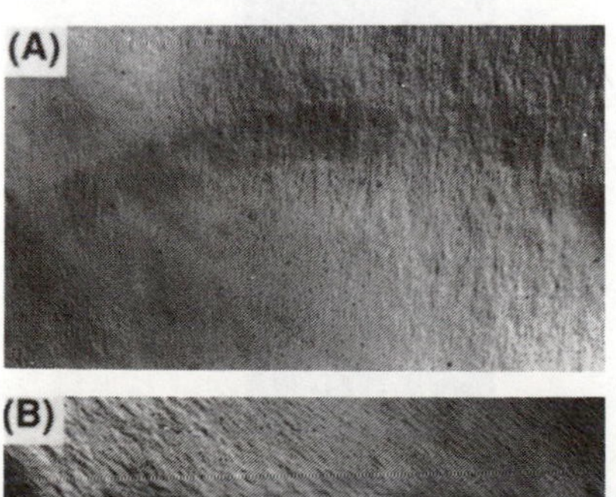

The behavior of the Si_3N_4 pin was in general the same as that of the steel pin, Fig. 15. The initial friction coefficients were the same indicating a surface contamination such as an oxide film having a role at the onset of the tribological process. The steady state values of the friction coefficients are lower in the case of the Si_3N_4 pin; 0.7 and 0.06 on the uncoated substrate and diamondlike film, respectively. Moreover, in the case of the Si_3N_4 pin on the diamondlike film the difference between the initial and the steady state friction is considerably larger than in the case of the steel pin.

Fig. 16. Wear scars on a i-C film. A steel pin (A), and Si_3N_4 *pin (B).*

The wear scars on the diamondlike film were viewed in the optical microscope and in Fig. 16 corresponding micrographs are shown. In the case of the steel pin the wear track shows signs of the minimal adhesive interaction and the path of the pin is just visible. The wear track in the Si_3N_4 case possesses more details and even morphology changes or slight wear can be observed. According to Fig. 16 deposition of the pin material onto the diamondlike film also seems possible, though this could not be verified with EDS.

More information about the sliding interaction of steel and Si_3N_4 against the diamondlike film on WC-Co cemented carbide and also on the uncoated substrate was obtained examining the changes on the pin after the test. SEM micrographs of the steel pins are shown in Fig. 17. The pin used on the uncoated WC-Co has worn mainly by the adhesive mechanism and the material has been transferred to the counterpart. EDS revealed, however, also the tungsten signal on the wear scar of the steel pin. The steel pin rubbed against the diamondlike film has not worn and original surface structure is still visible but a thin ring-shaped deposition has grown during sliding. This deposition originates in the film, although the corresponding wear scar does not reveal signs of wear, Fig 16. This is due to the small amount of material required for the deposition in Fig. 17. It seems evident that this deposition is formed at the beginning of the sliding, and the subsequent sliding occurs between the diamondlike film and the deposition.

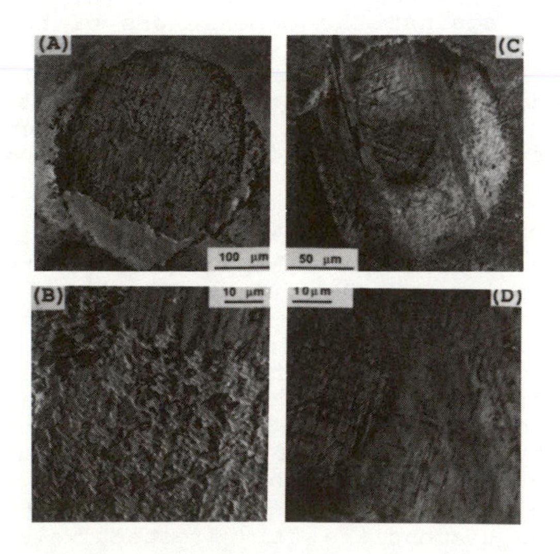

Fig. 17. SEM micrographs of the steel pins after 10^4 revolutions against WC-Co (A,B) and i-C (C,D).

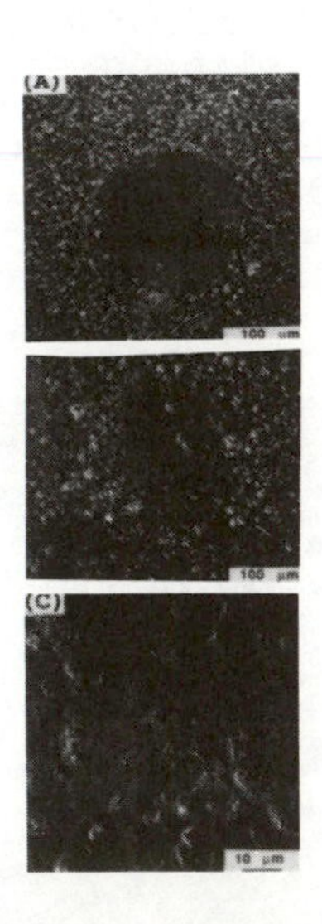

Fig. 18. SEM micrographs of the Si_3N_4 pins after 10^4 revolutions against WC-Co (A) and i-C (B,C).

The SEM micrographs of the Si_3N_4 pins are shown in Fig. 18. Again high wear rate during sliding against the uncoated WC-Co cemented carbide can be noticed. However, contrary to the corresponding steel pin the surface of the wear scar is very smooth. Moreover, no material transfer was detected. All these facts confirm that the main wear mechanism in this case has been the abrasive one. The Si_3N_4 pin used on the diamondlike film shows minimal wear and no deposited film can be observed either.

The sliding of the steel and Si_3N_4 pins on the hydrogen free diamondlike film results in the different friction coefficients. Moreover, the scanning electron and optical micrograps revealed also differences on the rubbed surfaces of the pins and the corresponding wear tracks. These facts indicate slight but obvious differences between these two sliding systems.

Fig. 17 showed some evidence about the transfer film on the steel pin. Without a doubt this transfer film originates in the diamondlike coating and the sliding finally occurred between materials consisting of the same element, i.e. carbon. One interesting feature of the transfer film is its ring-like shape. No deposition can be found in the center part of the pin. This same region, in fact, is the contact area on the substrate material under the normal force used here (the calculated diameter of the contact area is 56 μm and the deposition free area in Fig. 10 is 40 μm in diameter, respectively). Thus it seems apparent that no transfer film is formed on this high surface pressure region. This may be due to the weakness of the adhesion of the transfer film to the steel surface under the high pressure. The transfer film formation may alter the wear behaviour of the diamondlike film in the conditions where the virgin surface of the counterpart constantly slides on the diamondlike film.

The width of the wear tracks on the diamondlike film corresponds well to the calculated diameters of the contact areas. As mentioned above the diameter of the contact area in the steel pin case is 56 μm. The observed width of the wear track is 75 μm. In the case of the Si_3N_4 pin the calculated diameter of the contact area and the observed width of the wear track on the diamondlike film are excatly the same, 52 μm. This is also consistent with the observed width of the sliding region of the Si_3N_4 pin.

The Si_3N_4 pin caused visible changes in the surface morphology of the wear track as compared to the steel pin. This may be due to the higher surface pressure in the Si_3N_4 pin case. A trasfer film on the pin was not observed either, indicating the different sliding mechanism from that of the steel pin - diamondlike film system. Although Fig. 17 reveals some signs of the depositions on the wear track, it could not be identified with EDS. It is, however, unlikely that this possible deposion originates in the Si_3N_4 pin, because the friction coefficient remained very low throughout the test. It has been observed [48] that the friction coefficient of the Si_3N_4 - Si_3N_4 contact is much higher, μ = 0.67.

The friction coefficients obtained in this work are comparable to those measured on crystalline diamond. Depending on the crystal direction friction coefficients from 0.05 to 0.15 have been obtained in the sliding of the diamond rider on the single crystal diamond [45]. The wear of the pins sliding on the diamondlike film was below the detection limit. This zero wear can be compared to the wear of pins on the uncoated WC-Co cemented carbide. The wear coefficients of steel and Si_3N_4 can be determined (ignoring the run-in wear) as 2.4×10^{-16} and 5.0×10^{-16} m^3/Nm, respectively, on the basis of the present measurements. For example the wear coefficient of Si_3N_4 on steel is somewhat higher, about 10^{-14} m^3/Nm [48].

3. ARC-DISCHARGE DEPOSITION

3.1. GENERAL

Although with the mass-analyzed ion beam system *i*-carbon films can be produced very controllably, a serious difficulty is the lack of the effective sourse. Carbon has a high boiling point and it reacts with almost all elements at high temperature. Thus the high ion current needed for product purposes destroy conventional sources in short time.

One type of the carbon ionization method with which high quantities of carbon can be converted to ion form is based on the electric arc-discharge technique, i.e. the strong electric current between the carbon anode and cathode creates the carbon plasma stream. With this system energetic carbon ions can be produced directly without any additional acceleration of ions. This system is rather straightforward if the presence of neutral carbon atoms and microparticles in the plasmastream is accepted. However, if high quality films are desired the microparticles and neutral atoms should be eliminated and thus the system will be somewhat more complicated.

Many Russian groups have used this plasma evaporation system and developed it for the preparation of the *i*-C films [49-56]. They have achieved very promising results, e.g. films which were harder and more wear resistant than natural diamond have been prepared.

3.2.THEORY

There is not yet generally accepted theory of the formation and behavior of arc-disharge plasmas. Therefore the optimation of the arc device must be performed experimentally. Due to the many parameters it is a difficult task. According to Dorodnov [54] the acceleration of plasma occurs with the Ampere force. Because in the plasma acceleration case the Ampere force is proportional to the square of the current, this would mean, a priori, that by increasing the current higher plasma energies could be achieved. However, Aksenov et al. [53] have pointed out experimentally that the increase in the discharge current reduces the energy of the plasma ions. It may be true that Dorodnov's theory is correct but explosion effects in the cathode have optimum values as regards the plasma energy and quantity. Possibly the most physically justifiable theory is that presented by Lyubimov [57]. According to that theory the plasma has a gas-dynamic character in the region adjacent to the arc cathode spot.

3.3. DEVICE CONSTRUCTION

The Russian groups have presented many differerent plasma devices for carbon plasma production. However, only Kharkov group [53] used the system in which the microparticles and neutrals are eliminated . For the elimination they used a curvilinear plasma-optical separator. The separation is performed with the stationary current magnet system. The system is straightforward and reliable but for the curving of the ions a magnet system a several kilowatt power consumption is needed.

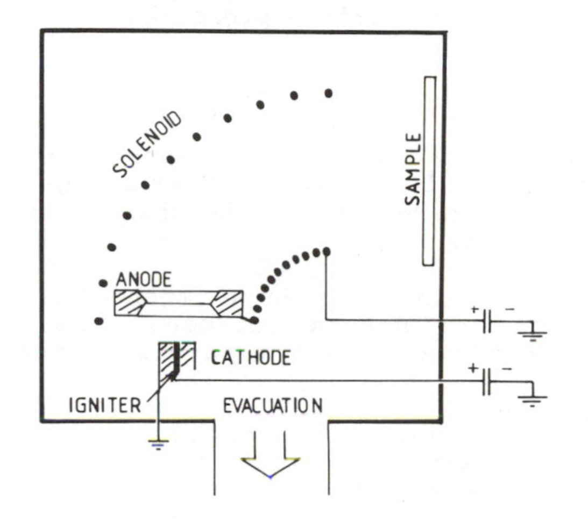

Fig. 19. *A schematic picture of an arc discharge deposition system.*

In Fig. 19 the type of plasma arc-disharge system used in our laboratory is shown. It is a modified combination of those presented by Aksenov et al. [53] and Maslov [55]. The main principle is as follows: In the ingniter current circuit the capacitor bank of some tens of pF is charged with a voltage of some tens of kV. Between the anode and cathode there is a much lower voltage, some hundreds or thousands of volts, but the capacitance of the capacitor bank is at least a thousand times higher than that of the ignitation circuit. In order to achieve a reasonable efficiency a high current, 3 kA or more, is necessary. It is also important for the curving of the high energy plasma-ions, e.g. for the solenoid 10 cm in diameter and for the 500 eV carbon ions a magnetic field of 2200 G is needed [57].

As to the practical use of the plasma evaporation system an essential feature is that the influence of the magnetic field decreases rapidly outside the solenoid. The compressing effect of the solenoid decreases rapidly and the plasma stream will expand. In addition, the ions follow the force lines of the magnetic field but the high energy ions go rather linearly. If it is needed, this effect can be used for the separation of the low and high energy ions.

3.4. PROPERTIES OF THE FILMS

Fig. 20. A wear scar on a i-C film deposited using the arc discharge method.

Although the developement of the plasma devices is still in progress some properties of the films have been studied. The properties have been observed to correspond closely to those of the films prepared with the mass-analyzing system. For example, the initial friction coefficient against steel of a *i*-C film deposited onto hardened tool steel was 0.2 and it decreased to 0.13±0.01 after the sliding distance of 1 130 m (180 000 wear cycles). In Fig. 21 the corresponding SEM micrograph of the wear track is shown. The normal load in this test was 250 g and the diameter of the pin 6 mm, which gives a high surface pressure.

4. CONCLUSIONS

The direct mass-separated ion beam method was demonstrated to be a powerful method in the investigation of *i*-C films. The advantages are the fine control of the deposition parameters and the ability to produce almost impurity-free carbon films.

The i-C layers prepared using C^+-ion beam have mechanical properties which are close to those of natural diamond. Their hardness and abrasive wear resistance is far better than with other hard materials.

The properties of the *i*-C layers depend strongly on the deposition ion energy. An *i*-C film with the most diamondlike qualities can be produced using C^+-ion energy of 100 - 500 eV. Using various hydrocarbon beams it was possible to produce *a*-C:H layers. A profound difference in the structure of *i*-C and *a*-C:H was indicated by the hardness, wear and density measurements. The hydrogen containing films are softer and less wear resistant.

The diamondlike film fabricated using the mass analyzed ion beam possesses very low friction in dry sliding against Si_3N_4 (μ=0.06) and reasonable low against steel (μ=0.14). The sliding mechanism was slightly different in these two cases. In the steel case a transfer film was found on the pin. The wear of the pins sliding on the diamondlike film was unmeasurable in the conditions which caused severe wear of the same materials sliding on the uncoated WC-Co. The slight wear of the diamondlike film was found only in the Si_3N_4 case.

Both methods presented in this review are still in developement but it can be foreseen that they will have their own advantages. Without doubt the mass-analyzing method is the most accurate way to prepare the diamondlike or even diamond films and the great challenge of the diamond preparation technology, i.e. the heteroepitaxy of diamond films will probably be performed with this method. The great advantage of the plasma deposition is that it is an effective low-cost method, with which high quality adhesive *i*-C films, on whatever vacuum-resistant substrates, can be prepared. In addition, both methods are low temperature processes.

5. REFERENCES

1. R.F. Bunshah ed., *Deposition Technologies for Thin Films and Coatings*, Noyes Publications, New Jersey, 1982.
2. J.J. Cuomo and S.M. Rossnagel, Nucl. Instr. and Meth.

1987, **B19/20**, 963.
3. S. Aisenberg and R. Chabot, J. Appl. Phys. 1971, **42**, 2953.
4. C. Weissmantel, K. Bewilogua, D. Dietrich, H.-J. Erler H.-J. Hinneberg, S. Klose, W. Nowich, and G. Reisse, Thin Solid Films 1980, **72**, 19.
5. S. Shanfield and R. Wolfson, J. Vac. Sci. and Technol. 1983, **A1**,323.
6. C. Weissmantel, *Thin Films from Free Atoms and Particles*, ed. K.J. Klabunde, Academic Press, Orlando, 1985, p.153.
7. R.J. Wedlake, *The Properties of Diamond*, ed. J.E.Field, Academic Press, London, 1979, p.501
8. J.E. Field ed., *The Properties of Diamond*, Academic Press, London, 1979, p.641.
9. R.M. Baum, Chemistry and Engineering, December 23, 1985, p.20.
10. R. Berman, *The Properties of Diamond*, ed. J.E.Field, Academic Press, London, 1979, p.1.
11. A. Matsumoto, Y. Sato, M. Kamo, and N. Setake, Jpn. J. Appl. Phys. 1982, **21**, L183.
12. N. Fujimori, T. Imai, and A. Doi, *Proc.* IPAT 85, ed. H.Oechsner, Munich, 1985, p.307.
13. M.W. Guinan and J.H. Kinney, J. Nucl. Mater. 1981, **103/104**, 1319.
14. R. Webb, D. Harrison Jr, and M. Jakas, Nucl. Inst. Meth. 1986 **B15**, 1.
15. K. Roessler and G. Eich, *Proc.* E-MRS 1987, Strasbourg, Les Editions de Physique, Paris.
16. Manufacturers' statements.
17. O. Almen and G. Bruce, Nucl. Instr. and Meth. 1961, **11**, 279.
18. E. Hechtl, J. Bohdansky, and J. Roth, J. Nucl. Mater. 1981, **103/104**, 333.
19. R. Kelly, Nucl. Instr. and Meth. 1978, **149**, 553.
20. F. Seitz and J.S. Koehler, *Solid State Physics*, **vol. 2**, Academic Press, New York, 1956, p.305.
21. D.A. Thompson, Rad. Effects **56**, 105 (1981).
22. E.G. Spencer, P.H.Schmidt, D.C.Jay, and F.J.Sansalone, Appl. Phys. Lett. 1976, **29**, 118.
23. F. Jansen, M.Machonkin, S.Kaplan, and S.Hark, J. Vac. Sci. and Technol. 1985, **A3**, 605.
24. T. Miyazawa, S. Misawa, S. Yoshida, and S. Gonda, J. Appl. Phys. 1984, **55**, 188.
25. J. Koskinen, J-P. Hirvonen, and A. Anttila, Appl .Phys. Lett. 1985, **47**, 941.
26. A. Anttila, J. Koskinen, M. Bister, and J. Hirvonen, Thin Solid Films, 1986, **136**, 129.
27. J.-P. Hirvonen, J. Koskinen, A. Anttila, D. Stone, and C. Paszkiet, Mater. Sci. and Eng. 1987, **90**, 343.
28. A. Anttila, J. Koskinen, R. Lappalainen, J-P. Hirvonen, D. Stone, and C. Paszkiet, Appl. Phys. Lett. 1987, **50**, 132.
29. J. Koskinen, J. Appl. Phys. 1988, **63**, 2094.
30. J. Ishikawa. Y. Takeiri, K. Ogawa, and T. Takagi, J. Appl. Phys. 1987, **61**, 2509.
31. J. Fink, T. Müller-Heinzerling, J. Pflüger, A. Bubenzer, P. Koidl, and G.Crecelius, Solid State Commun. 1983, **47**, 687).
32. B. Dischler, *Proc.* E-MRS 1987, Strasbourg, Les Editions de Physique, Paris.
33. F. Jansen and M. Machonkin, Thin Solid Films 1986, **140**, 227.
34. T. Moravec and T. Orent, J. Vac. Sci. and Technol. 1981, **18**, 226.
35. A. Anttila, J. Koskinen, J.Räisänen, and J. Hirvonen, Nucl.

Instr. and Meth. 1985, **B9**, 352.
36. A. Khan, D. Mathine, and J.Woollam, Phys. Rev. 1983, **B28**, 7229.
37. S. Fujimori and K. Nagai, Jpn. J. Appl. Phys. 1981, **20**, L194.
38. M. Sokolowski, A. Sokolowska, B. Gokieli, A. Michalski, A. Rusek, and Z.Romanowski, J. Crystal Growth 1979, **47**, 421.
39. D.S. Whitmell and R. Williamson, Thin Solid Films 1976, **35**, 255.
40. A. Fontell and E. Arminen, Canadian J. Phys. 1969, **47**, 2405.
41. S. Kasi, H. Kang, and J. W. Rabalais, Phys. Rev. Lett. 1987, **59**, 75.
42. J.H.Freeman, W.Temple, and G.A.Gard, Vacuum 1984, **34**, 305.
43. K.-H.Müller, J. Appl. Phys. 1986, **59**, 2803.
44. C.J.McHargue, Nucl. Instr. and Meth. 1987, **B19/20**, 797.
45. J.E. Field (ed.): *The Properties of Diamond,* Academic Press, London, 1979. p. 647.
46. K. Enke, H. Dimigen, and H. Hübsch, Appl. Phys. Lett. 1980, **36**, 291.
47. S.M. Ojha and L. Holland, Thin Solid Films 1977, **40**, L31.
48. K. Holmberg, P. Andersson, and J. Valli, Paper presented at *The 14th Leeds-Lyon Symposium on Tribology Interface Dynamics*, INSA de Lyon, September 1987.
49. I.I. Aksenov, V.A. Pelows, V.G. Padalka, and V.M. Khoroshikh, Sov. J. Plasma Phys. 1978, **4**, 425.
50. V.E. Strelnitskii, I.I. Aksenov, S.I. Vakula, V.G. Padalka, and V.A. Pelows, Sov. Tech. Phys. Lett. 1978, **4**, 546.
51. V.E. Stelnitskii, v.g. Padalka, and S.I. Vakula, Sov. Phys. Tech. Phys. 1978, **23**, 222.
52. I.I. Aksenov, S.I Vakula, V.G. Padalka, V.E. Strelnitskii, and V.M. Khoroshikh, Sov. Phys. Techn. Phys. 1980, **25**, 1164.
53. A.M. Dorodnov, Sov. Phys. Techn. Phys. 1978, **23**, 1058.
54. A.M. Dorodnov, S.A. Muboyadzhyan, Ya. A. Pomelov, and Yu.A. Strukov, J. Appl. Mech. and Tech. Phys. 1981, **22**, 28.
55. A.J. Maslov, G.K. Dmitriev, and Yu. D. Chistyakov, Instrum. and Exp. Tech. 1985, **28**, 662.
56. G.A. Lyubimov, Dokl. Akad. Nauk. SSSR 1975, **225**, 104.
57. N.A. Khizhnyak, Sov. Phys. Tech. Phys.1965, **10**, 655.

Materials Science Forum Vols. 52 & 53 (1989) pp. 217-236

ION BEAM DEPOSITED DIAMONDLIKE CARBON FILMS

M.J. Mirtich
National Aeronautics and Space Administration
Lewis Research Center, Cleveland, OH 44135, USA

ABSTRACT

Single and dual ion beam systems were used to generate hard, durable, chemically inert, optically transparent diamondlike films.

INTRODUCTION

There have been many publications reporting efforts to produce thin carbon films with diamondlike properties. A variety of plasma and ion beam techniques have been employed to generate diamondlike carbon (DLC) films. DLC films can be deposited by rf plasma decomposition of a hydrocarbon gas [1,2] or other alkanes [3], by low energy carbon ion beam deposition [4,5] or by ion plating and dual beam techniques [6]. Weissmantel, et al. [6] referred to these films as i-carbon (i-c) implying that some type of ion bombardment or ion beam is involved in the film preparation.

Because DLC films have been generated using a wide variety of techniques, the physico-chemical properties of such films vary considerably. In general DLC films have characteristics that are desirable in a number of applications. These characteristics can be changed by varying the deposition technique and potentially tailored to meet the needs of the particular application. With this in mind, diamondlike films were produced at NASA Lewis Research Center by using methane (CH_4) and argon in either single or dual ion beam sources, or by using an ion source to sputter deposit carbon from a graphite target. Many film properties were measured, and a variety of applications evaluated. Some of the applications include using DLC films as protective coatings on transmitting windows, power electronics or as insulating gates, and corrosion resistant barriers. The process techniques and the results of these measurements and evaluations are presented herein.

SINGLE AND DUAL BEAM ION SOURCE AND DEPOSITION PROCEDURE

A 30-cm-diameter ion source with its optics masked to 10 cm in diameter was used to directly deposit DLC films. The ion source, developed for electric propulsion technology, uses argon gas in the hollow cathode located in the main discharge chamber, as well as in the neutralizer [7]. After a discharge is established between the cathode and anode, methane (CH_4) is introduced through a manifold into the discharge chamber. For the depositions presented in this paper the molar ratio CH_4 to argon was 0.28. This ratio was found to be ideal for generating films. If the CH_4/Ar ratio was too large the discharge extinguished. No films were observed at low a CH_4/Ar ratios since this condition did not allow a net deposition of C atoms due to the more dominant sputtering effects of the Ar ions [8]. The ideal energy level for deposition of DLC films has been reported to be between 100 and 150 eV [3,4]. In these experiments the total ion beam energy is the sum of the discharge voltage and the screen grid voltage. Therefore, for a discharge voltage of about 50 V the screen grid voltage was approximately 50 V. At this low value of screen grid voltage it was necessary to increase the accelerator voltage more negative than usual (to around -500 V) to extract a beam. Typically current densities at these conditions were 1 ma/cm^2 at a distance 2.5 cm axially downstream of the grids [9].

Knowing the current density and the knowledge that divergent field ion sources can be operated at low ion extraction voltages (normal conditions usually utilize screen grid voltages of 1000 V) it was possible to design an experiment using a single ion source (30 cm) with CH_4/Ar at 100 eV and current densities of 0.2 ma/cm^2. Films were deposited at these conditions on Si and SiO_2 at deposition rates as high as 71 Å/min to film thicknesses as great as 1.5 um.

It is believed that the amorphous carbon films are produced under conditions where both growth and sputtering occur simultaneously increased sputtering may decrease the number of graphite precursors incorporated in the films and hence improve film quality.

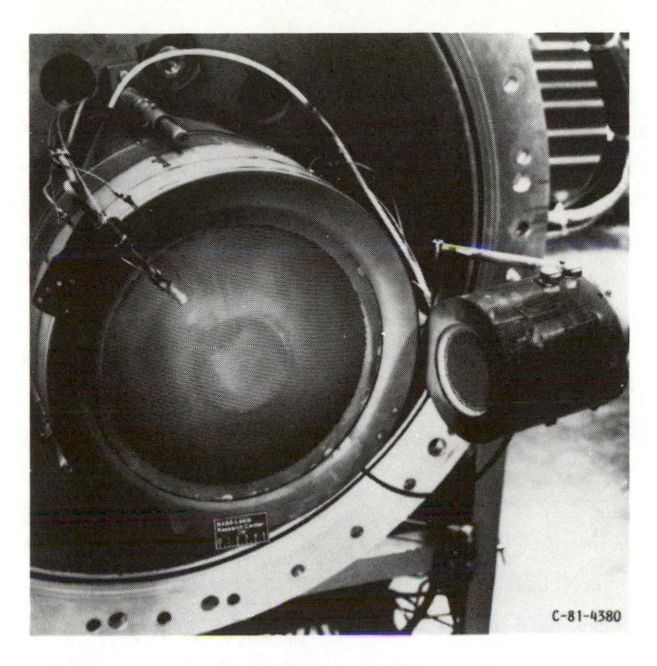

FIGURE 1. - DUAL BEAM ION SOURCE FOR DEPOSITION OF FILMS WITH DIAMOND-LIKE PROPERTIES.

In addition Marinow and Dobrew [10] have found that active sites for nucleation are created and the growth and coalesence of the nuclei enhanced due to an increased mobility of the condensing atoms when film structures are bombarded by inert gas beams. With these factors in mind, a dual beam system was created by adding an 8-cm diameter argon ion source. This system, shown in figure 1, was used to generate another set of diamondlike carbon films. The 8-cm source, using a filament cathode, was located at a 12° angle with respect to the 30-cm source and 25 cm from the substrate. There was no observed interaction between the two sources or the ion beams during operation.

The 8-cm ion source was used to direct a beam of energetic (200 to 600 eV) argon ions at a current density of 25 $\mu a/cm^2$ on the substrates while the deposition from the 30-cm ion source was taking place. When the ion energy of this second beam was greater than 550 eV, no net film formation was found. The beams were approximately monoenergetic; however no mass selection was attempted.

Arrival rates at surfaces. - The species arriving at the surface can be divided into two categories: 1) Primary ions including Ar^+, C^+ and hydrocarbon fragments of the general form $C_mH_N^{+Z}$, and 2) a flux of unchanged species arising sputtering. Contaminants from the residual gases in the source were analyzed by residual gas analyzer and found to be low in concentration thus verifying only small presence of reactive sputtering. A temporarily large C/Ar ratio was used in some points in figure 2, and in many other cases to inititate film growth, and then the ratio was reduced and deposition continued on the already

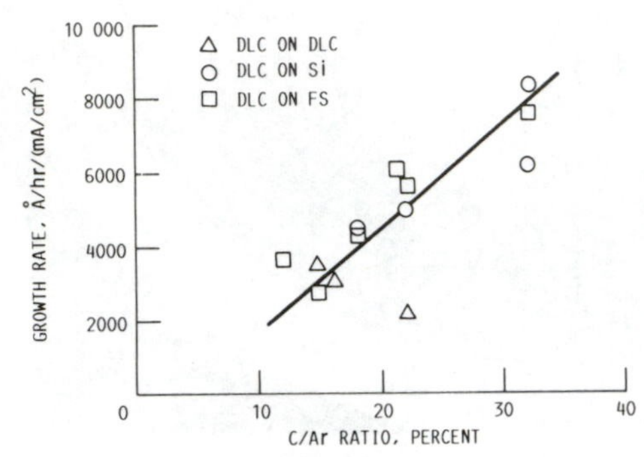

FIGURE 2. - GROWTH RATES OF DLC FILM IN FAVORABLE DEPOSITION CONDITIONS PLOTTED VERSUS THE CH_4/Ar RATIO. THE SUBSTRATE MATERIALS WERE Si, FUSED SILICA, AND DLC FILM.

existing DLC film. Most films in figure 2 were deposited with a source which had been used for long periods of time to make DLC films, thus its internal surfaces were covered with carbonaceous material layers from decomposition of the gas in the source. Depositions under these favorable conditions exhibited only small dependence of the growth rate on the substrate material. Film deposition was obtained at much lower C/Ar ratios, and the dependence of the growth rates on the ratio was linear in this region too. Only at considerably lower values of the C/Ar ratio were the growth rates dependent on the substrate material again.

The effect of the carbonaceous coating within the source can be probably explained with figure 3, which shows the residual gas analyzer mass spectrum when the source operated on mixture of methane and argon. The spectrum shows hydrocarbon fragments of the forms CH_x, C_2H_x, and C_3H_x. There are singly and doubly ionized argon ions, and ions of an argon isotope with mass 36. There are also oxygen containing ions like OH_x and CO. It is likely that the carbonaceous fragments came from the ionized methane, however the ratio of the heavier fragments to the light ones was changing with the voltages of the grids. The heavy fragments were very sensitive to the voltages of both grids, and their yields were not reduced much even without any methane flow into the discharged chamber. The carbonaceous coated source included more heavy carbonaceous fragments with lower hydrogen fractions, and part of them were probably coming from resputtering of the materials on the walls. One could speculate that these heavy fragments were probably depositing more readily on any substrate than the methane fragments, and thus initiating fast film growth.

The Direct Ion Source DLC deposition processes have some properties which were desirable for the study of growth rates. The source of carbon for the deposition was hydrocarbon gas whose flow rate was about one quarter of that of the Ar in terms of flow rates. The ratio of carbonaceous ions to Ar ions was considerably smaller as shown in figure 3 by the intensities of the appropriate peaks. Thus the material removal from the substrate was believed

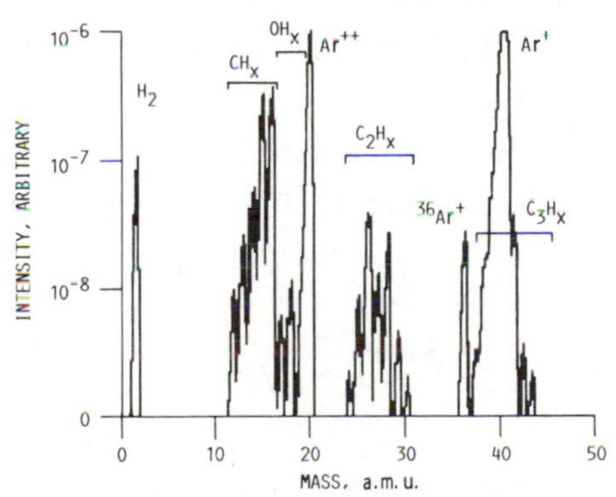

FIGURE 3. - MASS SPECTRUM OF THE GASEOUS SPECIES DURING DEPOSITION. THE PEAKS ARE MARKED WITH FRAGMENT IDENTIFICATIONS.

to be done mainly by Ar ions and only a small fraction by all other carbonaceous ions. The discharge plasma was predominantly sustained by the argon gas and was very stable. Therefore there appeared to be a separation between the deposition process, which was done by carbonaceous ions, and the removal process, which was done mainly by Ar. A constant flow rate of Ar meant roughly a constant removal rate. This was a very different experimental situation from the cases of rf glow discharges, in which CH_4 and Ar pressures and all currents and voltages were linked together and showed strong mutual interdependence [11-15].

Compositional characteristics. - Auger spectra of the DLC films showed no evidence of any elements other than carbon and small amounts of argon and oxygen. The films were remeasured after argon ion sputter profiling with an ion energy of 2000 V and a current density of 25 μa/cm. The oxygen signal disappeared, but the argon signal was enhanced. High resolution Auger spectra obtained from a single crystal of pyrolytic graphite, an ion beam deposited carbon film sample, and natural diamond are shown in figure 4. The lineshape for the ion beam deposited carbon film lies somewhere between those of pyrolytic graphite and diamond. The shoulder in the graphite spectrum at 250 eV is present in the DLC film spectrum, but not as pronounced. This result was consistently observed on numerous samples. The spectrum for natural diamond shows no shoulder and the main peak is shifted to higher energies. This latter effect may be caused by some charging of the sample.

After sputtering, the spectra from the ion deposited carbon film and the natural diamond became identical to that obtained from the pyrolytic graphite. This result is in agreement with earlier work which showed ion beam sputtering causes the surface of diamond to give a "graphitic" Auger signal [16,17]. One should also note that the escape depth of the 270 eV Auger electrons is only 7 Å [18]. Consequently, in all cases only the outermost surface layers were

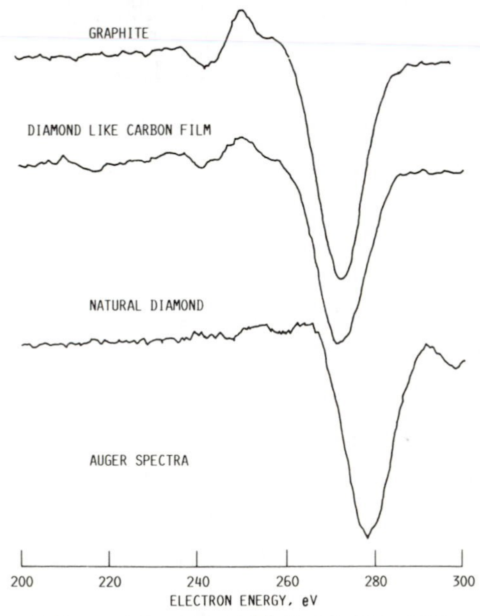

FIGURE 4. - AUGER SPECTRA OF PYROLYTIC GRAPHITE, ION DEPOSITED CARBON FILM AND NATURAL DIAMOND.

sampled. These would be the most prone to graphitization. It is, therefore, significant that differences between the ion beam deposited films and graphite were observable.

An example of a Secondary Ion Mass Spectroscopy (SIMS) spectrum from a dual beam deposited film on a silicon substrate is shown in figure 5. There are a cluster of peaks at 12, 13, 14, and 15 AMU from the hydrocarbon fragments C^+, CH^+, Ch_2^+, and CH_3^+. The peak at 14 AMU could also be assigned to N; even though the lower sensitivity AUGER analysis indicated no nitrogen present in the films. There is a strong H^+ peak at 1 AMU and a cluster of hydrocarbon peaks at 26, 27, 28, and 29 AMU. It has been noted by Benninghoven [19] that when multiatom ionic clusters are emitted from the surface, e.g., CH_2^+, these atoms were bond together in the original solid. It has been shown that these films [20], using semiquantitative infrared spectroscopy, the ratio of chemically bonded hydrogen to carbon is between 0.03 and 0.44. However, nuclear reaction and combustion analysis [21], which are in good agreement, indicate that the H/C atom ratio is close to unity. This difference between the two measurement techniques may arise from the presence of nonbonded hydrogen in the films.

The electron diffraction pattern of the films generated using either the single ion source or dual beam method was found to be characteristic of an amorphous solid. Angus, suggested that these films consist of a structure that is a random network of methylene and double bonded carbon linkages and tetrahedrally coordinated carbon atoms [20].

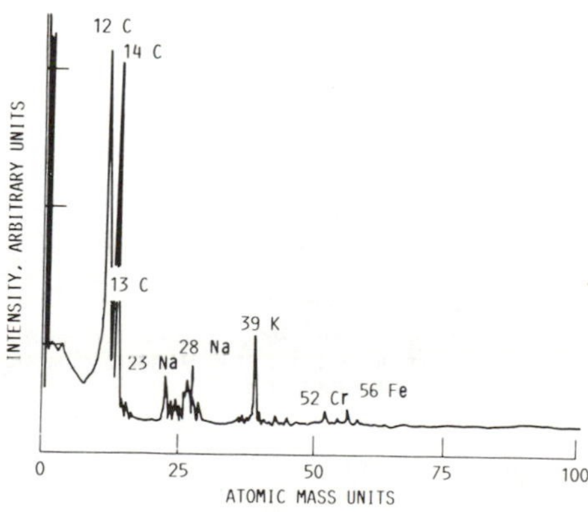

FIGURE 5. - SECONDARY ION MASS SPECTRUM OF ION-DEPOSITED CARBON FILM ON SILICON AFTER 17 MINUTES OF ARGON ION SPUTTERING.

OPTICAL PROPERTIES OF FILMS

The spectral transmittance, reflectance and absorptance of the films deposited on fused silica were obtained using the Gier-Dinkle integrating sphere and the techniques described in reference 22.

Shown in figure 6 is the spectral transmittance for DLC films varying in thickness from 800 Å to 3300 Å. These films were obtained using the dual beam ion source. Figures 7 and 8 show the corresponding spectral reflectance and absorptance for these same films. At short wavelengths the films all show a large decrease in transmittance with a corresponding large increase in absorptance. For film thicknesses between 800 and 1500 Å there are only small differences in transmittance at all wavelengths. Increasing the DLC film thickness to 3300 Å has only a small effect on the transmittance for wavelengths greater than 8000 Å, but reduces the transmittance to as low as 10 percent at 4000 Å. Most of this transmittance loss for the 3300 Å thick film is due to the corresponding increase in absorptance (figure 8). The thinner films (800 to 1500 Å) look clear to yellowlike in appearance and the thicker 3300 Å film brown.

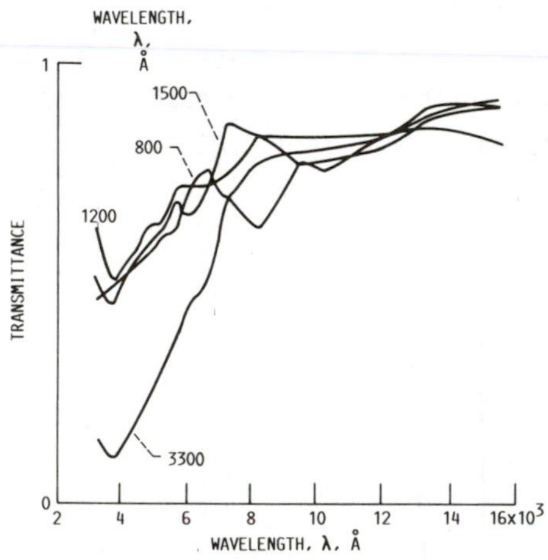

FIGURE 6. - TRANSMITTANCE VERSUS WAVELENGTH FOR VARIOUS THICKNESS DLC FILMS GENERATED USING CH_4 IN THE DUAL BEAM SYSTEM.

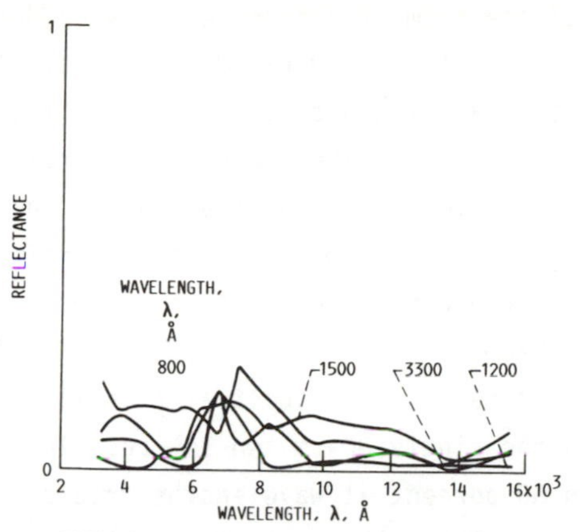

FIGURE 7. - REFLECTANCE VERSUS WAVELENGTH FOR DLC FILMS GENERATED USING CH_4 IN THE DUAL ION BEAM SYSTEM.

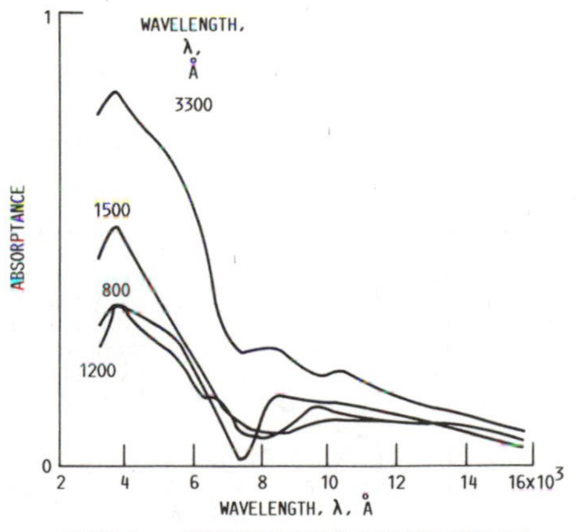

FIGURE 8. - ABSORPTANCE VERSUS WAVELENGTH FOR DLC FILMS GENERATED USING CH_4 IN THE DUAL BEAM ION SOURCES.

The spectral transmittance, reflectance and absorptance for the films generated using the single ion source are similar to those generated using the dual beam source, but have a lower spectral transmittance for films greater than 1200 Å thick. This is evident in figure 9 where the spectral transmittance is shown for two DLC films of similar thickness (1500 Å) generated using CH_4 in the dual beam and single ion source. Also shown in figure 9 are spectral data for DLC film 1700 Å thick, obtained by ion beam sputtering of a carbon target [5]. The transmittance was measured only between 4000 and 8000 Å for this film and is very low when compared to the CH_4 deposited films. The low transmittance of the film of reference 5 may be due to its low hydrogen content, in comparison to the CH_4-derived DLC films. The 1500 Å thick dual beam film has greater transmittance at all wavelengths when compared to the 1500 Å thick single beam film. The increased absorption most likely arises from the presence of systems of conjugated double bonds within the film although the presence of oxygen could also play a role. Both the graphitic precursors and oxygen would be expected to be reduced by the increased sputtering from the second beam.

Also shown in figure 9 is the spectral transmittance for a 500 Å film generated using CH_4 in the single 30-cm ion source. This film has transmittance values greater than 90 percent at wavelengths greater than 7000 Å. Since the transmittance is increasing with wavelength, these films could have use in high energy laser applications as a chemically inert encapsulant for coated

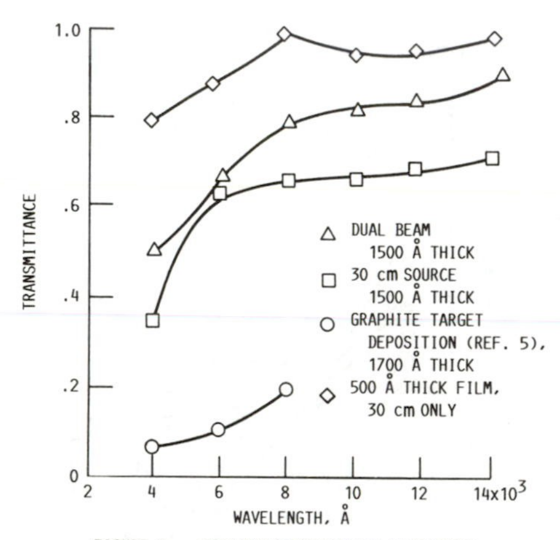

FIGURE 9. - TRANSMITTANCE VERSUS WAVELENGTH FOR DLC FILMS USING CH_4 IN DUAL BEAM OR SINGLE ION SOURCES AND GRAPHITE TARGET DEPOSITION OF REF. 5.

optics. If the transmittance of the films with high hydrogen constant could be improved further, they could find use as integral cover slides for solar cells. This might be accomplished by adding hydrogen to the films during the deposition process or by increasing the energy level of the second ion source to obtain better sputter removal of oxygen and conjugated, graphitic structures.

The IR transmittance of the DLC films was measured by depositing the films on an IR transmitting material, ZnSe. Shown in figure 10 is the IR transmittance of 1000 Å DLC film plus an intermediate layer of 300 Å of Ge, which allows for good DLC adherence on ZnSe; this is presented along with uncoated ZnSe. Although there is a reduction in transmittance at shorter wavelengths, only a 1 percent loss occurs at 10 μm. This reduction could be eliminated by picking the proper DLC and Ge thicknesses to allow the combination to become an antireflective coating.

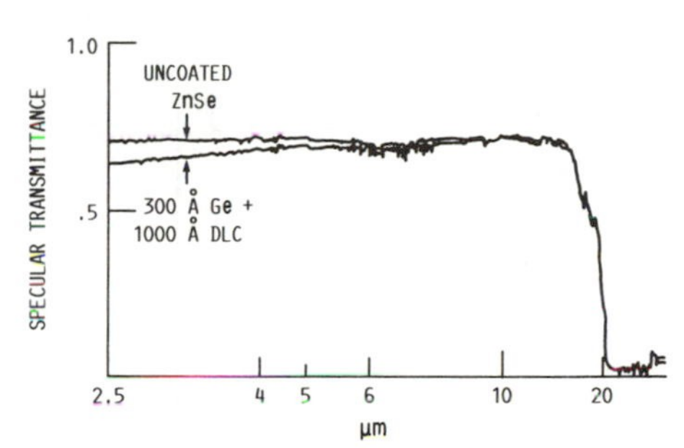

FIGURE 10. - INFRARED TRANSMITTANCE OF 1000 Å DLC + 300 Å Ge ON ZnSe AND UNCOATED ZnSe.

Listed in table I are values of H/C ratio, optical band gap, index of refraction, absorption coefficient and solar transmittance for CH_4-derived films generated using the single or dual beam ion sources, and, sputter deposited films of reference 5. Certain properties are enhanced by using a dual beam system, mainly the solar transmittance which is higher, as is the index of refraction, and a lower absorption coefficient, when compared to the single ion source films. The optical band gap and density for the single and dual beam system films are similar. The films obtained by Banks and Rutledge [5] have a higher band gap, but a lower solar transmittance (by a factor of two), when compared to the CH_4 single or dual beam films. There differences between the films of reference 5 and single or dual beam films are probably due to the difference in hydrogen content of the films.

TABLE I. - OPTICAL, CHEMICAL AND PHYSICAL PROPERTIES OF DIAMONDLIKE FILMS GENERATED USING CH_4 WITH SINGLE OR DUAL BEAM ION SOURCES, AND SPUTTER-DEPOSITION OF GRAPHITE

	30 cm CH4/A~ = 28 percent	Dual beam CH4/A~ = 28 percent	Graphite target
H/C ratio	1.0	1.0	Low
Resistivity, Ω-cm	8.66×10^6	3.35×10^6	6.29×10^6
Optical band gap (eV)	0.382	0.343	0.909
Density, gm/cm^3	1.8	1.8	2.2
Index of refraction	2.0	2.46	--------
Absorption coefficient/cm at 5000 Å	5.15×10^4	4.26×10^4	7.26×10^4
Solar transmittance	0.519	0.648	0.134
$t = \frac{\int Q(\lambda)T(\lambda)d(\lambda)}{\int Q(\lambda)d\lambda}$	Film = (1500 Å thick)	(1500 Å)	(1700 Å)
Adherence (Quaztz)	$>5.5 \times 10^7$ N/m2 >8000 psi	$>5.5010^7$ N/m^2 >8000 psi	

Chemical and physical properties. - The DLC single and dual beam films were subjected to solution of 3 parts H_2SO_4 and 1 part HNO_3 (concentrated acids, by volume) at 80 °C for periods up to 20 hr. The films on silicon were unaffected by the reagent. The films on fused silica showed varied behavior. In some cases they were unaffected by the reagent and in others the film was removed from the substrate, but not fully dissolved. These results clearly indicate that the best films, especially those on silicon, are far more resistant to chemical etching than normal polymeric hydrocarbons or graphite. This suggests and the potential use of the DLC films as chemical and/or a diffusion barrier for microelectronic or optical components.

Transmission electron microscopy at 30 000 X showed the films to be smooth and essentially free of features. No pinholes or other defects were observed.

The adherence of the films on quartz was measured followed the procedure used by Mirtich [23]. The adherence of the films generated with either the single or dual beam systems were as good as the maximum adherence of the Sebastian Adherence Tester used in the measurement (~$5.5x10^7$ N/m^2 or 8000 psi). The film adherence was so good for some films, portions of quartz gave way with the film still intact.

Some films deposited with the dual beam on Si have been kept for four years and show no visible signs of deterioration. However, the thickest film, 1.75 μm, deposited using the single ion source, spalled from the substrate within several weeks.

DLC films were deposited on Si and intrinsic stress measurements were

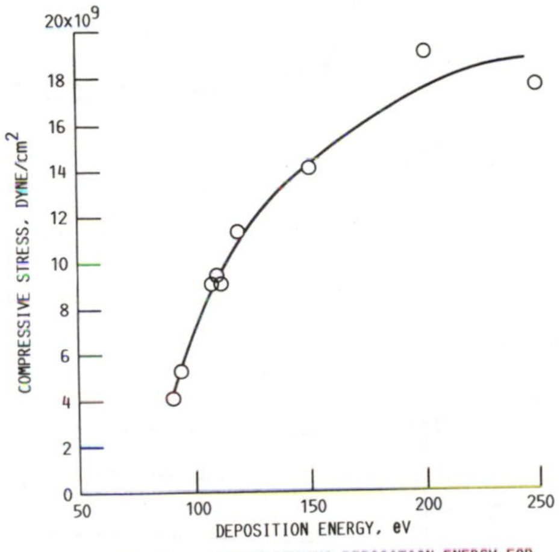

FIGURE 11. - STRESS VERSUS DEPOSITION ENERGY FOR DLC FILMS MADE WITH CH_4 IN THE SINGLE BEAM ION SOURCE.

made using an Ionic System's intrinsic stress gauge. All of the DLC films exhibited a compressive stress which varied depending on the deposition method, hydrocarbon gas, and energy of deposition. Figure 11 shows that the stress in the CH_4 single beam films can be reduced to values as low as $4x10^9$ dyne/cm^2 by decreasing the deposition energy to 90 eV.

Microhardness tests on DLC deposited films were measured using the Vickers diamond indentation method. This method consists of loading a small diamond-tipped stylus with a known weight and allowing it to penetrate freely the surface of the material being measured [24]. Figure 12 is a plot of microhardness versus load for a DLC film 4100 Å thick deposited on silicon. From the data shown in figure 12 the microhardness is seen to increase with decreasing load for the DLC films but not for the silicon. More data of microhardness versus load is shown in figure 13 for RF-sputtered DLC films [25]. For the geometry of the diamond indenter used here [24], the ratio of stylus penetration to film thickness can be shown to vary from ≈10:1 for a 1200 Å film with a 200 gm load to ≈1.21:1 for a 4600Å film with a 25 gm load.

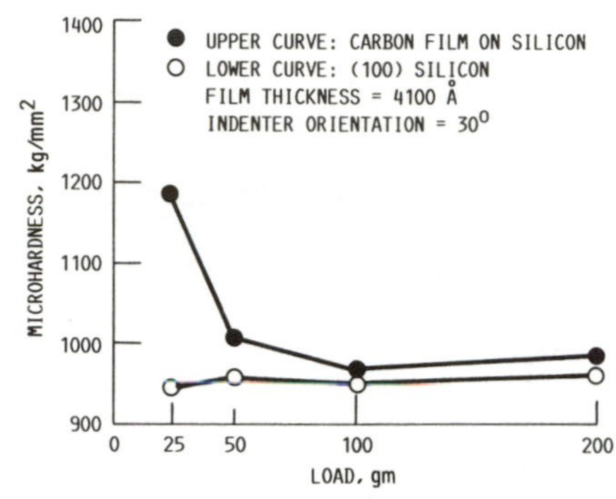

FIGURE 12. - MICROHARDNESS VERSUS LOAD FOR DLC FILM 4100 Å THICK DEPOSITED ON SILICON.

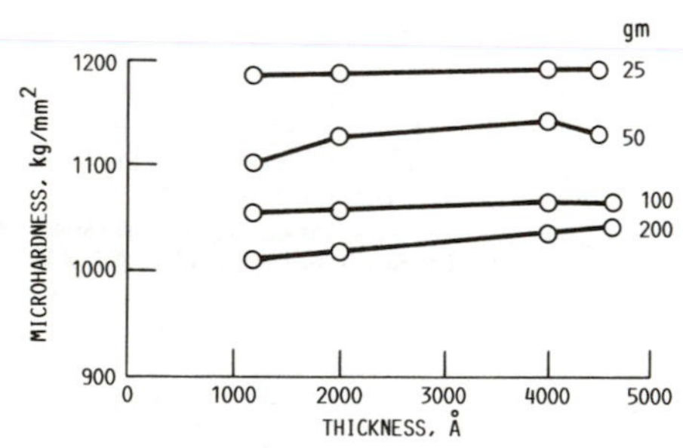

FIGURE 13. - MICROHARDNESS AS A FUNCTION OF LOAD AND THICKNESS.

It is clear from these two figures and the results of reference 24 that the stylus penetrates the DLC films. However, it has been shown in reference 26 that using the data of figures 12 and 13 and the following equation [26]:

$$H_{Film} = H_s \left[1 + \frac{D_s^2 - D^2}{tD4\sqrt{2}\tan(11°)\cos(22°)} \right] \quad (1)$$

where

t film thickness

D_s indent diagonal of substrate

D indent diagonal of film coated substrate

H_s hardness of substrate alone

H_{Film} hardness of film

it is possible to calculate the hardness of the DLC films. Using equation 1 to calculate the DLC film Vickers hardness, a value of 4,600 kgm/mm^2 was obtained. The Vickers hardness of diamond is 10,060 kgm/mm^2. So, DLC films do not have the hardness of diamond, but they are twice as hard as corundum, which has a Vickers hardness of 2,085 kgm/mm^2, a material considered hard. John Angus of Case Western Reserve University in an independent measurement of the dual ion beam DLC films found the microhardness of the films to be around 4 to 5000 kgm/mm^2. This shown in figure 14 along with the microhardness of SiC, BN and diamond.

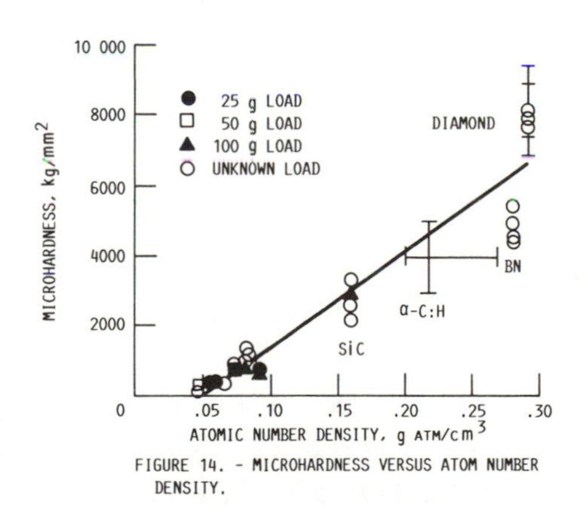

FIGURE 14. - MICROHARDNESS VERSUS ATOM NUMBER DENSITY.

Dr. Angus, using RBS spectroscopy, also found that the diffusion coefficient of Argon trapped in a dual ion beam DLC film deposited on Si, several years old, was less than 10^{-18} cm^2/sec. This diffusion coefficient is 10 orders of magnitude less than for conventional hydrocarbon polymers. This infers these films can be used as a good diffusion barrier.

Dr. John Angus of Case Western Reserve University has defined these dense "diamondlike" hydrocarbon films are a new class of solid. Their very unusual nature is shown in figure 15, in which the atom number density is plotted versus the atom fraction hydrogen. The "diamondlike" solids have atom number densities greater than any other hydrocarbon and fall between crystalline diamond and the adamantanes, which are hydrogen saturated molecular diamonds with 10 and 14 carbon atoms.

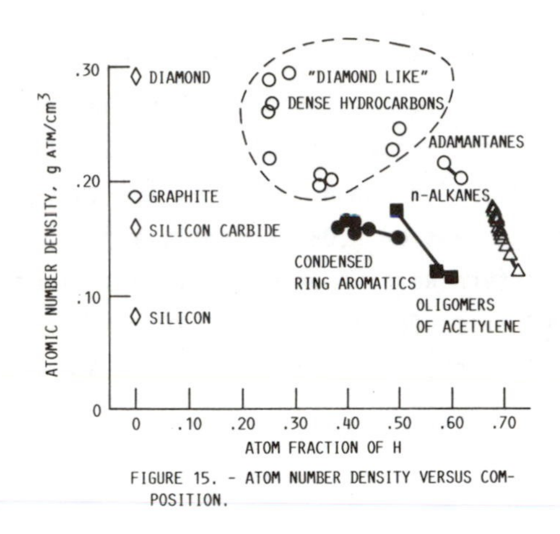

FIGURE 15. - ATOM NUMBER DENSITY VERSUS COMPOSITION.

Electronic Properties. - A systematic attempt to study the electrical characteristics of the DLC films on compound semiconductors and explore their potential as a gate dielectric for insulated - gate technology for very high (1 to 100 GHz) speed integrated circuits was performed. DLC films were deposited on InP, GaAs, and Si and evaluated as potential new electronic material in metal-insulator-semiconductor (MIS) device configurations for solid state power devices for space application [27].

After deposit of the DLC films using dual beam deposition and ion energies between 133 and 200 eV, circular 5000 Å aluminum gate electrodes were deposited on the carbon films from a resistance heated boat at 1×10^{-6} torr to form the metal-insulator (carbon)-semiconductor (MIS) sample. Ohmic contacts were formed by depositing 5000 A of aluminum (100 percent) on the Si back surface and 5000 A of Au-Zn alloy (5 percent) on the InP abd GaAs back surfaces. This was done to minimize resistance with the external circuitry. The contact metallization was sintered at 375 °C for 5 min in forming gas.

The fixed insulator charge density of the MIS system was evaluated by high frequency 1 MHz C-V measurements [27]. The analysis of the C-V data is shown in figure 16 for carbon films on InP, GaAs, and Si substrates. This figure displays the variation of fixed insulator charge density as a function of the ion beam energy at which the diamondlike carbon film was deposited on the substrates. The lower x-axis shows the energy of the single ion beam deposition (30 cm ion source). The upper x-axis of figure 16 shows the energy of the 30 cm ion source beam deposition with 25 mA beam current while simultaneously a beam of 600 eV argon ions with a beam current of 6 mA (dual beam process)

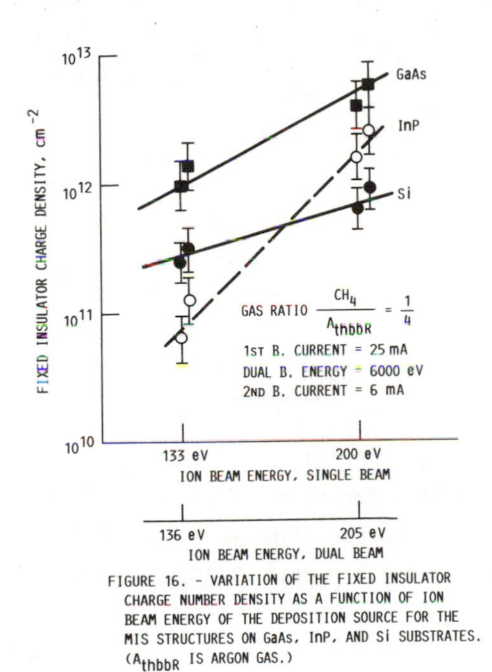

FIGURE 16. - VARIATION OF THE FIXED INSULATOR CHARGE NUMBER DENSITY AS A FUNCTION OF ION BEAM ENERGY OF THE DEPOSITION SOURCE FOR THE MIS STRUCTURES ON GaAs, InP, AND Si SUBSTRATES. (A_{thbbR} IS ARGON GAS.)

was used to sputter the film. The results show that the fixed insulator charge number density increases with ion beam deposition energy for all of the InP, GaAs, and Si substrates.

Analysis of the quasistatic C-V data yielded a U-shaped distribution of interface state density for InP, GaAs, and Si substrates. Figure 17 shows a minimum interface state density in the middle of the band gap of the substrate as a function ion beam energy of the deposition source. The lower and the upper x-axis are the same as for figure 16 and have been described above. The results of figure 17 show that the interface state density as a function of

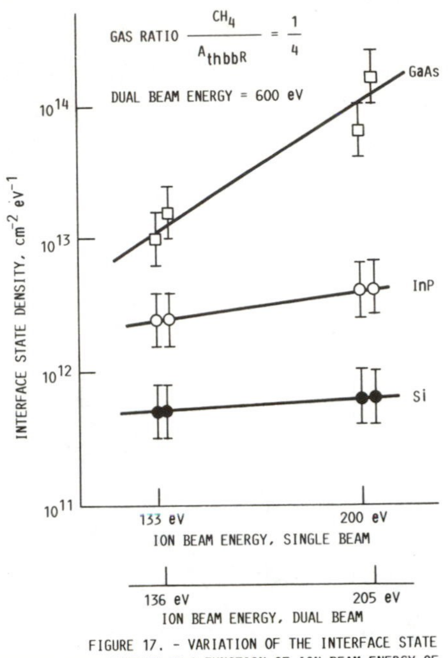

FIGURE 17. - VARIATION OF THE INTERFACE STATE DENSITY AS A FUNCTION OF ION BEAM ENERGY OF THE DEPOSITION SOURCE FOR THE MIS STRUCTURES ON GaAs, InP, AND Si SUBSTRATES. (A_{thbbR} IS THE ARGON GAS.)

ion beam deposition energy increases dramatically for carbon films deposited on GaAs, increases slightly for carbon films deposited on InP, and remains essentially unchanged for carbon films on Si substrates. This increase may be due to damage created by the ion beam at the film-semiconductor interface at higher energies.

These results indicated that these ion beam deposted diamondlike carbon films may not be a promising new electronic material as a gate dielectric for insulated-gate technology application in microelectronics. This is due to low optical band gap (1 eV), low resistivity (8×10^7 Ωcm), very high fixed insulator charge number density (10^{11} to 10^{13} cm^{-2}), and high interface state density (10^{12} to 10^{14} CM^2 $E\bar{V}^1$). In addition, these films generally decomposed at temperatures above 450 °C and, thus, are not suitable for use in microelectronics processing.

SUMMARY

Single or dual ion beam systems using CH_4/Ar gases were used to deposit amorphous diamondlike carbon film on various substrates. Auger spectra of the films show no evidence of any elements other than carbon and small amounts of argon and oxygen. Nuclear activation and combustion analysis indicate that the hydrogen to carbon ratio of the films is close to unity. This high value of H/C, it is felt, allowed the films to have good transmittance from visible wavelengths through infrared and up to 20 μ. The films were impervious to reagents which dissolve graphitic and polymeric carbon structures and suggest potential use as chemical or diffusion barriers. Film adherence on quartz was so good that portions of the quartz gave way with the film still intact. Evaluation of these films as new electronic material show the films to be deficient in electrical properties conducive to microelectronics. These diamondlike films have atom number densities greater than other hydrocarbons and are considered a new class of solid.

REFERENCES

1. Holland, L.; and Ojha, S.M.: Thin Solid Films, 1978, 48, 21.
2. Berg, S.; and Andersson, L.P.: Thin Solid Films, 1979, 58, 117.
3. Spencer, E.G.; Schmidt, P.H.; Joy, D.C.; and Sansalone, F.J.: Appl. Phys. Lett., 1976, 29, 118.
4. Aisenberg, S.; and Chabot, R.: J. Appl. Phys., 1971, 42, 2953.
5. Banks, B.A.; and Rutledge, S.K.: NASA TM-82873, 1982.
6. Weissmantel, C., et al.: Thin Solid Films, 1979, 63, 315.
7. Sovey, J.S.: AIAA Paper 76-1017, 1976.
8. Nir, D.; and Mirtich, H.M.J.: J. Vac. Sci. Technol. A, 1986, 4, 560.
9. Mirtich, M.J.: AIAA Paper 81-0672, 1981.
10. Marinov, M; and Dobrev, D.: Thin Solid Films, 1977, 42, 265.
11. Ojha, S.M.; Norstrom, H.; and McCuluch, D.: Thin Solid Films, 1979, 60, 213.
12. Andersson, L.P.: Thin Solid Films, 1981, 86, 193.
13. Anderson, L.P.; and Berg, S.: Vacuum, 1978, 28, 449.
14. Berg, S.; Gelin, B.; Ostling, M.; and Babulanam, S.M.: J. Vac. Sci. Technol. A, 1984, 2, 470.
15. Berg, S.; Gelin, B.; Svardstrom, A.; and Babulanam, S.M.: Vacuum, 1984, 34, 969.
16. Lurie, P.G.; and Wilson, J.M.: Diamond Research, 1976, 26.
17. Haas, T.W.; Grant, J.T.; and Dooley G.J., III; J. Appl. Phys., 1972, 43, 1853.
18. Jacobi, K.; Surf. Sci., 1971, 26, 54.

19. Benninghoven, A.: Surf. Sci., 1975, 53, 596.
20. Angus, J.C.; Mirtich, M.J.; Wintucky, E.G.: Metastable Materials Formation by Ion Implantation, S.T. Picraux and W.J. Choyke, eds., North-Holland, New York, 1982, pp. 433-440.
21. Angus, J.C.; Stultz, J.E.; Schiller, P.J.; MacDonald, J.R.; Mirtich, M.J.; and Domitz, S.: Thin Solid Films, 1984, 118, 311.
22. Bowman, R.L.; Mirtich, M.J.; and Weigand, A.J.: NASA TM X-52687, 1969.
23. Mirtich, M.J.: J. Vac. Sci. Technol., 1981, 18, 186.
24. Petty, E.R.: Technique of Metals Research, Vol. V, Part 2, Measurement of Mechanical Properties, R.F. Bunshah, ed., Interscience Publishers, 1971, pp. 158-215.
25. Kolecki, J.C.: NASA TM-82980, 1982.
26. Jonsson, B.; and Hogmark, S.: Thin Solid Films, 1984, 114, 257.
27. Kapoor, V.J.; Mirtich, M.J.; and Banks, B.A.: J. Vac. Sci. Technol. A, 1986, 4, 1013.

Materials Science Forum Vols. 52 & 53 (1989) pp. 237-290

CARBON (sp^3) FILM GROWTH FROM MASS SELECTED ION BEAMS: PARAMETRIC INVESTIGATIONS AND SUBPLANTATION MODEL

Y. Lifshitz *, S.R. Kasi ** and J.W. Rabalais

Department of Chemistry, University of Houston
Houston, TX 77204-5641, USA

ABSTRACT

Carbon (sp^3 diamondlike) film growth from mass selected ion beam deposition (MSIBD) is discussed. It is shown that MSIBD is capable of controlling and separating all the relevant carbon deposition parameters unlike other techniques such as plasma deposition, CVD or non-selected ion beam deposition that have a complex phsyical-chemical nature. Experimental results of parametric investigations of carbon MSIBD, emphasizing those obtained from the unique Houston facility (a UHV, MSIBD system with in situ surface analysis instrumentation), are presented. These results are discussed in terms of a new subplantation (subsurface shallow implantation) model supported by computer simulations of the carbon ion-target interaction. The role of ion energy, nature of bombarding species (C, C_nH_m, Ar, H), substrate material, and target temperature in carbon film evolution is evaluated. The capability of MSIBD to grow and tailor films with desired properties (e.g. dense, hard, transparent, metastable films) is emphasized.

*On sabbatical leave from Soreq NRC, Yavne 70600, Israel

**Present address: IBM T. J. Watson Research Center, Yorktown Heights, NY 10598

1. INTRODUCTION

Carbon layers have been extensively deposited during the past three decades in an attempt to fabricate diamond films [1-4]. The work was motivated by the unique set of physical properties of diamond [1-7]: hardest material known, excellent electrical insulator, best thermal conductor, high dielectric strength, highly transparent in the UV, visible and IR regions, chemically inert, resistant to oxidation and corrosion, and compatible with body tissues. The deposited carbon films, however, had properties varying between those of diamond and those of graphite, within a range of many orders of magnitude [3,4,8]. The unique possibility of "tailoring" desired properties for a specific purpose (by a proper choice of deposition technique and parameters) is advantageous for many applications, including [7-11]: optical coatings suitable for hazardous environments, protective films for magnetic recording materials, heat sinks, solid state devices, moisture barriers, low friction coatings, hard coatings for mechanical tools, and protective coatings compatible with body tissues for medical applications.

A variety of techniques have been developed for the deposition of carbon films resulting in extensive data covered by excellent recent reviews and literature surveys [3,4,8,10,12]. Nevertheless, the fundamental understanding of the nature of these films is still limited, due to the complex chemical-physical nature of the deposition techniques that have been used, where intrinsic deposition parameters to be defined later are difficult to define and control. Most of the work has concentrated on characterization of the final properties of the carbon films and very little is known about film growth processes and the relationship between the intrinsic deposition parameters, the structure, and the bulk properties of the deposited films.

The purpose of the present work is to:

- introduce the mass selected (carbon) ion beam deposition (MSIBD) technique which is capable of controlling the intrinsic deposition parameters;
- describe an experimental methodology for investigation of carbon film growth;
- present a subplantation model for carbon film growth from

hyperthermal (~10 eV - 1 keV) carbon species that is relevant to most of the present carbon deposition techniques;

- detail parametric investigations of carbon film deposition (most of them conducted by the MSIBD facility in Houston) that are interpreted in terms of the subplantation model;
- demonstrate the possibility of deposition of films with the short range order of diamond.

2. CATEGORIZATION OF CARBON FILMS, DEPOSITION TECHNIQUES AND INTRINSIC DEPOSITION PARAMETERS.

2.1 Nomenclature and Categorization of Carbon Films

The broad spectrum of properties and the variety of methods for preparation of carbon films has resulted in ambiguous and inconsistent nomenclature [3,13] that does not reflect the true physical structure of these films. Different names have been used for similar films while the same name has been used for entirely different films. It is an often overlooked fact that the field of carbon films covers several pure phases of carbon as well as hydrocarbon compounds. Graphite, the stable hexagonal form, and cubic diamond, the metastable form (stable at high temperature-high pressure conditions) are very well known. Four other metastable carbon phases have however been identified, including hexagonal diamond (Londsdalite)[2,14], a hexagonal phase - Chaoite [15,16], and two other cubic phases [17,18], all stable at the proper high pressure, high temperature conditions. While the crystalline structure of these four additional carbon phases was identified, very little is known about their properties, and present characterizations of carbon films just refer to the known properties of graphite and cubic diamond.

The properties of the different phases of carbon are related to the nature of the carbon bond, i.e. the hybridization state of the carbon [8,19,20]. Cubic diamond has a sp^3 tetrahedral structure where each atom is bonded to four different carbon atoms and no "dangling bonds" exist. Graphite has a sp^2 structure where each C atom is bonded only to three carbon atoms in a two dimensional arrangement, while the remaining p-type orbital forms a "dangling bond" (or a π electron band).

"Amorphous carbon" refers to a carbon matrix with any possible mixture of sp, sp^2, or sp^3 hybridized carbon that has no crystalline long range order. The term "diamond" usually refers to films that possess a true short range sp^3 electronic configuration along with the known cubic or hexagonal long range order. Unlike silicon, the possibility of "amorphous diamond" (i.e. sp^3 short range order with no diamond long range order) is overlooked. "Diamondlike" is a term that is widely and vaguely used to describe carbon films that possess properties (i.e. hardness, optical transmission, electrical resistivity, density) that are similar to those of diamond. A quantitative definition of the range of properties of "diamondlike" carbon (DLC) does not exist. Other terms often used are:

"a-C" - amorphous carbon;

"a-C:H" - amorphous carbon hydrogenated; a dense hydrocarbon film with a hydrogen concentration that varies between 10%-70%;

"i-C" or "i-carbon" - carbon films made by deposition from accelerated ions;

"dense carbonaceous" - films with density ρ>0.24 g/cm^3 that can be either "dense carbon" (pure carbon) or "dense hydrocarbon" (with significant amounts of hydrogen);

"hard carbon(aceous)" - films with hardness that approaches that of diamond.

To summarize, this nomenclature of carbon films reflects the present status of the field: (i) Carbon films have a variety of properties, (ii) Different preparation methods exist; and (iii) There is little fundamental understanding of film growth mechanisms.

2.2 Characterization of Deposition Methods

2.2.1. Basic Approach

Thermal evaporation of carbon results in films with properties of graphite [3,8,21-23] (i.e. high electrical conductivity, high absorption coefficient in the visible and the IR region). Since the dense metastable carbon phases (e.g. cubic diamond) require high temperature-high pressure conditions for their formation [2,7], two basic approaches have been used for deposition of dense carbon films: (a) Energetic species (~10-1000 eV) impinge on the substrate to form carbon films [3,4,9,22]. These species can be carbon containing, or other energetic species (e.g. Ar) that impinge on the substrate simultaneously with another flux of carbon species. The energy associated with these species is expected to form the dense, metastable phases. (b) Chemical reactions involving hydrocarbons and hydrogen, typically on a substrate held at elevated temperatures [1,23-30].

2.2.2. Ion Beam Deposition (IBD)

This term includes all the techniques [3,4,9,10,30] that use low energy (~10 eV - 1 keV) ions for deposition of carbon films. Direct beam deposition involves carbon containing ions that impinge on the target. Mass selected ion beam deposition (MSIBD) of carbon involves the use of mass selected, energy controlled, carbon containing ions as will be fully described in Sect. 3. Sputter ion deposition involves deposition of carbon containing species that are sputtered from a graphite target bombarded with ions (usually Ar). Dual ion deposition or ion assisted deposition involve a source of carbon containing species (either thermal or hyperthermal) that impinge on the target simultaneously with energetic ions (usually Ar) emerging from another source.

2.2.3. Plasma Deposition

Energetic species formed in a plasma decomposition of various hydrocarbon gases bombard substrates placed on a (negatively or positively) biased electrode [3,4,31-35]. RF, DC and pulsed plasma systems have been used. A relatively high hydrogen content in the film is typical for these techniques.

2.2.4. Chemical Vapor Deposition (CVD)

Chemically active hydrocarbon fragments react [1,23-30] on

the substrate to spontaneously grow diamond films under rather unstable conditions. A typical arrangement involves a mixture of methane (<1%) and hydrogen (>99%) where hydrocarbon fragments and atomic hydrogen species are generated by means of an excitation source (hot filament, RF or microwave). Diamond films are deposited on substrates maintained at elevated temperatures (~800-1000°C).

2.3. Intrinsic Deposition Parameters

Rigorous research and fundamental understanding of the growth processes of carbon films require the proper determination and control of the intrinsic deposition parameters. In most practical deposition systems these parameters are very difficult to define, measure, and control. Practical deposition parameters typical for a specific technique (e.g. gas pressure, bias voltage and RF power in a plasma discharge process [3,4,31-35]) are used but they do not reflect true intrinsic physical quantities from which film growth mechanisms can be deduced. This section details different intrinsic deposition parameters relevant to carbon film deposition in most of the present techniques. Detailed experimental results and discussion of the influence of these parameters on carbon film growth will be given in Sect. 5 in terms of the subplantation model.

2.3.1. Source Parameters

2.3.1.1. Impinging Parent Species

Species used for carbon deposition include [3,4,10] carbon ions (negative and positive) and atoms, hydrocarbon radicals and ions, carbon cluster ions (C_n^+), and non-carbon species such as hydrogen and argon. Most carbon deposition processes have an unidentified mixture of several species. Deposition from hydrocarbon species results in a high concentration of hydrogen in the films (a-C:H) [3,4,8]. Argon ions impinge on the targets in most ion assisted deposition processes and in some processes that use Ar sputtered carbon species as the carbon source [3,4,9,10,22,36]. Ar is often believed to preferentially sputter graphitic and amorphous carbon constituents enriching the diamond constituent of the film [3,36]. Hydrogen is believed to stabilize the carbon sp^3 arrangement by saturating "dangling bonds" and by preferentially etching graphitic constituents [1,3,13,37].

2.3.1.2. Ion Energy

Ion energy is the primary parameter that modifies the properties of the carbon films [3], inducing the formation of dense, adhering, hard carbon films of one or more of its metastable phases. The vague notation of "thermal spike" [4,38-40] (i.e. a localized high-pressure, high-temperature region) is often used to explain the modification of carbon film properties by the energetic species. Most carbon deposition systems are characterized by a broad and sometimes unknown energy distribution.

2.3.1.3. Incident Flux

The incident flux is known to have a major influence on the structure of films deposited from thermal evaporation [40-41]. No similar investigation has been made for carbon deposition from hyperthermal species, though the RF power of plasma deposited carbon films has been found to affect their optical properties. Apart from its influence on the film growth mechanism the incident flux can be associated with secondary effects such as: substrate temperature, ambient pressure, charging of insulating targets, and ratio of primary atoms to residual impurity atoms.

2.3.1.4. Angle of Incidence

Several phenomena that affect carbon growth are sensitive to the angle of incidence of the impinging species. The sputtering yield of the target material increases at glancing incident angles [43,44]. The range of the impinging species decreases with the angle of incidence and the reflection coefficient of the impinging species increases (lower trapping efficiency). Plasma deposition processes have a wide range of incident angles while glancing angles are used for Ar impingement in ion assisted or dual beam deposition. Direct ion beam deposition techniques usually use normal incident angles.

2.3.2. Substrate Parameters

2.3.2.1. Nature of Substrate

The features of the substrate may affect the growth of carbon films in different ways. The chemistry of the carbon-substrate system affects the adhesion and the evolution of the films. Carbide forming substrates are more likely to form adherent films and usually have a high reaction probability for carbon. The carbon-substrate solubility and diffusion rates determine the possibility of achieving a pure carbon film rather than a carbon-substrate alloy. Surface roughness induces nucleation of diamond crystallites in CVD processes [23]. The crystalline nature of the substrate (amorphous, polycrystalline, or single crystal, crystal orientation, phase) affects the possibility of crystalline oriented or epitaxial growth of the carbon films. Surface cleanliness (determined by the preparation techniques) is also of importance.

2.3.2.2. Substrate Temperature

IBD and plasma deposition of carbon are usually performed at room temperature while CVD processes typically necessitate elevated temperatures (~800-1000°C) [23]. IBD and plasma deposition on substrates held at elevated temperatures (i.e. 400°C) are often associated with low (<0.1) sticking probability and graphitization of films [3,4]. Substrate temperature is expected to influence the structure and the phase of the films by determination of diffusion rates (of carbon and substrate substitutional and interstitial atoms and of vacancies) and the thermodynamic equilibrium of the carbon-substrate system.

2.3.3. Environment

Most of the carbon depositions are performed at pressures $>10^{-6}$ torr [3,4]. In most cases this is due to the process itself that necessitates a gas pressure in the range ~1 millitorr - 1 torr (e.g. plasma deposition and CVD). Very few systems work under UHV conditions ($p<5x10^{-10}$ torr) needed for preparation of high purity carbon films. Since some of the residual gases are incorporated in the evolving carbon films, a high ratio of deposition flux to residual gas flux is needed. Poor, uncontrollable vacuum conditions may be one of the reasons for poor reproducibility of carbon films in many deposition schemes.

3. MASS SELECTED ION BEAM DEPOSITION (MSIBD)

3.1. Advantages and Disadvantages of MSIBD

Limited understanding of deposition phenomena due to the complex chemical-physical nature of many deposition systems is a general feature not related only to carbon film deposition. Mass selected ion beam deposition has been suggested [48] as an effective technique for fine control of all the intrinsic deposition parameters that were previously discussed. When combined with a UHV deposition chamber and in situ diagnostics it offers many advantages, particularly for carbon film deposition.

- Control of ion source parameters: selection of only one type of carbon containing ion from various possibilities at a specific deposition stage, control of ion energy (e.g. 10 eV - 1 keV), ion flux, angle of incidence and beam size over a wide range.
- Dual ion beam deposition of different species (e.g. carbon ions and hydrogen or argon ions) and simultaneous doping during deposition are possible.
- Target parameters (temperature, in situ pre-deposition and post-deposition treatments, nature of target) are controlled with great flexibility.
- UHV environment provides deposition of pure films on atomically clean substrates and controlled admission of gases.
- In situ parametric investigation of film growth becomes feasible using surface analysis instrumentation.

The technique has several drawbacks, especially for some practical applications:

- The MSIBD equipment is much more expensive than CVD, plasma deposition, or many ion beam deposition systems.
- Current densities are lower than in some other methods, resulting in longer deposition periods.
- The beam size of present instruments is limited to a few cm^2 at most.
- Line of sight is needed for deposition, eliminating the possibility of deposition on substrates of a complex geometry.

Mass selected carbon beam deposition can thus serve several purposes:

- Parametric investigations of carbon film deposition including simulation of other existing deposition methods.
- Development of processes with specific deposition parameters resulting in tailored film properties for specific applications. These processes can be attempted by other deposition methods (CVD, plasma deposition or IBD) depending on commercial considerations.
- Actual deposition of carbon films. Controlled, microfocused dual beams on atomically clean surfaces provide in situ flexibility of changing deposition parameters and capability of doping and/or successive use of different ions for multilayer deposition. These features are most advantageous for microelectronic applications and/or small size optical devices.

3.2. Short Review of Carbon Beam Deposition Using MSIBD

The pioneering work of Aisenberg and Chabot [9,22,49] has demonstrated the possibility of depositing hard, transparent, insulating carbon films from a mixture of 50-100 eV C^+ and Ar^+ ions. A diamond constituent in these films was identified by TEM analysis [22,36]. This work initiated several other groups to use pure, low energy, mass selected carbon ions in an attempt to deposit pure diamond films. Freeman et al. [50] deposited diamond films on existing diamond crystals held at 700°C, using mass selected 900 eV carbon ions in a vacuum of ~10^{-6} torr. Similar results were obtained with high energy C^+ beams (~10-100 keV) by Nelson et al. [51] (at these energies internal growth of diamond is obtained). Chaikovskii et al. [52-54] deposited diamond films in UHV using a mass selected high intensity ion beam system. They obtained small crystalline (~10-100 Å) diamond films on different substrates with ion energies of ~30-100 eV at temperatures of 170-293°K. The true diamond nature of these films was established by ex-situ Auger analysis and by TEM. Large crystalline inclusions (up to 50 μm size) were chaotically arranged in a finely dispersed small crystalline base. Deposition on substrates held at 360°K resulted in graphitic films. The groups of Miyazawa et al. [55] and Antilla et al. [56-59] used higher energies (C^+, 300-600 eV, Miyazawa; C^+ and CH_3^+, 500-1000 eV, Antilla) at a vacuum of ~10^{-6} torr to obtain amorphous films with diamondlike properties. Ishikawa et al. [60-64] used low energy (10-1000 eV) C^- and C_2^- beams in a vacuum of 10^{-6} torr to obtain amorphous diamondlike films. The optimal resistivity and optical properties of the films deposited from C^- ions were achieved in the energy region of ~100-200 eV.

Rabalais et al. in Houston [10,64-73] have developed a unique facility that combines a mass selected carbon ion beam, UHV (≤$5x10^{-10}$) deposition environment, and several in situ surface analysis tools [10,67,73]. In situ parametric studies of carbon film deposition performed with this facility [10,64-73] resulted in a profound understanding of the carbon film growth processes and will be described in the following sections. Complementary thick film carbon deposition was performed in Soreq NRC, Israel using the MEIRA facility, under conditions determined in Houston, but at only 10^{-6} torr. Table 1 lists the properties of some mass selected carbon beam facilities.

3.3 The Houston MSIBD Facility

A mass selected ion beam deposition system consists of an ion source, acceleration system for ion beam transport, mass selection capability for the transmission of only the desired species, deceleration system for achieving and controlling the low ion energies needed for (carbon) film deposition, target substrate assembly, deposition chamber maintained at low pressure (preferably with facilities for in situ measurements) and controlled gas admission.

The Houston facility (Fig. 1) is briefly described as an example. A more detailed description of the same system is available elsewhere [10,67,73]. Low energy ions are produced in a Colutron ion source with the ionization region maintained at 1-300 eV above the grounded sample. This is a plasma type source which produces small size (2 mm diameter) beams of microampere intensity with narrow (0.1-0.2 eV) energy spread. The ions are extracted from the ionization region by means of a drawout plate after which they are accelerated to 1500 eV by a two tube accelerating lens (the drawout plate and accelerating lens are located in region (1) of Fig. 1). All of the beam components following the accelerator except the final deceleration stage float at -1.5 kV. An electrostatic quadrupole doublet (Fig. 1 (2)) receives the beam from the accelerator lens and projects it through an entrance slit to the 60° sector electromagnet mass analyzer (Fig. 1 (5)) that selects the desired m/e species by varying the current through the magnet. The beam emerging from the exit of the mass analyzer is then reshaped and afterwards bent by 6° (Fig. 1 (7)) in order to eliminate the line of sight high energy (up to 1.5 keV) neutral beam that is created by charge exchange with residual gases. This neutral beam could change the properties of the carbon films once it impinges on the target. The final deceleration (Fig. 1 (9)) takes place directly in front of the sample (Fig. 1 (15)), maximizing ion current and minimizing space charge dilation. The beam energy is determined by the source voltage and is variable over the range of 1-300 eV. Typical C^+ current densities of 500 nA/cm^2 for a 0.1 cm^2 beam size are obtained (at ~100 eV). This corresponds to a deposition rate of only 7 Å/hour (assuming a sticking probability of unity and a carbon film density of 3 g/cm^2). Such current densities are impractical for thick film deposition but are ideal for in situ film evolution investigations. The base pressure in the deposition chamber is $3x10^{-11}$ torr, although the pressure in the ion source is ~10^{-5} torr. This is obtained by a four stage (e.g. Fig. 1 (8)) differential pumping scheme that mates the ion beam

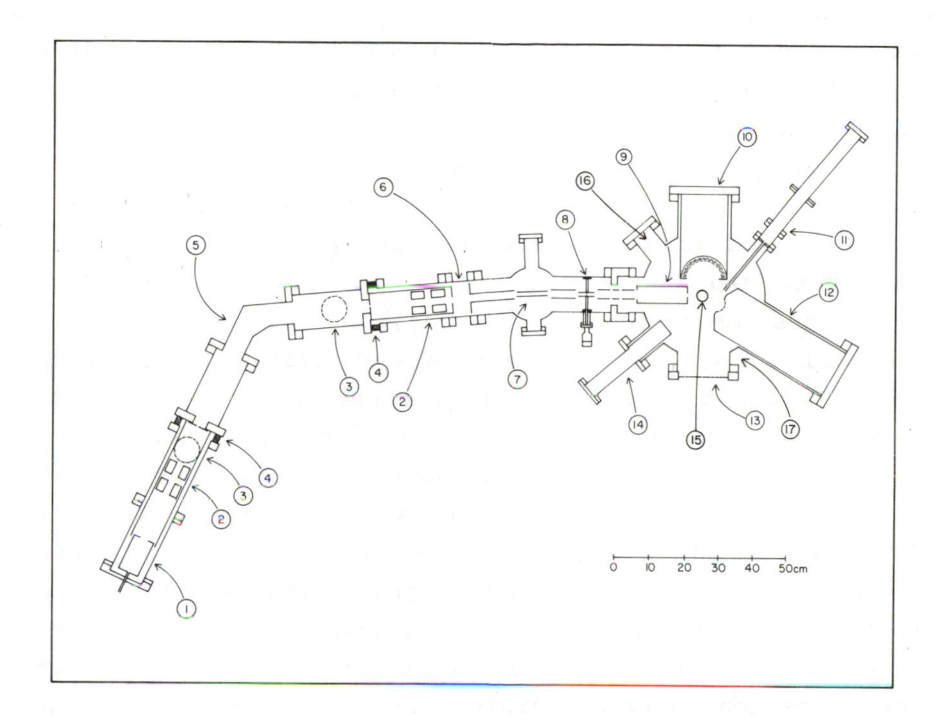

Fig. 1: Houston ion beam facility - Schematic drawing of ion beamline and analysis chamber - top view. The components include: (1) ion source and gas inlet, extraction and acceleration stage, (2) electrostatic quadrupole doublets in entrance and exit of mass analyzer, (3) vacuum pumping port, (4) ceramic insulator for flight tube, (5) 60° sector electromagnet - mass selection stage, (6) gate valve, (7) 6° deflector plates (to eliminate fast neutrals) and turbomolecular pumping stage, (8) rotatable flap serving as a differential pumping buffle and beam apperture, (9) decelerator lens, (10) LEED/AES hemispherical grid analyzer, (11) UPS source, x-ray XPS source located above this source, (12) CMA analyzer and electron gun for AES, (13) viewport, (14) RGA mass spectrometer, noble gas ion source located above this RGA, (15) sample holder and manipulator, (17) gas inlet sources at bottom of chamber. From [10] with permission.

assembly to the deposition chamber. This chamber is an ultra high vacuum (UHV) stainless steel bell jar that is pumped by a combination of a 500 l/sec turbomolecular pump, a 250 l/sec ion pump, and a titanium sublimator along with liquid nitrogen cooled cryobaffle. Additional gases (e.g. hydrogen) can be introduced through UHV compatible leak valves (Fig. 1 (11)). The deposition chamber also provides target annealing (to 1000°C) and Ar^+ sputtering facilities for in situ cleaning of substrates. Annealing is performed by electron bombardment from the back of the substrate and Ar^+ sputtering is performed by means of a commercial gas ion source (Fig. 1 (14)).

Several in situ surface analysis facilities are available inside the deposition chamber (Fig. 1 (10-14)), including Auger electron spectroscopy (AES), x-ray and UV photoelectron spectroscopy (XPS and UPS), low energy electron diffraction (LEED), residual gas analysis (RGA), electron energy loss spectroscopy (ELS), and ionization loss spectroscopy (ILS). Thus, dynamic monitoring of the film growth stages is feasible.

The Houston facility is a basic research system that is not well suited for thick film production due to its low current density. The ion current sharply decreases when the ion energy is reduced due to space charge repulsion resulting in beam profile dilation. Specially designed beam configurations are thus needed for the production of intense ion beams at energies in the range of 10-100 eV. For more intense mass selected carbon beam facilities, the reader is referred to Table 1 and the references therein.

4. EXPERIMENTAL METHODOLOGY FOR CARBON THIN FILM GROWTH

4.1. Methodology of In-Situ Measurements With Surface Analysis Techniques

Most of the carbon film deposition work includes two stages: (i) Preparation of films under different conditions and (ii) ex-situ characterization of these films using numerous characterization techniques. A comprehensive recent review of carbon film characterization is now available [8]. Nevertheless,

Table 1: Mass Selected Carbon Ion Beam Deposition Facilities

System	Ion Energy	Ion Current Density	Beam Size	Pressure During Deposition	In Situ Diagnostics
Freeman [50,74]	100-900 eV	200 $\frac{\mu A}{cm^2}$ (500 V) 100 $\frac{\mu A}{cm^2}$ (100 V)	3 cm^2	10^{-6}	No
Chaikovskii [52-54]	30-100 eV	1-10 $\frac{mA}{cm^2}$	?	$\leq 5x10^{-9}$	No
Miyazawa [55]	100-1000 eV	200 $\frac{\mu A}{cm^2}$ (600 V) 60 $\frac{\mu A}{cm^2}$ (300 V)	1 cm^2	$3x10^{-6}$	No
Anttila [56-59]	500-1000 eV	1 $\frac{mA}{cm^2}$	0.03 cm^2 1 cm^2 at low fluxs	10^{-6}	No
Ishikawa, Takagi [60-63,75]	10 eV-20 keV Negative	100 µA (100 V)	?	10^{-6}	No
Rabalais (Houston Facility) [10,67,73]	1-300 eV	500 $\frac{nA}{cm^2}$	0.1 cm^2	$\leq 5x10^{-10}$	surface analysis techniques (AES, XPS, UPS, LEED)
Soreq NRC	10 eV-20 keV	1 $\frac{mA}{cm^2}$ (100 V)	1 cm^2	$10^{-7}-10^{-6}$	No

the investigation of the growth mechanism in such a two stage approach is very difficult. Rigorous research should include preparation of films with different thicknesses (10 Å-1 μm) for each set of deposition parameters and ex-situ characterization of these films. Such an approach is practically tedious and can be performed only on a limited scale. An excellent example of this approach is the recent CVD work of Kobashi et al [23]. Apart from the tedious work involved in the large number of samples needed, ex-situ measurements may be affected by exposure of the samples to atmospheric pressure, especially when surface analysis is considered. An alternative approach, very widely used in thin film physics, is that of in situ measurements of the evolution of a single film with increasing deposition fluence for each set of deposition parameters [42,76]. This approach, not used before for carbon films, necessitates analysis techniques that can be coupled to the deposition chamber and are compatible with the deposition environment (sometimes UHV conditions). Surface analysis techniques fit very well to these demands and are capable of layer by layer analysis of film elemental composition, chemical nature, short and long range order, and phase. A variety of surface analysis techniques is, however, needed for a complete measurement of all these properties (e.g. Auger Electron Spectroscopy (AES), x-ray photoelectron spectroscopy (XPS), ultraviolet electron spectroscopy (UPS), low energy electron diffraction (LEED), reflection high energy electron diffraction (RHEED), electron energy loss spectroscopy (EELS), ionization loss spectroscopy (ILS), and x-ray induced Auger electron spectroscopy (XAES)). Previous establishment of data from known standards is also necessary. For the case of carbon films establishment of surface analysis data for the different carbon phases as well as specific carbon-substrate compounds is needed.

4.2. Example I - Establishment of Diamond sp^3 Short Range Order of Carbon Films Deposited in the Houston Facility

Differences in the electronic structure of diamond, graphite and amorphous carbon are reflected in their spectroscopic surface analysis features. The surface analysis data for diamond, graphite and amorphous carbon have been established as standards for further identification of the nature of deposited carbon films using AES, EELS, UPS, ILS, XPS, and XAES. Published results [77-80] were compared to those obtained in Houston and missing data were complemented. Fig. 2 shows the surface analysis data obtained with AES, XPS, UPS, and EELS for diamond, graphite, thermally evaporated amorphous carbon, and carbon films ~100 Å thick deposited on Ni(111) at room temperature from 75-

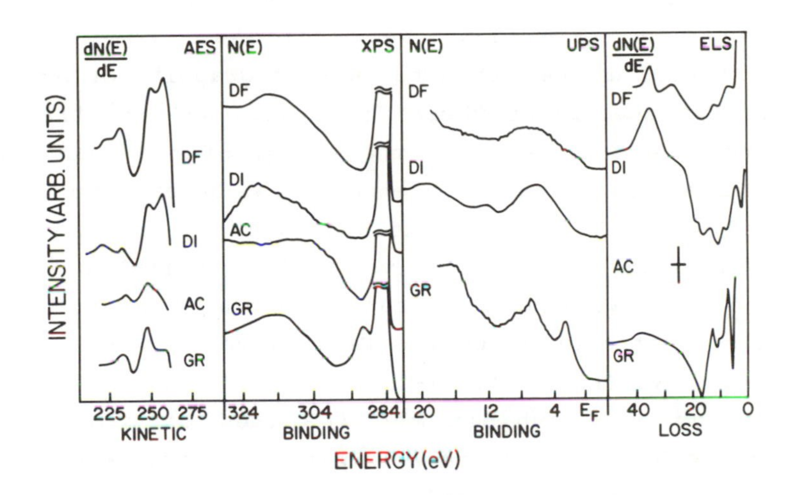

Fig. 2: Evidence for sp^3 (diamond) short range order of films deposited at room temperature by low energy (~60-180 eV) C^+ bombardment. Comparison of surface spectroscopic signatures of carbon in graphite (GR), amorphous carbon (AC), diamond (DI) and diamond films deposited using the Houston facility (DF), measured by AES, XPS, UPS and EELS. No information is presented for the UPS spectrum of AC. The vertical line drawn for AC in EELS indicates the energy loss position of the bulk plasmon peak. The data for GR, AC and DI in AES, and DI in EELS are from [77] while DI in UPS is from [80]. From [69] with permission.

150 eV C^+ ions using the Houston facility. EELS and UPS of amorphous carbon are not presented. These data have been discussed in detail elsewhere [68]. The features of the three different forms of carbon are different in all of these techniques, each of them being easily distinguishable. The differences lie in the lineshapes (e.g. in AES), the appearance or disappearance of peaks corresponding to π electrons (e.g. XPS,

UPS, and EELS), and the energy shift of corresponding loss peaks (e.g. XPS and EELS). It is therefore possible to identify the carbon form of the films deposited in the Houston facility.

In all four surface analysis methods, the spectra of the deposited films correspond to that of diamond and are different from that of graphite or amorphous carbon. In the AES spectra mild current conditions are needed to avoid electron beam damage typically observed for diamond surfaces [68]. Under routine experimental conditions, spectra of electron beam damaged diamond films are measured (Fig. 3). Nevertheless, these are more similar to that of diamond than to that of graphite or amorphous carbon. The XPS spectra of these films show no π-π^* transition at 6-7 eV from the C 1s peak and the energy of the bulk plasmon loss peak is similar to that of diamond (i.e. 34 eV from the C 1s peak compared to 28 eV for graphite and 23 eV for a-C). The UPS spectra of the films are indicative of the typical sp^3 hybrid orbitals of diamond with no contribution in the π band region and a recession of states close to the Fermi level. No UPS spectra were, however, available for a-C. The EELS spectra of the deposited films also exhibit the features of diamond, are different from graphite, and the bulk plasmon energy loss peak is at higher energy than that of a-C.

This example demonstrates the possibility of in situ conclusive determination of the diamond-sp^3 short range order of carbon films deposited on various substrates at room temperature in the Houston facility. This result is different from previously reported "DLC's" that exhibited surface analysis features of amorphous carbon [3,8,81].

4.3. Example II - Determination of Stages of Evolution of Carbon Films

The establishment of surface analysis reference data was the first step in the study of the carbon film growth process. Fig. 3 exhibits carbon AES and EELS lineshapes of Ni substrates bombarded with 75 eV C^+ ions at different fluences [68,69,73]. The carbon film evolves from a carbidic via a graphitic to a final diamond stage. The graphitic stage, associated with the disappearance of the Ni EELS line (the bump at ~65 eV in the EELS spectra), is more obvious in the EELS data since the AES lineshapes at this stage still show some contribution from the carbide layer. The significance of this three stage evolution sequence, along with additional experimental results, will be discussed in Sect. 5 in terms of the subplantation model.

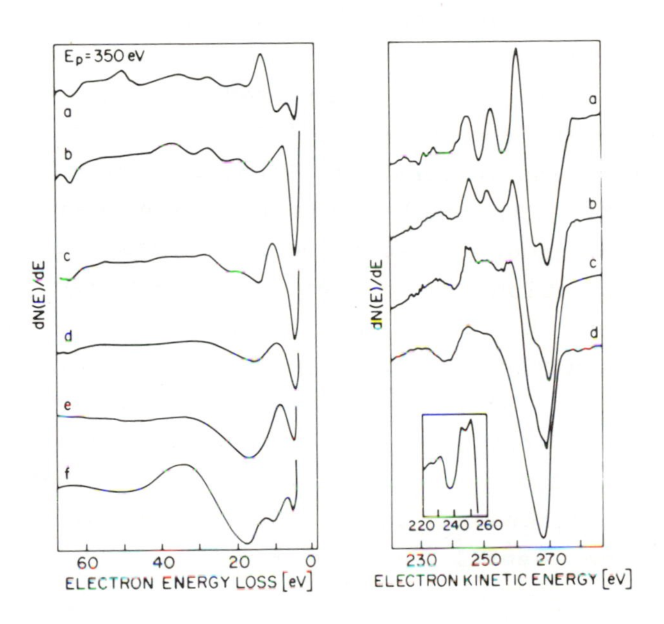

Fig. 3: Evolution of the EELS (left) and AES (right) lineshapes for 75 eV C^+ deposition on Ni(111). The doses in the EELS sequence are: (a) clean Ni, (b) 3 x 10^{15}, (c) 5 x 10^{15}, (d) 7 x 10^{15}, (e) 9 x 10^{15}, and (f) >2 x 10^{16} C^+ ions cm^{-2}. The doses in the AES are: (a) 2 x 10^{15}, (b) 6 x 10^{15}, (c) 9 x 10^{15}, and (d) >2 x 10^{16} ions cm^{-2}. The inset shows an Auger electron spectrum of a diamond film subject to minimal electron beam damage. From [69] with permission.

5. SUBPLANTATION MODEL OF FILM GROWTH FROM HYPERTHERMAL CARBON SPECIES

5.1. Introduction and Basic Assumptions

Carbon film growth mechanisms from hyperthermal species are presently poorly understood. Two main notations are used to explain the formation of dense, hard, transparent, insulating carbon films that are so different from those obtained from thermal evaporation: (i) "preferential sputtering" [3,36] of graphitic or amorphous carbon constituents, enriching the diamond component of the film and (ii) "thermal spikes" [4,38-40], localized, highly excited regions (due to the projectile penetration into the evolving matrix), that provide the high temperature-high pressure conditions necessary for the formation of a carbon metastable phase. Conventional surface growth mechanisms and notations are also used [42,76]. This section presents a recently proposed model [70,82,83] that discusses the carbon film evolution in terms of a shallow implantation process ("subplantation"). The model is based on experimental data (with emphasis on surface analysis measurements) and classical trajectory simulations of the projectile-target interaction (using TRIM, the Biersack Ziegler Monte Carlo program for classical trajectories of atoms) [44,84]. For a more detailed treatment of the model, the reader is referred elsewhere [70,82,83]. This section first describes a general outline of the model. Following, the role of different intrinsic deposition parameters is discussed in terms of the proposed model. Supporting experimental results and TRIM calculations are presented.

5.2. Outline of Subplantation Model

It is suggested that carbon film growth from hyperthermal species is a subsurface shallow implantation process that progresses along the following steps:

(a) Penetration of the impinging species into subsurface layers of the bombarded target. The penetration depth, trapping efficiency, and distribution of the species in the target depend on the subplantation scheme, i.e. type and energy of the impinging species and nature of target material. Some of the impinging species may be reflected and do not contribute to the net deposition process.

(b) Stopping of the energetic species in the target matrix by three principal energy loss mechanisms: atomic displacements, phonon excitations, and electron excitations.

(c) Occupation of a site in the host matrix that serves as a mold for the structure of the film to be formed.
(d) Increase in the concentration of the penetrating species in the host matrix resulting in the formation of an inclusion of a new phase and outward expansion of the subsurface layer (internal subsurface growth).
(e) During the early stages of film growth, the surface is mainly composed of substrate atoms while the penetrating carbon species mainly occupy subsurface sites. The substrate surface atoms are gradually sputtered and diluted by ion mixing until a surface consisting of only projectile species evolves. This is in contradiction to surface deposition processes where the substrate atoms are simply covered by the impinging thermal species.
(f) The phase and structure of the evolving film are determined by several effects: (i) The "mold" effect of the host matrix that determines the possible site occupancies of the penetrating species and places constraints on the initial evolution of the new phase (the structure of an inclusion embedded in a surrounding matrix is highly influenced by the boundary conditions imposed by that matrix); (ii) Preferential displacement of atoms with low displacement energies (LE_d) leaving atoms with high displacement energies (HE_d) in their more stable positions; and (iii) diffusion rates of vacancies and interstitials created in the deposition process. At low enough temperatures the carbon interstitials are frozen in their positions, a high concentration of interstitials is formed, and an athermal spontaneous transformation to a new phase occurs.
(g) Epitaxial growth and/or preferred orientation of films on crystalline materials is expected to result from: (i) the mold effect of the host matrix discussed in (c) and (f); (ii) the angular dependence of the displacement probability due to different E_d's needed for recoil along different crystal directions [85]; and (iii) the sharply defined incident angle of the impinging species that may result in channeling along specific crystalline directions.
(h) The surface features of the evolving film and the efficiency of the deposition process depend on the sputtering yield by the impinging ions. Low sputtering yields are needed for efficient

deposition and are necessarily less than unity for film growth to occur at all.

(i) Evolution of a pure carbon film on a non-carbon substrate is possible only when C-substrate collisional ion mixing and diffusion processes are small. In this case the "deposition" process has two distinct stages: (i) "heterodeposition" and evolution of a pure layer as described in (a)-(e) and (ii) "homodeposition" of energetic carbon species impinging on a pure carbon film.

5.3. TRIM Calculations of C-Target Interaction

Calculations of hyperthermal carbon penetration into the different substrates are needed for the analysis of a specific carbon deposition scheme. TRIM [84] (the Monte Carlo simulation program of Ziegler and Biersack) calculations were performed with the following assumptions: (a) an amorphous target material; (b) a binary collision approximation; (c) each trajectory is a single event, uncorrelated with trajectories of other species; (d) chemical effects (e.g. chemical reaction, diffusion) are neglected. Most of the calculations were performed assuming a low carbon fluence where the dynamic evolution of the composition of the target during bombardment was neglected. These calculations yield several physical quantities relevant to the deposition process including [70,71,83]: (i) range of carbon species; (ii) carbon distribution profile; (iii) backscattering (BS) yield (sticking probability equals 1-BS); (iv) sputtering yield; (v) damage (no. of atomic displacements per impinging C); and (vi) relative importance of the three different energy loss channels. Experimental values of physical parameters needed for the calculations were used when available (e.g. displacement energy, surface binding energy). Some dynamic calculations [86] were also performed [70,83], taking into account the changes in the composition of the target. These calculations yield the actual carbon concentration profile in the evolving film.

Fig. 4 shows results of TRIM calculations [71] of impingement of C^+ ions on C (diamond, graphite, or amorphous

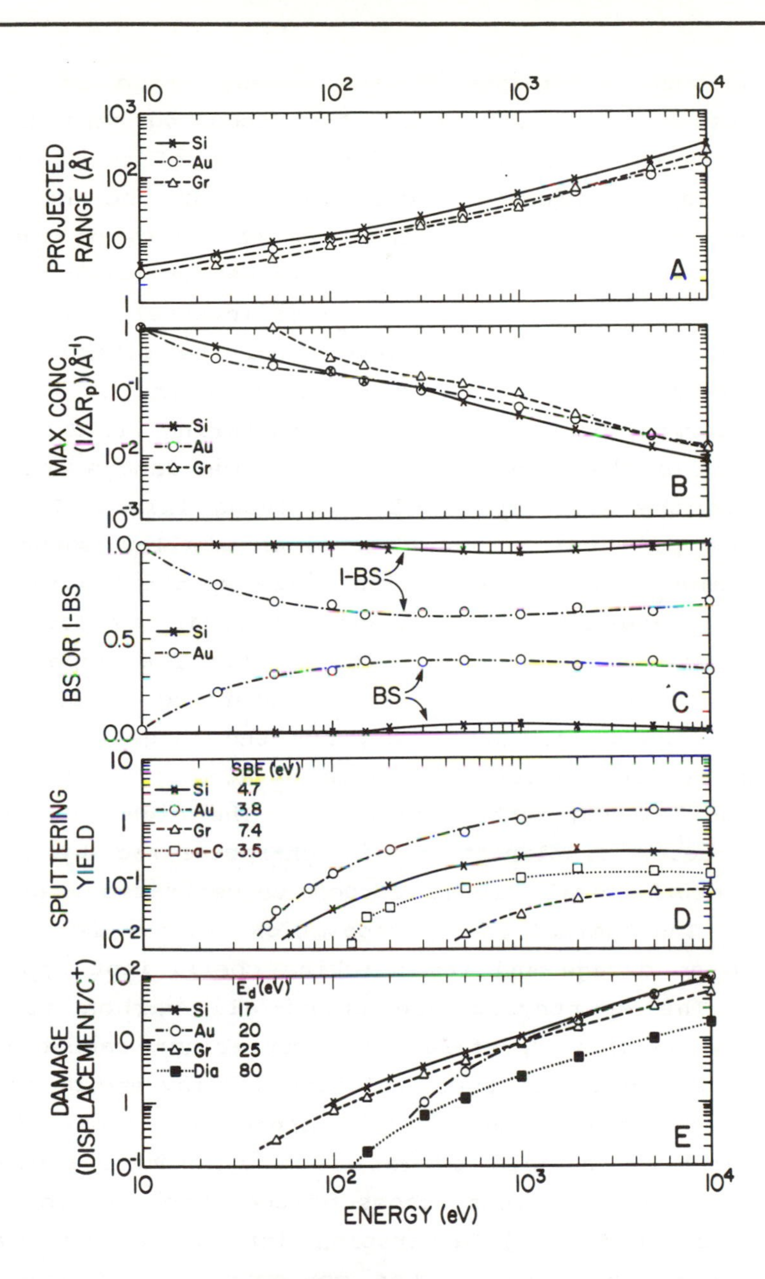

Fig. 4: TRIM [84] calculations of physical phenomena involved in deposition from the hyperthermal species 10 eV-10 keV C^+ ions on C, Si and Au: (A) range, (B) maximum local concentration ($1/(\Delta$ Rp); Δ Rp - range straggling), (C) backscattering efficiency or trapping efficiency (1 - backscattering efficiency), (D) sputtering yield, (E) damage (number of displacements per impinging C^+ ion). GR = graphite; a-C = amorphous carbon; Dia = diamond; SBE = surface binding energy; Ed = displacement energy. From [71] with permission.

carbon), Si, and Au targets in the energy range of 10 eV - 10 keV. These results will be briefly discussed here and more specifically in the following sections. Each of the calculated quantities has a different significance for the deposition process. The range (R_p) determines the thickness of substrate atoms that should be removed before a pure carbon layer is formed on the substrate. The straggling (ΔR_p) is inversely proportional to the local concentration and determines the minimum fluence needed for the formation of a carbon inclusion in the substrate. The backscattering coefficient and the sputtering yield determine the efficiency of the deposition process. The damage determines the phase and the structure of the evolving film. Successful deposition of pure carbon films involves shallow ranges, low straggling (high local concentration), low sputtering yield and backscattering, and controlled damage. Several conclusions can be immediately drawn from Fig. 4: (i) The hyperthermal (~50 eV - 500 eV) energy region is characterized by penetration to <u>subsurface</u> layers (subplantation) but the range is still low and the local concentration obtainable is high. In this region the damage is low and controllable; (ii) The high (~10 keV and up) energy region (implantation) is characterized by a large penetration depth, small local carbon concentration, and high damage; (iii) The thermal energy deposition (<1 eV) is a surface process with no damage and its sticking coefficient approaches unity; (iv) The sputtering yields of all carbon forms is extremely low so that "preferential sputtering" cannot play a significant role in carbon film evolution for low energy ions.

Fig. 5 shows the evolution of a carbon film from 150 eV C^+ ions impinging on a gold target from dynamic calculations [44,70,83,86]. The different stages of the subplantation growth are clearly observed: (i) Penetration into subsurface layers; (ii) Increase of the local carbon concentration of subsurface layers leaving a "skin" of gold surface atoms; (iii) Sputtering and dilution of the gold surface atoms until a pure carbon layer is formed; and (iv) Growth of a pure carbon film.

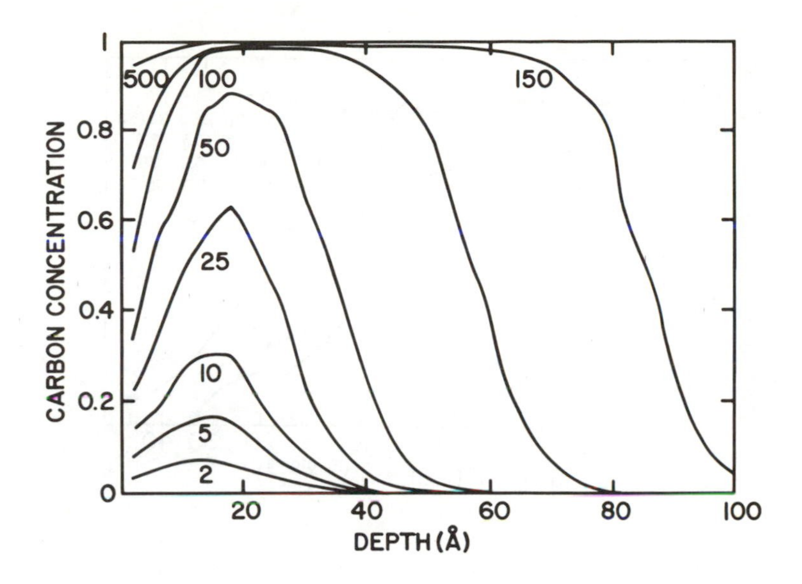

Fig. 5: Dynamic TRIM calculations [86] of C evolution with fluence from 150 eV C^+ impingment on Au. The fluence is given in $x10^{15}$ C^+/cm^2. Note the subsurface entrapment of the C^+ and the high fluence needed to remove the Au surface layer.

5.4. Evolution of Carbon Film With Fluence in a Subplantation Process

Surface analysis techniques are useful in the determination of the carbon film growth processes through detection of several features: (a) the intensities of the signals from the substrate and the deposited carbon and (b) the chemical nature of the carbon derived from the carbon lineshapes.

Fig. 6 shows the evolution of the substrate (Si, Ni, Au) and the C AES intensities with increasing 150 eV C^+ fluence [22,82,83]. The substrate AES intensities remain almost constant for a C^+ fluence equivalent to a layer thickness of more than twice the escape depth of the AES substrate electrons. Conventional surface deposition coverage of the same surface would result in an initial rapid decrease of the substrate AES intensities, as also indicated in Fig. 6 for Ge thermal evaporation on Si [81].

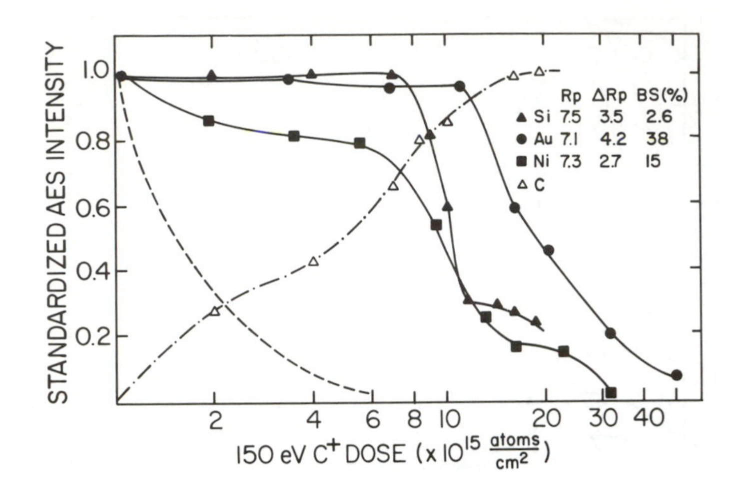

Fig. 6: Substrate AES peak intensities (solid lines) as a function of 150 eV C^+ fluence at normal incidence. Ni (61 eV), Si (92 eV), and Au (69 eV) AES transitions are used. The peak intensities are normalized to that of the clean surface. The dashed line corresponds to Si AES intensity as a function of thermally evaporated Ge dose [87]. The dashed-dotted line corresponds to the carbon KLL AES peak intensity specifically for the Si host. TRIM results include: R_p - Projected range (10^{15} atoms/cm^2), ΔR_p - Range straggling, and BS - Backscattering yield. From [82] with permission.

The C KLL AES intensity, however, increases consistently during the deposition, indicating that the carbon ions are trapped in the substrate. These observations can only be explained by penetration of C^+ ions into subsurface layers (R_p ~10 Å) and internal growth of the new carbon phase until the surface substrate atoms are sputtered and diluted, in accord with the subplantation model and the TRIM calculations.

Another question to be addressed is the phase evolution of the carbon film with carbon fluence. It was demonstrated in a previous section that the film evolves from a carbidic through a graphitic to a final diamond phase. Fig. 7 [82] schematically correlates this phase evolution with the corresponding carbon profile evolution in the subplantation process. The initial

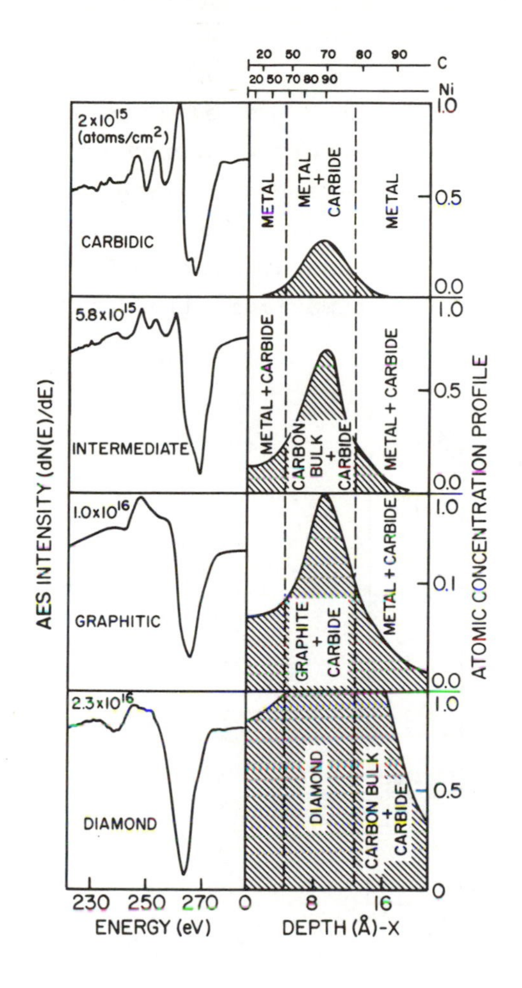

Fig. 7: Stages of subplantation growth. Left column - C KLL AES lineshapes for different C^+ fluences for 150 eV C^+ on Ni (111). Right column - subsurface entrapment of energetic carbon and buildup of carbon deposits. The top scale indicates the relative contribution to the AES intensity (%) derived from a layer of depth x (lower scale).

carbide phase is due to the penetration of carbon into the substrate matrix (M), forcing C-M bonds (a carbide) even when no stable carbides are known to be formed (e.g. Au). A further increase in the local carbon concentration at higher C fluence forms a two dimensional carbon layer (graphite). Higher local carbon concentration forms a three dimensional bulk carbon phase - graphite or diamond, depending on the deposition conditions, as will be discussed in the next section.

5.5. Role of Energy in Carbon Subplantation - Preferential Displacement

Ion energy is the crucial parameter that affects the evolution of carbon films with modified properties compared to thermally evaporated carbon films. Experimental data from several laboratories show that diamond or diamondlike films are formed from impingement of C^+ ions in the energy range of ~50-200 eV [3,22,52-54,64-73] (Fig. 8) and that maximum film density

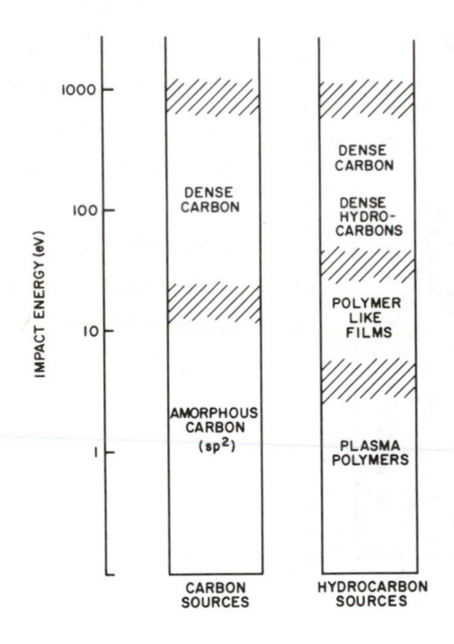

Fig. 8: Influence of impact energy on type of film produced. From [3] with permission.

and electronic band gap are achieved for E ~ 100-200 eV (Fig. 9) [60-63]. Fig. 10 [68] shows a "phase diagram" of the fluence needed for the evolution of a carbidic (A), an intermediate (B), and a graphitic (C) stage for C^+ impingement on Ni(111) at room temperature. Much higher fluences are needed for the formation of the graphitic stage for $E < 30$ eV and $E > 180$ eV while the diamond stage was not reached for the practical fluences obtainable in the experiment. TRIM calculations (some

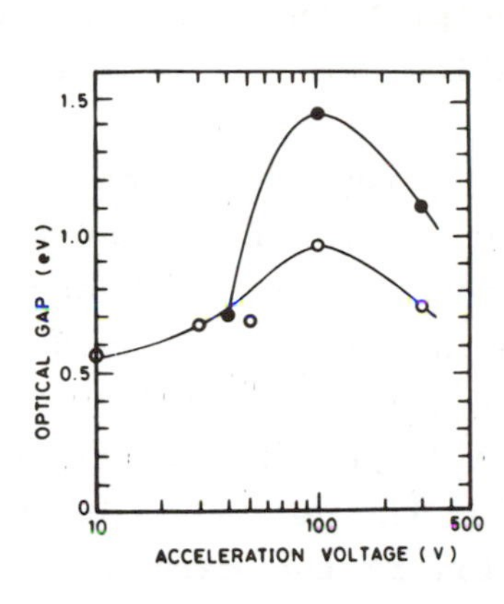

Fig. 9(a): Optical gap of the carbon film deposited by the C^- (O) and C_2^-(●) ion beams as a function of the acceleration voltage. From [60] with permission.

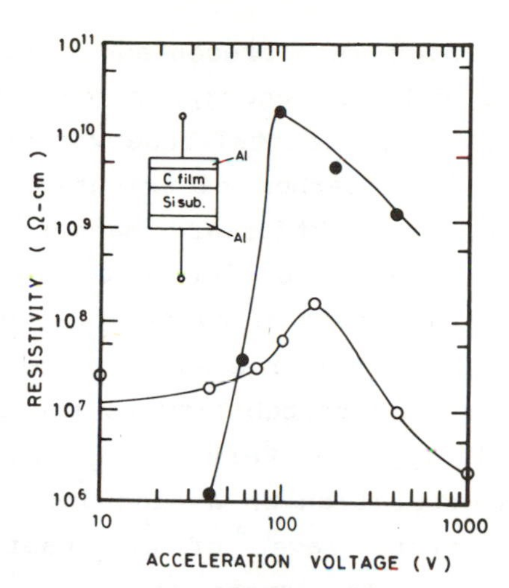

Fig. 9(b): Electrical resistivity of the carbon film deposited by the C^-(O) and C_2^-(●) ion beams as a function of the acceleration voltage. From [60] with permission.

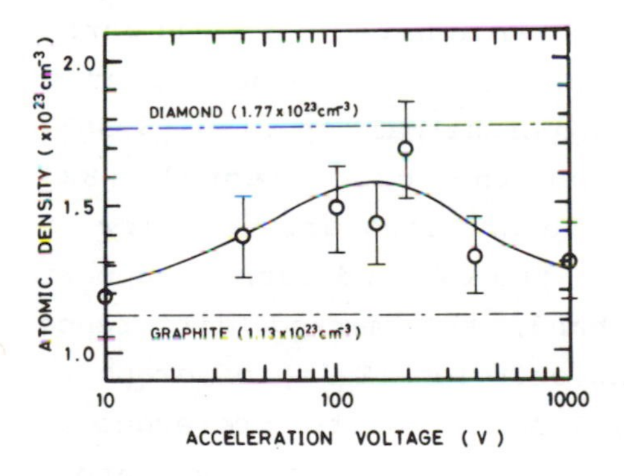

Fig. 9(c): Atomic density of the carbon film deposited by the C^- ion beam as a function of the acceleration voltage. The atomic densities of diamond and graphite are indicated in the figure. From [60] with permission.

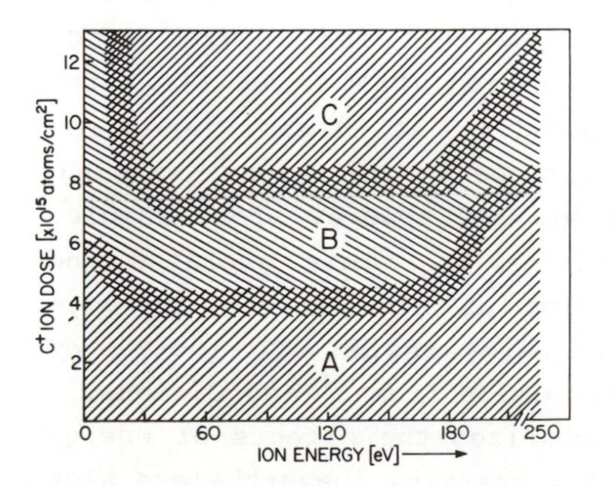

Fig. 10: Ion dose/ion energy phase diagram of C^+ ions impinging on Ni(111). (A) carbidic phase, (B) intermediate phase, (C) graphitic phase. Transition to (D) - diamond phase occurs at higher doses. From [68] with permission.

of which were presented in Fig. 4) are helpful in evaluation of the role of energy in carbon subplantation and in explaining these experimental observations.

The carbon distribution profile is energy dependent in a very straightforward manner: the carbon range increases with energy and the distribution profile broadens. A minimal energy (several tens of eV or more, depending on the target involved) is needed for a true subsurface entrapment of the carbon and a small migration probability to the surface. For high energies (~1 keV and up), a very high carbon fluence is needed for the incorporation of a high local concentration of the carbon and a substrate layer of at least ~30-50 Å should be sputtered or diluted to expose the pure carbon layer. This explains why higher fluences are needed for the formation of the graphitic stage in Fig. 10 at higher C^+ energies.

Of special interest to the structure evolution of the carbon film are the sputtering yield and damage. Preferential sputtering of amorphous carbon and graphite constituents, enriching the diamond component of the film was suggested as a mechanism for diamond growth from hyperthermal species [3,36]. Both the calculated [55,70,71] and the experimental [88] sputtering yields of graphite by C^+ and Ar^+ ions are very low at normal incidence angles (≤ 0.12 for $E \leq 500$ eV) and cannot support such a mechanism. On the other hand, the marked difference between the experimental displacement energies (E_d's) of graphite (25 eV) [3] and diamond (80 eV) [1,3,89] suggests preferential displacements of low E_d (graphitic or amorphous carbon) atoms leaving the high E_d (diamond, sp^3) atoms in their more stable positions. TRIM calculations indicate that for 100-200 eV C^+ ions, the number of displacements per ion (N_d) is 0.75-1.7 for graphite and 0.03-0.3 for diamond. At higher energies N_d (graphite) $\approx 3N_d$ (diamond), but both are high, causing significant damage and amorphization. At 1000 eV, for example, N_d (graphite) = 8.4 and N_d(diamond) = 2.6. This explains the deposition of amorphous carbon films at high carbon energies (e.g. $E > 500$ eV) and the optimal energy region of ~100-200 eV for diamond or diamondlike film deposition from C^+ and C^- ions.

Fig. 11 [82] schematically summarizes the effects of energy on carbon film deposition. At low energies (nevertheless high

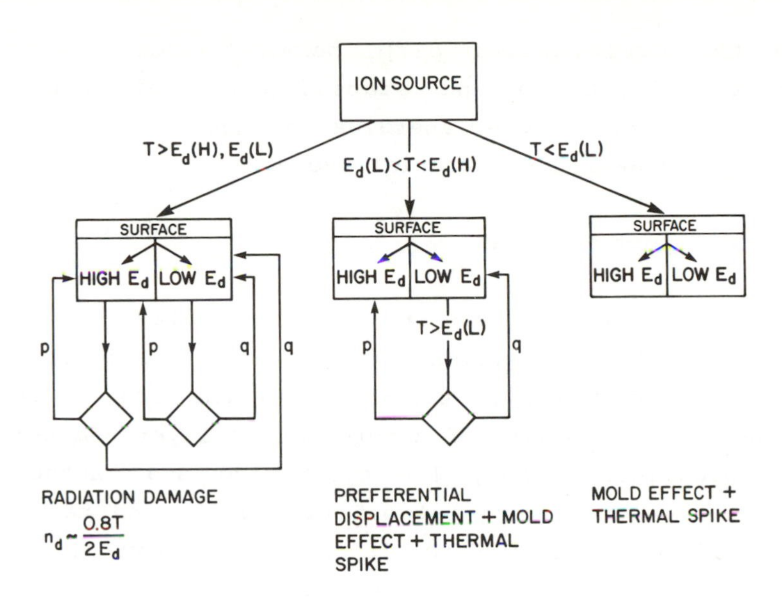

Fig. 11: Schematic illustration of dense matrix formation for three E ranges:
(I) $T < E_d(L)$ - only "mold effect",
(II) $E_d(L) < T < E_d(H)$ - preferential displacement,
(III) $T > E_d(H)$ - radiation damage and amorphization.
E_d - displacement energy; H,L - high and low E_d atom components; T - energy transferred in collision by primary ion; n_d - number of displacements per incident primary particle; p,q - probabilities of high and low E_d atomic site occupancies. From [82] with permission.

enough to trap carbon atoms in <u>subsurface</u> layers), no displacement occurs and the carbon species become trapped as interstitials. A high carbon interstitial concentration may be followed by an athermal, spontaneous precipitation of a new phase (similar to the "self-trapping" observed for He implantation) [93]. The formation of a diamond sp^3 constituent (or another dense phase) may be favored by the "mold" effect (constraints imposed by the environment of the host matrix) and by the highly excited environment due to the electronic and phonon excitations ("thermal spikes") [4,38-40]. A denser phase formation is also favorable since it reduces the volume expansion and stress involved in the transformation. At intermediate energies the preferential displacement becomes effective, favoring the formation and stabilization of diamond sp^3 inclusions. At still higher energies the damage is high for all carbon forms and amorphous material evolves.

5.6 Role of Target Material

The target material (type and structure) plays a significant role in carbon film deposition from hyperthermal species, since the first step of the film evolution is the projectile-target interaction. The formation of a pure carbon layer necessitates a shallow implantation range, low sputtering and backscattering yields, and low intermixing between projectile and target atoms (either collisional or due to diffusion). Several substrate materials covering a wide range of masses from Li to Au were investigated and will be considered in this section.

Many different substrates (Si, Cu, Ni, Au) exhibited similar film growth characteristics under C^+ bombardment, namely an evolution from a carbidic via a graphitic to a final diamond sp^3 stage [68,69,72,73]. The fluence dependence of C^+ bombardment of Ni and Si was similar (similar fluences were needed for phase evolution and for the decrease of intensity and disappearance of the substrate AES lines). Much higher (~60% higher) fluences were needed for carbon evolution on gold (Fig. 6) [72,73,82]. TRIM calculations show [70,71,82] that while the carbon ranges in Si, Ni, and Au (in terms of atoms/cm^2) are similar, the backscattering (BS) of 150 eV C^+ ions from an Au target is about 40%, leading to the higher fluences needed for carbon film evolution on this material (The BS yield of 150 eV C^+ from Ni and Si are 15% and 3%, respectively).

Carbon bombardment of a Li surface (a surface enriched CuLi alloy) exhibits different results. For 30 eV C^+ bombardment of Li, [71-73] formation of a Li-C alloy only, and no evolution of a pure carbon layer even for 10^{17} C^+/cm^2 (ten times the fluence needed for pure carbon evolution on Si and Ni, see Fig. 12) is observed. This behavior cannot be explained by pure subplantation arguments, since the range of 30 eV C^+ in Li is similar (in terms of atoms/cm^2) to that of 150 eV C^+ in Si and

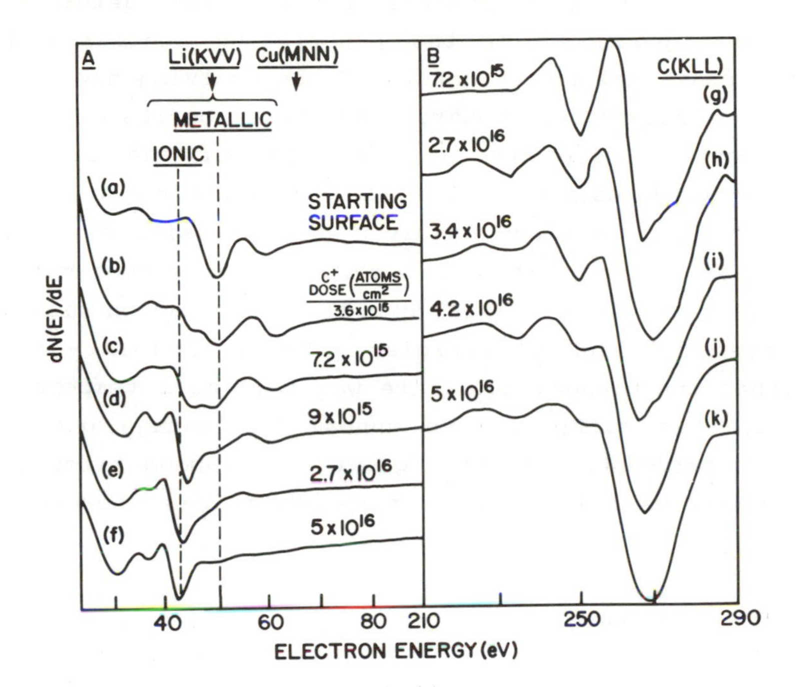

Fig. 12: Stages of film evolution on Li (surface enriched CuLi alloy) following 30 eV C^+ bombardment. Evolution of Li(KVV) (left) and C KLL(right) AES lineshapes indicate formation of Li-C and no graphite or diamond evolution. From [71] with permission.

Ni, the distribution profile is narrower, and the sputtering and backscattering yields for the 30 eV C^+ on Li are negligible. The only possible explanation is C-Li intermixing due to a combination of collisional and diffusion effects, resulting in the evolution of a C-Li compound rather than a pure carbon film, as indicated by the C and Li AES lineshapes.

When the diffusion rate of the carbon species in the target material is high compared to the C^+ flux, the carbon species diffuse into the bombarded surface forming a low concentration carbon solution in the substrate. This is the case for C^+ impingement on a hot (400°C) Ni substrate [10,71-73].

Another effect of the substrate is related to its

crystallinity. A specific crystalline environment determines the possible final positions of the penetrating species and also affects the phase and the structure of the evolving new inclusion by imposing constraints (boundary conditions) on its evolution (a "mold" effect). While the initial stage of the film growth involves "heterodeposition" of a carbon phase on a non-carbon substrate, the second stage involves "homodeposition" of carbon on a pure carbon layer. This is a more favorable host matrix for epitaxial growth. Homoepitaxial growth of diamond on diamond was reported even for high C^+ energies [50-51] (900 eV - 50 keV), provided that the diamond substrate was held at a temperature of ~700°C, which is sufficient to anneal the damage due to the 900 eV C^+ bombardment. Oriented growth of diamond films on non-diamond substrates held at room temperature was, however, also reported [52].

5.7 Role of Temperature

Two different problems are associated with substrate temperature effects on carbon film deposition: (i) The thermal stability of the final film and (ii) the evolution of the carbon film upon C impingement on a hot substrate.

Pure carbon films (amorphous, DLC, and diamond) and hydrogenated carbon films tend to graphitize upon annealing to temperatures of 400°C and up (graphitization of a-C:H is associated with hydrogen release) [3,4,8,72,73]. This tendency is accelerated when interdiffusion between the carbon layer and the substrate exists. Diamond films on Ni are thus unstable upon annealing while for similar films on Si and Au, no carbon diffusion into the substrate was detected even at 700°C [69,72,73].

The final evolution of the carbon phase is, however, much more sensitive to the temperature of the substrate during the subplantation process. At substrate temperatures of 100°C and higher, graphitic sp^2 films evolve (Fig. 13) [52-54,72,73]. These are different from the sp^3 films deposited at 70°C and lower, as previously discussed. This behavior is associated with

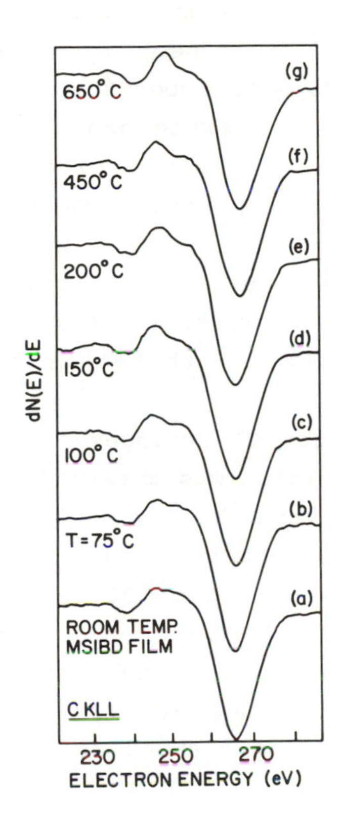

Fig. 13: C KLL AES signatures for carbon deposition onto carbon surfaces at varying temperatures. The C^+ fluence in each case was 1.8×10^{16} atoms/cm^2. The starting surface was a room temperature MSIBD film as shown in (a). Stages (b) - (g) correspond to carbon deposition onto (a) maintained at the specified temperatures.

the mobility of the carbon interstitials [90-92]. At temperatures slightly above room temperature, these interstitials become mobile while at lower temperatures they are immobile. Low temperature C^+ impingement on a surface thus creates immobile C interstitials whose concentration increases with carbon fluence, inducing strain in the matrix. At a certain stage spontaneous athermal transformation to a dense sp^3 (diamond) phase occurs, similar to that found for helium interstitials [93]. At substrate temperatures above 100°C, the interstitials become mobile and migrate to the surface to form graphitic (the stable carbon allotrope) layers.

This temperature dependence is in contradiction to the "thermal spike" model [4,38-40] where the dense sp^3 phase is presumed to evolve during the short period (~10^{-11} sec) of local excitation of the matrix resulting from projectile penetration. Since the energy (~0.01 eV) associated with substrate annealing from room temperature to 100°C is negligible compared to that associated with the "thermal spike" (~10-100 eV), it is expected from the "thermal spike" model that diamond sp^3 film evolution will be independent of substrate temperature as long as these films are stable upon post-deposition annealing (i.e. T < 400°C). The above experimental results thus invalidate the "thermal spike" model of diamond film evolution.

An entirely different behavior exists for C^+ impingement on a diamond surface. Successful diamond deposition was observed at a diamond temperature of 700°C [50-51] where the vacancy mobility (very low at T < 400°C) is high enough to account for the annealing of the damage induced by the impingement of 900 eV C^+ ions. Internal diamond growth thus occurs, similar to that reported for even higher C^+ energy (~50 keV) [51]. The driving force for the metastable diamond evolution is the mold effect of the diamond matrix that favors the inclusion of diamond to that of graphite.

5.8 The Role of Argon and Hydrogen

Ar hyperthermal species are often used (simultaneously with C species) for carbon deposition. They are supposed to preferentially etch amorphous carbon and graphitic constituents, enriching the diamond sp^3 constituent of the films [3,36]. TRIM calculations show [70] that Ar induces similar collisional effects to those of carbon species. Ar, however, contributes no carbon interstitials to the evolving film (essential for the evolution of the diamond layers according to the subplantation model) and is entrapped in the evolving matrix contributing to stress, damage, and undesirable changes of properties induced by impurity atoms. As for preferential sputtering, the sputtering yield of C by Ar ions is extremely low, especially in the low energy region (50-100 eV) used by Aisenberg and Chabot [9,22,49] (<0.05 for all C forms). An Ar flux at least 100 times more intense than the C flux is needed to efficiently sputter the amorphous carbon and graphitic constituents. Ar entrapment and damage is expected under these conditions. More efficient preferential sputtering may be achieved at glancing angles of Ar incidence, where the sputtering yields of C from Ar become significant. Impingement of 100 eV Ar^+ at an angle of 10° from the surface results in sputtering yields of 1.3, 0.7, and 0.2 for a-C, graphite and diamond, respectively. Direct pure C^+ impingement thus seems to be highly preferable to Ar^+ impingement for diamond sp^3 film formation.

Hydrogen plays an essential role in CVD processes [1,23-30] where it preferentially etches non-diamond constituents (very high hydrogen concentrations and substrate temperatures are needed). The possibility of a similar contribution to carbon film deposition is sometimes discussed. The role of hydrogen in stabilization of diamond surfaces by terminating "dangling bonds" is also emphasized [3,80]. TRIM [70] calculations of H^+ bombardment of C [70] show low displacement efficiency, low sputtering yields, and high backscattering efficiency. The main energy loss mechanism is ionization. Hyperthermal hydrogen species are thus inefficient in preferential displacement and preferential sputtering processes. The hydrogen is, however, expected to be entrapped in the carbon layer, forming films with high hydrogen concentration.

Fig. 14 shows an experiment conducted to demonstrate the Ar and H role in carbon film electron formation from hyperthermal species [70,72]. Ar impingement on a graphitic layer does not produce a diamond layer and the graphitic AES lineshape does not change. The Ar is, however, trapped in the film as is evident from the appearance of the Ar AES lines. Successive bombardment with 35 eV D_2^+ ions resulted in amorphization of the graphite (due to the high ionization efficiency) and chemical etching of the carbon layer exhibited by the disappearance of the Ar line that was used as a marker. This chemical sputtering by hydrogen ions is relatively efficient (sputtering yield $\sim 10^{-2}$-10^{-1}) at low energies and increases with temperature [94-95].

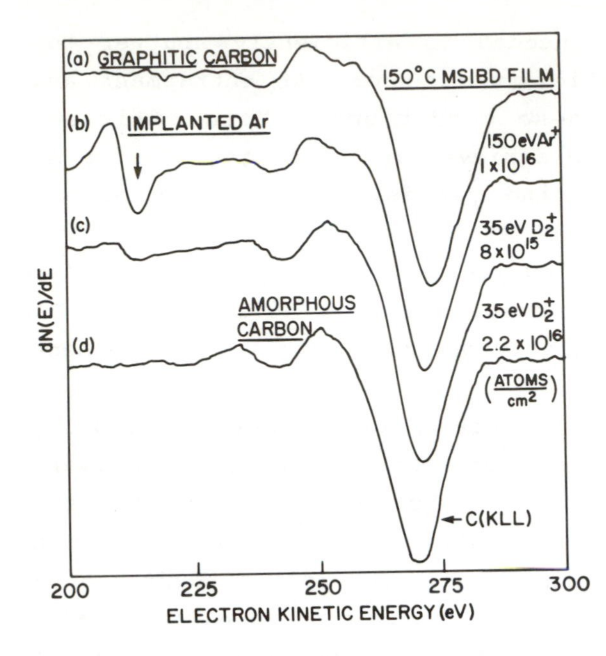

Fig. 14: Illustration of the role of low energy Ar and H bombardment on diamond film evolution: (a) as deposited graphitic carbon film (150°C MSIBD deposition on Au); (b) same film exposed to 1 x 10^{16} 150 eV Ar^+/cm^2; (c) film (b) exposed to 8 x 10^{15} 35 eV D_2^+/cm^2; (d) film (b) exposed to 2.2 x 10^{16} 35 eV D_2^+/cm^2. Note: (i) Ar incorporation and no diamond evolution due to Ar bombardment, (ii) amorphization and chemical sputtering (revealed by Ar disappearance) due to D_2 bombardment.

5.9 Role of Hydrocarbon Species

Impingement of hydrocarbon species is a complex phenomenon under controlled, well-defined conditions. Under practical hydrocarbon deposition conditions, a wide distribution of all physical parameters exists [3] and rigorous treatment of film growth phenomena is practically impossible.

The following arguments can be, however, deduced from TRIM calculations and subplantation notations: (a) Impingement of low energy hydrocarbons where the energy is insufficient for dissociation, results in hydrocarbon polymer films; (b) Higher energy impingement of C_nH_m molecules results in complete dissociation to carbon and hydrogen species where each atom energy is proportional to its mass (i.e. the carbon energy is $12E/(12m + n)$ and the hydrogen energy is $E/(12m + n)$; (c) The trajectories of all of these species that move in the target at exactly the same time are correlated. Since each projectile moves in an environment excited by several other species, displacement of atoms becomes possible at much lower energies than that for single C^+ impingement; (d) The low energy hydrogen can: (i) be trapped (forming a-C:H), (ii) be reflected, or (iii) chemically etch the evolving carbon layer; (e) The projected range of the penetrating hydrogen species is much lower than that of the carbon species (whose energy is 12 times higher). At high enough energies (e.g. > 200 eV for methane), the distribution profiles of the hydrogen and the carbon separate resulting in a shallow a-C:H layer and a deeper dense carbon layer. In the shallow region where the hydrogen is implanted it reaches saturation levels and a significant amount of hydrogen is released. In addition, if the etching efficiency of the low energy hydrogen ions is sufficient, this shallow a-C:H layer is etched, resulting in a dense carbon film with very little hydrogen content [3]. This also explains the very low sticking probability reported for plasma deposition from hydrocarbon species [3]. The decrease of this sticking probability with temperature (i.e. the increase of the etching with temperature) gives additional support to this proposed mechanism; and (f) Preferential bond breaking of C-H compared to C-C ($\Delta H = 3.5$ eV and 7.4, respectively) [3], followed by diffusion and release of the hydrogen atoms, may also contribute to low hydrogen concentration of the films.

These phenomena explain the energy dependence of carbon film deposition from hyperthermal hydrocarbons [3] (Fig. 8), namely: (i) polymer hydrocarbon film formation at low energies; (ii) amorphous carbon hydrogenated films at intermediate energies; and (iii) dense carbon formation (with low hydrogen concentration) at higher energies.

6. BULK CHARACTERIZATIONS OF MSIBD CARBON FILMS

Bulk characterizations are a necessary step in the investigation of MSIBD carbon films. Some characterizations of structure and bulk properties were reported from the few works that were conducted and are listed in Table 3. Most of the results were obtained from films deposited under only medium vacuum conditions ($p \geq 10^{-6}$ torr). Thick, large samples needed for these characterizations are difficult to deposit in presently available systems. The present characterizations also cover a very limited range of deposition parameters (some not clearly identified), usually restricted by the system capabilities. The influence of different substrates, crystalline orientations,

Table 2: Summary of Bulk Characterizations of MSIBD of Carbon Films on Non-Diamond Substrates

Technique	Reference	Deposition Conditions	Results
TEM	Chaikovskii et al. [52-54]	-100 eV C^+, room temp. <5.10^{-9} torr	small polycrystalline(-50Å) cubic diamond films in which large (>1 m) crystals are chaotically dispersed in preferred (111) orientation.
	Miyazawa et al. [55]	300, 600 eV C^+, room temp. -10^{-6} torr	amorphous material
Ellipsometry	Houston Facility [68,69]	150 eV C^+, room temp. <10^{-9}torr	n=2.2-2.6
	Soreq Facility [69]	120 eV C^+, room temp. -10^{-6} torr	n-3
	Ishikawa, Takagi et al. [60-63]	10-1000 eV C^-,C_2^-, room temp. -10^{-6} torr	n-3.1
	Miyazawa et al. [55]	300, 600 eV C^+, room temp. -10^{-6} torr	n-2.4
Resistivity	Anttila [56-58]	300, 600 eV C^+, room temp. -10^{-6} torr	1.7x10^6 cm
	Ishikawa, Takagi et al. [60-63]	150 eV C^-, room temp. -10^{-6} torr	2x10^8 cm
		100 eV C_2^-, room temp. -10^{-6} torr	1.5x10^{10} cm
Density	Ishikawa, Takagi et al. [60-63]	150 eV C^-, room temp. -10^{-6} torr	-3.2 g/cm^3
	Anttila	500-1000 eV C^+, room temp. -10^{-6} torr	3.2-3.5 g/cm^3

Table 2 (Cont'd): Summary of Bulk Characterizations of MSIBD of Carbon Films on Non-Diamond Substrates

Hardness	Anttila [56-58]	500-1000 eV C^+, room temp. -10^{-6} torr	18000 Kgf/mm^2 for 100 g weight 9000 for 500 g
Thermal Conductivity	Ishikawa, Takagi et al. [63]	-100 eV C^-, C_2^-, room temp. -10^{-6} torr	7.7 W.cm^{-1}.K^{-1}
RAMAN	Soreq NRC [69]	120 eV C^+, -0.7 m film	very broad peak at 1560 cm^{-1} typical for disordered carbon, similar to that found as base on CVD diamond films [23]
IR Transparency	Houston Facility [69]	150 eV C^+, room temp. <10^{-9} torr	transparent 2.5-25 m
	Soreq NRC [69]	120 eV C^+, room temp. -10^{-6} torr	transparent 2.5-25 m
	Ishikawa, Takagi et al. [60-63]	100 eV C^-, room temp. -10^{-6} torr	transparent 2.5-25 m
		100 eV C^-, 400°C -10^{-6} torr	transparent 4-25 m some absorption for <4 m
		100 eV C^-, 800°C 10^{-6} torr	absorption
		300 eV C^+, room temp. 10^{-6} torr	transparent 3-20 m
Visible Region	Miyazawa et al. [55]	300, 600 eV C^+ room temp., -10^{-6} torr	extinction coef. k=0 for >6500Å k=0.1 for 5000Å
	Ishikawa, Takagi et al. [60-63]	150 eV C^- 100 eV C_2^- room temp., -10^{-6} torr	optical gap 0.94 eV optical gap 1.44 eV
	Soreq NRC [69,70]		color fringes of varying thickness film indicate transparency

deposition conditions, etc. is usually not considered. One of the most difficult problems regarding the use and interpretation of these characterizations is the lack of established standard data for thin (<1 μm) diamond films with small (<1000 Å) crystalline size. This also holds for other carbon allotropes that are much less common than diamond. An important example of this situation, the use of Raman spectroscopy for analysis of thin diamond films, will be discussed.

Raman spectroscopy of bulk cubic diamond is well known [8,23,97-99]. The first order Raman scattering spectrum of bulk cubic diamond consists of a sharp 1332 cm^{-1} line. The first order Raman spectrum of bulk graphite is also known and consists of a sharp single line at 1580 cm^{-1} [8,23,98-100]. For small crystalline (<100 Å) graphitic material, however, a second line at 1360 cm^{-1} appears and is attributed to the disorder of the graphite [8,23,98-100]. Very little is however known about the Raman spectra of small diamond crystallites [98-99], and yet Raman is considered by many scientists as the best established identification technique for diamond films. Recent characterization work of CVD diamond films by Kobashi et al. [23] clearly demonstrates (Fig. 15) that the intensity of the cubic diamond 1332 cm^{-1} Raman line is inversely proportional to the crystalline size (determined by SEM). Furthermore, the Raman 1332 cm^{-1} signal of diamond films with a clear XRD cubic diamond pattern and a crystalline size of ~2000 Å is very weak and almost indistinguishable. This behavior can be explained in terms of the ratio between the number of surface and bulk atoms as demonstrated in Fig. 16. The surfaces of small diamond crystallites may be reconstructed or disordered and exhibit a Raman spectrum similar to amorphous carbon. Since the intensity of the Raman scattering from graphite and amorphous carbon is ~60 times stronger than that of diamond [101], it is expected that the contribution of the microcrystalline diamond surface atoms will eliminate the possibility of the detection of the 1332 cm^{-1} diamond line in diamond films with a grain size <500 Å. Careful consideration of the characterization techniques and establishment of the relations between structure (especially thickness and grain size) and characterizations of the different carbon allotropes is needed.

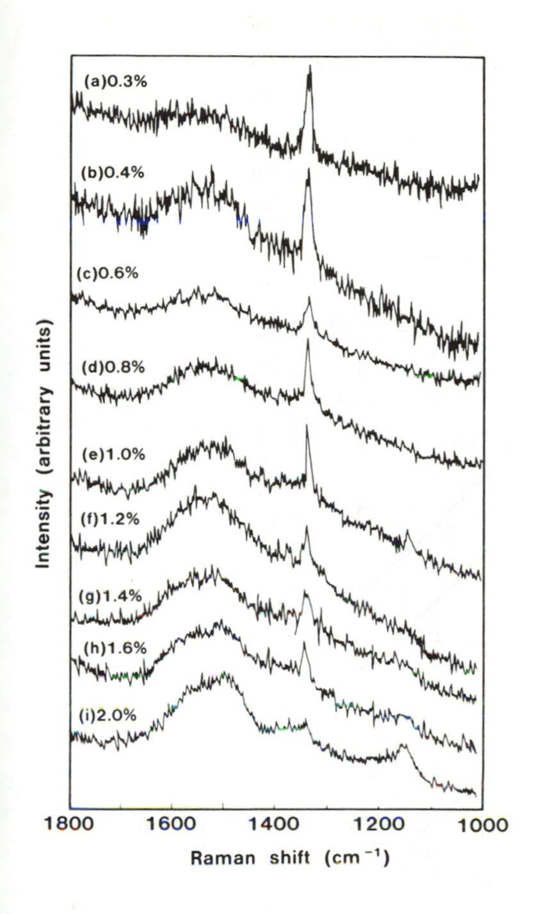

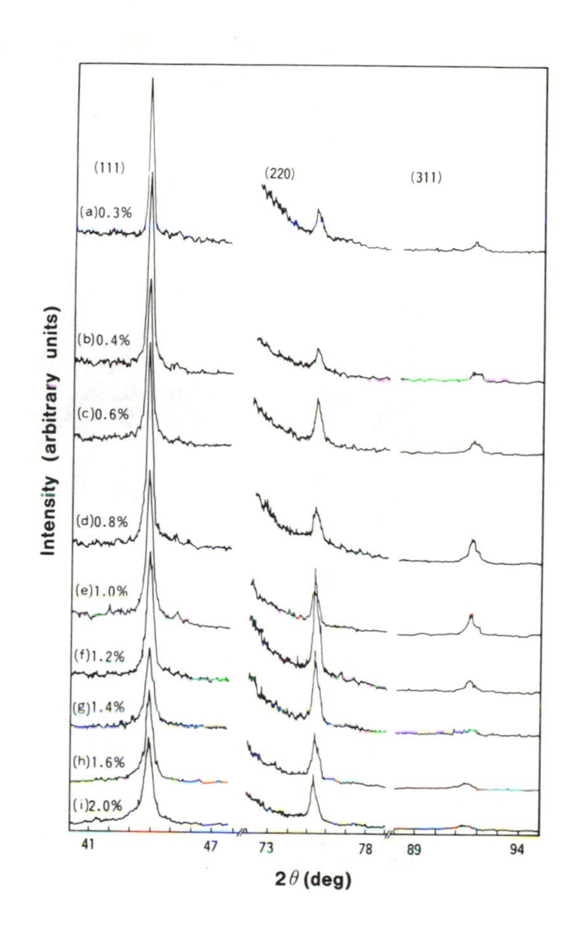

(a) (b)

Fig. 15: Comparison of XRD (a) and Raman (b) data of polycrystalline diamond CVD films [23]. The crystalline size was ~1 μ m for (a)-(e) and ~0.2 μ m for (h).

(a): X-ray diffraciton spectra of diamond films deposited on Si(100) substrates at different CH_4 concentrations: (a) 0.3%, (b) 0.4%, (c) 0.6%, (d) 0.8%, (e) 1.0%, (f) 1.2%, (g) 1.4%, (h) 1.6%, and (i) 2.0%. A weak diffraction of (400) is seen at 119.6° for c = 1.0% and 1.2%. From [23] with permission.

(b): Raman spectra of diamond films deposited on Si(100) substrates for 7 h. For bulk diamond only a single sharp line is observed at 1332 cm^{-1}. From [23] with permission.

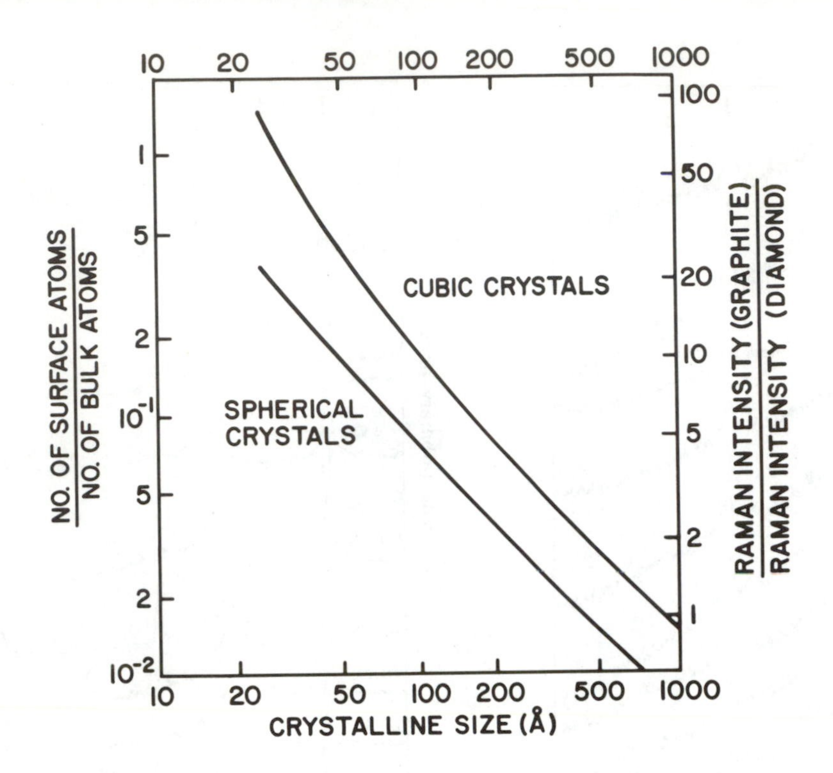

Fig. 16: Effect of crystalline size on relative Raman intensity, assuming surface atoms contribute to graphitic (or amorphous carbon) spectra and bulk atoms to diamond spectra.

8. APPLICATION OF THE SUBPLANTATION MODEL TO CARBON PLASMA DEPOSITION AND CARBON ION BEAM DEPOSITION TECHNIQUES

The previous discussion focused on mass selected carbon beam deposition where the intrinsic deposition parameters are well-defined and controlled. Plasma deposition systems and most ion beam deposition systems are, however, characterized by a large spread of the deposition parameters for the same process [3,4,10]. They usually include a mixture of different types of energetic ions and neutrals with a large energy distribution ranging from thermal species to sometimes 1 keV species. The angles of incidence vary between normal to glancing angles and the vacuum conditions are usually poor.

Irreproducibility of these conditions, variations over the deposition area, and fluctuations within a single deposition cycle are common. Since the carbon film growth may be very sensitive to some of the deposition parameters, it is difficult to rigorously treat these complex systems. It should also be noted that in most practical systems the deposition process involves a superposition of (i) subplantation due to hyperthermal species; (ii) conventional surface deposition processes due to thermal species and sometimes (iii) surface chemical reactions. Study of these systems involves the separate consideration of the surface deposition processes and the subplantation processes and evaluation of the possible effects of their superposition. As far as subplantation phenomena are concerned, evaluation of the species involved, and the distributions of the energy of the species and their angles of incidence upon the target are needed. TRIM calculations of ranges, distribution profiles, sputtering and backscattering yields, and damage for the different components of the impinging species should be performed. Surface deposition processes are better known and the film properties due to surface deposition from the thermal species component of the plasma should be evaluated. Analysis of the superposition of these two processes (surface deposition and subplantation) also requires some knowledge about synergistic effects, i.e. effects related to the two processes being performed simultaneously rather than successively. The general trends of subplantation may, however, be applicable also for these techniques under appropriate conditions. It seems, nevertheless, unlikely that reproducible, pure, crystalline carbon films of semiconductor quality will be produced from such complex and uncontrolled conditions. However, films of practical value for many other applications are obtainable from such processes. Most of the current and past work in this area concerns preparation and characterization of films, with little effort in acquiring data for understanding the fundamental film growth mechanisms. We believe that rigorous work along the lines sketched herein is essential for the future development of the field.

9. SUMMARY AND CONCLUSIONS

This chapter has presented an experimental and theoretical approach for the investigation of carbon film deposition from hyperthermal species. Mass selected ion beam deposition under UHV conditions with in situ surface analysis diagnostics capable of controlled deposition and analysis of fundamental growth mechanisms was presented. The different intrinsic parameters involved in the deposition process were discussed. The possibility of deposition of adherent carbon films with the sp^3 diamond short range order on different substrates held at room temperature was established.

Results of in situ parametric investigations of carbon film deposition from hyperthermal species were presented including the effect of fluence, energy, substrate nature, substrate temperature, and type of impinging species (carbon, argon, and hydrogen). Among the most important results are:

(1) When C-substrate diffusion and ion-mixing are low, a pure carbon film forms. The film evolution at room temperature advances from a carbidic via a graphitic to a final diamond stage.

(2) The optimal energy range for the diamond evolution is ~60-180 eV.

(3) sp^3 films are deposited on substrates held at $T \leq 70°C$ while graphitic films evolve on substrates held at $T \geq 100°C$.

These results along with those from other laboratories were discussed in terms of a recently proposed subplantation (subsurface shallow implantation) model. A dense sp^3 matrix at the proper temperature conditions evolves from hyperthermal carbon impingement due to preferential displacement of LE_d (graphitic) atoms, leaving the HE_d (sp^3) atoms in their positions, and due to the "mold" effect of the host matrix that imposes constraints on the carbon inclusion to be formed. Trajectory calculations (TRIM) were proven valuable for the analysis of the experimental data and rigorously support the subplantation model.

The present work has established guidelines for future work to follow. More in situ parametric investigations are required. The structure of the final films should be determined by bulk techniques (XRD, TEM, SEM) and their properties (optical, electrical, mechanical) should be investigated. The complex relations between intrinsic deposition parameters, film structure, and properties could then be properly determined. Further establishment of the bulk characterization of carbon thin films, e.g. Raman spectra of small diamond crystallites, is also necessary for that purpose. Additional theoretical work is needed, including improved trajectory calculations and atomistic calculations of the inclusion of carbon phases in specific host matrices (from a high concentration of interstitials). As for experimental work, the development of processes for deposition of many forms of carbon films that may be suitable for many applications, still remains a challenge.

10. FUTURE TRENDS

MSIBD has been shown as an excellent technique for controlled deposition of a variety of carbon films ranging from graphite to diamond on different substrates. The possibility of controlling and separating the different deposition parameters (that do not exist in other techniques that have a complex physical-chemical nature) is advantageous for several purposes:

- Parametric investigations of carbon (diamond) film formation, including simulation of other existing deposition methods.
- Development of processes for specific applications. These processes can be later adapted by other techniques, depending on economic considerations.
- Actual fabrication of (sp^3) films for special applications

This chapter focused specifically on carbon (sp^3) film deposition. Subplantation with mass selected ion beams has a general nature, and was actually applied for deposition of dense, hard metastable phases of other materials and for epitaxial growth of films at lower temperatures than needed for other deposition methods [4,46,102-108]. This new technique which bridges the existing gap between surface deposition and implantation, has the potential for development of films of a variety of materials with unique and beneficial properties.

11. ACKNOWLEDGMENT

This material is based on work supported by the National Science Foundation under Grant No. DMR-8610597. The authors are grateful to W. Eckstein for collaboration on the TRIM calculations.

REFERENCES

1) Angus, J. C. and Hayman, C. C. Science, 1988, 241, 913.

2) Bundy, F. P., Strong, H. M., Wentroff, Jr., R. H. in "Chemistry and Physics of Carbon", P. L. Walker, Jr. and P. A. Thrower, Eds., Dekker, NY, 1973, Vol. 10, 213.

3) Angus, J. C., Koidl, P., Domitz, S., in "Plasma Deposited Thin Ion Films", J. Mort and F. Jansen, Eds., CRC Press, Boca Raton, FL. 1986, chapter 4.

4) Weissmantel, C. in "Thin Films From Free Atoms and Particles", K. J. Klabunde, Ed., Academic Press Inc., Orlando, FL. 1985, chapter 4.

5) Berman, R., Ed., "Physical Properties of Diamond", Claredon Press, Oxford, 1965, 295.

6) Chrenko, R. M. and Strong, H. M., "Physical Properties of Diamond", G. E. Corporate Research Report No. 75 CRD 089.

7) Field, J. E., Ed., "The Properties of Diamond", Academic Press Inc., London, 1979.

8) Tsai, H. and Bogi, D. B., J. Vac. Sci. Technol. A, 1987, 5(6), 3287.

9) Aisenberg, S., J. Vac. Sci. Technol. A, 1984, 2(2), 369.

10) Lifshitz, Y., Kasi, S. R., and Rabalais, J. W., Adv. Mater. Manu. Process., 1988, 3(2), 157.

11) Meissner, R., Spear, K. E., Badzian, A. R., and Roy, R., J. Metals, September 1987, 8.

12) Woollam, J. A., Chang, H., and Natarajan, V., Appl. Phys. Comm., 1985-86, 5(4), 263.

13) Angus, J. C., Thin Solid Films, 1986, 142, 145.

14) Bundy, F. P. and Kasper, J. S., J. Chem. Phys., 1967, 46, 3437.

15) El-Goresy, A. and Donnay, G., Science, 1968, 161, 363.

16) Whittaker, A. G. and Kintner, P. L., Science, 1969, 165, 589.

17) Aust, R. B. and Drickamer, H. G., Science, 1963, 140, 817.

18) Fujii, K., Shohata, N., Mikami, M., and Yonezawa, M, Appl. Phys. Lett., 1985, 47, 370.

19) Robertson, J., Adv. Phys., 1986, 35(4), 317.

20) Robertson, J. and O'Reilly, E. P., Phys. Rev. B, 1987, 35(6), 2946.

21) Grigorovici, R., Devenyi, A., Gehorghiu, A., and Belu, A., J. Non-Cryst. Solids, 1972, 8-10, 793.

22) Aisenberg, S. and Chabot, R., J. Appl. Phys., 1971, 42(7), 2953.

23) Kobashi, K., Nishimura, K., Kawate, Y., and Horiuchi, T., Phys. Rev. B., 1988, 38(6), 4067.

24) Derjaguin, B. V. and Fedosev, D. B., Carbon, 1973, 11, 299.

25) Derjaguin, B. V. and Fedosev, D. B., Sci. Amer., 1975, 233, 102.

26) Swabe, A. and Inuzuka, T., Appl. Phys. Lett., 1985, 46, 146.

27) Sato, Y., Matsuda, S. and Nogita, S., J. Mat. Sci. Lett., 1986, 5, 565.

28) Matsumoto, O. and Katagiri, T., Thin Solid Films, 1987, 146, 283.

29) Frenklach, M. and Spear, K. E., J. Mat. Res., 1988, 3(1), 133.

30) Namba, Y., Wei, J., Mohri, T., and Heidarpour, E. A., J. Vac. Sci. Technol. A, 1989, 7(1), 36.

31) Ojha, S. M. and Holland, L., Thin Solid Films, 1977, 40, L31.

32) Nir, D., Thin Solid Films, 1984, 112, 41.

33) Nir, D., Kalish, R., and Lewin, G., Thin Solid Films, 1984, 117, 125.

34) Natarajan, V., Lamb, J. D., and Woollam, J. A., J. Vac. Sci. Technol. A, 1985, 3, 685.

35) Warner, J. D., Pouch, J. J., Altarovitz, S. A., Liu, D. C., and Landford, W. A., J. Vac. Sci. Technol. A, 1985, 3, 900.

36) Spencer, E. G., Schmidt, P. H., Joy, D. C., and Sansalone, F. J., Appl. Phys. Lett., 1976, 29(2), 118.

37) Angus, J. C. and Jansen, F., J. Vac. Sci. Technol. A, 1988, 6(3), 1778.

38) Seitz, F. and Koehler, J. S., Progress in Solid State Physics, Academic Press, New York, 1957, Vol. 12, 30.

39) Seitz, F. and Koehler, J. S., Solid State Phys., 1956, 2, 305.

40) Weissmantel, C., Thin Solid Films, 1982, 92, 55.

41) Chopra, K. L., "Thin Film Phenomena", McGraw-Hill, New York, 1969.

42) Reichelt, K., Vacuum, 1988, 38(12), 1083.

43) Gautherin, G., Bouchier, D., and Schwebel, C., in "Thin Films From Free Atoms and Particles", K. J. Klabunde, Ed., Academic Press, Inc., Orlando, 1985, chapter 5.

44) Biersack, J. P. and Eckstein, W., Applied Physics A, 1984, 34, 73.

45) Takagi, T., J. Vac. Sci. Technol. A, 1984, 2(2), 382.

46) Harper, J.M.E., Cuomo, J. J., Gambino, R. J., and Kaufman, H. R., in "Ion Bombardment Modification of Surfaces: Fundamentals and Applications", O. Auciello and R. Kelly, Eds., Elsevier, Amsterdam, 1984, chapter 4.

47) Greene, J. E. and Barnett, S. A., J. Vac. Sci. Technol. 1982, 21(2), 285.

48) Ref. 3, pp. 96-97, ref. 45, p. 388, ref. 46, p. 159, ref. 47, p. 299-301.

49) Aisenberg, S. and Chabot, R., J. Vac. Sci. Technol., 1973, 10, 104.

50) Freeman, J. H., Temple, W., and Gard, G. A., Vacuum, 1984, 34(1-2), 305.

51) Nelson, R. S., Hudson, J. A., Mazey, D. J. and Piller, R. C., Proc. R. Soc. London Ser. A, 1983, 386, 211.

52) Chaikovskii, E. F., Puzikov, V. M., and Semenov, A. V., Sov. Phys. Crystallogr., 1981, 26(1), 122.

53) Chaikovskii, E. F. and Rozenberg, G. Kh., Sov. Phys. Dokl., 1984, 29(12), 1043.

54) Chaikovskii, E. F., et al., Archinum Nauki O Materialach, 1986, t.7,2.2, 187 (in Russian).

55) Miyazawa, T., Misawa, S., Yoshida, S., and Gonda, S., J. Appl. Phys., 1984, 55(1), 188.

56) Anttila, A., Koskinen, J., Bister, M., and Hirvonen, J., Thin Solid Films, 1986, 136, 129.

57) Anttila, A., Koskinen, J., Lappalainen, R., Hirvonen, J. P., Stone, D., and Paszkiet, C., Appl. Phys. Lett., 1987, 50, 132.

58) Anttila, A., Koskinen, J., Räisänen, J., and Hirvonen, J., Nucl. Inst. Meth. Phys. Res., 1985, B9, 352.

59) Koskinen, J., Hirvonen, J. P., and Anttila, A., Appl. Phys. Lett., 1985, 47(9), 941.

60) Ishikawa, J., Takeiri, Y., Ogawa, K., and Takagi, T., J. Appl. Phys., 1987, 61(7), 2509.

61) Ishikawa, J., Ogawa, K., Miyata, K., Tsuji, H., and Takagi, T., Nucl. Inst. Meth. Phys. Res. B, 1987, 21, 205.

62) Ogawa, K., Murayama, H., Tsuji, H., Ishikawa, J., and T. Takagi, Proc. of 10th Symp. on Ion Sources and Ion-Assisted Technology, Tokyo, June 2nd-4th, 1986, 30.

63) Yoshida, M., Ogawa, K., Tsuji, H., Ishikawa, J., and Takagi, T., Proc. of 11th Symp. on Ion Sources and Ion-Assisted Technology, Tokyo, June 1st-3rd, 1987, 58.

64) Kasi, S. R., Kang, H., and Rabalais, J. W., Phys. Rev. Lett., 1987, 59(1), 75.

65) Kasi, S. R., Kang, H., and Rabalais, J. W., J. Vac. Sci. Technol. A, 1988, 6(3), 1788.

66) Rabalais, J. W. and Kasi, S. R., Science, 1988, 239, 623.

67) Kang, H., Kasi, S. R., and Rabalais, J. W., J. Chem. Phys., 1988, 88(9), 5882.

68) Kasi, S. R., Kang, H., and Rabalais, J. W., J. Chem. Phys., 1988, 88(9), 5914.

69) Kasi, S. R., Lifshitz, Y., Rabalais, J. W. and Lempert, G. D., Angewandte Chemie, Adv. Mater. Sec., 1988, 100(9), 1245.

70) Lifshitz, Y., Kasi, S. R., Rabalais, J. W. and Eckstein, W., submitted to Phys. Rev. B.

71) Lifshitz, Y., Kasi, S. R., and Rabalais, J. W., Proc. of MRS Meeting, Nov. 28-Dec. 3, 1988, Boston, Mass. Vol. 129 (in press).

72) Kasi, S. R., Lifshitz, Y., and Rabalais, J. W., submitted to Phys. Rev. B.

73) Kasi, S. R., Ph.D. thesis, University of Houston, 1988.

74) Freeman, J. H., Temple, W., Beanland, D., and Gard, G. A., Nucl. Instrum. Meth., 1976, 135, 1.

75) Ishikawa, J., Takeiri, Y., and Takagi, T., Rev. of Sci. Inst., 1986, 57(8), 1512.

76) Rhead, G. E., Barthes, M. G., and Argile, C., Thin Solid Films, 1981, 82, 201.

77) Lurie, P. G. and Wilson, J. M., Surface Sci., 1977, 65, 476.

78) Pepper, S. V., Surface Sci., 1982, 123, 47.

79) Pate, B. B., Oshima, M., Silberman, J. A., Rossi, G., Lindau, I., and Spicer, W. E., J. Vac. Sci. Technol. A, 1984, 2(2), 957.

80) Pate, B. B., Surface Sci., 1986, 165, 83.

81) Green, A. C. and Rehn, V., J. Vac. Sci. Technol. A1, 1983, 1, 1877.

82) Lifshitz, Y., Kasi, S. R., and Rabalais, J. W., Phys. Rev. Lett., 1989, 62, 1290.

83) Lifshitz, Y., Kasi, S. R., Roux, C., Rabalais, J. W., and Eckstein, W., 13th International Conference on Atomic Collisions in Solids, Aarhus, Denmark, August 7-11, 1989 (to be published in Nucl. Inst. Meth. in Phys. Res. B)

84) Ziegler, J. F., Biersack, J. P., and Littmark, U., "The Stopping and Ranges of Ions in Matter", Pergamon, Oxford, 1985, Vol. 1.

85) Dearnaley, G., et al., "Ion Implantation", North Holland Pub. Co., Amsterdam, American Elsevier Pub. Co., Inc., NY, 155, 1973.

86) Möller, W. and Eckstein, W., Nucl. Inst. Meth. Phys. Res. B, 1984, 2, 814.

87) Feldman, L. C. and Mayer, J. W., "Fundamentals of Surface and Thin Film Analysis", North Holland, Elsvier, New York, NY, 1986, chapter 6.

88) Andersen, H. H. and Bay, H. L., in "Sputtering by Particle Bombardment I", R. Behrisch, Ed., Springer Verlag, NY, 1981, chapter 4.

89) Burgermeister, E. A., Ammerlaan, C.A.J., and Davies, G., J. Phys. C., 1980, 13, L691.

90) Prins, J. F., Phys. Rev. B., 1985, 31(4), 2472.

91) Prins, J. F., Rad. Effects. Lett., 1983, 76, 79.

92) Prins, J. F., Derry, T. E., and Sellschop, J.P.F., Phys. Rev. B., 1986, 34(12), 8870.

93) Wilson, W. D., Rad. Eff., 1983, 78, 11.

94) Yamada, R., J. Vac. Sci. Technol. A, 1987, 5(4), 2222.

95) Roth, J. and Bohdanski, J., Nucl. Inst. Meth. Phys. Res. B, 1987, 23, 549.

96) Hsu, W. L., J. Vac. Sci. Technol. A, 1988, 6(3), 1803.

97) Solin, S. A. and Ramdas, A. K., Phys. Rev. B., 1970, 1, 1687.

98) Nemanich, R. J., Glass, J. T., Lucovsky, G., and Shroder, R. E., J. Vac. Sci. Technol. A, 1988, 6(3), 1783.

99) Plano, L. S. and Adar, F., SPIE Vol. 822, Raman and Luminescence Spectroscopy in Technology, 52, 1987.

100) Nemanich, R. J. and Solin, S. A., Phys. Rev. B, 1979, 20, 392.

101) Wada, N. and Solin, S. A., Physica B, 1981, 105, 353.

102) Gautherin, G., Bouchier, D., and Schwebel, C., in "Thin Films From Free Atoms and Particles", K. J. Klabunde, Ed., Academic Press, Inc., Orlando, FL, 1985, chapter 5.

103) Miyake, K. and Tokuyama, T., Thin Solid Films, 1982, 92, 123.

104) Zalm, P. C. and Beckers, L. J., Appl. Phys. Lett., 1982, 41(2), 167.

105) Greene, J. E., et al. J. of Crystal Growth, 1986, 79, 19.

106) Morishita, Y., Maruno, S., Isu, T., Nomura, Y., and Ogala, H., J. Crystal Growth, 1988, 88, 215.

107) Rossnagel, S. M. and Cuomo, J. J., Vacuum, 1988, 38(2), 73.

108) Zuhr, R. A., Pennycook, S. J., Noggle, T. S., Herbots, N. Haynes, T. E. and Appleton, B. R., Nucl. Inst. Meth. Phys. Res. B., 1988

Materials Science Forum Vols. 52 & 53 (1989) pp. 291-300

SUBSTRATE SPUTTER EROSION EFFECTS DURING a-C:H FILM FORMATION IN A GLOW DISCHARGE

S. Berg and B. Gelin
Institute of Technology, Uppsala University
Box 534, S-751 21 Uppsala, Sweden

ABSTRACT

Energetic carbon containing ions will be created in a glow discharge during a-C:H film deposition. These energetic ions will cause sputter erosion of the substrate surface and the growing film. The growth rate of the carbon film formation will be reduced somewhat due to this effect. Furthermore, at the initial stage of film growth part of the original substrate surface will be exposed to the incoming energetic ions. Sputter erosion of atoms from the substrate material will therefore also take place. This substrate sputter erosion will continue until the carbon film has grown to a thickness that completely prevents sputtering of atoms from the underlaying substrate.
Etching depths as high as 1000 Ångström has been observed when depositing carbon films from butane onto thick gold films but minor etching effects were observed when depositing carbon on glass or Si substrates.
The physics of this unavoidable and substrate dependent etching effect will be explained in this article.

INTRODUCTION

One of the techniques to fabricate a-C:H films uses a traditional rf-diode sputtering chamber as the processing equipment [1-11]. However, the target electrode is normally not used. Only the substrate electrode is excited by rf-power. Also the gas is of course no longer argon as usually used in the sputtering process. The argon gas is replaced by, or mixed with some hydro-carbon gas. The glow-discharge pressure is normally in the 5-50 mTorr region. The thickness growth obtained in such a system for different hydrocarbon gases are shown in figure 1 [7]. As expected the most carbon rich gases causes the largest deposition rates but there is no simple growth rate relationship for different gases.

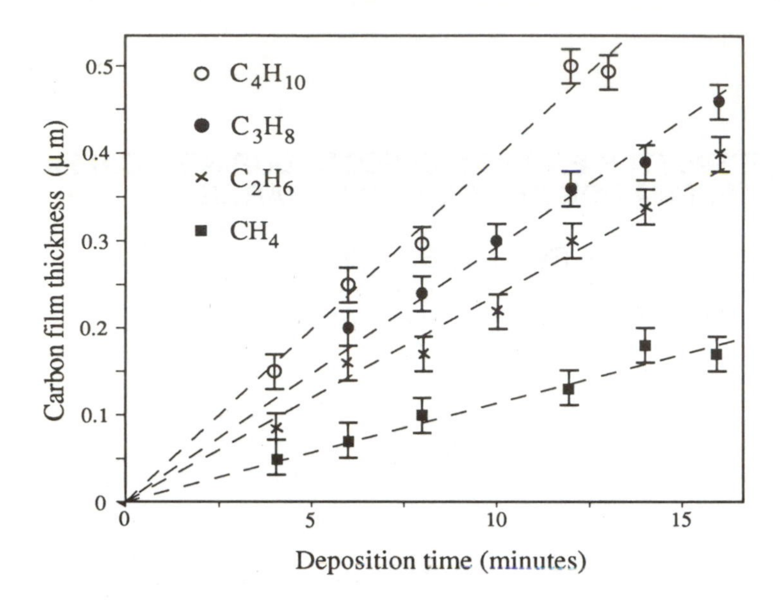

Figure 1. Thickness growth of carbon film onto glass substrates, obtained from deposition in four different gases. (-900 V and 10 mTorr.)

In the glow discharge system described above only a fraction of the hydrocarbon gas (CH_4, C_2H_2, C_2H_6 etc.) will be ionized. During the impact ionization the molecules will dissociate into some cracking pattern. The ions in the glow discharge will therefore be built up by a spectrum of these cracking fragments [7]. The positively charged fragments will be accelerated towards the negatively self biased rf-excited substrate electrode. These carbon containing ions will be the carrier of the carbon that will form the a-C:H film at the substrates resting on the rf-excited substrate table electrode. Normal negative bias voltages at this electrode range from 500-1000 Volts.

The energy of the incoming ions is high enough to cause significant sputtering at the substrate surface. The incoming ions will thus both carry the carbon to the substrate surface and also cause sputtering of the formed carbon film. A competing process will take place between deposition and sputter etching.

This is a situation that closely reminds of primary ion beam deposition. In the primary ion beam technique an ion beam of some material (e.g. carbon) may be directed towards a substrate. When low energy carbon ions strikes the substrate a carbon film will be formed at the substrate surface. However, if the energy of the carbon ions is increased the ions will be energetic enough to cause significant self-sputtering of the formed carbon film. Thus deposition of and sputtering by carbon ions will be competing processes taking place simultaneously. In fact, by increasing the energy of the incoming carbon ions the sputtering rate may be larger than the deposition rate. Instead of film deposition a net substrate etching will occur.

This will hardly happen in a glow discharge built up by a pure (100%) hydro-carbon gas but likely if an Ar/hydro-carbon gasmixture is used. In a glow discharge system the applied voltage can not be varied independently with respect to ion current density. Also, due to charge exchange between ions and molecules there will be a wide distribution of energies of the incoming ions. Some will have quite low energy and will therefore not markedly contribute to sputter erosion of the deposited film. These low energy carbon containing ions will only supply carbon to the surface and assist in carbon film formation. Therefore, in a 100% hydro-carbon glow discharge, deposition almost always dominates over etching and a net film deposition results.

There are exceptions, however, as shown in figure 2. A silicon substrate and a gold substrate are processed simultaneously in a CH_4-glow discharge. While the Si surface is coated with a carbon film, no carbon is found on the Au surface. In fact, the sputter etching effect is dominating on the Au surface. Based on that the Au- and Si-surfaces behave so differently in an Ar/CH_4 glow discharge a selective sputter etching/deposition technique, without the use of chemical active gases, has been developed [10-12].

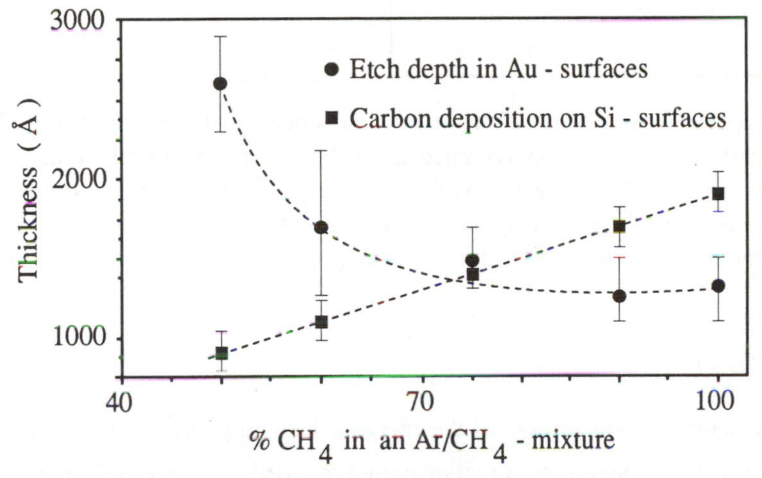

Figure 2.
Experimental results for carbon film formation on Au and Si substrates processed simultaneously for 20 min in a glow discharge of Ar/CH_4 at - 500V rf self bias on the substrate table.

In the following we will clearify the unavoidable substrate etching effect that always will take place at the initial stage of carbon film growth in a hydrocarbon glow discharge. This will also explain the observed difference in initial growth on different substrates.

ION /SUBSTRATE INTERACTION

A schematic drawing of the carbon coverage at the substrate surface at the initial stage of film growth by energetic ion bombambardments onto a substrate is shown in figure 3. This figure illustrates in a simplified way the conditions at the surface before the original substrate has been totaly covered by a carbon film. The energetic incoming carbon containing ions (F) will cause both carbon deposition and sputter erosion of carbon atoms from the carbon film (yield S_c) and also erosion of substrate atoms from the non carbon covered fraction ($1-\theta$) of the surface (yield S_s).

F represents the flux of incoming ions onto the substrate surface. θ is the fraction of the original substrate surface still not covered by a carbon film. When an energetic incoming ion bombards the ($1-\theta$) fraction of the surface, sputter erosion of the original substrate surface will take place. This sputter erosion will continue until the substrate is totally covered by the carbon film. This will happen when θ approaches unity. Thus, the original substrate surface will be sputter eroded during the time when θ passes through the interval $0 < \theta < 1$.

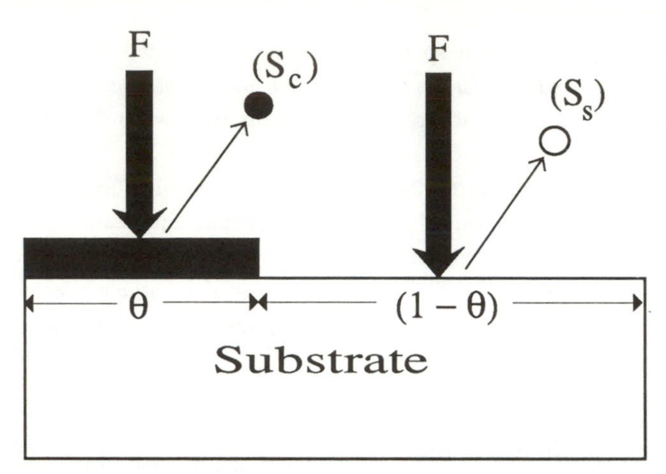

Figure 3.
Schematic illustration of the initial stage of deposition of a carbon film by energetic carbon containing ion bombardment.

As a consequence of this sputter erosion, some of the original substrate material will be removed at the initial stage of the carbon film deposition. As much as 1000Å of substrate material removal has been observed before a carbon film has been fully formed on a gold substrate processed an a butane plasma [6]. In some applications this may not be tolerable. This is especially the case in laminated multilayer structures where the a-C:H film constitutes one of the layers.

SPUTTER EROSION MODEL

To calculate the amount of sputter erosion caused by the energetic incoming carbon ions we must find an expression for $\theta(t)$. The first order approximation earlier reported by Berg and Andersson [6] will be recalled.

We will investigate the situation shown in figure 3. The fraction θ of the surface is covered by a carbon layer. If the flux of incoming ions is denoted F, the rate of carbon growth R_c [carbon atoms /unit area and time] at the substrate surface may be written

$$R_c = k F \tag{1}$$

where k is a constant that depends on the ionized hydrocarbon gas, the surface conditions (the sticking coefficient of carbon to the surface) and on the geometry of the equipment.
Simultaneously with the carbon growth the flux F of energetic ions will sputter erode some of the deposited carbon from the carbon deposited fraction θ of the surface. This sputter erosion rate R_i [carbon atoms /unit area and time] may be written

$$R_i = \theta F S_c \tag{2}$$

where S_c is the sputtering yield of carbon.
The mass m_c per unit area of the net deposited carbon onto the substrate surface at the time t will be the difference between the amount of deposited and sputter eroded carbon

$$m_c = \int_0^t (R_c - R_i) M_c \, dt \tag{3}$$

where M_c is the mass of the carbon atom.

The substrate surface will be assumed to be totally covered by a carbon layer when its average thickness is x_o. The surface coverage coefficient θ may be defined as

$$\theta = \frac{m_c}{\rho_c x_o} \tag{4}$$

where ρ_c is the density of the carbon film. From eq. (1) - (4) we can solve for $\theta(t)$ if $\theta(0) = 0$

$$\theta(t) = \frac{k}{S_c}\left[1 - \exp\left(- \frac{S_c F M_c}{\rho_c x_o} t \right) \right] \tag{5}$$

This equation gives the time dependence of $\theta(t)$. It is now possible to calculate the sputter erosion of substrate atoms from the non covered fraction $(1-\theta)$ of the substrate surface. The flux of sputter eroded substrate atoms F_s from the fraction $(1 - \theta)$ of the substrate surface at the time t may be written

$$F_s = F (1 - \theta) S_s \tag{6}$$

where S_s is the sputtering yield of the substrate material. The mass m_s of the sputter eroded substrate material per unit area at the time is

$$m_s = \int_0^t F_s M_s \, dt \tag{7}$$

where M_s is the mass of a single atom of the substrate material. This mass corresponds to an etch depth thickness d of

$$d = \frac{m_s}{\rho_s} \tag{8}$$

where ρ_s is the density of the substrate material. From eq. (5) - (8) we can obtain the time dependence of d(t)

$$d(t) = \frac{F S_s M_s}{\rho_s}\left[(1 - \frac{k}{S_c}) t + \frac{k}{S_c} \tau \left[1 - \exp(-\frac{t}{\tau}) \right] \right] \tag{9}$$

where

$$\tau = \frac{\rho_c x_o}{F M_c S_c} \tag{10}$$

If the condition

$$(k - S_c) > 0 \tag{11}$$

is satisfied, the contribution from the deposition will always be larger than the sputter erosion and a net carbon film formation will start to grow. θ (t) will gradually increase and approach $\theta = 1$ at $t = t_o$. Solving for t_o in eq.(5) gives

$$t_o = \frac{\rho_c x_o}{F M_c S_c} \ln\left[\frac{k}{(k - S_c)}\right] \tag{12}$$

At $t=t_o$ the original substrate material will be totally covered by the carbon film of thickness x_o.This carbon film will prevent the incoming ions from further sputter eroding of the original substrate material. This is also seen in eq.(6). For $\theta = \theta(t_o) = 1$

$$F_S = F\left[1 - \theta(t_o)\right] S_S = 0 \tag{13}$$

$F_s = 0$ implies that the sputter erosion of the substrate material ends at $t = t_o$.

The value of d(t) will saturate at $t = t_o$ since the sputter erosion of the original substrate ends when a fully protecting carbon film has been formed on the substrate. The saturation value $d(t_o)$ can be obtained from eq.(9). Note that eq.(9) only is valid in the time interval $0 < t < t_o$.

In figure 4, experimental values are compared to calculated d(t) values for depositing a-C:H films on Al and Au substrates [6]. This figure serves to illustrate the fact that the unavoidable substrate etching can be considerable, but approach different saturation values for different substrates.

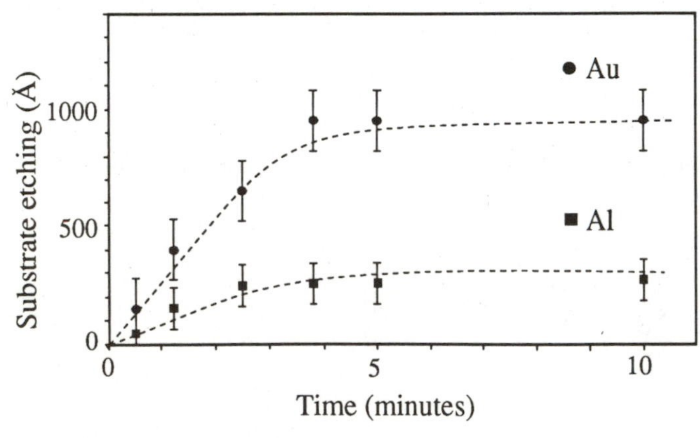

Figure 4.
Etch depth d(t) in gold and aluminium as a function of time for substrates procssed in an rf butane glow discharge with -1.2 kV rf self bias and butane pressure 0.8 Pa.

The formulas derived above may only serve as to give a general understanding of the ion/film/substrate interactions. The effect of several complex phenomena have been neglected. A more detailed description can be found in ref. [10-12].

SELF LIMITING SUBSTRATE ETCHING

The etch depth saturation derived above may find technological applications. For $t > t_o$ the value of d(t) saturates at $d = d(t_o)$. Thus the etch depth into the original substrate material for $t > t_o$ has became independent of the time t. This effect has been described as the Self Limiting Etch Depth effect (SLED-effect) [11,12]. This effect can be applied to situations where a predetermined etch depth is required. Possible applications may be in situations where normal plasma etch selectivity can not be used to determine a critical desired etching endpoint.
It is important to realize that the ion/substrate interaction in glow discharge deposition of a-C:H films may cause considerable substrate etching. Mixing the hydrocarbon gas with argon, to increase the ion bombardment during film formation, will of course even more pronounce this effect. As an illustration to this, experimental results of initial etching in Ar/CH_4 for some substrate materials [11] are shown in figure 5.

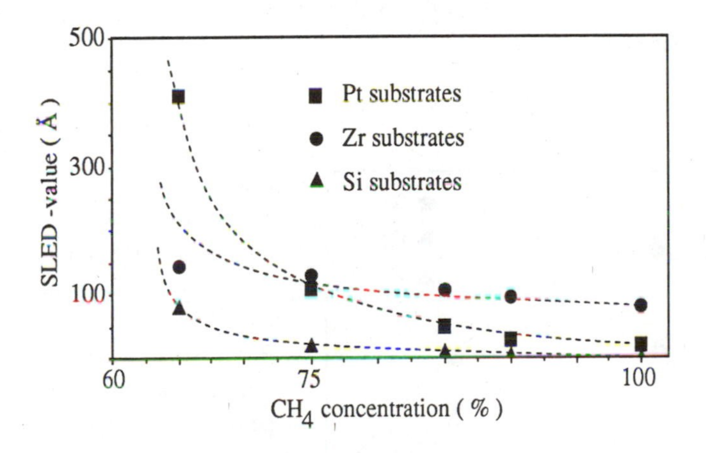

Figure 5.
Saturated etch depths (= SLED-values) in Pt, Si and Zr surfaces as a function of CH_4-concentration in an Ar/CH_4-gas mixture at 900 V rf peak value and 10 mTorr total gas pressure.

The unavoidable initial substrate etching effect may also cause a soft interface between the substrate and the carbon film. This has been observed using a very similar technique to selectively deposit titanium on different substrates. [13].

INITIAL CARBON FILM GROWTH

The time t_o derived above may be different for different substrate materials. It will thus take different times before a fully protecting carbon layer is formed on different substrate surfaces. Therefore the carbon film thicknesses may differ at different substrates even if they were processed together in the same glow discharge. The carbon growth rate will not be identical on different substrates until a fully protecting carbon layer has been formed. During the initial time interval, $0 < t < t_o$, the growth rate of carbon is nonlinear and substrate dependent [10,11]. In figure 6 are shown the carbon film thickness on Si, Pt, Al and Zr as a function of the methane concentration [10]. It can be observed that the minimum methane concentration needed to form a carbon film is different for different substrate materials. In figure 7 are shown the carbon film thicknesses on Si, Pt and Zr as a function of rf bias voltage [10] . The decrease in film thicknesses for increasing rf voltage are due to the fact that the sputtering yield values increase with increasing energy of the impinging ions.

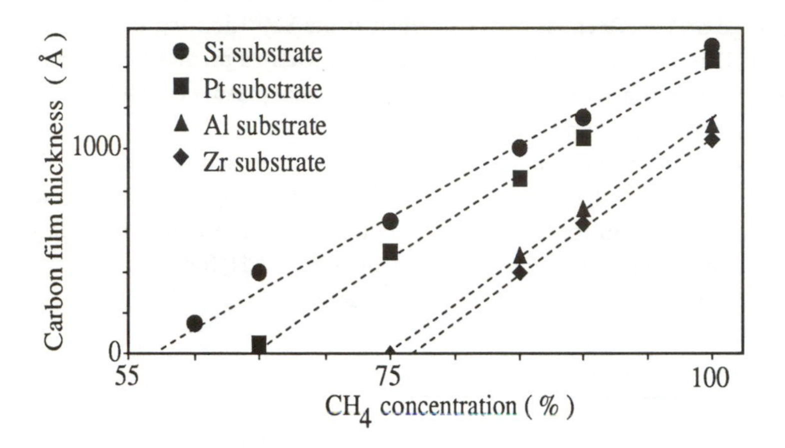

Figure 6.
Carbon film thickness on different substrates as a function of the CH_4 concentration in an Ar/CH_4 glow discharge. Processing time 20 min. Total pressure 10 mTorr and -900V rf peak voltage.

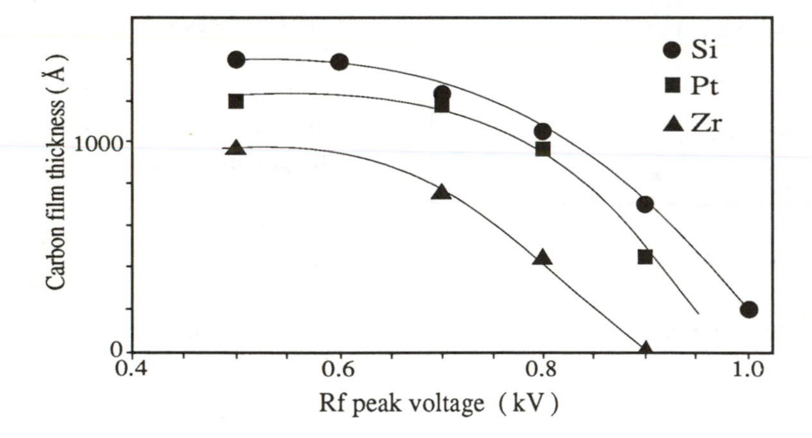

Figure 7.
Carbon film thickness on 3 different substrates as a function of the rf peak voltage. Processing time 20 min in a glow discharge of 75 % CH_4 in an Ar / CH_4 mixture with a total pressure of 10 mTorr.

COMPUTER SIMULATIONS OF CARBON RESPUTTERING

Sputtering of monoatomic carbon film overlayers is not identical to the sputtering of bulk carbon. The value S_c of the sputtering yield of carbon atoms introduced in eq. (2) will not be constant during the glow discharge deposition process. S_c may vary dramatically as the carbon film thickness increases. This can be illustrated by introducing the partial sputtering yield value S_{cc} defined as

$$S_{cc} = \frac{R_i}{F} \tag{14}$$

An example of the expected variation in this partial sputtering yield value as calculated by simulating the collision cascades in the film/substrate structure is shown in figure 8. These curves were obtained by using the Monte Carlo based program TRIM.SP [14-16].

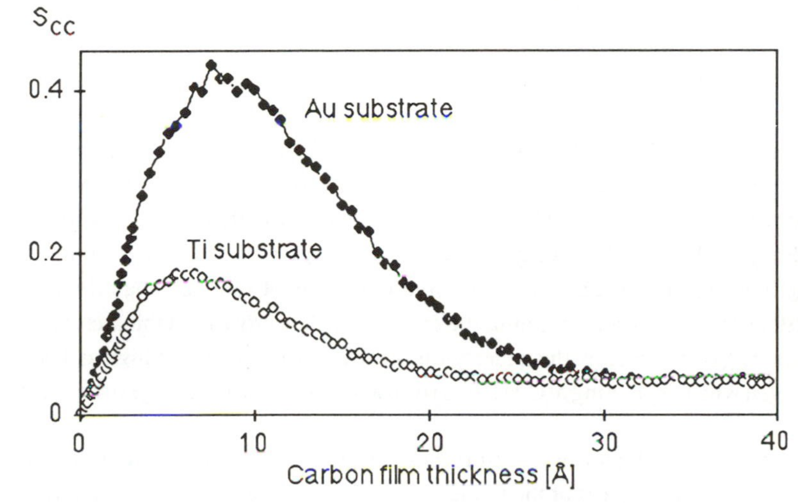

Figure 8.
Variation in carbon partial sputtering yield, S_{cc}, as a function of carbon film thickness on two different substrates. Energy of the incoming cabon ions = 1.0 keV.

These calculations clearly indicate that the behaviour of the sputter erosion of a thin carbon film from a gold substrate is quite different from what is expected from eroding the same film on a titanium substrate. However, as d > 30 Å both curves coincide. This is the thickness at which the carbon film partial sputtering yield approaches the bulk value, S_c.

With the TRIM.SP. program it is also possible to calculate the expected sputter erosion of bulk substrate atoms through the deposited carbon layer. Results from such calculations are shown in figure 9. This figure show the number of sputter eroded substrate atoms , S_{ss}, per incoming energetic (1.0 keV) carbon ion as a function of the average carbon film thickness. The situations on both a gold and a titanium substrate are shown.

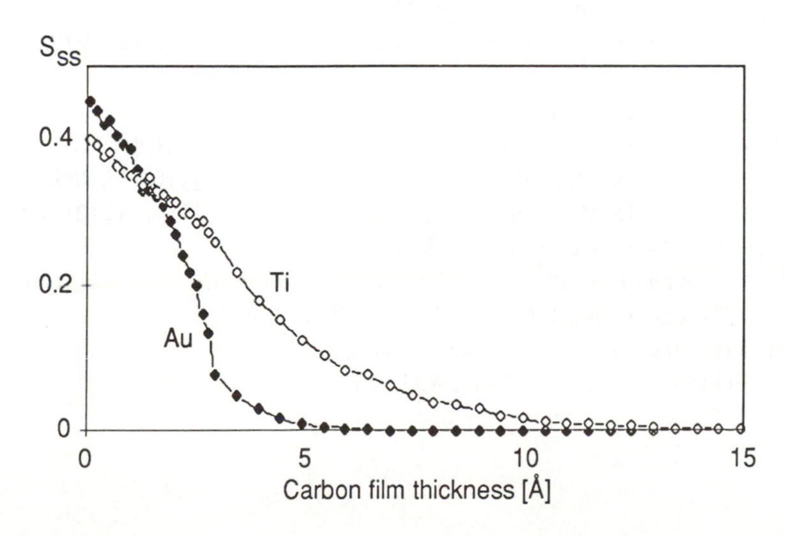

Figure 9.
Sputter erosion of substrate bulk atoms / incoming (1.0 keV) carbon ion as a function of carbon film thickness on two substrates. At thicknesses > x_o no further substrate bulk atoms will be sputter eroded.

From the results of figure 9 it is possible to evaluate the values x_o, as defined in eq.(4). The value of x_o can be obtained as the thickness at which the sputter erosion of substrate atoms drops to zero. It should be observed that x_o depends on the film / substrate combination.

CONCLUSIONS

Energetic ions will be created in a rf glow discharge during a-C:H film deposition. The formed carbon containing energetic ions will cause sputter erosion of the substrate surface. At steady state conditions the growth rate of the carbon film will be reduced somewhat due to this effect. Moreover, at the initial stage of film growth, part of the original substrate surface will be exposed to these incoming energetic ions. It is thus unavoidable that some sputter erosion of substrate material also will take place. This sputter erosion will continue until the carbon film has grown to a thickness that completely prevents sputtering of atoms from the underlying substrate. Etch depths as high as 1000 Ångström has been observed when depositing a-C:H films from a butane glow discharge onto thick gold films.
This etching effect may be a limiting factor in the usefulness of the glow discharge technique in fabricating very thin multilayers, since parts of the predeposited film will be removed at the initial stage of carbon film growth.

REFERENCES

1) A.Bubenzer, B.Dischler,G. Brandt and P.Koidl: J .Appl. Phys., 1983, 54 (8), 4590.
2) B.Meyerson and F.W.Smith.: J. Non. Cryst. Solids, 1980, 35/36 , 435.
3) K.Enke.: Thin Solid Films, 1981, 80 , 227.
4) L.Holland and S.M.Ohja.:Thin Solid Films, 1979, 58 , 107.
5) I.Watanabe.: Japanese Journal of Applied Physics, 1984, 23 (4), 487.
6) L.P.Andersson and S.Berg.: Vacuum, 1978, 28 (10/11), 449.
7) L.P.Andersson, S.Berg, H.Norström, R.Olaison and S.Towta.: Thin Solid Films, 1979, 63 , 155.
8) S.Berg and L.P.Andersson.: Thin Solid Films, 1979, 58 , 117.
9) H.Norström, R.Olaison, L.P.Andersson and S.Berg.: Le Vide, 1979, Suppl.196 , 11.
10) S.Berg, B.Gelin, A.Svärdström and S.M.Babulanam.: Vacuum, 1984, 34 (10/11), 969.
11) S.Berg , B.Gelin, M.Östling and S.M.Babulanam.: J.Vac.Sci.Technnol., 1984, A2 (2), 470.
12) S.Berg, C.Nender and B.Gelin.: Vacuum, 1988, 38 (8-10), 621.
13) C.Nender, S.Berg, B.Gelin and B.Stridh.: J.Vac.Sci.Technol., 1987, A5 (4), 1703.
14) J.P.Biersack and L.G.Haggmark.: Nucl..Instr. and Meth., 1980, 174, 257.
15) J.P.Biersack.: Nucl..Instr. and Meth. in Phys.Res.,1987, B27 , 21.
16) J.P.Biersack and W.Eckstein.:Appl.Phys., 1984, A34 , 73.

Materials Science Forum Vols. 52 & 53 (1989) pp. 301-322

NOVEL FORMS OF CARBON FROM POLY(ACRYLONITRILE): FILMS AND FOAMS

C.L. Renschler and A.P. Sylwester
Sandia National Laboratories
Albuquerque, NM 87185, USA

ABSTRACT

A number of interesting carbon forms can be prepared by the carbonization of polyacrylonitrile (PAN). We have expanded on the well known PAN-derived carbon fiber work to produce fully dense thin carbon films, porous carbon films, and microcellular carbon foams. These carbon materials can be prepared with a wide range of tailored properties. We report mechanistic studies of the PAN carbonization process together with potential useful device applications for these novel materials.

INTRODUCTION

Nearly any organic material will carbonize, or char, at sufficiently high temperatures in the absence of oxygen. Carbon precursors include lignin [1], cellulose [2,3], novolacs [4], specialty polymers [5], and polyacrylonitrile (PAN) [6-10]. PAN has been found to be particularly useful, especially as a precursor to carbon fibers [6,11]. At relatively low temperatures (200-300°C) in the presence of oxygen, PAN undergoes a cyclization process, resulting in the formation of the so-called polyimine ladder polymer [9]:

PAN —Δ, 200-300°C→ LADDER POLYMER

This material is dimensionally stable and will not melt during subsequent high temperature (600-3000°) carbonization in an inert atmosphere during which the ladder polymers coalesce into an extended aromatic structure [9]:

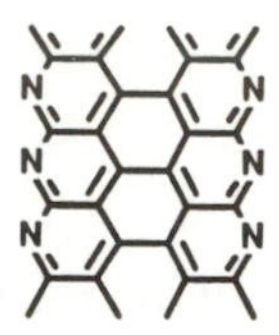

The degree of ordering achieved in this carbonized phase is dependent on the carbonization temperature, with higher temperatures leading to progressively smaller heteroatom content and greater extended aromaticity. At sufficiently high temperatures (2500-3000°C), nearly pure graphite is formed [12]. The exact chemical structure of the ladder polymer is in dispute, but, in any case, the extended ring structure appears to assist in the ordering which must take place to form the more highly graphitized structures realized at high temperatures. In addition to dimensional stability and ordering mentioned above, PAN has the advantage of relatively low weight loss during carbonization, with half the loss shown by rayon [13]. In this chapter, we will outline the techniques by which the carbonization of PAN, long applied to the production of carbon fibers, can be extended to other carbon forms. Specifically, we will describe the use of PAN as a precursor to carbon films and foams. In order to more completely describe the utility of films and foams prepared in this way, a brief description of alternate means for production of these materials is in order.

Carbon films can be prepared by a wide variety of techniques, including decomposition of acetylene in an arc discharge [14], plasma deposition [15], laser-initiated organic decomposition [16], sputtering [17], and evaporation [18]. Advantages and disadvantages accrue to each of the above deposition methods. For example, evaporation and sputtering each produce nearly pure carbon and can be monitored directly to give in situ thickness control. However, both of these methods are line-of-site techniques and cannot be used to coat blocked or otherwise inaccessible surfaces. Laser-initiated decomposition can be used to precisely control the area of carbonization and can produce very small carbonized structures. Unfortunately, such rastor-controlled carbonization has inherently low throughput. Plasma deposition will be reviewed in other chapters of this book and will not be further discussed here. In general, a solvent-based coating method, such as is used for PAN deposition, can uniformly coat almost any surface for subsequent conversion to carbon. A number of surfaces coated in this way and associated device fabrication will be discussed later in the chapter.

Carbon foams have been prepared using a wide variety of approaches. Without reviewing these in detail, it is important to recognize that each approach has inherent advantages and disadvantages as well as unique capabilities. Commercially available carbon foams include reticulated vitreous carbon (RVC) [19]. Reticulated carbon foam results from the carbonization of a selectively oxidized polyurethane foam which has been swollen with a reactive monomer [20]. The resultant foam is open-celled with large pore sizes (ca. 100 micron). Phenolic microballoons can be stabilized with an appropriate binder and carbonized to produce syntactic carbon foams [21]. These low-density, closed-cell carbon foams can be prepared with high compressive strengths [21]. Substrate replication processes have also been utilized to produce carbon foams. Here a sacrificial pore-former (e.g. sintered salt) [22] is coated with a carbonizable polymer or resin. After the resin is carbonized, the pore-former is dissolved away resulting in a carbon foam.

A number of high energy physics applications for carbon foams (e.g., inertial confinement fusion) simultaneously require properties such as 1) low densities (0.04 to 0.07 g/cc), 2) small cell or pore sizes (less than 20 micron), 3) an open-celled morphology with a continuous void, and 4) dimensional stability (e.g., low coefficient of thermal expansion). Most of the approaches described above do not yield carbon foams which would be useful in these applications. We have described the preparation of low-density, microcellular, carbonized PAN foams using a solution process [23] which makes it possible to provide carbon foams with a wide range of attainable properties that fulfill these high energy physics requirements. As we will discuss in this chapter, our PAN process is very versatile as demonstrated by the capability to produce numerous structures from thin solid films to thin porous films to structural foams which are then carbonized.

CARBON FILMS FROM PAN

Film Formation

Thin films or coatings of PAN are produced in several different ways depending on the thickness required and the type of surface to be coated. In each of the solvent-based coating techniques described below, a solution of PAN in dimethylformamide (DMF) is used, with a typical concentration between 3 and 6 g/100 ml. For very thin films (a few hundred Å to ca. 1 micron) on flat surfaces, spin casting is appropriate. Typically, the substrate is held in place via a vacuum chuck, which can be spun at any rate up to 10,000 rpm. The casting solution is dispensed on the substrate either before or after the substrate has begun spinning. This is the technique almost universally employed to apply photoresist for semiconductor patterning, so spin casting equipment is commercially available [24]. Any nominally flat substrate which can be held in place via vacuum can be coated in this way. Our most common substrate is a quartz plate, 25 mm by 25 mm by 1 mm thick. The film thickness can be varied by changing the concentration of polymer in the casting solution or by changing the spin speed. Once these parameters are determined, uniform and reproducible films can be produced.

For thicker films (one to several microns), solvent casting is the preferred coating technique. By this method, casting solution is allowed to stand in a quiescent layer over the substrate until the solvent evaporates, leaving a polymer film behind. The relatively thick layers produced in this way act to planarize roughened or semi-porous surfaces.

Finally, for completely non-planar substrates, a dip-coating method may be employed. However, one must take care when removing the coated part from the solution that the solvent drains evenly so that a uniform film is produced.

We have discovered that atmospheric moisture can cause phase separation of PAN in DMF, producing films which have a foam-like morphology. In some instances this may be employed to advantage (vide infra), but it is generally undesirable and can be prevented by coating films in an atmosphere with no more than 20 per cent relative humidity. For solvent casting and dip-coating, where the solvent is removed slowly, the coating process is best done in a dryroom (<3 per cent humidity).

After the coating process is completed, the PAN is thermally pretreated in a forced air convection oven at 220°C. Thin samples can be placed directly into a pre-heated oven. However, thick samples (> 1 micron) have a tendency to crack under thermal stress, so such films should be brought to final temperature at a slow ramp (ca. 5°C/min), held at 220°C for 16 hr, and then ramped back down to room temperature. The pretreated films are then carbonized under inert atmosphere at a final temperature of at least 600°C. A typical thermal cycle involves a ramp to the final temperature, a hold at that temperature (soak) for 8 hr, and a ramp back to room temperature. In our laboratory, argon is continuously flowed through the furnace during carbonization. The argon is passed through an oxygen scrubber to reduce oxygen levels to < 3 ppm. Higher levels of oxygen will partially or completely combust the carbon during carbonization, especially in the case of thinner films.

Characterization and Spectroscopy

A number of techniques can be applied to the characterization of these films during various stages of processing. UV-visible absorbance spectra of thermally pretreated and carbonized films are shown in Fig. 1. Untreated PAN is water-white, and has no absorbance at wavelengths greater than ca. 225 nm. After pretreatment, the development of conjugated carbon-nitrogen double bonds leads to the formation of a longer wavelength absorption. Since this absorption tails into the visible region, the pretreated films take on a brown coloration. The carbonized films have a lustrous, mirror-like appearance, and the absorption spectrum is nearly structureless. It should be noted that the spectrum shown (for a film carbonized at 1020°C) is a transmission spectrum plotted in absorbance units and includes both absorption and reflection effects. The increased absorbance with decreasing wavelength to ca. 240 nm, with increased transparency at shorter wavelengths is characteristic of free-electron-like metals, and the shape of this curve is similar to that observed with 1000 Å flakes of graphite [25]. The apparent absorptivity of PAN carbon films is ca. 2 x 10^5 cm^{-1} at 400 nm.

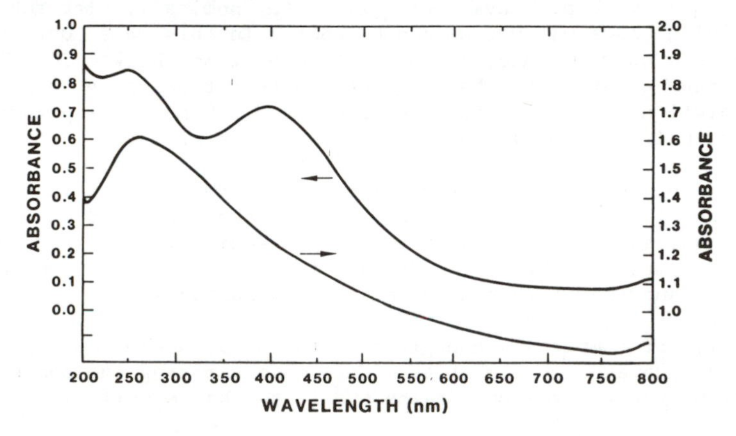

Fig. 1: UV-visible absorbance spectra: (left scale) pretreated PAN film, 16 hr at 220°C, and (right scale) carbonized PAN film, 1020°C.

Transmission infrared (IR) spectroscopy is a useful probe of chemical changes taking place in these films during pretreatment. Fig. 2 shows the IR spectra of PAN films before and after thermal pretreatment. Peaks in the untreated PAN film spectrum at 3440 and 1660 cm^{-1} were assigned to absorbed water and residual solvent respectively, and were verified by comparison with the spectrum of a film dried under vacuum for 2 hr. The sharp band at 2240 cm^{-1} is due to the nitrile stretch frequency. This band is almost completely lost during pretreatment, as expected. Methylene groups of the PAN mainchain give rise to peaks at 1430 and 2940 cm^{-1}. These peaks are greatly reduced by oxidative pretreatment. One also observes the growth of peaks at 1380 and 1590 cm^{-1}, which are assigned to carbon-carbon or carbon-nitrogen double bonds. The broad peak at 3400 cm^{-1} could be due to hydroxyl group (- OH) formation. These results are in agreement with literature results on thicker films [26]. The IR spectra of carbonized PAN showed little discernable structure.

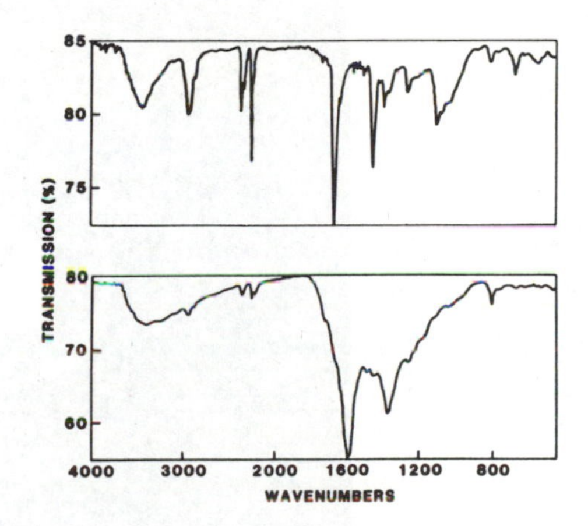

Fig. 2: IR transmittance spectra of (a) untreated PAN film and (b) PAN film thermally pretreated at 220°C for 16 hr.

Raman spectroscopy was used to monitor the degree of ordering (graphitization) which occurs at different pyrolysis temperatures. Peaks observed at 1350 and 1580 cm^{-1} shift were assigned to the disordered and ordered phases, respectively, and are shown in Fig. 3. Similar peaks were reported for 500 Å thick carbon films formed by arc-condensation [27]. The size of the 1580 cm^{-1} band increases relative to the 1350 cm^{-1} band with increasing carbonization temperature. The arc-condensed films showed similar trends due to increased ordering with higher temperature treatment. Quantitative comparison of our spectra with those of the arc-condensed films is difficult, since significant heteroatom concentration exists in our lower temperature carbons (see foam elemental analysis below). However, since the spectra in Fig. 3 are nearly identical to those of Ref. 27, and since the peak positions of our spectra do not shift with carbonization at temperatures high enough to eliminate most heteroatoms, the 1580 and 1350 cm^{-1} bands can reasonably be assigned to more ordered and less ordered phases, respectively.

Both electron diffraction and dark-field transmission electron microscopy were performed on films carbonized at 1220°C. The electron diffraction pattern is shown in Fig. 4. These data indicate that while considerable crystallinity exists, the average domain size is very small (< 20 Å). The number and relative sharpness of the electron diffraction rings indicates the presence of crystallinity with the intense, innermost ring corresponding to a combination of the 1 0 $\bar{1}$ 0 and 1 0 $\bar{1}$ 1 reflections from graphite. Since no basal planes are observed, one can conclude that these planes are parallel to the surface. The full rings, as opposed to dot patterns, indicate there is no preferred orientation relative to the c-axis. This is in agreement with literature data on carbon black particles and carbon fibers, in which the graphitic basal planes were found to be parallel to the local surface [28,29].

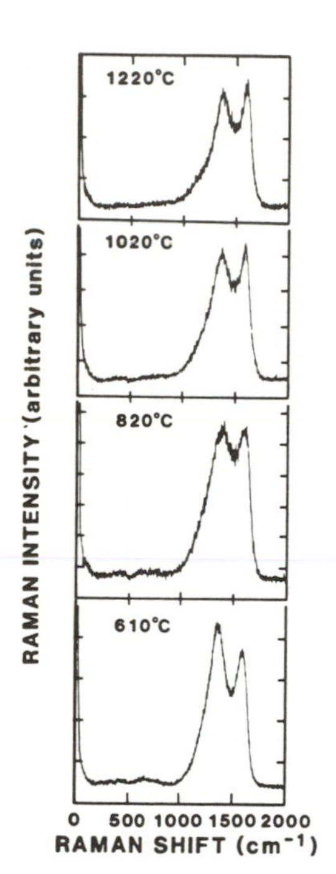

Fig. 3: Raman spectra of PAN films carbonized at the indicated temperatures.

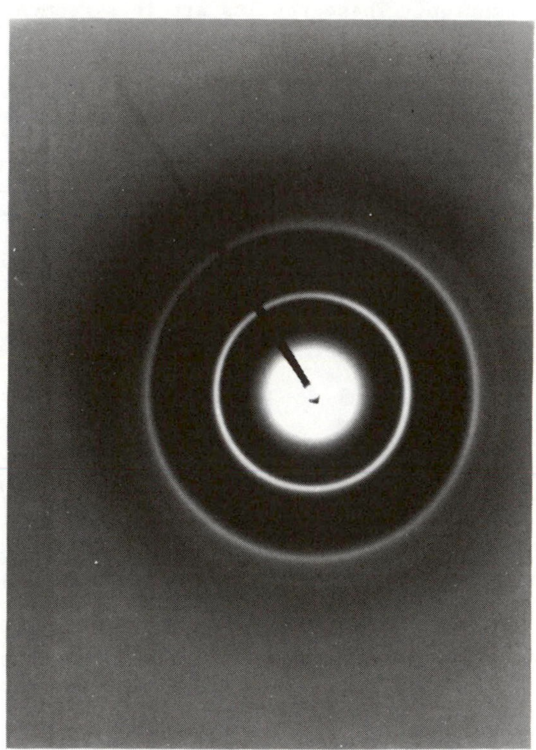

Fig. 4: Electron diffraction pattern of PAN film carbonized at 1220°C.

One of the advantages of producing carbon films from PAN is that the electrical conductivity can be varied by changing the carbonization temperature. Table I shows the conductivities (four-point probe measurement) for carbon films carbonized between 610 and 1220°C. The increase in conductivity with temperature is consistent with increased structural ordering. The conductivity of PAN-derived films at any given temperature is relatively high compared to other organic precursors. For example, the conductivity of PAN-derived films carbonized at 1220°C is approximately as high as that of poly(p-phenylene-1,3,4-oxadiazole) carbonized at 1800°C [5].

The conductivity of the 1220°C PAN-derived film is within a factor of 30 of the basal-plane value for pure graphite [30].

TABLE I

ELECTRICAL CONDUCTIVITY OF PAN FILMS

CARBONIZATION TEMP. (°C)	CONDUCTIVITY ($S\ CM^{-1}$)
610	0.12
820	160
1020	500
1220	830

It appears reasonable that the high conductivity is the result of the ease with which the ladder polymers coalesce into extended condensed ring systems. Therefore, we also produced films from other polymer types in which this ordering process would be disrupted. Fig. 5 shows conductivity data for carbon derived from PAN homopolymer, poly(4-vinylpyridine) (PVP), and PAN-co-PVP copolymer. IR analysis was used to determine that the copolymer was ca. 35 per cent by weight PAN. Of course, no cyclization takes place in PVP during thermal pretreatment. As expected, at higher temperatures of carbonization, the conductivity of each film increased with PAN content. The change in conductivity in going from pure PAN to pure PVP is equivalent to lowering the carbonization temperature from 1200 to 800°C. However, a more pronounced effect is the ease with which quality films can be produced, i.e., the dimensional stability of cyclized PAN greatly facilitates the fabrication of smooth, unbroken films. An alternate means of stabilizing other carbon precursors, reported in the literature [4], involves the use of UV flood exposure to induce cross-linking.

Oxidation and Processing

Oxidation and other etch processes are of interest both from a mechanistic standpoint and for device applications. For example, the rate at which thermal oxidation occurs has significant implications in the use of carbon films in high temperature devices. Fig. 6 is an Arrhenius plot for the rate of thermal oxidation of carbon films, carbonized at 640°C, in air. Rates were measured as the loss of film thickness per unit time. Gravimetric methods could not be applied here, as each film typically contained significantly less than 1 microgram of material. The other advantage of measuring rate by thickness loss is that rates could be measured at temperatures much lower than can be used with gravimetric techniques. Rates as low as about 1 Å/hr could be measured in this way, at temperatures as low as 300°C. In contrast, gravimetric measurements typically require temperatures of at least 500°C. The use of this method was justified by previous reports of surface-limited oxidation [31,32]. An activation energy of 33 ± 7 kcal/mol was observed for these films, in good agreement with gravimetric data on carbon-carbon composites at temperatures below 600°C, which gave 35-45 kcal/mol [31]. The 1-3 Å/hr loss at 300°C is quite low, so that films of sufficient thickness exposed to such a continuous operating temperature would be fairly stable. Of course, in the absence of oxygen, devices could be operated at any temperature up to the carbonization temperature without degradation.

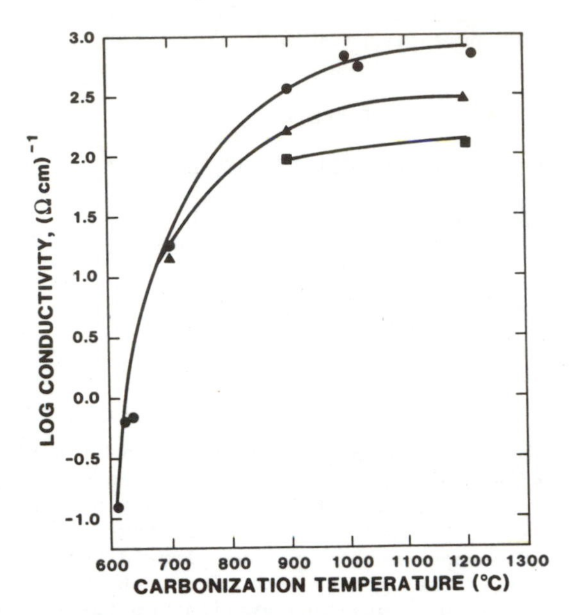

Fig. 5: Electrical conductivity of films as a function of carbonization temperature: PAN homopolymer (circles), PAN-PVP copolymer (triangles) and PVP homopolymer (squares).

The mechanistic aspects of carbon oxidation were studied by Auger analysis, as shown in Fig. 7, for films carbonized at 900°C. The spectra are of films as produced and after oxidation in air at 320°C for 65 hr. An approximate doubling of surface-bound oxygen (relative to other elements) is observed after heating in air. This is consistent with the formation of an intermediate surface-bound oxygen, as opposed to direct oxidation to CO or CO_2. The broadening of the carbon 1s line, as seen on high resolution scans, is consistent with the formation of carbon-oxygen bonds.

Obviously, the carbonization of these films must take place on a high temperature substrate. In those instances where the film must then be moved to another substrate, at least two techniques are available to accomplish the transfer. The first of these techniques is a wet etch process and is useful for transfer of carbon films from quartz substrates. The film/substrate combination is either immersed in or floated on a solution of aqueous hydrofluoric acid, (HF, 0.9-12 per cent by weight, depending on substrate size and type). The carbon is inert to attack by HF, but the underlying quartz is partially dissolved, allowing the film to float off and rest on the fluid surface. The new substrate is then placed under the film and the liquid level made to recede until the film rests on the new surface. This technique works best with porous surfaces and grids, which allow the receding solution to drain, rather than become entrained between the film and substrate. With care, films to at least 4 in. diameter can be removed and transferred intact.

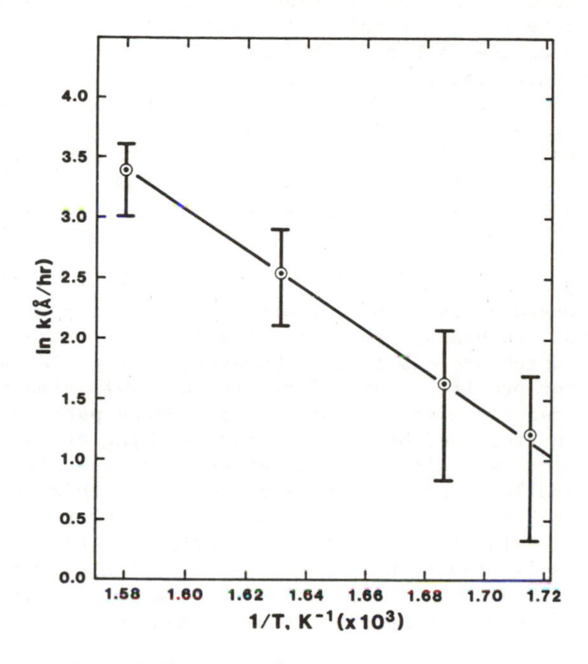

Fig. 6: Arrhenius plot for the oxidation of carbon film in air. Film carbonized at 640°C.

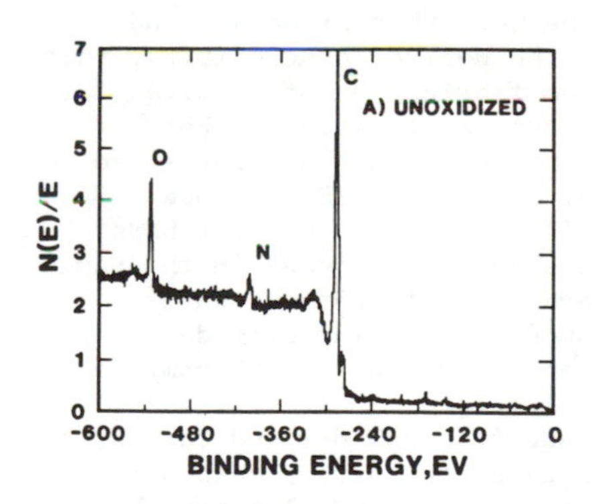

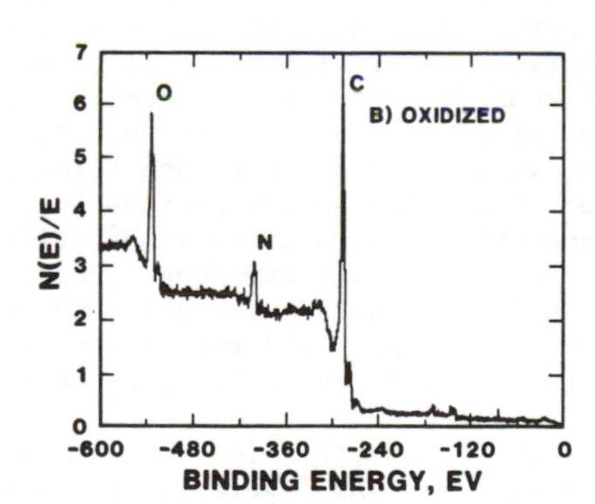

Fig. 7: XPS spectra of films carbonized at 900°C: A) without air oxidation, and B) after 65 hr in air at 320°C.

The alternate transfer method is more straightforward. In a dry bond transfer, the film is affixed to the second substrate with an appropriate adhesive. After cure, the transfer is completed with a simple peel off. Of course, the carbon/adhesive and adhesive/second substrate bonds must be stronger than the carbon/original substrate bond, so the technique requires fairly poor adhesion of the carbon to the original substrate. With quartz, low temperature carbonization results in poor adhesion. As an example, we have successfully transferred 640°C carbon films to polyimide using type 3109 unsupported epoxy film adhesive (3M Co., Minneapolis), cured at 125°C for 1-2 hr, as the bonding agent. However, if high conductivities (and, therefore, high carbonization temperatures) are required, dry bond transfer from quartz is difficult. In such a case, pyrolysis on a substrate to which PAN-derived carbon has an inherently low adhesion, such as lithium tantalate, is preferred.

As will be discussed below, applications frequently arise in which films need to be patterned. Carbon's high chemical and thermal stability limit the types of patterning which are realistic. However, oxygen plasma etching provides an option for precise, rapid patterning of small dimension structures. We have demonstrated the ability to produce pattern masks from photoresist on carbon, followed by oxygen plasma etching of features as small as several tens of microns. Although not yet demonstrated, it appears likely that significantly smaller features can be patterned in this way. Typical plasma conditions are 0.5 torr oxygen and 50 W plasma power. This gives carbon etch rates of 70-150 Å/min. After completion of the etch process, the photoresist mask is solvent-stripped. Care must be taken when etching thick films already transferred to low temperature substrates, as long uninterupted etches can cause surface heating and substrate damage effects.

Applications

Although these films are not semiconductors in the sense of having non-linear current-voltage characteristics, the conductivities produced over a wide range of carbonization temperatures fall between that of metals and insulators. Coupled with the ability to tailor the desired conductivity, this makes carbon films attractive candidates for the fabrication of passive devices for microelectronic and hybrid microcircuit applications. For example, we have produced carbon films on alumina substrates and patterned resistor bars using AZ1350 photoresist as an etch barrier. Resistance measurements were made after evaporative deposition of Cr-Au contact bars. In this way, resistors have been produced with sheet resistances in the range 100-400,000 ohm/square. The 500-1000 ohm/square range is not currently accessible with thin film resistor materials such as tantalum nitride. Temperature coefficient of resistance measurements gave ca. - 1000 ppm/°C.

Patterned conductor elements can be fashioned from PAN-derived carbon, as shown in Fig. 8. The figure shows a quartz plate coated with a carbon film which has been patterned to yield a series of interdigitated electrode bars. The interelectrode spacing is 300 microns. These electrode materials were produced for electrochemical measurements relating to battery research.

Since remote surfaces can be coated by solvent-based deposition, PAN-derived carbon is finding application in protective barrier coatings. A cathode current collector consisting of an open-celled nickel foam was dip-coated with PAN, pretreated, and carbonized to yield a carbon-coated foam. The carbon was required to protect the nickel from attack by thionyl chloride, an extremely agressive electrolyte. Nickel passivation (by conversion to nickel chloride) was monitored by measuring the electrode capacitance initially and after 16 months immersion in thionyl chloride. An

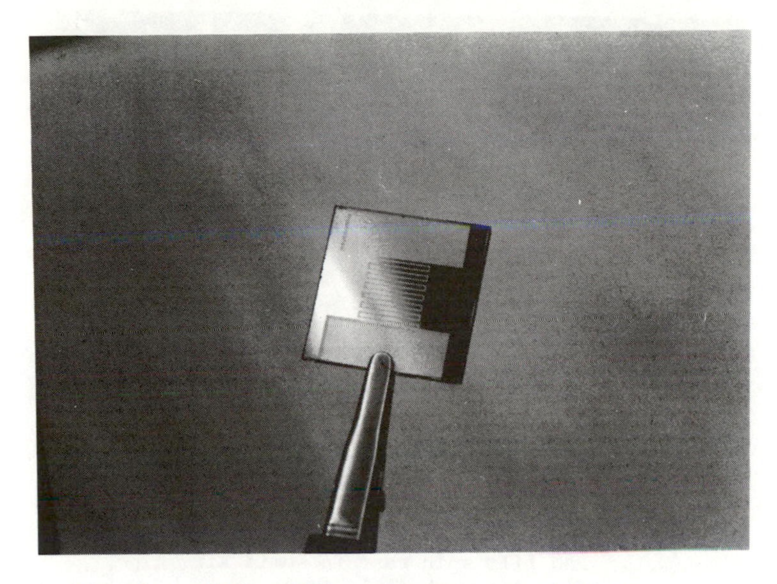

Fig. 8: Carbon film on quartz patterned as interdigitated electrodes. Interelectrode spacing is 0.3 mm.

approximate 2 per cent change in capacitance indicated the carbon was quite effective as a barrier in this case. Other metals, such as stainless steel, have also been carbon-coated in this way.

Finally, it is possible to coat carbon-on-carbon, for the purpose of modifying surface morphology. A semi-porous graphite cylinder was carbon-coated in order to planarize the surface. In this way, a similar chemical structure was maintained while smoothing the topography for subsequent coating operations.

POROUS FILMS

As mentioned above, PAN film deposition in a humid environment leads to phase separation and a film with a foam-like morphology. This structure can be maintained through the pretreatment and carbonization steps. Fig. 9 shows a PAN film spin cast on quartz in a humid environment. Since such films have a high surface area, possible applications might include microsensors, in which a porous, electrically conductive film must be loaded with an active sensing element. These films bridge the gap between fully dense films and the carbon foams discussed in the next section. In effect, one can think of these foam-like films as macroscopic foams with the thickness dimension collapsed. However, as will be described, the control one has over foam morphology is much more sophisticated in the case of the bulk foams.

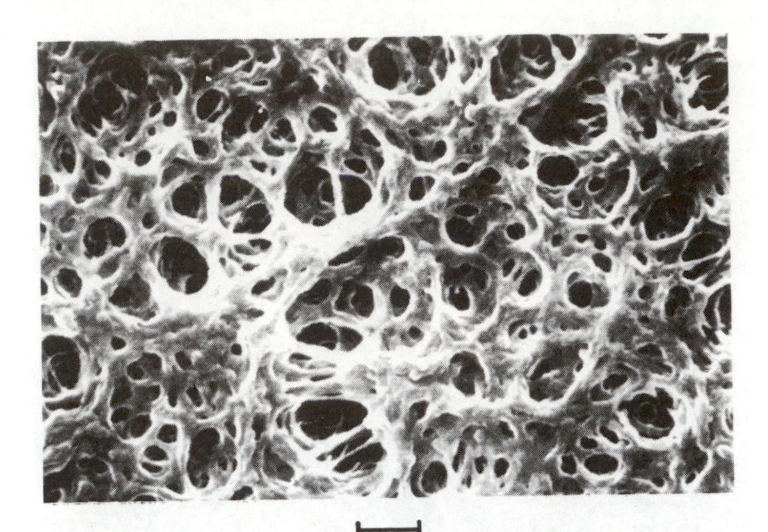

Fig. 9: Scanning electron photomicrograph of PAN film spin cast in humid air. Fiducial mark is 1 micron.

CARBON FOAMS FROM PAN

Preparation of PAN Carbon Foams

Carbonized PAN foams are prepared in a three-step process [23]. In the first step, a low-density PAN foam is prepared by thermally-induced phase separation (TIPS) of a solution of PAN or a PAN-based copolymer (e.g., acrylonitrile-co-maleic anhydride). The use of TIPS for the preparation of low-density polymer foams has been reported for polymers such as poly-4-methyl-1-pentene (TPX) [33] and polystyrene [34]. The TIPS process is illustrated schematically in Figure 10. To prepare a PAN foam, a homogeneous solution of PAN (typically 1.5 to 10% by weight) in a solvent (e.g., maleic anhydride, MA) is prepared above the solution critical temperature (T_c). The solution is then vacuum degassed and poured into a mold. The mold consists of a high thermal conductivity base of copper and insulating sides made of RTV silicone [34]. The mold is transferred to a copper plate on a temperature controlled bath to effect a thermal quench. Upon cooling the solution, phase separation occurs with the formation of two interpenetrating phases: one rich in solvent and one rich in polymer. Once the solution is cooled below its freezing point (T_f), the solvent can be removed by vacuum sublimation. Removal of the solvent affords a low-density microcellular PAN foam. Alternatively, the solvent can also be removed by extraction [23].

In the second step of the process, the PAN foams are thermally pretreated in a forced-air convection oven as described previously for thin PAN films. The white PAN foam takes on a brown coloration of the "ladder polymer" structure. In order to maintain dimensional control and retain the microcellular morphology of the PAN foam during subsequent high temperature carbonization, the pretreatment conditions were optimized to complete the cyclization of PAN and remove the associated exotherm as revealed by differential scanning calorimetry (DSC). Optimized pretreatment conditions

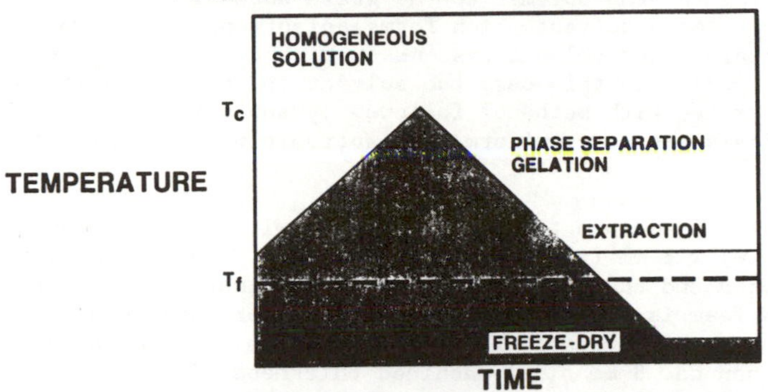

Fig. 10: Schematic representation of the TIPS process. T_c and T_f refer to the critical and freezing temperatures of the polymer solution, respectively.

for the low-density PAN foams were found to be a slow ramp (ca. 5°C/min) to 180°C for 6 hr followed by a slow ramp (ca. 5°C/min) to a final temperature of 220°C for 16 hr and finally a slow ramp (ca. 5°C/min) to room temperature. Additionally, the cyclization reaction has been monitored by infrared spectroscopy for loss of the nitrile stretch band at 2240 cm^{-1} as reported for thin PAN films.

In the final process step, the pretreated PAN foams are carbonized at high temperatures in a purified argon atmosphere. The high temperature treatments were carried out in a programmable (Honeywell controller/programmer) three-zone tube furnace (Lindberg, 1200°C) with a quartz retort. Programmed cycles consisted of a 5°C/min ramp to an intermediate temperature 100°C below the final temperature for 1 hr followed by a 3°C/min ramp to the final temperature for 8 hr and a slow ramp (ca. 3 to 5°C/min) back to room temperature.

The result of this three-step process is a rigid, low-density, microcellular carbonized foam which retains the same shape and microstructure as the initial PAN polymer foam. It should be noted that these low-density foams can be readily machined to close tolerances at each stage of processing.

Characterization of Foams

In general, the properties of carbonized PAN foams are dependent upon the foam morphology, density and final carbonization temperature. The density of the initial PAN foam is determined by the PAN solution concentration. Using TIPS to prepare PAN foams permits the preparation of a wide range of PAN foam densities and morphologies. The PAN foam morphology is determined by the mechanism by which phase separation occurs. If liquid-liquid phase separation occurs prior to the solvent freezing, the resulting PAN foam morphology is very uniform and isotropic as shown in Figure 11 for a carbonized PAN foam prepared with MA as the solvent. Anisotropic or directional PAN foams are produced when the solvent freezes prior to phase separation forcing the PAN polymer to the grain boundaries. Methyl sulfone (MS) is an example of a solvent which forms anisotropic PAN foams as shown in Figure 12. Certain mixed solvent systems such as 84:16 DMF/water (v/v%) form stable PAN gels [23]. In this case the solvent is not frozen and the PAN gels can be extracted with methanol followed by sub-critical CO_2. The resulting PAN foams are very uniform and isotropic with a very small cell size.

Carbonized PAN foams are typically prepared with final dimensions of 120 mm (l) x 30 mm (w) x 6 mm (t). In solvent systems, such as MA, where the solution is frozen and the solvent removed by sublimation, the thickness for the initial PAN foam is limited due to heat transfer considerations during the thermal quench. As a result it is difficult to prepare carbonized PAN foams greater than ca. 8 mm final machined thickness with MA as the solvent. The other aerial dimensions are limited only by processing time and equipment constraints. Solvent systems such as DMF:water which yield stable gels can be cast to greater thicknesses and dimensions which are limited only by processing constraints [35].

The PAN foams undergo both mass loss and shrinkage during pretreatment and carbonization. The overall shrinkage is uniform in all dimensions and is typically 20 to 30% linear. Greater mass loss and greater shrinkage accrue with higher processing temperatures. Overall densification of the foams ranges from ca. 30% at 600°C to ca. 50% at 1200°C. After machining the low-density PAN foams to parallel dimensions, the bulk foam densities are determined by metrology (i.e., mass/volume). Low-energy (6 KeV helium-flight-path) X-ray radiography is used to detect non-uniformities on the order of several percent. In general, the carbonized PAN foams exhibit excellent radiographic homogeneity.

Examination of foam cross-sections by scanning electron microscopy (SEM) demonstrates that the initial PAN foam morphology is preserved through carbonization. Of merit for various applications is a carbon foam with a uniform, isotropic, open-celled microstructure with no cell-size gradient through the thickness. Using MA as the solvent, carbonized PAN foams can be prepared with densities ranging from 0.03 to 0.5 g/cc with a uniform isotropic morphology (Figure 11) and cell sizes of ca. 10 to 2 micron respectively. PAN gel-derived carbon foams can be prepared with a uniform, isotropic morphology and cell size of ca. 0.5 to 1 micron as shown in Figure 13. Using a mixed solvent system composed of 55:45 MA:succinonitrile (SN), which also gels, carbonized PAN foams can be prepared with uniform cell sizes ranging form 0.1 to 0.5 micron as shown in Figure 14. Carbonized PAN foams with anisotropic morphologies with 50 to 100 micron pores oriented through the thickness (Figure 12) can be prepared with a density range of 0.04 to 0.5 g/cc.

Due to their very small cell sizes, microcellular foams exhibit high surface areas [36]. The surface areas of carbonized PAN foams are determined primarily by the foam morphology. In general, smaller cell sizes afford greater surface areas at a fixed foam density. For MA-derived PAN carbon foams, the surface areas determined by the BET method are typically 5 to 80 m^2/g. As shown in Table II, the BET surface area for MA-derived PAN carbon foams decreases with increasing carbonization temperature for a foam density of ca. 0.05 g/cc. This is likely due to annealing of microporosity within

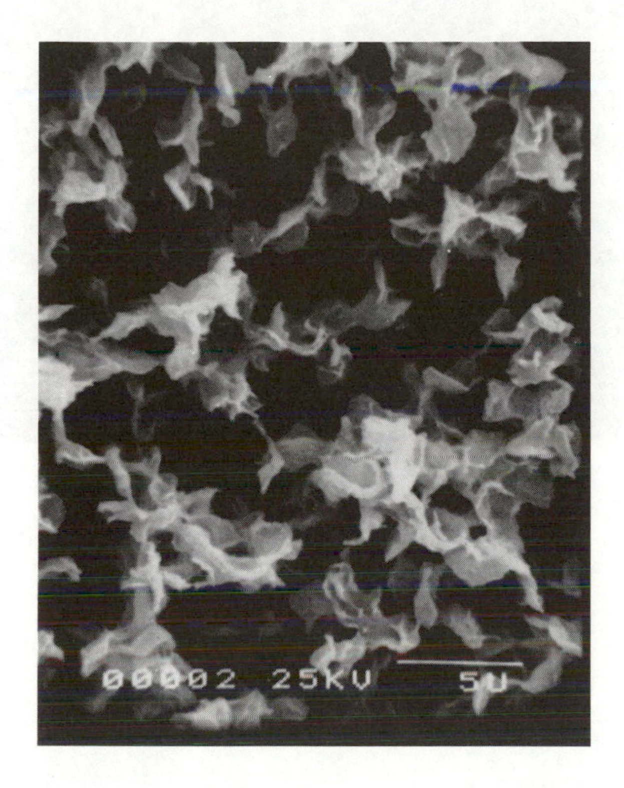

Fig. 11: Scanning electron (SEM) photomicrograph of a carbonized PAN foam prepared using MA as the solvent. The foam was carbonized at 1200°C and has a density of 60 mg/cc. The fiducial mark is 5 micron.

the struts of the carbonized foam with increasing processing temperature.

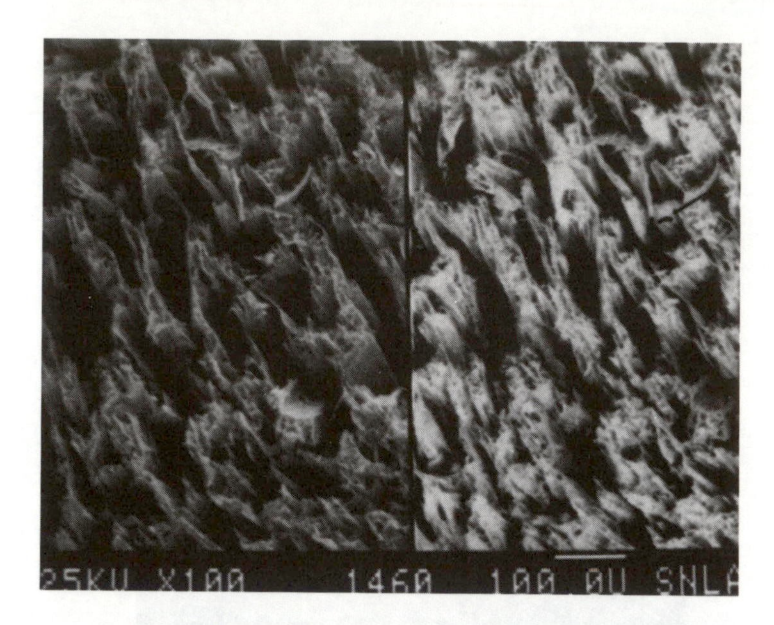

Fig. 12: SEM photomicrograph (left) of a carbonized PAN foam prepared using MS as the solvent. Backscatter image (right) of the same region demonstrates the uniform incorporation of Pd (10% w/w) onto the carbon foam. The image is a fracture surface parallel to the mold bottom. The fiducial mark is 100 micron.

Helium pycnometry varifies that the density of the carbon struts increases with increased carbonization temperature as shown in Table II.

The elemental composition of low-density carbonized PAN foams is summarized in Table III. Increasing the carbonization temperature increases the carbon content of the foams. The small residual oxygen and nitrogen contents at 1000 and 1200°C are consistent with data obtained from PAN-derived carbon fiber work [6,10]. The residual nitrogen content is likely due to the difficulty in removing nitrogen from heterocyclic aromatic structures such as the aromatic ladder polymer. Carbonization of PAN foams at 1400°C is sufficient to produce elemental carbon. The presence of nitrogen or oxygen at the surface of the foam may be useful for improved wetting characteristics or for coordination of catalysts.

For applications where carbonized PAN foams must be machined to close tolerances (e.g., hemispherical targets for inertial confinement fusion experiments) it is desirable to have a material with good dimensional stability and a low coefficient of thermal expansion (CTE). The CTE's for carbonized PAN foams are summarized in Table II. The CTE's were measured over a wide temperature range (-55 to 150°C) using a Harrop fused-silica pushrod dilatometer. Foams carbonized at temperatures greater than 1000°C showed a linear response and low CTE's characteristic of glassy carbon (typically 3-4 ppm/°C) [37]. Foams carbonized at 600 to 900°C showed non linear (sigmoidal) response with regions of larger expansion (e.g., 6-8 ppm/°C) followed by regions of contraction and then expansion with increasing temperature. This non-linear behavior is attributed to desorption of gases or adsorbed water from the more porous struts of these foams.

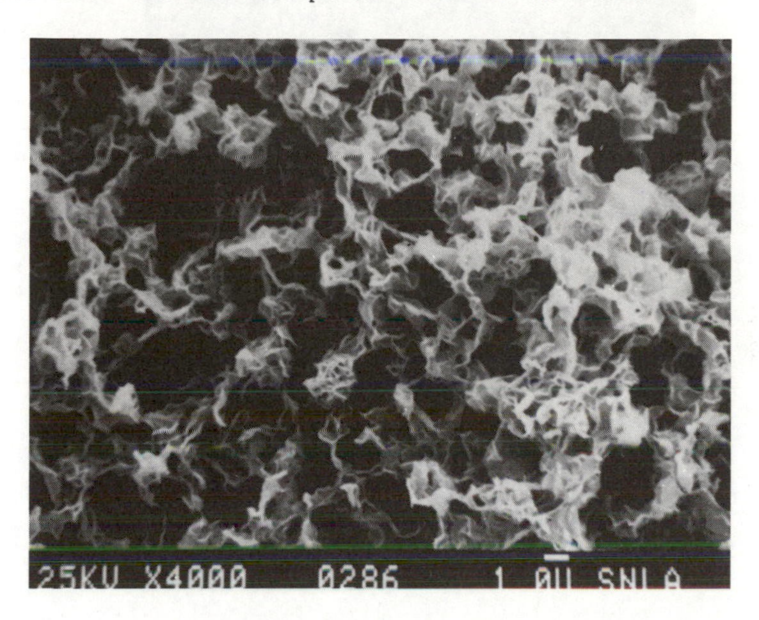

Fig. 13: SEM photomicrograph of a carbonized PAN foam prepared using a DMF/water gel. The foam was carbonized at 1200°C and has a density of 110 mg/cc. The fiducial mark is 1 micron.

TABLE: II

PROPERTIES OF CARBONIZED PAN FOAMS*

CARBONIZATION TEMP. (°C)	FOAM DENSITY (mg/cc)	CARBON DENSITY (g/cc)	BET SURFACE AREA (m^2/g)	CTE (ppm L/L°C)
800	54.4	1.52	76.6	6 (Sigmoidal)
1000	58.8	1.59	55.7	4.0
1100	56.5	1.73	48.9	3.7
1200	55.0	1.76	28.2	3.2

*Maleic anhydride solvent
* 220°C Pretreatment

Fig. 14: SEM photomicrograph of a carbonized PAN foam prepared from a MA:SN gel. The foam was carbonized at 1100°C and has a density of 120 mg/cc. The fiducial mark is 1.5 micron.

Potential Applications

In addition to their use as materials for the fabrication of high-energy physics targets, carbonized PAN foams have other potential applications as porous electrodes, catalyst supports, filters, high temperature insulation materials and in the preparation of electrically conductive reticulated carbon composites.

The continuous void of carbonized PAN foams can be filled with an epoxy resin or a monomer (e.g., styrene) polymerized *in-situ*. The result is a novel type of electrically conductive composite [38]. The bulk electrical conductivity of these "reticulated carbon composites" can be controlled over a wide range through adjustment of the carbonized PAN foam processing parameters. Figure 15 illustrates that the electrical conductivity of carbon foam/epoxy composites smoothly increases with increasing foam carbonization temperature. The electrical conductivity can be controlled from approximately 10^{-7} to 10^{-1} S cm^{-1} simply through adjustment of the carbonization temperature for foams at a constant density of 0.06 g/cc. Adjusting the PAN carbon foam density at a fixed carbonization

TABLE III

ELEMENTAL ANALYSIS OF CARBONIZED PAN*

CARBONIZATION TEMP.(°C)	%C	%H	%N	%O
600	70.13	0.85	18.38	10.64
800	75.41	0.23	13.43	9.93
1000	90.60	<0.05	5.03	2.90
1200	96.93	<0.05	1.75	0.61
1400	99.99	-	-	-

* 220°C Pretreatment
* Weight %
* Oxygen content by neutron activation

temperature of 1200°C allows the conductivity to be controlled further from approximately 10^{-2} to 10^{2} as shown in Figure 16. Using this approach, greater electrical conductivity is achieved at a much lower volume percent carbon (when compared to carbon powder filled composites) presumably due to the "built-in" conductive framework of the carbon foam. Most of the volume in these composites is occupied by the organic resin of polymer filler, hence the bulk mechanical properties and environmental stability of these composites resemble those of the chosen filler polymer

The high surface areas and the open porous microstructure of the carbonized PAN foams make them attractive as "flow-through" catalyst supports. We have demonstrated that catalytic metals such as Pd, Pt and Ni can be incorporated onto the carbon foam surfaces [23]. This can be accomplished through the incorporation of a soluble organometallic compound (e.g., palladium acetylacetonate) into the initial PAN solution prior to TIPS. A carbonized PAN foam with highly dispersed Pd deposited on the foam structure (Figure 12) results. The selective hydrogenation of terminal acetylenes has been demonstrated in solution with Pd on a carbonized PAN foam support [23]. Soluble organometallic compounds can also be incorporated after the foams are carbonized. Subsequent thermal decomposition can be utilized to form immobilized active catalysts [39].

In preliminary experiments, carbonized PAN foams show nearly ideal theoretical porous electrode behavior [40]. The ability to control the density, surface area and morphology (i.e., cell size) over a very wide range make the carbonized PAN foams attractive for investigation as novel porous electrodes. Additionally, the ability to incorporate redox metals onto the carbon foam surfaces and the ability to control the carbonized foam elemental composition presents opportunities for the development of improved battery electrodes.

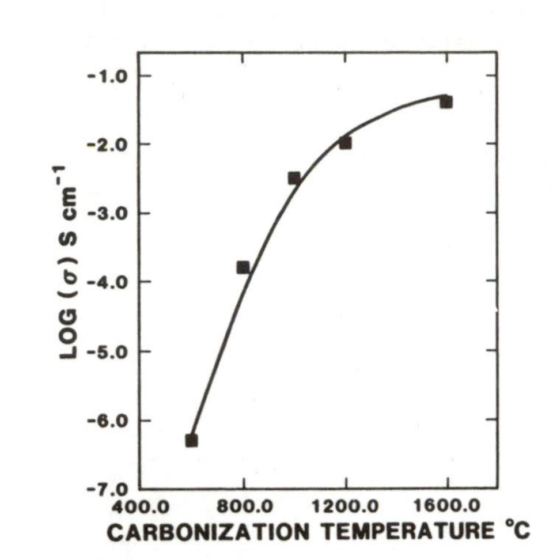

Fig. 15: Electrical conductivity of carbon foam/epoxy composites as function of carbonization temperature for foams of a constant density of 60 mg/cc.

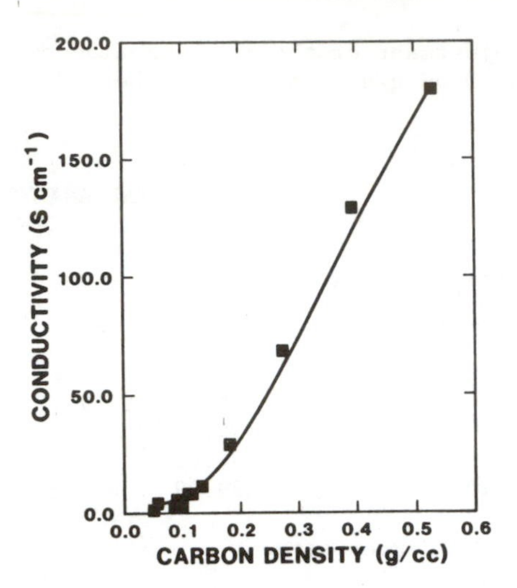

Fig. 16: Electrical conductivity of carbon foam/epoxy composites as a function of density for foams carbonized at 1200°C.

Acknowledgments

The authors wish to thank coworkers J. Aubert, P. Rand, R. Clough, C. Arnold (all of SNLA) for collaborative efforts. We thank F. Delnick for collaborative efforts on the battery work. L. Maestas and D. Husskinson provided the SEM assistance. J. Aubert and E. Russick provided the BET surface area analysis. We gratefully acknowledge R. Martinez for assistance with film bonding and transfer, D. Norwood, K. Schubert and E. Mueller for assistance with electrical measurements. We also wish to thank M. Keenan for the helium pycnometry, A. Galuska for XPS spectroscopy, D. Tallant for spectra Raman and C. Hills for the electron diffraction work. We thank W. Morgan and A. Sanchez for the patterning work. We are grateful to L. Salgado, J. Miller, E. Perea, D. Strall and M. Malone for the general technical assistance. This work was performed at Sandia National Laboratories supported by the U.S. Department of Energy under contract number DE-AC04-76DP00789.

REFERENCES

1. Bacon, R. in "Chemistry and Physics of Carbon", vol. 9, ed. by P. L. Walker and P. Thrower (Marcel Dekker, New York) 1973

2. Ross, J. H.: Appl. Polym. Symp., 1976, 29, 151

3. Shindo, A.; Nakanishi, Y. and Sema, I.: Appl. Polym. Symp., 1969, 9, 271

4. Lyons, A. M.: J. of Non-Crystalline Solids, 1985, 70, 99

5. Yasujima, H.; Murakami, M. and Yoshimura, S.: Appl. Phys. Lett., 1986, 49(9), 499

6. Henrici-Olivé, G. and Olivé, S.: Adv. Polym. Sci., 1983, 51, 1

7. Shindo, A.: Report Osaka Industrial Research Institute, 1961, No. 317

8. Johnson, J. W.: Appl. Polym. Symp., 1969, 9, 229

9. Goodhew, P. J.; Clarke, A. J. and J. E. Bailey: Mater. Sci. Engin., 1975, 17, 3

10. Watt, W.: Nature, 1972, 236, 10

11. Ko, T. -H.; Ting, H. -Y. and Lin, C. -H.: J. Appl. Polym. Sci., 1988, 35, 631

12. Fitzer, E. and Miller, D. J.: Chem. Ztg. Chem. Appel., 1972, 96, 20

13. Ezekiel, H. M. and Spain, R. G.: J. Polym. Sci., 1967, C 19, 249

14. Anderson, D. A.: Philos. Mag., 1977, 35, 17

15. Staryga, E.; Lipinski, A.; Mitura, S. and Has, Z.: Thin Solid Films, 1986, 145, 17

16. Ohdaira, H.; Suzuki, H. and Saito, M.: Int. J. for Hybrid Microelectronics, 1983, 6(1), 276

17. Hauser, J. J.: Solid State Commun., 1975, 17, 1577

18. Morgan, M.: Thin Solid Films, 1971, 7(5), 313

19. ERG Corp., Oakland, CA 94608.

20. Geer, H. C. in "Encyclopedia of Polymer Science and Technology", ed. by H. F. Mark, N. G. Gaylord and N. M. Bikales (Interscience, New York) 1970, p. 102.

21. Benton, S. T. and Schmitt, C. R.: Carbon, 1972, 10, 185

22. Pekala, R. W. and Hopper, R. W.: J. Mat. Sci., 1987, 22, 1840

23. Sylwester, A. P.; Aubert, J. H.; Rand, P. B.; Arnold, Jr., C. and Clough, R. L.: Proc. Am. Chem. Soc., Div. Polym. Mat. Sci. Engin., 1987, 57, 113

24. Model 1-EC101D-R485 Photo-resist Spinner, Headway Research, Garland, TX.

25. Zhang, J. M. and Ekland, P. C.: J. Mater. Res., 1987, 2(6), 858

26. Chung, T. -C.; Schlesinger, Y.; Etamed, S.; MacDiarmid, A. G. and Heeger, A. J.: J. Polym. Sci:Polym. Phys., 1984, 22, 1239

27. Ronz, J. N.; Oberlin, A. and Beny-Bassey, C.: Thin Solid Films, 1983, 105, 75

28. Kinetlso, E. A.: Proc. First and Second Conf. on Carbon, 1956, 21

29. Knibbs, R. H.: J. Microscopy, 1971, 94, 273

30. Murakami, M.; Watanabe, K. and Yoshimura, S.: Appl. Phys. Lett., 1986, 48(23), 1594

31. McKee, D. W.: Carbon, 1987, 25(4), 551

32. Walker, Jr., P. L.; Rusinko, Jr., F. and Austin, L. G.: Adv. in Catalysis, 1959, 11, 164

33. a) Young, A. T.; Moreno, D. K. and Marsters, R. G.: J. Vac. Sci. Technol. 20, 1982, 1094; and b) Young, A. T.: J. Cellular Plastics, 1987, 23, 55

34. Aubert, J. H. and Clough, R. L.: Polymer, 1985, 26, 2047

35. Aubert, J. H.: private communication

36. Aubert, J. H.: J. Cellular Plastics, 1988, 24, 132

37. Grayson, M. in "Encyclopedia of Composite Materials and Components", (Wiley, New York), 1983

38. Sylwester, A. P. and Clough, R. L.: Proc. Am. Chem. Soc., Div. Polym. Sci. Engin., 1988, 58, 1049

39. Sylwester, A. P.: unpublished results

40. Delnick, F. M. and Sylwester, A. P.: unpublished results

Materials Science Forum Vols. 52 & 53 (1989) pp. 323-340

OPTICAL PROPERTIES AND LOCAL ATOMIC BONDING IN HYDROGENATED AMORPHOUS CARBON AND SILICON-CARBON ALLOYS

F.W. Smith
Department of Physics
City College of the City University of New York
New York, NY 10031, USA

ABSTRACT

A simple microstructural model has been presented for a-C:H and a-$Si_{1-x}C_x$:H alloy films which is based on the presence of four components in the films (amorphous diamond-like or tetrahedral, graphitic, polymeric, and void). Values for the volume fractions of these components present in the films have been determined from an effective medium approximation analysis of the measured film optical dielectric function and density. A wide range of film properties can be understood on the basis of this microstructural model. In addition, a ternary phase diagram for alloys of tetrahedral carbon, C(sp^3), trigonal carbon, C(sp^2), and hydrogen is presented and is suggested to represent a useful way of demonstrating the compositional relationships of these a-C:H alloy films.

I. INTRODUCTION

Films of hydrogenated amorphous carbon (a-C:H) have received considerable attention due to their interesting "diamond-like" properties (high hardness, high transparency, high resistivity, and chemical inertness) and also due to the fact that, as amorphous semiconducting films, it has been demonstrated that they can be doped *n* or *p* type. Thus, these films have potential applications as hard, transparent optical coatings, wear-resistant coatings, and also as a novel semiconducting material. A strong indication of the interest in this area is demonstrated by the fact that several very useful reviews of amorphous carbon films have appeared recently [1-4].

The microscopic origins of these interesting properties have been to a large degree clarified. Both threefold (trigonal) and fourfold (tetrahedral) coordinated carbon atoms are present in the films and play important roles in film structure and properties. Hydrogen is also incorporated in the films and plays a crucial role in helping to stabilize the tetrahedral coordination of carbon atoms.

It is the goal of this presentation to propose a simple microstructural model [5] for a-C:H and to determine the degree to which such a model based on four components, amorphous diamond-like, graphitic, polymeric, and void, can explain macroscopic film properties such as the measured optical dielectric function ε and film density ρ when used along with the effective medium approximation (EMA). Use of the EMA has allowed the determination of the volume fractions υ_i of the four proposed components both for as-deposited and annealed films. On the basis of this simple microstructural model, a wide range of a-C:H film properties can be understood, including not only ε and ρ, but also film hardness, IR absorption, optical energy gap, electrical resistivity, valence band and carbon K near-edge absorption, doping, and thermopower.

A similar microstructural model [6] is also presented for a-$Si_{1-x}C_x$:H alloy films. These alloys have been studied in order to determine the degree to which the incorporation of Si in a-C:H can stabilize the tetrahedral bonding of carbon.

II. ATOMIC BONDING AND THE EFFECTIVE MEDIUM APPROXIMATION

We will first discuss the local atomic bonding units expected to be present in amorphous carbon and carbon-based alloy films. The effective medium approximation which will be used to model the measured optical constants and density of these films in terms of their microstructure will then be presented.

A. ATOMIC BONDING

For the a-C:H films studied, it is proposed that the local atomic bonding units will be formed from (1) carbon atoms with tetrahedral coordination, C(sp^3), (2) carbon atoms with trigonal coordination, C(sp^2), and (3) hydrogen atoms, H. There are obviously many possible atomic bonding units which can be constructed from these three atoms and which could therefore be present in the films. In order to keep the model simple, we will focus on only three fundamental atomic bonding units, the minimum number from which a useful model for film properties can be obtained.

1. DIAMOND-LIKE BONDING UNIT

The bonding unit C(sp^3) - C(sp^3)$_4$, consisting of a central carbon atom tetrahedrally coordinated to four other C(sp^3) atoms (figure 1A), is a clear choice for the diamond-like bonding unit. In a-$Si_{1-x}C_x$:H alloys, the corresponding tetrahedral bonding unit will consist of either a C or Si atom surrounded by four other tetrahedrally coordinated C(sp^3) or Si(sp^3) atoms.

2. GRAPHITIC BONDING UNIT

The graphitic bonding unit C(sp^2)- C(sp^2)$_3$ π consists of a central carbon atom trigonally coordinated to three other $C(sp^2)$ atoms, with an additional π-electron present (figure 1B). In crystalline graphite this π - electron is delocalized, while in amorphous carbon films it can be expected to be localized to some extent, depending on the nature of the adjacent bonding units.

3. POLYMERIC BONDING UNIT

Although a wide variety of hydrogen-containing bonding units may be present, we choose for simplicity to consider only the $C(sp^3)$ -$C(sp^3)_2H_2$ unit (figure 1C), which is equivalent to the fundamental unit of the polyethylene chain polymer, $(CH_2)_n$. We note that another possible polymeric bonding unit would be $C(sp^2)$ -$C(sp^2)_2H\ \pi^*$, equivalent to the $(CH)_n$ unit of the polyacetylene chain polymer. This latter unit is interesting in that a "localized" π - electron (indicated as π^*) can be considered to be present in the bonding unit.

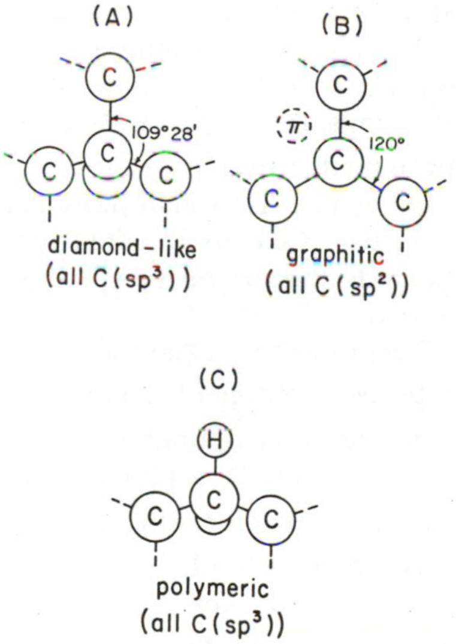

Fig. 1. Three bonding units proposed to be present in a-C:H films: A. diamond-like $C(sp^3)$ - $C(sp^3)_4$; B. graphitic, $C(sp^2)$ - $C(sp^2)_3\ \pi$; C. polymeric $C(sp^3)$ - $C(sp^3)_2H_2$.

B. THE EFFECTIVE MEDIUM APPROXIMATION

In order to model the optical constants of these a-C:H films, we will use the Bruggemann effective medium approximation [7] in which we assume that the film is a composite heterogeneous medium consisting of amorphous diamond-like, graphitic, polymeric, and void components [5]. The first three components mentioned correspond to the three local atomic bonding units described in the previous section. The EMA expression for ε, the dielectric function of the composite medium, is

$$\sum_i v_i \frac{\varepsilon_i - \varepsilon}{\varepsilon_i + 2\varepsilon} = 0 \qquad \sum_i v_i = 1, \tag{1}$$

where υ_i is the volume fraction and ε_i the dielectric function of component i. For the EMA to be valid, the scale of heterogeneity in the film (i.e. the size of regions consisting of a single component) should be much smaller than the wavelength of light in the medium, yet large enough so that the individual regions have "bulk-like" dielectric properties. Thus while these films are proposed to have a heterogeneous microstructure, we suggest that they are nevertheless macroscopically homogeneous.

Specific questions concerning the scale of heterogeneity present, i.e. the sizes of the regions consisting of a single component, are most naturally discussed in terms of the medium range order (MRO) present in the film. This issue has been addressed in detail by Robertson and O'Reilly for the case of the sp^2- bonded C atoms. Robertson and O'Reilly [8-10] have carried out electronic structure calculations for a variety of atomic structures. Briefly, they find that planar 6-member rings are the lowest-energy configuration for sp^2-bonded clusters and propose further that these π -bonded clusters are separated from each other by sp^3 bonds. They also show that E_{opt} is determined by the size of these clusters, i.e. by the medium range order and that E_{opt} is inversely proportional to the size of the sp^2 -bonded cluster. We will return to this question below.

In order to use the EMA expressed by equation 1 to determine the volume fractions υ_i of the proposed components from the measured $\varepsilon = \varepsilon_1 + i\,\varepsilon_2$ of the film, the ε_i for the individual components must first be known. As outlined previously [5] for the case of a-C:H, we have used "smoothed" values of ε_1 and ε_2, derived from measurements on crystalline diamond, for the ε_i of the amorphous diamond-like component (with $\upsilon_i = \upsilon_{ad}$). The same type of smoothing procedure has been carried out for the amorphous polymeric component ($\upsilon_i = \upsilon_{ap}$) where ε_1 and ε_2 data for a polyethylene film have been used. For the amorphous graphitic component ($\upsilon_i = \upsilon_{ag}$) the measured ε_1 and ε_2 for an evaporated carbon film have been used. The medium range order [8] present in such a film corresponds to sp^2-bonded aromatic clusters of average size about 15 Å and with an E_{opt} of about 0.7 eV. A void component ($\upsilon_i = \upsilon_v$) has also been included, with $\varepsilon_1 = 1$ and $\varepsilon_2 = 0$.

An additional equation which can be used to help determine the υ_i involves the measured film density ρ, and is expressed by

$$\rho = \sum_i \rho_i \upsilon_i \,, \tag{2}$$

where $\rho_{ad} = 3.2$, $\rho_{ap} = 0.92$, and $\rho_{ag} = 2.0$ gcm^{-3}. The measured film densities have in this way been very important for setting constraints on the possible υ_i. This is due in large part to the fact that the four proposed components have quite different densities.

For the a-$Si_{1-x}C_x$:H alloys, we have used an amorphous tetrahedral component based on Si- and C- centered tetrahedra in place of the amorphous diamond-like component. The development of this approach and of the calculation of ε for this component have been discussed in detail previously [11], and will not be discussed further here.

III. EXPERIMENTAL RESULTS FOR OPTICAL PROPERTIES

The experimental results obtained for ε_1, ε_2, and the optical energy gap E_{opt} for a-C:H and a-$Si_{1-x}C_x$:H alloy films will now be presented. These results for ε_1 and ε_2, obtained from measurements of film transmittance and reflectance [5,6], will be used in the EMA in section IV in order to derive values for the volume fractions υ_i of the various components proposed to be present in these films.

A. a-C:H FILMS

a-C:H films have been prepared [5] via dc glow discharge of pure C_2H_2. ε_1 and ε_2 have been determined as functions of photon energy E in the range 1.5 to 4.75 eV, and are shown below for the as-deposited film (T_d= 250°C) and for the same film following anneals at T_a = 350, 450, 550, 650, and 750°C.

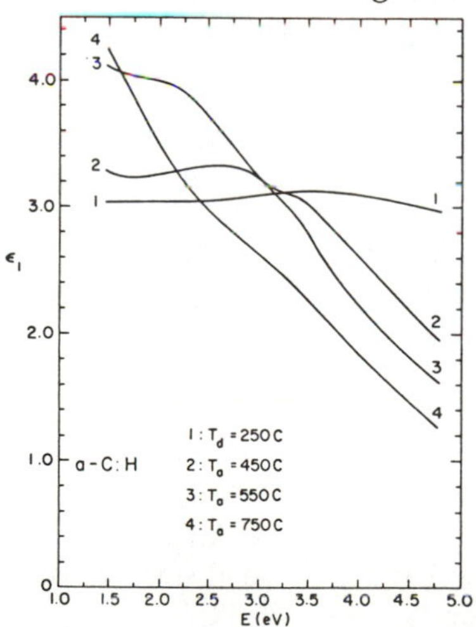

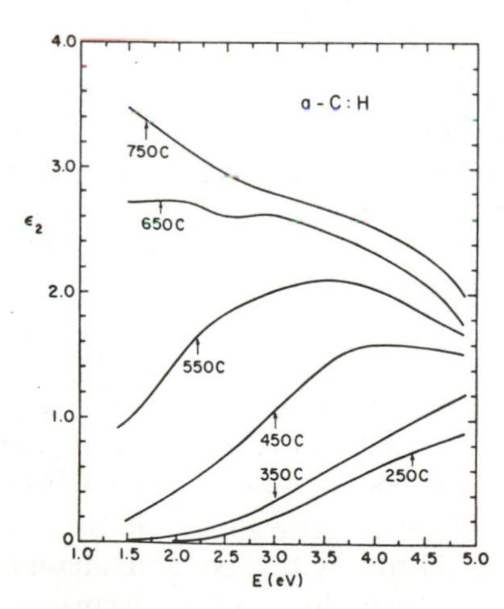

Fig. 2. Real part ε_1 of the dielectric constant vs energy for an a-C:H film as deposited (T_d = 250°C) and following anneals at 450, 550, and 750°C.

Fig. 3. Imaginary part ε_2 of the dielectric constant vs energy for an a-C:H film as-deposited (T_d=250°C) and following anneals at 350, 450, 550, 650, and 750°C.

It can be seen that $\varepsilon_1 = n^2 - k^2$, and hence n, the real part of the index of refraction, are essentially constant for the as-deposited film in the energy range studied, with dispersion in ε_1 increasing with increasing T_a up to 750°C. This increasing dispersion, with ε_1 decreasing with energy, is characteristic of glassy carbon and evaporated carbon films where graphitic short range order is expected to dominate. ε_2 is observed to be low for the as-deposited film and increases rapidly with T_a, as hydrogen is evolved and graphitic short range order develops.

The density of the as-deposited film is low (1.35 gcm^{-3}) and increases with T_a, as shown in the table below.

Table 1 Variation of film density with annealing temperature

T_a (°C)	ρ (gcm^{-3})
250	1.35
300	1.37
350	1.31
400	1.31
450	1.42
500	1.47
550	1.47
600	1.47
650	1.50
700	1.56
750	1.64

The first increase observed for ρ at T_a =450°C can be seen in figure 3 to coincide with a strong increase in the optical absorption of the film and also with the beginning of a significant decrease in E_{opt}, see below. It is important to point out that measurements of film thickness as a function of T_a have indicated that the film loses 35% of its mass while being annealing between 400 and 650°C, thus indicating that not only hydrogen but also a significant amount of carbon has been volatized.

With annealing, the optical energy gap E_{opt} decreases from a value of 2.2 eV at T_d=250°C to zero following the anneal at T_a = 650°C, as shown in figure 4. These values of E_{opt} have been obtained from the intercepts of plots of $E (\varepsilon_2)^{1/2}$ vs E. Also shown in figure 4 are results [12] obtained for an a-$Si_{1-x}C_x$:H film with x equal to about 0.9. The enhanced thermal stability of this latter film is clear, as E_{opt} decreases to zero for T_a = 800 - 850°C.

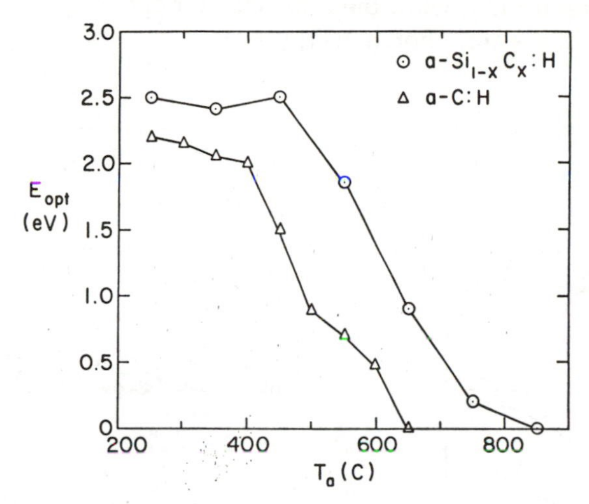

Fig. 4. E_{opt} vs annealing temperature for a-C:H and for a-$Si_{1-x}C_x$:H (x~0.9).

For completeness, we show in figure 5 the dependence of E_{opt} on deposition temperature for a-C:H [13]. E_{opt} extrapolates to zero for values of T_d greater than about 400°C.

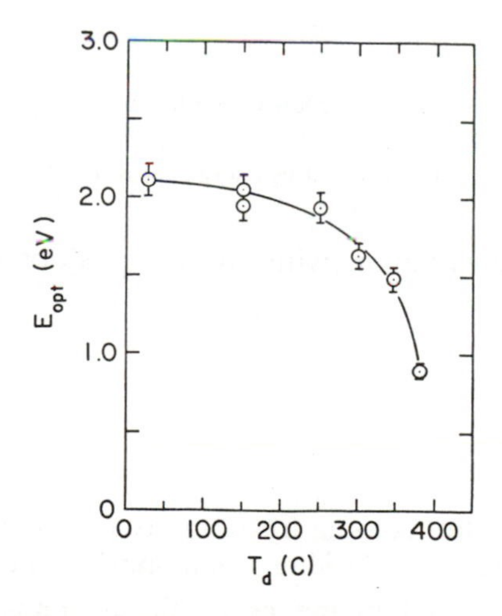

Fig. 5. E_{opt} vs deposition temperature for a- C:H.

B. a-$Si_{1-x}C_x$:H FILMS

A series of a-$Si_{1-x}C_x$:H films (from x = 0, a-Si:H, to x = 1, a-C:H) have been prepared via rf glow discharge of mixtures of C_2H_2 and SiH_4 (in Ar) at

T_d = 250°C. In figures 6 and 7 below the results [6] for ε_1 and ε_2 are shown as functions of composition x and photon energy E.

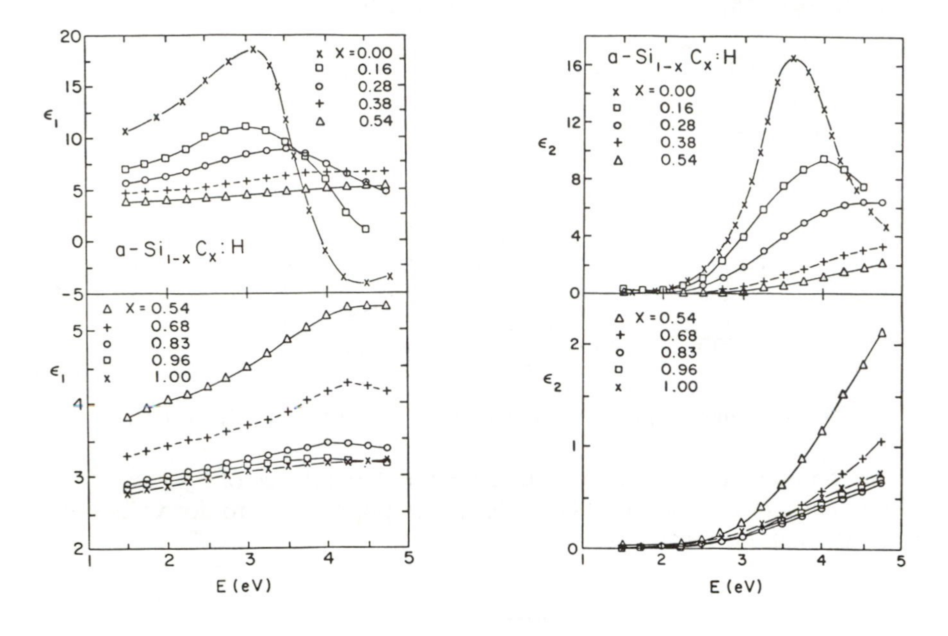

Fig. 6. Real part ε_1 of the dielectric constant vs energy for nine a-$Si_{1-x}C_x$:H films.

Fig. 7. Imaginary part ε_2 of the dielectric constant vs energy for nine a-$Si_{1-x}C_x$:H films.

ε_1 and ε_2 show systematic variations with film composition from a-Si:H (x = 0) to a-C:H (x = 1).

IV EMA ANALYSIS

A. a-C:H FILMS

In order to determine the volume fractions υ_i for the four components proposed to be present in these a-C:H films (diamond-like, polymeric, graphitic, and void), the measured ε_1 and ε_2 for the as-deposited and annealed film have been substituted into equation 1, along with the ε_i spectra used for each of the four components. A least-mean-square fitting procedure consisting not only of a minimization of the real and imaginary parts of equation 1, but also of the difference between the measured density and the calculated density $\rho = \rho_{ad}\,\upsilon_{ad} + \rho_{ap}\,\upsilon_{ap} + \rho_{ag}\,\upsilon_{ag}$ (equation 2) has yielded the values of υ_i listed in table 2 and shown in figure 8 as functions of T_a.

Table 2. Results of EMA for volume and atomic fractions, fraction of sp^3-bonded carbon, and average coordination <Z> as functions of T_a.

<Z>	T_a(C)	υ_{ad}[a]	υ_{ag}[b]	υ_{ap}[c]	υ_v[d]	f (H)	f(C,sp^3)	f(C,sp^2)	F(C,sp^3)
2.45	250	0.138	0.108	0.754	0.000	0.486	0.424	0.088	0.828
2.39	350	0.088	0.176	0.736	0.000	0.489	0.363	0.148	0.710
2.68	450	0.101	0.360	0.410	0.129	0.321	0.321	0.357	0.473
3.05	550	0.048	0.634	0.034	0.275	0.036	0.120	0.844	0.124
3.0	650	0.000	0.751	0.000	0.249	0.0	0 .0	1.0	0.0
3.0	750	0.000	0.819	0.000	0.181	0.0	0.0	1.0	0.0

[a] Volume fraction of amorphous diamond-like component.

[b] Volume fraction of amorphous graphitic component.

[c] Volume fraction of amorphous polymeric component.

[d] Volume fraction of void component.

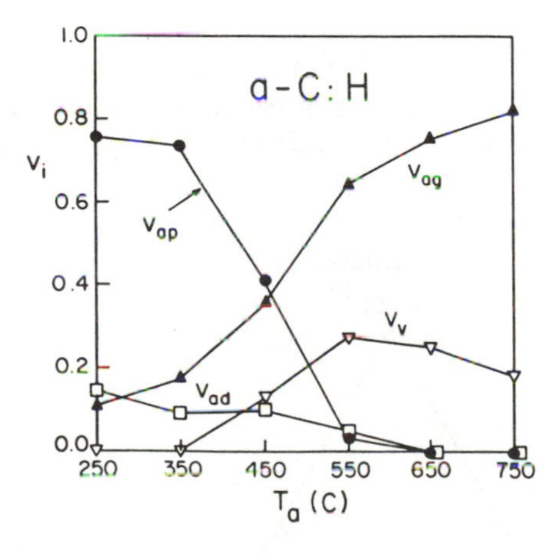

Fig. 8. Volume fractions υ_i of the various components of the a-C:H film obtained from the EMA (see text) vs annealing temperature T_a. υ_{ad}, amorphous diamond-like; υ_{ag}, amorphous graphitic; υ_{ap}, amorphous polymeric; υ_v, void.

It can be seen from table 2 and figure 8 that the amorphous polymeric component is dominant in the film up to the 450°C anneal, at which point the amorphous graphitic component rapidly increases. In the as-deposited film the amorphous diamond-like component comprises only 14% of the volume and this contribution falls to zero following the 650°C anneal. The void component has zero volume fraction up to and including the 350°C anneal, after which υ_v begins to increase. It is likely that voids are created as the polymeric material either leaves the film or is converted to the graphitic component via loss of hydrogen. This void component begins to decrease following the 650°C anneal as the film begins to densify.

We can make use of these results for the υ_i to obtain estimates for the atomic fractions of hydrogen, sp^3-bonded carbon $C(sp^3)$, and sp^2-bonded carbon $C(sp^2)$ atoms present in the film. For this purpose we assume that 1) the amorphous diamond-like component ($\rho_{ad} = 3.2 gcm^{-3}$) is composed of only sp^3-bonded carbon, yielding $N_{ad}(C) = 1.60x10^{23}$ atoms cm^{-3}, 2) the amorphous polymeric component ($\rho = 0.92$ gcm^{-3}) has a nominal composition CH_2 and is composed of sp^3-bonded carbon (with $N_{ap}(C) = 3.93x10^{22}$ atoms cm^{-3}) and hydrogen (with $N_{ap}(H) = 2 N_{ap}(C) = 7.96x10^{22}$ atoms cm^{-3}), and 3) the amorphous graphitic component ($\rho_{ag} = 2.0 g/cm^3$) is composed of only sp^2-bonded carbon, so that $N_{ag}(C) = 9.98x10^{22}$ atoms cm^{-3}.

The atomic fractions of the three atomic species proposed to be present in the film satisfy the relationship $f(H) + f(C,sp^3) + f(C, sp^2) = 1$, where, for example, $f(C,sp^3)$ is given in terms of the volume fractions υ_i by

$$f(C,sp^3) = \frac{N_{ad}(C)\,\upsilon'_{ad} + N_{ap}(C)\,\upsilon'_{ap}}{\sum_i N_i \upsilon'_i}, \qquad (3)$$

with $\upsilon'_i = \upsilon_i/(1-\upsilon_v)$ and $\Sigma N_i\upsilon'_i = N_{ad}(C)\,\upsilon'_{ad} + N_{ap}(C)\,\upsilon'_{ap} + N_{ag}(C)\,\upsilon'_{ag} + N_{ap}(H)\,\upsilon'_{ap}$. These atomic fractions are presented in table 2 and are shown in the ternary phase diagram for the $C(sp^3)$-$C(sp^2)$-H system presented in figure 9.

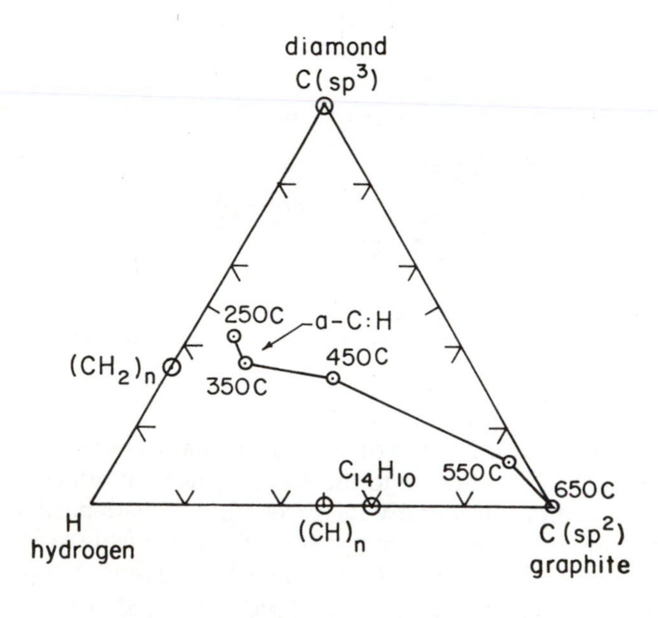

Fig. 9. Phase diagram for the $C(sp^3)$-$C(sp^2)$-H system. Shown are points corresponding to the a-C:H films studied, to polyethylene $(CH_2)_n$, to polyacetylene $(CH)_n$, and to anthracene $C_{14}H_{10}$.

The three vertices of this phase diagram correspond to diamond, C(sp^3), (upper vertex), graphite, C(sp^2), (lower right vertex) and hydrogen (lower left vertex). In addition, also indicated are points representing polyethylene $(CH_2)_n$, polyacetylene $(CH)_n$, and anthracene ($C_{14}H_{10}$). We note that the carbon chars formed from anthracene lie in this phase diagram on the line between anthracene and the graphite vertex (lower right).

The fraction of all carbon atoms in the film which are sp^3-bonded, given by

$$F\,(C,sp^3) = f\,(C,sp^3) / \left(f\,(C,sp^3) + f\,(C,sp^2)\right), \tag{4}$$

is also shown in table 2. The fraction F (C,sp^2) which are sp^2-bonded is given simply by 1-F (C,sp^3). In figure 10 the resulting fraction F (C,sp^3) for this a-C:H film is plotted as a function of T_a, along with results obtained by Dischler et al [14] from infrared absorption measurements on a film prepared from C_6H_6. It can be seen that the two quite different analyses give very similar results.

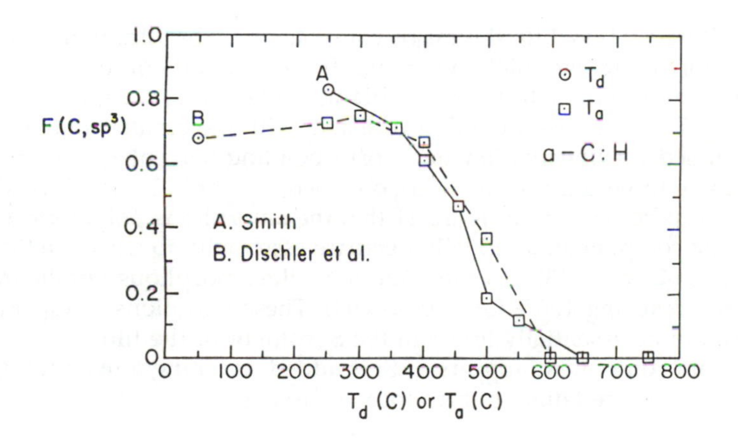

Fig. 10. Fraction F(C,sp^3) of carbon atoms in a-C:H films which are sp3-bonded vs annealing temperature. A. Smith (from results obtained in ref. 5), B. Dischler et al (ref. 14).

B. a-$Si_{1-x}C_x$:H FILMS

An EMA analysis [6] similar to that performed for the a-C:H film has been carried out for the a-$Si_{1-x}C_x$:H films and has yielded the volume fractions shown below in figure 11.

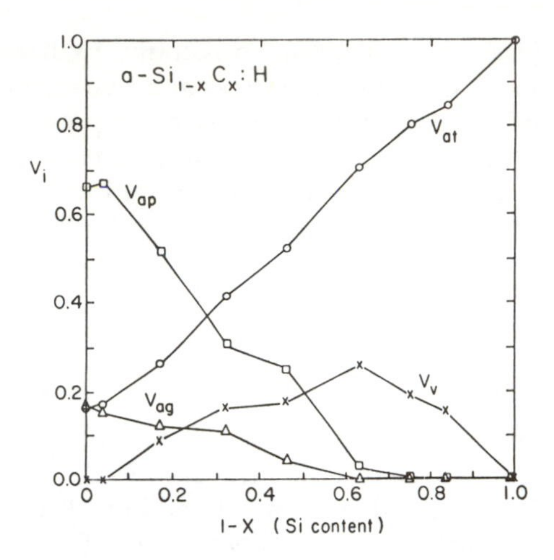

Fig. 11. Volume fractions of the various components of the a-$Si_{1-x}C_x$:H films as functions of 1-x (Si content). υ_{at}, amorphous tetrahedral, υ_{ap}, amorphous polymeric, υ_{ag}, amorphous graphitic, and υ_v, void.

Beginning with a-C:H (x=1.0), the major component in the film is found to be the amorphous polymeric one (67%), with significant amounts of amorphous graphitic (17%) and amorphous tetrahedral or diamondlike (16%) components also found to be present. These results are quite consistent with the results obtained for the a-C:H film discussed previously [5], where corresponding percentages of 75%, 11%, and 14% for these three components were obtained, see table 2. As Si is added to the a-C:H film, it can be seen from figure 11 that the amorphous polymeric and amorphous graphitic components gradually decrease essentially to zero at a Si fractional content of 0.62 (x = 0.38). At the same time, the amorphous tetrahedral component increases, reaching 100% for x=0 (a-Si:H). These variations of υ_{ap}, υ_{ag} and υ_{at} are observed to be essentially linear in the Si content of the films (see figure 11). A void component appears as Si is added, reaching a maximum value of 26% for x=0.38 before falling to zero for x=0 (a-Si:H).

V. DISCUSSION

A. a-C:H FILMS

The extent to which the interesting properties of these a-C:H films can be understood in terms of the proposed microstructural model consisting of four components, amorphous diamond-like, amorphous graphitic, amorphous polymeric, and void will now be discussed.

1) Optical Dielectric Function

As discussed in Section IV, the optical dielectric function $\varepsilon = \varepsilon_1 + i\varepsilon_2$ measured in the range 1.5 to 4.75 eV has been used in conjunction with the EMA and the measured film density to determine the volume fractions v_i of the four components proposed to be present in the films [5]. The agreement between the measured ε and the ε predicted using the EMA with the derived υ_i has been shown to be quite satisfactory. The amorphous diamond-like and polymeric components are essentially transparent in the energy range investigated. Therefore the film absorption which grows with annealing can be explained in terms of the increase in the volume fraction of the amorphous graphitic component, υ_{ag}, with increasing T_a. The low absorption measured for the as-deposited film can be attributed to the dominance of υ_{ad}, and especially υ_{ap}, over υ_{ag}. Nevertheless, υ_{ag} in the as-deposited film (about 11%) does limit the attainable transparency. It has been very difficult to reduce υ_{ag} below 10% without adding much more hydrogen and hence increasing υ_{ap}. This, of course, leads to a very soft polymeric film.

2. Density

The measured film density has also been very important in the EMA for determining the υ_i. The low density, 1.35 g/cm^3, measured for the as-deposited film (T_d = 250°C) can be attributed to the dominance of υ_{ap} and the increase of the density with increasing T_a up to 750°C, correlates quite well with the loss of hydrogen and the growth of υ_{ag} at the expense of υ_{ap}. The presence and growth of the void component with increasing T_a (450°C and above) is also demonstrated clearly by the EMA.

3.) Hardness

Although the hardness of this film was not measured, the high hardness typically observed for a-C:H films is consistent with the proposed microstructure. It is quite possible that an a-C:H film with 14% diamond-like and 75% polymeric components, assuming significant cross-linking, could indeed have a significant hardness. The amorphous diamond-like component serves to bind the polymeric and graphitic components together in a very hard composite material with the high hardness typically observed.

4.) IR Absorption

The polymeric component in the film is clearly the source of the strongly-absorbing C-H modes typically observed in IR absorption spectra. The fact that the measured fraction F (C,sp^3) of tetrahedrally coordinated C atoms in the as-deposited and annealed film agrees very well with the results obtained by Dischler et al [14] from their IR absorption study lends additional support to the film microstructural model proposed here. The fact that Dischler et al found that a significant amount of hydrogen was bonded to sp^2-bonded carbon indicates that the amorphous polymeric component can contain both sp^2- and sp^3-bonded carbon, instead of only sp^3-bonded carbon as assumed here.

5.) **Optical Energy Gap**

The value of E_{opt} = 2.2eV obtained here for the as-deposited film is limited by the presence of the amorphous graphitic component. In addition, the rapid decrease in E_{opt} which occurs for T_a = 450°C and above correlates very well both with the strong increase for υ_{ag} and decrease in υ_{ap} observed for increasing T_a. E_{opt} goes to zero at T_a = 650°C, the point at which both υ_{ap} and υ_{ad} also go to zero.

To explore this issue further, the dependence of E_{opt} on F (C, sp^2) the fraction of C atoms in the film which are in the amorphous graphite component (and hence sp^2-bonded) is shown below in figure 12 for this film and also for films prepared by other workers [15-18].

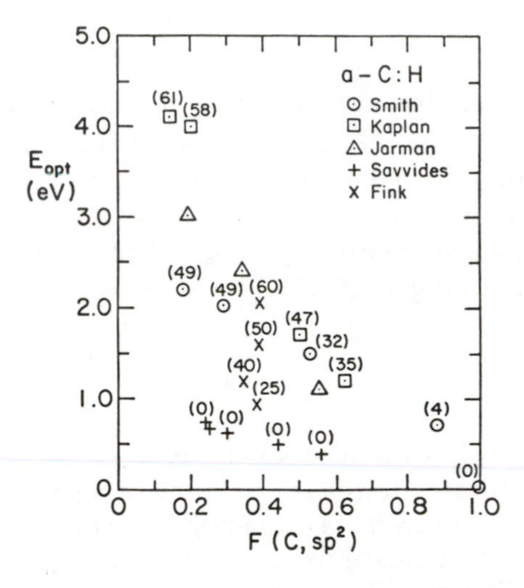

Fig. 12. E_{opt} vs fraction F (C, sp^2) of carbon atoms in a-C:H films which are sp^2-bonded. Circles, Smith (ref. 5); triangles, Jarman et al (ref. 16); squares, Kaplan et al (ref. 15); x, Fink et al (ref. 17); +, Saviddes (ref. 18). Shown also are hydrogen atomic percentages in films (in parentheses), where available.

The fact that there is a considerable amount of scatter present in the results shown in figure 12 can be attributed to the presence of widely-varying amounts of hydrogen bonded in the films. For example, consider that for films with F (C, sp^2) = 0.20 ± 0.05, values of E_{opt} from 0.75 up to 4.1 eV are obtained, with E_{opt} increasing with increasing H content. Specifically, the 0.75 eV result [18] corresponds to 0 at % H, the 2.2 eV result [5] to about 50 at % H, and the 4.1 eV result [15] to 61 at % H. In addition, the results of Fink et al [17] for four films with F (C, sp^2) = 34 to 39% yield E_{opt} values increasing from 0.95 to 2.05 eV as the H content increases from 25 to 60 at %. The simplest explanation for these results is that hydrogen bonded in the film in some way prevents the sp^2-bonded C atoms from absorbing at low energies.

This is perhaps not surprising as there exist hydrocarbon polymers such as polyacetylene, $(C(sp^2)H)_n$, in which all the C atoms are essentially sp^2-bonded and yet which have a non-zero energy gap. It is very possible that the network - terminating effect of H atoms in the amorphous graphitic component leads to a tendency for localization of the π-electrons and hence a decrease in their low energy optical absorption. This, of course, would lead to an increase in E_{opt} for a given $F(C, sp^2)$ as the H content increases.

This discussion is consistent with the results of Robertson and O'Reilly [8-10] who found that it is not only the fraction of sp^2-bonded carbon but also the size distribution of sp^2-bonded clusters which determine E_{opt}. It is clear from figure 12 that hydrogen can thus play an important role in determining E_{opt}.

6) Electrical Resistivity

The very high electrical resistivity [13], about 10^9 ohm cm at room temperature for T_d=250°C, can be attributed on the basis of the microstructural model proposed here to the dominance of the polymeric component. As T_d (or T_a) increases, the resistivity drops rapidly, to about 10^5 ohm cm for T_d = 350°C, which can be attributed to the increasing υ_{ag} of the more highly conducting amorphous graphitic component.

7) Photoelectron Spectroscopy

The transition from a-C:H to graphitic carbon with annealing has been carefully studied [19] via valence-band photoemission and near-edge absorption at the carbon K edge. Both of these measurements show clear graphitic features developing, at the top of the valence-band and in the near-edge region respectively, as T_a is increased. This transition is in very good agreement with the microstructural model presented here, including its evolution with annealing. In particular, there is evidence from the near-edge absorption for a small graphitic component even in a film deposited at 100°C, in essential agreement with the results obtained here from the EMA analysis.

8) Doping

The successful *n* and *p* type doping of a-C:H films via the incorporation, respectively, of P and B atoms during deposition has been previously accomplished [20]. The fact that doping is possible in a-C:H can be attributed to the substitutional incorporation of B and P in the amorphous diamond-like component, where tetrahedral coordination is possible. The relative inefficiency of the doping, ie. the weak shift of the Fermi level, can be attributed to the high density of electronic states extending from the tails of the valence and conduction bands into the energy gap. This high density of states can be attributed to the localized π– electron states associated with the amorphous graphitic component. It will probably be necessary to reduce υ_{ag} to essentially zero before more efficient substitutional doping can be achieved in these a-C:H films.

9) Thermopower

The measured sign of the thermopower S for doped a-C:H films has confirmed [21] that successful doping has occurred. In addition, the observed temperature dependences of S and of the electrical conductivity have provided strong evidence that the Fermi level E_F has been shifted via doping from near the middle of the energy gap into broad bands of localized tail states at the appropriate band edges, where conduction via variable range hopping at or near E_F dominates. As stated previously, these localized tail states can be associated with the π– electron states of the amorphous graphitic component.

The microstructure of a-C:H films can also be characterized by the average coordination number <Z> of the atoms in the network. Using the notation a-C_yH_z with y + z =1, the carbon fraction can be written as the sum of tetrahedral or diamond-like (y_t) and graphitic (y_g) parts, i.e. $y = y_t + y_g$. Using this notation, we have $<Z> = 4 - y_g - 3z$ or, using our previous notation, $<Z> = 4 - f(C,sp^2) - 3f(H)$. The resulting values for <Z> for the as-deposited and annealed films are given in table 2, where it can be seen that <Z> increases from about 2.4 to 3.0 as the film is annealed. The result for the as-deposited a-C:H film is very close to the value $<Z> = \sqrt{6} = 2.45$ given by Phillips [22] as the average coordination number for an alloy in which all bond-length and bond-angle constraints are satisfied, thereby reducing the strain in the alloy.

The ternary phase diagram shown in figure 9 for the $C(sp^3)$-$C(sp^2)$-H alloy system is a very useful way of representing the compositional relationship of these a-C:H alloy films to other closely-related materials, namely diamond, graphite, polyethylene, polyacetylene and anthracene. It is clear from our previous discussion that important film properties such as E_{opt} depend not only on the $F(C,sp^2)$ and $F(C,sp^3)$ fractions but also on the H content of the film. This ternary phase diagram is a very useful way of illustrating such dependences.

B. a-$Si_{1-x}C_x$:H FILMS

We have found that the microstructure proposed for these films, consisting of amorphous tetrahedral, graphitic, polymeric, and void components, gives a successful description [6] of the variations of both the optical dielectric function and film density as functions of film composition x. In addition, it is clear from this analysis that it is the appearance of the amorphous graphitic component in the films which limits the attainable value of E_{opt} in this alloy series as the carbon content increases. The model also provides strong evidence that complete chemical ordering with homogeneous dispersion exists within the amorphous tetrahedral component across the entire alloy series.

Our use of the EMA to describe the optical spectra of these a-$Si_{1-x}C_x$:H films has provided useful information on the variations of the volume fractions of the four proposed components as functions of film composition. Although clearly an oversimplification for the structure of these films, our model is a successful first approximation to the local bonding and has provided a useful framework for understanding the properties of this technologically important alloy series. Improvements to the model should focus on a better treatment of the optical absorption due to the localized π - like electrons associated with the amorphous graphitic component and explicit calculations of the effects of hydrogen incorporation on the optical response of the amorphous polymeric and tetrahedral components.

There exists additional experimental evidence supporting our use of amorphous polymeric, graphitic, and tetrahedral components in the EMA. IR studies [23], including our own results, have indicated that the majority of the hydrogen in these films is bonded to C, as opposed to Si, thus supporting the use of an amorphous polymeric component with approximate composition CH_2 and an amorphous tetrahedral component with a much lower H content, ~ 10 at %. We note here that an additional improvement to our model would be to allow the hydrogen content of the amorphous tetrahedral component to increase as the C content of this component increases.NMR studies [24] also point to the existence of two hydrogen-containing phases in a-$Si_{1-x}C_x$:H alloys: heavily hydrogenated C clusters (the amorphous polymeric component) and a weakly hydrogenated a-Si lattice (the amorphous tetrahedral component).

The EMA indicates that a void component appears in these a-$Si_{1-x}C_x$:H films from the Si-rich end as C is added to a-Si:H and also from the C-rich end as Si is added to a-C:H. The increased compositional disorder in the films as a result of alloying leads to the generation of voids due the inability of the various components to fit together perfectly. The maximum void volume fraction of 0.26 occurs for a C fraction of 0.38, the composition at which the volume fraction of the amorphous polymeric component drops to a very low value (0.03). The amorphous polymeric component is thus proposed to be effective in helping to relax the amorphous network so that the components fit together better, with less of a tendency for the presence of voids.

The goal of enhancing the tetrahedral coordination of C in an a-C:H film via incorporation of Si has been only partially successful in these a-$Si_{1-x}C_x$:H films. Adding Si does decrease the optical absorption in a-C:H as can be seen in figure 7 where ε_2 is observed to decrease as x decreases from 1 (a-C:H) to 0.83. Nevertheless, the volume fraction υ_{ag} of the amorphous graphitic component falls to zero only for x = 0.38, i.e. for a Si-rich film.

VI. CONCLUSIONS

We have proposed that there exist three local atomic bonding units in a-C:H films and that these three units are the basis of the amorphous diamond-like, amorphous graphitic, and amorphous polymeric components present in the films. When these three components are used in the EMA along with a fourth void component, the proposed microstructural model can not only provide a basis for understanding the measured ε_1 and ε_2 and film density, but also the observed film hardness, IR absorption, optical energy gap, electrical resistivity, valence band and carbon K near-edge absorption, doping, and thermopower. Thus the proposed local atomic bonding units provide, to a first approximation, a sufficient basis for describing the microstructure of these interesting and technologically-important films.

A key issue which requires further research, both theoretically and experimentally, concerns the nature of the localized (or partially-localized) π - electron states and their contribution to the measured film properties.

ACKNOWLEGEMENTS

The author wishes to thank B. Meyerson, K. Mui, M. Strongin, and other collaborators for their contributions to the research on which this chapter is based. This research has been supported by the PSC-BHE Research Award Program of the City University of New York and by the U.S. Department of Energy under grant DE-FG02-84ER45168 and contracts DE-AC02-76CH00016 and DE-AC02-80ER10750.

REFERENCES

1) Angus, J.C., Koidl, P., and Domitz, S., in "Plasma Deposited Thin Films," ed. J. Mort and F. Jansen (CRC Press, Boca Raton, 1987) Ch. 4.
2) Robertson, J., Advances in Physics, 1986, 35, 317.
3) Tsai, H. and Bogy, D.B., J.Vac. Sci. Technol., 1987, A5, 3287.
4) Angus, J.C. and Hayman, C.C., Science, 1988, 241, 913.
5) Smith, F.W., J. Appl. Phys., 1984, 55, 764.
6) Mui, K., Basa, D.K., Smith, F.W., and Corderman, R., Phys. Rev., 1987, B35, 8089.
7) Bruggemann, D.A.G., Ann. Phys. (Leipzig), 1935, 24, 636.
8) Robertson, J., and O'Reilly, E.P., Phys. Rev., 1987, B35, 2946.
9) O'Reilly, E.P., J. Non-Cryst. Solids, 1987, 97/98, 1095.
10) Robertson, J., Phil. Mag. Lett., 1988, 57, 143.
11) Mui, K., and Smith, F.W., Phys. Rev., 1987, B35, 8080.
12) Mui, K., Basa, D.K., and Smith, F.W., J. Appl. Phys., 1986, 59, 582.
13) Meyerson, B., and Smith, F.W., J. Non-Cryst. Solids, 1980, 35/36, 435.
14) Dischler, B., Bubenzer, A., and Koidl, P., Solid State Comm. 1983, 48, 105.
15) Kaplan, S., Jansen, F., and Machonkin, M., Appl. Phys. Lett., 1985, 47, 750.
16) Jarman, R.H., Ray, G.J., Standley , R.W., and Zajac, G.W., Appl. Phys. Lett., 1986, 49, 1065.
17) Fink, J., Müller-Heinzerling, Th., Pflüger, J., Scheerer, B., Dischler, B., Koidl, P., Bubenzer, A., and Sah, R.E., Phys. Rev., 1984, B30, 4713.
18) Savvides, N., J. Appl. Phys., 1986, 59, 4133.
19) Wesner, D., Krummacher, S., Carr, R., Sham, T.K., Strongin, M., Eberhardt, W., Weng, S.L., Williams, G., Howells, M., Kampas, F., Heald, S., and Smith, F.W., Phys. Rev, 1983, B28, 2152.
20) Meyerson, B. and Smith, F.W., Solid State Comm., 1980, 34, 531.
21) Meyerson, B. and Smith, F.W., Solid State Comm., 1982, 41, 23.
22) Phillips, J.C., Phys. Rev. Lett, 1979, 42, 153.
23) Catherine, Y., Zamouche, A., Bullot, J. and Gauthier, M., Thin Solid Films, 1983, 109, 145.
24) Reimer, J., Vaughan, R.W., Knights, J.C., and Lujan, R.A., J. Vac. Sci. Technol., 1981, 19, 53.

Materials Science Forum Vols. 52 & 53 (1989) pp. 341-364

REAL TIME ELLIPSOMETRY CHARACTERIZATION AND PROCESS MONITORING FOR AMORPHOUS CARBON DEPOSITION

R.W. Collins

The Pennsylvania State University
Materials Research Laboratory and Department of Physics
University Park, PA 16802, USA

ABSTRACT

This article reviews a recent investigation of amorphous carbon (a-C:H) thin film growth, surface modification, and etching, undertaken with real time and spectroscopic ellipsometry. In the first part of the article, ellipsometry instrumentation, data interpretation, and substrate characterization are summarized. In the second part, more detailed discussion focuses on materials and process characterization in real time. For a-C:H prepared from CH_4 by plasma-assisted chemical vapor deposition (PACVD) and by direct ion beam deposition, accurate values of the film thickness, deposition rate, and bulk optical functions have been deduced from the measurements in real time during film growth. The optical functions provide information on the bonding configurations in a-C:H. Substrate modification and film-substrate reactions that occur in the initial stages of growth upon igniting the deposition plasma or ion beam have been detected and characterized at the monolayer level. This information is not accessible to *ex situ* measurements. Real time ellipsometry has also allowed monolayer detection of changes in the near-surface bonding and microstructure generated by inert and reactive gas plasma exposure. Transitions between polymer-like and graphitic bonding at the surface as well as increases in microscopic surface roughness for graphitic films have been observed. The importance of real time ellipsometry in thin film deposition technology is emphasized by showing the ability to map PACVD preparation parameter space according to specific regimes of deposition and etching.

I. INTRODUCTION

A wide range of analytical techniques, some capable of real time application, have been developed to characterize the near-surface chemical bonding and physical structure of *in-situ*-prepared materials. For the particular example of epitaxial growth of crystalline materials in ultrahigh vacuum (uhv), the application of *in situ* probes for real time monitoring and control has been closely coupled with advances in processing. Crystalline materials are amenable to the use of diffraction techniques, and reflection high energy electron diffraction is one that has provided valuable information on growth and structure at the monolayer level.[1]

For amorphous thin films, *in situ* Auger electron spectroscopy[2] and x-ray and ultraviolet photoelectron spectroscopy[2,3] have been used to study the chemical nature of films and their interfacial regions, but these probes are insensitive to the structural aspects of nuclei formation, coalescence, and roughness development. These *in situ* spectroscopies tend to be applied to model systems, and are not generally used as routine tools for passive real time monitoring and control.

Real time scanning electron microscopy has provided very important insights into the physics of the nucleation and growth of metal films[4] and has also been applied to amorphous thin films of high atomic number.[5] In particular, the geometry of nuclei and their remarkable coalescence behavior can be directly observed.[6] One drawback results from the complexity of the apparatus, rendering this technique unsuitable other than for model studies, as well. In addition, there is a lack of sensitivity to monolayer-scale phenomena such as Stranski-Krastanov behavior[7] and an absence of information on chemical interactions at the substrate-film interface. Real time scanning tunneling microscopy (STM) may provide solutions to some of these problems,[8] but not the complexity issue.

It is clear that none of the above-mentioned real time techniques is a panacea. Each one has its strengths and any one of them cannot provide answers to all relevant questions concerning the structural and chemical characteristics of film growth from the monolayer scale upwards. However, one major problem, relevant for the particular case of hydrogenated amorphous carbon (a-C:H) thin film preparation, is that most of the real time measurements described so far require uhv or high vacuum deposition. None can be employed in real time in the severe environments of high pressure plasmas, ion beams, and chemical vapor deposition. Such deposition techniques are scalable to large areas, suitable for mass production and, as a result, are often chosen for a given application unless property considerations demand the uhv processing required for single crystal films.

In contrast to the other techniques, *in situ* ellipsometry is a monolayer-sensitive characterizational tool that can be used for real time monitoring, control, reproducibility and end-point detection in these adverse environments.[9] Ellipsometry involves determining the change in polarization state that occurs upon reflection of a weak polarized light beam from the surface of a growing film at a non-normal incidence angle.[10] As a result, it is passive, non-destructive, non-perturbing, and can be employed at atmospheric pressure or with high density plasmas. Because the technique allows detection of $< 0.01°$ changes in the phase of light upon reflection, submonolayer sensitivity to changes in film or roughness layer thickness is achieved. Because it is sensitive to the polarizability per unit volume of thin film material,[11] it can detect bulk voids in absorbing materials at relative volume fractions as low as 0.001. The main drawback of ellipsometry is the lack of chemical specificity, particularly for amorphous materials in which the electronic transitions are very broad. However, for films to be used in optical applications such as for scratch-resistant antireflection coatings,[12] it can provide a complete description of the sample properties.

The requirements for ellipsometry are quite simple in comparison to many of the real time probes mentioned above. To perform ellipsometry experiments one needs two basic requirements: (1) optical access to the sample surface at a 60°-70° angle of incidence during growth and (2) the ability either to align the sample surface with respect to the fixed optical path or to align the optical path with respect to the fixed sample surface, with a known angle of incidence in both cases. Care must also be taken to mount and align the windows of the vacuum chamber to minimize the influence of strain-induced birefringence if high accuracy absolute (rather than differential) measurements are to be obtained. As the use of automatic ellipsometry becomes more commonplace through problem solving at the research and development level, such equipment may eventually be incorporated into advanced industrial thin film deposition for process and quality control.

Earlier real time ellipsometry studies have provided information on the nucleation, growth, and interfacial properties of thin film hydrogenated amorphous silicon (a-Si:H) prepared by different methods.[13-15] In this review, the capabilities of the technique are demonstrated for thin film a-C:H, prepared both by plasma-assisted chemical vapor deposition (PACVD) and by direct ion beam deposition with CH_4 source gases. For a-C:H, the experiments provide real time information and a level of detail inaccessible to *ex situ* optical measurements. Comparable monolayer sensitivity in the initial stages of film growth has been achieved only recently with *in situ* surface analysis in ultrahigh vacuum.[2,3]

Applications for a-C:H materials rely on their mechanical hardness, chemical inertness, and infrared transparency, attributed to a large number of tetrahedral (sp^3) bonds (thus referred to as "diamond-like" a-C:H).[16,17] However, many questions concerning the details of film structure remain unanswered. Because tetrahedral bonding appears to result when film precursor impact energies are well above thermal (> 30 eV), high plasma power or negative dc bias magnitude (> 100 V) and low pressure (< 0.1 Torr) are required to prepare diamond-like films by PACVD from hydrocarbon gases.[12,17-23] At lower impact energies, films are softer with polymer-like bonding and properties.[12,17,24,25] Films of the latter type have been prepared by methods analogous to those used to prepare a-Si:H[26,27] and doping of such films has been studied.[28,29] A number of ion beam techniques have been employed to obtain hard, diamond-like films, including single and dual ion beam sputtering using C targets and direct ion beam deposition from pure C and CH_4 sources.[2,3,17,24,30-39] A hybrid technique combining sputtering of a C target and concurrent plasma decomposition of hydrocarbon gas has also been used.[40]

In Section II, a brief review of the automatic ellipsometry apparatus will be presented, with the primary emphasis placed on the rotating analyzer configuration. Section III will provide an overview of the fixed wavelength real time and spectroscopic ellipsometry data analysis and interpretation techniques that have been applied in the a-C:H research. The dielectric functions for different forms of disordered carbon will be discussed in Section IV. These serve as reference data to assist in the interpretation of the real time and spectroscopic results on complex multilayer structures. In Section V, a detailed spectroscopic ellipsometry analysis of the Mo thin film substrate material used in the depositions described in Sections VI and VII will be provided. Such analysis is desirable in order to determine the substrate native oxide thickness and to ensure that the substrate is structurally homogeneous. Earlier studies have suggested that, for depositions on substrates with extensive surface roughness and surface-connected microvoids, it is difficult to detect monolayer scale nuclei coalescence and substrate-film reactions owing to the loss of phase coherence of the light beam in the experiment.[41] Experimental results on the growth, etching, and surface modification of a-C:H prepared by PACVD, electron beam evaporation, and ion beam deposition are presented in Sections VI and VII. To date, the most studied process is PACVD.

II. EXPERIMENTAL NOTES

Two types of automatic ellipsometers have been most commonly employed for real time studies of thin film deposition, and both have been applied to study amorphous semiconductor preparation. For the polarization modulation ellipsometer (PME), the active element consists of a piezobirefringent quartz crystal driven by an ac voltage.[42] For the rotating element ellipsometer (REE), one of the polarizer prisms is rotated continuously, for example, by mechanically mounting it within the hollow shaft of a motor.[13,43,44] The PME can operate at data collection rates higher than 100 kHz, whereas the typical REE is operated in the 10-100 Hz range.

In the real time studies of amorphous thin film growth presented to date, ellipsometric data were obtained at a fixed (but selectable) photon energy with time intervals between successive data points of 0.2 s for the PME and a minimum of 0.9 s for the REE.[44] Thus, both systems provide monolayer resolution during film deposition at rates of 1 Å/s. Because these systems typically rely on combined Xe lamp/monochromator sources and photomultiplier tube (PMT) detectors, several minutes are required to step through a range of photon energies to obtain spectroscopic ellipsometry (SE) data. Thus, SE measurements are time consuming and can be performed *in situ* with the deposition suspended, but not in real time. The latest advances in ellipsometric instrumentation employ either the REE configuration with optical multichannel detection[45] or the PME configuration with rapid photon energy scanning,[46] in each case to achieve real time SE capability.

The only real time ellipsometric study of a-C:H deposition reported thus far was performed with an REE ellipsometer.[47] As a result, some additional details will be presented on its design, operation, and coupling to the deposition system. The configuration of optical elements was the following: source-monochromator-polarizer-sample-analyzer-detector, with the analyzer as the rotating element. This rotating analyzer ellipsometer (RAE), based on a popular design by Aspnes and Studna, was used because it provides exceptional precision, reasonable accuracy, and is achromatic, requiring no spectral calibration.[48] In addition, it is easier to approach the requirement of an ideal, polarization insensitive detector for the RAE than an ideal, unpolarized source, the requirement if the polarizer was used as the rotating element.

With a 54 Hz rotation rate for this RAE, the minimum time interval between two measurements was 0.4 s. This included 0.2 s for real time data collection from 10 analyzer rotations, and 0.2 s for the following operations: (1) real time data transfer, (2) averaging of the acquired waveforms from each of the 20 half-rotations, and (3) Fourier analysis of the averaged waveform. The signal output from the PMT has the following mathematical form:[9,48]

$$I(t) = I_0 \{ 1 + \alpha \cos 2A(t) + \beta \sin 2A(t) \} \qquad \text{(II.1)}$$

with the instantaneous analyzer position, A, given by: $A(t) = 2\pi f_m t + \phi$. Here f_m is the frequency of the analyzer, ϕ is an arbitrary constant phase factor for the analyzer position at $t = 0$, and α and β are the normalized Fourier coefficients to be determined. The Fourier coefficients provide the conventional ellipsometry angles ψ and Δ:

$$\begin{aligned} \tan\psi &= \{(1+\alpha)(1-\alpha)^{-1}\}^{1/2} \tan P \\ \cos\Delta &= \beta(1-\alpha^2)^{-1/2} \end{aligned} \qquad \text{(II.2)}$$

where P is the fixed polarizer angle. ψ and Δ are related to the sample structure and properties through the relationship $r_p/r_s = \tan\psi \exp(i\Delta)$, where r_p and r_s are the complex reflectivities of the sample for pure p- and s-polarization states.

For real time measurements with a RAE, the data are susceptible to errors owing to the detection of stray light emitted from sources associated with the deposition. A "dark cycle", obtained with a shutter blocking the incident beam, corrects the data for all ambient light sources and dc level drifts, but more than doubles the minimum data cycle time from 0.4 s to the 0.9 s value quoted above. In monitoring processes at a slower rate, data from a greater number of analyzer rotations can be averaged and/or an adjustable delay can be inserted between successive data cycles.

To obtain the results reviewed here, two different bakable deposition systems, each 6" in diameter with optical access at a 70° angle of incidence, could be positioned at the axis of the RAE. Fused quartz windows, mounted to minimize stress-induced birefringence, were centered with respect to the incident and reflected beams and aligned normal to them. The first deposition system was a capacitively coupled rf PACVD reactor, turbo-pumped to a base pressure of 10^{-7} Torr. The substrates were mounted on the rf active cathode plate (28 cm^2 in area), and the cathode bias could be measured and controlled independent of the rf power. The area of the grounded anode surfaces was ~ 50 cm^2. The second system housed a 3 cm Kaufman ion source which was aimed directly at the substrate surface, using a preparation method similar to that of Ref. 39. Other deposition details will be noted when the results are presented.

III. ANALYSIS OF REAL TIME AND SPECTROSCOPIC DATA

A. REAL TIME ANALYSIS

The experimental data can be expressed as a trajectory in the (ψ,Δ) plane that is swept out as a function of time during film deposition, etching, or surface modification. For the work on a-C:H, the raw data were plotted in the plane of the real and imaginary parts of the pseudo-dielectric function, ($<\varepsilon_1>$,$<\varepsilon_2>$). The relation connecting the two sets is:

$$<\varepsilon_1> + i <\varepsilon_2> = \sin^2\theta \,[\, 1 + \{(1-\rho)/(1+\rho)\}^2 \tan^2\theta \,] \qquad \text{(III.1)}$$

where $\rho \equiv \tan\psi \exp(i\Delta)$, and θ is the angle of incidence. For an opaque sample having a surface which is an abrupt termination of the bulk structure, ($<\varepsilon_1>$,$<\varepsilon_2>$) is equal to the actual dielectric function, (ε_1,ε_2). Thus, such a sample must be atomically smooth and clean to obtain its characteristic bulk (ε_1,ε_2) spectrum from the raw data.

The computational techniques used to fit the experimental ($<\varepsilon_1>$,$<\varepsilon_2>$) data for a-C:H growth have been described in similar studies of hydrogenated amorphous silicon (a-Si:H).[44,49] Basically, one starts with a idealized model of layer-by-layer growth on an atomically smooth and abrupt substrate surface with no roughess or void density evolution vs. film thickness and no substrate modification or substrate/film reaction (henceforth referred to as "uniform growth"). Thus, in the uniform growth model thickness-independent (ε_1,ε_2) values are assumed for the growing film. If there is a homogeneous oxide on the starting substrate surface and its thickness and dielectric function are known (eg. from a separate set of measurements as in Sect. V), then this layer can be included in the model as well.[50] The model trajectory for uniform growth is computed with standard multilayer optical analysis,[10] and the (ε_1,ε_2) values for the growing film are adjusted to provide the best fit to the data. When the film is grown to opacity, the (ε_1,ε_2) values used in the uniform growth model are established trivially by the ($<\varepsilon_1>$,$<\varepsilon_2>$) convergence point obtained at the end of the experimental trajectory.

Figures 1(a) and 1(b) provide examples of calculated trajectories for an incident photon energy of 3.2 eV which simulate the first 1000 Å of a-C:H growth on Mo metal substrates. In Fig. 1(a), trajectories are presented for film growth with $\varepsilon_1 = 3.5$ and four different values of ε_2. In Fig. 1(b), the trajectories are presented for $\varepsilon_2 = 0.25$ and different values of ε_1. In no case do the simulated films reach opacity after 1000 Å since trajectory convergence points are not attained. Such models can be used to establish graphical fits to experimental trajectories. In Fig. 1(b), all curves tend to intersect near ($<\varepsilon_1>$,$<\varepsilon_2>$) = (-3, -7). (For each calculated trajectory, the film thickness at the point of intersection is different.) Thus, comparison of the data with Fig. 1(a) in this regime can provide an estimate for ε_2 for the data. Once ε_2 is established, ε_1 can be approximated by comparison with Fig. 1(b). The process can be iterated to minimize the deviation between the calculation and the data. Alternatively linear regression may be applied, but the technique may be troublesome when the experimental trajectory crosses itself.

In most cases, no layer-by-layer growth model can be found to fit the experimental trajectory over the full range of thicknesses. As a result, inhomogeneities must be introduced into the model structure in an attempt to account for the discrepancies. These may include effects such as island nucleation, void structure or surface roughness evolution, reaction between the film and substrate, or chemical/structural modification of the near-surface of the substrate. For example, a coalescing island microstructure can be modeled as a composite film consisting of bulk material and void with volume fractions that depend on the accumulated film thickness. (ε_1,ε_2) of such composite materials can be determined with the Bruggeman effective medium approximation (EMA) given by:[11]

$$0 = \sum_{i=1}^{n} f_i \,(\varepsilon_i - \varepsilon)(\varepsilon_i + 2\varepsilon)^{-1} \qquad \text{(III.2)}$$

where ε_i and f_i are the dielectric function and volume fraction of the ith component of the n-component composite and ε is the effective dielectric function of the composite. Details concerning the validity of this approach are treated elsewhere.[51]

For very thin (<100 Å) weakly absorbing a-C:H films on c-Si and metal substrates, it is difficult to determine unique values of both ε_1 and thickness. In practical terms, this means that model trajectories for a-C:H films of different void content nearly superimpose in the initial stages of growth, making it somewhat difficult to identify nucleation effects definitively. In contrast, the

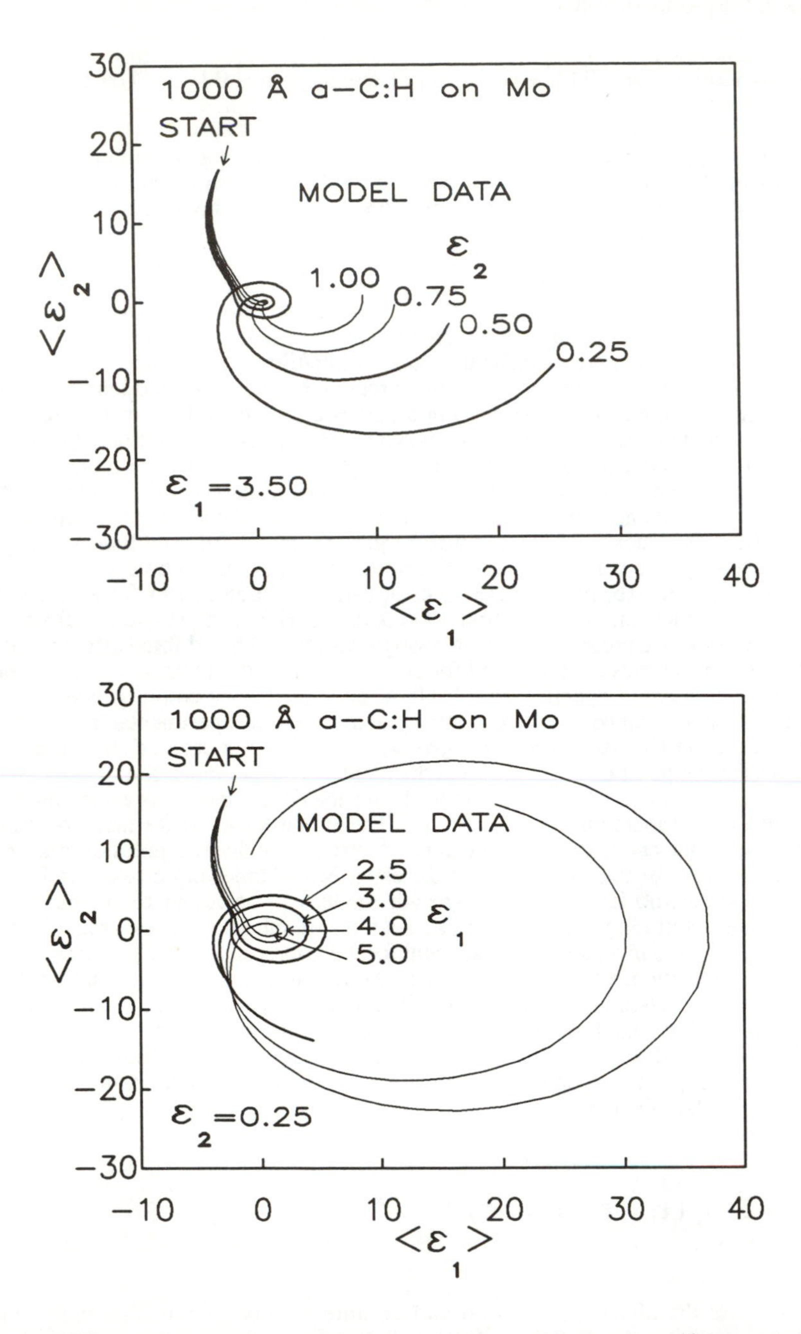

FIGURE 1. Model pseudo-dielectric function trajectories simulating the uniform growth of 1000 Å a-C:H layers on Mo substrates monitored at 3.2 eV. For each of the eight trajectories a different dielectric function for a-C:H is chosen. In (a) the real part is fixed (3.5), and in (b) the imaginary part is fixed (0.25).

($<\varepsilon_1>,<\varepsilon_2>$) trajectories for a-C:H are extremely sensitive to absorbing layers that may be generated at the substrate interface owing to plasma- or film-substrate reactions. However, one is confronted with the difficulty of developing appropriate models for the dielectric functions to characterize these interfacial layers. In work reviewed here, these dielectric functions were simulated with the EMA, using physical mixtures of the substrate and film materials. This provided the thickness and a compositional ratio for the layer. So far, this approach has been found to lead to consistent results when both SE and real time data on the interface structure has been analyzed. Future SE studies of thicker reacted layers formed by C ion bombardment or implantation for various substrate materials should provide better dielectric functions for more detailed characterization of the interfacial bonding and structure.

B. SPECTROSCOPIC ANALYSIS

Analysis of spectroscopic (ψ,Δ) or ($<\varepsilon_1>,<\varepsilon_2>$) data on multilayer structures employs multilayer optical computation, linear regression analysis, the EMA, and a library of reference data for the dielectric functions of the expected components of the sample.[52] The reference data, carrying the photon energy dependence in the calculated spectra, are manipulated in the analysis in an attempt to deduce (1) the structure of the sample (eg. number of layers, number of components in each layer) and (2) a set of best fit values of photon-energy-independent structural variables (layer thicknesses and component volume fractions) which minimize the unbiased estimator of the mean square deviation between the calculated model and experimental data. This technique has found extensive use in the analysis of complex samples, and its validity has been established from comparisons with other structural measurements.[53-57] A variation of this method eliminates the necessity of a complete reference dielectric function set by parameterizing one or more unknown dielectric functions with energy-independent variables that are determined in the analysis along with the structural parameters.[58] With both techniques, one must calculate confidence limits on all deduced variables to prevent overparameterization.

In the course of a single energy real time ellipsometry experiment, the deposition, etching, or surface modification process can be suspended and *in situ* spectroscopic measurements obtained. These data can be analyzed using the techniques just described in order to verify consistency with the structural model and the heterogeneities found to fit the real time data.

A second SE analysis technique instead *relies on* structural model parameters (eg. thicknesses, volume fractions) deduced from real time ellipsometry trajectories at a single photon energy to determine one unknown dielectric function from the experimental spectra. The dielectric function is typically that of a bulk thin film grown in the real time experiment but, in addition, may be that of a component of composite layers such as those simulating surface and interface roughness. The analysis technique performs the exact inversion ($<\varepsilon_1>,<\varepsilon_2>$) -> ($\varepsilon_1,\varepsilon_2$) using Newton's method with all other quantities in the multilayer optical problem known. The validity of the analysis can be checked by insuring that there are no spurious artifacts in the ($\varepsilon_1,\varepsilon_2$) data which arise from spectral features either in a second known dielectric function of the structure (eg. that of the substrate)[59] or in an interference pattern,[53] if present. These artifacts would be present in the deduced ($\varepsilon_1,\varepsilon_2$) if the "known" structural parameters were in error or provided an incomplete definition of the multilayer system.

IV. DIELECTRIC FUNCTIONS OF CARBON FILMS

Many workers have studied the optical properties of differently-prepared disordered carbon-based films as part of larger research programs to deduce their chemical bonding and physical structure, and to understand the underlying process/property relationships.[16,17] The intent of this article is not to advance the current level of understanding in this particular area but, rather, to build upon the wealth of existing information. *In doing so, the dielectric response will be employed as a fingerprint of C-bonding and structure which can be identified with monolayer resolution by ellipsometry during film growth (ie. in real time).* This approach can provide *both* unique insights into the process/property relationships *and* instantaneous feedback for process control.

Previous research has suggested that the dielectric function of rf PACVD a-C:H, prepared as a function of annealing temperature, can be simulated using an EMA with physical mixtures of disordered C(H) components including: (1) tetrahedral (sp^3) or diamond-like C (2) polymer-like $(CH_2)_x$, (3) trigonal (sp^2) or graphitic C, along with void.[60] In a similar vein, in a more recent study, a-C:H dielectric functions have been modeled using two Lorentz oscillators representing the π-π^* and σ-σ^* optical transitions associated with trigonal and tetrahedral bonding.[61]

In general, for weakly absorbing a-C:H films prepared from hydrocarbon gas sources, a strong correlation has been noted between the physical and optical properties.[17] An index of refraction (n) of 1.8 (or $\varepsilon_1 \sim 3.2$) at 1 µm provides an approximate dividing line between materials with predominantly polymer-like (n<1.8) and diamond-like (n>1.8) structure.[25] As noted earlier, a high average impact energy for the $C(H_x)$ precursors tends to generate the metastable diamond-like structure with the transition near 30 eV.[17] Thus, the index of refraction for the film is normally a monotonically increasing function of the impact energy in the deposition process.

The magnitude of the extinction coefficient (k) in the optical photon energy range appears to provide a gauge of the fraction of graphitic bonding.[17,60,62] k also tends to be a monotonically increasing function of impact energy for the polymer- and diamond-like films, and k continues to increase as n approaches or exceeds 2.4 ($\varepsilon_1 \sim 5.6$), the value for diamond. This increase in graphitic bonding reduces network strain by suppressing the increase in average coordination number that would otherwise accompany a transition to a pure diamond-like structure.[16] Because n for disordered graphitic C is greater than n for diamond when $h\nu < 2.0$ eV, the EMA approach suggests that the observed increase in n (1 µm) with impact energy also has significant contributions owing to an enhancement in trigonal bonding. Of course, quoting n ~ 2.4 alone, is not sufficient to demonstrate completely tetrahedral bonding.

Corresponding concepts hold for a-C prepared without hydrogen, for example, by electron beam evaporation or sputtering a C target. In this case the structure obtained at very low impact energy appears to be completely trigonally bonded. Predominantly tetrahedral bonding has been proposed on the basis of optical data obtained on sputtered a-C with higher ion impact energies in the range of 10-20 eV.[62]

Although a rudimentary understanding of the relationships between the optical properties and structure of a-C(:H) materials has been developed along these lines, more detailed quantitative analysis is required. One problem which may occur in some optical analyses is the potential for misinterpretation of graphitic bonding as diamond-like owing to a π-π^* gap which may be as wide as 5 eV.[16] This gap is thought to be a sensitive function of the medium range order in the film.

Fig. 2 shows the dielectric functions of three forms of disordered carbon thin films characteristic of the real time studies reported below. The data for crystalline diamond is also presented for reference. The photon energy, 3.2 eV, where most of the real time measurements were performed is highlighted. The electron beam evaporated film was prepared from pure C with low impact energy and shows a dielectric function typical of trigonally bonded a-C.[62] The PACVD film was prepared on c-Si from pure CH_4 at a substrate temperature (T_s) of 250 °C, rf power (P) of 20 W, a negative dc bias (V_{dc}) of -90 V, and a pressure (p) of 0.25 Torr. The weak absorption of this film throughout the visible and the value n(3.2 eV) = 1.69 suggests a polymeric structure.[25] The direct ion beam deposited film, prepared on c-Si at ambient T_s, from a 1.5:3.5 flow ratio of CH_4:Ar, and a beam voltage (V_B) of 200 V, when compared with the PACVD film, clearly shows diamond-like characteristics but at the expense of a greater fraction of graphitic bonding. The latter gives rise to the larger ε_2.

The wide range of dielectric responses obtained for different a-C-based thin films, demonstrates clearly that a real time measurement sensitive to the optical properties of the growing film will provide a distinct technological advantage in establishing regions of preparation parameter space most appropriate for optical applications.[12,21]

V. NOTES ON SUBSTRATE CHARACTERIZATION

Before presenting the experimental results, a review is appropriate concerning the choice of substrates for the a-C:H growth studies. Single crystal Si (with 18 - 24 Å native oxide) was chosen in many initial studies owing to the quality of its surface and to the desire to achieve reproducibility from deposition to deposition. Thin films of ion beam sputtered Mo (typically 2000 Å thick) on c-Si have also been used as substrates. However, for these thin film substrate materials, extra care should be taken to minimize complexities resulting from surface roughness or surface-connected microvoids. For example, conformal coverage of the substrate by the growing

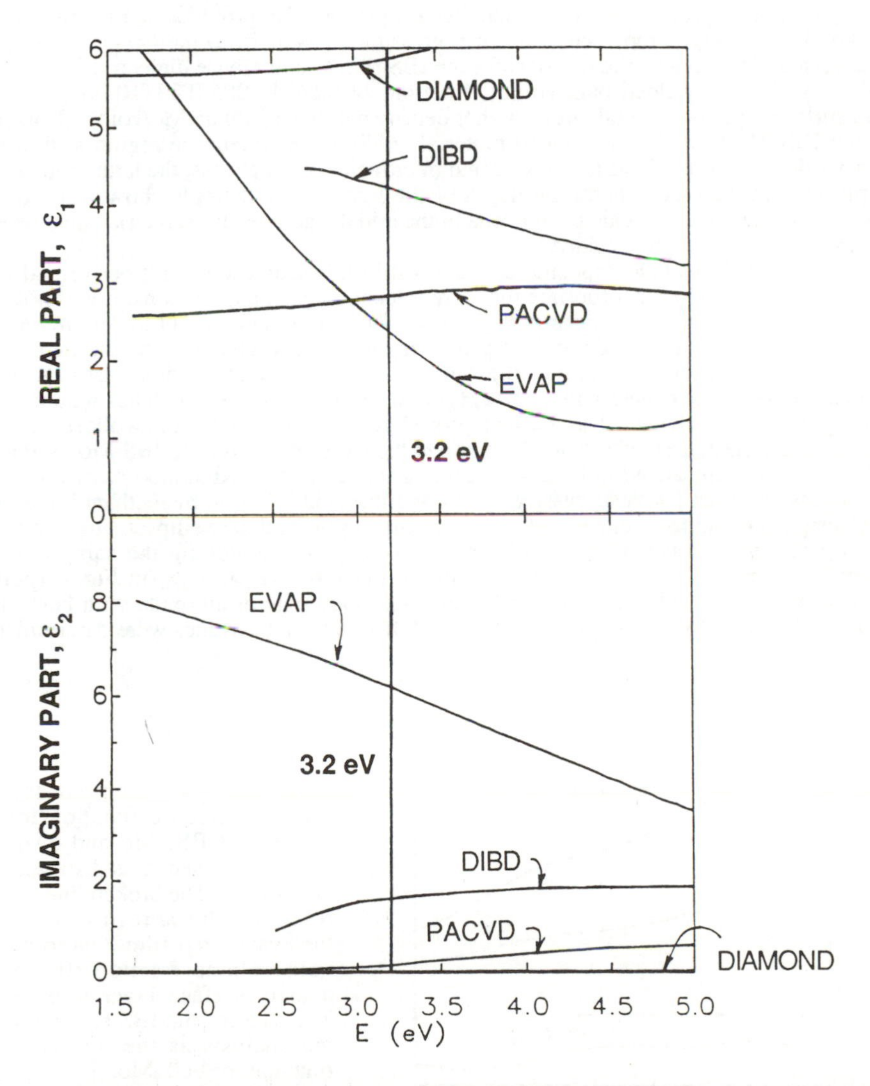

FIGURE 2. Dielectric functions of three forms of a-C(:H) films discussed in this review along with diamond for reference. DIBD = direct ion beam deposited a-C:H from 1.5:3.5 CH_4:Ar; PACVD = plasma-assisted chemical vapor deposited a-C:H from pure CH_4; EVAP = electron beam evaporated a-C.

film and penetration of the film into the substrate void structure will lead to a complex substrate/film composite interface layer which can modeled with the EMA.[41] This non-planar interface degrades the phase sensitivity of ellipsometry and makes detection of any monolayer-scale nucleation or film/substrate reaction phenomena extremely difficult. Furthermore microvoids in the substrate may introduce voids in the film, giving a film microstructure which is dominated by the substrate, masking finer trends as a function of the deposition parameters.

To demonstrate the potential of spectroscopic ellipsometry (SE) to assess these problems, Figure 3 shows a comparison of pseudo-dielectric function data for both evaporated and 1.4 keV ion beam sputtered (IBS) Mo thin films. For the IBS film, the $\langle\varepsilon_2\rangle$ data match (within 10% of the maximum value) those obtained on bulk Mo mirrors, Mo-ion implanted at 150 keV and annealed to achieve high density and microscopically smooth surfaces.[63] This comparison, along with other probes,[64] shows that the IBS process also provides dense, smooth films. The broken lines in Fig. 3 represent a linear regression analysis fit to the ($\langle\varepsilon_1\rangle$,$\langle\varepsilon_2\rangle$) data for the evaporated Mo film with the raw data for the IBS film as a reference dielectric function (see Sect. III). This best fit yielded bulk volume fractions of 0.665±0.005/0.335±0.005 Mo/void, and a thin surface layer of material with a higher density deficit (0.35/0.65 Mo/void). Thus, in contrast to the IBS Mo film, the evaporated material exhibits a near surface region with a significant density deficit caused by surface-connected microvoids or roughness, the latter with a modulation depth greater than the penetration depth of the light. This example shows that, for optimum sensitivity to monolayer-scale phenomena in the initial real time ellipsometric data, the evaporated film substrates should be avoided.

Because the a-C:H depositions described in this article were not performed in uhv, but rather under conditions appropriate for current technological applications, it was not possible to ensure oxide-free substrates prior to deposition. Thus, it was important to determine the starting thickness of the native oxide and its properties in order to characterize the deposition at the monolayer level with the real time ellipsometry probe. Although in some cases substrate oxides can be reduced in thickness with *in situ* H_2 or Ar plasma or ion beam etches without significant damage to the underlying substrate, such procedures were not used in the work reviewed here.

Real time and SE efforts to characterize the native oxide for the IBS Mo used as thin film substrates are summarized in Figs. 4-6. First, a substrate exposed to laboratory air for a number of months was rinsed with successive cycles of dilute acid followed by methanol in a windowless cell designed for surface studies. After each etching cycle, real time ellipsometry at a fixed photon energy of 3.5 eV was performed with a flow of dry N_2 enveloping the sample to minimize reoxidation.[65] These data are plotted in the plane of ($\langle\varepsilon_1\rangle$,$\langle\varepsilon_2\rangle$) in Fig. 4 (points). The broken line was calculated assuming homogeneous removal of an oxide with best fit (ε_1,ε_2) = (3.74, 0.84) on Mo with (ε_1,ε_2) = (18.37, -4.42). The latter values were determined from the

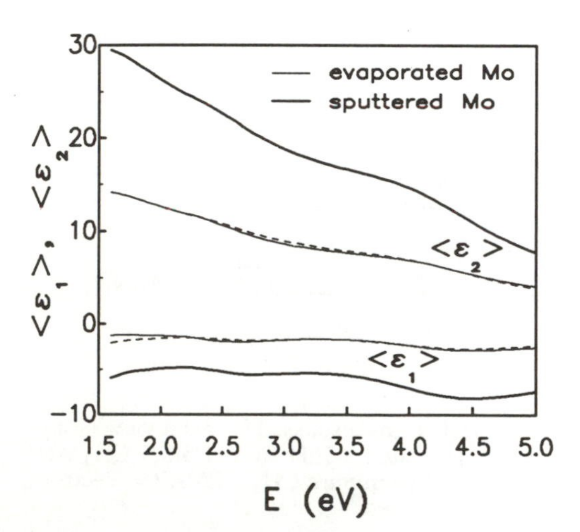

FIGURE 3. A comparison of the pseudo-dielectric function for ion beam sputtered (IBS) Mo and evaporated Mo thin films used as substrates for a-C:H deposition. The broken line represents the results of a linear regression analysis fit to the evaporated film data to determine its void volume fraction (0.335) and low density surface layer thickness (10 Å). The data for the IBS film were applied in the analysis as the reference dielectric function for bulk Mo.

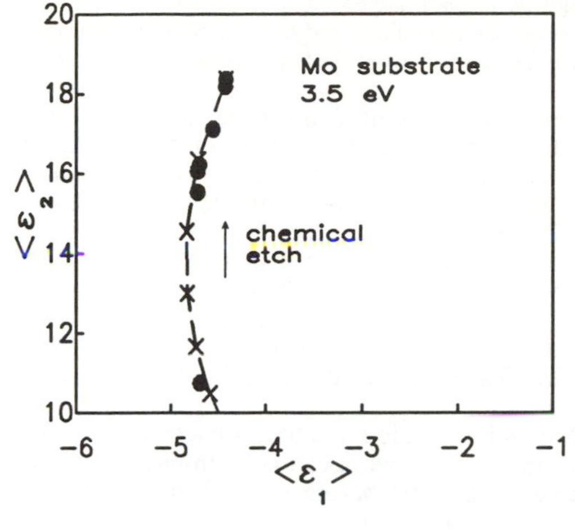

FIGURE 4. Pseudo-dielectric function at 3.5 eV during chemical etching of a thin film IBS Mo substrate (solid points). The broken line (markers at 10 Å increments) is the best fit trajectory for the uniform removal of oxide with $(\varepsilon_1,\varepsilon_2) = (3.74, 0.84)$.

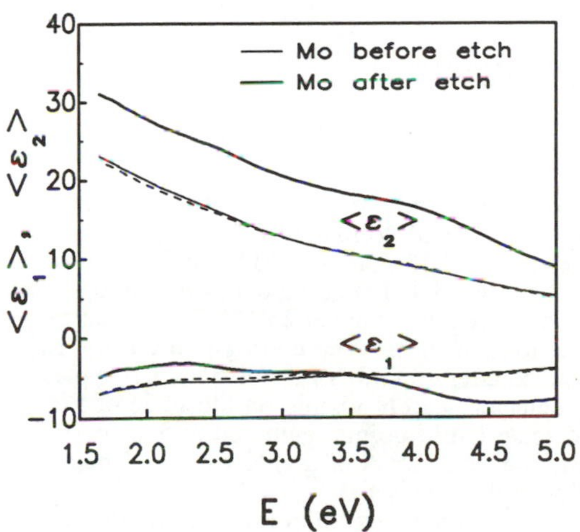

FIGURE 5. A comparison of the pseudo-dielectric function for an IBS Mo substrate film determined before and after the chemical etching treatments of Fig. 4. The broken line represents the results of linear regression analysis to determine the oxide thickness (48 Å) and the composition which simulates the optical properties of the oxide (0.17/0.83 Mo/dielectric). The "after etch" data were applied in the analysis as the reference dielectric function for bulk Mo.

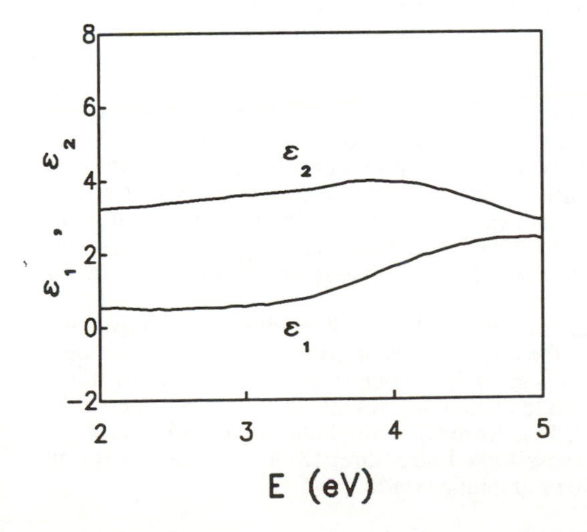

FIGURE 6. Dielectric function of the Mo substrate oxide obtained by assuming a 48 Å thickness and a three phase model (air/oxide/metal) and inverting the "before etch" pseudo-dielectric function of Fig. 5.

stable endpoint of maximum $<\varepsilon_2>$ in Fig. 4 which was unchanged with further oxide etching treatments. Thus, from the markers at 10 Å increments on the calculated trajectory, it is clear that the first etch cycle removes 32 Å of oxide and the following five cycles remove less than 6 Å each.

SE data obtained before and after the treatments are presented in Fig. 5. Using the "after etch" ($<\varepsilon_1>,<\varepsilon_2>$) data as the reference dielectric function for bulk Mo, the data for the oxidized sample were analyzed by two methods. First, the linear regression analysis technique was applied using a composite of materials of known dielectric functions to simulate the dielectric function of the native oxide. The broken line in Fig. 5 was obtained for a best fit 48±1 Å layer having a 0.17±0.01/0.83±0.01 Mo metal/dielectric composition. The dispersion equation for fused silica was used for the dielectric component of the oxide.[66] The 48 Å oxide thickness deduced from the SE data matched that from the real time etching data of Fig. 4. The second technique uses the 48 Å value from Fig. 4 to invert ($<\varepsilon_1>,<\varepsilon_2>$) for the oxidized structure and obtain ($\varepsilon_1,\varepsilon_2$) for the oxide as described in Sect. III. The results of this calculation are given in Fig. 6.

Now that the dielectric function of both Mo and its oxide have been deduced, the oxide thickness for any other Mo substrate prepared in the same process can be computed from the starting ($<\varepsilon_1>,<\varepsilon_2>$) point of the real time ellipsometry trajectory before a-C:H film growth. The importance of this will be made clear in the following sections.

VI. RESULTS AND DISCUSSION: RF DISCHARGES

A. THIN FILM PREPARATION BY PACVD

Figure 7 presents an experimental ($<\varepsilon_1>,<\varepsilon_2>$) trajectory collected at 3.2 eV during the growth of an rf PACVD a-C:H film from pure CH_4 (bold solid line). The deposition was performed with the following preparation parameters: p = 0.25 Torr, P = 20 W, and V_{dc} = -90 V. The starting point at the upper left near ($<\varepsilon_1>,<\varepsilon_2>$) = (-3.4, 16) corresponds to the ion beam sputtered thin film Mo substrate with a 15 Å native oxide layer, held at 250°C. The initial oxide thickness was estimated as described in the previous Section. The ending point at the right corresponds to a semitransparent a-C:H film obtained after ~33 min of deposition. Upon plasma ignition, a slight increase in ($<\varepsilon_1>,<\varepsilon_2>$) was detected, barely visible on the scale of Fig. 7. Figure 8 shows the same data on an expanded scale (solid points, gathered at 5 s intervals), highlighting this initial rapid increase along with the first 125 Å of a-C:H growth. The initial increase is clearly faster than the 5 s time resolution of this data set.

The short broken line segment (markers at 2 Å increments) increasing from the starting point [(-3.4, 16)] in Fig. 8 was calculated to fit the initial rapid movement in ($<\varepsilon_1>,<\varepsilon_2>$) by assuming that 7 Å of the initial 15 Å of native oxide on the Mo was converted to Mo metal in the first 5 s after plasma ignition. Oxide reduction is reasonable, apparently originating upon exposure of the substrate to atomic H from the CH_4 plasma.

Returning to the data of Fig. 7, the full ($<\varepsilon_1>,<\varepsilon_2>$) trajectory could now be fit assuming uniform growth of a-C:H with a dielectric function, ($\varepsilon_1,\varepsilon_2$), of (2.84, 0.425) on the reduced Mo substrate surface. These ($\varepsilon_1,\varepsilon_2$) values correspond to (n,k) of (1.69, 0.13). As in Sect. IV, on the basis of n, it was concluded that this rf PACVD a-C:H is polymer-like.[25] The model fit (broken line with markers at 20 Å increments) virtually coincides with the data on the scale of Fig. 7 and, thus, could be used to determine the thickness at any point along the trajectory. In particular, the endpoint corresponds to ~ 1230 Å, allowing an average deposition rate of 37 Å/min to be calculated. The instantaneous deposition rate could also be determined as will be described in Part C of this Section.

The dominant deviation of the uniform growth model (broken line with markers) from the experimental data (solid line) in Fig. 7 occurs in the initial stages of growth. This is shown most clearly in Fig. 8 where the uniform growth model is given by the short-dashed trajectory (markers every 5 Å). Two types of models were considered to explain the slightly larger $<\varepsilon_1>$ values in the data trajectory: (1) n for the near-substrate a-C:H is lower than the bulk in the initial stages of growth, as for a higher void density or less cross-linked structure; (2) k is greater as for an interfacial roughness layer, or a layer of carbidic or graphitic bonding.

For the long-dashed trajectory in Fig. 8 (markers at 5 Å increments), an initial void density was assumed whereby hemispherical nuclei on a 50 Å square grid increase in radius and coalesce. As discussed earlier, the EMA was used to determine the dielectric function of the nuclei/void composite. This overall structural model has been successfully applied to the problem of a-Si:H nucleation on Mo substrates.[67] However, for a-C:H, the nucleation model predicted lower $<\varepsilon_1>$ values than the uniform growth model. As a result, although the nucleation model cannot not be ruled out because of its slight perturbation to the uniform growth model, other complications must be dominant.

For the solid line trajectory in Fig. 8 (markers at 5 Å increments), a 6 Å layer, modeled as a 0.5/0.5 volume fraction mixture of the film/substrate materials, was assumed to form at the substrate interface in the first minute of deposition (first 10 data points). The 6 Å thickness provided the close fit to the experimental data. Because the interfacial layer is so thin, it was difficult on the basis of the data of Fig. 8 to determine its $(\varepsilon_1,\varepsilon_2)$ values and distinguish from among the possibilities of graphitic and carbidic bonding, and microscopic roughness layers. In

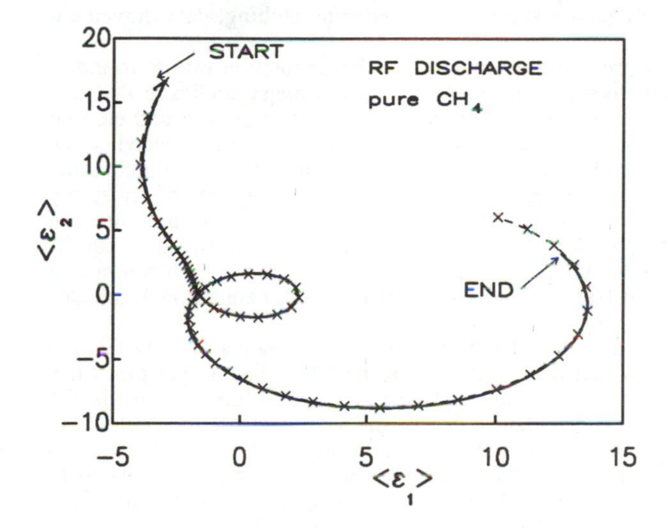

FIGURE 7. Experimental pseudo-dielectric function trajectory at 3.2 eV for the growth of a-C:H on Mo by PACVD from CH_4 (solid line). The broken line, virtually coincident with the data on this scale, was determined assuming uniform growth with $(\varepsilon_1,\varepsilon_2)$=(2.84, 0.425). Markers on the model trajectory denote 20 Å increments. A final film thickness of 1230 Å was deduced.

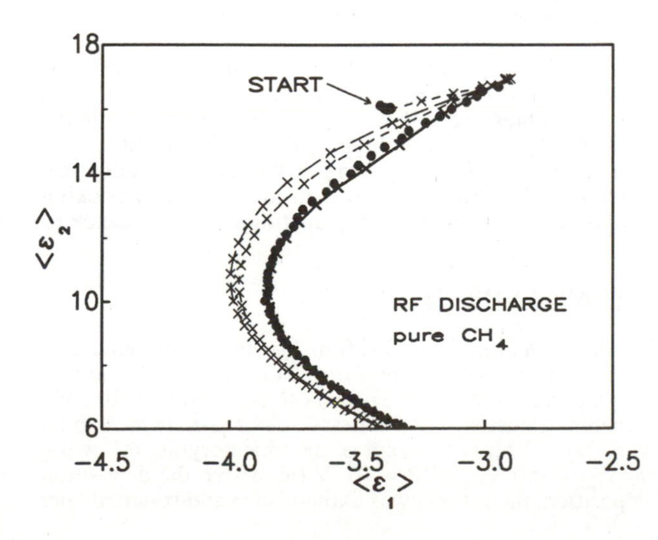

FIGURE 8. Data of Fig. 7 on an expanded scale to highlight the first 125 Å of a-C:H growth (data points, at 5 s intervals). The lines with markers represent models depicting (1) Mo oxide reduction to metal (first short-dashed segment, increasing from 'start'), followed by the uniform growth model of Fig. 7, (2) island nucleation (long dashed line), and (3) formation of a 6 Å interface layer, interpreted as a carbide, in the first minute of growth (solid line). Markers denote 2 Å increments for oxide reduction and 5 Å increments for all other models.

Part C, further experiments are reported which resolve this ambiguity, favoring an interfacial carbide layer. The presence of such interfacial layers has also been deduced by *ex situ* Raman spectroscopy,[68] as well as by *in situ* (but not real time) surface analysis techniques.[2,3]

B. SCANNING PARAMETER SPACE FOR RF DISCHARGES OF CH_4

The results of the previous section demonstrate quite clearly the capabilities of real time ellipsometry. The optical functions, thickness, deposition rate, and also substrate modification and interface effects can be established. Owing to space limitations, it is not possible to cover the wide range of applications of the technique for studying a-C:H film growth and properties. In this section, the use of real time ellipsometry as a deposition rate monitor to characterize a-C:H growth and etching regimes in preparation parameter space will be reviewed. This is one particular area of importance which is extremely time-consuming to study with *ex situ* techniques.

Other workers have also noted regimes of PACVD a-C:H growth and etching and developed models for such behavior.[69-72] It should be clear that real time ellipsometry can be used to delineate these regimes quite easily and accurately. For the trajectory of Fig. 7, the data spiral in a counterclockwise sense during deposition. When plasma deposition and etching processes are in balance, however, the data are stationary, and during etching, data movement reverses to the clockwise sense.

As expected, among the a-C:H preparation parameters, the deposition rate is found to depend most sensitively on the rf plasma power. From real time ellipsometry studies of PACVD a-C:H growth from pure CH_4, an rf power threshold is found to separate growth and etching regimes. Above this threshold, continuous film growth occurs. As the rf power is reduced below the threshold, the deposition rate decreases to a negative value with etching processes dominating. In Figure 9, this power threshold, as determined by real time inspection of ellipsometry trajectory movement, is depicted as a surface in the preparation parameter space of CH_4 pressure and substrate electrode dc bias. For example, at a pressure and dc bias of 0.08 Torr and -100 V, deposition will occur at any power at which the plasma can be sustained (>0.5 W). In contrast, at 0.8 Torr and 0 V bias, the rf power must be increased above 40 W before deposition dominates over etching.

The power threshold was observed to increase by a factor of ~2 when a 1:1 CH_4:H_2 gas ratio was used. This supports the view that it is atomic H in the CH_4 plasma that plays the dominant role in the etching reactions.[72] As the pressure is increased, the increase in the probability of gas phase recombination of C_mH_n radical precursors to form stable molecules may be greater than that for H atom recombination in order to explain the observed transition from growth to etching with pressure. Under such circumstances, the flux of C-containing radicals to the surface may be suppressed relative to the H-flux as the pressure is increased. This supposition is consistent with the observation that if a W-filament is lit (to 1500 K) well outside the plasma zone, the power threshold drops dramatically. The role of the filament then is to enhance the flux of C-containing radicals relative to H at the surface.

A more detailed analysis of the data of Fig. 9 shows that the power threshold for deposition is linearly related to pressure with a prefactor that is independent of dc bias. The constant in the linear relationship between power threshold and pressure is in turn approximately linear with the negative dc bias. Deposition may be enhanced relative to etching by the impingement of ions which supply energy to the surface layers of the film. This may increase the surface reaction rate for the incorporation of precursors such as CH_3 which are expected to have a low sticking coefficient. This mechanism may also provide an alternative explanation of the influence of pressure on the deposition threshold.

C. THIN FILM ETCHING IN RF DISCHARGES OF CH_4

In this part, the ($\langle\varepsilon_1\rangle$,$\langle\varepsilon_2\rangle$) trajectories obtained for a-C:H film growth/etching sequences will be reviewed in greater detail. These experiments were performed to characterize the thin interface layer between PACVD a-C:H prepared from pure CH_4 on both c-Si and thin film Mo substrates. Figure 10 shows an experimental trajectory for the first 45 Å of a-C:H growth on c-Si heated to 250°C (solid circles, every 3s). For this first part of the trajectory the following parameters were used: p = 0.09 Torr, V_{dc} = -50 V, and P = 10 W (ie. above the deposition threshold in Fig. 9). After 80 min of deposition, the plasma was extinguished and restarted later

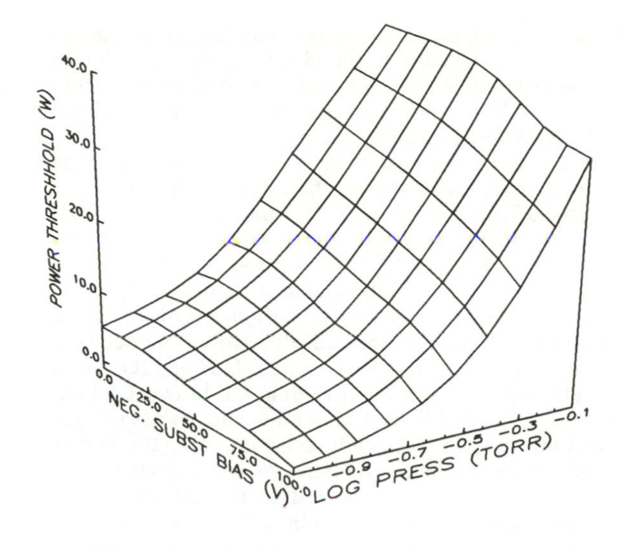

FIGURE 9. Rf power threshold separating deposition (above surface) and etching (below surface) regimes as a function of substrate dc bias magnitude and CH_4 gas pressure. The gas pressure range was 0.08 to 0.8 Torr. The results were determined from real time ellipsometry studies.

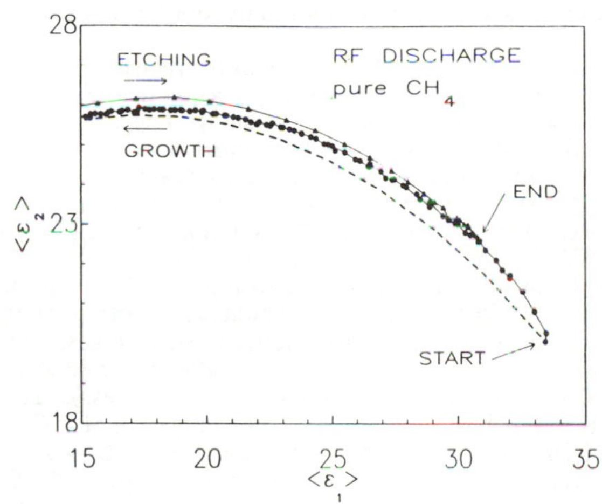

FIGURE 10. Experimental pseudodielectric function trajectory at 3.2 eV for the first ~45 Å of PACVD a-C:H deposition on c-Si using CH_4 (solid circles, at 3 s intervals). After 80 min (~ 850 Å, data not shown), the CH_4 pressure was increased from 0.09 to 0.7 Torr, entering the etching regime. Only the last 40 Å of etching is shown here (solid triangles, at 9 s intervals). The broken line is the uniform growth model with $(\varepsilon_1,\varepsilon_2)=(2.90,\ 0.45)$, found to fit the deposition data for thicknesses > 50 Å.

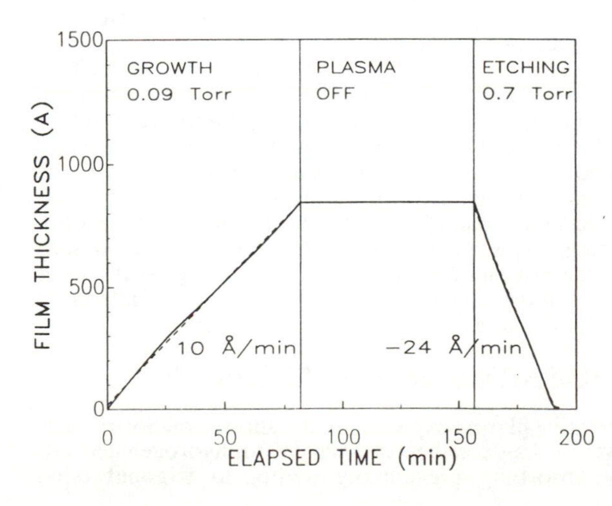

FIGURE 11. Film thickness vs. elapsed time derived from a comparison of deposition-etching data and the model of Fig. 10 (solid line). The broken lines are least-squares fits to constant rates for the deposition and etching regimes.

at p = 0.7 Torr with all other parameters the same. The increase in gas pressure forced the plasma into an etching regime (see Fig. 9), and the ($<\varepsilon_1>$,$<\varepsilon_2>$) data were found to return to values close to the substrate starting point (solid triangles, every 9 s). The broken line provided the best fit to the data in the growth regime for thicknesses > 50 Å (not shown in Fig. 10) and was calculated assuming uniform growth with (ε_1,ε_2)=(2.90, 0.45) [or (n,k)=(1.71, 0.13)].

This model allowed an estimate of the thickness at every point along the growth/etching trajectories. Since the times associated with the data were known, the thickness vs. time and instantaneous deposition rate could be estimated. Such data for the full experiment of Fig. 10 is presented in Figure 11 (solid lines). Least squares fitting was applied to obtain the average rates during growth at 0.09 Torr, 10 Å/min, and etching at 0.7 Torr, -24 Å/min (broken lines).

It should be noted that there are minor errors in the absolute thickness values in Figure 11 owing to the use of the uniform growth model. These will be discussed only briefly here. First, the experimental growth data in Fig. 10 show deviations from the uniform growth model as in Fig. 8. These could be attributed to a 7 Å layer at the interface between the a-C:H and the native oxide of the c-Si substrate. This layer has different optical properties than the bulk a-C:H and could be fit using a silicon-carbon composite and the EMA, as discussed earlier in association with Fig. 8. The etching trajectory shows that this layer is more etch-resistant, being left behind on the substrate surface. Thus, the etching endpoint is displaced along the growth trajectory from the starting substrate point. Second, the etching trajectory is slightly different from the growth trajectory, and converges to it only at the etching endpoint. This can be attributed to inhomogeneous etching with a stable roughness layer which is only removed after underlying material has been removed.

The corresponding interface analysis for the case of a-C:H on Mo will be treated in greater detail next. Figure 12 depicts experimental ($<\varepsilon_1>$,$<\varepsilon_2>$) data for a two step plasma treatment of a 23 Å layer of a-C:H deposited on Mo held at 250°C, exactly as in Figs. 7 and 8. The long-dashed curve traversing the figure (markers every 5 Å of a-C:H) is the model trajectory found to best fit the growth data for the 23 Å thick a-C:H (data points omitted for clarity). This model calculation also included a 6 Å interface layer, modeled as a 0.5/0.5 mixture of Mo/a-C:H, as in Fig. 8.

In the first part of the trajectory, the thin a-C:H at 250°C was exposed to a pure He plasma with p = 0.25 Torr, P = 10 W, and V_{dc} = -50 V. The data (solid points) were collected every 14 s for the duration of the 23 min He plasma exposure. The data shift to higher $<\varepsilon_1>$ values, indicating the formation of a strongly absorbing near-surface layer, then gradually tail off to lower $<\varepsilon_2>$. Two model calculations were performed in an attempt to characterize these changes. The short broken line in Fig. 12 was calculated assuming that polymer-like a-C:H was converted to an equal thickness of low density graphitic a-C (markers every 5 Å of converted thickness). The solid line in Fig. 12 used a conversion ratio of 2 Å of graphitic a-C for every 1 Å of a-C:H (markers every 5 Å of graphitic a-C). A dielectric function for the graphitic a-C of (ε_1,ε_2)=(1.94, 2.01), estimated from the damaged surface layer on plasma-etched, electron-beam evaporated a-C (see Part D), was used owing to the expected damage to the surface layer by He ion bombardment. The data fit the former model in the first few minutes of exposure; the later time deviations may be attributed to a gradual increase in porosity and expansion of the layer.

After He exposure, the thin film was exposed for 23 min to a CH_4 plasma under etching conditions as described earlier (p = 0.4 Torr, P = 10 W, and V_{dc} = -50 V). The data points in this regime were collected at 12 s intervals. Although quantitative modeling was not performed for this part of the trajectory, it is clear that the graphitic overlayer regains transparency in the first minute, presumably by converting back to a lower density polymeric form owing to plasma atomic H exposure. More gradually, this layer and and any underlying a-C:H are etched away. As in Fig. 10, however, the ($<\varepsilon_1>$,$<\varepsilon_2>$) data do not return to the initial substrate values near (-3, 17), but rather to a point displaced 7-8 Å along the original a-C:H growth trajectory (long-dashed line). Again, this behavior suggests an etch resistant layer at the interface to the a-C:H 6-8 Å thick. Because of its resistance to etching, the layer cannot be graphitic or microscopic interfacial roughness. Thus, it has been tentatively identified as a carbide layer.

D. SURFACE MODIFICATION OF EVAPORATED a-C IN AN RF DISCHARGE

In this final section, the effect of reactive plasma exposure on the surface microstructural evolution of a-C will be briefly discussed. As described in Sect. IV, unhydrogenated a-C prepared by evaporation is strongly absorbing presumably owing to trigonal (sp^2)

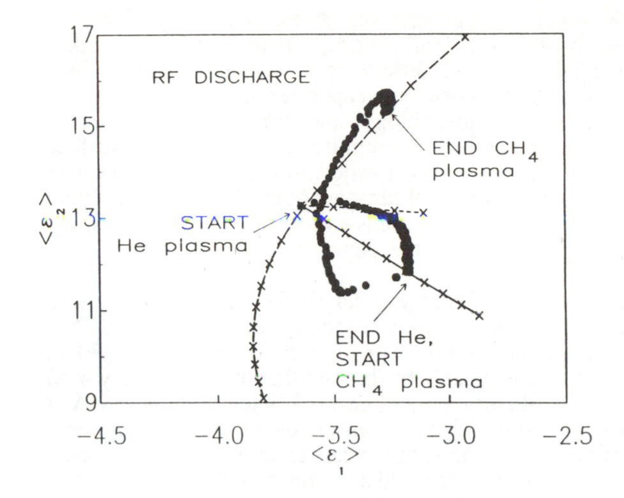

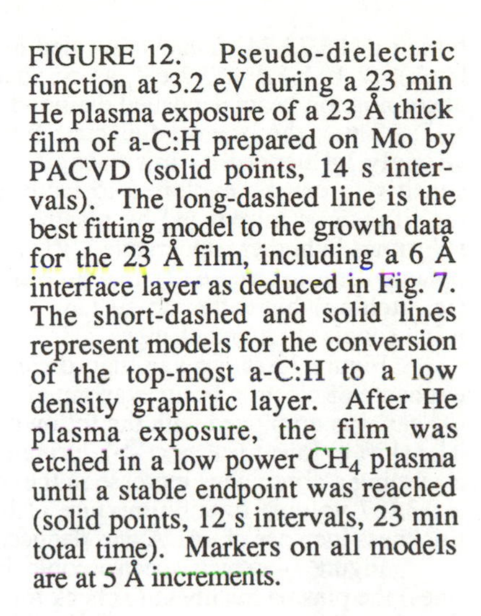

FIGURE 12. Pseudo-dielectric function at 3.2 eV during a 23 min He plasma exposure of a 23 Å thick film of a-C:H prepared on Mo by PACVD (solid points, 14 s intervals). The long-dashed line is the best fitting model to the growth data for the 23 Å film, including a 6 Å interface layer as deduced in Fig. 7. The short-dashed and solid lines represent models for the conversion of the top-most a-C:H to a low density graphitic layer. After He plasma exposure, the film was etched in a low power CH_4 plasma until a stable endpoint was reached (solid points, 12 s intervals, 23 min total time). Markers on all models are at 5 Å increments.

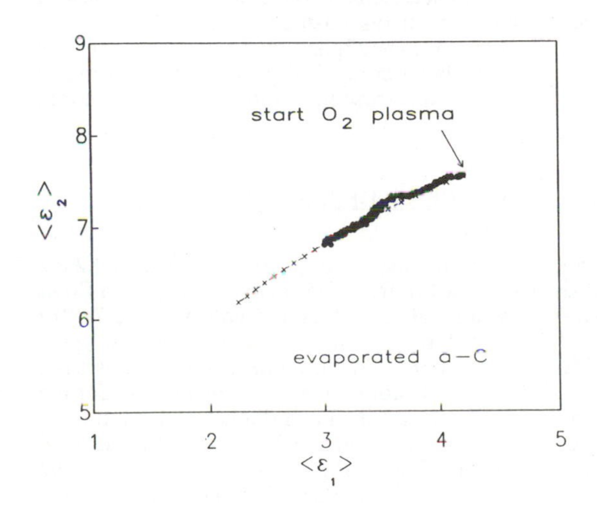

FIGURE 13. Pseudo-dielectric function trajectory at 2.4 eV for a 20 min O_2 plasma exposure of a-C prepared by electron beam evaporation (points, at 5 s intervals). The broken line was calculated assuming an increase in surface roughness, modeled as a 0.3/0.7 volume fraction mixture of bulk a-C/void (markers at 4 Å increments). A 41 Å roughness increase is deduced from the endpoint.

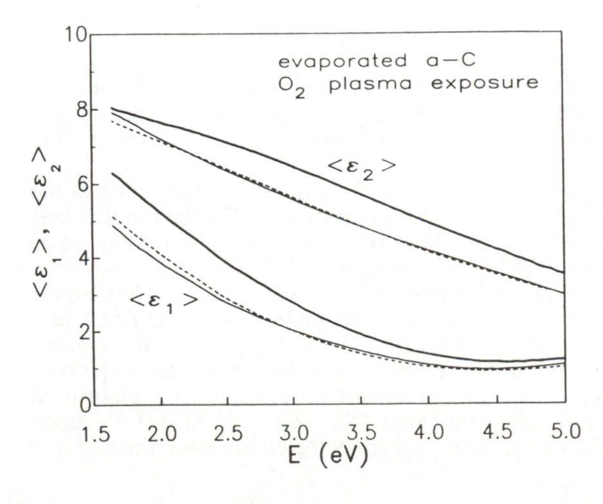

FIGURE 14. Pseudo-dielectric functions obtained before (bold solid) and after (light solid) the O_2 plasma exposure of Fig. 13. The broken lines are the results of a best fit linear regression analysis of the 'after' data using the 'before' data as a bulk reference dielectric function. A roughness layer of 43±2 Å was deduced, best fit by a 0.3/0.7 volume fraction mixture of bulk a-C/void, supporting the real time data of Fig. 13.

bonding.[16,73,74] This is revealed in the a-C dielectric function, $(\varepsilon_1,\varepsilon_2)$, which is ~(2.4, 6.2) [(n,k)=(2.1, 1.5)] at 3.2 eV, as compared to (4.1, 1.2) [(n,k)=(2.1,0.3)] for the more weakly absorbing ion beam deposited diamond-like a-C:H of Fig 2 (also to be described in detail in Sect. VII). Thus, the pseudo-dielectric function data obtained on evaporated a-C films are more strongly influenced by microscopic voids, surface morphology, and oxide layers.[75] This sensitivity can be exploited to study (1) the nucleation of graphite films or (2) the surface modifications introduced by plasma exposure of a-C. The former may also occur on some substrates in higher temperature CVD regimes where diamond films are obtained. When studying opaque films as in case (2), real time ellipsometry data are not sensitive to film thickness changes (ie. etching) but rather to surface and bulk (throughout the penetration depth of the light) microscopic structural changes.

Figure 13 shows a pseudo-dielectric function trajectory at 2.4 eV for a 20 min O_2 plasma exposure of electron beam evaporated a-C held at room temperature (points at 5 s intervals). Pure O_2 gas was employed with the following parameters: p = 0.4 Torr, P = 20 W, and V_{dc} = -60 V. The shift to lower $(<\varepsilon_1>,<\varepsilon_2>)$ during exposure was closely fit (broken line, markers every 4 Å) assuming a continuous increase in the thickness of surface roughness, modeled with the EMA as a 0.3/0.7 volume fraction mixture of bulk a-C/void. After 20 min exposure, a total roughness thickness increase of ~41 Å was deduced from a comparison of the data endpoint with the model.

Figure 14 depicts spectroscopic data collected before (bold solid lines) and after (light solid lines) the plasma treatment to check for consistency at all accessible photon energies. The broken line fit to the 'after' data was obtained by linear regression analysis using the 'before' data as that of the bulk dielectric function of a-C. This, in effect, assumes negligible roughness on the sample before plasma treatment. The analysis gave a best fit layer composed of a 0.3/0.7 mixture of a-C/void with a 43±2 Å thickness value. Thus, the spectroscopic and single wavelength interpretations are consistent within the confidence limits.

VII. RESULTS AND DISCUSSION: a-C:H BY ION BEAM DEPOSITION

Figure 15 depicts an experimental pseudo-dielectric function trajectory obtained at 3.2 eV during the preparation of an ion beam deposited a-C:H film, from 1.5 sccm and 3.5 sccm flows of CH_4 and Ar (solid line).[39] The operating pressure and beam flux and voltage were 7×10^{-4} Torr, 1.5 mA/cm^2, and 200 V, respectively. As in Fig. 7, the starting point at the upper left near $(<\varepsilon_1>,<\varepsilon_2>) = (-4.5, 8)$ corresponds to the Mo substrate, unheated during deposition in this case. The ending point at center is reached after ~ 27 min of deposition time. The lower $<\varepsilon_2>$ for the starting point in Fig. 15 in comparison to that in Fig. 7 was ascribed to a thicker native oxide on the Mo substrate. Chemical etching experiments as in Sect. V revealed a 90 Å oxide thickness for the substrate of Fig. 15 with $(\varepsilon_1,\varepsilon_2) = (3.70, 0.65)$ at 3.2 eV. Fig. 15 shows that, when the deposition is initiated, an increase in $(<\varepsilon_1>,<\varepsilon_2>)$ occurs. This increase is more extensive than the corresponding effect observed in Fig. 8 for the PACVD deposition. The data obtained in this initial growth stage are reproduced for clarity on the expanded scale of Figure 16 (solid points, gathered at 3 s intervals).

In Figs. 15 and 16, the long-dashed line with the markers (virtually indistinguishable from the data on the scale of Fig. 15) represents the simplest model calculation that reproduces the overall experimental results. The two growth stages required to model the data are described in detail in the following two paragraphs.

As in Fig. 8, the upward shift in $(<\varepsilon_1>,<\varepsilon_2>)$ in the first 30 s indicates a sharpening of the dielectric discontinuity at the surface in the initial stage of the deposition. The data in this first stage were closely fit assuming conversion of the top 32 Å of the native oxide into an equal thickness, not of pure Mo metal as for the PACVD deposition, but of a metal-rich composite. This model is given by the long-dashed trajectory to increasing $(<\varepsilon_1>,<\varepsilon_2>)$ in Fig. 15 (markers every 8 Å) and Fig. 16 (markers every 4 Å). The composite was modeled with the EMA as a 0.7/0.3 volume fraction mixture of bulk Mo metal and a-C:H. The initial positive slope of the data in the $(<\varepsilon_1>,<\varepsilon_2>)$-plane provided the composition estimate, and the magnitude of this data movement provided the converted thickness. Figure 17 presents a linear regression analysis fit (broken lines) to spectroscopic data (solid lines) gathered from 2.75 eV to 5.0 eV upon suspending the deposition after 75 s for ~ 20 min. This analysis verified the structure deduced

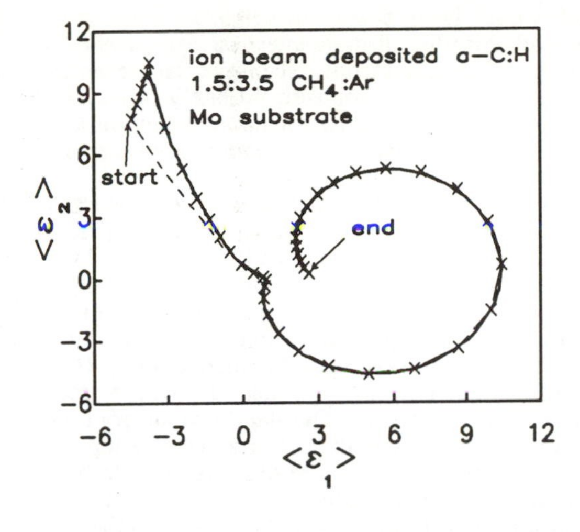

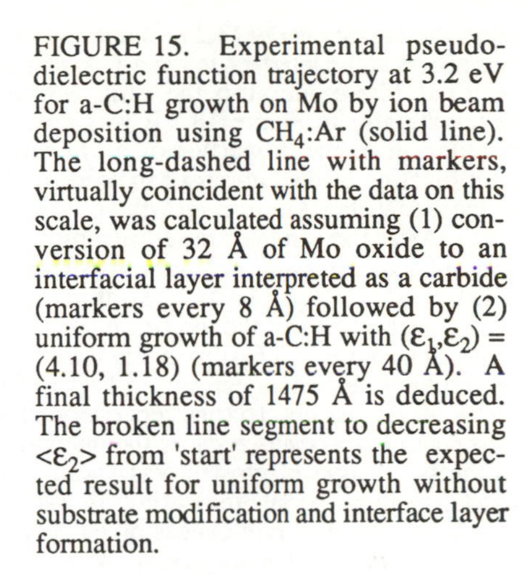
FIGURE 15. Experimental pseudo-dielectric function trajectory at 3.2 eV for a-C:H growth on Mo by ion beam deposition using CH_4:Ar (solid line). The long-dashed line with markers, virtually coincident with the data on this scale, was calculated assuming (1) conversion of 32 Å of Mo oxide to an interfacial layer interpreted as a carbide (markers every 8 Å) followed by (2) uniform growth of a-C:H with $(\varepsilon_1,\varepsilon_2)$ = (4.10, 1.18) (markers every 40 Å). A final thickness of 1475 Å is deduced. The broken line segment to decreasing $\langle\varepsilon_2\rangle$ from 'start' represents the expected result for uniform growth without substrate modification and interface layer formation.

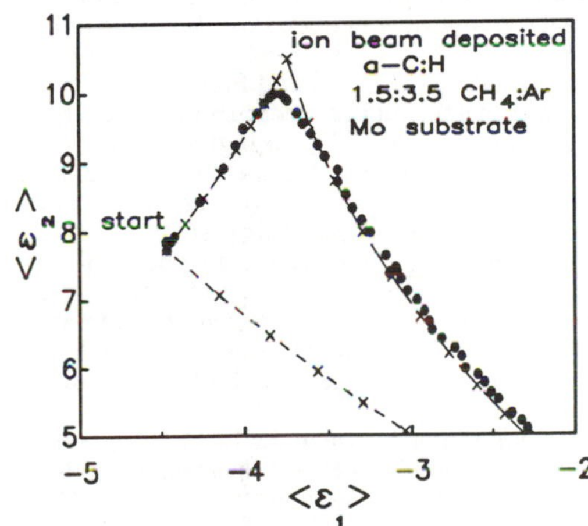

FIGURE 16. Data (solid points, 3 s intervals) and models of Fig. 15 shown on an expanded scale to highlight the first 80 Å of a-C:H growth. Markers for interface layer formation and the two uniform growth models are at 4 Å and 10 Å increments, respectively.

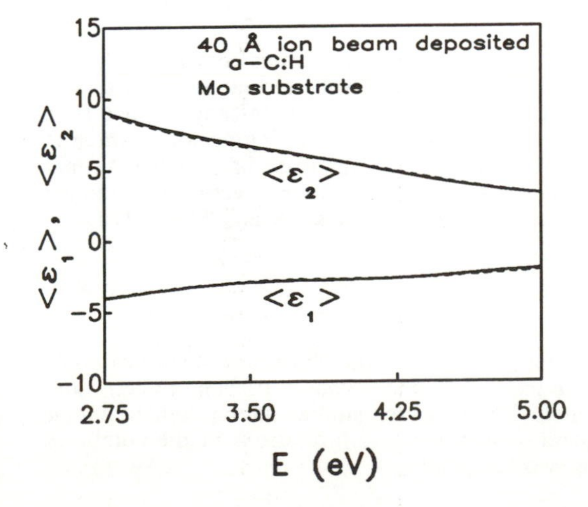

FIGURE 17. Pseudo-dielectric function obtained upon suspending the deposition of Figs. 15 and 16 after 75 s (solid line). The broken lines represent a linear regression analysis fit which supported the real time analysis and gave the structure shown schematically in Fig. 18.

from the single energy (3.2 eV) data: (1) a 34±4 Å thick layer, best fit as a 0.7/0.3 mixture of Mo/a-C:H at the interface to the a-C:H and (2) beneath this film/substrate interface layer, a 59±2 Å residual Mo oxide atop the bulk Mo substrate. The full sample structure is presented schematically in Figure 18. It was noted that the Mo/a-C:H composite probably provides a simulation of the properties of a *chemically-mixed* layer which forms upon reduction and carbidization of the top-most Mo oxide under the influence of CH_n, H, and Ar ion bombardment in the initial stages of film growth.

After formation of the film/substrate interface layer in the first 30 s, the decrease in $<\varepsilon_2>$ and subsequent spiral behavior was fit closely assuming uniform growth of semitransparent a-C:H with $(\varepsilon_1,\varepsilon_2)$=(4.10, 1.18) atop the interface layer. This implies (n,k) values of (2.05, 0.29), suggesting that the ion beam deposited film can be categorized as diamond-like,[25,62] in contrast to the PACVD film of Figs. 7 and 8. As noted above, this a-C:H growth model is given by the long-dashed trajectories in Fig. 15 (indistinguishable from the data on this scale; markers every 40 Å) and Fig. 16 (markers every 10 Å). The slope discontinuity in the model in Fig. 16 at maximum $<\varepsilon_2>$ occurs because the interface between the 32 Å carbide layer and the a-C:H formed in the two stages of growth was assumed to be atomically abrupt. The data deviate from the model in this region, indicating an 8-12 Å interface width. Finally, the trajectory endpoint provided a final film thickness of 1475 Å and an average deposition rate of 55 Å/min. In both Figs. 15 and 16, the trajectory with the short dashes to decreasing $<\varepsilon_2>$ from 'start' was calculated assuming a uniform growth model for a-C:H with $(\varepsilon_1,\varepsilon_2)$=(4.10, 1.18) atop the Mo oxide without incorporating oxide modification and the composite interface layer (markers in Fig. 16 every 10 Å).

The differences between the results of modeling of Figs. 7 and 8 and Figs. 15 and 16 are consistent with the two deposition techniques. It is clear that the higher precursor impact energies of ion beam deposition in comparison with PACVD at low bias (< 100V) or high pressure (> 0.1 Torr) favor diamond-like over polymer-like bonding. As noted above, this accounts for the large difference in the indices of refraction (1.69 vs. 2.05) at 3.2 eV. In addition, the larger extinction coefficient for the sputtered film (0.29 vs. 0.13) indicates a higher fraction of trigonal bonding which relieves internal strain associated with tetrahedral bonding, again as noted earlier.[16] The fact that the top-most Mo oxide is completely reduced for the plasma-assisted CVD film interface, but is converted to a carbide for the sputtered film, probably reflects differences in the relative fluxes atomic H and C-containing species. Finally, the thickness of the interfacial carbide layers (6 Å vs. 32 Å) is another indication of the different impact energies for the two deposition techniques.

To expand on this latter point, Fig. 19 shows a correlation between the index of refraction at 3.2 eV, deduced from the fit to uniform growth, and the carbide interface layer thickness for a number of depositions, deduced as in Figs. 7, 8, 12, 15, and 16. The squares denote PACVD depositions with the designated gas sources: either CH_4 or 1:1 CH_4:H_2. The triangles denote direct ion beam deposited films with the designated gases and beam voltages. The open and solid markers represent depositions on thin film Mo (those of Figs. 7, 8, 15 and 16) and c-Si substrates, respectively.

It is clear that PACVD and ion beam depositions on the two substrates obey a trend which supports a common origin for the variations. Impact energy controls the formation of metastable tetrahedral bonding which increases n. Thus, the increase in interface layer thickness with n for n > 1.95, suggests that the layer is established by the penetration of impinging ions, which is sensitive to the ion energy and species. For n < 1.95, the carbide interface layer is 5-8 Å, only weakly dependent on n. In this regime, the ion penetration may be a monolayer or less, and the interface layer may be established by a slower reaction between the substrate and the a-C:H.

VIII. SUMMARY

Real time ellipsometry at a fixed photon energy, supplemented by selected spectroscopic measurements, have been applied to characterize the growth of amorphous carbon (a-C:H) films, as well as the structure of the starting substrates. The real time studies have provided unique insights into deposition processes. The general capabilities and limitations of the combined techniques will be summarized in the following two paragraphs.

In the later stages of a-C:H film growth (thicknesses > 100 Å), the real time data [expressed as a trajectory in the plane of the pseudo-dielectric function, ($\langle\varepsilon_1\rangle$,$\langle\varepsilon_2\rangle$)] can be fit to obtain the bulk dielectric function (or the index of refraction, n and extinction coefficient, k) of the film. Once such a fit is obtained, the instantaneous deposition rate for the a-C:H can be established from a comparison of the thicknesses for the model to the elapsed time associated with each data point.

In the initial stages of film growth (thickness < 50 Å), the real time data provide information on deviations from uniform, layer-by-layer growth models. For weakly absorbing films, the data are most sensitive to substrate surface modification, such as oxide reduction owing

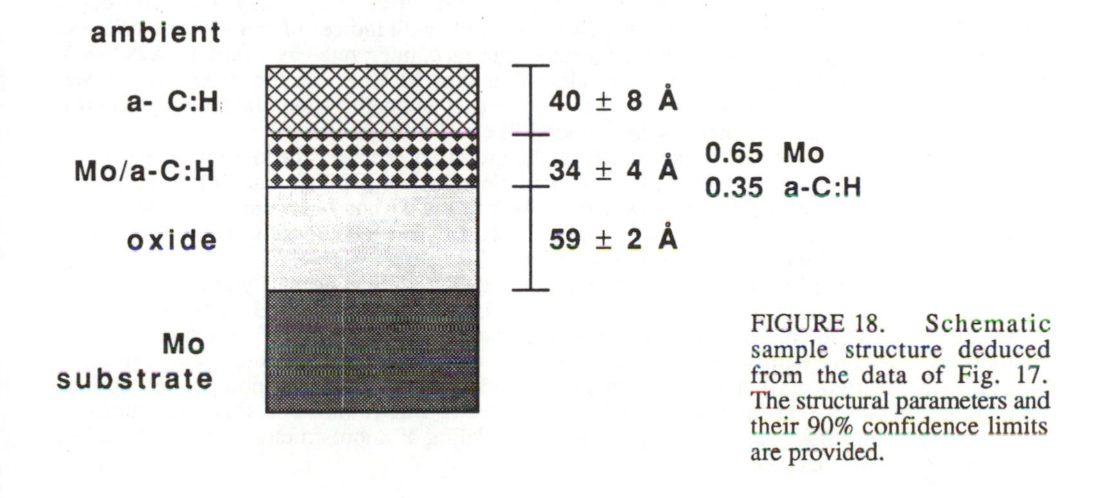

FIGURE 18. Schematic sample structure deduced from the data of Fig. 17. The structural parameters and their 90% confidence limits are provided.

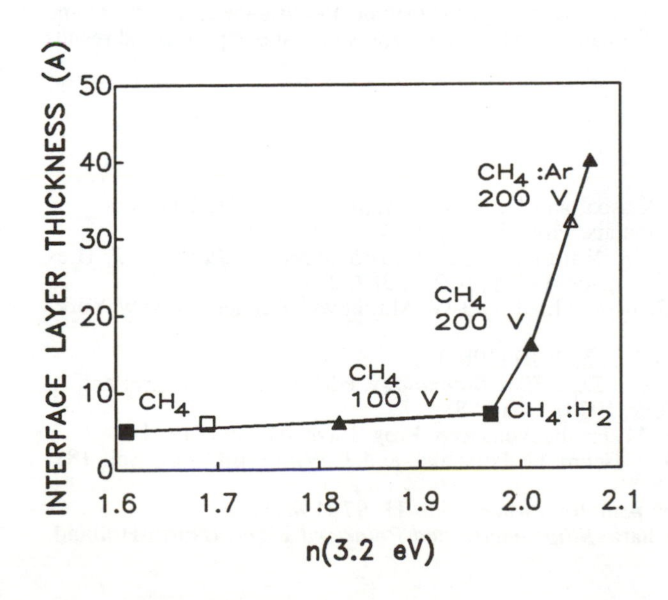

FIGURE 19. Thickness of the interface layer, interpreted as a carbide, between a-C:H films and c-Si (solid markers) and Mo (open markers) substrates plotted for films of different index of refraction. Squares and triangles denote PACVD and direct ion beam deposited films, respectively, with the designated gas supplies and beam voltages.

to atomic H generated in the growth process, and to strongly absorbing interface layers, such as carbides formed by substrate-film reactions. These phenomena can be detected with monolayer sensitivity. Plasma etching has been found to be useful in removing a-C:H films to expose underlying modifications or reactions with the substrate surface. In particular, from this one can distinguish chemical reactions at the substrate interface from physical mixtures characteristic of microscopic interfacial roughness. Limitations do exist in characterizing the reacted interface layers by real time ellipsometry owing to (1) the lack of appropriate comparison standard dielectric functions for the carbides and (2) the use of fixed photon energy measurements. Ultimately, advances in instrumentation including a real time spectroscopic capability employing optical multichannel detection may be required for more detailed interface characterization. In the spectroscopic analyses of this work, the thickest reacted layers (30-40 Å) appear to be best simulated using a 0.7/0.3 physical mixture of substrate/a-C:H.

Next, specific results pertaining to deposition processes will be summarized. First, real time determination of n and k allowed categorization of a series of rf PACVD and ion beam deposited films as polymer-like or diamond-like. For films with indices of refraction < 1.95, (at 3.2 eV) the chemically-reacted layer at the substrate interface, interpreted as a carbide, was 5-8 Å thick. The corresponding layer for higher index films was 30-40 Å. In the latter case, it was suggested that the layer is generated by energetic ion impact and penetration. It is the high energy impact of these ions that also generates the diamond-like bulk film structure.

The influence of deposition parameters on the growth process has been established in real time using stepwise successive depositions on a single substrate. In this way, a plasma power threshold separating etching from growth for PACVD a-C:H has been characterized. This threshold was found to increase with increasing CH_4 pressure, with decreasing substrate negative dc bias magnitude, and with increasing CH_4 dilution in H_2.

For a-C prepared by electron beam evaporation, the bonding is primarily graphitic, and, as a result, ε_2 is much larger than for PACVD and ion beam deposited a-C:H. For these films, ellipsometry measurements are more sensitive to surface roughness. As a final example of the capabilities of real time ellipsometry, the increase in the thickness of surface roughness during a-C etching in a reactive plasma was shown to be detectable with sub-monolayer resolution. Consistency between real time and spectroscopic results was found, and this has improved confidence in using effective medium appromation modeling of microstructure for future studies of C-based thin film nucleation.

Ackowledgements

With pleasure, the author wishes to acknowledge the technical assistance J.M. Cavese and Dr. H. Windischmann of BP America Research and Development, where the experimental results were obtained.

REFERENCES

1) J.H. Neave, B.A. Joyce, P.J. Dobson, and N. Norton, Appl. Phys. A **31**, 1 (1983).
2) S. Kasi, H. Kang, and J.W. Rabalais, Phys. Rev. Lett. **59**, 75 (1987).
3) D. Ugolini, P. Oelhafen, and M. Wittmer, *Proc. E-MRS Meeting, June 1987*, (Les Editions de Physique, Les-Ulis, France, 1987) pp. 267, 287, 297.
4) D.W. Pashley, in *Epitaxial Growth*, edited by J.W. Matthews (Academic, New York, 1975), Part A, p. 2.
5) P.B. Barna, J. Vac. Sci. Technol. A **3**, 2610 (1985).
6) C.A. Neugebauer, in *Handbook of Thin Film Technology*, edited by L.I. Maissel and R. Glang (McGraw-Hill, New York, 1970), Chapt. 8, p. 7.
7) J.A. Venables, G.D.T. Spiller, M. Hanbucken, Rep. Prog. Phys. **47**, 399 (1984).
8) See, for example, E. Ritter, R.J. Behm, G. Potschke, and J. Wintterlin, Surf. Sci. **181**, 403 (1987).
9) J.B. Theeten and D.E. Aspnes, Ann. Rev. Mater. Sci. **11**, 97 (1981).
10) R.M.A. Azzam and N.M. Bashara, *Ellipsometry and Polarized Light*, (North-Holland, Amsterdam, 1977).

11) D.E. Aspnes, Thin Solid Films **89**, 249 (1982).
12) T.J. Moravec and J.C. Lee, J. Vac. Sci. Technol. **20**, 338 (1982).
13) J.B. Theeten and F. Hottier, J. Cryst. Growth **48**, 644 (1980).
14) B. Drevillon, Thin Solid Films **130**, 165 (1985).
15) R.W. Collins, J. Vac. Sci. Technol. A **4**, 516 (1986).
16) J. Robertson, Adv. Phys. **35**, 317 (1986).
17) For an excellent review with comprehensive references see: J. Angus, P. Koidl, and S. Domitz in: *Plasma Deposited Thin Films*, edited by J. Mort and F. Jansen, (CRC, Boca Raton, FL, 1986) p. 89.
18) D.S. Whitmell and R. Williamson, Thin Solid Films **35**, 255 (1976).
19) L. Holland and S.M. Ojha, Thin Solid Films **38**, L17 (1976).
20) S. Berg and L.P. Andersson, Thin Solid Films **58**, 117 (1979).
21) A. Bubenzer, B. Dischler, G. Brandt, and P. Koidl, J. Appl. Phys. **54**, 4590 (1983).
22) A.R. Nyaiesh and W.B. Nowak, J. Vac. Sci. Technol. A **1**, 308 (1983).
23) S.S. Khan, D. Mathine, and J.A. Woollam, Phys. Rev. B **28**, 7229 (1983).
24) C. Weissmantel, K. Bewilogua, K. Breuer, D. Dietrich, U. Ebersbach, H.-J. Erler, B. Rau, and G. Reisse, Thin Solid Films **96**, 31 (1982).
25) B. Dischler, R.E. Sah, P. Koidl, W. Fluhr, and A. Wokaun, in *Proc. 7th Int. Symp. Plasma Chemistry*, edited by C.J. Timmermans (The Technical University at Eindhoven, Eindhoven, The Netherlands, 1985) p. 45.
26) D.A. Anderson, Philos Mag. **35**, 17 (1977).
27) B. Meyerson and F.W. Smith, J. Non-Cryst Solids **35&36**, 435 (1980).
28) B. Meyerson and F.W. Smith, Solid State Commun. **34**, 531 (1980).
29) D.I. Jones and A.D. Stewart, Philos Mag. B **46**, 423 (1982).
30) S. Aisenberg and R. Chabot, J. Appl. Phys. **42**, 2953 (1971).
31) E.G. Spencer, P.H. Schmidt, D.C. Joy, and F.J. Sansalone, Appl. Phys. Lett. **29**, 118 (1976).
32) J.J. Hauser, J. Non-Cryst. Solids **23**, 21 (1977).
33) C. Weissmantel, C. Schurer, R. Frohlich, P. Grau, and H. Lehmann, Thin Solid Films **61**, L5 (1979).
34) B.A. Banks and S.K. Rutledge, J. Vac. Sci Tech. **21**, 807 (1982).
35) T. Mori and Y. Namba, J. Vac. Sci. Technol. A **1**, 23 (1983).
36) T. Miyazawa, S. Misawa, S. Yoshida, and S. Gonda, J. Appl. Phys. **55**, 188 (1984).
37) F. Jansen, M. Machonkin, S. Kaplan, S. Hark, J. Vac. Sci. Technol. A **3**, 605 (1985).
38) J. Koskinen, J.-P. Hirvonen, and A. Anttila, Appl. Phys. Lett. **47**, 941 (1985).
39) M.J. Mirtich, D.M. Swec, and J.C. Angus, Thin Solid Films **131**, 245 (1985).
40) J. Zelez, J. Vac. Sci. Technol. A **1**, 305 (1983).
41) R.W. Collins and J.M. Cavese, Superlattices and Microstructures **4**, 729 (1988).
42) B. Drevillon, J. Perrin, R. Marbot, A. Violet, and J.L. Dalby, Rev. Sci. Instrum. **53**, 969 (1982).
43) J.B. Theeten and F. Hottier, J. Electrochem. Soc. **126**, 450 (1979).
44) R.W. Collins in *Advances in Amorphous Semiconductors*, edited by H. Fritzsche, (World Scientific, Singapore, 1989, in press).
45) A commercial grade OMA-based spectroscopic ellipsometer is presently available from SOPRA, 68 rue Pierre Joigneaux, 92270 Bois-Colombes, France.
46) S.T. Mayer and R.H. Muller, J. Electrochem. Soc. **135**, 2133 (1988).
47) R.W. Collins, Appl. Phys. Lett. **52**, 2025 (1988). See also R.W. Collins, J. Vac. Sci. and Technol. (1989, in press).
48) D.E. Aspnes and A.A. Studna, Appl. Opt. **14**, 220 (1975).
49) R.W. Collins and J.M. Cavese, J. Appl. Phys. **61**, 1869 (1987).
50) The computed trajecory is insensitive to the assumption that a thin surface oxide contributes to the starting ($<\varepsilon_1>$,$<\varepsilon_2>$) values.
51) G.A. Niklasson, C.G. Granqvist, and O. Hunderi, Appl. Opt. **20**, 26 (1981).
52) D.E. Aspnes, Proc. Photo-Opt. Instrum. Eng. (SPIE) **276**, 188 (1981).
53) D.E. Aspnes, A.A. Studna, and E. Kinsbron, Phys. Rev. B **29**, 768 (1984).
54) K. Vedam, P.J. McMarr, and J. Narayan, Appl. Phys. Lett. **47**, 339 (1985).
55) J. Narayan, S.Y. Kim, K. Vedam, and R. Manukonda, Appl. Phys. Lett. **51**, 343 (1987).

56) R.W. Collins, B.G. Yacobi, K.M. Jones, and Y.S. Tsuo, J. Vac. Sci. Technol. A **4**, 153 (1986).
57) J.A. Woollam, P.G. Snyder, A.W. McCormick, A.K. Rai, D. Ingram, and P.P. Pronko, J. Appl. Phys. **62**, 4867 (1987).
58) S.Y. Kim and K. Vedam, Thin Solid Films (1989, in press).
59) H. Arwin and D.E. Aspnes, Thin Solid Films **113**, 101 (1984).
60) F.W. Smith, J. Appl. Phys. **55**, 764 (1984).
61) S. Orzesko, J.A. Woollam, D.C. Ingram, and A.W. McCormick, J. Appl. Phys. **64**, 2611 (1988).
62) N. Savvides, J. Appl. Phys. **59**, 4133 (1986).
63) P.G. Snyder, M.C. Rost, G.H. Bu-Abbud, J. Oh, J.A. Woollam, D. Poker, D.E. Aspnes, D. Ingram, and P. Pronko, J. Appl. Phys. **60**, 779 (1986).
64) H. Windischmann, J. Appl. Phys. **62**, 1800 (1987).
65) D.E. Aspnes and A.A. Studna, Appl. Phys. Lett. **39**, 316 (1981).
66) I.H. Malitson, J. Opt. Soc. Am. **55**, 1205 (1965).
67) R.W. Collins and J.M. Cavese, J. Non-Cryst. Solids **97&98**, 269 (1987).
68) M. Ramsteiner, J. Wagner, Ch. Wild, and P. Koidl, J. Appl. Phys. **62**, 729 (1987).
69) H. Yasuda and T. Hsu, Surf. Sci. **76**, 232 (1978).
70) Z. Has, S. Mitura, M. Clapa, and J. Szmidt, Thin Solid Films **136**, 161 (1986)
71) S. Mitura, L. Klimek, and Z. Has, Thin Solid Films **147**, 83 (1987).
72) S. Nishikawa, K. Kakinuma, H. Fukuda, T. Watanabe, K. Nihei, Jap. J. Appl. Phys. **25**, 511 (1986).
73) N. Wada, P.J. Gaczi, and S.A. Solin, J. Non-Cryst. Solids **35&36**, 543 (1980).
74) D.F.R. Mildner and J.M. Carpenter, J. Non-Cryst. Solids **47**, 391 (1982).
75) Here a comparison is given concerning the influence of surface roughness on the pseudo-dielectric function of strongly and weakly absorbing amorphous thin films. With endpoint roughness layers of 10 Å, for example, pseudodielectric function values of (13,25) and (4,1) for a-Si:H and a-C:H, respectively, would imply bulk dielectric functions of (16.5,26.5) and (4.04,0.93).

Materials Science Forum Vols. 52 & 53 (1989) pp. 365-386

RAMAN SPECTRA OF DIAMONDLIKE AMORPHOUS CARBON FILMS

M. Yoshikawa

Toray Research Center, Inc., Otsu, Shiga 520, Japan

ABSTRACT

Raman spectra of diamondlike amorphous carbon (a-C) films prepared under atmosphere with various hydrogen gas content have been measured as a function of excitation wavelength. The Raman spectral profiles vary with excitation wavelength depending on electronic absorption spectra associated with π-π^* electronic transitions. Dependence of Raman spectra on excitation wavelength is explained in terms of π-π^* resonant Raman scattering from aromatic rings with various sizes. From the results on Raman and infrared measurements, a-C films are confirmed to consist of a mixture of tetrahedral (sp^3) and trigonal (sp^2) bonding structures. The relative Raman intensity is found to decrease with increase of sp^3/sp^2 bonding ratio in a-C films. It is shown that Raman spectroscopy is a powerful tool for the characterization of a-C films.

1. INTRODUCTION

Diamondlike amorphous carbon (a-C) films have recently attracted attention because of their hardness, chemical inertness and optical transparency. Typical a-C films prepared by a variety of methods such as chemical vapor deposition (CVD), sputtering and pulsed laser evaporation possess refractive index of 1.8-2.2, optical gap of 1-2 eV and electrical resistivity of 10^4-10^8 Ω-cm(1-3). A number of analytical techniques have been employed to characterize the microstructure of a-C films including Raman spectroscopy(4-6), infrared spectroscopy (IR)(7), electron energy loss spectroscopy (EELS)(8) and other optical measurements(1,2).

Raman spectra of a-C films have been studied by several authors(3-6). It is reported that the Raman spectrum consists of a band centered at around 1530 cm^{-1} and a shoulder band at around 1400 cm^{-1}, and that the spectral profile varies with the deposition conditions. However, it is not clear whether those bands are vibrations originating from carbon atoms with sp^3 or sp^2 configuration. Consequently, the relation between the film properties and the Raman spectral profile is ambiguous.

Ramsteiner and Wagner have measured Raman spectra of a-C films as a function of excitation wavelength. They have found that position of a broad band at around 1600 cm^{-1} shifts to lower energy with increase of the excitation wavelength(9). We have measured Raman spectra of a-C films prepared by plasma CVD. In a previous paper(10), we demonstrated that the spectra of the a-C films could be well resolved to two bands with Gaussian line shapes, and proposed that these bands originated from carbon clusters with a sp^2 configuration from a comparison between the spectra of sp^3 and sp^2 bonded materials. Furthermore, it was found that the positions of these bands shifted to lower energy with increase of the excitation wavelength.

We measured Raman spectra of several samples prepared by changing the hydrogen gas content in the sputtering atmosphere in order to study the cause of the frequency-shift and to elucidate the microstructure of a-C films. It was found that the Raman spectral profiles changed sensitively reflecting variation of absorption spectra associated with π-π^* electronic transitions with the hydrogen content. From a comparison between Raman and electronic absorption spectra, it was confirmed that the frequency-shift was due to the π-π^* resonant Raman scattering from sp^2 carbon clusters(11). However, there remains some uncertainty concerning detailed structure of sp^2 carbon clusters.

The hydrogenation introduces many sp^3 sites in the carbon films(12) and absorption spectra vary with the hydrogen content in the films(11). The Raman spectral profiles of the a-C films change, reflecting the variation of absorption spectra. Hence, Raman spectroscopy can be used as a tool to estimate the sp^3/sp^2 bonding ratio in a-C films. In this chapter, we discuss about the microstructure of a-C films, based on the Raman spectral variation with the excitation wavelength, and try to elucidate the relation between film properties and Raman spectral profile.

This chapter is organized as follows. In sec.2 we describe Raman spectroscopy used to analyze the microstructure of a-C films. A comparison between Raman spectra of trigonally and tetrahedrally bonded materials is made in sec.3. Variations of Raman spectra with the excitation wavelength and deposition condition are discussed in sec.4. These results are then used to interpret the microstructure of a-C films. The schematic model for the microstructure of a-C films is represented in sec.5.

2. RAMAN SPECTROSCOPY

When the crystal is illuminated with monochromatic light of frequency w_L, a series of much weaker scattered light with frequencies $w_L \pm w_j(q)$, where $w_j(q)$ are optical phonon frequencies, is radiated from the crystal in addition to a very strong light at the frequency w_L(13,14). The strong line centered at w_L is due to elastic scattering of phonons in the spectrum of the radiation and is known as Rayleigh scattering. The series of weak lines at $w_L \pm w_j(q)$ originates from inelastic scattering of photons by phonons which induce the electronic dipole moment arising from the modulation of the electronic polarizability by the vibration of the atoms(lattice or molecular vibrations) and constitutes the Raman spectrum. The Raman bands at frequencies $w_L - w_j(q)$ are called Stokes lines, those at frequencies $w_L + w_j(q)$ are known as anti-Stokes lines. Since the intensities of the anti-Stokes line are considerably weaker than those of the Stokes lines, the Stokes lines are usually detected. Since the wave vector of visible light is still extremely small compared to the extension of the Brillouin zone, first-order Raman scattering yields information about phonons at q=0. If two or more phonons are involved in the scattering process, it is also possible to obtain informations about phonons with $q \neq 0$. However, the Raman lines associated with these higher order process are weak compared with those associated with first-order processes.

Since the number of the observed Raman lines, the peak frequencies, the intensities or half width depends on the crystal structure and crystal states, we can inversely obtain informations about the crystal structure and the degree of the crystallization by measuring the Raman spectrum of the materials. Nowadays, Raman spectroscopy is used as a powerful tool for the characterization of carbon materials, semiconductors, ceramics, and organic materials.

Photons in the infrared and far-infrared region of the spectrum have energies comparable with phonon energies. Such photons can be strongly absorbed by certain phonons at q=0. In one-phonon process a photon can be absorbed only by those phonons which induce a non-vanishing dipole moment. Since the selection rules in infrared absorption and Raman scattering are generally different, the two techniques provide complementary information.

The principle type of instrumentation is illustrated in figure 1. The monochromatic and polarized light of the Ar or Kr laser passes through an interference filter that rejects spurious

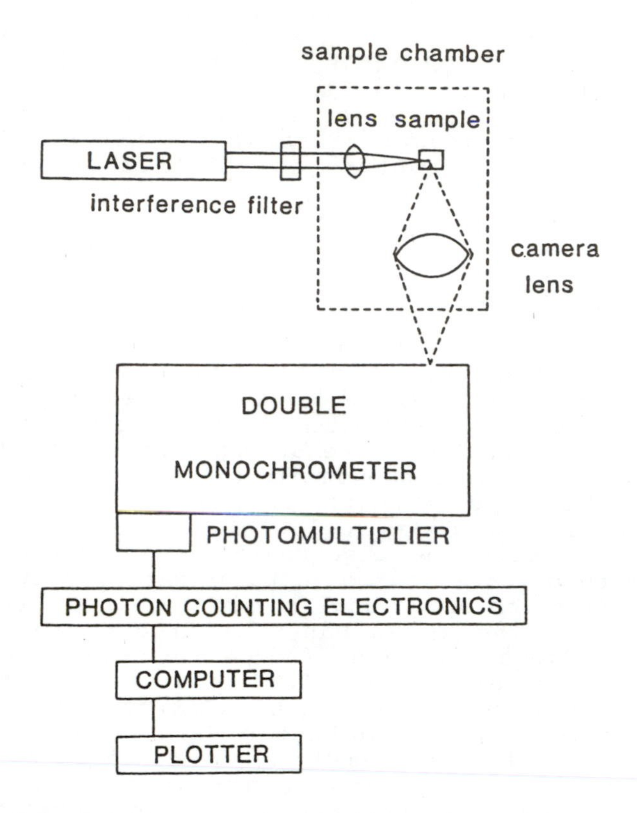

Figure 1. Raman scattering apparatus.

lines from the laser sources. The light beam then enters the polarization rotator and is focused by the lens onto the sample. Light scattered from the sample is focused by the camera lens onto the entrance slit of the double-grating spectrometer. Light leaving the final exit slit of double-grating monochromator is focused onto the cathode of the photomultiplier, whose output is processed with photon counting electronics which include the preamplifier and the discriminator.

3. RAMAN SCATTERING IN CARBON MATERIALS

3.1. TRIGONALLY BONDED MATERIALS

Graphite crystallizes according to the $D_{6h}{}^{4}$ space group and has twelve vibrational modes at q=0(15). The frequencies of the two in-plane Raman-active E_{2g} modes have been measured directly by Raman spectroscopy yielding $w(E_{2g2})=1581\ cm^{-1}$ for the high frequency mode and $w(E_{2g1})=42\ cm^{-1}$ for the low frequency mode.

The atomic displacement for E_{2g2} mode is shown in figure 2-(a).

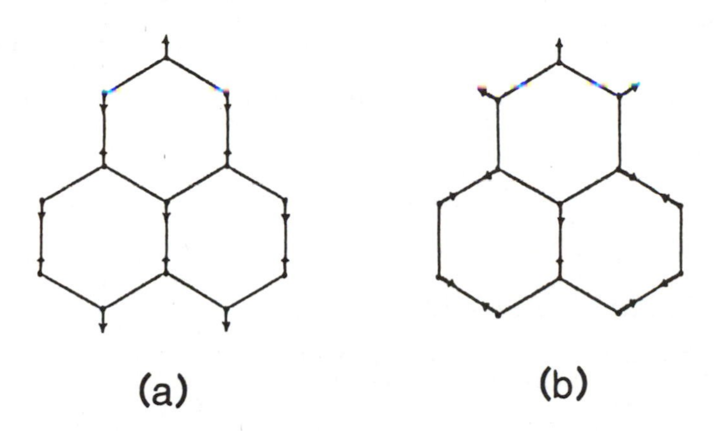

Figure 2. Atomic displacements for Raman active modes;
(a) E_{2g2} mode (b) The disorder mode

The highly oriented pyrolytic graphite (HOPG) is highly oriented along the c-axis, but in the layer planes it consists of a randomly ordered collection of crystallites of ~1um average diameter. The E_{2g2} band for HOPG is observed at 1581 cm^{-1}. Raman spectra of HOPG, pyrolytic graphite (PG) and glassy carbon (GC) are shown in figure 3.

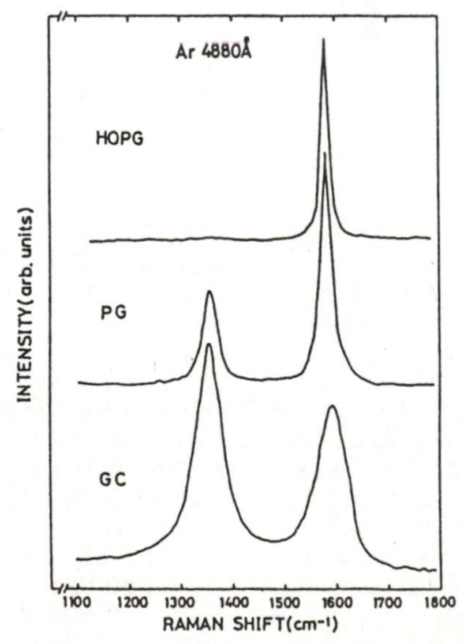

Figure 3. Raman spectra of HOPG, PG and GC.

PG with small crystallites relative to HOPG exhibits two Raman bands, the E_{2g2} mode near 1581 cm^{-1} and an additional broad band at about 1360 cm^{-1} (the disorder mode) which is assigned to phonon associated with finite crystallite size(16). The atomic displacement for the 1360 cm^{-1} band is shown in figure 2-(b). The intensity and bandwidth of the 1360 cm^{-1} band increase with decreasing the crystallite size. In addition, the bandwidth of the E_{2g2} band increases as the crystallite size decreases, and the peak frequency of the E_{2g2} band upshifts since the breakdown in the wave vector selection rule brings in a range of modes away from the zone center which are upshifted relative to the zone-center mode. In GC with smaller crystallites than PG(~25Å), the E_{2g} band is observed at about 1590 cm^{-1}.

3.2. TETRAHEDRALLY BONDED MATERIALS

Diamond has the same sp^3 bonding structure as that of silicon. A single peak is observed at 1333 cm^{-1} in diamond and at 520 cm^{-1} in silicon(10). Raman spectra of diamond, silicon and amorphous silicon are shown in figure 4. In amorphous materials,

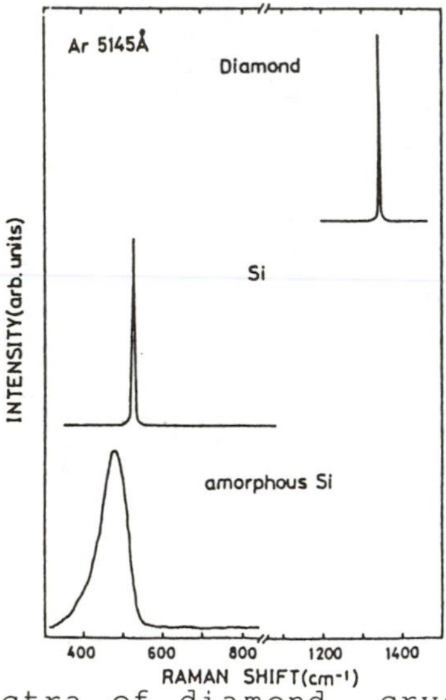

Figure 4. Raman spectra of diamond, crystal silicon and amorphous silicon.

observed Raman spectrum reflects the vibrational density of states which is similar to that of the crystal. There is a maximum of the calculated state density near 480 cm^{-1} in amorphous silicon and a broad band is actually observed at 480 cm^{-1}(10). Although the Raman spectrum in amorphous diamond has not been reported yet , a broad band is expected to be observed at around 1200 cm^{-1} from the calculation of the density of states(6).

4. RAMAN SCATTERING IN DIAMONDLIKE FILMS

4.1. SPUTTERING FILMS

4.1.1. EXPERIMENTAL

The a-C films were prepared by sputtering a graphite target with 1 kW power and depositing carbon on silicon and glass substrates held at about 280K in a mixture gas of Ar and H_2. Several samples were prepared by changing the hydrogen gas content ($H_2/(H_2+Ar)$) from 0-70 %. The total pressure was kept at about 10^{-3} Torr.

Raman spectra were measured at room temperature with several lines of an argon ion laser and a krypton ion laser. The Raman spectra were recorded by a Jobin Yvon Ramanor U-1000 double monochromator equipped with a photomultiplier and photon-counting electronics. Raman spectra were measured in a backscattering geometry and recorded at a low power of 40-60 mW to avoid thermal decomposition of the samples. The frequency of the observed Raman bands was calibrated using a 320 cm^{-1} band in CaF_2.

4.1.2. RESULTS AND DISCUSSIONS

A. RESONANT RAMAN SCATTERING

Figure 5 shows Raman spectra of a-C films deposited on silicon substrates excited at various wavelengths. A relatively sharper Raman band at 1530 cm^{-1} and a broad shoulder band at around 1400 cm^{-1} are observed in the spectra excited by 5145 Å line. The Raman spectral profile varies with the excitation wavelength. Dependence of the frequencies of the two Raman bands on the excitation wavelength was plotted after the Gaussian line shape analysis in figure 6(17). The calculated lines agree well with the experimental data points. The experimental results for HOPG, PG and GC are also plotted in figure 6 for comparison. As seen in figure 6, the position of the high-frequency band for samples A and C varies from 1558 to 1515 cm^{-1} and from 1563 to 1516 cm^{-1} between 4579 and 6471 Å excitations, respectively. The frequency-shift of the shoulder band for samples A and C is from 1388 to 1351 cm^{-1} and from 1393 to 1356 cm^{-1}, respectively.

In HOPG which consists of aromatic rings with infinitesize, the E_{2g2} band is observed at 1581 cm^{-1}(15,16), independently of the excitation wavelength. On the other hand, the positions of the E_{2g2} bands of PG and GC, which have finite aromatic ring structures, shift to higher frequency by about 2-5 cm^{-1}(10). The low-frequency band which is assigned to phonon associated with finite crystallite size is observed at about 1360 cm^{-1} in PG and GC, and the position shifts to lower frequency with increase of the excitation wavelength(16). The wavelength dependence of the E_{2g2} bands observed for HOPG, PG and for GC suggests that the variation of the high-frequency bands for PG and GC is related to the crystallite size. The frequency-shift of the shoulder bands for a-C films agrees with that of the 1360 cm^{-1} band in PG and

GC(10). In addition, the frequency variation for the high-frequency bands is also similar to that for the 1360 cm^{-1} band in PG and GC. From the comparison of Raman spectral features of a-C films, PG, GC and HOPG, the frequency variation of Raman bands in a-C films is considered to be related to the graphite crystallite size(10).

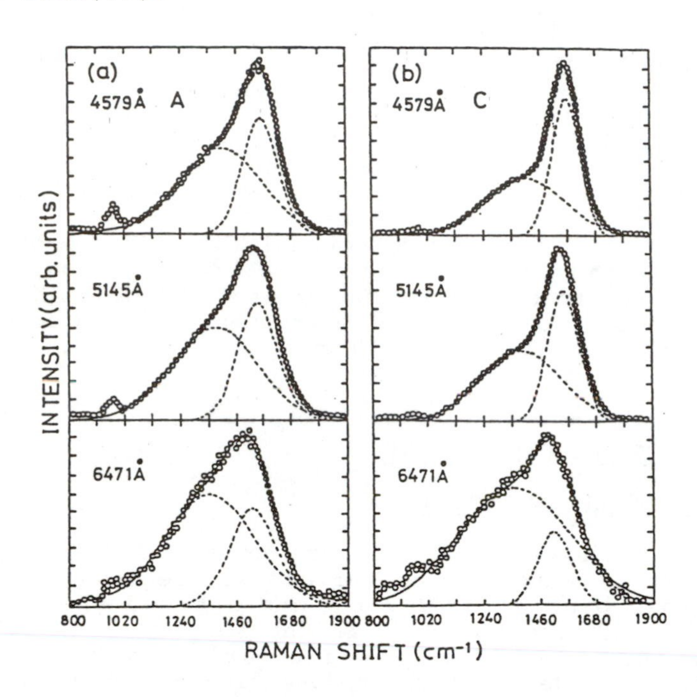

Figure 5. Raman spectra of a-C films obtained by various excitation wavelengths. Figure 5-(a) and -(b) show Raman spectra of samples A (the hydrogen content 0%) and C (30%), respectively. The Raman spectra were given after subtracting the back ground due to the luminescence. The open circles are experimental data points and solid lines are the Gaussian fit to the data. The decomposed bands are also shown in Figs.5-(a) and -(b) in broken lines. The Raman band at about 970 cm^{-1} corresponds to the second order line of the Si substrates.

A single peak is observed at 1333 cm^{-1} in diamond(10). The position of the Raman band does not depend on the excitation wavelength. In addition, the position of Raman bands in amorphous sp^3 bonded materials such as amorphous silicon does not depend also on the excitation wavelength. From the comparison between Raman spectral features for sp^2 and sp^3 bonded materials and expectation of appearance of an amorphous diamond band in a much lower wavenumber region as discussed in sec.3.2, we conclude that Raman bands at around 1400 and 1530 cm^{-1} in a-C films originate from carbon clusters with sp^2 configuration. Considering that the Raman scattering cross-section of diamond

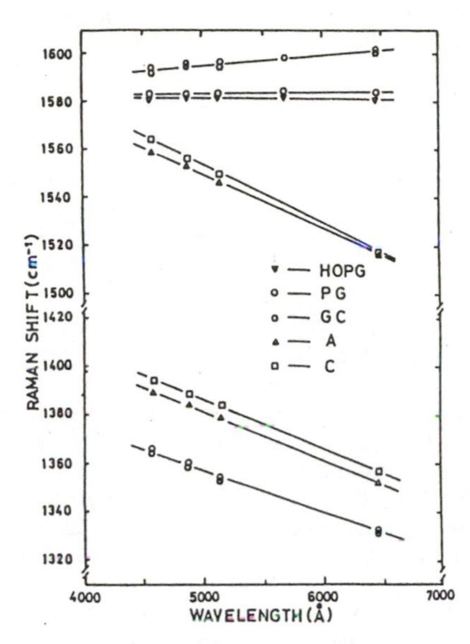

Figure 6. Variation of the peak frequencies of the major Raman bands as a function of excitation wavelength for a-C films (samples A and C), HOPG, PG and GC.

($9.1 \times 10^{-7} cm^{-1} sr^{-1}$) is much lower than that of graphite ($5 \times 10^{-5} cm^{-1} sr^{-1}$)(20), it seems to be very difficult to detect Raman bands due to the sp^3 carbon clusters.

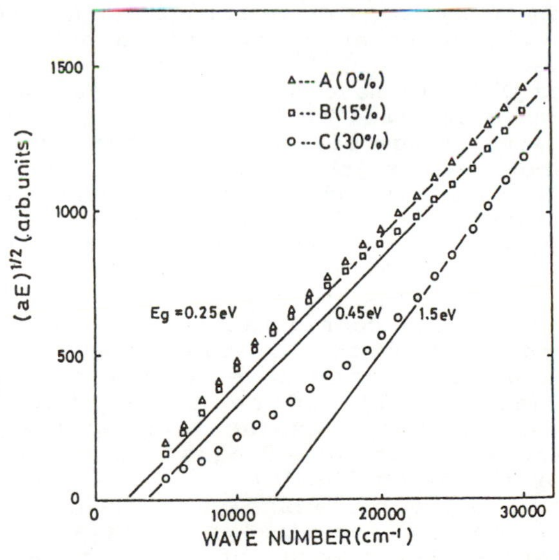

Figure 7. The optical gaps for samples A(0%), B(15%) and C(30%) deposited on the glass substrates.

Absorption spectra of a-C films deposited on glass substrates were measured and replotted in figure 7 by the following relation:

$$(aE)^{1/2} \sim (E-E_{opt})$$

where a and E are absorption coefficient and energy, respectively. The relation holds not only for the absorption edge of a-C films but also for that of a-Si:H and chalcogenide glasses(18). Extrapolation of the straight line to the abscissa gives the " optical gap " E_{opt}. The optical gaps of a-C films increase from 0.25 to 2.0 eV with increase of the hydrogen content(17). The values of the optical gap for various films are listed in table I. The absorptions associated with π-π^* transition in aromatic compounds are observed between 10000 and 45000 cm^{-1} (for example, at about 16000 and 43000 cm^{-1} for transpolyacetylene and benzene, respectively)(12,19). The absorption at around 20000 cm^{-1} decreases with increase of the hydrogen content. The spectral variation here with the hydrogen content agrees with the results reported by several authors(8,12,17).

As seen in Figs.5-(a) and 5-(b), the intensity of the shoulder band becomes stronger with increase of the excitation wavelength, but the intensity of the high-frequency band inversely becomes weaker(11). Consequently, a bend at about 1400 cm^{-1} in the Raman spectra becomes dull with increase of the excitation wavelength. The bend at about 1400 cm^{-1} becomes sharper with increase of the hydrogen content. This spectral variation will be explained by the change of $(aE)^{1/2}$ spectra with the hydrogen content in figure 7. In the previous paper(11), we have interpreted the spectral variation with the excitation wavelength by the π-π^* resonant Raman scattering from sp^2 carbon clusters, based on a comparison between Raman and absorption spectra.

Robertson and O'Reilly have calculated electronic structure of a-C films and have asserted that the width of the optical gap varies inversely with change of the sp^2 carbon cluster size(12). Considering the above result, the variations of Raman spectra and the absorption edge with the hydrogen content are explained by a change of distribution of the sp^2 carbon cluster size with the hydrogen content. A comparable downshift with increase of excitation wavelength has been observed for Raman bands of C=C and C-C stretching vibrations (v_1 and v_3 modes) with A_g symmetry in polyenes with the trans configuration(21). The polyenes have chain-length-dependent absorption wavelength and forceconstants, and the downshift has been attributed to the resonant coupling with longer chains with increase of the excitation wavelength(21). An analogous explanation by the wavelength-dependent coupling with sp^2 carbon clusters with various sizes has been given for the a-C films(10).

The v_1 and v_3 modes for transpolyacetylene are observed at about 1170 and 1490 cm^{-1}, respectively, in the spectra excited by the 5145 Å(21) whereas Raman bands of a-C films are observed at about 1400 and 1530 cm^{-1}. The frequencies of v_1 and v_3 modes for transpolyacetylene do not agree with those of the Raman bands for a-C films. Smith et al. measured a Raman spectrum of the damaged layer of diamond after ion bombardment, and observed an asymmetric broad band which is centered at about 1530 cm^{-1} and superimposed with a sharp Raman line of diamond at 1333 cm^{-1}(22). An asymmetric broad band is also observed on the damaged layer of graphite after ion bombardment. The spectral profiles of the damaged layers of graphite and diamond coincide well with those of a-C films except the sharp line of diamond. These observations and comparison among the Raman spectral features of a-C films, PG, GC and HOPG suggest that Raman bands in a-C films originate from aromatic rings with various sizes rather than polyene chains. The interpretation here is consistent with the result on an electron-diffraction analysis by McKenzie et al.(23).

As seen in figure 5, the high-frequency band is enhanced by high energy excitation whereas the shoulder band is enhanced by low energy excitation. Referring to the result that the electronic absorption maximum shifts to a lower energy with increase of ring size in aromatic molecules (i.e., naphthalene →anthracene→tetracene)(19), we assume that the Raman bands at around 1400 and 1530 cm^{-1} originate mainly from aromatic rings with large and small sizes, respectively.

The hydrogenation introduces more sp^3 sites in the films, and results in change of the absorption spectra and increase of the optical gap(12). Considering that Raman spectral profiles vary sensitively reflecting change of absorption spectra, Raman spectroscopy is expected to be used as an indicator for the sp^3 content of a-C films. In the next section, we will discuss the relation between the sp^3 content and Raman spectra.

B. CHARACTERIZATION

Refractive index and thickness of the a-C films deposited on silicon substrates were determined by ellipsometry. The thickness of samples is about 400-1000 Å and the refractive index at 6328 Å was listed in table I.

Figure 8 shows Raman spectra of various films excited at 5145 Å. Luminescence becomes intense with increase of the hydrogen content and Raman spectra superimposed on it are observed in a-C films. The relative intensity of the shoulder band against that of the high-frequency band decreases with increase of hydrogen content in the range 0-30 %. The relative intensity obtained by the Gaussian line shape analysis is listed in table I(17).

For a-C films deposited under the hydrogen content in the range 0-30 %, the relative intensity decreases from 2.3 to 1.2 with increase of the hydrogen content. The optical gap and the refractive index varies from 0.25 to 1.5 eV and from 2.54 to 2.01, respectively. It is reported that optical gaps of a-C films increase with increase of the sp^3/sp^2 ratio(1,2). It is also reported that the refractive index of a-C films decreases with increase of the sp^3/sp^2 ratio(1,2). Combining the results on Raman and optical measurements, it is considered that the decrease of the relative intensity is related to increase of the sp^3/sp^2 ratio in a-C films(17).

Table I. The relation between the relative intensity of Raman bands and optical constants. The relative intensity shows the intensity of the shoulder band normalized by that of the high-frequency band. E_g and n are optical gap and refractive index, respectively.

Sample	Gas ratio ($H_2/(H_2+Ar)$) (%)	Relative intensity	n (6328Å)	E_g (eV)
A	0	2.3	2.54	0.25
B	15	1.4	2.47	0.45
C	30	1.2	2.01	1.5
D	50	*	1.83	2.0
E	70	*	1.81	2.0

In figure 8 and figure 3 of ref.11, the peak frequencies of the shoulder and high-frequency bands do not appreciably vary with the samples. Since a change in the sizes of the sp^2 carbon clusters is expected to cause the frequency-shift of the Raman bands, it is considered that the sample dependent change in the relative intensity is not due to the change of the size but due to a change in the size distribution. The decrease of the intensity of the shoulder band assigned to the large sp^2 carbon clusters with increase of the hydrogen content suggests that the sp^2 carbon clusters with large sizes are converted into the sp^3 structures or a merely decrease of content of sp^2 carbon clusters with large sizes. The decrease of content of the sp^2 clusters with large sizes or the conversion into sp^3 structures is considered to cause an increase of the optical gap, that is, an increase of the sp^3/sp^2 ratio in a-C films.

Fink et al.(8) measured electron energy loss spectra of a-C films. From a Kramers-Kronig analysis of the loss functions, they concluded that 2/3 of carbon atoms were sp^3 bonded for a typical a-C film having an optical gap of ~1.5 eV and that1/3 were sp^2 bonded. Their estimation agrees with the result from IR measurement by Dischler et al.(7). On the basis of their results on EELS and IR measurements, we estimate that 2/3 of the carbon atoms are sp^3 bonded for our sample C.

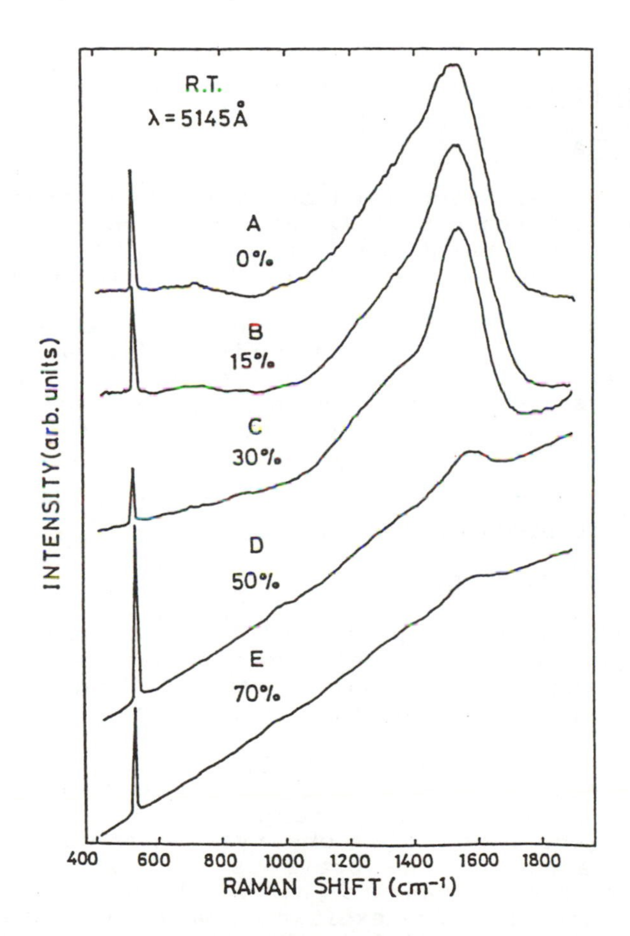

Figure 8. Raman spectra of several samples excited at 5145 Å; A(0%), B(15%), C(30%), D(50%) and E(70%). The Raman bands at about 520 and 970 cm^{-1} correspond to the first and second order lines of the Si substrate, respectively.

For samples D and E, luminescence becomes more intense and obscures Raman spectra due to sp^2 carbon clusters. The optical gap increases to 2.0 eV and the refractive index decreases to 1.8

with increase of the hydrogen content. The samples D and E decompose easily even at laser power less than 10 mW. It is known that excessive increase of the hydrogen content results in not only extreme increase of the optical gap but also decrease of hardness and increase of the polymeric structure in the film(4). McKenzie et al. also obtained an evidence of the polymeric structure in the a-C films from an electron-diffraction analysis(23). In addition, it is well-known that the refractive index of most polymers is in the range 1-2. From these results, we conclude that polymeric structures are formed for samples D and E.

4.2. PLASMA CVD FILMS

4.2.1. EXPERIMENTAL

The a-C films are deposited on silicon substrates held at room temperature. Mixture of C_2H_4 and H_2 was used. Two types of sample were obtained by changing the gas ratio (A:C_2H_4/H_2=1/0,B:C_2H_4/H_2=1/50). The total pressure was kept at about 1 Torr.

The thickness and the refractive index of the a-C films were determined by ellipsometry. The refractive index was about 2.1 at 6328Å in agreement with an earlier work on a-C films deposited by the pulsed laser evaporation(3). The optical gap and hardness of the films was about 1 eV and 9 on the Mohs scale, respectively.

4.2.2. RESULTS AND DISCUSSIONS

A. RESONANT RAMAN SCATTERING

Figure 9 shows Raman spectra of the a-C films excited at various wavelengths. A relatively sharp Raman band centered at about 1530 cm^{-1} and a broad shoulder band at about 1400 cm^{-1} are observed in the spectra excited at 5145Å line(10). The spectral profile agrees with that of the a-C films deposited by the sputtering method. As seen in figure 9, the position of the high frequency band both for samples A and B varies from 1550 to 1480 cm^{-1} and from 1565 to 1500 cm^{-1}, respectively, with increasing the excitation wavelength. This observation is consistent with the results obtained by the previous section. The relative intensity of the shoulder band to that of the high-frequency band increases with increasing the excitation wavelength. Furthermore, the relative intensity of the shoulder band is found to become larger with increase of the amount of H_2 gas.

The Raman spectra can be decomposed into two bands with Gaussian lineshapes(10). The calculated lines agree well with the experimental data points. The dependences of the frequencies of the two decomposed Raman bands on the excitation wavelength were plotted in figure 10. The experimental results for HOPG, PG and GC are also plotted in figure 10 for comparison.

The position of the high-frequency band in a-C films shifts to lower energy with increase of the excitation wavelength; from 1554 to 1511 cm^{-1} and from 1570 to 1544 cm^{-1} for samples A and B,

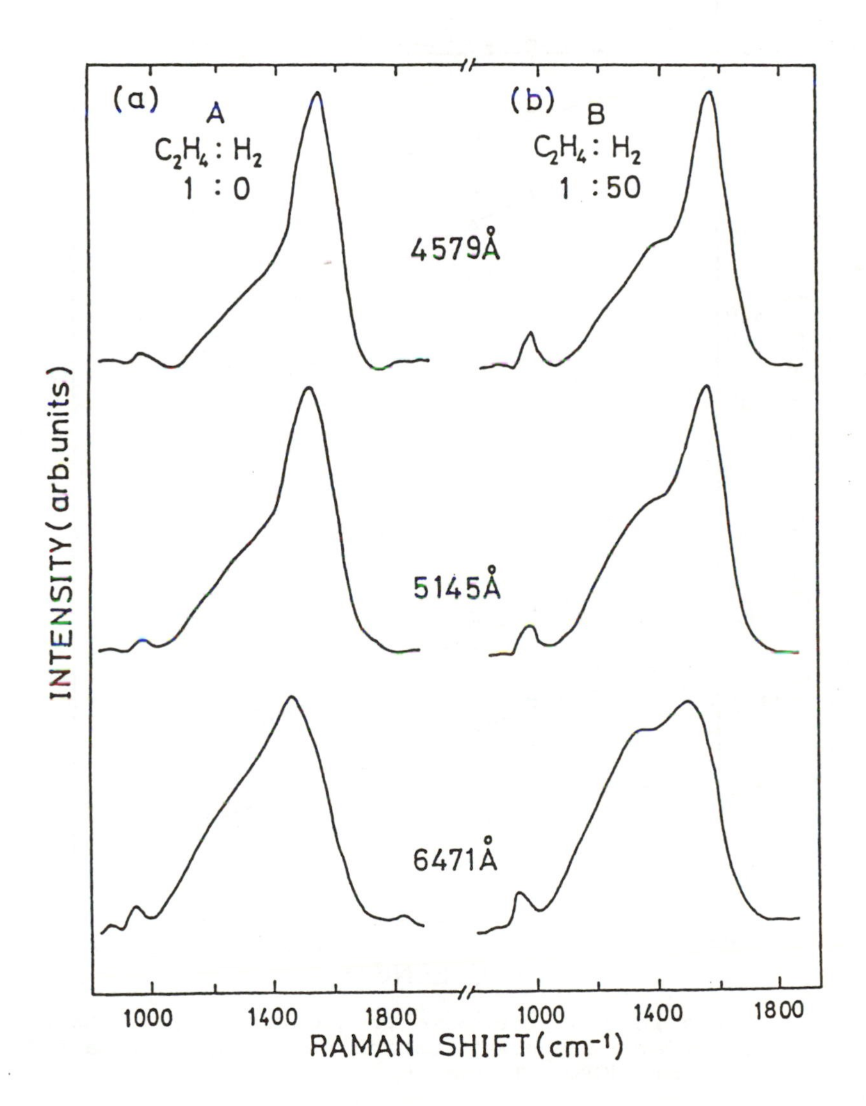

Figure 9. Raman spectra of a-C films prepared by plasma CVD method as a function of various excitation wavelengths. Figure 9-(a) and -(b) show Raman spectra of sample A (C_2H_4/H_2=1/0) and sample B (C_2H_4/H_2=1/50), respectively. The thickness of samples A and B is 700 and 500 Å, respectively.

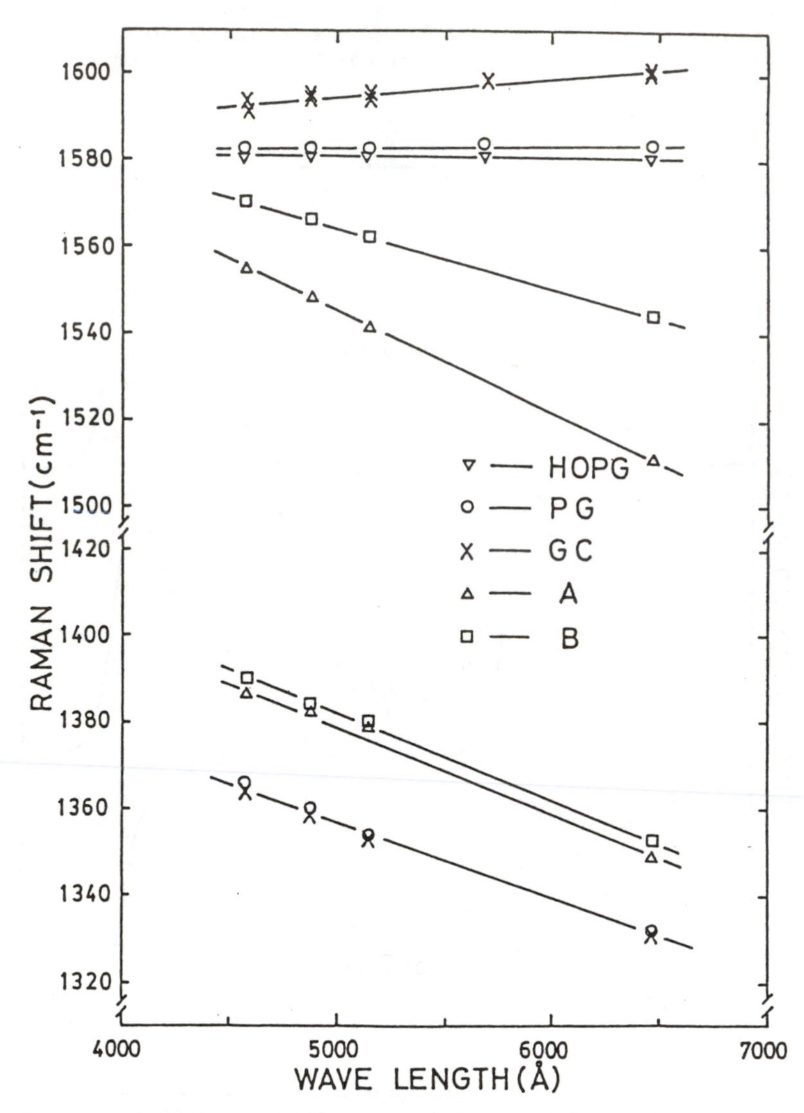

Figure 10. Variations of the peak frequencies of the major Raman bands as a function of the excitation wavelength for a-C films (samples A and B), HOPG, PG and GC.

respectively, between 4579Å and 6471Å excitation. The decrease of the peak frequency by increasing the excitation wavelength is observed for the shoulder band of a-C films, too. The decrease for samples A and B is from 1386 to 1350 cm^{-1} and from 1390 to 1353 cm^{-1}, respectively. As discussed in the previous section, the dependence of the peak frequency of Raman band on excitation wavelength is interpreted in terms of π-π^* resonant Raman scattering from aromatic rings with various sizes(10).

B. CHARACTERIZATION

In sec. 4.1, we suggested that the relative intensity of Raman bands decreased with an increase of sp^3/sp^2 ratio in a-C films and that the relative intensity could be used as a parameter for sp^3 content in a-C films. In this section, we will measure infrared absorption spectra of a-C films and try to elucidate the relation between sp^3/sp^2 ratio in a-C films and Raman spectral profiles from a comparison between Raman and infrared absorption spectra.

Figure 11 shows Raman spectra of a-C films prepared by plasma CVD method(24). A relatively sharper Raman band and a broad shoulder band is observed at around 1530 cm^{-1} and 1400 cm^{-1}, respectively and the spectral profile agrees with that of a-C films prepared by sputtering method. In addition, a weak shoulder band is observed at about 1200 cm^{-1} in the spectra of a-C films prepared by plasma CVD method. Raman spectra of the a-C films are decomposed into two bands with Gaussian line shapes. A typical result of the lineshape analysis for samples A and B is also shown in figure 11. The relative intensity of the 1400 cm^{-1} band against 1530 cm^{-1} band for samples A and B is 0.9 and 1.2, respectively(24).

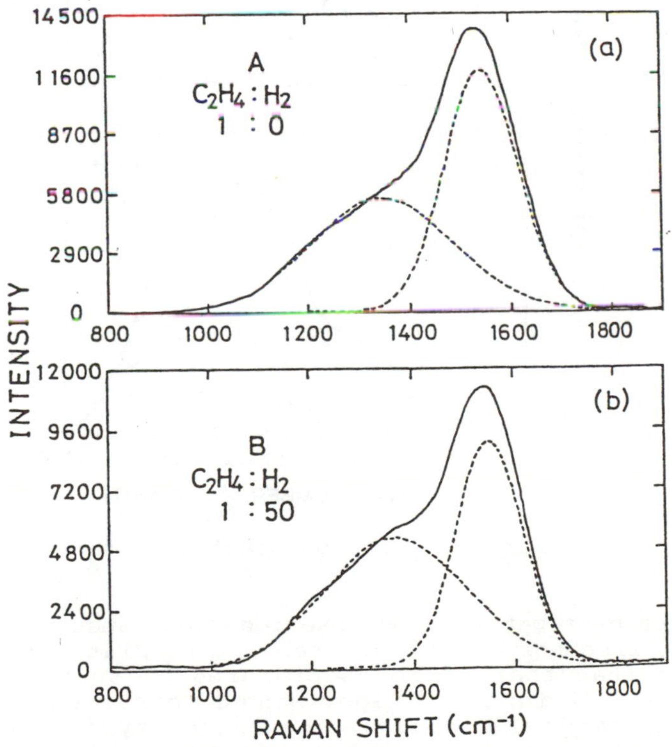

Figure 11. Raman spectra of a-C films prepared by plasma CVD method; (a) sample A(C_2H_4/H_2=1/0) (b) sample B(C_2H_4/H_2=1/50)

Figure 12 shows infrared absorption spectra of a-C films for samples A and B. The C-H stretching absorption bands due to hydrogens bonded to sp^2 and sp^3 carbons (C_{sp2}-H and C_{sp3}-H bands) are observed at around 3000 and 2900 cm^{-1}, respectively. Absorption spectra are resolved to six bands with Gaussian line shapes. From the relative intensity between C_{sp3}-H and C_{sp2}-H bands, we estimate that 72% and 69% of the carbon atoms are sp^3 bonded for samples A and B, respectively(24). Combining the results on Raman and IR measurements, the decrease of the relative intensity of Raman bands is considered to be caused by an increase of sp^3/sp^2 ratio in a-C films. It is confirmed that the relative intensity of 1400 cm^{-1} band against 1530 cm^{-1} band can be used as a parameter for sp^3/sp^2 ratio.

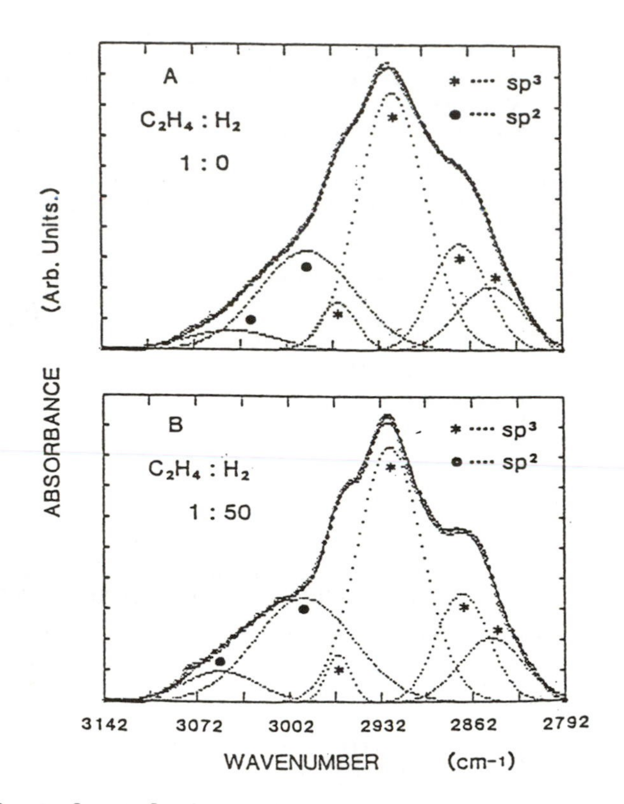

Figure 12. Infrared absorption spectra of samples A and B.

An agreement between the observed Raman spectrum and calculated line for a-C films prepared by plasma CVD method is not better than that for ones prepared by sputtering method, because the additional shoulder band is observed at about 1200 cm^{-1} in the former films(24). The intensity of the additional shoulder band for sample B is stronger than that for sample A. The v_3 mode for transpolyacetylene is observed at about 1170 cm^{-1} in the spectra excited by the 5145Å line(21) whereas a broad

Raman band for amorphous diamond is expected to be observed at around 1200 cm^{-1} from the calculation of the density of states(6). Referring to the results that the sp^2/sp^3 bonding ratio for sample B is larger than that for sample A and that the Raman scattering cross-section of diamond is much lower than that of sp^2 bonded materials(20), the shoulder band at about 1200 cm^{-1} is considered to originate from polyene structures such as polyacetylene rather than amorphous diamond. Based on this assignment, we conclude that a-C films prepared by plasma CVD method contain much polyene structures than those by sputtering method. Considering that a-C films prepared by plasma CVD method are generally harder than those prepared by the sputtering method(25), polyene structures may play an important part in the determination of hardness of a-C films.

5. MICROSTRUCTURE OF a-C FILMS

Based on the obtained results, a schematic model for the microstructure of a-C films is drawn in figure 13. The structure of a-C films consists of sp^3 and sp^2 carbon clusters. Its ratio depends on the deposition conditions. Most of sp^2 carbon clusters are aromatic rings and the others perhaps polyenes or polymers. The connection of those sp^2 clusters determines hardness of a-C films. The hydrogen atoms bond to the termination of these bonds. For a-C films having E_g=~1.0 eV, 2/3 % of carbon atoms are considered to be sp^3 bonded.

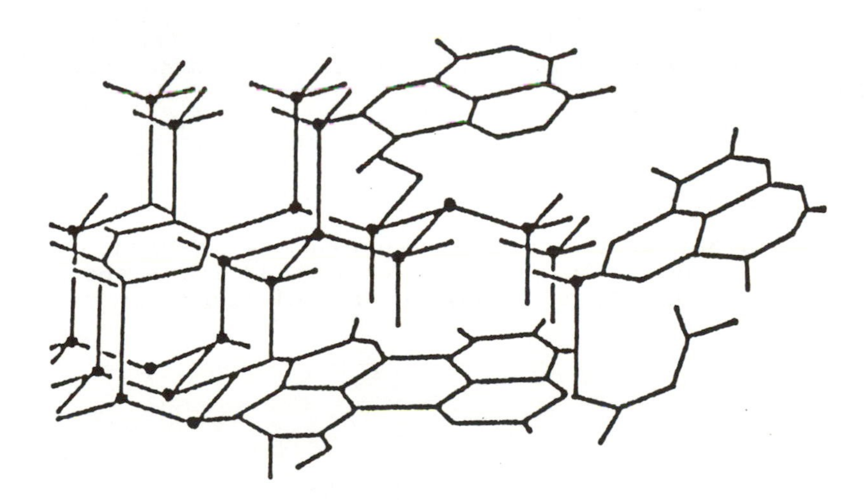

Figure 13. The schematic model for the microstructure of a-C films.

In GC with crystallite size L_a = ~25Å, two Raman bands are observed at 1590 and 1360 cm^{-1}(25) whereas asymmetric broad Raman bands assigned to the sp^2 carbon clusters with various sizes are observed at around 1530 cm^{-1} in a-C films. The comparison between Raman spectra of GC and a-C films speculates that sp^2 cluster size of a-C films is smaller than 25Å in diameter. From the calculation of the electronic structure of a-C films, Robertson and O'Reilly(12,25) assert that the absorption edge of a-C films is determined by the sp^2 carbon cluster size and that the optical gap varies inversely with the sp^2 cluster size. Their interpretation suggests that the a-C films having an optical gap of ~1eV consist of sp^2 clusters with the crystallite size of ~15Å in diameter.

6. CONCLUSIONS

We have measured Raman spectra of a-C films prepared under atmosphere with various hydrogen content. It is found that the Raman spectral profiles vary sensitively, reflecting the change of the electronic absorption spectra associated with π-π^* electronic transitions. Raman spectral variation with excitation wavelength is explained in terms of the π-π^* resonant Raman scattering from aromatic rings with various sizes. From Raman and infrared absorption measurements, a-C films is confirmed to consist of sp^3 and sp^2 carbon clusters such as aromatic rings with various sizes and polyenes. The Raman bands at 1400 and 1530 cm^{-1} are considered to originate mainly from aromatic rings with large and small sizes, respectively. The relative intensity of 1400 cm^{-1} band against 1530 cm^{-1} band is found to decrease with increase of the sp^3/sp^2 bonding ratio in a-C films. It is shown that Raman spectroscopy can be used as a parameter for the sp^3 content. It is confirmed that an excessive hydrogen content results in increase of the polymeric structure in the a-C films.

ACKNOWLEDGMENTS

The author would like to express his gratitude to his co-workers G.Katagiri, H.Ishida, N.Nagai, A.Ishitani and T.Akamatsu. The author thanks Miss.Y.Nishibori for ellipsometry measurement.

REFERENCES

1) N.Savvides, J.Appl.Phys.58,518(1985).
2) N.Savvides, J.Appl.Phys.59,4133(1986).
3) T.Sato, S.Furuno, S.Iguchi and M.Hanabusa, Jpn.J.Appl.Phys.26,L1487(1987).
4) P.Couderc and Y. Catherine, Thin Solid Films 146,93(1987).
5) R.O.Dillon, J.A.Woollam and V.Katkanant, Phys.Rev.B29,3482(1984).
6) D.Beeman, J.Silverman, R.Lynds and M.R.Anderson, Phys.Rev.B30,870(1984).
7) B.Dischler, A.Bubenzer and P.Koidl, Solid State Commun.48,105(1983).
8) J.Fink, T.Müller-Heinzerling, J.Pflüger, A.Bubenzer, P.Koidl and G.Crecelius, Solid State Commun.47,687(1983).
9) M.Ramsteiner and J.Wagner, Appl.Phys.Lett.51,1355(1987).
10) M.Yoshikawa, G.Katagiri, H.Ishida, A.Ishitani and T.Akamatsu, Solid State Commun.66,1177(1988).
11) M.Yoshikawa, G.Katagiri, H.Ishida, A.Ishitani and T.Akamatsu, Appl.Phys.Lett.52,1639(1988).
12) J.Robertson and E.P.O'Reilly, Phys.Rev.B35,2946(1987).
13) P.Brüesch, Phonons:Theory and Experiments II, p.65. Springer-Verlag, Berlin Heidelberg New York(1986).
14) M.H.Brodsky, Light Scattering in Solids I, p.1. Springer-Verlag, Berlin Heidelberg New York (1975).
15) M.Cardona and G.Guntherodt, Light Scattering in Solids III, p.3. Springer-Verlag, Berlin Heidelberg New York(1982).
16) F.Tuinstra and J.L.Koenig, J.Chem.Phys.53,1126(1970).
17) M.Yoshikawa, G.Katagiri, H.Ishida, A.Ishitani and T.Akamatsu, J.Appl.Phys.64,6464(1988).
18) B.Dischler, A.Bubenzer and P.Koidl, Appl.Phys.Lett.42,636(1983).
19) J.B.Birks, Photophysics of Aromatic Molecules (Wiley, New York, 1970),p.56.
20) N.Wada, P.J.Gaczi and S.A.Solin, J.Non-Cryst.Solids, 35-36, 543(1980).
21) I.Harada, Y.Furukawa, M.Tasumi, H.Shirakawa and S.Ikeda, J.Chem.Phys.73,4746(1980).
22) J.E.Smith, Jr., M.H.Brodsky, B.L.Crowder and M.I.Nathan, J.Non-Crystalline Solids 8-10,179(1972).
23) D.R.McKenzie, L.C.Botten and R.C.McPhedran, Phys.Rev.Lett.51,280(1983).
24) M.Yoshikawa, N.Nagai, G.Katagiri, H.Ishida and A.Ishitani, First International Conference on The New Diamond Science and Technology,p.222 .Japan New Diamond Forum, Japan(1988).
25) J.Robertson, Advances in Physics 35,317(1986).

Materials Science Forum Vols. 52 & 53 (1989) pp. 387-406

NUCLEAR MAGNETIC RESONANCE STUDIES OF AMORPHOUS HYDROGENATED CARBON

M.A. Petrich
Department of Chemical Engineering
Northwestern University, Evanston, IL 60208, USA

ABSTRACT

Amorphous hydrogenated carbon is a technologically important material because of its wide range of useful and interesting properties. These properties are a result of the atomic bonding configurations and microstructural features formed during deposition. Understanding and improving a-C:H requires improvements in fundamental understanding of this class of materials. This article discusses characterization of a-C:H by solid state nuclear magnetic resonance. Nuclear magnetic resonance can quantitatively observe the hybridization of carbon atoms, the amount of hydrogen in a sample, and the distribution of hydrogen atoms in the sample. The observations made in nuclear magnetic resonance experiments correlate well with measured properties of a-C:H films.

INTRODUCTION

The chemical behavior of carbon atoms has occupied chemists for many years. The appeal of carbon is a result of its versatile bonding. Carbon atoms bond with each other in a variety of ways, reflective of the possible molecular orbital hybridizations. Carbon bonding orbitals can hybridize in sp^3, sp^2, and *sp* configurations, corresponding to single, double, and triple bonds. Additionally, aromatic rings can be formed, as in the benzene molecule, in which six carbon atoms form a planar hexagonal molecule, with a sharing of bonding electrons among all of the carbon atoms in the molecule.

The properties of solid forms of carbon are of current interest to materials scientists and engineers. Important materials for applications range from very strong carbon fibers [1] to friction resistant carbon films [2]. In the solid state, carbon bonding takes two extremes. When carbon hybridizes completely in the sp^3 configuration, it forms diamond. When the hybridization is exclusively sp^2, the resulting material is graphite. The low natural abundance of pure diamond or graphite suggests that the production of either of these materials will be a difficult task. The unique and important properties of diamond and graphite suggest that synthesis efforts are worthwhile. Amorphous carbon generally contains carbon atoms bonded in both diamond-like and graphitic configurations [3]. In amorphous hydrogenated carbon, the situation is further complicated by the presence of hydrogen, which terminates network bonding. The types of carbon bonding and the relationships between these bonding environments and hydrogen determine the properties of the material.

Hydrogen microstructure in amorphous materials is a topic of great importance because of the effect hydrogen has on properties [4]. Hydrogen may be randomly distributed throughout the bulk of a material, or it may segregate in heavily hydrogenated microscopic regions or near surfaces. Hydrogen is a serious problem for would-be producers of diamond films, since a hydrogen atom bonded to a carbon atom introduces a defect into the diamond lattice. Although the sp^3 character of the molecular orbitals is retained, the diamond lattice is terminated. The properties of a-C:H are certainly dependent upon the amount and location of hydrogen. Hydrogenation of sp^2 environments produces sp^3 hybridization [5], but hydrogen must be eliminated if diamond-like films are the desired product. The electronic properties of amorphous hydrogenated silicon are dependent upon the hydrogen microstructure [6,7], and it seems that the electronic properties of amorphous carbon will also depend upon hydrogen microstructure. Properties such as wear resistance and optical band gap are a function of the carbon bonding configurations in amorphous hydrogenated carbon [3], but hydrogen is also a significant component in these materials, with reported hydrogen contents ranging from 0 to 50 atomic % [3]. Hydrogen affects the carbon bonding configurations, and hydrogen directly impacts properties such as optical band gap, hardness, and electrical conductivity.

Thorough characterization is helpful in ascertaining the chemical effects of processing changes and understanding the chemistry of amorphous carbon. Combined with property measurements, characterization data can be used to identify promising process modifications and "ideal" chemical features. The questions that must be answered to characterize the local bonding configurations and microstructure of amorphous hydrogenated carbon are: What are the relative amounts of hydrogen and carbon in the material? What carbon bonding configurations are present, and what are the relative amounts of those configurations? How are the carbon bonding environments affected by hydrogen?

This article focuses on a particularly useful experimental method to answer these characterization questions— nuclear magnetic resonance spectroscopy (NMR). NMR has been used to great advantage to study a-C:H and other amorphous materials such as the amorphous hydrogenated silicon alloys [8]. The advantages of nuclear magnetic resonance over other spectroscopies are that NMR observes local bonding configurations in solids in a way that is both quantitative and selective. The experimenter has tremendous control over what is observed, and these observations can be interpreted quantitatively. Despite the author's enthusiasm for this technique, it is important to remember that any analytical technique benefits from supporting evidence obtained by other methods. NMR can play an important part in any materials characterization effort. The remainder of this article is devoted to descriptions of NMR spectroscopy, and applications of its use in studying the atomic-level structure and chemical bonding of amorphous hydrogenated carbon.

NUCLEAR MAGNETIC RESONANCE SPECTROSCOPY

Nuclear magnetic resonance spectroscopy (NMR) exploits the interactions between radio-frequency radiation and atomic nuclei to reveal information about the composition and structure of molecules and solids. NMR experiments provide information about chemical bond lengths, the local chemical environment of an atomic species, the number of nuclei in each environment, the chemical identity of the sample, and spin interactions within the sample. The spin interactions which are important in solid state NMR experiments are typically expressed by a quantum mechanical operator, the total Hamiltonian for the system [9]—

$$\mathbf{H}_{total} = \mathbf{H}_{Z} + \mathbf{H}_{d,ii} + \mathbf{H}_{d,is} + \mathbf{H}_{cs} + \mathbf{H}_{Q} + \mathbf{H}_{rf}(t) + \mathbf{R}(t) \qquad (1)$$

where

$\mathbf{H}_{Z}$ represents the interaction between the observed nucleus and an external magnetic field. Only nuclei with non-zero spin quantum number are "NMR active". Examples of NMR active nuclei are ^{13}C, ^{1}H, and ^{29}Si. This interaction is manifested by each nucleus in the periodic table having its own resonance frequency. (Zeeman Hamiltonian)

$\mathbf{H}_{d,ii}$ represents the interaction between the observed nucleus and the magnetic field resulting from a similar neighboring nucleus. (homonuclear dipolar Hamiltonian)

$\mathbf{H}_{d,is}$ represents the interaction between the observed nucleus and the magnetic field resulting from a different neighboring nucleus. (heteronuclear dipolar Hamiltonian)

$\mathbf{H}_{cs}$ represents the interaction between the nucleus of interest and nearby chemical bonding electrons. (chemical shift Hamiltonian)

$\mathbf{H}_Q$ represents the interaction of the nucleus with electric field gradients. This is only present when I (nuclear spin) > 1/2. (quadrupolar Hamiltonian)

$\mathbf{H}_{rf}$(t) represents the experimenter's control over the other Hamiltonians. This Hamiltonian describes the radio frequency pulses that perturb the nuclear system during the NMR experiment.

R(t) describes the coupling of nuclear spin angular momenta to fluctuating magnetic fields in the lattice. This spin-lattice relaxation is usually due to atomic motion such as phonon modes, diffusion, or hopping. The time constant for this process is orders of magnitude larger than that for the spin-spin processes described above, and thus forms a separate class of NMR experiments. In the interest of brevity we shall ignore these phenomena in this paper. An excellent description of the role of spin-lattice relaxation in amorphous hydrogenated silicon is found in reference 10.

The Zeeman Hamiltonian is the dominant interaction. The other Hamiltonians are generally treated as perturbations on the Zeeman term. The perturbations convey most of the desired information about a spin system. Experimentally and theoretically, the perturbations are studied by adopting a reference frame which eliminates the Zeeman term [11]. One of the advantages of NMR as an experimental technique for the study of solids is that the experimenter can selectively suppress any of the interactions represented by these Hamiltonians. This suppression is achieved by exposing the sample to carefully timed pulses of radio-frequency radiation at precisely known frequencies and phases described by $\mathbf{H}_{rf}$(t). We obtain an NMR spectrum described by the resultant total Hamiltonian. By selectively eliminating sub-Hamiltonians, we simplify calculation of molecular and solid-state parameters from the spectral data [9,11]. Another significant feature of NMR is that it is quantitative. Experiments can be performed such that the signal from each observed nucleus has exactly the same magnitude. Thus, NMR can be used to measure atomic concentrations, such as the hydrogen content of amorphous hydrogenated carbon. While the width of a spectral peak may vary, the integral of the spectral peak from a given sample remains constant. This feature is particularly significant in several of the experiments described below.

The magnetic resonance spectra of solids are characterized by large linewidths caused mainly by the dipolar interactions ($\mathbf{H}_{d,ii}$ and $\mathbf{H}_{d,is}$ in equation 1). These interactions depend upon the angle between the internuclear vector and the direction of the applied magnetic field. In liquids, molecular motions cause this angle to be averaged over all possible values, effectively erasing the effect of dipolar interactions. To overcome the broadening effect of fixed nuclear spins in solids and thus obtain high resolution solid spectra, we must use sophisticated techniques to prevent the dipolar interaction from broadening the spectrum. Advances in NMR equipment and experimental techniques in recent years allow the study of atoms in a solid with liquid-like resolution [11-13]. However, high resolution experiments are not always advantageous. In narrowing the spectral linewidths, we discard information about those interactions (and atomic configurations) that caused the broadening of the line.

Consider the line broadening interactions which affect the NMR spectra of amorphous hydrogenated carbon. The dominant interactions are the heteronuclear dipole interaction ($\mathbf{H}_{d,is}$) between hydrogen and carbon atoms and the homonuclear dipole interaction ($\mathbf{H}_{d,ii}$) between hydrogen nuclei. These dipolar Hamiltonians can be manipulated experimentally with dipolar decoupling. Heteronuclear dipolar decoupling is achieved by the irradiation of hydrogen at its magnetic resonance frequency while simultaneously observing the carbon nuclei [13]. This irradiation produces a rapid rotation of the hydrogen nuclei in spin space, mimicking the rapid isotropic motion that molecules experience in liquids. Experiments which use proton decoupling to narrow carbon magnetic resonance spectra of amorphous hydrogenated carbon are described below and are illustrated in figure 1.

The heteronuclear dipolar interaction is the basis for another technique, cross-polarization, which uses the heteronuclear dipolar interaction to transfer energy from an abundant spin system (such as hydrogen) to a dilute spin system (such as carbon-13). This transfer of energy results in an enhancement of the intensity of the dilute spin signal [12,13]. However, because the signal enhancement decreases with the cube of the distance from a hydrogen nucleus, the quantitative nature of the NMR experiment can be lost. We shall discuss this important detail later in the context of specific studies of a-C:H. Most of the reported NMR studies of a-C:H employ cross-polarization. Figure 1 shows the details of this experiment.

With the heteronuclear dipolar interaction removed, the chemical shift Hamiltonian dominates the carbon NMR spectrum. Chemical shifts are changes in the resonance position of a nucleus caused by coupling of electron wave functions to the applied external magnetic field [11]. These shifts can be measured quite accurately and have been the primary source of information in NMR studies of liquid systems. Chemical shifts can provide valuable information such as the ratio of sp^3 to sp^2 bonding environments. However, in solid samples there are two broadening mechanisms which obscure these data. The first is dispersion in the chemical shift caused by bond length and bond angle distributions in the sample. Second, the chemical shift spectra of powdered or amorphous samples are broadened with a characteristic shape by chemical shift anisotropy [12]. Analysis of the line shape can reveal information about the electron distribution around a nucleus. Unfortunately, when more than one chemical shift line shape is present, the overlap may render the spectrum featureless. Spinning the sample at the "magic angle" (54.7°) with respect to the applied magnetic field eliminates the chemical shift anisotropy and narrows the spectrum, provided the spinning rate is greater than the linewidth [12]. Magic angle spinning does not, however, remove line broadening due to disorder.

A general application of line narrowing techniques is the difference experiment. Because the spin Hamiltonians can be experimentally controlled, we can perform sets of experiments comparing the effect of allowing an interaction to occur to the effect of removing that interaction.

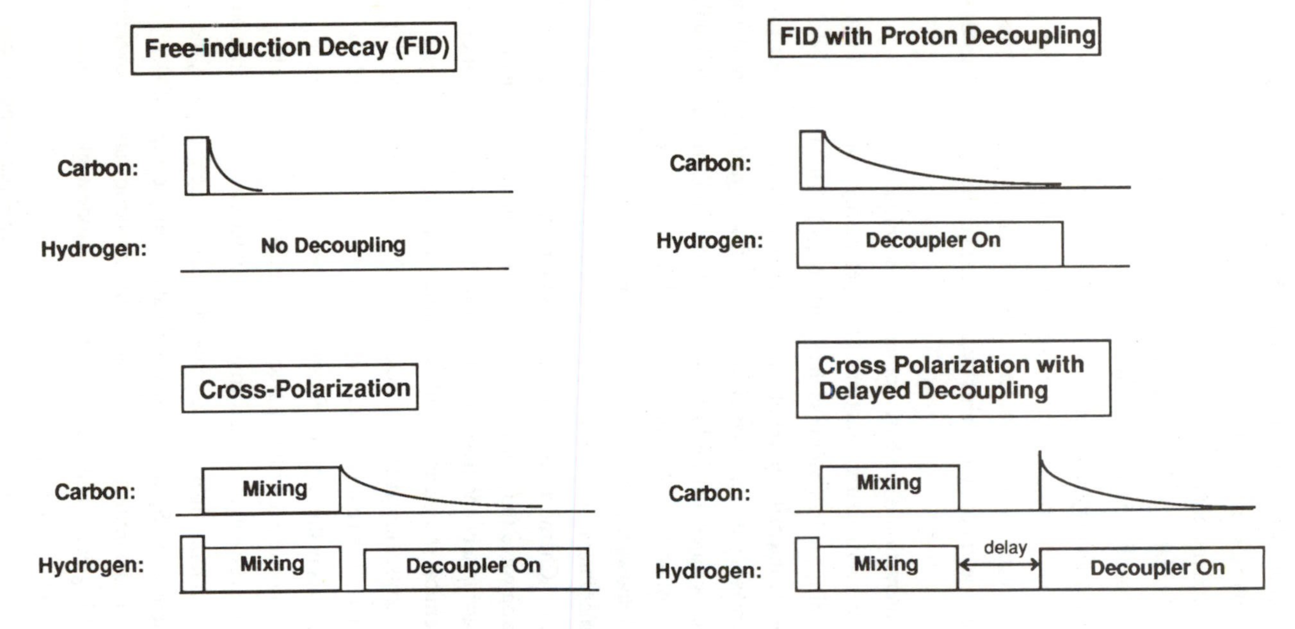

Figure 1 — Timing diagrams for the NMR pulse sequences described in the text. The x-axis represents real time during the experiment. The rectangular regions indicate that the sample is being irradiated during that time at the resonance frequency of carbon or hydrogen. The curved lines represent the signal from the decaying carbon magnetization. ***Free Induction Decay:*** *The free induction decay is simply the observation of the decay of the carbon magnetization with no signal enhancement. The linewidth of the Fourier transformed signal is inversely proportional to the rate of decay of the FID.* ***FID with Proton Decoupling:*** *The irradiation of the sample at the hydrogen resonance frequency decreases the decay rate of the carbon magnetization and increases resolution.* ***Cross Polarization:*** *Magnetization is transferred from the normally abundant hydrogen nuclei to the usually rare carbon-13 nuclei. This allows data to be acquired much more rapidly than in the first two experiments. However, sufficient hydrogen must be present for this experiment to work quantitatively. Hydrogen decoupling is employed during observation of the signal to improve resolution.* ***Cross Polarization with Delayed Decoupling:*** *Same as cross polarization except that hydrogenated nuclei are erased from the spectrum during a hydrogen decoupling delay period immediately after the cross polarization. During this delay, the heteronuclear dipolar interaction causes a rapid relaxation of carbon nuclei that are directly bonded to hydrogen. The delay can be varied to probe the spectrum of distances between carbon and hydrogen atoms.*

For example, to study hydrogenated and non-hydrogenated regions of amorphous carbon, we first observe the carbon spectrum without manipulating any of the Hamiltonians. All interactions are manifested in the spectrum. Then, a proton decoupling experiment removes the heteronuclear dipolar interaction between carbon and hydrogen. Carbon nuclei which are distant from hydrogen are unaffected while those close to hydrogen contribute to the spectrum differently than in the first experiment. Examples of this type of experiment are described in the next section.

Although narrowed linewidths provide higher resolution spectra, it is often useful to retain the dipolar interaction in our spectra. A particularly good example is hydrogen NMR of amorphous semiconductors where the homonuclear dipolar Hamiltonian can be used to measure distances between nuclei and to observe nuclei arranged in clustered environments. Because the spectral linewidth due to this dipolar interaction varies inversely with the cube of the distance between nuclei [11], linewidth measurements can provide information about internuclear spacing. In solids, the dipolar interaction is usually not analyzable as a two-body interaction, and the method of Van Vleck [14] must be used to predict linewidths for various atomic configurations. The advantage of this method is that it allows for gross discrimination between different local bonding configurations. The disadvantage of the method of Van Vleck is that a large number of microscopically different atomic configurations yield the same predicted linewidth. The situation is improved by the development of multiple quantum NMR [9,15,16] techniques which use the homonuclear dipolar interaction to "count" the number of nuclei within a spin system. This technique has not yet been used to study amorphous carbon, but has been used successfully to study the hydrogen microstructures of amorphous silicon [6,17] and amorphous silicon carbide [18].

A limitation of magnetic resonance as a tool for thin film studies is its relative insensitivity as compared to other spectroscopic and beam techniques. The practical lower limit for detection of hydrogen at room temperature is about 10^{18} nuclei. Ten milligrams (~1 μm of amorphous carbon deposited on a 62 cm^2 electrode) is ample material for hydrogen magnetic resonance experiments, although we can obtain NMR spectra on as little as 1 μm of film on a 1 cm^2 substrate. The detection limit is even higher for other nuclei because of the lower inherent sensitivities (gyromagnetic ratio) and the lower natural abundances of their NMR active isotopes. For carbon spectra, 150 to 200 mg of material are required, which corresponds to roughly 15 - 20 μm thick films on a 62 cm^2 electrode. If the deposition area is limited, sample thicknesses must be increased to yield enough nuclei for NMR observation. NMR parameters have been shown to be independent of film thickness for plasma-deposited amorphous silicon films thicker than 0.1 μm [19,20]. To prepare them for NMR, films are removed by flexing the substrate, by scraping, or by dissolution of the substrate material. It is common to deposit films on aluminum foil, and to recover the films by dissolving the aluminum in dilute hydrochloric acid. The removed powder and flakes are rinsed thoroughly with deionized water and dried prior to weighing and packing into NMR sample holders. The removal of material from the substrate has a negligible effect on the observed NMR parameters [21].

Another disadvantage of NMR is the heavy reliance on signal averaging techniques to improve signal-to-noise ratios. The experimenter must wait for at least 3 spin-lattice relaxation time constants (T_1) between spectral acquisitions [13]. In materials such as gem-quality diamond this relaxation time constant can be as long as several hours [22]. Amorphous polymers generally have T_1 values of several seconds. Spin-lattice relaxation time constants can be measured [9-13], although there is little reported data for ^{13}C or ^{1}H in a-C:H. The most serious problem in studies of materials like a-C:H is that T_1 may vary throughout the sample. If the delay between signal averages is too short, regions of the film with longer T_1 will not be observed. This problem is particularly extreme if the long T_1 regions are a small (<10 %) fraction of the sample. In that case, the "missing" signal may not be noticed, and the spectrum will provide an erroneous picture of the material. In the cross polarization experiment, the carbon T_1 value is not important [12,13]. The experiment can be repeated at a rate determined by the hydrogen T_1 value. In materials with large amounts of hydrogen this T_1 value is usually quite short (<10 seconds) at room temperature.

Before starting an NMR experiment, the experimenter must first answer the following questions:

1. Is the sample large enough? Remember that at least 10^{18} NMR active nuclei are required to run a spectrum. With 10^{19} or more nuclei, experimental set-up and acquisition of data is considerably easier.

2. What do I expect the value of T_1 to be? This controls the repetition rate you use for signal averaging. Generally, start with an estimated, or literature reported value for a similar material. However, it is best to measure T_1 for each sample.

3. What aspect of the sample do I want to investigate? This determines the experiments you must run.

4. Will the experiment I've chosen provide all of the information I want? If not, it may be necessary to run a series of experiments, or a difference experiment.

CHARACTERIZATION OF A-C:H

A representative example of the application of ^{13}C NMR is illustrated by the three spectra of a-C:H films shown in Figure 2 [23]. These films were prepared from methane plasmas at different rf powers and gas pressures to produce materials with optical band gaps ranging from 1.1 eV to 3.0 eV. The two broad peaks in these spectra arise from sp^2 and sp^3 carbon bonding configurations as labelled in the figure near 150 ppm and 30 ppm. The reasons that the spectra are so broad relative to spectra from other carbon-based systems such as organic polymers [24] are that a number of chemically distinct sp^3 and sp^2 carbon environments exist in a-C:H films, and many of these environments are perturbed due to the disordered nature of the film microstructure. This "chemical shift dispersion" and large number of carbon bonding environments make these spectra potentially quite rich in information, but unfortunately lacking in visible detail due to the overlapping of signals from each environment. The spectra in figure 2 were acquired with cross polarization and magic angle spinning (CPMAS). If enough hydrogen is present to efficiently cross polarize all of the carbon nuclei, the two peaks in this spectrum quantitatively represent the amounts of sp^2 and sp^3 carbon in the sample [9]. Integration of these peaks is a fairly simple way to quantitatively measure the sp^3/sp^2 carbon bonding ratio in a-C:H films. It is interesting to note that the sp^3/sp^2 ratios determined from these spectra agree well with ratios determined by core-electron energy loss spectroscopy [23].

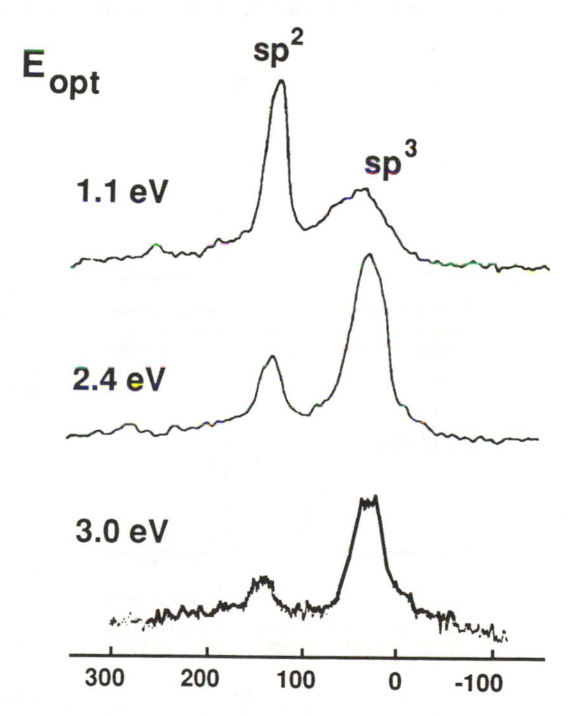

Figure 2: *^{13}C NMR spectra of three amorphous carbon films. Spectra are acquired with cross-polarization and magic angle spinning (2 msec mixing time; 0.5 second delay between signal averages; 400,000 signal averages). Carbon resonance frequency is 15.1 MHz. Optical gaps shown beside the spectra are determined from Tauc plots. (Taken from reference 23 and reprinted with the permission of the American Institute of Physics).*

Figure 3 is an example of how NMR techniques can be used to study the effects of plasma chemistry on film microstructure [25]. ^{13}C NMR spectra of amorphous carbon films prepared from ethane, ethylene, and acetylene each show absorption regions characteristic of sp^2 and sp^3 carbon atoms. However, the

sp^3/sp^2 ratio decreases from approximately 5 in the film deposited from an ethane plasma to approximately 1.6 in the film deposited from an acetylene plasma. As is well known, the sp^3/sp^2 ratio is affected by all of the plasma processing parameters. It is also well established that the sp^3/sp^2 ratio is an important predictor of film properties. NMR is an outstanding analytical tool for determining the sp^3/sp^2 ratio of a-C:H films. Studies of carbon bonding in deposited films are an essential adjunct to studies of the deposition chemistry itself.

Although the spectra in Figure 3 were acquired with the cross-polarization technique described above, the authors explain that the large hydrogen contents (H:C ~ 1:1) and other factors allow all carbon environments to be sampled equally. In addition to the requirement of a high hydrogen-to-carbon ratio, another criterion reported for the acquisition of quantitative cross polarization spectra is that the time required for cross polarization (see figure 1) must be much smaller than the proton rotating frame spin lattice relaxation time constant, $T_{1\rho}$ [25].

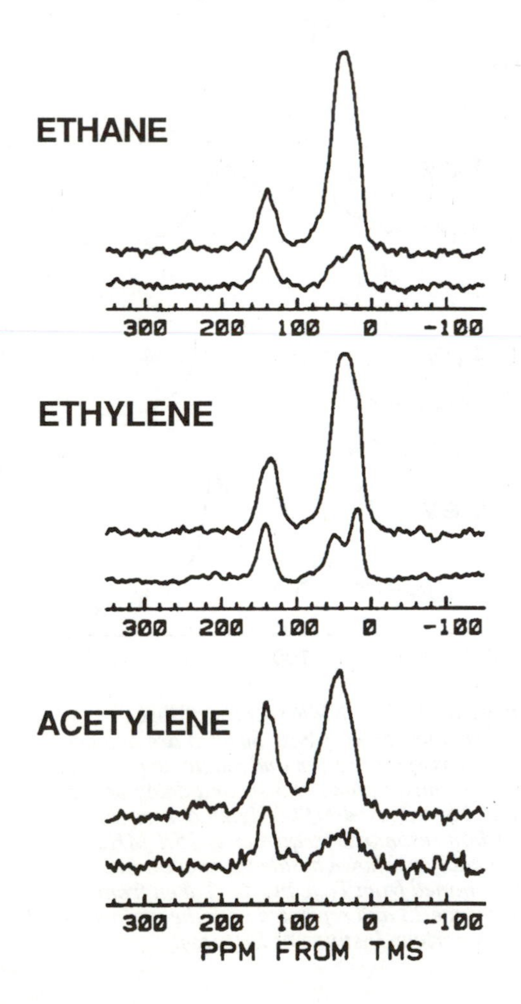

Figure 3: *^{13}C NMR spectra of a-C:H films prepared from ethane, ethylene, and acetylene. Spectra labelled (a) are acquired with cross-polarization and magic angle spinning (1 msec mixing time; 1 second delay between signal averages for ethane and ethylene-based samples, 0.4 second delay for acetylene-based sample; 25,000 signal averages, 20,000 signal averages, and 100,000 signal averages for the ethane, ethylene, and acetylene-based samples respectively; 2.2 kHz MAS rate). All samples are 20 mg. Spectra labelled (b) are acquired with cross-polarization with delayed decoupling and magic-angle spinning. A 40 μsec delay is inserted between the cross-polarization time and signal acquisition as illustrated in figure 1. Carbon resonance frequency is 22.63 MHz. (Adapted from an original figure in reference 25, by permission of John Wiley & Sons, Inc.).*

The relationship between the hydrogen atoms and particular carbon bonding environments is important information. For example, in the deposition of diamond, it is important that hydrogen be eliminated from sp^3 carbon environments to allow the formation of sp^3 carbon-carbon bonds. However, it appears equally important to hydrogenate sp^2 carbon environments, to promote sp^3 hybridization [26]. Regardless of the application, the location and amount of hydrogen in a-C:H films is as important as the sp^3/sp^2 carbon ratio. Hydrogen has a strong effect on the electronic properties of other plasma-deposited amorphous materials [8], and affects the optical [27] and mechanical properties [28] of a-C:H films. Knowing the location of hydrogen in a-C:H films allows the researcher to postulate a deposition mechanism, and also to evaluate the relative success of process modifications.

The amount of hydrogen, and information about the distribution of hydrogen, in a-C:H can be determined with hydrogen NMR [29]. The typical ^{1}H-NMR spectrum of a-C:H has a linewidth of ~60 kHz, which indicates that the distances between hydrogen nuclei are small. Typical configurations giving linewidths of this order are polymeric $(CH_2)_n$ regions. Linewidths for particular configurations can be calculated from the Van Vleck equation [14]. Since there are many arrangements of nuclei which result in the same observed linewidth, a certain amount of information from other sources, and some insight, is required to determine exactly which configuration is actually in a given material. The amount of hydrogen is measured by comparing the integrated intensity of the spectrum with the integrated intensity of the spectrum of a reference compound. Consideration of the average distance between hydrogen atoms (as determined by the bonding configuration) and the atomic percentage of hydrogen allows the determination of the relative homogeneity or heterogeneity of the hydrogen distribution in a-C:H.

The simplest way to observe carbon-hydrogen relationships in a-C:H with NMR is with a dipolar decoupling experiment. This experiment, depicted schematically in figure 1, involves the observation of the ^{13}C free induction decay while simultaneous applying continuous irradiation at the proton resonance frequency. The effect of this irradiation of the protons is to "decouple" the hydrogen nuclei from the carbon nuclei, thus prolonging the decay of the carbon magnetization and narrowing the Fourier transformed carbon resonance signal. By also acquiring the carbon free induction decay signal without irradiating the protons, one can observe the spectral differences caused by the hydrogen nuclei, and determine the extent of hydrogenation of the various carbon bonding environments.

The results of a dipolar decoupling experiment are shown in Figure 4 [30]. The top spectrum, acquired with proton decoupling, reflects all of the carbon atoms in the sample. The middle spectrum, acquired without the resolution enhancement of proton decoupling, shows only those carbon atoms which are not bonded to hydrogen. The bottom spectrum, a subtraction of the first two, shows only the hydrogenated carbon atoms. Although this information appears similar to that provided by figures 2 and 3, there is an important distinction. The spectra in figure 4 were acquired *without* cross polarization. The relationship between hydrogen and carbon does not affect the top spectrum. All carbon atoms are observed, regardless of their proximity to hydrogen. The middle spectrum is affected by hydrogen, in that the hydrogenated carbon atoms are not observed. However, the carbon atoms contributing to the spectrum are not hydrogenated, nor are they closer than two or three bond distances to hydrogen atoms. This experiment demonstrates that cross polarization techniques are not required to study a-C:H, and that such techniques may be detrimental to an accurate determination of the chemical nature of a film. Non-hydrogenated or low hydrogen content films can only be studied in experiments which directly observe the carbon nuclei.

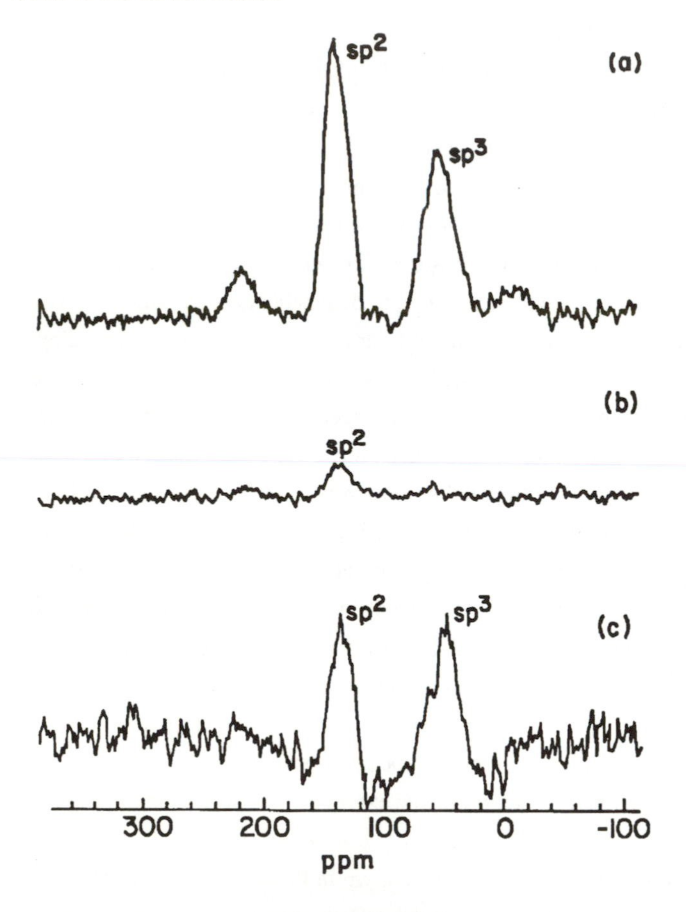

Figure 4: *^{13}C NMR spectra of a-C:H films prepared from acetylene. Spectrum (a) is a Fourier transformed free induction decay, acquired with proton decoupling and magic angle spinning. All carbon nuclei in the sample contribute to this spectrum. Spectrum (b) is a Fourier transformed free induction decay, acquired with magic angle spinning and no further signal enhancement. Only non-hydrogenated carbon nuclei contribute to this spectrum. Spectrum (c) is the difference of spectra (a) and (b), and represents the hydrogenated carbon nuclei. Carbon resonance frequency is 48.29 MHz. Magic angle spinning rate is 3.5 kHz. Spectra are plotted on the same vertical scale. (Taken from reference 30 and reprinted with the permission of the American Institute of Physics).*

Amorphous hydrogenated carbon films can be deposited by a variety of methods. Relationships between film properties and process conditions can be elucidated by considering characterization data from NMR. Figure 5 is a ^{13}C NMR spectrum of an ion beam sputter deposited sample containing 35 atomic % hydrogen [31]. The sp^3/sp^2 ratio, determined by integrating the absorption peaks, is 0.67. The optical band gap of this material is 1.2 eV, which is significantly larger than that of non-hydrogenated amorphous carbon. The authors postulate that the hydrogenation of the film causes the sp^2 carbon bonding to convert to sp^3 hybridization. This is in agreement with other observations that increasing hydrogenation increases the optical band gap and sp^3/sp^2 ratio of a-C:H films. Given the sp^3/sp^2 ratio, the trend in the optical band gap with increasing hydrogen content can be explained. In this type of study, where the effect of a process variable (hydrogen content of a sputtering atmosphere) on a film property (optical band gap) is being investigated, NMR provides the intermediate information required to explain the effect of the process variable on the property.

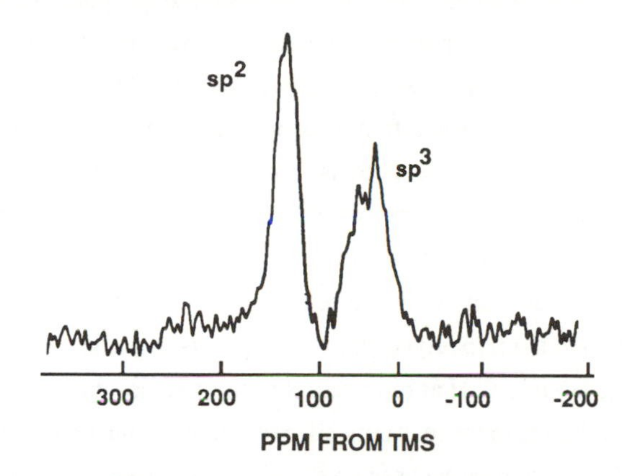

Figure 5: *^{13}C NMR spectrum of ion beam sputter deposited a-C:H film. Spectrum is acquired with cross polarization and magic angle spinning. No details of the experiment are provided in the source. The sample has a Tauc gap of 1.2 eV. (Taken from reference 31 and reprinted with the permission of the American Institute of Physics).*

Both hydrogen and carbon affect the properties of a-C:H, and it is the distribution of hydrogen among the carbon bonding environments which may be most important. An interesting variation of the cross polarization experiment, cross polarization with delayed decoupling (CPDD), can also be used to determine the relationship between hydrogen and carbon in a-C:H films [25]. This experiment, depicted schematically in figure 1, involves the same basic steps as a cross-polarization experiment, but instead of detecting the carbon magnetization immediately after the mixing period, the experimenter waits a certain amount of time between the end of cross-polarization and the beginning of detection. During this delay the dipole coupling between hydrogen and carbon nuclei causes a decay of the carbon magnetization. This decay is most severe for those carbon atoms bonded directly to hydrogen. After the delay the hydrogen decoupler is turned on and the decay of the carbon magnetization is recorded. The spectra observed in a delayed decoupling experiment are thus devoid of signals from hydrogenated carbon nuclei. Comparison with spectra observed with standard decoupling techniques allows an interpretation of which carbon bonding environments are hydrogenated and which are not. The relative amounts of hydrogenated and non-hydrogenated carbon nuclei can also be determined.

Figure 3 displays spectra acquired with standard cross polarization (CPMAS), and with the CPDD technique [25]. The spectra labelled (a) are the cross polarization spectra discussed earlier. Those labelled (b) are the result of CPDD experiments. The relative amounts of various carbon bonding environments can be determined by comparing the two groups of spectra. The large difference in intensity between the sp^3 peaks in the CPDD spectra and the sp^3 peaks in the standard cross polarization spectra shows that the sp^3 environments are more heavily hydrogenated than the sp^2 environments. The splitting observed in the CPDD spectrum from the ethylene deposited film illustrates an interesting point. This spectrum supposedly shows only non-hydrogenated carbons. Since the samples consist of only carbon and hydrogen, it would seem that non-hydrogenated sp^3 carbon atoms can only exist in the quaternary carbon configuration. The chemical shift of diamond and other quaternary carbons is about 45 ppm [32]. The spectrum has a feature near 45 ppm, but also has a peak near 15 ppm. This "extra" peak is a signal from methyl carbons. It appears in the spectrum because the rapid rotation of a methyl group in a-C:H at room temperature reduces the ^1H-^{13}C dipolar coupling strength. The methyl carbon resonance is only slightly affected during the delay period in the CPDD sequence. The CPDD experiment can provide helpful information about the distribution of hydrogen in a material, but is subject to the same limitations as CPMAS.

The NMR spectroscopist can focus on particular structural features by using isotopic enrichment. The selective substitution of "NMR active" isotopes, e.g. ^{13}C, for "inactive" isotopes, e.g. ^{12}C, enhances the signal from the enriched environment. This technique has been used to great advantage in organic chemistry, biochemistry, and polymer science. In these fields, highly developed synthetic capabilities make it possible to selectively enrich any desired molecular site or group of sites. Deposition of thin films, especially by plasma enhanced chemical vapor deposition, does not yet allow such careful molecular level surgery. Despite this restriction, isotopic enrichment can be used to great advantage in thin film studies. Deuterium enrichment of plasma deposited films, and the subsequent isotope shift of infrared absorptions, has been used to identify hydrogenated (deuterated) bonding environments in amorphous silicon [33] and amorphous carbon [34] with FTIR techniques. Deuteration of amorphous silicon films has also been used to facilitate deuterium NMR studies of trapped molecular species and of the distribution of hydrogen (deuterium) in these materials [35]. The appreciable control that the NMR spectroscopist normally has is enhanced by isotopic enrichment and labelling of samples.

Isotopic enrichment can be used to study deposition chemistry by labelling a particular molecule or molecular configuration. The fate of the molecule can be traced by observing the NMR signal from standard films (no isotopically labelled precursors) and films prepared with the isotopically enriched precursor. This is illustrated by an experiment studying the plasma deposition of a-C:H from toluene [36]. The toluene molecules are ^{13}C labelled at the methyl carbons. The carbon NMR spectra of films prepared from toluene and ^{13}C enriched toluene are shown in figure 6. The top pair of spectra come from a film prepared with unenriched toluene. The bottom pair of spectra come from a film prepared with the methyl group labelled toluene. Standard CPMAS (A) and CPDD (B) experiments are shown for each film. While these spectra are similar to those from the materials discussed above, certain features deserve further mention. A sharp peak appearing near 128 ppm in the unenriched spectrum is indicative of intact rings

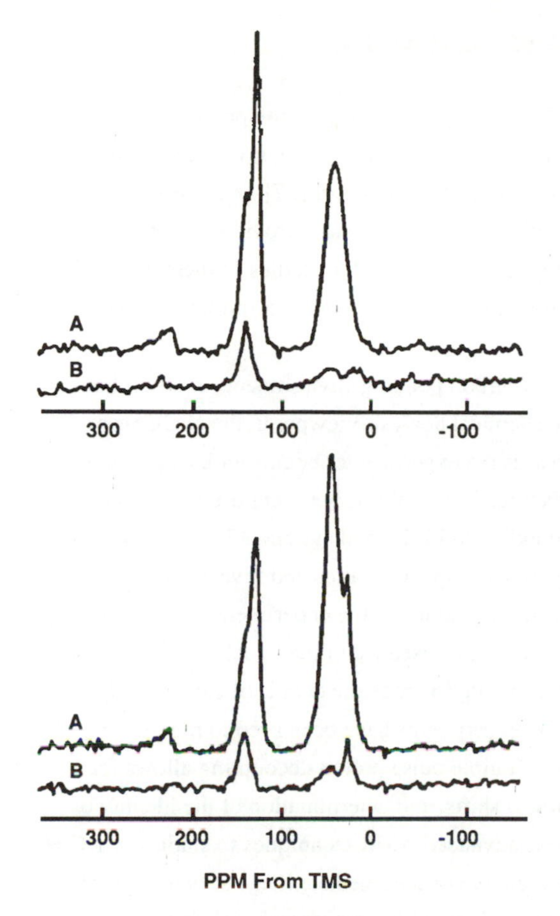

Figure 6: *^{13}C NMR spectra of a-C:H films deposited from toluene plasmas. The bottom set of spectra is deposited from toluene in which 10% of the methyl carbons are ^{13}C enriched. The top set of spectra is deposited from a toluene plasma containing the natural abundance (1.1%) of ^{13}C. Spectra labelled* **A** *are acquired with cross-polarization and magic angle spinning. (1 msec mixing time; 1 second delay between signal averages; 25,000 signal averages; 2.2 kHz MAS rate). Spectra labelled* **B** *are acquired with cross-polarization with delayed decoupling and magic angle spinning. A 40 μsec delay is inserted between the cross-polarization time and signal acquisition as illustrated in figure 1. Carbon resonance frequency for all spectra is 22.63 MHz. See figure 1 for an illustration of these experiments. (Adapted from reference 36).*

being incorporated into the film. And, the peak at 141 ppm in the CPDD spectrum is typical of phenyl C-1 carbons in benzyl groups. Isotopic enrichment provides further information on the fate of the precursor molecules. The sharply defined methyl resonance (~15 ppm) in the enriched spectrum shows that one outcome of the plasma chemistry is the formation of phenyl and methyl radicals, with subsequent incorporation of methyl groups in the film. However, the large resonance peak at 43 ppm is assigned to benzyl methylene carbons, providing evidence for a plasma process which generates benzyl radicals, rather than phenyl radicals. Evidence that the deposition chemistry is not entirely explained by these two processes is provided by the increase in absolute intensity of the sp^2 carbon absorption. Some of the methyl carbons (carrying the ^{13}C label) must participate in reactions whose products are unsaturated carbon bonds. This experiment is an elegant demonstration of the role that characterization of solid products can play in deciphering deposition mechanisms.

THE FUTURE OF NMR AND A-C:H

Although NMR has proved to be a useful technique for studying the processing and structure of a-C:H, the full potential of NMR has not been exploited. A review of the NMR literature reveals numerous studies of analogous materials such as coal [37], polymers [24], and diamonds [22,38], in which many powerful experimental schemes have provided important information. I will briefly describe some key applications of NMR to studies of these materials, and suggest how similar experiments could be used to learn more about amorphous hydrogenated carbon.

Many similarities are evident between the carbon bonding environments in amorphous hydrogenated carbon and those in coal. From a chemical analysis viewpoint, the challenges of studying these materials are the same. In both cases the experimenter begins with a system of unknown stoichiometry and non-crystalline structure. The coal science literature provides many excellent articles describing studies of carbon bonding in ill-defined systems [37,39]. As with a-C:H, NMR studies of coal began with determinations of sp^3/sp^2 ratios and investigations of hydrogen bonding. However, as more sophistication is added to the experimentation, the amount and quality of information improves. Coal scientists have used techniques such as two-dimensional NMR and multiple pulse proton decoupling to probe the details of carbon-hydrogen relationships in their materials. Two-dimensional experiments have been used to measure distances between hydrogen and carbon nuclei. Multiple pulse proton decoupling allows for higher resolution observation of hydrogen chemical shifts, and determination of the identity of hydrogenated carbon atoms. Application of these advanced NMR techniques to studies of a-C:H will allow researchers to develop a more detailed picture of intermediate range structures such as non-hydrogenated sp^2 carbon clusters [3,40] and "polymer-like" regions of hydrogenated sp^3 carbon.

Organic polymers have probably been studied more than any other class of materials using solid state NMR. The polymer scientist has the advantage of knowing the composition and basic molecular structure of the sample prior to performing any spectroscopic characterization. Despite the fact that a-C:H is not as well understood as organic polymers, the similarities between the two classes of materials make the polymer science literature a useful resource for a-C:H researchers. NMR has been used to advantage in studies of the chemical bonding environments in polymers [24]. However, measurements of NMR relaxation times have provided even more information about molecular configurations and microstructure of polymers [24,41]. These experiments reveal the vibrational characteristics of polymer side-groups and chain "backbones". This information is used to interpret physical property measurements, and also helps in deciding on possible modifications of the material.

Relaxation time measurements have been used sparingly in a-C:H studies [25], where the primary investigation has been of proton rotating frame relaxation times ($T_{1\rho}$). These experiments were used to study the relationship between hydrogen atoms and paramagnetic defects in a-C:H, and also to verify that the cross-polarization method generated quantitative spectra in this case. Further experiments which study the relaxation time distribution among the carbon nuclei should prove helpful in determining the exact nature of both the sp^2 and the sp^3 environments.

A powerful technique in NMR spectroscopy is the observation of spectra as a function of temperature. Polymeric materials, for example, are often studied by observing the spectrum at temperatures ranging from well below the glass transition temperature to well above the glass transition. These experiments, which probe the molecular motions in the material, are indirect measures of relaxation phenomena. In general, molecular configurations which are vibrating freely (such as a polymer molecule at high temperatures) are observed as narrow absorption lines in the NMR spectrum. Those configurations whose motion is hindered, or which are rigid, are observed as broad lines. Certain motions that are present at high temperature "freeze out" at lower temperatures, with our experimental observation being a broadening of the NMR absorption line. Such observations provide clues to the nature of the material, and can be used in the same context as direct measurements of relaxation time constants. Spectra as a function of temperature have been used to observe the relative flexibility of polymer chains [24], and to demonstrate the existence of mobile hydrogen in amorphous semiconductors [42]. In a-C:H studies, temperature dependent spectra will be valuable in studying methyl group incorporation, hydrogen mobility at elevated temperatures, and the nature of the carbon lattice. It is important in many protective coating applications to observe any structural changes which may occur in a material exposed to harsh environments. Recently developed NMR hardware allows for the observation of spectra during thermal treatment of materials at temperatures exceeding 700 °C [43].

Most of the NMR data used as examples in this paper were acquired with cross-polarization techniques. It is important to remember that, in general, cross-polarization spectra are not quantitative. This technique is best suited to materials with large hydrogen contents and even hydrogen distributions, such as organic polymers. For a given cross-polarization time, those carbon atoms that are far removed from a hydrogen nucleus will not develop magnetizations as strong as carbons which are directly bonded to hydrogen. For long cross-polarization times, relaxation effects become evident as described in various references [9,11-13]. Relaxation of the carbon atoms directly bonded to hydrogen causes a decay in the observable signal from these nuclei even as the observable signal from other carbon nuclei is increasing. When hydrogen contents are modest (<20 atomic %), or when it is suspected that the hydrogen distribution is extremely irregular, it is best to avoid cross-polarization experiments. Planned NMR studies of plasma-deposited diamond films will not use cross polarization because of the very low (<5 atomic %) hydrogen content. The quantitative accuracy of cross-polarization experiments can be checked by observing changes in the spectrum as a function of cross-polarization contact time [13].

Another drawback of the cross-polarization method is that carbon atoms near paramagnetic centers such as carbon free radicals cannot be observed due to the rapid decay of nearby spin-locked hydrogen nuclei. As shown in figure 1, the carbon signal develops during simultaneous irradiation of carbon and hydrogen nuclei. Paramagnetic centers cause the hydrogen magnetization to decay before it can be transferred to the carbon nuclei. Thus, carbon nuclei in these environments will be invisible in this experiment. This can be averted by direct observation of carbon spectra using the free-induction decay pulse sequences shown in figure 1. In these cases, the decay of the carbon magnetization will be rapid, but detectable.

A criticism of NMR is that even though all of the bonding environments should be sampled, some of them escape detection because of extremely rapid decay leading to extremely broad lines. Lines may be broadened beyond detectability in spectra such as those shown in this article, and become indistinguishable from the baseline. In fact, this broadening is used to great advantage in the delayed decoupling and direct detection experiments already described. To insure that spectra are representative of the entire sample, it is possible to count the number of NMR active nuclei in a sample by comparing the integrated intensity of the spectrum with the intensity of the spectrum obtained from a sample with a known number of NMR active nuclei. The number of carbon nuclei in an a-C:H sample can be determined by measuring the hydrogen concentration and mass of the sample. Subtracting the hydrogen mass from the total mass provides the mass of carbon atoms. Once we know the number of carbon atoms in a particular sample, comparison of the integrated intensity of the a-C:H spectrum with the intensity of the spectrum from a known reference material, such as adamantane, allows a check on whether or not all of the carbon nuclei in the a-C:H sample are observed in the spectrum.

CONCLUSIONS

The properties of a-C:H are directly related to the carbon and hydrogen microstructure in a particular film. Microstructural features, in turn, are determined by processing conditions. Thorough characterization of chemical bonding and microstructure of a-C:H is an integral part of any research program attempting to gain engineering control of film properties via adjustments in processing parameters. Solid state nuclear magnetic resonance is a quantitative and selective technique for the observation of chemical bonding environments and microstructural features in a-C:H. NMR data provide necessary information for the understanding of the relationships among processing conditions, structure, and properties of amorphous hydrogenated carbon films. Despite its demonstrated utility, NMR spectroscopy has not been used to its full capability in a-C:H studies. More intensive application of NMR, and more sophisticated experiments will contribute greatly to the further understanding of a-C:H, and a consequent improvement in materials performance.

REFERENCES

1) T. Hamada, T. Nishida, Y. Sajiki, M. Matsumoto, and M. Endo: J. Mater. Res., 1987, **2**, 850

2) K. Enke, H. Dimigen, and H. Hubsch: Appl. Phys. Lett., 1980, **36**, 291

3) J. Robertson: Adv. Physics, 1986, **35**, 317

4) Gust Bambakidis and Robert C. Bowman, Jr., eds., *Hydrogen in Disordered and Amorphous Solids*, (Plenum Press, New York, 1986).

5) Michael Frenklach and Karl E. Spear: J. Mater. Res., 1988, **3**, 133

6) Karen K. Gleason, Mark A. Petrich, and Jeffrey A. Reimer: Phys. Rev. B, 1987, **36**, 3259

7) Enakshi Bhattacharya and A.H. Mahan: Appl. Phys. Lett., 1988, **52**, 1587

8) Jeffrey A. Reimer and Mark A. Petrich: in *Advances in Disordered Semiconductors*, edited by Hellmut Fritzsche, (World Scientific, Singapore, 1989), p. 1.

9) Richard R. Ernst, Geoffrey Bodenhausen, and Alexander Wokaun, *Principles of Nuclear Magnetic Resonance in One and Two Dimensions*, (Oxford, New York, 1987).

10) P.C. Taylor, in *Semiconductors and Semimetals*, Vol. 21C, edited by J.I. Pankove, (Academic Press, Orlando, 1984), p. 99.

11) C.P. Slichter, *Principles of Magnetic Resonance*, (Springer-Verlag, Berlin, 1980).

12) Michael Mehring, *Principles of High Resolution NMR in Solids*, (Springer-Verlag, Berlin, 1983).

13) B.C. Gerstein and C.R. Dybowski, *Transient Techniques in NMR of Solids*, (Academic Press, Orlando, 1985).

14) A. Abragam, *Principles of Nuclear Magnetism*, (Oxford, New York, 1961).

15) Y.S. Yen and A. Pines: J. Chem. Phys., 1983, **78**, 3579

16) J. Baum and A. Pines: J. Am. Chem. Soc., 1986, **104**, 7447

17) J. Baum, K.K. Gleason, A. Pines, A.N. Garroway, and J.A. Reimer: Phys. Rev. Lett., 1986, **56**, 1377.

18) Mark A. Petrich, Karen K. Gleason, and Jeffrey A. Reimer: Phys. Rev. B., 1987, **36**, 9722

19) Jeffrey A. Reimer, Robert W. Vaughan, and John C. Knights: Phys. Rev. B., 1981, **24**, 3360

20) J.B. Boyce and M.J. Thompson: J. Non-Cryst. Solids, 1984, **66**, 115

21) E.D. VanderHeiden, W.D. Ohlsen, and P.C. Taylor: J. Non-Cryst. Solids, 1984, **66**, 115

22) P. Mark Henrichs, Milton L. Cofield, Ralph H. Young, and J. Michael Hewitt: J. Mag. Res., 1984, **58**, 85

23) R.H. Jarman, G.J. Ray, R.W. Standley, and G.W. Zajac: Appl. Phys. Lett., 1986, **49**, 1065

24) Richard A. Komoroski, ed., *High Resolution NMR Spectroscopy of Synthetic Polymers in Bulk*, (VCH Publishers Inc., Deerfield Beach, Florida, 1986).

25) A. Dilks, S. Kaplan, and A. Van Laeken: J. Polym. Sci.: Polym. Chem. Ed., 1981, **19**, 2987

26) John C. Angus and Cliff C. Hayman: Science, 1988, **241**, 913

27) A. Bubenzer, B. Dischler, G. Brandt, and P. Koidl: J. Appl. Phys., 1983, **54**, 4590

28) S. Kaplan, F. Jansen, and M. Machonkin: Appl. Phys. Lett., 1985, **47**, 750

29) J.A. Reimer, R.W. Vaughan, J.C. Knights, and R.A. Lujan: J. Vac. Sci. Technol., 1981, **19**, 53

30) A. Grill, B.S. Meyerson, V.V. Patel, J.A. Reimer, and M.A. Petrich: J. Appl. Phys., 1987, **61**, 2874

31) F. Jansen, M. Machonkin, S. Kaplan, and S. Hark: J. Vac. Sci. Technol., 1985, **A3**, 605

32) T.M. Duncan: J. Phys. Chem. Ref. Data, 1987, **16**, 125

33) M.H. Brodsky, Manuel Cardona, and J.J. Cuomo: Phys. Rev. B, 1977, **16**, 3556

34) B. Dischler, R.E. Sah, P. Koidl, W. Fluhr, and A. Wokaun, Proc. 7th Int. Symp. on Plasma Chemistry, C.J. Timmermans, Ed. (IUPAC Subcommittee of Plasma Chemistry, Eindhoven, 1985), p.45.

35) D.J. Leopold, J.B. Boyce, P.A. Fedders and R.E. Norberg: Phys. Rev. B, 1982, **26**, 6053

36) S. Kaplan and A. Dilks: Thin Solid Films, 1981, **84**, 419

37) David E. Axelson, *Solid State Nuclear Magnetic Resonance of Fossil Fuels*, (Multiscience Publications Ltd., Canada, 1985).

38) H.L. Retcofsky and R.A. Friedel: J. Phys. Chem., 1973, **77**, 68

39) Randall E. Winans and John C. Crelling, editors, *Chemistry and Characterization of Coal Macerals (ACS Symposium Series #252)*, (American Chemical Society, Washington, D.C., 1984).

40) M. Ramsteiner and J. Wagner: Appl. Phys. Lett., 1987, **51**, 1355

41) Jacob Schaefer, E.O. Stejskal, T.R. Steger, M.D. Sefcik, and R.A. McKay: Macromolecules, 1980, **13**, 1121

42) J.A. Reimer, R.W. Vaughan, and J.C. Knights: Solid State Commun., 1981, **37**, 161

43) Leo J. Lynch, David S. Webster, and Wesley A. Barton: Adv. Mag. Res., 1988, **12**, 385

Materials Science Forum Vols. 52 & 53 (1989) pp. 407-426

DIAMONDLIKE THIN FILMS AND THEIR PROPERTIES

N. Savvides
CSIRO Division of Applied Physics
Sydney, 2070, Australia

ABSTRACT

Diamondlike carbon thin films were synthesized by low-energy ion-assisted deposition, and their optical and electrical properties were studied. The deposition technique and the properties of the films are reviewed with special emphasis placed on the role of ion bombardment in modifying these properties, in altering the electronic band structure, and in developing the tetrahedral bonding of the amorphous carbon network.

I INTRODUCTION

The development of ion-assisted deposition (IAD) techniques over the last 15 years has been accompanied by the synthesis of numerous new metastable materials (1). Among these, amorphous carbon thin films containing a high degree of sp^3 carbon-carbon bonding have been produced that are not available by other thin film processes. The properties of such films are close to those of diamond and hence they are known as diamondlike carbon (DLC) films.

The driving force for the development of DLC films has been the combination of the many outstanding properties of crystalline diamond. DLC films are in part an outgrowth of the continuing efforts to synthesize diamond under metastable conditions from the vapour phase. These efforts have largely succeeded and both crystalline diamond films and DLC films can now be produced by a variety of IAD techniques based primarily on the use of plasma discharges and low-energy ion beams. A common feature of these techniques is the supply of superthermal energy to the growing film by bombarding inert or reactive ions, and by ionized species from which the film is condensed.

Diamondlike carbon (DLC) is an amorphous material with spectacular metastability and extraordinary properties. As a metastable state between the naturally occuring carbons, graphite and diamond, DLC displays a wide range of physical properties including extreme hardness, chemical inertness, high dielectric strength, variable optical constants, IR transparency, and optical gap up to 3 eV. Films of DLC are generally completely amorphous and homogeneous, and consist mainly of a mixture of tetrahedral (sp^3) and trigonal (sp^2) carbon-carbon bonds. The sp^3/sp^2 bonding ratio determines the properties of films, and for bonding ratios greater than unity the films possess properties closer to those of diamond than to those of graphite. Under certain deposition condition DLC films can be prepared which have diamond microcrystals dispersed in the amorphous phase. DLC includes both pure amorphous carbon (a-C), and hydrogenated amorphous carbon (a-C:H) which can contain up to 30 at.% of hydrogen bonded in the amorphous carbon matrix. The presence of hydrogen generally increases the optical gap, and by stabilizing sp^3 bonding it can lead to improved diamondlike properties.

Amorphous carbon films possessing diamondlike properties were first reported in 1971 by Aisenberg and Chabot (2), and later by Spencer et al (3) who obtained hard and transparent films by the direct condensation of C^+ ions of about 40 eV. Weissmantel et al (4) used dual ion-beam deposition to obtain similar films, whereas in the absence of ion bombardment graphitic films were obtained. Subsequently many workers used the various IAD techniques to produce hard a-C and a-C:H. It is now well established that a transition from graphitic to diamondlike material takes place when the film is deposited under concurrent inert ion bombardment, is condensed from accelerated ionized hydrocarbon species, or from C^+ and C^- ions (1,5). The optimum energy for obtaining DLC films with high sp^3/sp^2 ratio is in the range 50 - 100 eV per deposited carbon atom. Higher ion energies tend to reverse this trend with graphitization taking place. There is considerable accumulated literature on the synthesis, characterization and applications of DLC films and the reader is referred to recent reviews (6-12). Table I gives a brief summary of some general properties of the various forms of carbon.

TABLE I
General Properties of Various Forms of Carbon

	Conductivity ($\Omega^{-1}cm^{-1}$)	Optical gap (eV)	Density ($g\ cm^{-3}$)	Hardness Vickers ($kg\ mm^{-2}$)	Refractive index n
Diamond	10^{-16}-10^{-18}	5.45	3.515	10000	2.4
Graphite	$2.5\times10^4(\perp c)$	-0.04	2.26	-	3.0
Evaporated carbon	10^{-2} - 1	0 - 0.4	1.5 - 1.8	-	
Plasma DLC	10^{-7} - 10^{-16}	1.2 - 2.2	1.5 - 1.8	1200 - 4500	1.8-2.2
Ion-beam DLC	10^{-2} - 10^{-16}	0.4 - 3.0	1.7 - 2.8	1200 - 9500	2.1-2.4
Diamond film	$\gtrsim10^{-12}$	5.0	3.5	9500	2.38

In general, DLC films deposited by different IAD techniques and from different carbon sources (solid carbon or hydrocarbons) can have different bonding networks with sp^1, sp^2 and sp^3 carbon-carbon bonding which pose problems both in characterizing and classifying these materials. For the present study both a-C and a-C:H films were prepared by a low-energy IAD technique (ion energies 30 - 75 eV per C atom) based on dc magnetron sputtering. A single graphite target was used as the source of carbon atoms, and ion beams of Ar^+ and H_2^+ arising from the sputtering plasma were used to bombard the growing film. Amorphous carbon films produced by sputtering from a graphite target and using inert Ar^+ ions to bombard the depositing film offer a pure amorphous carbon network that does not contain hydrogen. Also, since the material is produced by condensing single carbon atoms, its structure is not influenced by the structure of the source or precursor material. The backbone of such a network is found to consist of a mixture of diamondlike (sp^3) and graphitic (sp^2) bonds. Hydrogenation of this material is made possible by sputtering the graphite target in an argon-hydrogen plasma and bombarding the film with H_2^+ ions.

This chapter deals with the deposition and characterization of DLC films, and emphasizes the role of ion bombardment in influencing the bonding configuration of the carbon network and the hydrogen content of films. This is done not only to illustrate the broad range of material properties which are influenced by low-energy ion bombardment, but also to provide some understanding of how the growth environment affects the material properties. An attempt is also made to bridge the gap between DLC and diamond by including results which show that a continuum of materials from DLC to quasi-crystalline diamond is produced by the same low-energy IAD technique.

II LOW-ENERGY ION-ASSISTED DEPOSITION OF FILMS

A FILM DEPOSITION

The DLC films were synthesized by low-energy IAD from a graphite target using an unbalanced magnetron (UM-gun) sputter source. Films of thickness from 20 nm to several μm were deposited on a variety of substrates at a substrate temperature of 300 K. The effective sputtering area of the target was approximately 20 cm^2. The a-C films were deposited by sputtering the graphite target in ultrahigh purity (99.999%) argon gas at a pressure of 1 Pa. The a-C:H films were deposited by sputtering in an argon-hydrogen mixture with the argon partial pressure fixed at 1 Pa and at partial pressures of hydrogen p_H=0.1 and 0.5 Pa (Table II).

The characteristics of the sputtering source and the deposition apparatus have been described in detail previously (1,13,14). A schematic diagram of the deposition apparatus is shown in Figure 1. Briefly, the unbalanced magnetron source provides, simultaneously, sputtered species from which the film is condensed and high fluxes of impinging ions whose energy is controlled in order to modify the film structure and properties. The operation of the UM-gun source is similar to that of conventional planar magnetrons but it differs in one important aspect by using an unbalanced magnetic field configuration designed to produce a column of plasma which is directed at the substrate position. The substrates are electrically isolated from the deposition chamber and, when exposed to this plasma, develop a negative floating potential with respect to the plasma. Ions from the plasma drift towards the sheath that surrounds the substrates and are accelerated across it to bombard the growing film. The principal ion species impinging on the films during growth are Ar^+ ions in the case of a-C, and mixed Ar^+ and H_2^+ ions in the case of a-C:H films.

To carry out measurements of optical properties one requires very thin films deposited on transparent substrates that are invariably insulating. Insulating substrates are also required for other measurements such as electrical conductivity. Since the films are also insulating, external biasing could not be used, and so the energy of bombarding ions was varied by controlling the sputtering power to produce the required floating potentials. The ion energy is then the sum of the plasma and floating potentials.

All films for optical measurements were 20 nm thick and were deposited onto high internal transmittance fused silica (Heraeus Suprasil I) substrates. Similar substrates were used to deposit films of thickness 0.5-1.0 μm for electrical measurements, while films 2 - 3 μm thick were deposited onto crystalline silicon wafers for IR absorption and microhardness measurements. The substrate temperature during deposition was kept at about 300 K. The deposition rate varied linearly with sputtering power, and was approximately 1.0 μm per hour when the magnetron was operated at 100 W of power.

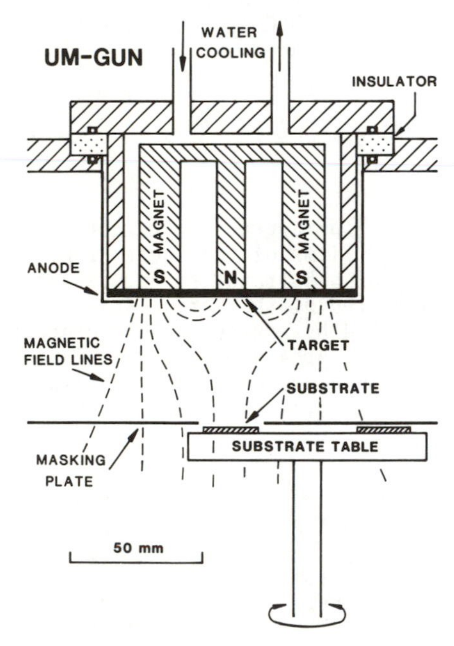

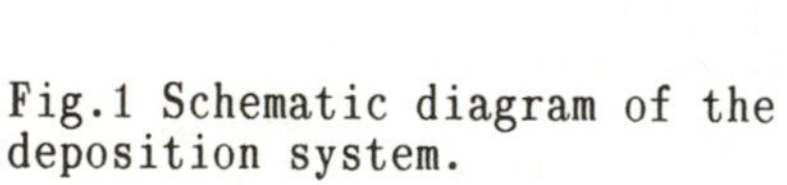

Fig.1 Schematic diagram of the deposition system.

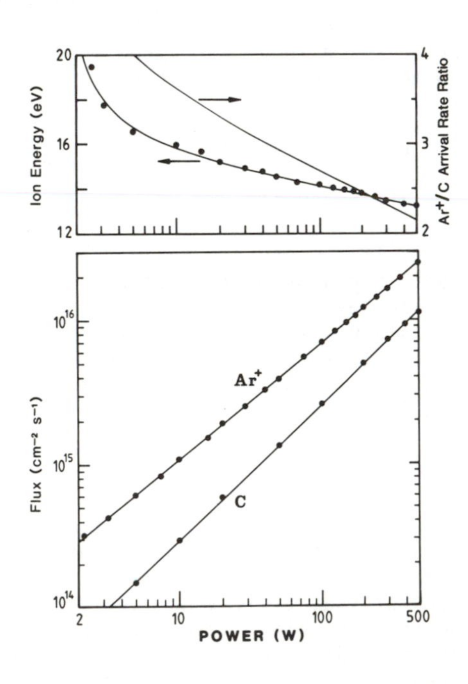

Fig.2 Dependence on sputtering power of the flux of Ar^+ ions and C atoms, the ion energy and arrival rate ratio.

B INERT ION BOMBARDMENT

The purpose of using energetic ions is to supply superthermal energy and chemical activity to the film growth process. The interaction of ions with the growing film depends on the incident energy E_i of ions and their mass, the arrival rate ratio or flux ratio of ions to condensing atoms, and whether inert or reactive ion species are used. The UM-gun source operates essentially as a dc magnetron and so the ion and atom fluxes depend on sputtering power, gas pressure and gas composition. Figure 2 illustrates the dependence on sputtering power of the flux of Ar^+ ions and C atoms, the ion energy and Ar^+/C arrival rate ratio when sputtering a graphite target in pure argon gas at 1 Pa. A summary of the deposition conditions of films is given in Table II. Since the ion energy E_i and the arrival rate ratio Ar^+/C change with sputtering power they have been combined to yield the ion energy per C atom, E_t/C. The values of E_t/C are in the range of those used in the work of Aisenberg and Chabot (2) and Spencer et al (3). The flux of H_2^+ ions is not included in the ion-to-atom arrival rate ratios. For the energies used, the energy transferred in an elastic head-on collision between a H_2^+ ion and a carbon atom (1.9-2.8 eV) is less than that transferred in the argon-carbon collision (9.2-14.2 eV). Therefore, the displacement caused by hydrogen bombardment can be assumed to be less significant, particularly since the transferred energy is below the threshold for carbon-carbon bond dissociation, e.g. sp^3-type bonds have energy of 6.3 eV while in both diamond and graphite the bond energies are 7.4 eV.

TABLE II

Deposition parameters. p_H, partial pressure of hydrogen gas; E_i, ion energy; Ar^+/C, ion to condensing carbon atom arrival rate ratio; E_t/C, ion energy per C atom. The partial pressure of argon gas, p_{Ar} = 1 Pa.

Specimen	Power (W)	p_H (Pa)	E_i (eV)	Ar^+/C	E_t/C (eV)
1	500	0	13	2.2	29
2	50	0	14	2.9	41
3	10	0	16	3.7	59
4	500	0.1	13	2.2	29
5	50	0.1	16	2.9	46
6	10	0.1	18	3.7	67
7	500	0.5	14	2.2	31
8	50	0.5	18	2.9	52
9	10	0.5	20	3.7	74

C THE ROLE OF HYDROGEN IONS AND HYDROGENATION

When accelerated H_2^+ ions with energy greater than the binding energy of the hydrogen molecule (4.5 eV) impinge on the growing film they dissociate to yield atomic hydrogen. Thus bombarding the growing film with H_2^+ ions of energy 13-20 eV provides a flux of atomic hydrogen which imparts chemical activity to the film growth process.

The presence of atomic hydrogen creates unique conditions for stabilising sp^3 or diamondlike bonding in the amorphous carbon matrix. The Russian studies(15-17) of the synthesis of diamond from the vapour phase have shown that atomic hydrogen is a selective etchant of graphitic bonds. By continuously removing graphitic precursors from the film during deposition, atomic hydrogen allows a buildup of sp^3 bonds. If the concentration of atomic hydrogen over the growth surface of the film reaches supersaturation levels then diamond synthesis is possible.

The hydrogenation of amorphous carbon using atomic hydrogen has another advantage over the use of molecular hydrogen. Since the sp^3 C-H bond has a lower bond energy than sp^3 CH_2 and CH_3 bonds (4.3 and 4.8 eV), there is a higher probability of forming monohydride material if atomic hydrogen is available. We have used the IR absorption spectrum of films near 3.5 μm photon wavelength to gain information about the carbon-hydrogen bonding configuration, and to determine the hydrogen concentration from the integrated absorption of the C-H stretch bands (Table III). The dominant line in all the films was the 2920 cm^{-1} line assigned to the sp^3 C-H stretch band. Absorption bands that are characteristic of sp^1 or polymeric material were not observed.

IV RESULTS OF CHARACTERIZATION OF FILMS AND DISCUSSION

A OPTICAL PROPERTIES AND ASSOCIATED FUNCTIONS

1. Optical constants

Among the important properties of DLC films are their optical properties, such as high IR transmittance and variable refractive index, which are made use of to provide scratch-resistant antireflection coatings on IR optics, laser mirrors and compact optical disks. Diamondlike carbon is an amorphous semiconductor, and in the absence of crystalline inclusions of either diamond or graphite, is isotropic.

The optical properties of an isotropic solid are most conveniently described in terms of the complex refractive index $N = n + ik$ and the complex dielectric function $\epsilon = \epsilon_1 + i\epsilon_2$. The refractive constants (index n and extinction coefficient k) and the dielectric constants ϵ_1 and ϵ_2 are real functions of photon frequency ω. The two representations are related at any given frequency according to

$$\begin{aligned} \epsilon_1 &= n^2 - k^2 \\ \epsilon_2 &= 2nk \\ \alpha &= 4\pi k/\lambda \end{aligned} \qquad (1)$$

where α is the absorption coefficient and λ is the photon wavelength.

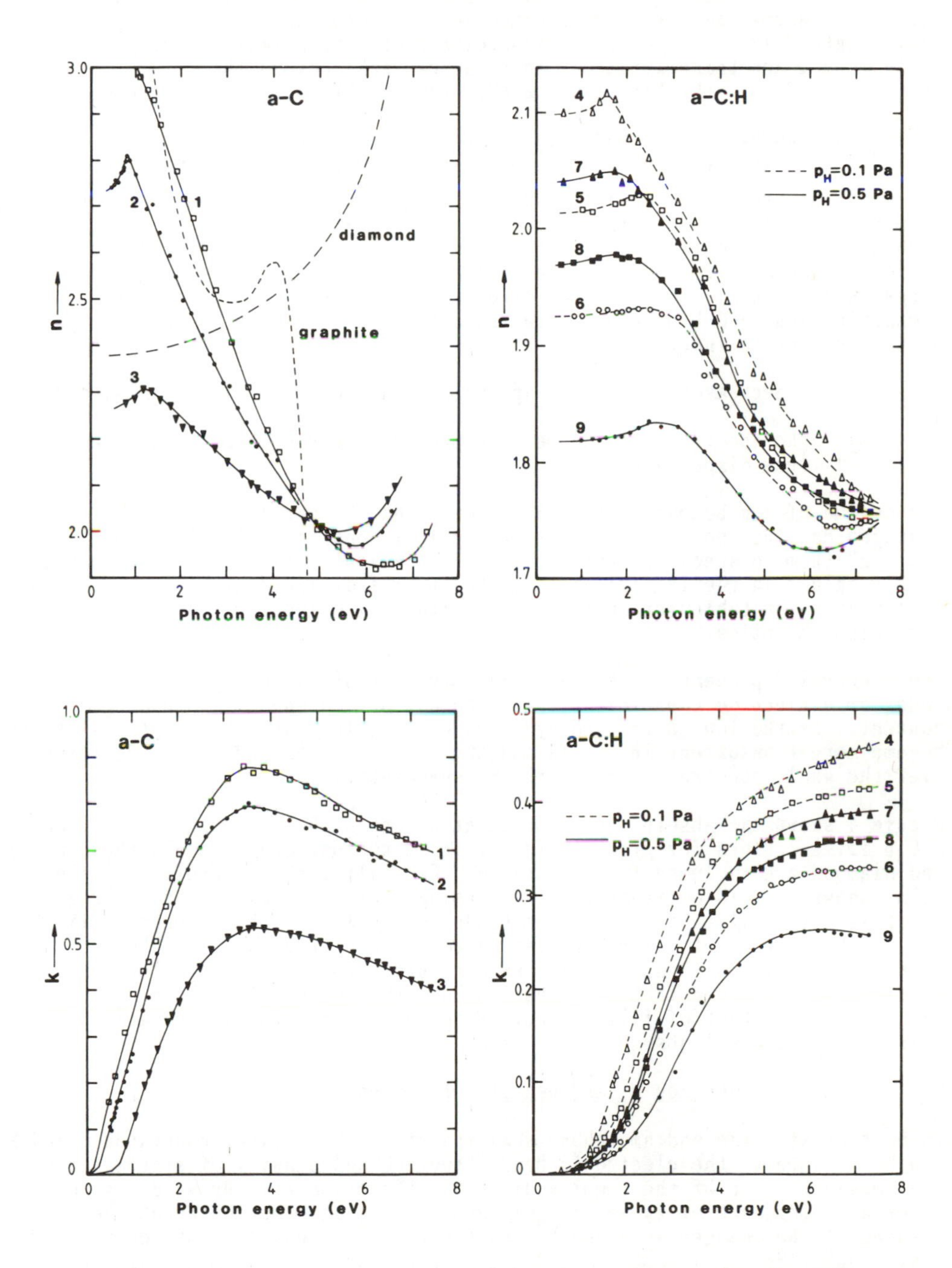

Fig.3 Refractive indices n and k vs photon energy for the DLC films 1-9 (Tables II and III), and n for graphite and diamond.

The refractive indices of thin films deposited on fused silica of high internal transmittance were determined in the photon energy range 0.5 - 7.3 eV (wavelength 0.18 - 2.5 μm) from measurements of near-normal incidence reflectance, and transmittance. This energy range covers the IR (0.5 - 1.6 eV), the VIS (1.6 - 3 eV), and the UV (3 - 7.3 eV) regions.

Figure 3 shows plots of n and k for a group of a-C, and for two groups of a-C:H films. The hydrogen partial pressures p_H used during deposition of a-C:H films are indicated in the plots. The figure also shows plots of n for diamond and graphite (electric vector perpendicular to the c-axis). Within each group of three films we see that n decreases with increasing ion energy (refer to Table II). Hydrogenation has a similar effect, producing rather large changes in n. The combined effects of increasing the energy of bombarding ions and hydrogen content have a great influence on the real part of the refractive index, n:

(a) n decreases from very high values in the IR to VIS regions, and

(b) the spectral dependence of n progressively weakens, i.e. n becomes less dispersive.

The degree of ion bombardment or ion energy and the partial pressure of hydrogen can thus be used to alter the optical constants of films thereby "tuning" them to specific applications. For example, films can be prepared with n values in the IR similar to those of crystalline diamond (n = 2.4), or with n value suitable for quarter wavelength antireflection coating on germanium IR optics.

The spectral dependence of k, the imaginary part of the complex refractive index or extinction coefficient, also depends on ion energy and hydrogen content. As the ion energy and hydrogen content increase, both types of films become more transparent in the IR and the total optical absorption decreases over the whole spectral range of the measurements.

Figure 4 shows the absorption coefficient α versus photon energy for a-C and a-C:H films. Plots for graphite (electric vector perpendicular to the c-axis) and diamond are included for comparison. Generally, the absorption of the films shows a rapid decrease with decreasing photon energy in the VIS region, which is a typical behaviour of semiconductors. At low photon energy (IR region) both types of films are transparent. This transparency makes DLC films attractive in IR application as hard protective and antireflection coatings. It is seen that α depends on both the energy of bombarding ions and the hydrogen content of films. Increasing ion energy and hydrogenation decrease the overall absorption of films.

2. Dielectric Function ϵ_2 and Density of States

In both crystalline and amorphous semiconductors the optical constants n and k can be related to the electronic band structure and density of states through the imaginary part of the complex dielectric function ϵ_2. By studying the spectral dependence of ϵ_2 we can gain substantial information about the meaning of the changes in n and k, and the way they relate to the density of states DOS (18).

Figure 5 shows schematically the DOS of a $\pi+\sigma$ electronic system appropriate to diamondlike a-C and a-C:H films (8,14). The π states, being more weakly bound, lie closer to the Fermi level E_F than the σ states, and thus the band edges are π-like. Consequently π states form the top of the valence band, empty or antibonding π^* states form the bottom of the conduction band, and so define the optical gap $E_0 = E_c - E_v$. Photon-assisted electronic transitions occur between the occupied bonding states (π or σ) and the empty antibonding states (π^* and σ^*) of the carbon-carbon covalent bond.

Electronic transitions requiring the absorption of a photon contribute to ϵ_2 since it represents the absorptive part of the complex dielectric function. In amorphous semiconductors, as in their crystalline counterparts, such transitions can be treated as one-electron excitations.

The dielectric constant ϵ_2 relates optical properties to the DOS as follows. The electronic contribution to $\epsilon_2(\omega)$ is proportional to the sum of all optical transitions between occupied valence-band states and empty conduction-band states. The product $\omega^2\epsilon_2$ is proportional to the convolution of valence and conduction-band states separated by energy $\hbar\omega$:

$$\omega^2\epsilon_2 = \text{const} \int_0^{\hbar\omega} g_c(E)\, g_v(E-\hbar\omega)\, dE \qquad (2)$$

where $g_c(E)$ and $g_v(E - \hbar\omega)$ are the one-electron density of conduction and valence-band states, respectively. It is customary to assume that g_c is a step function and hence the product $\omega^2\epsilon_2$ represents the integrated density of valence-band states.

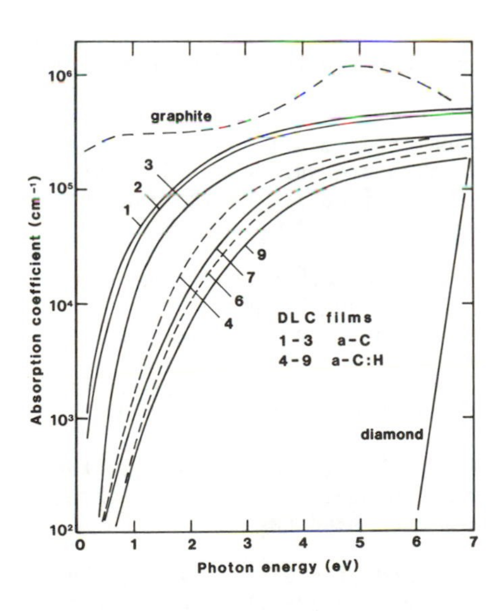

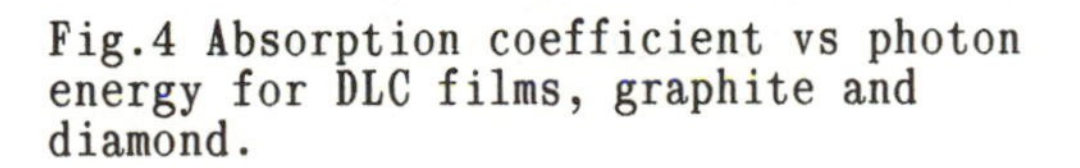
Fig.4 Absorption coefficient vs photon energy for DLC films, graphite and diamond.

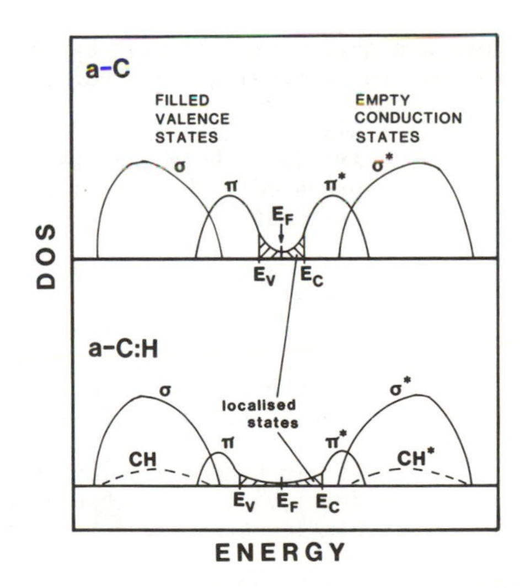

Fig.5 Schematic representation of the DOS of a $\pi+\sigma$ electronic system appropriate to DLC.

Figure 6 shows ϵ_2 vs photon energy for the DLC films. Due to the loss of long range order in amorphous semiconductors, the ϵ_2 spectra for DLC films show little resemblance to those for either diamond or graphite, i.e. the spectra are smooth and do not show the singularities that arise in crystalline materials where momentum conservation rules apply. Absorption at the fundamental absorption edge is due to transitions from the top of the valence band to the bottom of the conduction band. The results show that both ion bombardment and increasing hydrogen content cause the absorption edge to shift to higher energies. This indicates that states are removed from the top of the valence band and thus the band edge recedes leading to increased optical gap, as is indeed observed.

In the high energy, UV region, the optical properties are determined entirely by the band structure. Increasing the energy of bombarding ions and hydrogenation cause two significant changes in the ϵ_2 spectra:

(a) the peak or ϵ_{2max} in the spectra decreases or is suppressed, and

(b) the area under the curves, which is a measure of the total absorption, decreases dramatically.

The ϵ_{2max} seen in the a-C films is associated with the π electron. It is evident that ion bombardment with Ar^+ ion decreases the concentration of sp^2 bonding states. Bombardment with hydrogen ions completely suppresses ϵ_{2max}. Thus bombardment by H_2^+ ions and the ensuing dissociation into atomic hydrogen is very effective in "etching" graphitic (sp^2) bonds. Since there are a total of four valence electrons in carbon, the marked contraction of the area under the curves suggests substantial redistribution of valence band states from the band edge to deep inside the band, i.e. π-like states are removed and σ-like states are created.

There are three possible ways by which ion bombardment with Ar^+ and H_2^+ ions can lead to increasing sp^3 bonding:

(a) Diamondlike bonding forms in the localized impact zone of energetic Ar^+ ions where high-temperature and high-pressure conditions are produced. This mechanism would be more effective as the ion energy increases.

(b) Bombarding Ar^+ ions remove weak sp^2 carbon-carbon bonds leaving behind a higher concentration of stronger sp^3 bonds. The most likely mechanism is by preferential resputtering of weakly bonded graphitic precursors from the film surface. This mechanism can operate at low ion energies since the bonding energy of graphitic precursors to the film surface is about 0.8 eV.

(c) Atomic hydrogen selectively removes sp^2 bonds thereby raising the density of sp^3 carbon-carbon bonds by forming sp^3 CH bonds, whose bonding states lie well below the valence band edge.

These mechanisms deplete the top of the valence band of π electrons and thus fewer of them are available for optical transitions. Since the number of $\pi+\sigma$ electrons is fixed, electrons which are redistributed to low-lying σ states will make their contribution to optical absorption at higher photon energies.

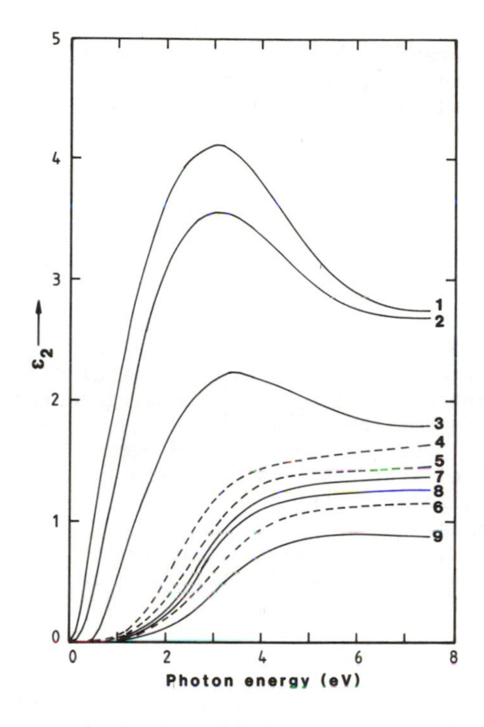

Fig.6 Dielectric constant ϵ_2 vs photon energy for DLC films.

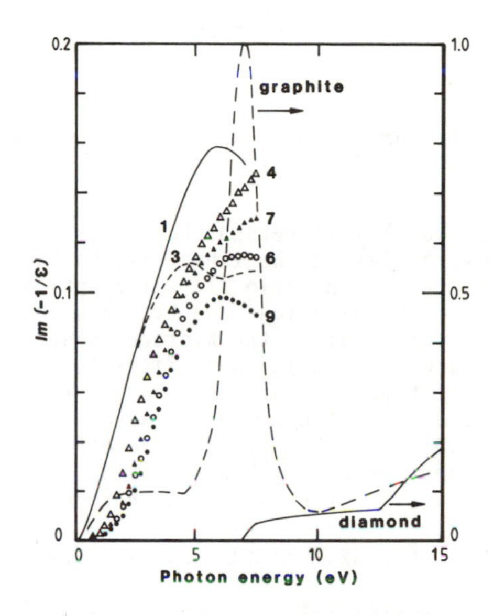

Fig.7 Energy loss function vs photon energy for DLC films, graphite and diamond.

TABLE III

Summary of film properties. C_H, hydrogen concentration; n, refractive index at 0.62 eV (λ = 2.0 μm); E_0, optical gap; n_{eff}, effective number of valence electrons per C atom at 7.5 eV; C_{sp^2} and C_{sp^3}, concentrations of C atoms in sp^2 and sp^3 configurations; sp^3/sp^2, ratio of sp^3 to sp^2 bonding. For graphite n_{eff} = 0.855 at 7.5 eV.

Specimen	C_H (at.%)	n	E_0 (eV)	n_{eff}	C_{sp^2} (%)	C_{sp^3} (%)	sp^3/sp^2
1	0	3.15	0.4	0.53	62	38	0.61
2	0	2.75	0.5	0.41	48	52	1.08
3	0	2.27	0.7	0.24	28	72	2.57
4	4	2.10	1.2	0.21	25	75	3.00
5	12	2.01	1.4	0.17	20	80	4.00
6	18	1.92	1.6	0.12	14	86	6.14
7	13	2.04	1.5	0.18	21	79	3.76
8	20	1.97	1.6	0.14	16	84	5.25
9	25	1.82	1.8	0.09	11	89	8.09

3. Optical Gap

In amorphous semiconductors it is normally assumed that the optical matrix element for interband transitions is independent of energy, and that the densities of states in both bands just beyond the mobility edges have a square-root dependence on energy. Thus equation (2) for optical transitions at the absorption edge reduces to

$$\omega^2 \epsilon_2(\omega) = B\,(\hbar\omega - E_0)^2, \qquad (3)$$

defining the optical gap E_0, where B is a constant proportional to the joint density of states.

Plots of $\omega\epsilon_2^{\frac{1}{2}}$ versus photon energy were used to obtain the optical gap by extrapolating the linear fit to the data onto the energy axis. The results are given in Table III. The optical gap increases with both increasing energy of bombarding ions and hydrogen content, and for the a-C:H films E_0 is proportional to the hydrogen content. Note, however, that the hydrogen concentration in these films is about half of that reported by other investigators for films having the same optical gaps but prepared by condensing ionized hydrocarbon species (rf glow discharge) of energy >500 eV (7,19). In the present study low-energy Ar^+ ions continuously resputter excess hydrogen from the film without causing graphitization or reduction of E_0 as is often seen when much higher ion energies are used.

4. Energy Loss Function and Plasmons.

The energy-loss spectra of DLC films can provide additional information on the density of valence band states (18). In a solid, the excitation of plasma oscillations (plasmons) is determined by the energy loss function, defined as the negative imaginary part of the reciprocal value of the complex dielectric function $\epsilon(\omega)$:

$$\mathrm{Im}\,(-1/\epsilon(\omega)) = \epsilon_2/(\epsilon_1{}^2 + \epsilon_2{}^2) = 2nk/(n^2 + k^2)^2. \qquad (4)$$

The energy loss function considers only electron excitations in the bulk of the solid and has maxima at plasma frequencies. For a well defined group of valence electrons (four in DLC), the loss function spectrum for DLC films would have plasmon peaks at frequencies in the vicinity of the π and σ valence electron free plasmon energy.

Graphite has a sharp plasmon peak at 7 eV, which is due to plasma oscillations of the π electron, and a second broad peak at 25 eV which involves $1\pi+3\sigma$ electrons. Diamond has just one broad peak due to 4σ electrons which is centred at about 34 eV with a low energy tail extending down to 7 eV. Core level excitations of the 1σ electrons in carbon occur at and above the K-edge threshold near 284 eV and are observed using electron energy loss spectroscopy (EELS). Figure 7 shows the loss function of some DLC films, of graphite, and the low energy tail of diamond. The plasmons in DLC films are centered close to the graphite plasmon and can be assumed to be due to π electrons. The films have rather weak plasmons compared with graphite. The plasmon peak value near 7 eV is relative to the number of π electrons. Increasing ion bombardment and hydrogenation tend to lower the peak maxima, providing further evidence that sp^2 bonding decreases with increasing ion energy and hydrogen content. The exact position of the maxima depends on the ratio of π-to-σ electrons and the polarisability of the σ electrons. In a-C films the effect of increasing ion energy causes a shift of the plasmons to lower photon energies. Similar shifts occur in a-C:H films with both increasing ion energy and hydrogen content.

5. The n_{eff} Sum Rule and Carbon Coordination

The optical constants are connected via the Kramers-Kronig dispersion relations which can be used to derive a number of "sum rules" useful in interpreting the optical properties. One such sum rule connects ϵ_2 with n_{eff}, the effective number of valence electrons per atom taking part in optical transitions in the photon energy from zero to some cut off E_M:

$$n_{eff} = \frac{0.766A}{\rho} \int_0^{E_M} E \, \epsilon_2(E) \, dE \tag{5}$$

where A is the mean atomic weight, ρ is the mass density in kg m^{-3}, and $E = \hbar\omega$ is the photon energy in eV.

In carbon there are four valence electrons per atom and the integral (5) should reach a plateau with $n_{eff} = 4$ as all photon-assisted interband transitions are exhausted at high photon energies. Diamond is essentially transparent below 7 eV so that $n_{eff} = 0$ below 7 eV but rises to 4 by about 25 eV. For graphite the integral (5) saturates near $n_{eff} = 1$ at about 9 eV, then rises with photon energy and saturates again near $n_{eff} = 4$ at about 25 eV. It is known that for graphite the interband transitions in the energy range 0-9 eV involve only the one π electron per atom (18). At about 9 eV $\pi \rightarrow \pi^*$ transitions are exhausted while $\sigma \rightarrow \sigma^*$ transitions, involving three σ electrons per atom, commence so that $n_{eff} = 4$ at high energies. This clear separation into π and σ contributions can be used to determine the sp^3-to-sp^2 ratio or the carbon coordination in DLC films.

Figure 8 shows n_{eff} for the DLC films, for graphite and diamond. Since the band edges in DLC are π-like, then all transitions below 8 V are $\pi \rightarrow \pi^*$. This fact can be used to determine the π-electron fraction per carbon atom from the experimental data. The fraction is simply given by the ratio of n_{eff} for films to n_{eff} for graphite. The relative concentration of carbon atoms with sp^2 configuration, C_{sp2}, and the sp^3/sp^2 bonding ratio are determined using n_{eff} values at 7.5 eV and the results are given in Table III. Figure 9 shows the variation of sp^3/sp^2 with the ion energy per deposited carbon atom. The combined effects of increasing the degree of ion bombardment and hydrogenation are to raise the sp^3/sp^2 bonding ratio and thus the diamondlike component in the films.

B ELECTRON ENERGY LOSS SPECTROSCOPY (EELS)

When high energy electrons pass through a thin film in a transmission electron microscope they suffer energy loss due to inelastic electron-electron scattering processes. The energy loss occurs in two different energy regimes: low-energy loss due to plasma excitations of valence or conduction band electrons, and high-energy loss associated with X-ray production by ejection or ionization of core electrons to states above the Fermi level.

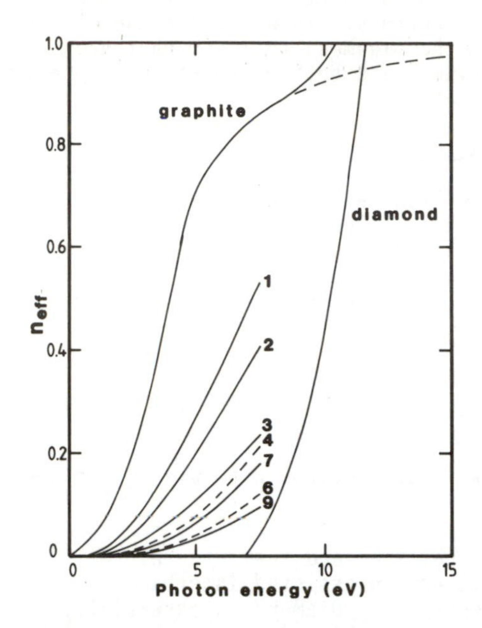

Fig.8 Effective number of valence electrons vs photon energy for DLC films, graphite and diamond.

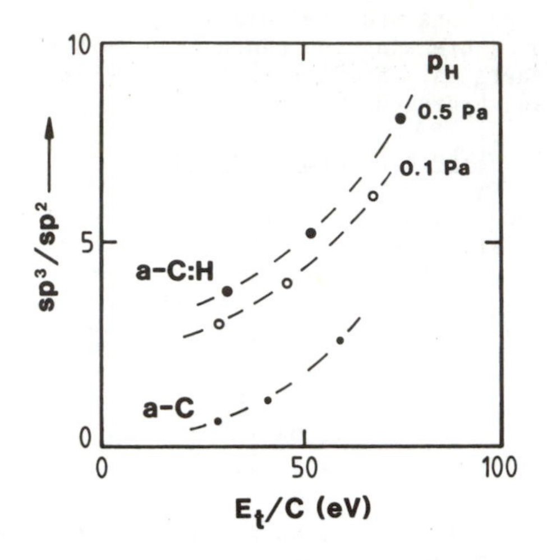

Fig.9 Variation of sp^3 to sp^2 bonding ratio with ion energy per condensing carbon atom E_t/C for the DLC films.

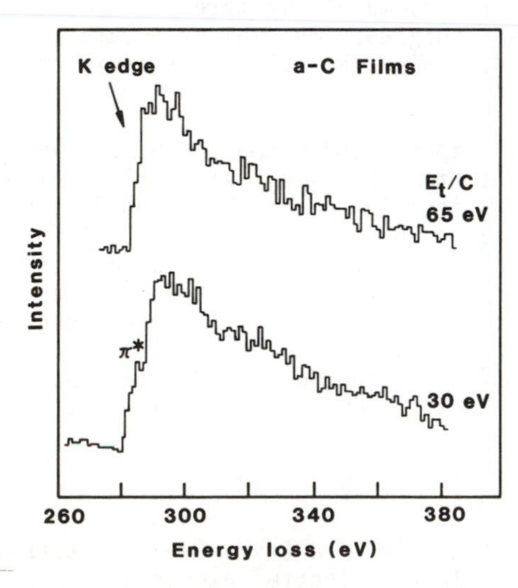

Fig.10 Electron energy loss spectra near the carbon K edge for two a-C films showing the effect of increasing the energy of ion bombardment from 30 to 65 eV per condensing carbon atom. (π^* indicates the peak due to excitation of 1s electrons into the π^* energy level).

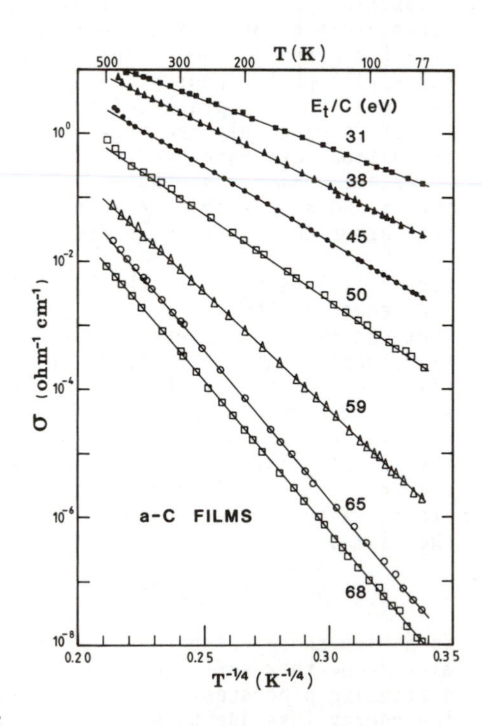

Fig.11 Temperature dependence of electrical conductivity σ for a-C films deposited at various values of ion energy per condensing carbon atom E_t/C.

In the various forms of carbon, plasma effects occur below about 40 eV while core excitations occur at and above 284 eV. Figure 10 shows the high energy-loss spectra near the carbon K edge for two a-C films. Detail around this edge is quite revealing of the band structure of the material. In graphite there is a distinct peak at the K edge at 285 eV which represents the transition of 1s electron excitation into π^* states. This structure is observed in the film deposited at the lower ion energy but not in the film deposited at the higher ion energy. Evidently increasing the energy of bombarding ions during film growth decreases the density of π states.

C ELECTRICAL PROPERTIES

Electron localization in amorphous semiconductors leads to zero conductivity at zero temperature. At finite temperatures the dc electrical conductivity is phonon-assisted or thermally activated. Although a variety of charge transport mechanisms are possible it is accepted that total transport can be split into band conduction due to electrons in extended states of the conduction band, and variable-range hopping between localized states at the Fermi level E_F. Band conduction requires high activation energies and is generally confined to the higher temperatures while variable-range hopping becomes dominant at the lower temperatures and/or at a large density of gap states.

The dc conductivity σ of films was measured as a function of temperature (T=100-500 K) using the standard 4-probe technique. Specimens were rectangular in shape, 16x6 mm^2, and contacts were made using top evaporated chromium electrodes. A Keithley model 616 digital electrometer sensed the voltage across the potential probes. The measurements were made with the specimen placed in a variable temperature cryostat (77 - 500 K), by first cooling the specimen and then taking measurements while the temperature was raised. A lower limit in conductivity of 10^{-8} ohm^{-1} cm^{-1} was possible with this apparatus. The a-C:H films had values below this limit and could not be measured. Thus only measurements for a-C films are reported.

For the a-C films the temperature dependence of the conductivity is found to be best described by $\log \sigma \sim T^{-1/4}$ (Figure 11), which indicates that the conductivity is dominated by hopping between localized states. Thus the mechanism of conduction in DLC films is very similar to that found in other tetrahedrally coordinated amorphous semiconductors and in carbon-ion-implanted diamond (20,21). Mott's treatment of variable-range hopping describes a mechanism of charge transport by hopping or tunnelling between distant but energy favourable localized states (20). If E_F lies in a region of constant density of states and the jump rate is thermally activated then at a finite temperature T the conductivity is given by

$$\sigma = \sigma_0 \exp\left[-(T/T_0)^{-1/4}\right] \quad (6)$$

where

$$T_0 = 18\ \alpha^3/kN(E_F). \quad (7)$$

$N(E_F)$ is the density of states at the Fermi level, α^{-1} is the extent of the localized wave function, and k is Boltzmann's constant. By fitting data to equation (6) we obtain $T_0 = 1.6\times10^6 - 1.4\times10^8$ K. The extent of the localized wave function $\alpha^{-1} \sim 10$ Å (22), so that to a first approximation $N(E_F)$ takes values between $1.3x10^{20}$ and $1.5x10^{18}$ cm^{-3} eV^{-1} as the ion energy per deposited carbon atom increases from about 30 to 70 eV. These values are similar to those for getter-sputtered a-C (22) and carbon-ion-implanted diamond (21), and are comparable to those of a-Si (20). Apart from increasing the optical gap and decreasing the density of states at the band edges, ion bombardment is seen to also decrease the density of gap states, thus leading to a "cleaner" optical gap.

V TRANSITION FROM DIAMONDLIKE CARBON TO DIAMOND

Although there can be only one true or pure diamond structure, in practice many different types of films can be produced ranging from DLC to amorphous or quasi-crystalline diamond and on to crystalline diamond. This diversity is a product of the many growth techniques and the methods used to characterize the films (11,23).

TABLE IV
Historical Review of the Developments of the Growth of Diamond and Diamondlike Carbon from the Vapour Phase

Year	Researchers	Development
1956 1958	Derjaguin & Spitsyn, Inst. of Physical Chem., Moscow, USSR. Eversole, Union Carbide, USA	Thermal pyrolysis of methane. Epitaxial growth of diamond on diamond powder at 900-2700°C. Very low growth rates due to co-deposition graphite. Graphite removed by etching in H_2 at high temperature.
1971 1976	Aisenberg & Chabot Spencer et al	Synthesise DLC films by direct ion beam deposition of C^+ ions. The films are extremely hard and show evidence of microcrystalline diamond.
1968 to 1984	Derjaguin, Spitsyn, Bouilov Fedoseev & coworkers	Study the growth kinetics of diamond. Establish the importance of atomic hydrogen in preferentially etching graphite. Develop enhanced CVD using catalytic, hot-filament and plasma methods. Attain growth rates ~1μm/hr. Grow diamonds on Cu, Si, W.
1981	Weissmantel & coworkers	Ion-beam deposition of hard DLC films.
1981 to 1987	Matsumoto & coworkers at NIRIM, Japan	Develop further the plasma-enhanced methods. Hydrocarbon/H_2 mixture (0.5-1% CH_4). Growth rates up to 5 μm/hr.
1984 1985	Mori & Namba Kitabatake & Wasa	Carbon ion-beam deposition of DLC. Films contain large diamond crystals.
1986	Savvides	Low energy ion bombardment of DLC films during growth causes graphitic to diamond transition.
1986	Messier, Badzian, Rustum Roy at Penn State Uni., USA	Confirm Russian and Japanese findings. Investigate microwave plasma CVD.
1987	Many research groups	Convergence of CVD, plasma and ion-beam techniques. Diamond growth rates ~ 1 μm/min.

Today's techniques for producing crystalline diamond films by metastable synthesis from the vapour phase owe their origins and success to work carried out in the Soviet Union by Derjaguin, Fedoseev, Spitsyn and co-workers (15-17), and to later advances made by Japanese workers led by Matsumoto (24,25). Table IV gives a brief historical review of these developments. The main features common to all the present methods of producing crystalline diamond films are:

(a) A high concentration of atomic hydrogen, generated by plasma or thermal means, is needed to prevent or remove non-diamond carbon deposits.

(b) The carbon atoms or carbon-containing species must be activated by plasma or thermal means.

(c) The substrate temperature must be in the range 500-1200°C.

Plasma-enhanced techniques, such as microwave-plasma and rf-plasma dissociation of hydrocarbon gases, require the presence of methyl groups, as well as atomic hydrogen, and this is accomplished by diluting methane (0.5 - 2 %) in hydrogen gas. Ion-beam and sputtering techniques use either hydrocarbon gases or solid carbon sources and hydrogen gas to provide ionized carbon species of energy 50-100 eV and atomic hydrogen.

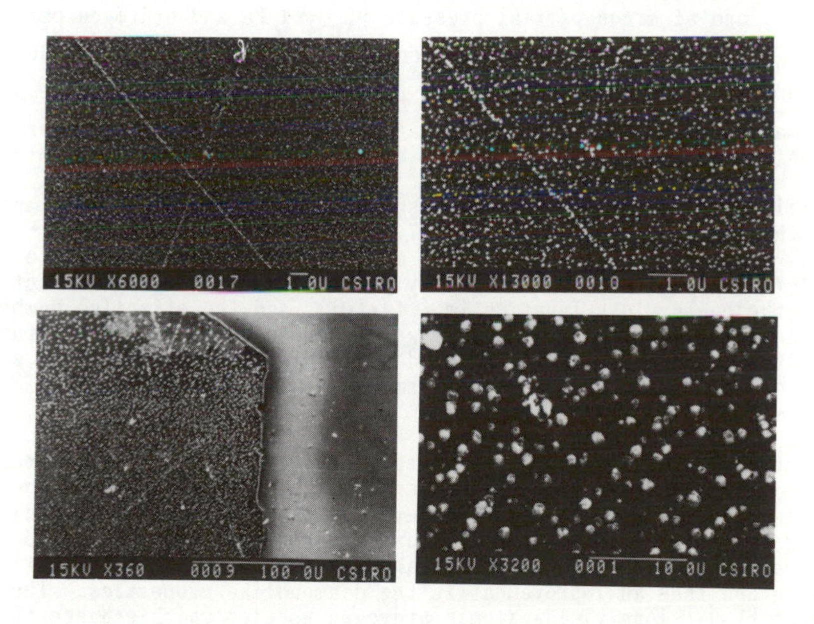

Fig.12 SEM micrographs showing the nucleation and growth of diamond microcrystals in a DLC film.

We investigated the nucleation and growth of diamond onto c-Si wafers held at relatively low substrate temperatures (300-600°C) by sputtering the graphite target in a plasma dominated by hydrogen (p_H = 1.5, p_{Ar} = 0.5 Pa) so as to produce the required high concentrations of atomic hydrogen over the film surface (11). The energy of bombarding Ar^+ and H_2^+ ions was 20 - 100 eV. The use of low substrate temperatures results in low growth rates (60 nm/h at a discharge power of 100 W). A variety of materials was produced ranging from homogeneous DLC films to films containing large concentrations of diamond microcrystals, and discontinuous crystalline diamond deposits of size up to 1 μm. Figure 12 shows the initial nucleation and growth of diamond microcrystals in a film which was deposited at 500°C. Both spontaneous nucleation and nucleation on scratches on the substrate's surface are evident. Deposition at higher temperatures would be expected to lead to higher growth rates and to fusion of the microcrystals.

VI SUMMARY

DLC thin films (amorphous carbon a-C and hydrogenated amorphous carbon a-C:H) were synthesized by low-energy IAD using dc magnetron sputtering of a graphite target in pure argon and argon-hydrogen plasma. Ion beams of Ar^+ and H_2^+ arising from the plasma were used to bombard the growing film. Thus a-C films were bombarded by Ar^+ ions while the a-C:H films were bombarded by a mixture of Ar^+ and H_2^+ ions. These films were deposited onto substrates held at room temperature, and at argon partial pressure p_{Ar} = 1 Pa and hydrogen partial pressure p_H = 0.1 and 0.5 Pa. Deposition at moderate temperatures (300 - 600°C) and at p_{Ar} = 0.5 Pa and p_H = 1.5 Pa produced quasi-crystalline and microcrystalline diamond, thus demonstrating that a continuum of materials from graphitic to DLC and on to diamond can be synthesized by low-energy IAD.

The DLC films were characterized by measurements of the optical constants n and k in the photon energy range 0.5 - 7.5 eV, temperature-dependent electrical conductivity, and EELS. The optical constants were used to determine the optical gap, the imaginary part ϵ_2 of the complex dielectric constant, the energy loss function Im $(-1/\epsilon(\omega))$, and the effective number of valence electrons n_{eff}. These results were then used to obtain qualitative and quantitative information on the band structure and the density of states at and below the band edge, and to determine the bonding coordination of the carbon atoms.

The properties of DLC films were found to depend primarily on the degree of ion bombardment and their hydrogen content. Ion bombardment by low-energy Ar^+ ions is considered to lead to the development of tetrahedral coordination. Increasing the Ar^+ ion energy per condensing carbon atom from 30 to 75 eV brings about a gradual trigonal-to-tetrahedral (or sp^2 to sp^3) transition in C-C bonding and thus an improvement in the diamondlike properties. The bombardment by H_2^+ ions yields atomic hydrogen species which enhance the selectivity toward tetrahedral coordination by the preferential removal of graphitic bonds. The combined effects of the bombardment by Ar^+ and H_2^+ ions lead to improved diamondlike properties and raise the sp^3/sp^2 ratio from 0.6 to 2.6 in a-C, and from 3 to 8 in a-C:H.

REFERENCES

1) Savvides, N.: Thin Solid Films, 1988, 163, 13
2) Aisenberg, S. and Chabot, R.: J. Appl. Phys., 1971, 42, 2953
3) Spencer, E.G., Schmidt, P.H., Joy, D.C. and Salsalone, F.J.: Appl. Phys. Lett., 1976, 29, 118
4) Weissmantel, C., Bewilogua, K., Breuer, K., Dietrich, D., Ebersbach, U., Erler,H.-J., Rau, B. and Reisse, G.: Thin Solid Films, 1982, 96, 31
5) Savvides, N.: J. Appl. Phys., 1985, 58, 518
6) Weissmantel, C.: in "Thin Films from Free Atoms and Particles", edited by K.J. Klabunde (Academic, New York, 1985) p.153
7) Angus, J.C., Koidl, P. and Domitz, S.:in "Plasma Deposited Thin Films", edited by J. Mort and F. Jansen (CRC Press, 1986) p.89
8) Robertson, J.: Adv. Phys., 1986, 35, 317
9) DeVries, R.C.: Ann. Rev. Mater. Sci., 1987, 17, 161
10) Tsai, H. and Bogy, D.B.: J. Vac. Sci. Technol. A, 1987, 5, 3287
11) Savvides, N.: Mater. Sci. Forum, 1988, 34-36, 487
12) Angus, J.C. and Hayman, C.C.: Science, 1988, 241, 913
13) Savvides, N. and Window, B.: J. Vac. Sci. Technol. A, 1986, 4, 504; J. Vac. Sci. Technol. A, 1985, 3, 2386
14) Savvides, N.: in "Amorphous Hydrogenated Carbon Films", edited by P. Koidl and P. Oelhafen (Les Editions de Physique, Paris, 1987) p.275
15) Spitsyn, B.V., Bouilov, L.L. and Derjaguin, B.V.: J. Cryst. Growth, 1981, 52, 219
16) Derjaguin, B. V. and Fedoseev, D. V.: "Growth of Diamond from the Gas Phase" (Nauka, Moscow, 1987)
17) Fedoseev, D. V., Varnin, V. P. and Derjaguin B. V.: Russ. Chem. Rev., 1984, 53 435
18) Savvides, N.: J. Appl. Phys., 1986, 59, 4133
19) Dischler, B., Bubenzer, A. and Koidl, P.: Appl. Phys. Lett., 1986, 42, 636; Solid St. Commun., 1983, 48, 105
20) Savvides, N.: J. Appl. Phys., 1984, 56, 2788
21) Hauser, J.J., Patel, J.R. and Rodgers, J.W.: Appl. Phys. Lett., 1977, 30, 129
22) Hauser, J.J.: J. Non-Cryst. Solids, 1977, 23, 21
23) Badzian, A.R., Bachmann, P.K., Hartnett, T., Badzian, T. and Messier, R.: in "Amorphous Hydrogenated Carbon Films", edited by P. Koidl and P. Oelhafen (Les Editions de Physique, Paris, 1987) p.63
24) Matsumoto, S., Sato, Y., Kamo, M. and Setaka, N.: Jap. J. Appl. Phys., 1982, 21, L183
25) Matsumoto,S., Hino, M. and Kabayashi, T.: Appl.Phys.Lett., 1987, 51, 737

Materials Science Forum Vols. 52 & 53 (1989) pp. 427-474

RADIATION EFFECTS IN AMORPHOUS HYDROGENATED CARBON

R. Kalish and M.E. Adel
Solid State Institute and Physics Department
Technion Israel Institute of Technology
Haifa, 32000, Israel

Abstract

The effects of energy deposition by photons, ions and electrons into amorphous hydrogenated carbon (a-C:H) films are reviewed. In all cases the treated films turn electrically conductive and optically opaque; nevertheless it is shown that different physical processes are responsible for the observed changes for the different modes of energy deposition. Results obtained by a variety of experimental techniques which probe the structural and compositional changes are described. Models which fit the experimental data on hydrogen loss and on changes in electrical and optical properties as a result of ion irradiation are proposed.

I Introduction

Amorphous hydrogenated carbon (a-C:H) films attract increasing interest both scientifically and technologically [1]. Their unique physical and chemical properties (extreme hardness, optical transparency, electrical resistivity and chemical innertness) make a-C:H attractive as protective antireflection films for windows and optical components and as coatings for the reduction of wear and corrosion. The ability of carbon to form both sp^3 (diamond-like) and sp^2 (graphite-like) bonds makes the study of the basic properties of a-C:H and their dependence on structure and hydrogen content very interesting.

The majority of papers published so far on a-C:H have concentrated on the techniques of producing the films and their as-grown properties, although a considerable number of thermal annealing studies have also been carried out [1]. The modification of film properties by methods other than heat treatment has not been extensively investigated. Only recently, some groups have turned their attention to the study of the post growth modification of a-C:H films caused by different kinds of energy deposition. Such studies are significant from both basic and practical aspects. The possibility to modify the films in a continuous and controlled manner by external means enables the researcher to follow gradual changes that the films undergo and to correlate them. Furthermore, the fact that in most cases the energy deposited into the film can be directed to any desired spot on the specimen offers the possibility of direct writing of patterns in a-C:H in which the properties differ markedly from those of the surrounding film; i.e. conductive or opaque patterns (with desired electrical or optical properties) can be produced in the films.

In the present review, we survey the work done on the post-growth modification by irradiation of a-C:H films. Three modes of energy deposition will be considered: (i) laser irrariation (ii) ion bombardment, (iii) electron irradiation. These processes show some common features: They all cause major structural changes, the most pronounced of which is the loss of hydrogen from the film which is accompanied by variations in a number of other physical properties. Common to all three is the threshold behaviour in deposited energy density required to induce these structural changes. Nevertheless, the physical mechanisms responsible for these alterations are different for the different modes of energy deposition, as will be described below.

In the following, we first briefly describe the various kinds of interactions with matter exhibited by the different forms of irradiation, emphasizing their uniqueness with regard to a-C:H, being an alloy composed of very light elements. We then describe the analytical tools most frequently used to evaluate the physical properties of the transformed films. Next, we describe the experimental data published to date in the literature on irradiated a-C:H, where most of the existing results are for ion bombarded films. Finally, we offer explanations for the observed phenomena whereby hydrogen effusion, bonding modifications and changes in electrical and optical properties are considered.

II Radiation Effects

Heat treatment of metastable materials is the most straightforward method of inducing changes towards more thermodynamically stable forms. The graphitization of hydrocarbons, such as a-C:H, by thermal annealing is well documented [1]. In particular, release of hydrogen acompanied by a transformation from sp^3 tetrahedrally bonded to sp^2 trigonally bonded carbon takes place for anneal temperatures between 400 and 600 °C. It has also been shown that an increase in medium range order, i.e. microcrystallite growth results from thermal treatment [2]. In contrast to the irradiation effects described below which are of a transient nature, thermal annealing is an equilibrium thermodynamic process.

II.1 Laser Irradiation Effects.

Focussed laser radiation may interact with matter resulting in a broad range of transient effects. At sufficiently high intensities, photon irradiation may lead to irreversible changes in target properties. In semiconductors the most important factors in determining the dominant photon absorption mechanism are the photon energy, the optical band gap of the target and the laser intensity. If the photon energy exceeds the optical band gap of the target, absorption is primarily due to band to band excitation i.e. electrons are promoted from the valence to the conduction band. In this case, the *damage threshold* is typically laser pulse energy dependent. If the photon energy is less than the optical band gap then at moderate laser intensities absorption is due mainly to the excitation of impurity or defect states within the gap. However, at sufficiently high intensities, non-linear mechanisms such as two-photon absorption may occur and quickly dominate as the radiation intensity is increased. Furthermore, once a significant density of carriers has been excited into the conduction band, free carrier absorption may occur irrespective of photon energy as in the case of optical absorption in metals. In this case the peak power density is critical in determining the damage threshold [3].

Energy is therefore absorbed by the target via photon-electron interactions however this energy is quickly transferred to the atomic lattice by electron-phonon interactions. These interactions occur on time scales much faster than typical pulse durations, which in turn are much shorter than characteristic thermal diffusion times.

Modification by laser irradiation may be due to a number of mechanisms. Firstly defects may be introduced by direct interactions of photons with the electronic structure as is the case with photosensitive materials. Electromagnetic breakdown may occur if the local field strength is sufficiently high. Local heating at the beam spot may also induce modifications similar to those resulting from thermal annealing. This is of particular importance in metastable materials such as a-C:H since, as was already described, hydrogen effusion is known to commence at relatively low temperatures. Finally, intense thermal gradients in the vecinity of the irradiated region may cause catastrophic damage such as blistering and ablation. Such effects are critically dependent on surface conditions so that discontinuities such as cracks, groves or pores greatly reduce the damage threshold by causing local hot spots.

II.2 Ion Irradiation Effects.

When an energetic ion impinges upon a solid the amount of energy lost per unit distance traversed depends on the mass, charge and energy of the projectile and on the density and composition of the target [4]. The specific energy loss $\frac{dE}{dx}$ per unit path length of the incident beam is related to the stopping cross-section, ε by

$$\varepsilon = \frac{1}{N}\frac{dE}{dx} \tag{1}$$

where N is the atomic density in the target. The two dominant energy loss processes are the interaction of the moving ion with the bound or free electrons in the target (ε_e) and the interaction of the moving ion with the screened nuclei of the target atoms (ε_n). The total stopping cross-section is the sum of the electronic and nuclear stopping, i.e.

$$\varepsilon_t = \varepsilon_e + \varepsilon_n \, . \tag{2}$$

Both processes may be viewed as Coulomb interactions between two charged particles. Whether the interaction is between ion and electron or ion and nucleus, in the vast majority of encounters the direction and speed of the incident ion are perturbed only slightly.

In both cases, the energy transferred per collision is, at high energies, very nearly [4]

$$E \approx E_{\perp} = \frac{M_1(Z_1 Z_2 e^2)^2}{M_2 E_k b^2}. \tag{3}$$

Here, Z_1 and Z_2 are the atomic numbers or the ion and target respectively ($Z_2 = -1$ for electrons), M_1 and M_2 are the ion and target atomic masses respectively, E_k is the ion kinetic energy and b is the impact parameter. From this expression it is evident that electrons with their small mass absorb much more energy per encounter than the target nuclei do. The inverse dependence on ion energy is also worth noting. For both nuclear and electronic energy loss this inverse dependence breaks down at low energies, but for different reasons. In the case of electronic stopping, as the ion velocity drops below the orbital velocity of the innermost target electrons, the cross-section for interaction with these electrons drops significantly. Furthermore, the above equation assumes that the projectile charge is $Z_1 e$, i.e. the ion is fully stripped, however, at lower energies the ion picks up electrons reducing its effective charge. Both these effects tend to reduce the electronic energy loss processes. At sufficiently low energies, nuclear energy loss due to Coulomb interactions between ion and target nuclei may become dominant, but eventually, as the ions slows further even these interactions become weaker as the nuclei are screened by their electrons. Therefore, immediately after penetration of the surface, energy transfer from ion to solid is predominantly due to excitation and ionization of target electrons. As the ion travels deeper into the solid, it loses energy and the cross-section for elastic collisions between ion and target nuclei increases so that towards the end of the ion's path energy deposition is primarily due to nuclear processes. The result is a different depth profile of deposited energy for the two processes and hence the radiation damage

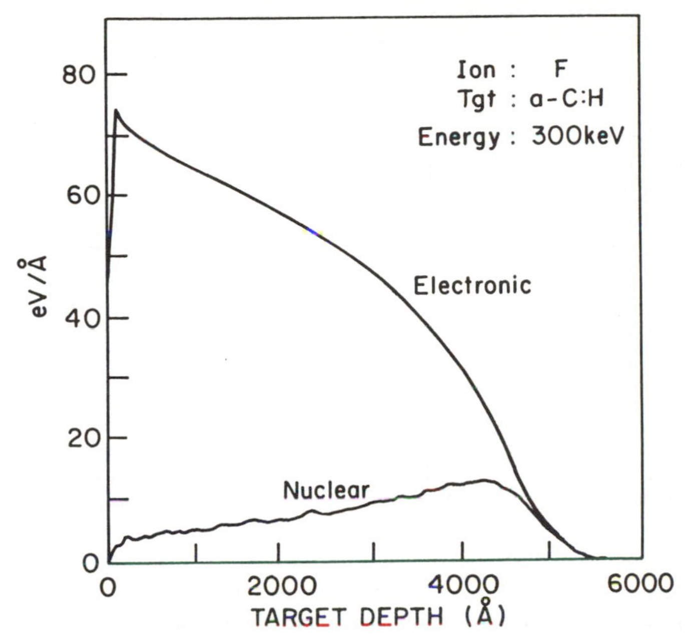

Figure 1: Energy deposited by electronic and nuclear processes for 300 keV Flourine ions in a-C:H, as simulated by computer program TRIM [5] with parameters: density 1.8 g/cm^3, composition 62 % C & 38 % H, displ. energy 20 eV & bind. energy 1 eV.

characteristic of each form of energy loss is also spatially separated, or at least not entirely overlapping, as demonstrated in figure 1.

The mechanism by which an energetic ion loses energy while penetrating a solid does not necessarily determine the ultimate destination of the deposited energy. For instance, although a significant amount of kinetic energy may be transfered by an ion in small impact parameter nuclear collisions, the recoiling nucleus will still lose a significant proportion of its energy to electronic processes. Ultimately most of the energy must end up as heat since the excited electrons will relax via electron-phonon interactions, although a number of radiative processes may also result in the production of photons from visible to X-ray energies. Other competing energy channels are re-emission of energetic target electrons or atoms (i.e. sputtering), backscattering of incident ions and finally changes in the free energy of the target solid due to phase changes or the creation of defects such as vacancies or interstitials.

This last process is probably the most significant when considering radiation damage induced by ion implantation. The elastic collisions which occur between ions and target nuclei result in significant energy transfer and providing the recoiling lattice atom receives energy in excess of about 20 eV (the displacement energy) it will become permanently displaced in the solid [6]. In many cases the transfered energy is far in excess of the displacement energy and a cascade of secondary collisions is innitiated in the vicinity of the primary collision event. This process is illustrated in figure 2 which shows the simulation

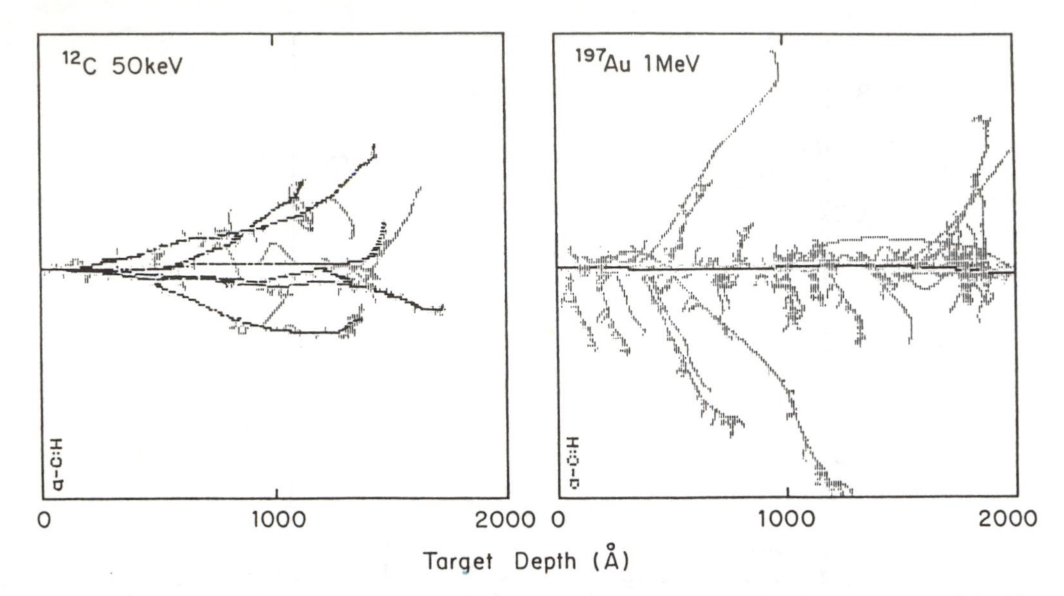

Figure 2: Left: Collision cascades (grey) produced by several 50 keV carbon ions (black) impinging on a-C:H, and right:a single 1 MeV gold ion. Simulations by computer program TRIM [5]. Parameters as in previous figure.

of collision cascades for several 50 keV carbon ions impinging on a-C:H, using the TRIM simulation program [5]. In crystalline semiconductors this results in the creation of local regions rich in vacancies and interstitials which eventually become amorphous at high enough implantation doses. In metals at room temperature, the high mobility of defects such as vacancies prohibits complete amorphization, but results in the production of high concentrations of extended defects such as dislocation loops. In an amorphous material structural modification may still result, such as local changes in bonding configurations or stoichiometry, or the introduction of dangling bonds. Under certain conditions, the passage of ions through an amorphous or partially amorphized layer may even cause annealing or epitaxial recrystallization [7, 8]. However, it is clear that if stoichiometric changes have resulted from the ion irradiation the modifications will be irreversible. As will be described, hydrogen evolution is perhaps the most prominant of ion-induced effects in a-C:H.

Electronic processes may also be responsible for radiation damage. In analogy to the above described nuclear collision cascade, energetic electrons produced by the passage of an ion through a solid may go on to excite other target electrons. The result is a dense cascade of secondary electrons, which may also be responsible for the breaking of bonds. Considering carbon based materials where different forms of bonding coexist, ion irradiation may result in bond restructuring. Simple kinematics, however, precludes the possibility of atomic displacement by electrons of energy typically in the tens or hundreds of electron volts range. In insulators catastrophic damage may result from enormous Coulomb forces produced by the disruption of local charge neutrality in the wake of an energetic ion. This so called Coulomb explosion effect has been suggested as one of the main radiation damage mechanisms in cases where electronic energy loss is very high during ion implantation into insulators.

Finally, it is worth noting that ion-solid interactions occur on time scales which do not neccesarily allow the system to relax to it's equilibrium thermodynamic state and hence the resultant structure may still be metastable.

II.3 Electron Irradiation Effects.

When an energetic electron impinges on a solid the processes by which it transfers energy to the solid may again be roughly categorized as electronic or nuclear. Looking at the electronic interactions, the nature of the energy transfer process is very similar to that of electronic stopping of ions in solids; i.e. excitation and ionization of both core and valence target electrons. Secondary electron cascades may also result with typical energies in the range ten to a hundred electron volts. For electron kinetic energies E_k around 5 keV the amount of energy lost in each scattering process relative to E_k is quite large such that the number of encounters between projectile and target electrons are fewer. Therefore, the statistical fluctuations in range (i.e. straggling) of the incident electrons are greater than in the case of ion implantation. Furthermore, the density and overlap of the secondary electron cascades are expected to be less for electron irradiation, since the energy loss per unit path length is much lower than in ion implantation.

In crystalline materials, a wide variety of defects may result from the above described processes including bond breaking, creation of colour centres and charge trapping.

The effects of electron irradiation in amorphous materials do not appear to have been researched extensively. However, photoluminescence studies of electron irradiated amorphous hydrogenated silicon have indicated the creation of a thermally annealable metastable defect which quenches the photoluminescence intensity around 1.5 eV [9].

With regard to elastic collisions with target nuclei, kinematic calculations show that electron energies in the kilo-electron volt range are sufficient to produce atomic displacement of hydrogen due to it's low mass and displacement energy. However the cross-sections for these collisions are expected to be very small ($\sim 10^{-20}$ cm^2) and have been ruled out in explaining defect production in a-Si:H since the measured damage threshold energies for a-Si:H and a-Si:D are about equal, in contradiction to the expectation of a higher threshold for a-Si:D due to the more massive deutereum nucleus [9].

III Analysis Techniques and Underlying Theory

III.1 Hydrogen Content

The most pronounced effect in a-C:H resulting from energy deposition is the depletion of hydrogen. It is therefore of prime importance to be able to determine the hydrogen content of the specimen at various stages of it's treatment. The techniques employed for evaluation of hydrogen content can be categorized into two groups. Those in which the amount of hydrogen *lost* from the sample is determined and those in which the amount of hydrogen *remaining* in the film is measured.

III.1.1 Hydrogen evolution.

Hydrogen loss measurements are best carried out by employing Residual Gas Analysis (RGA) techniques. In these, a mass spectrometer is tuned for the required atomic mass so as to allow the collection of the selected species emitted from the sample. With proper callibration, this technique can be made quantitative. It yields information not only on the amount of H lost, but also on the chemical form in which the hydrogen leaves the sample. (As will be discussed at a later stage, it has been verified that hydrogen effusion from a-C:H films is mainly in molecular form). Such techniques however, yield no information about the spatial distribution of hydrogen in the sample.

III.1.2 Hydrogen concentration.

Hydrogen concentration measurements are mostly carried out by the use of one of several Ion Beam Analysis (IBA) techniques. In these, the sample is bombarded by a beam of charged particles. The products resulting from the interaction between the probing projectile and the hydrogen (or deutereum) in the film are detected. These may be in the form of charged particle or gamma ray emission following a nuclear reaction, forward scattered hydrogen atoms or backward scattered projectiles resulting from Rutherford scattering events. The major disadvantage of these techniques is that during the ion beam probing process, the energy deposited in the film may itself alter the hydrogen content. Hence, most IBA techniques are destructive to a greater or lesser extent, and the measurement process itself may influence the results. This may be particularly severe when energetic heavy ions are used. Nevertheless, this has been turned to the advantage of the researcher in a number of cases in which the reaction inducing probing ion is the one that causes the changes which are to be investigated.

The major nuclear reactions applicable to hydrogen detection have been listed by Feldman [10], and are based on carrying out the "inverse" of well known proton or deutereum induced nuclear reactions, hence they often require bombardment by MeV light (< 20 amu) projectiles. If the nuclear reaction has a sharp resonance or threshold at a particular collision energy, then by varying the projectile energy and monitoring the reaction product yield, quantitative depth profiling with high depth resolution can be carried out. Among the most commonly used nuclear reactions are the $^{1}H(^{15}N, \alpha\gamma)^{12}C$ mainly developed by Lanford and coworkers [11], the $^{2}D(^{3}He,p)\alpha$, the $^{1}H(^{19}F, \alpha\gamma)^{16}O$ and the $^{1}H(^{18}O, \alpha)^{15}N$ reactions.

Ion scattering experiments have also been used for hydrogen profiling. In the Elastic Recoil Detection (ERD) technique MeV heavy ions are used to forward scatter light constituents out of the near surface layers of the target. The energy of the detected recoiling particles is converted into a depth from which they have been ejected by using known kinematics and stopping powers. An alternative ion scattering approach is to perform standard Rutherford Backscattering Spectrometry (RBS). In this case light target constituents are not directly observable, however their presence is reflected in the RBS spectra by variations in the backscattering yields from heavy constituents, resulting from the stopping power dependence on composition. This method is described in more detail in reference [12].

III.1.3 Hydrogen concentration by infrared absorption.

Information on concentration and bonding of hydrogen may be obtained from IR absorption measurements. This technique yields only relative information and therefore needs to be callibrated against another method, limiting the accuracy of the measurements. Furthermore, it is not depth sensitive. The advantages of this technique are that it is totally non-destructive and that it may yield information on the chemical bonding of the hydrogen in the matrix. Dischler et. al. [13] have well characterized a-C:H using this technique while Gonzalez-Hernandez and coworkers [14] have used IR to estimate hydrogen content in ion beam irradiated a-C:H.

III.2 Electron Spin Resonance

Electron Paramagnetic Resonance is a form of spectroscopy in which an oscillating electromagnetic field induces magnetic dipole transitions between the energy levels of a system of paramagnets. These energy levels are produced by a second static (or slowly varying) magnetic field via the Zeeman interaction. In the simplest case, the resonance condition is given by

$$h\nu = g\beta H \tag{4}$$

where H is the magnitude of the local (static + internal) magnetic field, β is the Bohr magneton and g is a dimensionless constant which is determined by the magnetic dipole moment. In ESR the paramagnets are of electronic origin, that is they result from the orbital or spin angular momentum of free or bound unpaired electrons for which g is very close to 2 ($g_{free} = 2.0023$).

The most important information obtained from ESR measurements on amorphous materials such as a-C:H are usually with regard to structural defects such as dangling bonds [15]. The intensity of the absorption peak may be related to the number (i.e. density) of such defects in the material by comparison with the signal from a standard sample with a known number of spins [16]. Furthermore, the shape and width of the signal may be related to the nature of interaction between the spins (dipolar or exchange) [17]. The interpretation of these interactions in terms of strutural information is non-trivial but may yield important results.

III.3 Optical Measurements

III.3.1 Optical Absorption.

Optical absorption is an important characterization method for amorphous materials, since the absorption edge is generally determined by optical transitions between defect states in the band tails, the density and position in energy of which are of great interest [18]. The optical gap may be estimated if a *Tauc plot* is applied to measurements of optical absorption versus photon energy [19].

The Tauc model of optical absorption in amorphous semiconductors assumes that due to disorder there is no conservation of electron momentum in optical transitions. Based on this assumption, and a parabolic density of states in the conduction and valence bands, Tauc has shown that the absorption dependence on photon energy E should be given by

$$\sqrt{\alpha E} = G(E - E_g) \tag{5}$$

where E_g is the energy difference between the conduction and valence band edges. However, if the bands are not entirely parabolic but posess tails, the densities of which diminish exponentially into the gap, then the optical gap E_g will generally be smaller than the mobility gap as determined by transport measurements.

III.3.2 Raman Scattering.

Raman scattering measurements provide a means of studying the lattice dynamics of disordered systems, giving insight into the nature of disorder. In the Raman effect, a photon is scattered inelastically by a crystal (solid) causing the annihilation or creation of a phonon. There are two important terms in the expression for the first-order Raman scattering intensity which may be related to the microstructure of the scattering solid. The first is the effective coupling parameter which represents the coupling of the light to the lattice vibrations of the solid. The second is the density of phonon states $\rho(\omega)$ and is determined primarily by the short range order in the solid. Hence it's coarse features are essentially the same in crystalline and non-crystalline materials. However, in amorphous materials the breakdown of momentum conservation means that $\rho(\omega)$ modulates the scattering intensity over it's whole energy range. It is the most important term with respect to sensitivity to structural modification since local variations and distortions in bond angles and lengths as well as the presence of specific microstructures such as rings alter the position or width of peaks in the density of states.

The interpretation of Raman spectra in terms of microstructure is by no means straightforward. However, extensive experimental as well as theoretical work has been carried out on a wide variety of materials including a-C:H so that new experimental results may be analysed by comparison with results from the literature. To illustrate this, figure 3 shows the Raman spectrum of a number of materials, including a-C:H, the most significant features being the G (graphite) peak at 1350 cm^{-1} and the D (damage) peak at 1550 cm^{-1}.

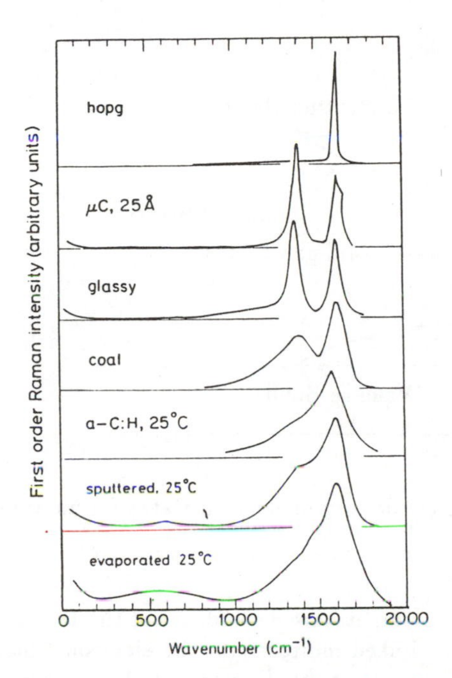

Figure 3: Raman spectra obtained from various carbon based materials; adapted from Robertson [20].

III.3.3 Photoluminescence.

When an incident photon has an energy greater than the band gap of a semiconductor, it is absorbed created an electron-hole pair which proceeds to recombine. The recombination process may be radiative, non-radiative or a combination of both. Spectroscopic analysis of the emitted radiation, in particular from amorphous materials, may yield information on characteristic defect states within the gap. Luminescence in hard a-C:H is of weak intensity, however stronger signals have been reported from more polymer-like a-C:H,due to the lower density of mid-gap states [21].

III.4 Electrical Measurements

Figure 4 shows a typical structure of the density of electron states in an amorphous semiconductor. In the case of a-C:H the above picture is considerably complicated by the ability of carbon to form both σ and π type bonding sites which, in principle, allows both σ and π bonding defects. However, the weaker π bonds reduce the energy separation between bonding and anti-bonding states as compared with σ bonding and so in reality it is the π defect states are those which predominate in the band tails [20].

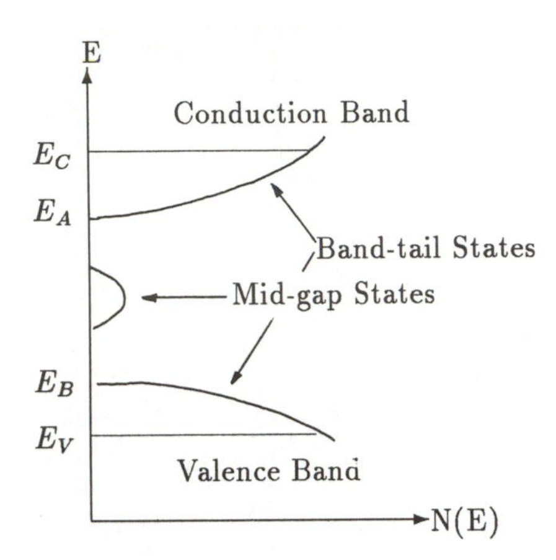

Figure 4: Typical density of electron states for an amorphous material.

III.4.1 Resistivity.

The measurement of the temperature dependence of the DC electrical resistivity is simple in principle and may be linked indirectly to the electronic band-structure. Assuming a density of states (DOS) similar to that described above, there are a number of different conduction paths which may dominate, depending on temperature [22].

a) At high temperatures, carriers may be excited beyond the mobility edges into *extended states*. The temperature dependence of the conductivity will be similar to that in crystalline semiconductors, i.e.

$$\sigma = \sigma_0 exp[-(E_C - E_F)/kT] \tag{6}$$

where σ_0 is given approximately by

$$\sigma_0 = eN(E_C)kT\mu_0. \tag{7}$$

In this expression $N(E_C)$ is the density of states at E_C and μ_0 is the mobility of carriers in extended states.

b) Conduction may be via charge carriers excited into localized states at the *band edges*, i.e. at E_A or E_B; see figure 4. The availability of thermal energy allows electrons (in the case of n-type conductivity) trapped in localized states to hop (tunnel) to nearby available states with an activation energy W_1. In this case the conductivity is given by

$$\sigma = \sigma_1 exp[-(E_A - E_F + W_1)/kT]. \tag{8}$$

σ_1 is typically $10^2 - 10^4$ times smaller than σ_0 primarily due to the much lower mobility of the localized states.

c) If there is a significant density of states *around the Fermi energy* then conduction may occur by hopping between localized states at or near the Fermi energy. In this case, at higher temperatures the electron (or hole) will jump to its nearest neighbour state and the temperature dependence of the conductivity is given by

$$\sigma = \sigma_2 exp(-W_2/kT), \tag{9}$$

where W_2 is the activation energy for hopping. At lower temperatures (or higher DOS) however, it may be favourable for carriers to jump to a further site which is closer in energy. A detailed calculation, due to Mott [22] shows that in this case (so called *variable range hopping*) the temperature dependence of the conductivity is given by

$$\sigma = \sigma_3 exp(-T_0/T)^{1/4}, \tag{10}$$

where

$$T_0 \approx 16/(\lambda^3 k N(E_F)). \tag{11}$$

Here, λ is the characteristic decay length of the localized wave function at the Fermi energy.

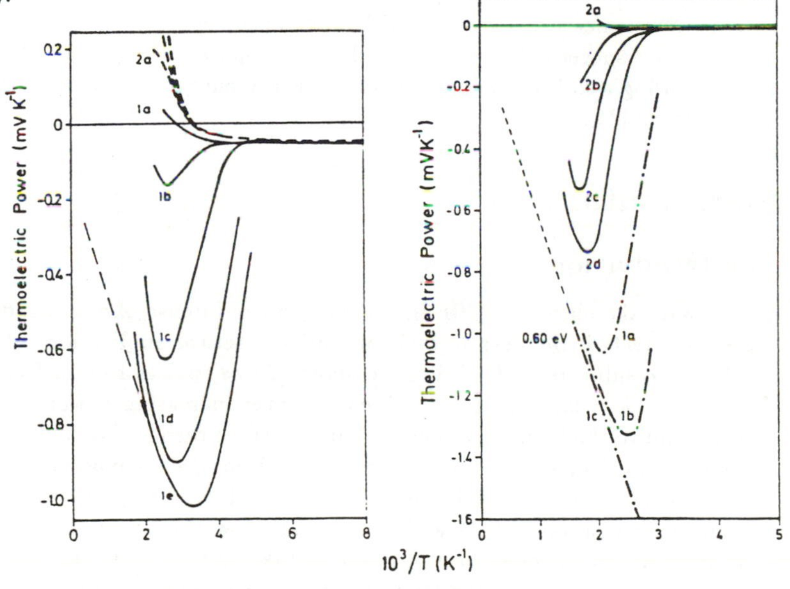

Figure 5: Inverse temperature dependence of thermopower S for a-Ge (left) and a-Si (right) for various annealing conditions, adapted from Beyer [23].

III.4.2 Thermopower.

In order to identify the dominant conduction mechanism or to decide whether two conduction processes participate in parallel, Hall effect or thermo-electric power measurements are usually carried out in addition to the conductivity measurements. Unfortunately, Hall effect measurements in non-crystalline materials are often difficult since Hall voltages are very small due to low mobilities and since the interpretation of the Hall voltage in terms of the sign of the charge carriers is often ambiguous. On the other hand thermopower measurements are more straight forward. The thermopower S is measured by the ratio $\frac{\triangle V}{\triangle T}$ of the voltage $\triangle V$ developed across two points in the material held at a small temperature difference $\triangle T$. If conduction is predominantly due to charge carriers in extended states above the mobility edge (i.e. $E > E_C$ for n-type or $E < E_V$ for p-type) then the thermopower is typically of the order ~ 1 mV/°K and is inversely dependent on temperature. When conduction is dominated by states near E_F (metallic conduction) it can be shown that thermopowers are much smaller ($\sim$ tens of μV/°K), are temperature independent and the sign is an indication of the slope of the density of states $N(E)$ at the Fermi energy E_F.

As an example figure 5 shows the temperature dependence of S for a-Si in which two distinct regimes are apparent, the temperature independent regime corresponding to hopping conductivity around the Fermi energy giving way to extended state conductivity at higher temperatures.

Very few thermopower measurements on a-C:H have been reported in the literature. Meyerson and Smith have succeeded in measuring the thermopower of doped a-C:H which was interpreted as showing a shift in the position of the Fermi energy as a result of doping by boron or phosphorous [24].

IV Experimental

IV.1 Laser Irradiation

The modifications that take place in a-C:H films as a result of intense photon irradiation have been studied relatively little, despite their potential practical applications. Prawer and coworkers [25] have subjected a-C:H films to short (70 ns) pulses from a frequency doubled Nd YAG laser ($\lambda = 0.53 \mu$m, E_{ph}=2.35 eV) at ever increasing power densities. The response of the film to the laser could be divided into three regimes. Below a threshold of about 0.2 J/cm^2, no changes could be detected in the films even if exposed to a large number of pulses. Hence it seems as if at low energy densities there is no observable interaction between the laser radiation and the a-C:H material. However, above this threshold a transformation from a-C:H to some form of graphite takes place. With increasing energy density, graphitization, accompanied by ablation proceeds until at sufficiently high energy density (1.5 J/cm^2 for a 1 μm thick layer) the material is totally ablated. Hence sharp trenches with some "spill over" shoulders can be created in the coating.

By subjecting the transformed material to some of the diagnostic methods described above, it was found that the irradiated material exhibits some physical properties which are characteristic of graphite: it has a low electrical resistivity of 0.2 Ωcm (as compared to 10^7 Ωcm for the untreated material) and it's Raman spectrum, displayed in figure 22 exhibits graphitic features, as will be discussed in section V. The infrared absorption in the 400 - 4000 cm^{-1} range is increased by 20 %, and the morphology and hardness of the film deteriorate. These features resemble very closely those reported [26] for a-C:H heated to 400 - 500°C. Hence it was concluded that the pulsed laser irradiation has locally heated the a-C:H to $\sim$ 500°C, a temperature at which partial graphitization takes place.

There are some technological implications to the laser irradiation induced modification of a-C:H. By judicious choice of the laser parameters it is possible to produce conducting pathways or well defined channels in the a-C:H layer. The former may have applications in device manufacture of micrometer scale thin-film resistors, while the latter may be used for high resolution lithography [27]. Rothschild and Ehrlich [28] have exposed a-C:H films to UV ($\lambda = .193\mu$m) radiation from an ArF excimer laser and have demonstrated that extremely fine lines can be dry etched in the film used as a "photo resist". As in the case for visible radiation, their results show that a threshold in energy density exists above which thin a-C:H films are totally ablated. At energy densities of 0.13 J/cm^{-2}, a single 20 ns pulse was found to be sufficient to totally remove a $\sim$ 1000 Å thick a-C:H layer. This energy density is lower than that required to induce similar effects with visible radiation, probably due to the higher photon energy and hence the smaller absorption depth for the radiation. a-C:H films have also been exposed to laser radiation of energy well below the absorption edge. Dischler [29] has used 10.6 μm radiation and has found that the damage threshold depends on the laser power density (damage threshold $\sim$ 300 MW/cm^{-2}). In this case it seems as if no simple direct absorption is responsible for the damage, but rather more complex power dependent processes must be considered.

IV.2 Ion Irradiation

The changes that occur in a-C:H films by ion irradiation have recently been studied by several groups. Hence in this field there now exists a rather clear picture of the effects that the irradiation has on the material structure and composition. The most relevant parameters in these studies are the ion mass and energy (both of which determine the dE/dx values), the ion fluence and the a-C:H film temperature during irradiation. Below, the currently available data will be briefly reviewed, concentrating on the results of the Technion group which has carried out an extensive set of measurements, employing most of the diagnostic techniques described above.

IV.2.1 DC Conductivity.

Resistivity changes as a result of ion irradiation are possibly the most spectacular of ion induced effects. A wide variety of ions have been implanted into a-C:H and the resulting changes in conductivity were measured. Figure 6 shows the dose dependence of the resistivity of a-C:H films irradiated with C^+ (50 keV), Ar^+ (110 keV), and Xe^+ (270 keV). The ion energies were chosen so that the projected range in each case was about 1000 Å. The experiments were carried out with the samples held at 100 °C so as to avoid blistering which otherwise occured for high dose noble gas implantations. The room temperature resistivity of the unirradiated films, ρ_0, was typically about 10^7 Ωcm.

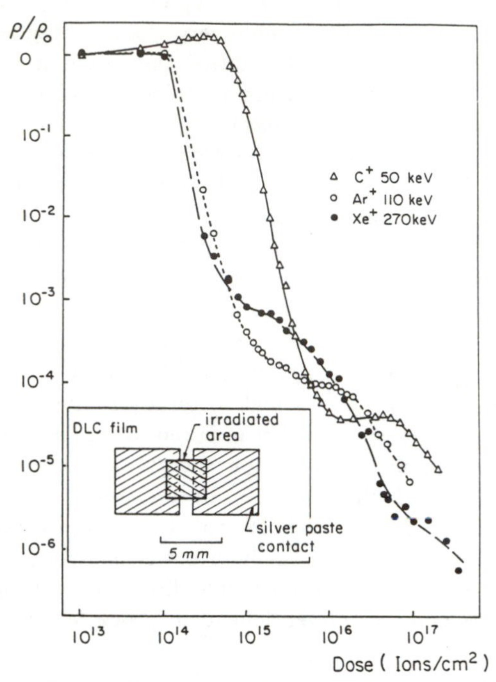

Figure 6: Ion dose dependence of resistivity for a) C^+, Ar^+ and Xe^+ irradiation of an a-C:H film. Inset : the irradiation geometry. Taken from [31].

The dose dependence of the resistivity is qualitatively the same for each ion studied, i.e. above a certain critical dose, the resistivity drops rapidly ($\rho \sim D^{-4}$) until it reaches an intermediate saturation level which appears as a knee in the curve. With still greater doses a further decrease in ρ is observed. It is also worth noting that for the C^+ implantation there is a small, yet definite resistivity *increase* for fluences below the threshold for the major resistivity drop.

Ingram and coworkers [30] have also measured the DC conductivity dependence on fluence in a similar geometry, but for 6.4 MeV flourine ions and 1 MeV gold ions. Their irradiation parameters were chosen so that both ions deposit roughly the same amount of energy via electronic processes, yet the gold ions deposited some 350 times more energy to nuclear processes than the flourine ions. Their results are shown in figure 7 and display a similar threshold behavior in the onset of the conductivity increase. However the resistivity drop following this onset differs drastically for the two cases; the flourine irradiation results in a drop in resistivity by some nine orders of magnitude while the resistivity changes for the gold irradiation is much less pronounced, leveling off after a reduction of only two or three orders of magnitude.

IV.2.2 DC Conductivity: Temperature Dependence.

The large increase in conductivity measured as a function of irradiation dose invites closer inspection. The study of it's temperature dependence may help in the understanding of the physics of the conduction mechanism. The temperature dependence of the conduc-

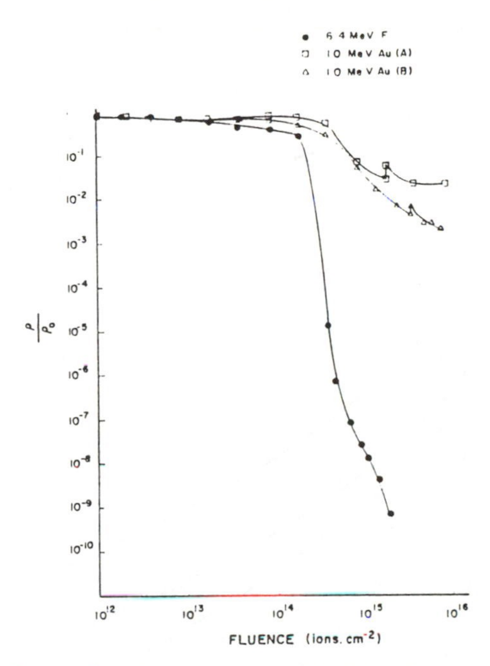

Figure 7: Ion dose dependence of resistivity for 6.4 MeV flourine and 1 MeV gold, taken from Ingram et. al. [30].

tivity [$\sigma(T)$] for samples irradiated at different doses is shown in Figure 8 where $\ln\sigma$ has been plotted versus $1/T$. Samples irradiated with doses of less than 1 $\times 10^{15}$ C^+ ions cm^{-2} have $\sigma(T)$ dependences virtually indistinguishable from the unirradiated film and are therefore not included in Figure 8. Also shown in the figure is the $\sigma(T)$ dependence for a sample irradiated at a higher temperature of 220 °C. This sample has, as indicated by Raman measurements, undergone significant graphitization.

IV.2.3 Hydrogen Evolution.

Many researchers have recently measured the effects of ion irradiation on hydrogen content. Below, the measurements of Prawer and coworkers [31] are presented in some detail since they can be best correlated to the other measurements done by the group. In these measurements, the hydrogen content of the as-grown and irradiated films was determined by Rutherford backscattering. To this end, a-C:H films were deposited on polished polygraphite or highly oriented pyrolytic graphite substrates. The probing was done using 320 keV protons backscattered into a surface barrier detector at 165°. Numerical estimates of the hydrogen content were obtained using the simulation and fitting program RUMP. Details can be found in references [32, 33]. In addition, in the absence of incorporated oxygen, the hydrogen content was also calculated by comparing the yield from a-C:H and that from pure carbon and good agreement with the values deduced from RUMP were obtained. By these methods it was possible to measure the hydrogen content of the films to ± 5 at. %. Figures 9 and 10 show RBS spectra from as-grown (36 at. % H) and

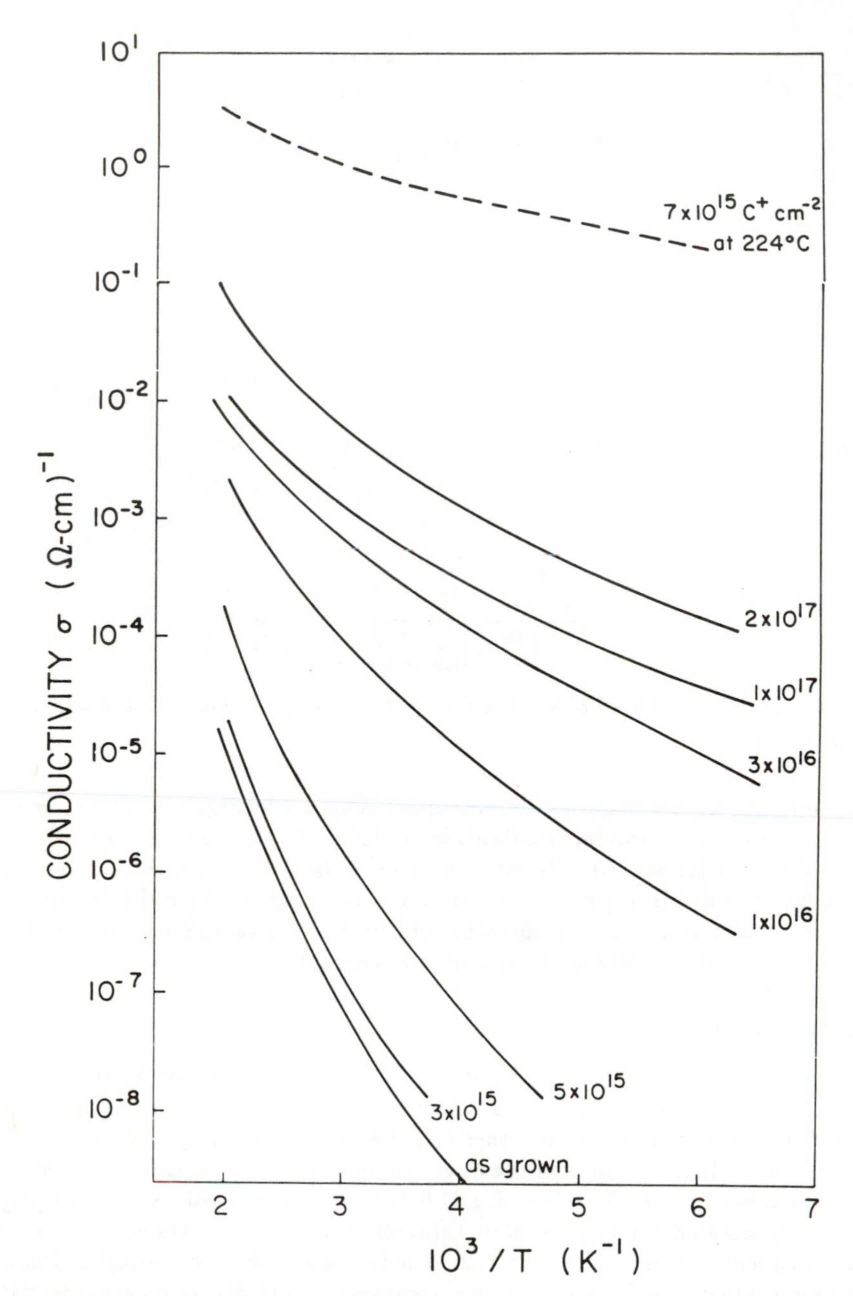

Figure 8: Inverse temperature dependence of the conductivity (σ) for samples irradiated with different doses of C^+ ions. The broken line is from a sample irradiated with 7×10^{15} C^+ cm^{-2} at 220 °C, taken from [31].

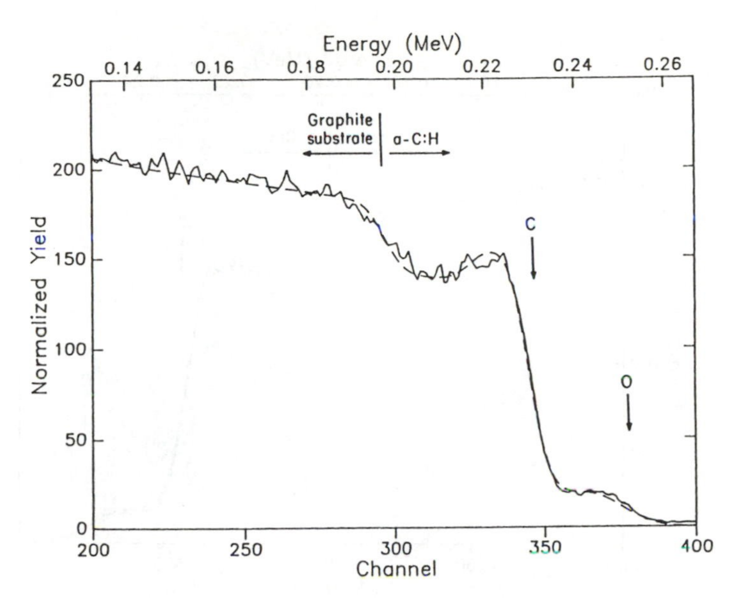

Figure 9: RBS spectrum (solid line) of as-grown a-C:H on a graphite substrate. The broken line is a simulation fit of the data. Adapted from [31].

irradiated (22 at. % H) a-C:H, including fits using RUMP. A comparison of the spectra demonstrates that the technique is indeed sensitive to hydrogen content.

In Figure 11 the hydrogen content of an a-C:H film is plotted as a function of irradiation dose with 50 keV C^+ ions at 100 °C. It is apparent that at a dose of about 10^{15} C^+ cm^{-2} hydrogen starts to effuse out of the film until above 10^{17} C^+ cm^{-2} nearly all the hydrogen has been removed.

Several other authors have carried out similar ion induced hydrogen evolution studies, covering a wide range of ion species and energies. As an example of the more precise hydrogen profiling techniques employing nuclear reaction analysis, figure 12 shows data from Baumann and coworkers as obtained by the use of the $^{1}H(^{15}N, \alpha\gamma)^{12}C$ nuclear reaction. In this case the probing ion was also responsible for the ion induced hydrogen release. The functional dependence of the remaining hydrogen content on the ion fluence is similar to that caused by the carbon irradiation. A model describing the ion induced hydrogen release process which summarizes and correlates all other available data will be given in the discussion.

IV.2.4 Optical Band Gap.

Concomitant with the increase in conductivity, a change in the colour of the film from light brown to shiny black was observed. This indicated a decrease in the optical band gap caused by the ion bombardment. In order to measure this decrease quantitatively, the photon energy dependence of the optical absorption coefficient, α, was measured for

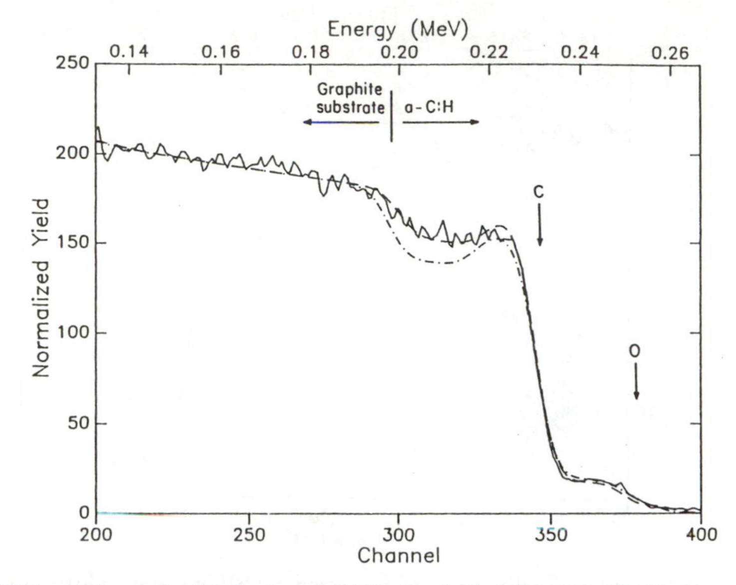

Figure 10: RBS spectrum (solid line) and RBS simulation (broken line) of an a-C:H film deposited on graphite which has been irradiated with 3×10^{15} C^+ cm^{-2} at 50 keV. For comparison, the simulation from the data above is also included. Adapted from [31].

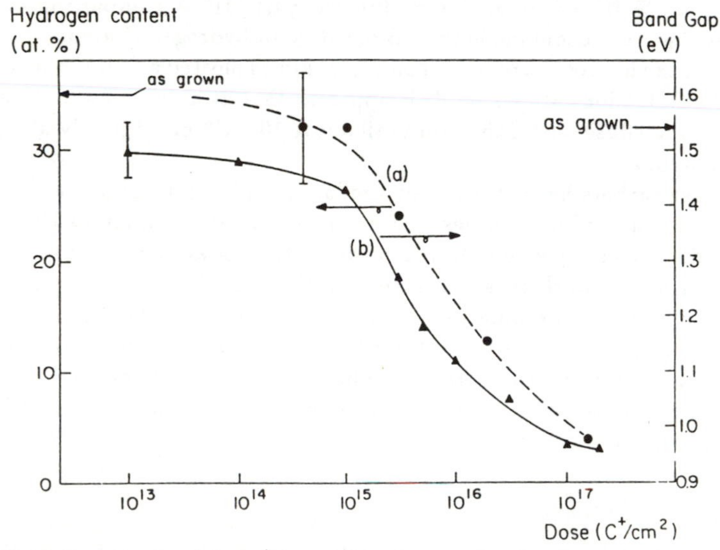

Figure 11: a) Hydrogen content as determined by RBS versus dose for a-C:H irradiated with 50 keV C^+ ions. b) Optical band gap as determined by Tauc plots versus dose for a-C:H irradiated with 50 keV C^+ ions. Adapted from [31].

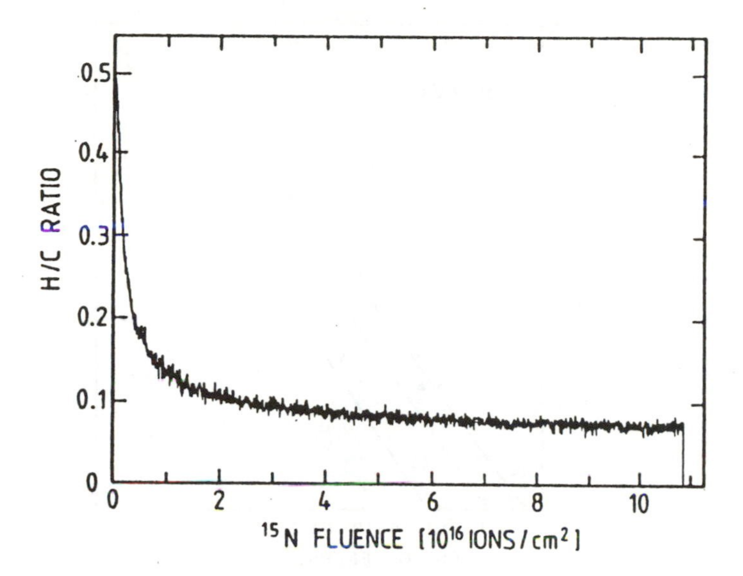

Figure 12: Hydrogen content as determined by NRA versus dose for a-C:H irradiated with 7 MeV N^+ ions. Taken from Baumann et. al.[34]

a-C:H samples irradiated with varying doses of 50 keV C^+ at 100 ± 20 °C. Representative data for three different doses is shown in Figure 13 where $(\alpha E)^{1/2}$ has been plotted versus E to show that, except for the lowest tail, the absorption edge follows the Tauc relation as described above. In figure 11 the band-gap determined from such plots is given as a function of irradiation dose.

A more detailed optical study of ion irradiated a-C:H has been carried out recently by Orzeszko, Woollam, Ingram and McCormick [35]. Using ellipsometric analysis, they have determined the real and imaginary parts of the dielectric function and their variation with ion fluence, for flourine and gold ions under the same conditions as for their conductivity measurements. In this case too a clear correlation between the reduction in band gap with ion dose and the increase in conductivity is evident.

IV.2.5 Raman Spectra.

Prawer and coworkers have carried out Raman spectroscopy in order to investigate modifications in microstructure resulting from ion irradiation of a-C:H. Representative Raman spectra for films irradiated at 100 °C are shown in figure 14. The as-grown films show a Raman peak at 1540 ± 6 cm^{-1} with a FWHM of about 150 cm^{-1}. The irradiated films do not show any significant shift of the G peak, however the D peak appears to be slightly enhanced as compared to the as-grown films. This is interpreted as showing that no large microcrystalline growth ($\sim$ 200 Å) has taken place, however subtle structural modification cannot be ruled out. Figure 15 shows Raman spectra of films irradiated to an intermediate dose of 50 keV C^+ ions at different substrate temperatures. As the temperature is raised above about 100 °C there is a clear upward shift and narrowing of the

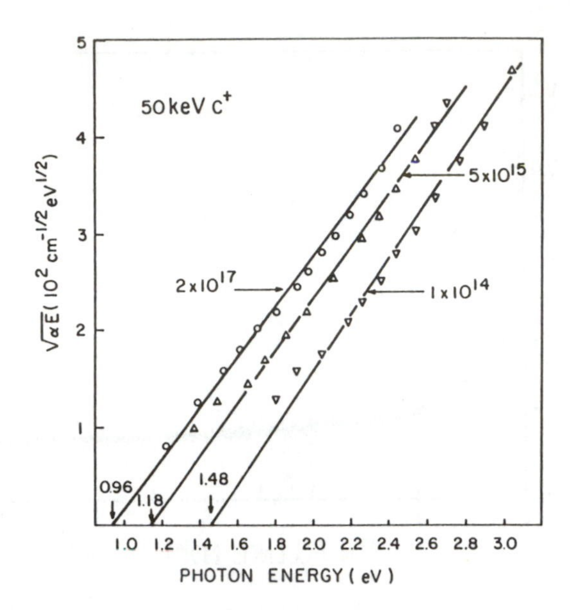

Figure 13: Tauc plots of the photon energy (E) dependence of the optical absorption coefficient (α) for three different doses (ions/cm^2) of C^+ irradiation of a-C:H films at 50 keV. Adapted from [31].

G peak together with a growth in the relative intensity of thge of the D peak compared with the G peak.

IV.2.6 Photoluminescence.

Photoluminescence in a-C:H has also been observed to be effected by ion bombardment. Gonzalez-Hernandez and coworkers investigated the PL intensity of a-C:H films irradiated with both hydrogen and nitrogen ions in the energy range beween 100 and 600 keV. Figure 16 shows the PL intensity dependence on photon energy for films irradiated with various doses of 100 keV hydrogen ions. It is interesting to note that for low fluences the PL intensity increases by a factor of 2 while for higher dose irradiations the signal is quenched, allowing the observation of the Raman peak.

IV.2.7 Electron Spin Resonance.

As was already described the nature of the ion irradiation process is such that an increase in structural defects is expected as a result of the high dose bombardments. It is well known that Electron Spin Resonance (ESR) can provide a measure of the total number of unpaired spins in the sample under study and, if the affected volume is known, the average spin defect density can be estimated. Furthermore, the large ion-induced conductivity increase was also suspected to be related to an increase in the dangling-bond density.

A set of representative derivative spectra are shown in figure 17, for samples deposited on silcon and implanted with 50 keV C^+ ions at 100°C, as described in reference [36].

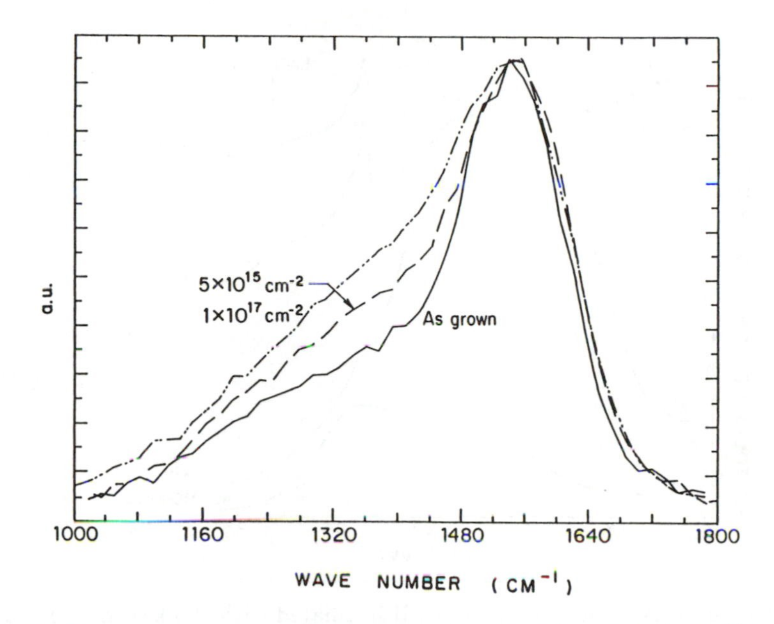

Figure 14: Raman spectra for a-C:H irradiated at 100 °C with 50 keV C^+ ions to various doses. (I. Sela, unpublished.)

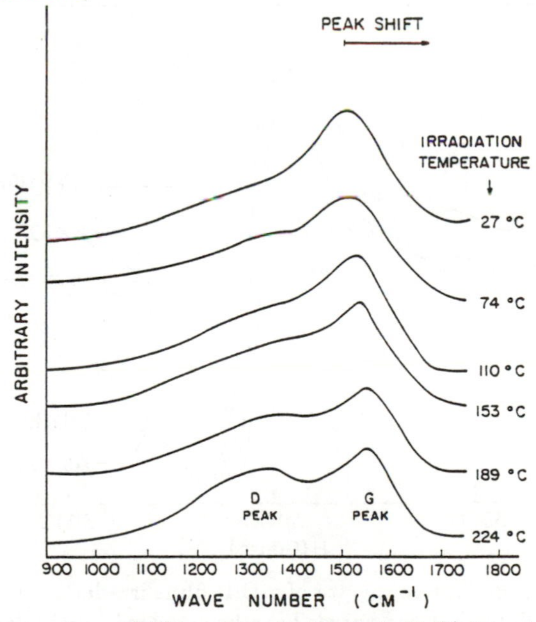

Figure 15: Raman spectra for a-C:H irradiated with 50 keV C^+ ions to a dose of 7×10^{15} ions cm^{-2} at 100 °C at various substrate temperatures. Adapted from [31].

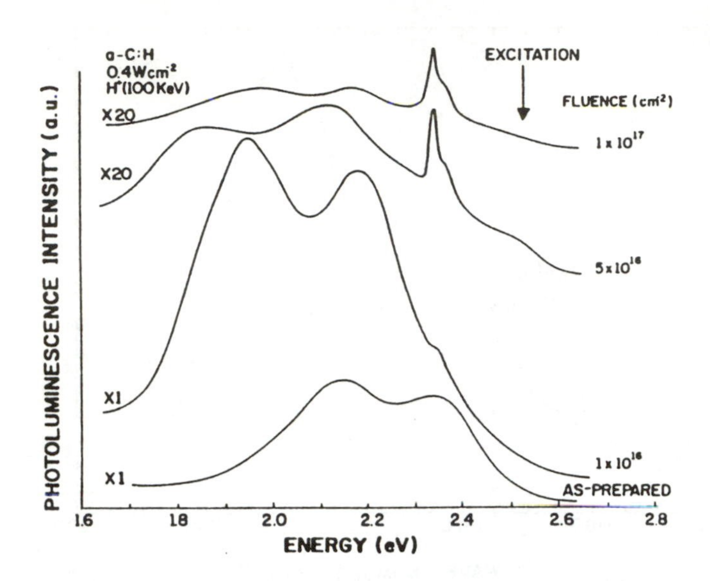

Figure 16: Photoluminescence spectra for a-C:H irradiated with 100 keV H^+ ions to three different fluences, taken from [14].

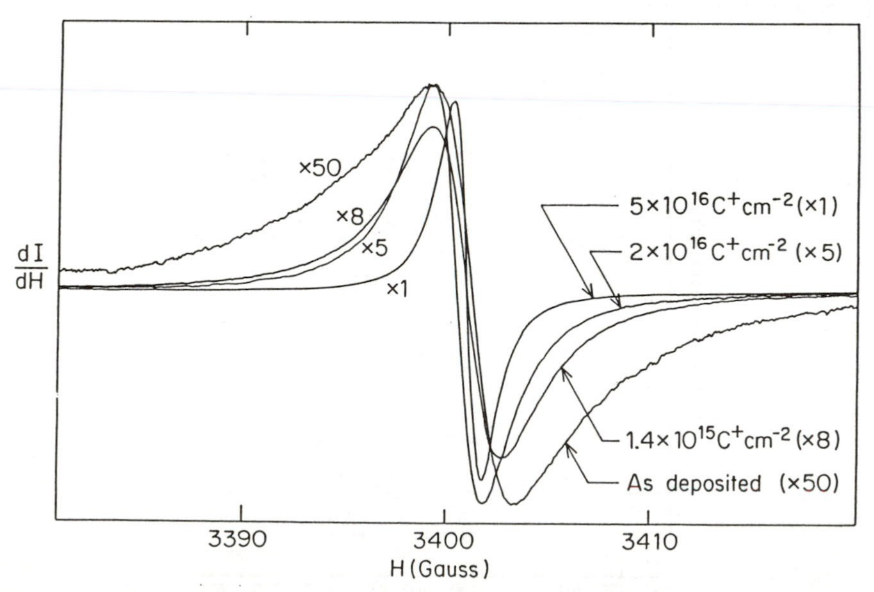

Figure 17: The ESR derivative spectra of a-C:H films irradiated with varying doses of 50 keV carbon ions. Note that the curves have been scaled by the factors indicated in the figure. Adapted from [31].

The slight horizontal displacement of the curves with respect to each other does not reflect any significant change in the g value but rather is due to minor variations in the sample position within the resosnant cavity. The resonace linewidth and spin density obtained from these data are plotted in Figure 18 together with the dose dependence of the resistivity and hydrogen content as described above. The close correlation between the dose dependence of the resistivity and hydrogen content (Fig. 18a), resonance linewidth (Fig.18b), and spin density (Fig.18c) is clearly evident. The scatter in the calculated spin density (Fig.18c) was mainly due to uncertainties inherent in the subtraction of the spin contribution attributable to the unaffected a-C:H layer and variations in the Q of the cavity for different sized samples. Obviously, there is much less error in the determination of the linewidth as this is directly measured from the spectra.

IV.2.8 Thermopower.

The changes in activation energy (i.e. changes in slope of Ahrenious plots) as a function of ion dose described in section IV.2.2 suggest that significant band structure modifications have occured as a result of ion irradiation. Thermopower measurements were therefore carried out by the Technion group in order to investigate further the conduction mechanism.

Reliable, reproducible results could be obtained only for samples with resistances of the order of 10^9 Ω or less, which, for the particular geometry and an effected layer thickness of about 3000 Å puts an upper limit on the material resistivity of about 10^4 Ωcm. To overcome this problem, films were grown with lower resistivities by raising the substrate temperature during deposition. To this end, a-C:H films were deposited on quartz substrates at a temperature of 260 °C, which yielded films with room temperature bulk resistivities around 10^6 Ωcm.

Figure 19 shows the inverse temperature dependence of the thermopower of two a-C:H samples, irradiated by 70 keV C^+ ions to intermediate doses using the same irradiation geometry and conditions as described above. The higher irradiation energy was chosen in order to increase the thickness of the irradiated layer hence reducing the sample resistance, and increasing the thermopower contribution of the irradiated material, relative to the unirradiated layer beneath. The figure clearly indicates the existence of two temperature regimes. Furthermore, the sign of the thermopower in the low temperature regime is different for each case. These results were reproducible, although minor changes in sample resistances resulted from thermal cycling. The as-grown sample was considerably more difficult to measure and no consistent reproducible result was obtained.

IV.3 Electron Irradiation

For the experiments described in this section, a-C:H films 7500 $\pm$ 500 Å thick were deposited on glass and diamond substrates (see ref. [37]). The films were subjected to electron bombardments in an ultra-high vacuum system, based on a scanning Auger microprobe. A beam energy of 5 keV was chosen corresponding to a range of 3000 Å in

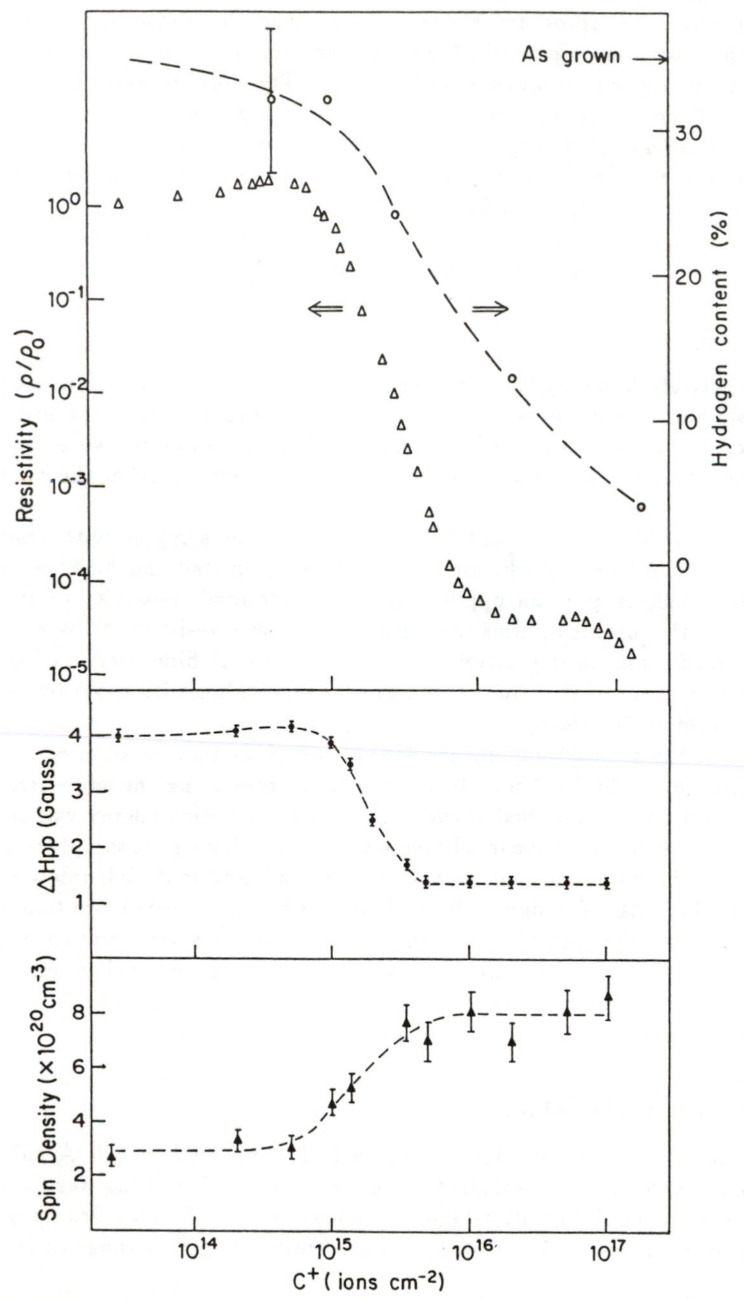

Figure 18: Variation with 50 keV C^+ ion dose of (a) resistivity and hydrogen content, (b) resonant linewidth and (c) spin density, from [31].

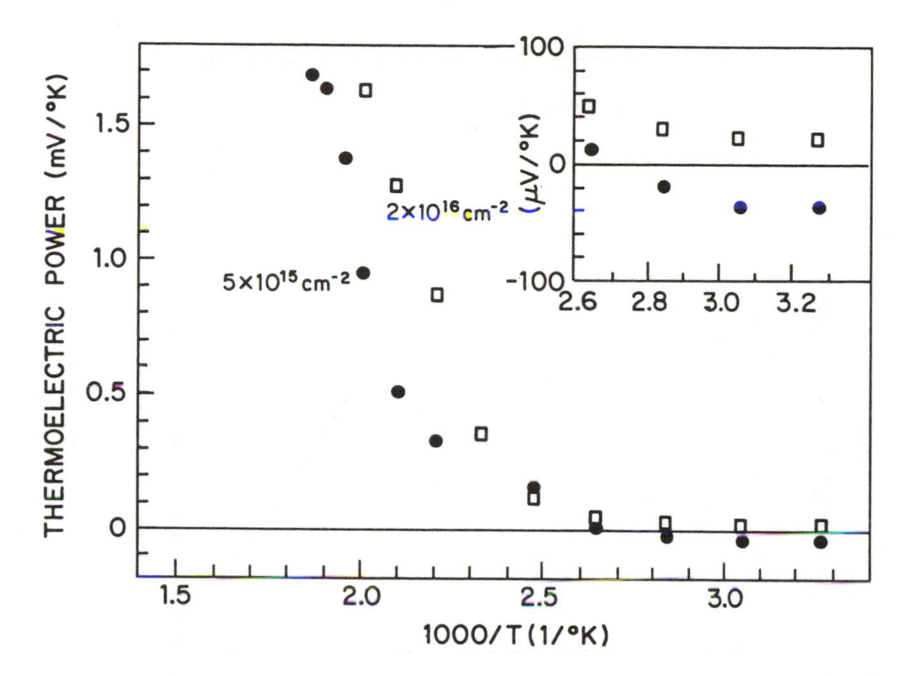

Figure 19: The thermo-electric power for a-C:H films deposited at 260 °C, irradiated to intermediate doses with 70 keV C^+ ions plotted against inverse temperature, Circles - 5×10^{15} ions cm^{-2}; Squares - 2×10^{16} ions cm^{-2}. Taken from [12].

a-C:H. This range is well short of the film thickness, so that any irradiation effects in the substrate may be precluded. Areas, 1×1 mm^2 were irradiated at a current of 5 μA to fluences ranging from 2×10^{17} to 5×10^{19} electrons cm^{-2}. Under these conditions, beam heating of the film is expected not to exceed about 20 °C. Due to this small irradiated area, a number of analysis techniques which require significantly larger irradiation areas (i.e. ESR, thermopower) have not been applied.

IV.3.1 DC Conductivity.

Figure 20a shows the resistivity dependence on electron fluence. The as-grown resistivity for the film was ($\rho_0 \sim 10^7$ Ωcm). No significant conductivity changes are observed below 10^{18} electrons cm^{-2}. Above this threshold, two distinct dose regimes in which the dose dependence of the resistivity is markedly different can be seen. In the first regime between 10^{18} and 10^{19} electrons cm^{-2} the resistivity follows quite closely an inverse exponential dependence [$\exp(-kD)$] on electron dose. In the second regime, from 10^{19} electrons cm^{-2} onwards the dose dependence can be described by a power law (D^{-3}).

IV.3.2 Optical Band Gap.

Optical absorption measurements were performed on the same samples as used for the conductivity measurements. As inferred from RBS measurements, the region of the film

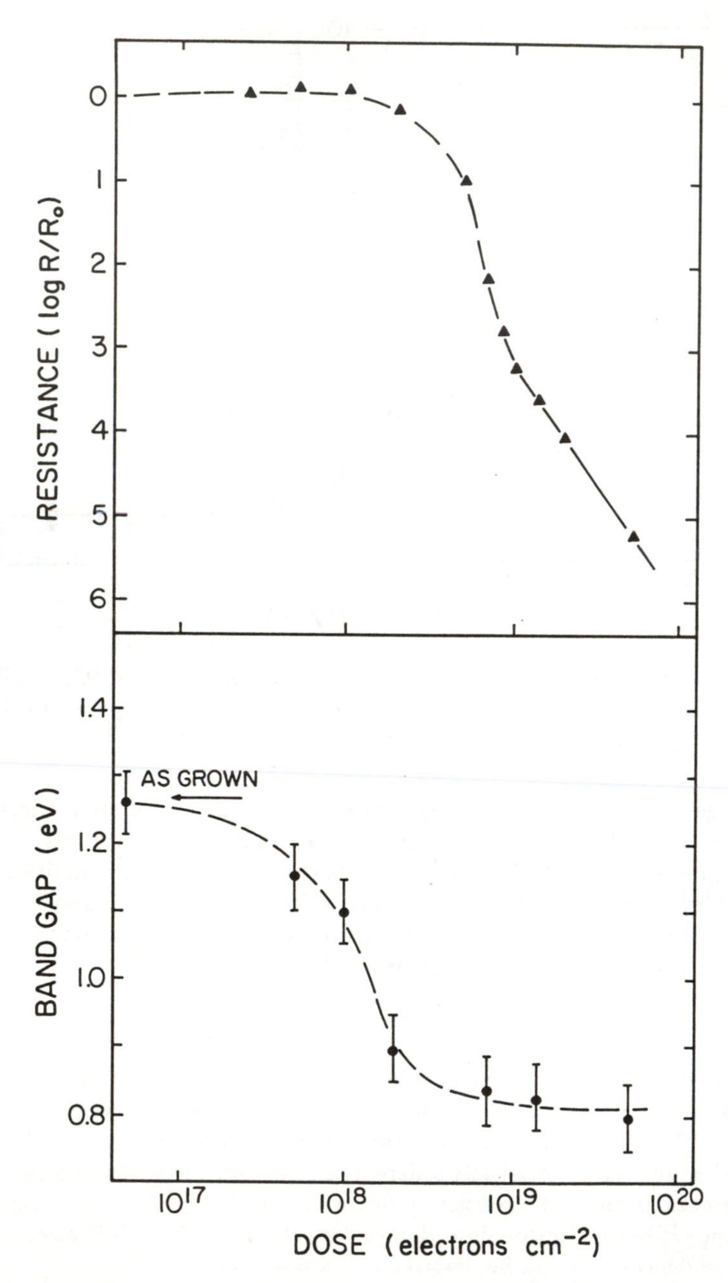

Figure 20: a) Dependence of the resistance R on electron dose irradiations of a-C:H on glass with 5 keV electrons. R_0 was 1.2×10^{11} Ω for the unirradiated a-C:H. b) Optical band gap as determined by Tauc plots versus electron dose for irradiated a-C:H films, from [37].

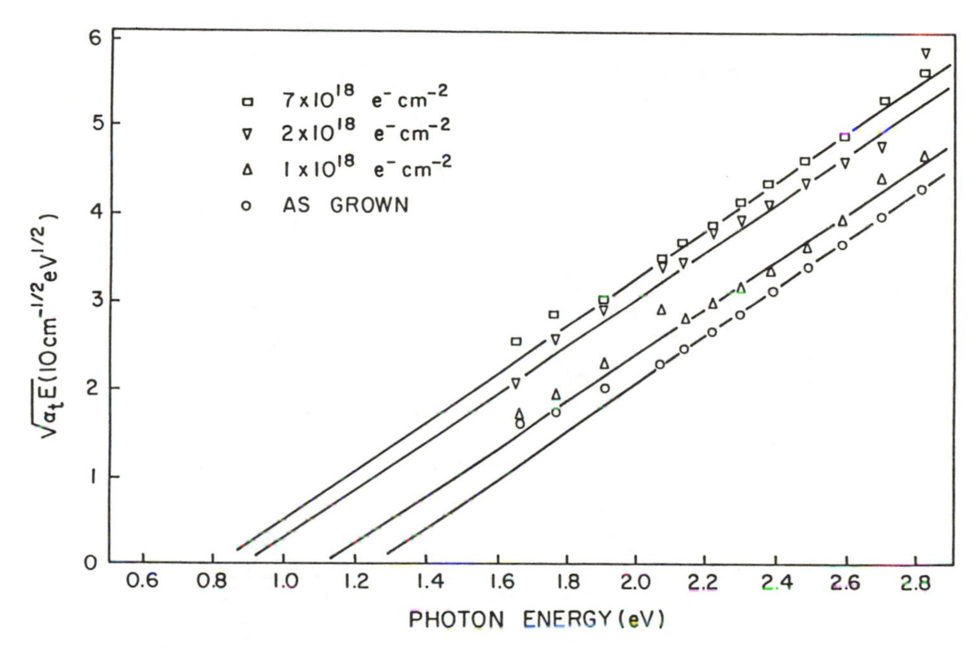

Figure 21: Tauc plots of the photon energy E dependence of the optical absorption coefficient α for a-C:H films irradiated to different electron doses at 5 keV, from [37].

affected by electron irradiation for all but very high doses is of the order of 2000 Å. Hence the net absorption coeffecient for the modified material was obtained by subtracting the contribution to the absorption of the pristine a-C:H beneath. Using this absorption coefficient $\alpha_t(E)$, the optical band gap could be assesed by plotting $(\alpha_t E)^{1/2}$ versus E. As can be seen in Figure 21, except for the low energy tail, the absorption appears to follow the the same Tauc relation as described above. Figure 20b shows the optical band gap obtained in this way, as a function of electron dose. It is quite remarkable that the most significant decrease in band gap occurs at doses below the threshold for conductivity increase.

IV.3.3 Raman Spectra.

Figure 22 demonstrates quite clearly that even after a dose of 2×10^{19} electrons cm^{-2} no significant change in the Raman spectrum from irradiated a-C:H is observable. For comparison, the Raman spectrum of an a-C:H film graphitized by laser treatment is also included. A side-effect observed during initial Raman measurements with excessive beam intensites is that some optical bleaching occured as a result of the probing laser radiation (4880 Å). This interesting observation requires further investigation.

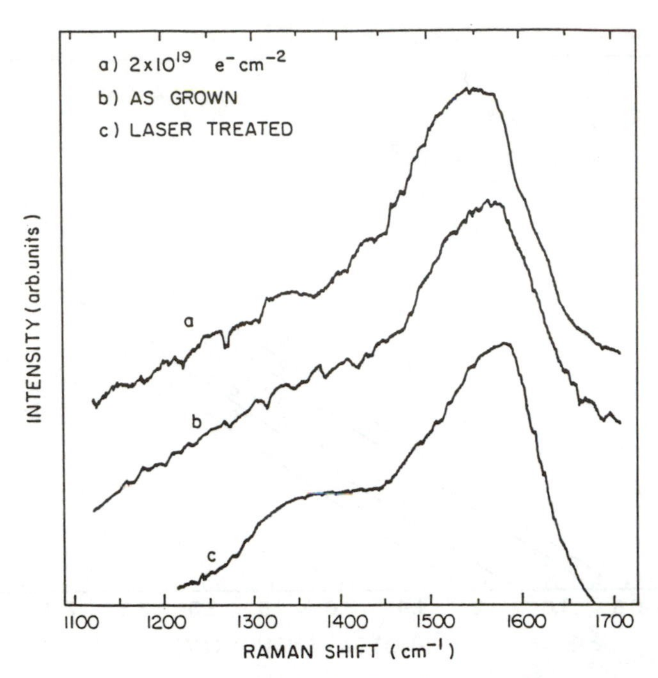

Figure 22: Raman spectra of as-grown, high dose irradiated and laser-treated a-C:H films. Adapted from [37]. See reference [25] or section IV.1 for laser treatment details.

IV.3.4 Hydrogen Evolution.

Rutherford backscattering experiments similar to those described in the previous section were carried out in order to assess the hydrogen content of electron irradiated films. The RBS spectra from a film irradiated with 5×10^{18} electrons cm^{-2} displayed no significant changes as compared with the RBS spectrum from an unirradiated layer. Direct comparison of the experimental spectrum from a sample irradiated to a dose of 1.5×10^{19} electrons cm^{-2} with the computer simulation indicated that less than a 1/4 of the original hydrogen content has been lost from the top 2000 Å. Upon further irradiation, more hydrogen is removed from the film, however well over half of the original hydrogen content is still present in the top layer. The hydrogren depleted region at this high dose appeared to be about 3000 Å thick, with about 25 atomic % of H remaining compared to the as grown value of 40 at. %.

V Discussion

The wealth of results presented above, in particular for the case of ion irradiation effects on a-C:H films, allows us, by comparison with theoretical predictions to draw conclusions regarding the mechanism which causes the transformations induced by energy deposition and their effects on film properties. Each energy deposition process will be discussed seperately however comparisons will be made between them.

V.1 Laser Irradiation

Laser irradiation leads to a transformation of a-C:H which is basically different from the transformations induced by charged particle irradiation. The experimental evidence given in section IV.1 above suggests that partial graphitization of the films takes place when the laser pulse energy is sufficient. The results of Raman scattering provide convincing evidence that the a-C:H layer has undergone a structural transition (turning graphitic), beyond the visible changes in morphology. Large single crystal graphite displays a single sharp Raman line at about 1580 cm^{-1} (the G line) [38]. However, small graphite crystallites (< 200 Å) show an additional disorder induced D peak at 1355 cm^{-1} [2], which is seen to be strongly enhanced by laser irradiation (see figure 22). Comparison with Raman data published on annealed a-C:H films [2] suggests that the laser treated a-C:H layer has reached a temperature of about 400-500 °C during the laser pulse.

Further evidence of a transformation temperature of 400-500 °C was provided by the IR spectrum of a laser treated area, which had a transformed layer approximately 0.5 microns thick. It showed an increase of about 20 % in absorption over the range 4000 - 400 cm^{-1} as compared to the as-grown layers. According to Nadler et al. [26] such an increase is typical of a-C:H films annealed in vacuum at 400-500 °C, whereas films annealed above 800 °C showed very strong absorption in the IR.

Dischler et al. [13] found from IR absorption measurements that films annealed at 400 °C retain a high proportion of diamond-like sp^3 bonds, which rapidly decreased with increasing anneal temperature until at 600 °C the film consisted solely of graphite-like sp^2 bonds. Hence at temperatures between 400 and 500 °C it would seem reasonable to describe the films as being partially graphitized whilst still retaining some of the original diamond-like properties.

V.2 Ion Irradiation

The discussion will be divided into four sections, the first two (H loss and bonding rearrangement) in which structural modifications are compared with theory, followed by two sections (band structure and conduction mechanism)in which electrical modifications are discussed.

V.2.1 Hydogen loss.

The most prominent observable structural modification is the loss of hydrogen as a result of ion bombardment or any other energy deposition mode. Since our original report [39], several studies of ion induced hydrogen release from a-C:H have appeared in the literature [34, 14, 30, 40, 41]. The results of these studies indicate that for films with a typical H concentration of 30 - 40 at. % , substantial H loss takes place at relatively low irradiation doses (10^{15} ions cm^{-2}), however a saturation level of about 5 at. % H is reached, at which H loss practically stops. This behaviour seems to be almost universal, and independent of irradiation conditions; MeV heavy ions, or sub-MeV light ions all exhibit a similar dependence for the H concentration on dose regardless of the dominant physical mode of energy transfer from the ion to the material. The question which one of the stopping mechanisms (nuclear or electronic) is responsible for the H loss is not entirely clear, however evidence will be given below that each may play a different role.

Baumann et al.[34] have fitted the measured hydrogen concentration versus dose curves with three different exponentials attributed to three different hydrogen release cross-sections. These may be interpreted as the result of different C–H bonding configurations in the amorphous carbon matrix. Sellschop and coworkers [42] proposed a similar interpretation for their data on ion-beam induced hydrogen release from natural diamond, suggesting that the changes in the slope in the logarithm of H concentration versus dose reflect different bonding states of H to C in diamond.

The mechanism by which H is lost from a-C:H films during thermal annealing has been illuminated in a series of experiments by Wild and Koidl [43]. By performing mass resolved measurments of the species effused from sandwiches of deuterated and hydrogenated carbon films, they have shown that hydrogen leaves the a-C:H films always in molecular form, i.e. molecular recombination occurs in the bulk, followed by diffusion as molecules through the layer until they are lost into the vacuum. A similar conclusion was also reached by Moeller who has shown that during D^+ irradiation of H^+ implanted graphite, recombination of atomic to molecular hydrogen occurs in the bulk, and the effusing species are H_2 or HD molecules [44].

An alternative description of hydrogen effusion from a-C:H films induced by ion irradiation has recently been proposed [45]. It is based on the above described experimental evidence for molecular effusion and does not require the introduction of different C–H bonding configurations. The model is statistical in nature and is based on a second order kinetic process, namely the recombination of atomic to molecular hydrogen. The basic assumption in the model is that in order for a hydrogen molecule to be created in the bulk, two C–H bonds within some characteristic recombination length of each other must be broken by the energy deposited by the passage of the same ion. As a consequence of this requirement the change in hydrogen concentration $d\rho$ due to a dose increase dD is proportional to the number of hydrogen *pairs* in a characteristic volume V, i.e.

$$\frac{d\rho}{dD} = -K\rho(\rho V - 1), \tag{12}$$

where K is an effective molecular release cross-section.

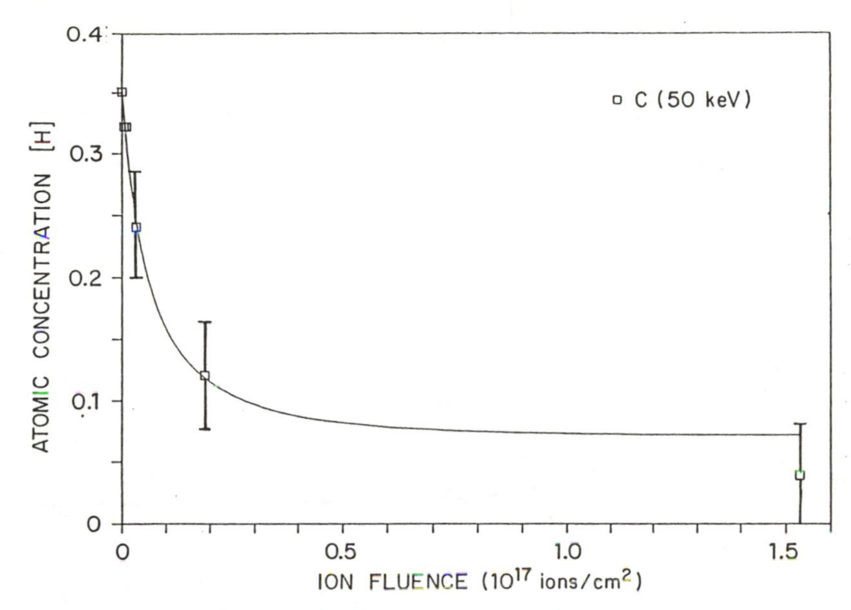

Figure 23: Hydrogen atomic concentration in a-C:H as a function of irradiation dose with 50 keV C^+ ions. Solid line indicates fit using functional dependence from bulk molecular recombination model [45].

This differential equation has the solution

$$\rho(D) = [1/\rho_f + (1/\rho_o - 1/\rho_f)exp(-KD)]^{-1}. \quad (13)$$

Here ρ_o is the initial volume hydrogen density and $V = 1/\rho_f$, i.e. the inverse of the final hydrogen concentration. The details of this calculation have been published elsewhere [45].

The model has been applied to all available published data on hydrogen effusion from a-C:H and two examples are presented here. Figure 23 shows the atomic hydrogen concentration dependence on 50 keV C^+ ion fluence, adapted from figure 11 including the fit given by the bulk molecular recombination model, using ρ_f and K as adjustable parameters. A second example is given in figure 24 which shows the atomic hydrogen concentration dependence on N^+ ion fluence, adapted from Baumann and coworker's data (figure 12). It is evident that the hydrogen concentration follows very closely the functional dependence given by the model; similar fits were obtained for all data analyzed, even for hydrogen release from diamond [42]. Table 1 summarizes the relevant experimental parameters for the various data including the release cross-section K and the final hydrogen concentration ρ_f used to generate the fits.

The physics governing the release mechanism resides in the parameter K and perhaps the final H concentration ρ_f. The correlation between hydrogen loss during ion irradiation and the electronic stopping power of the incident ion has already been demonstrated by Baumann et al. [34]. The approximate correlation between K and the electronic stopping powers observed in table 1 is illustrated in figure 25 where the energy loss due to electronic

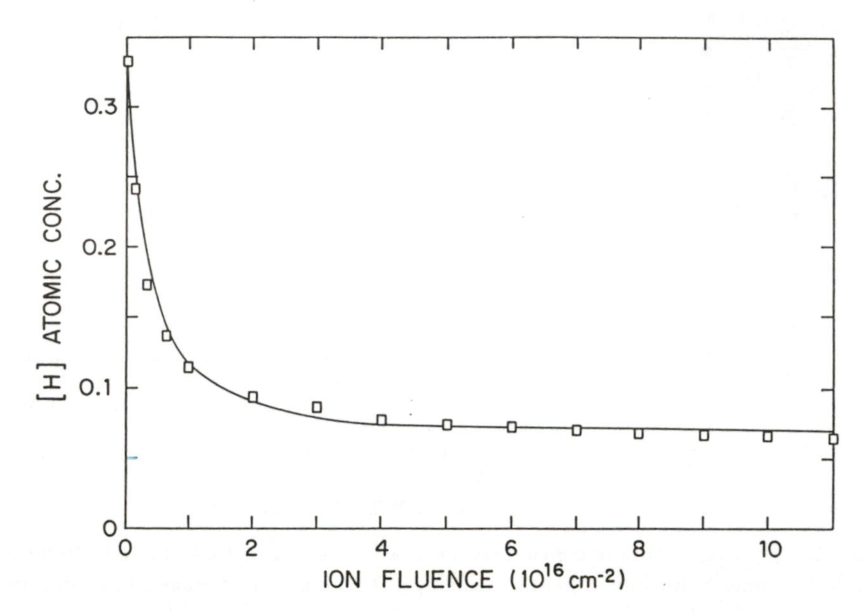

Figure 24: Hydrogen atomic concentration in a-C:H as a function of irradiation dose with 6 MeV N^+ ions, due to Baumann et.al. [34]. Solid line indicates fit using functional dependence from bulk molecular recombination model.

Table 1: Values of parameters used to fit bulk molecular recombination model for various experimental data on ion-induced hydrogen release from a-C:H. The electronic and nuclear energy loss for each case is also shown.

Author	reference	Ion	Energy (MeV)	release cross-section K (Å^2)	dE/dx electronic (eV/Å)	dE/dx nuclear (eV/Å)	final [H] ρ_f(at. %)
Baumann	[34]	^{15}N	7	0.5 ± 0.03	140	1.0	6.5
Baumann	[34]	^{15}N	3.5	4.5 ± 0.5	155	1.4	10
Baumann	[34]	^{20}Ne	1.6	4.5 ± 0.5	160	6	9
Baumann	[34]	^{40}Ar	2.15	4.5 ± 0.5	180	22	10
Zou	[41]	^{58}Ni	4	13 ± 3	192	17	7
Ingram	[35]	^{19}F	6.4	5	250	0.67	6
Ingram	[35]	^{127}Au	1	7	221	307	20
Fujimoto	[40]	^{12}C	12	5	100	0.14	9
Prawer	[31]	^{12}C	0.05	0.06 ± 0.02	15-27	18-14	5

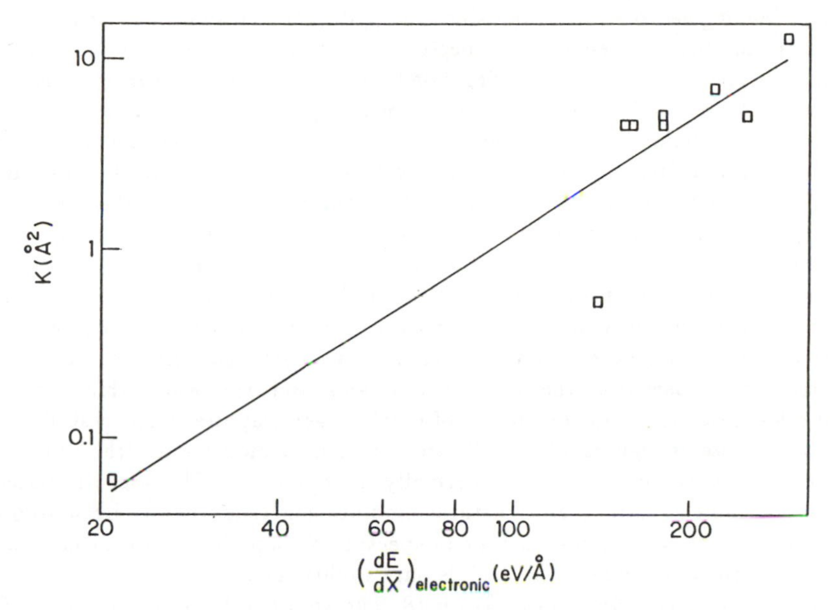

Figure 25: Log-log plot of molecular release cross-section K versus electronic energy loss $(dE/dX)_{electronic}$ as given in table 1. The straight line is of slope 2.

processes is plotted versus the effective cross-section K. Although the points are somewhat scattered, a least squares fit to this data produced a line of slope 2.0 with standard error 0.3. This is in fact the value expected for a molecular release cross-section governed by the probability of independently freeing two hydrogen atoms within a characteristic recombination distance.

In almost all cases reported the nuclear energy loss is neglible as compared to the electronic, with the exception of Ingram and coworkers' data for 1 MeV gold ions [30]. In this data, a final H concentration of $\sim$ 20 at. % was obtained, significantly larger than those observed to result from lighter ion induced effusion (typically 5-10 at. %, see table). As pointed out in refererence [30], the exceptionally high density of the collision cascade in this case results in a high density of H trapping centres. The characteristic volume (i.e. $1/\rho_f$) appears therefore to be governed by the density of retrapping centres which may compete with the molecular recombination process.

V.2.2 Bonding rearangement.

As a starting point, the as-grown samples will be described in terms of the two phase model along the lines suggested by McKenzie and coworkers [46], as well as Robertson and O'Rielly [47]. In this model, hard a-C:H is considered to consist of sp^2 clusters, typically planar aromatic ring structures, which are interconnected by randomly oriented tetrahedral sp^3 bonds. The hydrogen may be bonded on either the tetrahedral sites where they are required to reduce bond angle disorder or on the edges of the ring structures where they cause termination.

Based on this model it is clear that significant loss of hydrogen must be accompanied by major structural changes. Firstly, since a random tetrahedral phase is highly over constrained (analogous to a-Si:H), and since the hydrogen is responsible for reducing the co-ordination number, it's removal is expected to destabilize the sp^3 phase. This would either increase the number of unsatisfied bonds or result in restructuring whereby two previously sp^3 C–H bonds "recombine" to form an sp^2 bond. The latter effect would result in a net conversion from sp^3 type bonding to sp^2 type bonding. The growth of the sp^2 clusters would be further enhanced by the removal of hydrogen from the ring edges, so that a net increase in cluster size is expected. It is worth recalling that this restructuring is occuring during ion bombardment, which is a process expected to *increase* disorder. Upon inspection of figures 6 and 7 it is observed that at the intermediate saturation dose, the conductivity levels off at lower resistivities the lighter the ion. In fact it is suggested that the density of the collision cascade resultant from the passage of an energetic ion is the limiting factor which determines the extent of the cluster size increase. This is what sets ion irradiation apart from thermal annealing and laser irradiation which also induce hydrogen release but result in the growth of much larger graphite microcrystallites. The Raman data shown in figures 14 and 22 are useful in demonstrating the difference in microstructure between irradiated and thermally treated a-C:H. The laser irradiated film has undergone more drastic microstructural modification than the ion irradiated films, which for all doses show very similar Raman spectra. No significant shift or narrowing of the G peak is observed, while the D peak is only slightly enhanced.

Considering now the ESR data in figure 18, The spin density in the as-grown film is surprisingly high, considering the large hydrogen content (35 at. % or 4×10^{22} cm^{-3}). This is contrary to expectations for a random tetrahedral network such as a-Si:H, for which a much lower hydrogen content is effective in passivating dangling bonds, reducing the ESR signal from this particular spin active defect. However, in light of the two phase model unique to carbon structures, the spin signal could be due σ state dangling bonds, or to unpaired electrons in π states on the edges of sp^2 clusters, as proposed by Miller and McKenzie [48], the latter seeming more likely.

The functional form of the resonance curve may be revealed by integrating the measured differential spectrum. The dotted line in figure 26 shows the result of such an integration for an a-C:H film irradiated with a dose of 1×10^{15} C^+ cm^{-2} and is representative of the line shape obtained from all the measured spectra. It should be noted that the curve is highly symmetric and follows a Lorentzian rather than a Gaussian lineshape. This is indicative of exchange interactions between unpaired electrons [17], either σ state dangling bonds or π state defects. This effect is further enhanced by electron delocalization over a whole cluster, which is characteristic of π electrons resonating on aromatic ring structures. Miller & McKenzie have suggested this mechanism in their ESR study of as-grown and thermally annealed a-C:H. Figure 18b shows that as the irradiation proceeds the linewidth narrows indicating increased delocalization of π states, perhaps due to cluster size increase.

Figure 17 shows that regardless of ion dose the resonances remain centered about the same magnetic field strength, i.e. the g value remains constant. The ESR signal from graphite displays very marked anisotropy (g varies from 2.0026 to 2.0495) [49].

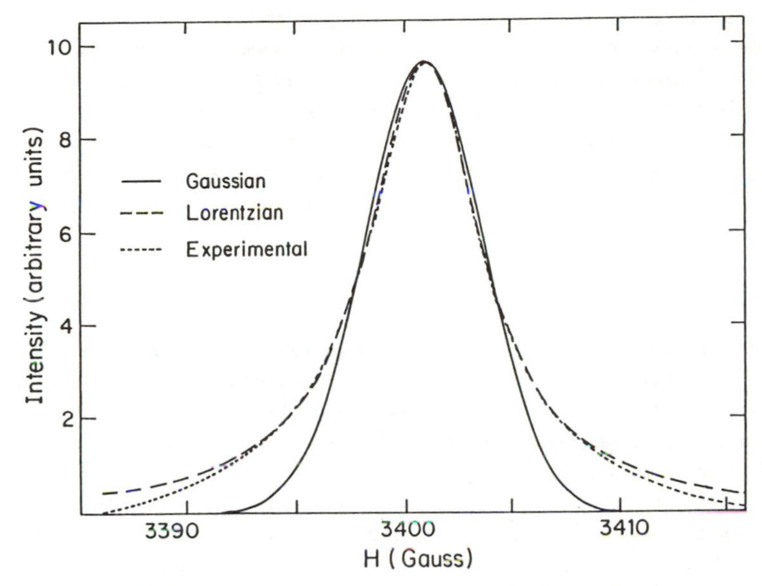

Figure 26: Numerically integrated spectrum of the ESR derivative spectrum of an a-C:H film irradiated with 1 $\times 10^{15}$ C^+ ions cm^2 together with Gaussian and Lorentzian curves adjusted to fit the Full Width Half Maximum and Peak Height of the experimental curve. Adapted from ref. [36].

Since the measured peak remains symmetric and with no such anisotropy observable even in the highly irradiated material, this is further conclusive evidence that macroscopic graphitization has not resulted from irradiation.

The measured g value is consistent with that obtained from crushed diamond powder [50] where the ESR signal is attributable to dangling bonds on the exposed diamond surfaces, however, it is worth noting that it is also consistent with that obtained from carbonized polymers where the signal is due to free radicals, i.e. unsaturated π states [48]. The recurrence of this g-value is not surprising since it is in both cases determined primarily by the spin orbit interaction with the carbon nucleus.

It is clear from figure 18 that as a result of ion irradiation, the density of spin active defects has increased by a factor of about 3.5 from the as-grown value of 2.5×10^{20} cm^{-3}, the increase again coinciding with the critical dose range for hydrogen effusion and conductivity increase. By comparing the number of hydrogen atoms removed from the film (2×10^{22} cm^{-3}) with the number of unsatisfied bonds (i.e. 6×10^{20} cm^{-3} spins) created over the same dose range, it is concluded that the vast majority of broken C–H bonds disappear, presumably recombining to form C–C bonds, with only some two percent remaining as either σ or π type dangling bonds. Since, as was previously argued, the sp^3 tetrahedral phase is expected to transform to the clustered ring phase, it is likely that the majority of these newly created defects are in the latter phase.

For the dose range below 5×10^{14} C^+ cm^{-2}, the spin density remains essentially unchanged from the as-grown value. This is remarkable in view of the fact that such a number of impinging ions is expected to generate about 10^{21} vacancies cm^{-3} [5]. The absence of any observable increase in spin density despite the considerable damage produced by the ion beam may indicate a dynamic annealing process in which reactive atomic hydrogen freed by ionization or collision cascades is trapped by newly created defects. This dynamic passivation may find support in the gradual initial resistivity *increase* observed for C^+ irradiation doses up to 5×10^{14} ions cm^{-2} at which short range hydrogen diffusion occurs but significant effusion has yet to commence. A simlilar idea was also put forward by Gonzalez-Hernandez and coworkers to explain the initial PL intensity enhancement observed for low dose H bombardment of a-C:H.

V.2.3 Band Structure Modification.

As seen above, irradiation of any form causes rather extreme changes in a-C:H. These are also reflected in changes in electrical properties. The important questions to address in this respect are i) in what way has the density of electron states been modified by irradiation, and ii) how does this modification affect the electrical conductivity, a question which will be adressed in the next section.

In the context of the structural model adopted in the previous section, the density of states of as grown a-C:H is expected to be qualitatively similar to that shown in figure 4. The electrical properties are determined primarily by the band-tail and mid-gap states. It has already been stressed that although both σ and π bonds are present, π states are expected to dominate at the band edges and in the gap due to their lower energy. Robertson & O'Rielly [20] have performed band structure calculations for a-C:H by neglecting σ states and considering only π states. A one electron tight binding model retaining only π orbitals and nearest neighbour interactions was used by them to calculate the energy eigenvalues of various simple ring structures. Figure 27 shows the energy spectra obtained for a number of structures of increasing size. A few important features resulting from these calculations should be noted. Firstly, an energy gap is opened, the energy difference between the last occupied and first unoccupied state being determined by the number of vertices, N. For $N = 6$, i.e Benzene-like rings, all states lie well away from E_F, producing the widest gap. From figure 27c, it is evident that as the number of rings in a cluster increases the band gap is diminished. This is shown more quantitatively in figure 28 where E_g is plotted against M, the number of rings in a cluster.

Regarding the band gap, the optical absorption data of figure 13 indicate that except for the lowest tail, the data for all samples appears to follow the Tauc relation described above. Figure 11 shows that the as grown films have an optical band gap of about 1.5 eV, typical of hard a-C:H [1]. However, with increasing C^+ irradiation above the threshold of 5×10^{14} ions cm^{-2}, the band gap shrinks sharply from 1.5 to 0.9 eV (at 1×10^{17} ions cm^{-2}). This reduction in band gap may be taken as an indication of an increase in cluster size from clusters of about 12 rings to clusters of perhaps 40 rings (see figure 28).

The Tauc plots of figure 13 also show evidence of a deviation from a parobolic energy dependence of the density of states at the lowest energies as inferred from the low energy absorption tails seen on all three spectra. This is indicative of band-tail states extending

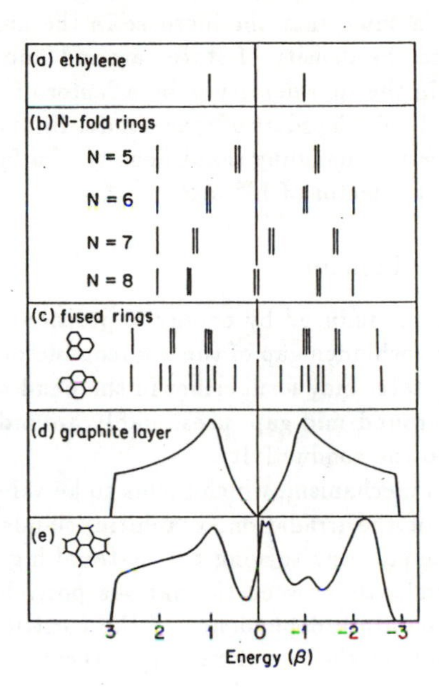

Figure 27: Energy eigenvalues as given by the Huckel Hamiltonian method for various ring structures. Calculations are due to Robertson & O'Reilly [47].

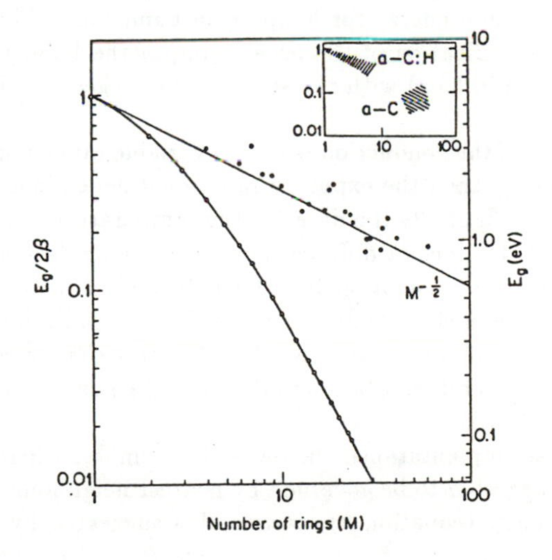

Figure 28: Band gap dependence on ring number (M) in cluster (upper curve) as deduced by the Huckel Hamiltonian technique. Calculations are due to Robertson & O'Reilly [47].

into the gap which may be important in determining the conduction properties of the films. It is reasonable to assume that the increase in the number of spin defects, sp^2 or sp^3 like, leads to a rise in the density of states around mid-gap. It is noteworthy that eventhough the increase in the spin density is by a factor of about 2.5, it is accompanied by changes in the resistivity of 4-5 orders of magnitude. Similar effects have been reported for a-Si:H where an increase in dangling bond density of a factor of about 7 corresponds to a resistivity decrease by a factor of 10^4 [51].

V.2.4 Conduction Mechanism.

The structural modifications induced by energy deposition into a-C:H films create electronic energy levels in the forbidden gap of the semiconductor, thus affecting it's electrical conductivity. These new states may form either in the band tails, causing them to extend further into the gap, or around mid-gap, presumably around the Fermi level (E_F), each contributing differently to the conductivity.

A different conduction mechanism, which seems to be valid for ion irradiated diamond [52] is that, as a result of the irradation, graphitic islands are formed which if dense enough form a percolative pathway turning the material highly conductive.

A clear way to discriminate between the various possible conduction mechanisms is by the study of the temperature dependence of the electrical conductivity for different irradiation conditions and by the investigation of their thermopower, as described in section III.4.

Figure 8, in which $\log \sigma$ is plotted vs $1/T$ for different irradiation doses, shows that the experimental data do not follow a straight line as expected for activated conductivity beyond the mobility edge. The observed curvature may be due to the temperature dependence of W_1, the activation energy for hopping in band tails. This behaviour was taken in reference [31] as an indication of the broadening of the band tails towards the Fermi energy which goes hand in hand with the structural changes (i.e. hydrogen loss and ESR signal increase).

On the other hand, if the conduction is due to variable range hopping of charge carriers around the Fermi energy then the expected functional dependence of σ on T is given by equation 10. As seen in figure 29 the data for the temperature dependence of a-C:H films irradiated with 50 keV C^+ ions at different doses (the same data as shown in figure 8) fit the functional dependence of equation 10 extremely well in the cases of highly irradiated a-C:H. The slight change in slope found for the different irradiation doses reflects a rather large change in T_0 from 10^9 to 5×10^7 °K. Whether this change results from an increase in λ due to cluster size increase, or an increase in $N(E_F)$ cannot be determined from the present data.

For the lower dose implantations, (below $\sim 10^{16}$ cm^{-2}), where the $T^{-1/4}$ fit is poor, the conductivity is suggested to be governed by nearest neighbour hopping amoung states around the Fermi energy (equation 9), as was also suggested by Jones and Stewart for as-grown material [53].

Further information on the conduction mechanism can be extracted from the ther-

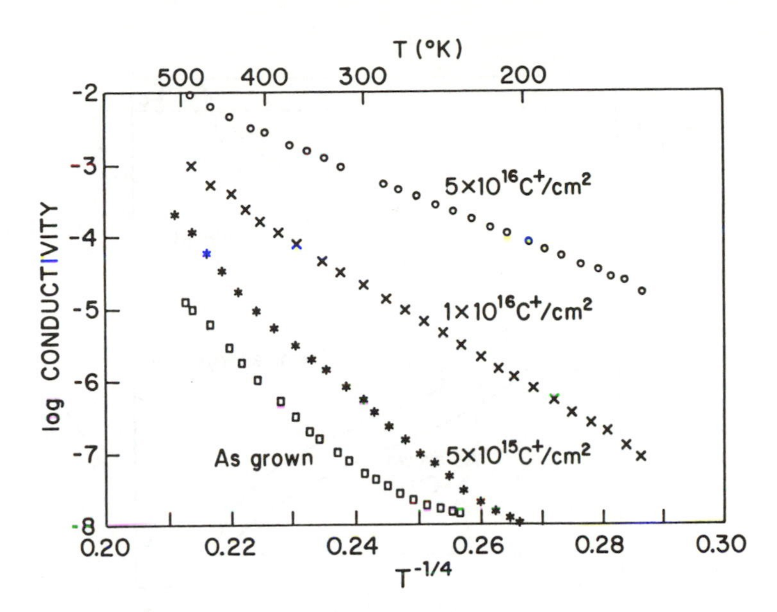

Figure 29: Log of the conductivity vs $T^{-1/4}$ for the conductivity data shown in figure 8 for irradiated films.

mopower measurements. Low values of the thermopower may indicate conduction around E_F while larger values are typical for conduction away from the Fermi level, either in band tails or extended states. In the latter case the sign of the thermopower indicates the sign of the charge carriers which dominate the conductivity.

With this in mind, the thermopower data in figure 19 are now reviewed. The low values of the thermopower (-40 $\mu V/°$K and +20 $\mu V/°$K for samples irradiated with 5×10^{15} and 2×10^{16} ions cm^{-2} respectively) and their temperature independence between room temperature and about 350 °K can be taken as confirmation of hopping conductivity between states around the Fermi energy. The fact that at lower temperatures the thermopower changes sign with increasing dose, if indeed significant, may indicate that while for unirradiated material, the Fermi energy is positioned below the peak of mid-gap states, as the irradiation progresses this peak expands and it's tail overlaps the Fermi energy causing an increase in $N(E_F)$ and changing the sign of the thermopower from negative to positive. These changes are schematically represented in figure 30 which shows the proposed density of states of as grown a-C:H and it's modification as a result of high dose ion irradiation.

At higher temperatures the thermopower is positive and increases dramatically, presumably as conductivity by holes beyond the mobility edge begins to contribute. Similar behavior was also observed in doped a-Si:H (see figure 5, taken from reference [23]). It should be noted that the thermopower, in contrast to the conductivity, is very sensitive to shifts in the conduction path from states around the Fermi energy to activated conduction since it does not only depend on the number of carriers in any particular conduction channel but also on their free energy.

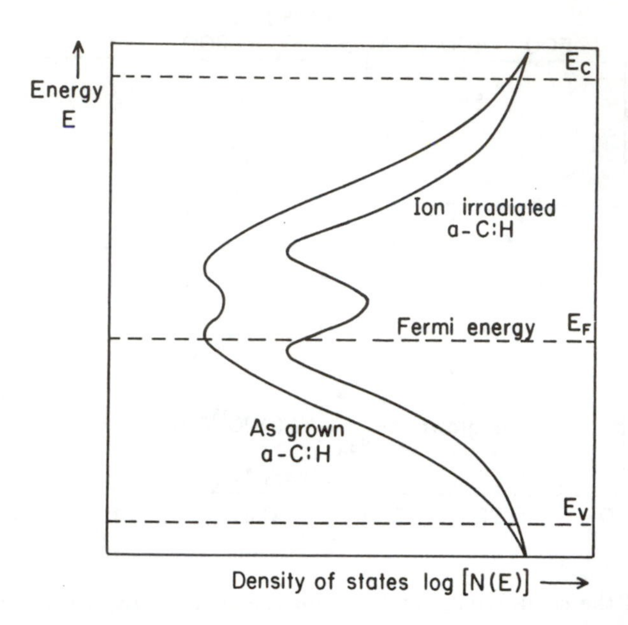

Figure 30: Schematic representation of the proposed as-grown and ion beam modified band structure of a-C:H.

V.3 Electron Irradiation

In contrast to the wealth of data available for ion irradiation effects in a-C:H, little is known about electron irradiated films. The experimental limitation of small electron treated regions regions has reduced the number of applicable analysis techniques, so that although some general conclusions may be drawn about the similarities and contrasts with ion irradiated a-C:H, certain aspects of electron irradiation have not been fully explored.

The relevant experimental results described in section IV.3 which must be explained are:

i) The optical absorption increases (i.e. band gap shrinks) at electron doses an order of magnitude below those required for significant conductivity enhancement and two orders of magnitude before hydrogen effusion begins. This is in stark contrast to ion irradiation in which the conductivity increase, band gap reduction and hydrogen effusion occur concomitantly over the same dose range.

ii) At high electron fluences (3×10^{19} electrons cm^{-2}) at which significant modifications in the electrical and optical properties have already occured, relatively little H loss is observed. This amounts to less than half of the initial hydrogen content in the top 3000 Å layer, which is smaller than the expected electron range of 5000 Å.

iii) Raman results indicate no induced graphitization.

iv) For low dose electron irradiation, the increased optical absorption was observed to be bleachable by high intensity laser irradiation.

V.3.1 Structural Modification.

To account for these observations the interactions between the electron beam and an insulating material should be considered. As mentioned in section II.3 above, energetic primary electrons interact with core and valence electrons of the material to produce a cascade of several secondaries. However, in contrast to ion irradiation the cascade is of much lower density so that the passage of an electron is much less likely to cause multiple bond breaking within a small volume, say 10 $Å^3$. This accounts for the extremely high doses ($> 10^{19}$ electrons cm^{-2}) required for the onset of hydrogen release, since, as was suggested in the ion induced hydrogen release mechanism, H effusion may occur only if two C–H bonds are simultaneously broken within some characteristic recombination distance. It is worth mentioning that on purely kinematic grounds (and assuming a C–H binding energy of about 4 eV) a 5 keV electron carries enough energy to displace a hydrogen atom in a small impact parameter ellastic collision. The classical Rutherford cross-section however, turns out to be of the order of 10^{-20} cm^{-2} so that this process is insignificant since, as two such events must occur simultaneously within a small distance, the molecular hydrogen release cross-section would be even smaller. The more likely alternative for H release is bond breaking by ionization during electron-electron interactions.

Electron irradiation is therefore different from ion irradiation in that the former causes mainly isolated bond-breaking without H loss, while in the latter the broken bonds must restructure in an environment of ever decreasing hydrogen concentration.

It may be speculated that the breaking of a single C–C bond may result in either five-fold or seven-fold ring structures whose presence in the as-grown material is rare due to their higher free energy. As seen in figure 27, they create π states in the tails which are closer in energy to E_F than the six-fold ring states and their proliferation with electron irradiation may be the explanation required for the band gap shrinkage at lower doses. No major conductivity enhancement results from this modification since the high hydrogen content remaining in the film allows passivation of the newly created π states. Such ring structures are also expected to be highly unstable and since no compositional changes are observed in films irradiated to moderate electron doses, it is reasonable that the structures are metastable and may revert back to the lower free energy six-fold rings if energy is made available to the system. The laser bleaching may be taken as an indication of this effect, although it is not clear whether the annealing is as a result of local heating or a more direct photon-defect interaction.

V.3.2 Band Structure and Conductivity.

In section V.2.4 the as-grown conductivity was suggested to be due to either activated hopping in band-tail states or hopping between states around the Fermi energy. The observation that following electron irradiation the band tails may shift towards mid-gap without any significant conductivity increase supports the proposition that the as-grown conductivity must be due to mid-gap states, otherwise the band gap reduction would have to be accompanied by a conductivity increase.

For the higher electron doses, where the conductivity dependence on dose follows a power law and where hydrogen release becomes significant, it is speculated that the main change in the nature of the film's structure is an increase in sp^2 cluster size as the hydrogen is removed in analogy to the case of ion irradiation. The weaker dose dependence of the conductivity increase for electrons (D^3) as opposed to ions (D^4) may indicate a slower, more gentle mechanism for hydrogen release.

VI Summary

The effects of energy deposition by photons, ions and electrons into amorphous hydrogenated (diamond-like) carbon have been reviewed and the results of a broad range of experimental methods have been described. The findings of these measurements show that the application of directed energy beams to a-C:H results in a number of alterations to the physical properties of the films.

Laser irradiation of the films results in a transformation similar to that due to heating to 500 °C, i.e. partial graphitization. The damage threshold was seen to be pulse energy dependent for photon energies above the band gap and power dependent for energies below the gap.

Ion bombardment above a certain threshold affects both electrical and optical properties of the material. Hydrogen release which goes hand in hand with an increase in the number of ESR active sites is acompanied by these changes. In contrast to laser irradiated a-C:H, ion implanted films do not show phenomena associated with significant graphitization.

In light of a two-phase model, ion irradiation is believed to modify the structure of a-C:H by causing hydrogen effusion which facilitates a conversion from sp^3 to sp^2 bonded carbon. This results in a net increase in the sp^2 graphite-like cluster size. The extent of the cluster size increase is limited by the ion bombardment process, allthough cluster growth is enhanced when irradiations are carried out at elevated temperatures. It is proposed that the neccesary condition for hydrogen release to occur during ion irradiation is that two C–H bonds are broken by the passage of a single ion within some characteristic recombination distance of each other.

The structural modification described above significantly modifies the density of electron states, the major manifestations of this alteration being the large conductivity increase and the narrowing of the optical band gap. In keeping with the two-phase structural model, the band structure and transport properties are considered to be determined by the nature of the π states since they lie closer in energy to the Fermi level than the σ states. Furthermore, it has also been demonstrated that the dominant conduction mechanism varies with irradiation dose, but in all cases is a form of hopping between states in the vecinity of the Fermi level.

Electron irradiation was also shown to modify the optical and conduction properties of a-C:H, however the structural modification seems to be of a different, more gentle nature, since these changes occur prior to significant irradiation induced hydrogen effusion.

Acknowledgements

This work was partially supported by the fund for the promotion of research at the Technion. The authors would also like to thank O. Amir for useful discussions.

References

[1] For a good review, see Angus J.C., Koidl P. & Dominitz S. in "Plasma Deposited Thin Films", Ch. 4, edited by Mort J. & Jansen F., CRC Press, Inc, (Florida, 1986).

[2] Dillon R.O., Woollam J.A. & Katkanant V., Phys. Rev., B15, 3482, (1984).

[3] Wood R.M., "Laser Damage in Optical Materials", Adam Hilger, (Bristol & Boston, 1986).

[4] Chu W.K., Mayer J.W. & Nicolet M.A., "Backscattering Spectrometry", Academic Press, (New York, 1978).

[5] Zeigler J.F., Biersack J.P., & Littmark U., "The Stopping & Range of Ions in Solids", Pergamon Press, (New York, 1985).

[6] Dearnaley G., Freeman J.H., Nelson R.S. & Stephen J. "Ion Implantation", North Holland Publishing Co. (Amsterdam, 1973).

[7] Adel M., Kalish R. & Richter V., J. Mater. Res. 1, 503, (1986).

[8] Elliman R.G., Johnson S.T., Pogany A.P. & Williams, J.S. Nuc. Instr. & Meth., B7, 310, (1985).

[9] Schade H., in "Semiconductors & Semimetals", 21, Part B, Chapter 11, edited by J. Pankove, Academic Press, (1984).

[10] "Fundamentals of Surface and Thin Film Analysis." Feldman L.C. & Mayer J.W., North Holland, (New York, 1986).

[11] Lanford W.A., Trantvetter H.P., Ziegler J.F. & Keller J., Appl. Phys. Lett. 28, 566, (1976).

[12] Adel M.E., Ph.D. thesis, Technion Israel Institute of Technology, (1989).

[13] Dischler B., Bubenzer A., & Koidl P., Solid State Commun. 48, 105, (1983).

[14] Gonzalez-Hernandez J., Asomoza R., Reyes-Mena A., Rickards J., Chao S.S. & Pawlik D., J. Vac. Sci. & Technol., A, 6, 1798, (1988).

[15] Taylor P.C. in "Semiconductors and Semimetals", 21, Part B, Chapter 3, sect. III, edited by J. Pankove, Academic Press, (1984).

[16] Poole C.P., "Electron Spin Resonance", John Wiley & Sons, (New York, 1967).

[17] Anderson P.W. & Weiss D.R., Rev. Mod. Phys., 25, 269, (1953).

[18] Cody G.D., in "Semiconductors and Semimetals", 21, Part B, Chapter 2, edited by J. Pankove, Academic Press, (1984).

[19] J. Tauc, "Amorphous and Liquid Semiconductors", Chapter 4, Plenum, (London, New York, 1974).

[20] Robertson J., Adv. Phys., 35, 317, (1986).

[21] Lin S. & Feldman J. Phys. Rev. Lett. 48, 829, (1982).

[22] Mott N.F., & Davis E.A., "Electronic Processes in Non-Crystalline Materials", second edition, Chapter 6, Clarendon Press, (Oxford, 1979).

[23] Beyer W. & Stuke J., Fifth International conf. on Amorphous and Liquid Semiconductors, 252, (1978).

[24] Meyerson B. & Smith F.W., Solid State Commun. 34, 531, (1980)

[25] Prawer S., Kalish R., & Adel M.E., Appl. Phys. Lett., 48, 1585, (1986).

[26] Nadler M.P., Donovan T.M., & Green A.K., Appl. Surf. Sci., 18, 10, (1984).

[27] Rothschild M., Arnone C. & Ehrlich D.J., J. Vac. Sci. & Technol. B, 4, 310, (1986).

[28] Rothschild M. & Ehrlich D.J., J. Vac. Sci. & Technol. B, 5, 389, (1987).

[29] Dischler B., Bubenzer A., Koidl P., Brandt G. & Schirmer O.F., NBS Special Publication, 669, 249, (1982).

[30] Ingram D.C. & McCormick A.W., Nucl. Inst. & Meth. B, 34, 68, (1988).

[31] Prawer S., Kalish R., Adel M.E., & Richter V., J. Appl. Phys., 61 4492, (1987).

[32] Doolittle L.R., Nucl. Inst. Meth. B9, 344, (1985).

[33] Doolittle L.R., Nucl. Inst. Meth. B15, 227, (1986).

[34] Baumann H., Rupp Th., Bethge K., Koidl P. & Wild Ch., Les Editions de Physique, EMRS conference proceedings, vol. 17, 149, (Strasbourg 1987).

[35] Orzeszko S., Woollam J.A., Ingram D.C. & McCormick A.W., J. Appl. Phys. 64, 2611, (1988).

[36] Adel M.E., Kalish R., & Prawer S., J. Appl. Phys., 62, 4096, (1987).

[37] Adel M.E., Brener R. & Kalish, R. Les Editions de Physique, EMRS proceedings, 334, (Strassbourg, 1987).

[38] Braunstein G., Steinbeck J., Dresselhaus M.S., Dresselhaus G., Elliman B.S., Venkatesan T., Wilkens B. & Jacobson D.C., MRS proceedings, symposium A, (Boston, 1985).

[39] Prawer S., Kalish R., Adel M., & Richter V., Appl. Phys. Lett. 49, 1157, (1986).

[40] Fujimoto F., Tanaka M., Iwata Y. Ootuka A., Komaki K. & Haba M., Nucl. Inst. & Meth. B, 33, 792, (1988).

[41] Zou J.W., Schmidt K., Reichelt K. & Stritzker B., J. Vac. Sci. & Technol. A, 6, 3103, (1988).

[42] Sellschop J.P.F., Madiba C.C.P. & Annegarn H.J. Nucl. Inst. & Meth. 168, 529, (1980).

[43] Wild Th. & Koidl P., Appl. Phys. Lett., 51, 19, (1987).

[44] Moeller W., Borgesen P. & Scherzer B.M.U., Nucl. Inst. & Meth., B19/20, 826, (1987).

[45] Adel M.E., Amir O., Kalish R. & Feldman L. J. Appl. Phys., accepted, (1989).

[46] McKenzie D.R., McPhedran R.C., Savvides N. & Cockayne D.J.H., Thin Solid Films, 108, 247, (1983).

[47] Robertson J. & O'Rielly E.P., Phys. Rev. B, 35 2946, (1987).

[48] Miller D.J. & Mckenzie D.R., Thin Solid Films, 108, 257, (1983).

[49] Wagoner G., Phys. Rev. 118, 647, (1960).

[50] Walters G.K. & Estle T.C., J. Appl. Phys. 321, 1854, (1961).

[51] Thomas P.A. & Flachet J.C., Philos. Mag. B., 51, 55, (1985).

[52] Kalish R., Bernstein T., Shapiro B., & Talmi I., Rad. Eff., 52, 153, (1980).

[53] Jones D.I. & Stewart A.D., Phil. Mag. B 46, 423, (1982).

Materials Science Forum Vols. 52 & 53 (1989) pp. 475-494

DIAMONDLIKE CARBON (DLC): ITS FABRICATION, ANALYSIS AND MODIFICATION BY ION BEAMS

D.C. Ingram

Department of Physics and Astronomy
Ohio University, Athes, OH 45701, USA

DIRECT ION BEAM DEPOSITION OF DLC

Diamondlike Carbon (DLC) is an unusual material because it can only be fabricated using plasma assisted chemical vapor deposition (PACVD) or by direct or indirect ion beam deposition (IBD). The latter really being specializations of PACVD since the ion beams used for this type of work can be considered as directed plasmas. This is particularly true of ion beams which are not mass analyzed, since they contain not only different ionized species but also neutral particles. They also carry sufficient electrons to ensure space charge neutrality of the beam. This chapter will concentrate on the use of ions beams in the deposition, analysis and modification of DLC. Conclusions will be drawn about the conditions which are present during the deposition of DLC from the changes produced in DLC films by ion irradiation.

The use of ion beams in the production of DLC originated with the work of Aisenburg and Cabot [1]. They deposited films from carbon ions extracted from a carbon arc discharge. Their depositions were performed at energies as low as 40 eV. In their systems no mass analysis was used to determine or restrict the masses of the arriving species. This is typical of the systems used in the fabrication of DLC by direct or indirect ion beam deposition. Reliance is placed on the purity of the gas used in the ion source and of the sputter target in the case of indirect IBD. The deposited film may thus contain contaminates from components of the ion beam system as well as the residual gas of the vacuum system. In fact the material depends on the presence of hydrogen at about 20 to 40 atomic percent to stabilize the sp3 bond states which give the material its diamondlike properties, e.g. electrical insulation, hardness, optical properties. The hydrogen can be introduced either from the residual gas of the vacuum system

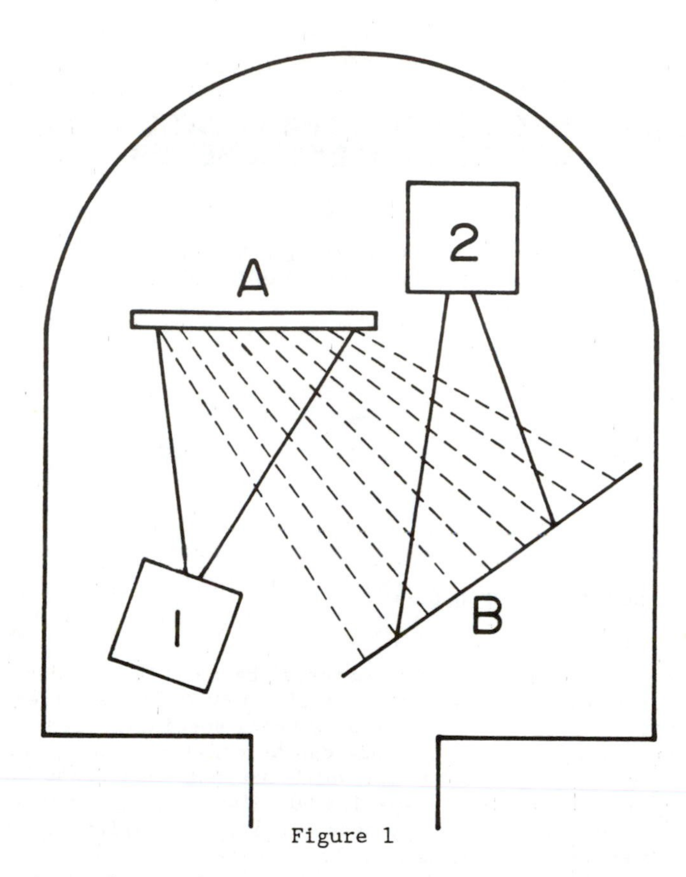

Figure 1

DLC deposition by ion beam assisted sputtering. Ion gun 1 sputters carbon from a carbon target A onto the substrate B. A second ion gun 2 may be used to modify the coating during deposition.

or through using, as ion source feed stock, hydrocarbon gases with methyl groups, e.g. methane, ethane etc. The work of Memming [2] has shown that in PACVD of DLC the use of double or triple bonded hydrocarbon precursor gases leads to graphitic and polymeric phases of carbon being deposited in addition to diamondlike material.

No work has been done to demonstrate the minimum energy and/or ion fraction of the arriving species necessary to fabricate DLC. For all of the PACVD techniques the energies of the arriving species will be well above a few eV. In the case of direct ion beam deposition the energies may be as high as 1 or 2 keV but are not likely to be lower than 10 eV. For indirect ion beam deposition where sputtering from a solid carbon target is undertaken then the energy of many of the deposited particles will be higher than several eV and there will be some scattered particles, dependent on the geometry of the system, which could have several hundred eV. Occasionally, a second ion beam is employed to assist with the deposition by simultaneously irradiating the sputtered coating, figure 1, [3].

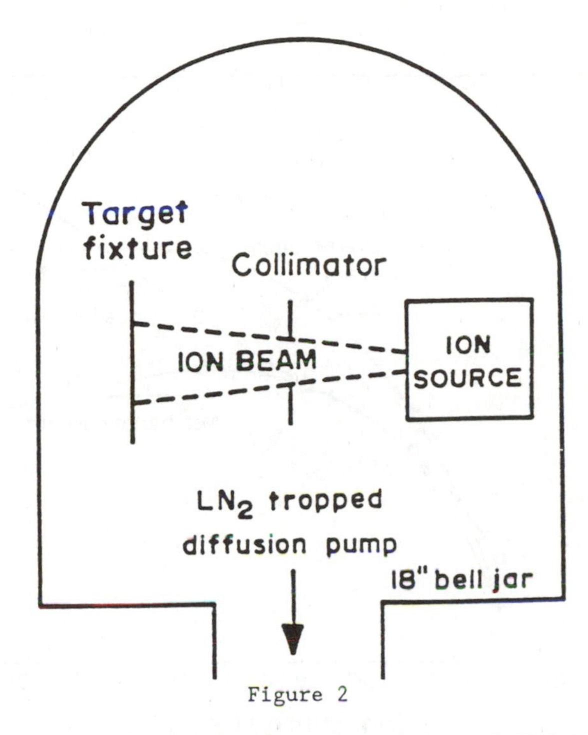

Figure 2

Typical system for direct ion beam deposition of DLC without mass analysis.

Direct ion beam deposition offers a way to control the species and the energy of the deposited species to a much greater extent than either ion beam sputter deposition or PACVD of DLC. Two groups [4,5] are currently working on mass analyzed low energy ion beam deposition with the aim to achieve depositions below 10 eV and perhaps as low as 1 eV. This technology is similar to the work of Freeman et al. [6,7] who achieved mass analyzed ion beam deposition of crystalline diamond with carbon ions from less than 100 eV up to 1000 eV.

To achieve direct ion beam deposition an ion beam is directed at the substrate to be coated, figure 2. The beam energy is maintained low enough so that the sputtering coefficient is less than unity i.e. fewer than one atom is sputtered from the target per incident ion. Sputtering is the process whereby energy, which is deposited by the incoming ion beam in nuclear displacements in the first or second layers of the target, causes atoms to be displaced with sufficient energy and momentum that they are lost from the target. For many target elements or compounds the sputtering coefficient exceeds unity at ion energies of only 100 eV for most ions. However, because carbon, in most forms, has a high surface binding energy and low nuclear stopping power it has a sputter coefficient which exceeds unity at several keV depending on the state of the target and the ion species [8]. Thus, if the sputtering coefficient as a function of energy for carbon on carbon is convoluted with the typical ion beam current as a function of energy for a Kaufman [9] ion source then a maximum will occur in the film growth rate at about 1 keV, figure 3.

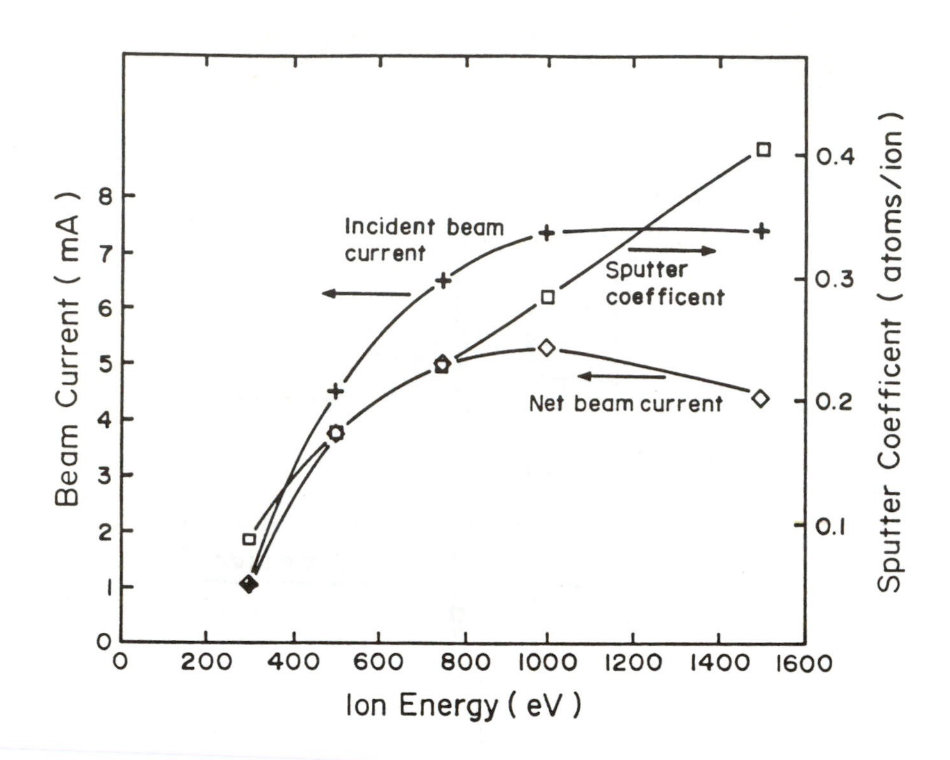

Figure 3

Incident ion beam current, carbon-carbon sputtering coefficient and net ion beam current arrival rate as a function of ion energy.

For any ion beam technique it is possible for the ions to be either mass analyzed or not mass-analyzed before they arrive at the target. The use of mass analysis, as compared to using no mass analysis, carries the penalty of higher capital cost and lower efficiency owing to reduced beam transmission but offers several orders of magnitude better purity and control of the residual gas environment of the target area and of the delivered ion beam. Indeed small areas (~cm^2) of a few tens of nanometers of thickness of isotopically pure single crystal films have been grown at high temperatures and under ultra high vacuum (uhv) conditions [4,5]. Recently, it was reported that the compound semiconductor β-SiC had been grown under these conditions by switching on first a carbon beam then a silicon beam in order to grow successive layers of carbon and silicon atoms required for this compound [5].

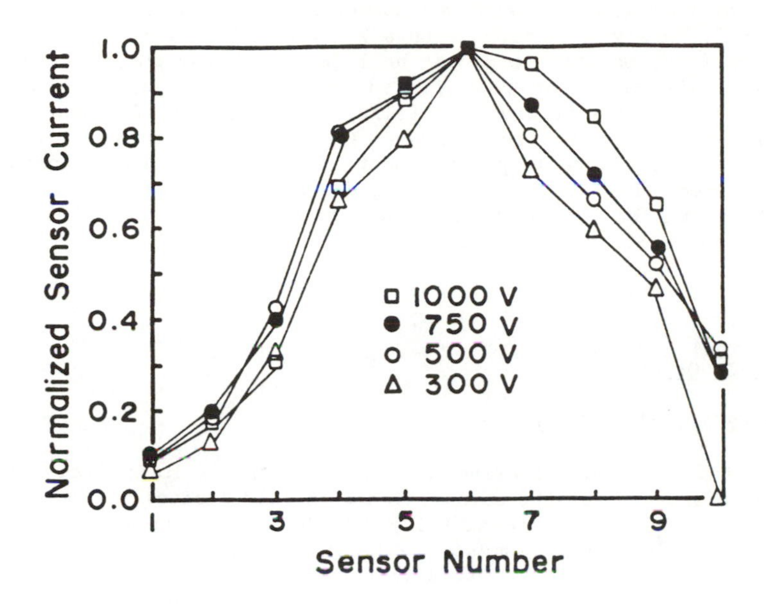

Figure 4

Typical ion beam profiles obtained from an ion beam deposition system similar to the one shown in figure 2 but without the collimator and with a special sensor plate at the target (see text). The lines joining the data points are to guide the eye.

For DLC it has been found that nonmass-analyzed ion beam deposition is practical and quite efficient at covering large (100 to 1000 cm^2) areas in reasonable times with thicknesses of the coating which are of practical use (100 to 1000 nm). Further since these films require carbon and hydrogen to be present it is convenient to run the ion source on pure methane and accept all the species of ions and neutrals produced. In the deposition process the ions probably dissociate on impact with excess hydrogen being lost to the vacuum system. Typically these films contain 70 % carbon atoms and 30 % hydrogen atoms. Thus, approximately, only one sixth of the arriving hydrogen is retained, assuming methane as the source gas. The precise mechanisms involved in the retention or loss of the hydrogen are not clear but some light will be shed on the processes present during deposition later in this chapter.

Without mass analysis the equipment for direct ion beam deposition is relatively simple, figure 2. An ion source capable of ionizing gaseous materials such as the Kaufman source [9] is run on methane and ions extracted at 1000 eV. They are directed onto the target which may be held on a fixture which can be scanned mechanically in front of the ion source to ensure a uniform coating is achieved. A collimator may be used to define the beam spot because this is usually not a perfect step function as can be seen in figure 4.

The DLC films, which were analyzed and modified by ions beams as described in later sections of this chapter were fabricated in a system similar to the one shown in figure 2. This was equipped with a 2.5 cm Kaufman type ion source [9] made by Ion Tech Inc., mounted in an 18" bell jar which was pumped by a liquid nitrogen trapped oil diffusion pump filled with 705 oil and backed, via a molecular sieve trap, by a rotary pump filled with TW7 oil. For the ion beam profile measurements shown in figure 4, the target plate was mounted 12 cm from the source. On the plate were 10 probes about 12.5 mm apart used to obtain the profile of the beam. The ion source was supplied with 99.99% research purity methane from Matheson Inc. The flow rate into the source was about 1.5 sccm and the pressure in the bell jar about $1x10^{-4}$ torr.

ION BEAM ANALYSIS OF DLC

RUTHERFORD BACKSCATTERING (RBS)

RBS is a well established quantitative technique for thin film analysis [10,11]. It can provide information on the composition of a film typically from 1 to 1000 nm in thickness. It has a sensitivity which varies from approximately 0.1 atomic % (at.%) for heavy elements in light element substrates to 10 at.% for light elements in heavy element substrates. It can also provide information on the crystalinity of single crystal targets and information about the crystallographic orientation of impurities in single crystal samples.

The utility of RBS in analyzing DLC is to provide a measure of the number of carbon atoms in the film and to establish whether and to what extent any heavy impurities have been deposited together with the film. RBS normally consists of scattering a beam of 2 MeV helium particles off the target to be investigated into a solid state barrier detector at some fixed angle to the incident beam. The detected particles are energy analyzed via a multichannel analyzer and the RBS spectrum or plot of the number of scattered particles as a function of their energy is obtained. As a helium ion moves through the target, and prior to its collision with a target atom, it loses energy to the electronic structure of the target. It also loses energy in the same manner after its collision as it leaves the target. The amount of energy it loses this way is dependent on the composition of the material. The probability or cross-section for collision is quite small and so the probability for multiple collisions is insignificant. In addition to losing energy while it moves through the target, the particle also loses energy at the collision, figure 5.

The collisional energy loss is unique for a given scattering angle and mass combination. Thus, if the atom from which the particle is scattered is at the surface of the target then it is possible to use the energy with which the helium particle was backscattered as a means to identify that atom on the surface. Also, the probability for scattering a helium particle from a given atom depends strongly on the square of its atomic number and a plot of yield versus surface energy from various element is shown in figure 6.

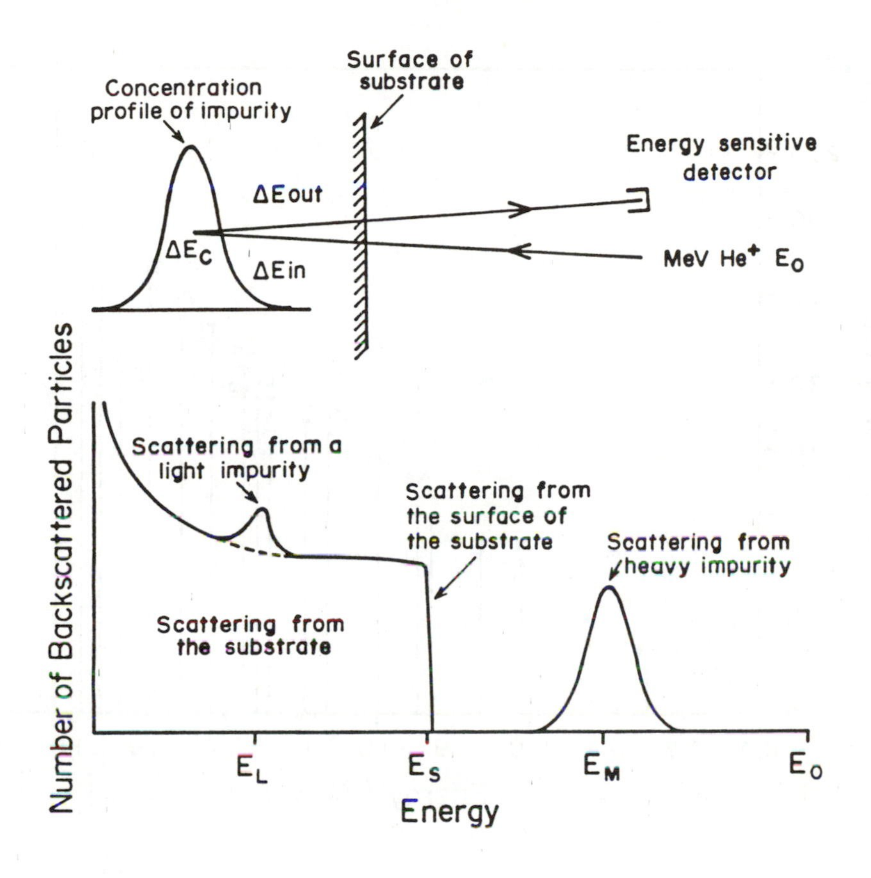

Figure 5

RBS analysis: upper figure shows geometrical arrangement of beam, target with impurity concentration profile, and detector; lower figure shows highlights key feature of an RBS spectrum.

RBS is able to provide information about the absolute number of atoms of an element per unit area of the target. Therefore, if there are N atoms per unit volume in a film of thickness t then the number of counts in the RBS spectrum from those atoms is given by:

$$A = Q\Omega\sigma Nt$$

where Q is the number of incident helium particles, Ω is the acceptance angle of the detector, σ is the Rutherford scattering cross-section. For this

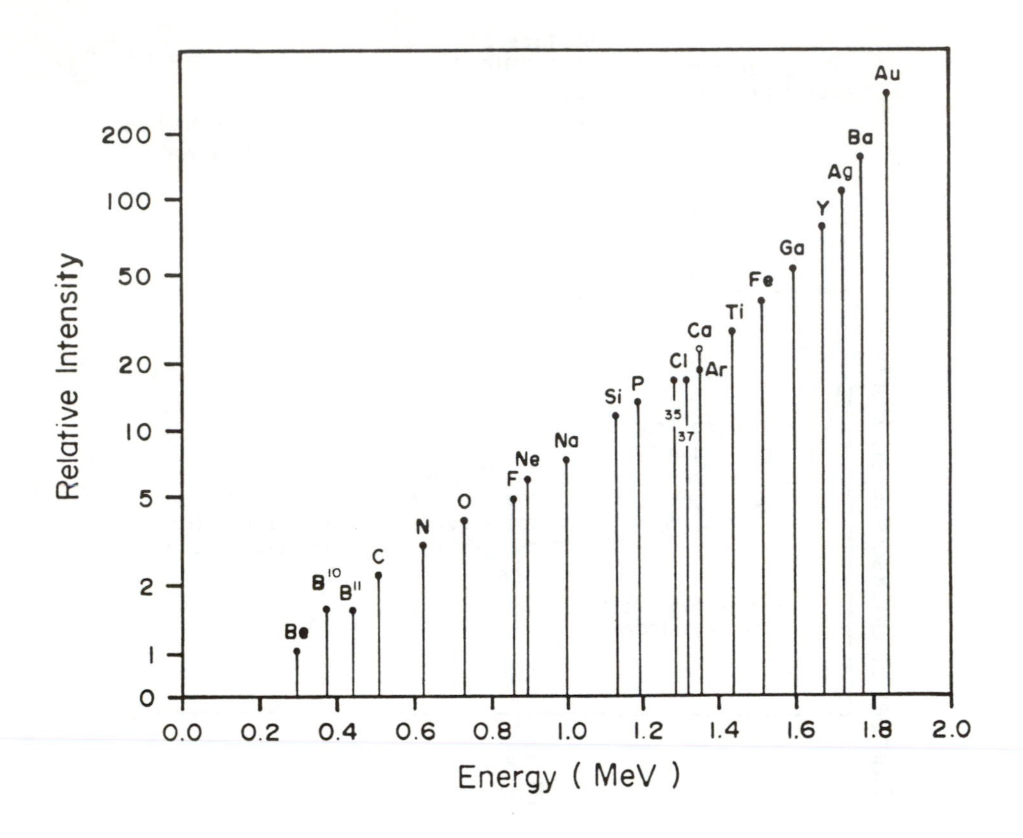

Figure 6

Relative scattering intensity of different elements as a function of the energy with which 2 MeV helium ions are backscattered from their surface at 168°. Note that isobaric elements scatter helium ions with the same energy and are therefore indistinguishable from each other e.g. Ca and Ar.

approximation it is assumed that the energy lost in traversing the film is negligible so that the scattering cross-section, which is a function of energy, does not change. For thick films or for determination of impurity concentration profiles with depth, the variation of cross-section with energy and the variation of energy loss rate with energy of the helium particles has to be taken into account and is this usually done through the use of simulation codes such as RUMP [12].

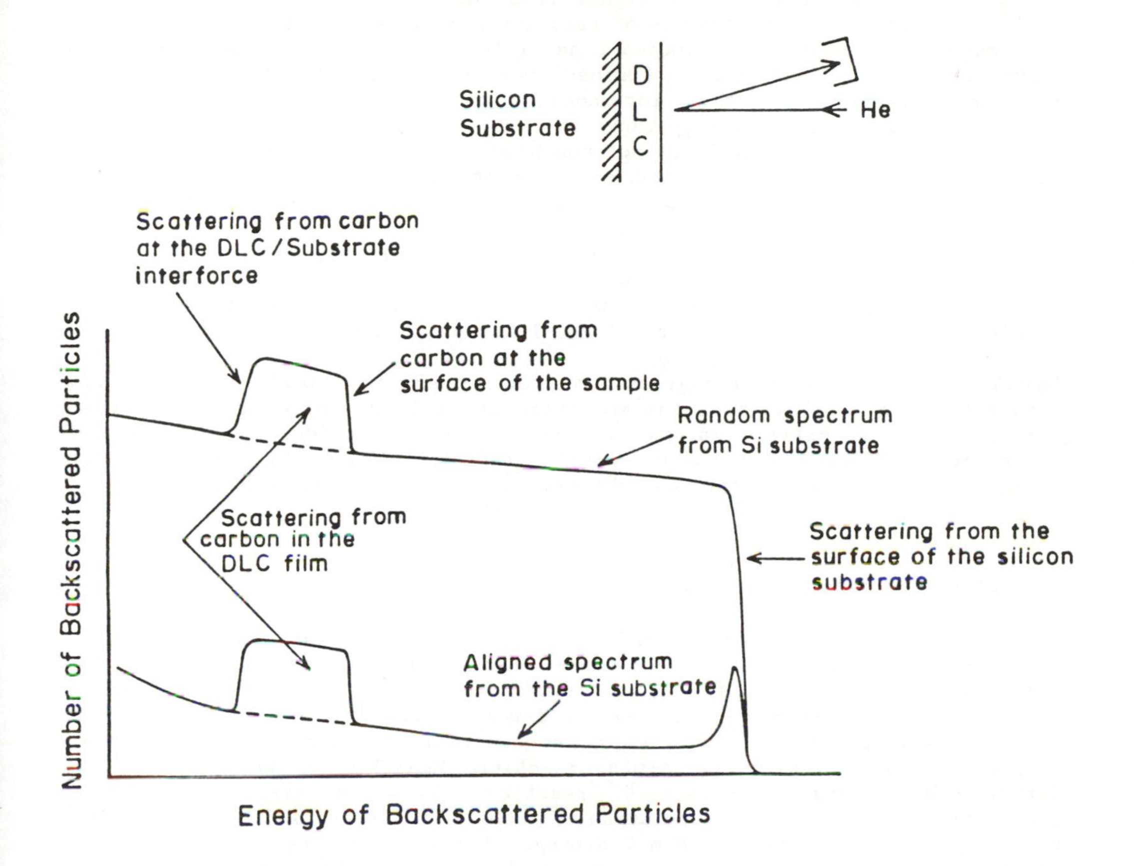

Figure 7

Two RBS spectra from a typical DLC film on a single crystal silicon substrate. The lower spectrum was obtained when the principle axis normal to the substrate surface was aligned with the beam. The upper spectrum was obtained while the sample was deliberately misaligned to avoid channeling in the substrate.

A typical spectrum from a DLC sample on silicon is shown in figure 7. It illustrates the property of RBS that low mass films on higher mass substrates can be difficult to analyze owing to the high background count from the substrate. It is possible subtract this background by fitting a curve to the silicon spectrum at lower and higher energies than the signal from the carbon film. This can be done such that the area under the carbon signal can be found to better than 1 %. Any higher accuracy is superfluous since the cross-section, the counting statistics, the beam current or the detector acceptance angle are usually not known to that accuracy.

It is possible to increase the signal from carbon relative to that from the silicon substrate through the use of resonant elastic scattering [10,11,13]. In this mode the energy of the incident particle and its mass are set so that a nonRutherford cross-section is obtained when the scattering event occurs at a magnitude greater than that for conventional Rutherford scattering. The disadvantage is that the energy range over which the cross-section is enhanced may be quite small so that it is not enhanced uniformly through the film to be analyzed. Also, the magnitude of the cross-section may not be known to the same accuracy as the Rutherford cross-section, which can be predicted reasonably accurately since it is derived from the Coulomb potential, which is itself an analytic function. With 1 MeV protons there is a fairly broad and uniform enhancement of the cross-section for scattering from carbon [13] this can be used to give a factor of 3.9 increase in the carbon signal relative to that predicted from the coulomb potential for 1 MeV protons scattering off carbon.

Another way to reduce the background from the substrate is to align the ion beam with a major crystal axis of the substrate if it is a single crystal such as electronic grade silicon wafers, figure 7. This can reduce the number of scattered particles from the substrate by a factor of 10 but this depends on the thickness the over lying film and the crystal quality of the substrate.

NUCLEAR REACTION ANALYSIS (NRA)

With nuclear reaction analysis (NRA) it is possible to analyze for specific elements and/or isotopes by using a high energy ion beam to excite particular states within the nucleus of the atom to be detected [13]. This causes various particles to be emitted depending on the ion atom combination and the energy of the ion. NRA is typically limited to concentration limits of 1 to 10 at.%, strongly dependent on the particular reaction. Examples of reactions used to detect hydrogen are the $F^{19}(p,\alpha\gamma)O^{16}$ reaction [14] which uses a fluorine ion energy of 6.4 MeV and has a resolution of about 20 nm in DLC and the $N^{15}(p,\alpha\gamma)O^{16}$ reaction [15] which uses a 6.4 MeV nitrogen ion and has a resolution of about 10 nm in DLC. For both of these reactions the same technique is used to obtain a profile of the hydrogen content of the film to be analyzed. A gamma ray detector is used to monitor gamma ray emission from the target. The ion energy is increased from just below the reaction energy in small steps. Each time it is increased a gamma ray spectrum is taken and the number of gamma ray counts from the particular reaction is plotted as a function of ion energy. As the energy is increased the depth at which the reaction takes place also increases such that it always coincides with the depth at which electronic energy loss of the ion to the target atoms causes the ion energy to fall to that just required for the reaction to take place. Both of these reactions have good depth sensitivity and sufficient concentration sensitivity to allow profiling of the hydrogen content of DLC fairly easily. However, for both it is necessary to have a standard available if total amounts of hydrogen are required. Also, care needs to be taken to ensure that exposure to such a high mass ion beam does not result in significant loss of hydrogen from the film during analysis. This is simple to check for by taking repeated spectra from the same position on the film with the same energy ion beam and ensuring that the gamma ray yield does not change. As will be seen later, a 6.4 MeV fluorine ion beam does apparently cause significant hydrogen loss which eventually stabilizes. It has been reported by Westerburg et al. [16] that the nitrogen reaction causes insignificant loss of hydrogen from a film deposited from pyrolysed methane. This reaction has also been used by Angus et al. [17] who found that the profile of hydrogen in sputtered deposited DLC was deficient in hydrogen in the first 50 nm., However, since they used the dual ion beam technique shown in figure 1 to make the film it is likely that the depletion was due to the low energy ion irradiation during deposition driving hydrogen either further into the film or out from the surface.

PROTON RECOIL DETECTION (PRD)

This is a specialization of elastic recoil detection (ERD). ERD has been used to analyze for light impurities in heavy substrates by using 35 MeV Cl ions to recoil the light nuclei out of the target with near Rutherford cross-sections [18]. PRD uses an MeV beam of helium particles to recoil hydrogen nuclei or protons out of the material to be analyzed [19]. The significant difference is that the cross-section is non-Rutherford and is known to be reasonably flat in the region of 3 MeV [19]. The analysis system is similar to RBS except that the silicon barrier detector, used to analyze the energy of the recoiled protons, is placed in a forward position, typically at 30° to the beam and 15° to the surface of the target, figure 8. Also, a thin foil of either mylar or aluminum is placed in front of the detector to prevent forward scattered helium particles from entering the detector.

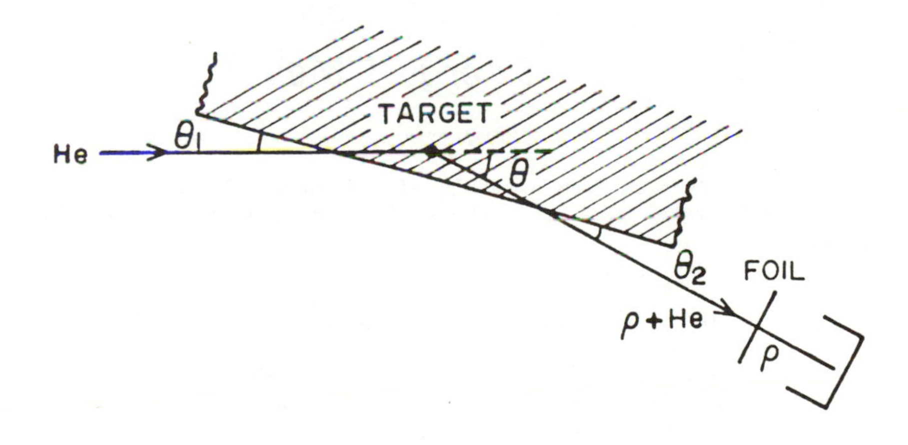

Figure 8

Geometric arrangement for PRD showing the helium ion beam, target with surface layer to be probed for its hydrogen content, foil to stop forward scattered helium particles from entering the detector, and silicon barrier detector used in the energy analysis of the recoiled protons or hydrogen nuclei from the target.

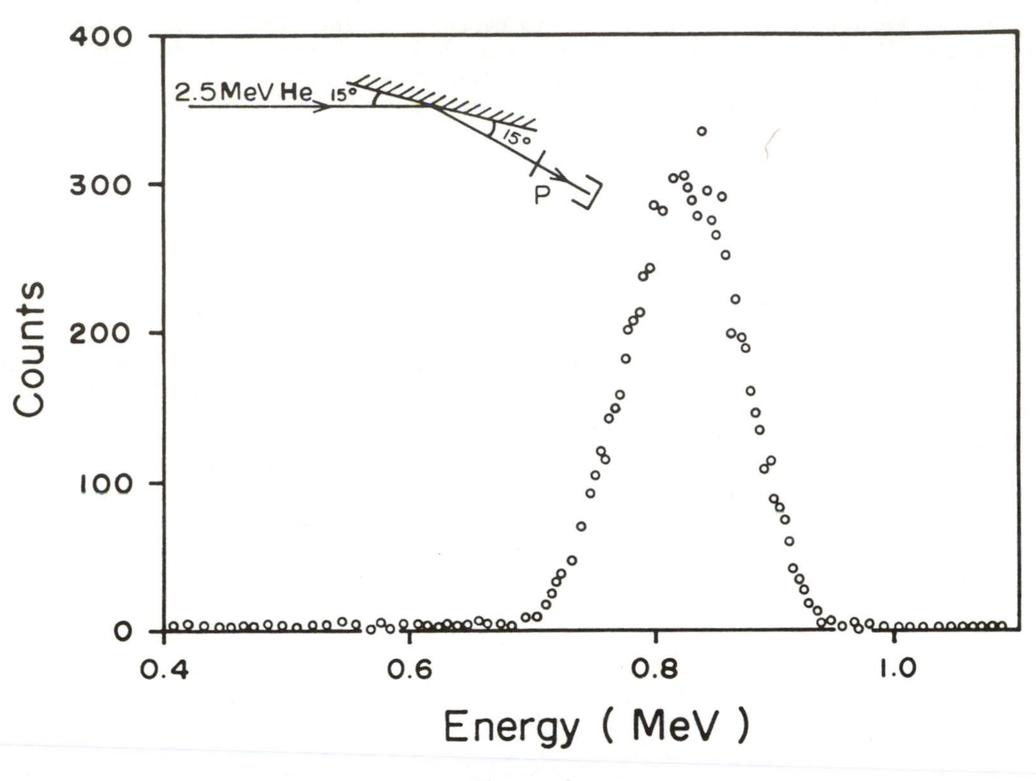

Figure 9

Typical PRD spectrum from a sample of DLC about 100 nm thick on a silicon substrate. The inset shows the recoil geometry for this spectrum.

Because helium nuclei have a significantly greater electronic energy loss rate than protons in the same material the energy lost in the target by the recoiled proton may be ignored when converting the energy spectrum of recoiled protons into a depth scale [20]. Thus, the energy of the recoiled proton may be considered to be made up of two components. The first is energy transferred to the proton in the recoil event from the helium particle. The energy of the helium particle will be less any energy it has lost as electronic energy loss in travelling to the depth at which the recoil event has occurred. The second significant effect is the energy lost by the proton the foil used to stop helium particles entering the detector. The energy loss of the protons in the foil leads to a significant energy straggle and loss of depth resolution.

If the cross-section for the recoil event is known then, like RBS, the amount of hydrogen in the target may be determined from the number of counts in the spectrum and a knowledge of the beam charge and detector acceptance angle. A typical spectrum from a 100 nm DLC film is shown in figure 9.

The depth resolution is not as good as for the nuclear reaction techniques (100 nm) but an entire spectrum can be obtained in less beam exposure time than a single data point for the two NRA techniques cited above. This is because data at all depths can be gathered simultaneously and the cross-section is very much higher. This has allowed the technique to be exploited in measuring hydrogen profiles in polymers which are very radiation sensitive [20].

ION BEAM MODIFICATION OF DLC

As an ion travels through a solid it loses energy through two separate mechanisms. It loses energy to the electronic structure of the atoms of the solid which is sometimes referred to electronic energy loss. It also loses energy through direct ion-atom collisions. This mechanism is referred to as nuclear energy loss or displacement energy loss. As displaced atoms move off their lattice sites, assuming they have received sufficient energy to cause an adiabatic displacement (about 25 eV), then they in turn lose energy to the other atoms of the target by both mechanisms.

Consideration must be given to the effect of these energy loss processes when materials are being analyzed by ion beams since it is possible that permanent radiation damage will be done to them which may affect the data being collected. Many hydrocarbon systems are very sensitive to radiation damage. Once atoms are rearranged or large amounts of energy are deposited in the electronic system of these materials then permanent chemical changes may occur and volatile species may be desorbed. Also, since energetic particles are present at the fabrication of DLC then these effects may influence the composition and properties of DLC. During ion beam analysis of DLC the most likely chemical change is the loss of hydrogen, which will also occur if the material is heated over 400°C [21]. Therefore, one effect which must investigated is the loss of hydrogen from DLC during exposure to MeV ion beams of the particles used in the ion beam analysis of DLC.

Experiments have been performed where DLC samples have been irradiated with ions of various energies and masses [22,23] . In [22] DLC films 200 nm thick were irradiated with various ions and changes in the electrical resistivity of the films were attributed to damage produced by nuclear energy loss. Using the TRIM simulation code [24] with ion target combinations used in [22] it is possible to make some observations about the reported effects. The nuclear energy loss rate for the 270 keV xenon ions is much higher than that for the 50 keV carbon ions, as shown in table 1 by the difference in the number of vacancies per ion, which is directly related to the nuclear energy loss rate. But, the films were sufficiently thin for the ions to deposit all their energy and come to rest within the film. Thus, the total electronic energy deposited by each 50 keV carbon ion will be about five times lower than that from each 270 keV xenon ion and it is dispersed over a greater depth. Therefore, it is possible that if the effects reported in [22] were due to electronic loss rather than nuclear energy loss, then the critical fluence for the observed changes would be much higher for the 50 keV carbon ions than for the 270 keV xenon ions, which is the case for the observations reported in [22]. Thus, it is unclear from this work which energy loss mechanism is responsible for the observed changes.

Table 1

Electronic energy loss fractions, vacancy production and mean ranges from TRIM 87 [24]: target was DLC from [22]; 73 at.% C, 22 at.% H, 5 at.% O; density 2 g cm^{-3}; displacement energy 20 eV; binding energy 1 eV.

Ion Species and energy	Electronic energy loss [%]	Vacancies per ion	Mean range [nm]
50 keV C	85	210	125
110 keV Ar	75	670	96
270 keV Xe	77	1650	86

Ingram and McCormick have performed ion beam modification experiments on direct ion beam deposited DLC [23]. The films were fabricated as described earlier in this chapter. The films were about 160 nm thick and were irradiated with either 1 MeV gold ions or 6.4 MeV fluorine ions. The fluorine beam was also used to obtain concentration profile as a function of depth of the hydrogen in the films in addition to performing in situ measurements of hydrogen loss from the films during irradiation. These two beams were chosen because they would deposit roughly the same amount of energy per ion in electronic energy losses but would differ greatly in the amount atomic displacements they would achieve, table 2. The gold ions producing about 350 times more displacements than the fluorine ions. Neither the gold nor the fluorine ions were likely to come to rest in the films, so the energy loss rates would be constant across the films and there would be no doping effects from the presence of the ions.

Table 2

Electronic energy loss rate and displacement rate from TRIM 87 [24]: target was DLC from [23]; 70 at.% C , 30 at.% H; density 2 g cm^{-3}; displacement energy 20 eV; binding energy 1 eV.

Ion Species and energy	Electronic energy loss [eV/ion/nm]	Displacements per ion per nm
1 MeV Au	3300	18
6.4 MeV F	2500	0.051

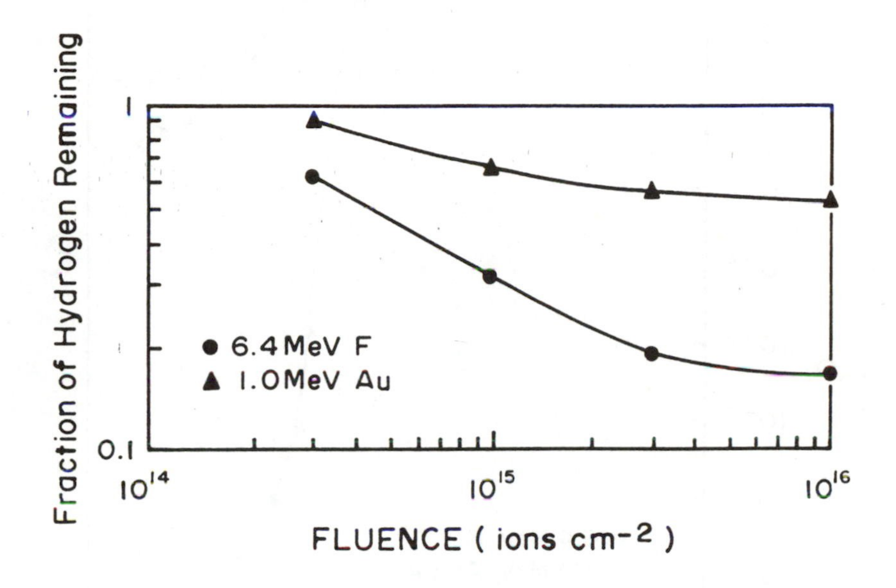

Figure 10

Hydrogen retained by DLC film as a function of fluence following bombardment by 6.4MeV fluorine or 1 MeV gold, as measured by PRD.

In [23] the hydrogen in the films as a function of ion fluence was monitored by PRD, figure 10. PRD had been determined not to cause significant hydrogen loss during irradiation [23], also using RBS no carbon loss was detected. From the measurements of electrical resistivity with ion fluence very large changes in the resistivity were seen for the fluorine ions once a critical fluence had been achieved figure 11.

In summary, both the hydrogen loss data and the change in resistivity data show a threshold for their onsets which occur at roughly the same fluences. The fluorine ion beam is much more efficient at removing hydrogen and permanently changing the resistivity of the DLC than the gold ion beam. If the gold ion beam was off the target several hours then some recovery of the resistivity occurred but this was lost once the beam was turned back on.

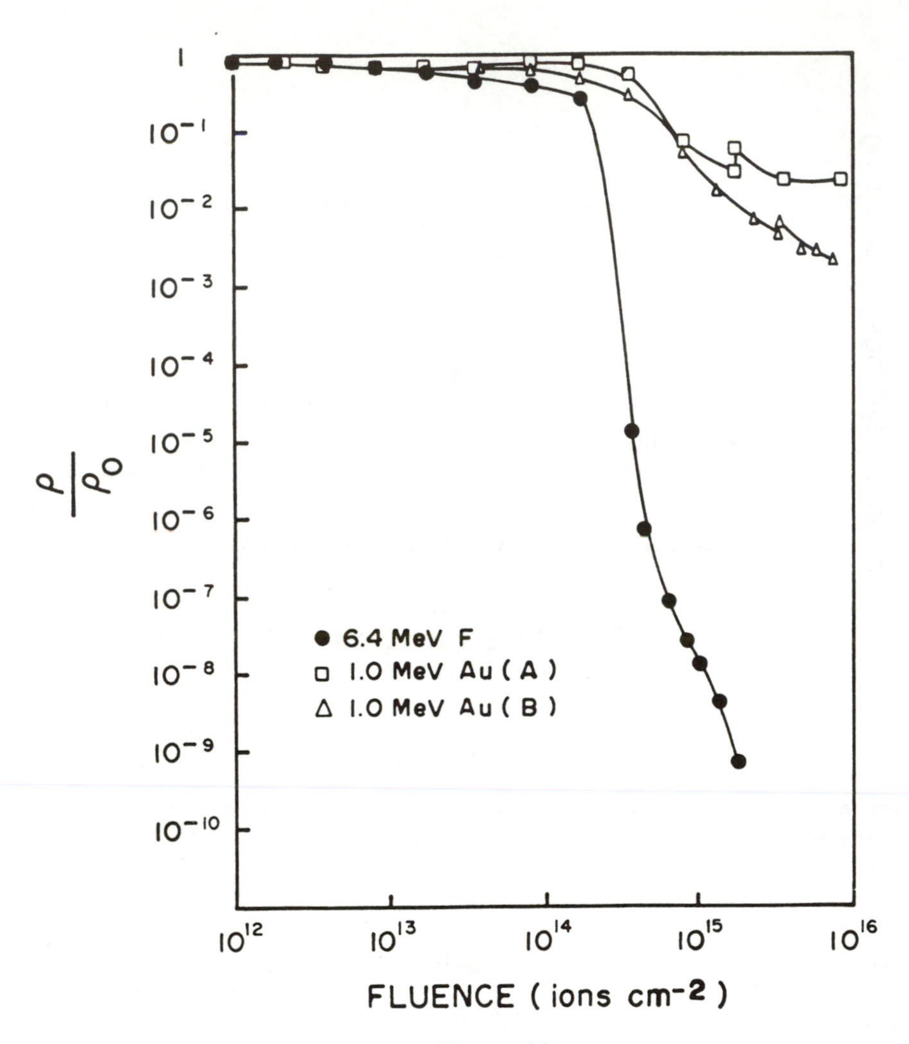

Figure 11

Change in resistivity of DLC films on fused silica as a function of fluence of 6.4 MeV fluorine or 1 MeV gold. Data from two samples is shown which were both irradiated with gold, A and B, both show some recovery of resistivity if the beam is off for an extended period (16 hours). This was not found for fluorine irradiated samples.

The conclusion of the work in [23] was that electronic energy lost to the films during ion bombardment was the cause of the hydrogen loss and the change in resistivity. The threshold effect is likely to be caused by some minimum fluence being required to set up low resistance and high speed diffusion paths in the films through consecutive overlap of ion tracks or cascades. The fact that the gold ions were apparently less able to remove hydrogen than the fluorine ions depositing similar amounts of electronic energy was because the gold ions also created many more retrapping centers for the hydrogen. Thus, although the gold ions may have detrapped the hydrogen many new trapping centers were immediately available for the hydrogen. This ability of the nuclear damage to help retain hydrogen also explains why hydrogen is retained at all by ion beam deposited DLC. It might be expected that a methane ion arriving at a surface with 1000 eV would dissociate on impact. But some of the hydrogen is retained because the damage created during the deposition is similar to the displacement damage created during the bombardment by 1 MeV gold ions. Using TRIM 87 it is possible to show that the median energy of the displaced atoms even in the cascade created by a 1 MeV gold ion is about 100 eV. Thus since ions of roughly the same energy are present during both deposition and modification then it is likely that similar defects with similar properties will also be present.

SUMMARY

Ion beams have been shown to be useful not in only fabricating and analyzing DLC but also in understanding the fabrication process. It has been shown that the radiation damage present during irradiation by heavy ions causes loss of hydrogen through electronic excitation and, concurrently, the resistivity of the films also changes. However, if high levels of displacement damage also occur during the irradiation then the loss hydrogen and the change in resistivity are both reduced. Thus, electronic excitation stimulates the loss of hydrogen and the change in resistivity and displacement damage helps to anneal the effect. Since the displacement damage occurs with particle moving with similar energies to those present during the deposition of the DLC it is likely that this environment also promotes hydrogen retention and high resistivity during deposition of DLC.

Ions beams have been shown to be useful in characterizing DLC as well as depositing and modifying it. For carbon and any heavier impurities, the total number of atoms in a film can measured by RBS. PRD may be used to determine the hydrogen content of the films. In using any ion beam technique care needs to be exercised to ensure that the ion beam does not cause sufficient radiation damage during the measurement so as to invalidate the measurement. This is particularly true for NRA where beams much heavier than helium are used and the cross-sections are not as high as for PRD.

REFERENCES

[1] S. Aisenburg, R. Chabot, J. Appl. Phys. 42 1971 2953.

[2] R. Memming, Thin Solid Films 142 1986 145.

[3] J.C. Angus, P. Koidl, S. Domitz, "Plasma Deposited Thin Films", ed. J. Mort, F. Jansen, pub. CRC Press 1986.

[4] D.G. Armour, P. Bailey, G Sharples, Vacuum 36 1986 769.

[5] S. Withrow, International Conference on Low Energy Ion Beams, Surrey University, UK 1989.

[6] J.H. Freeman, W. Temple, G.A. Gard, Nature 1978, vol 275 634.

[7] J.H. Freeman, W. Temple, G.A. Gard, Vacuum 1984, vol 34 305.

[8] P. Sigmund, "Sputtering by Particle Bombardment 1", ed. R. Behrisch, Topics in Applied Physics, vol 47, Springer Verlag, 1981.

[9] H.R. Kaufman, J. Vac. Sci. Technol. 15 1978 272.

[10] W-K Chu, "Backscattering Spectrometry", Academic Press 1972.

[11] L.C. Feldman, J.W. Mayer, "Fundamentals of Surface and THin Film Analysis", North-Holland, 1986.

[12] RUMP, for information contact Computer Graphic Service, 221 Asbury Rd. Lansing, NY 14882.

[13] J.W. Mayer, E. Rimini, "Ion Beam Handbook for Material Analysis", Academic Press, 1977.

[14] C.A. Barnes, J.C. Overley, Z.E. Switkowski, T.A. Tombrello Appl. Phys. Lett. 31 1977 239.

[15] W.A. Lanford, H.P. Trautvetter, J.F. Ziegler, J. Keller Appl. Phys. Lett. 1976 34 173.

[16] L. Westerburg, L.E. Svensson, E. Karlsson, M.W. Richardson, K. Lundstrom, Nucl. Inst. and Meths. in Phys. Res. 1985 B5 49.

[17] J.C. Angus, J.E. Stultz, P.J. Shiller, J.R. Macdonald, M.J. Mirtech, S. Domitz, Thin Solid Films 1984 118 311.

[18] J.L'Ecuyer, C. Brassard, C. Cardinal, J. Chabbal, L. Deschen, J.P. Labrie, B. Terreault, J.G. Martel, R. St-Jacques, J. Appl. Phys. 47 1976 381.

[19] D.C. Ingram, A.W. McCormick, Nucl. Inst. and Meths. in Phys. Res. B34 1988 68.

[20] J.D. Carlson, P.P. Pronko, D.C. Ingram, Materials Reasearch Society Annual Meeting, Symp. on Ion Implantation, Boston 1983.

[21] P. Bailey, D.G. Armour, J.B.A. England, N.R.S. Tait, D.W.L. Tolfree, Daresbury Laboratory Report DL/NUC/P169A, Daresbury, Warrington U.K. 1982.

[22] S. Prawer, R. Kalish, M. Adel, V. Richter, J. Appl. Phys. 61 1987 4492.

[23] D.C. Ingram, A.W. McCormick, Nucl. Inst. and Meths. in Phys. Res. B34 1988 68.

[24] J.F. Ziegler, J.P. Biersack, U. Littmark, "The Stopping and Range of Ions in Solids" Vol. 1, Pergamon Press 1980.

Materials Science Forum Vols. 52 & 53 (1989) pp. 495-514

MICROBEAM ANALYSIS STUDIES OF a-C:H FILMS

A.G. Fitzgerald and A.E. Henderson
Department of Applied Physics and
Electronic & Manufacturing Engineering
University of Dundee, Dundee DD1 4HN, Scotland

ABSTRACT

Carbon films prepared by sputtering, Fast Atom Beam (FAB), rf glow discharge, plasma assisted hybrid PVD and ion beam deposition methods have been studied by a range of microbeam analytical techniques. The presence of hydrogen and hydrocarbon components in these films, regardless of whether hydrocarbon gas is used in the deposition process, has been established. Crystalline components have also been identified in these films as α-carbyne or Ries crater carbon. Both are linearly polymerized forms of carbon. Polymerized forms of carbon have also been identified with larger crystal lattice parameters. The sensitivity of these carbon films to electron and ion irradiation has been investigated. A qualitative investigation of the degree of tetrahedral sp^3 bonding has been made by Auger electron spectroscopy, and scanning electron diffraction.

INTRODUCTION

Hydrogenated carbon (a-C:H) films can be produced by a wide variety of preparation methods. Some examples are magnetron sputtering in an argon-hydrocarbon gas mixture [1], ion beam deposition from an argon-hydrocarbon gas mixture [2] and r.f. glow discharge in a hydrocarbon gas [3]. There is also evidence that hydrogen is included in some carbon films when hydrocarbon gas is not involved in the deposition process. The source of hydrogen is believed to be residual water vapour in the high vacuum system used in the preparation process, contamination from the vacuum pumping system [4,5] or from hydrogen contamination in the graphite source.

To understand the role of hydrogen included in amorphous carbon films it is useful to consider the possible bonding arrangements for carbon atoms in various forms of carbon. Robertson [6] has summarized the bonding possibilities. Carbon atoms form bonds in three different configurations. These are the sp^1, sp^2 and sp^3 bonding configurations. The sp^3 configuration has the four carbon valence electrons assigned to tetrahedral sp^3 hybrid orbitals. This hybrid orbital forms a strong σ bond with an adjacent carbon atom. This is the bonding arrangement in the diamond form of carbon. In the sp^2 configuration, three of the four electrons are assigned to the trigonally directed sp^2 hybrids which form σ bonds and the fourth electron lies in a $p_z(p\pi)$ orbital lying normal to the plane containing the σ bonds. The $p\pi$ orbital forms weaker π bonds with adjacent $p\pi$ orbitals. At sp^1 sites, only two electrons form σ bonds, and the remaining two electrons lie in orthogonal p_y and p_z orbitals forming π bonds. A σ bond between two sites is called a single bond. This type of bond is represented by a single line, while a σ-π bond pair is labelled as a double bond and is represented by two lines. Hydrocarbons containing only single bonds are termed 'saturated'. Unsaturated systems can take the form of a system of separate double bonds in 'olefinic' systems such as ethylene, $H_2C{=}CH_2$, or as delocalized or conjugated π bonded systems such as the aromatic six-membered rings in benzene (C_6H_6) and graphite. There is some evidence [5] that minor amounts of -C≡CH groups are present in a-C:H.

Many disordered forms of carbon have structures based on the graphite lattice since this is the stable allotrope of carbon. The short range order in these amorphous carbon films is determined mainly by two parameters, the form of carbon bonding and the hydrogen content. These two parameters do not completely define the carbon film structure because there exists a substantial degree of medium-range order on an approximately 10Å scale. It is believed that the sp^2 sites of a-C tend to occur in warped graphite layer clusters and the sp^2 and sp^3 sites in a-C:H are segregated and clustered [5]. The structure of amorphous carbon is of fundamental importance, particularly because of the effect of the disordered arrangement on the π electron system and the electron energy band structure. Since the π states are weakly bound, they lie closer to the Fermi level than the σ states. The result is that the filled π states form the valence band and the empty π^* states form the conduction band and therefore determine the size of the energy gap.

The relation between bonding in a-C:H films and the bonding in the less well known crystalline allotropic forms of carbon, such as α-carbyne, β-carbyne and carbon VI [7] has not been investigated although these crystalline forms are often grown during deposition of hydrogenated carbon films. The α-carbyne allotrope is believed to be a polymerized form of carbon with -C≡C- chains, while β-carbyne consists of =C=C= chains [8].

A complete characterization of a-C:H films clearly requires the correlation of results from structural, microstructural and analytical techniques. The use of photoelectron spectroscopy and Auger electron spectroscopy gives information about the nature of the valence band structure in a-C:H, and techniques such as electron microscopy and electron diffraction give information about microstructure and atomic arrangement. Correlation with the results from composition analysis techniques such as SIMS must also be made to enable the effects of hydrogen to be assessed.

EXPERIMENTAL

The a-C:H films investigated in this work have been prepared by a number of techniques including the use of fast atom bombardment (FAB) sources, a d.c. triode sputtering gun, a dual ion beam deposition system, an rf glow discharge system and a physical vapour deposition system. Butane, butane-argon mixtures and propane have been used with the FAB source to produce hydrogenated films. Methane was used in depositions using the dual ion beam system. A high purity graphite sputtering target was used with argon as the sputtering gas for films prepared with the d.c. triode sputtering gun. A high purity graphite target was also used to prepare films by physical vapour deposition. The films were deposited on glass and aluminium foil for surface analysis work and on single crystal sodium chloride substrates for transmission electron microscopy and electron diffraction studies.

Characterization of the a-C:H films has been made by transmission electron microscopy (TEM), electron diffraction, scanning electron diffraction, electron energy loss spectroscopy (EELS), x-ray photoelectron spectroscopy (XPS), Auger electron spectroscopy (AES) and secondary ion mass spectrometry (SIMS).

Hydrogenated carbon films are mainly amorphous in form and therefore a technique that can give direct structural information about these films is important. A scanning electron diffraction system has been specially developed for this programme of characterization of a-C:H films. The system is shown diagramatically in figure 1. The system has been constructed by adapting an AEI transmission electron microscope. This instrument has a facility for imaging electron diffraction patterns at high resolution from thin films placed below the final (projector) lens. The modification of the instrument for scanning electron diffraction has involved placing coils below this specimen position. These coils are used to scan the high resolution electron diffraction pattern across the electron detector. The scan is controlled by an APPLE IIe microcomputer. The electron detector consists of an electron energy loss spectrometer, used to filter out inelastically scattered electrons from the electron beam followed by a scintillation screen and a photomultiplier tube. The EELS spectrometer is based on a design of Egerton [9]. The Apple IIe microcomputer scans the high resolution electron diffraction pattern across the entrance aperture to the EELS spectrometer in incremental steps, storing electron intensity for each step. The APPLE IIe microcomputer is interfaced to a PRIME mainframe computer and the file of the electron intensity distribution in the amorphous electron diffraction pattern forms the input data for radial distribution function analysis.

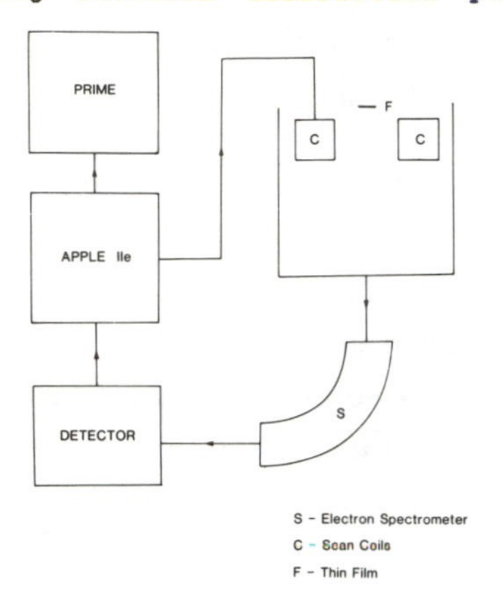

Figure 1 Schematic diagram of the scanning electron diffraction system.

A JEOL 100C STEM and an AEI EM6G TEM have been used for high resolution microstructural investigations of the a-C:H films. Surface analytical investigations have been made with a VG HB100 surface analytical system. This system consists of a scanning Auger electron microscope with a field emission electron gun, a CLAM 100 electron spectrometer used for Auger electron spectroscopy and x-ray photoelectron spectroscopy and an SQ300 mass spectrometer for SIMS. A FAB61 fast atom beam source is used for SIMS work and an AG60 ion gun was used for specimen cleaning for AES and XPS. Electron energy loss spectrometry studies were carried out at Glasgow University on a VG electron spectrometer fitted to a VG HB501 STEM.

STRUCTURE AND MICROSTRUCTURE

A number of techniques can give information about the atomic arrangement in a-C:H. Diffraction techniques are the most direct. McKenzie et al [10] and Sproul et al [11] have studied this material by electron diffraction. McKenzie et al [10] have interpreted the interference function obtained from these diffraction experiments in terms of both graphitic and hydrocarbon polymer domains. Hydrogen does not contribute to the interference function and the interference function measured in their work peaked at wave vectors characteristic of graphite. A peak at approximately 2Å^{-1} was believed to be associated with hydrocarbon domains.

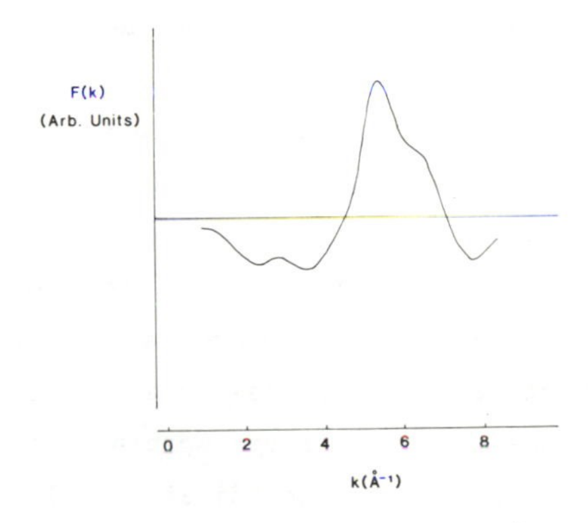

Figure 2 Interference function obtained by processing the electron intensity distribution from an a-C:H film prepared by rf glow discharge.

Preliminary structure studies have been made by scanning electron diffraction. Figure 2 shows an interference function obtained recently by scanning electron diffraction from a carbon film prepared by rf glow discharge. A micrograph of the corresponding electron diffraction pattern and the corresponding electron intensity distribution in reciprocal space obtained by the microcomputer controlled scanning electron diffraction system are shown

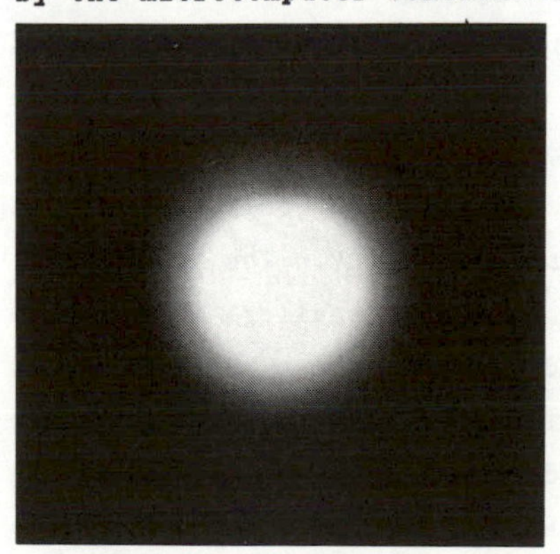

Figure 3 Electron diffraction pattern from an a-C:H film prepared by rf glow discharge.

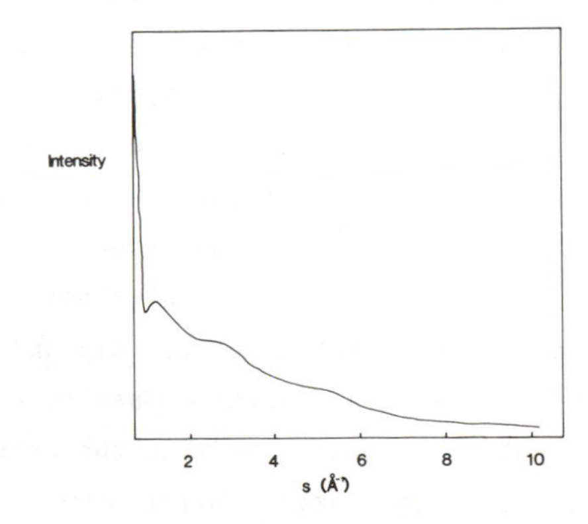

Figure 4 Intensity distribution in the electron diffraction pattern shown in figure 3 recorded with the scanning electron diffraction system.

in figures 3 and 4. The interference function has peaks at 1.1, 2.8 and 5.1Å^{-1} which are characteristic of a-C. The positions of these peaks agree well with the continuous random network model of Beeman et al [12] for amorphous carbon composed of 50% sp^3 and 50% sp^2 bonding. A plot of the radial distribution function from this film is shown in figure 5. The nearest neighbour spacing and nearest neighbour coordination obtained from this film are 0.15nm and 3.2 respectively. The values obtained for the nearest neighbour spacing and nearest neighbour coordination also agree well with the Beeman model [12].

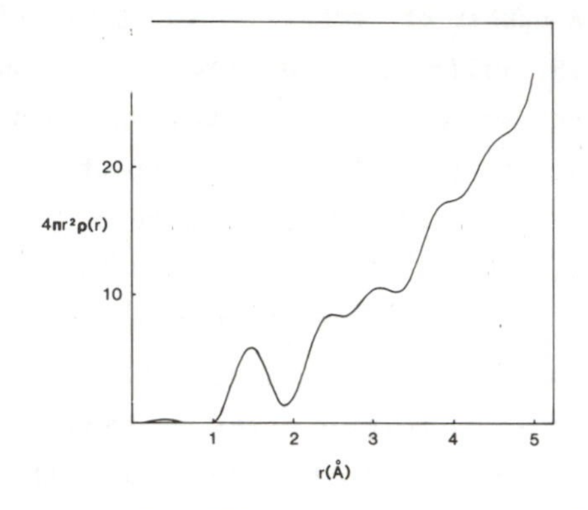

Figure 5 Plot of the radial distribution function obtained after processing the electron intensity distribution shown in figure 4.

High resolution scanning electron diffraction patterns originate from areas of the order of 200 μm. Structural and microstructural information from localised areas of the film require high resolution electron microscopy and selected area electron diffraction. High resolution studies of a-C:H films have revealed that the films are not entirely amorphous. Crystals are observed distributed randomly over the surface of the mainly amorphous film. Many of these crystals clearly belong to a layer-like crystal structure with dislocation networks resulting from slip between crystal layers. Figure 6 is an electron micrograph showing a layer structure crystal with dislocation networks grown in a film prepared by deposition from a FAB source using butane gas. A typical selected area electron diffraction pattern from one of these layer-like crystals is shown in figure 7. The pattern is composed of an arrangement of diffraction spots with hexagonal symmetry. The reciprocal lattice spacings arising in these patterns are much smaller than those expected from graphite or diamond. Detailed investigation of previous structural studies of carbon has revealed, as discussed earlier, that carbon does not exist exclusively in the graphite and diamond forms and can occur in at least eight other allotropes, some of which are complex structures [7,14,15]. The α-carbyne form of carbon is believed to be a carbon chain polymer possibly consisting of alternately single and triple bonded carbon atoms [15]. To index these electron diffraction patterns exactly in this work a computer program developed by Rhoades [16] has been used. This program

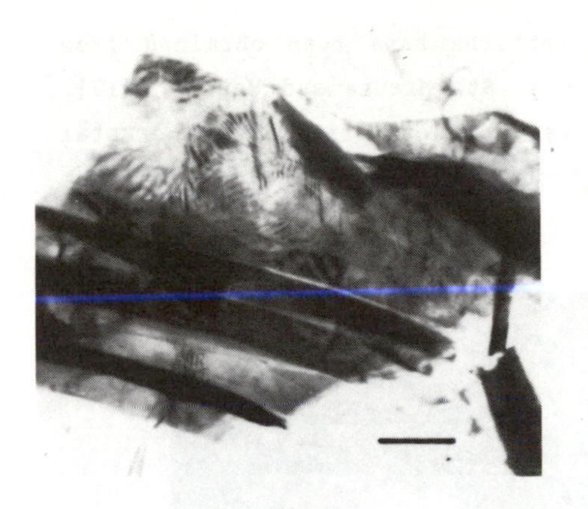

Figure 6 A crystal identified as either the α-carbine or Ries crater form of carbon (Fitzgerald et al[13]). Scale mark 0.2μm.

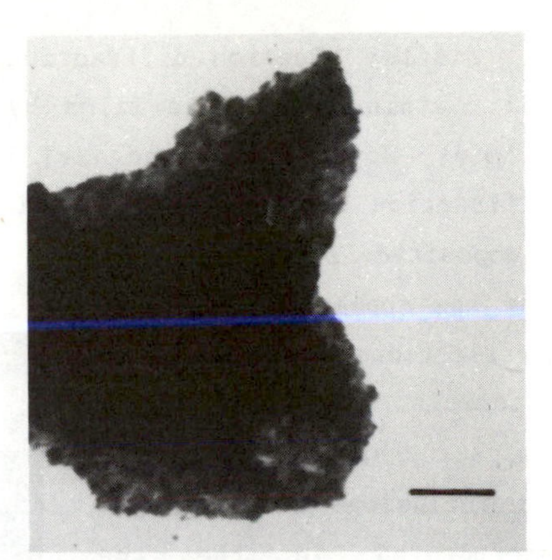

Figure 7 Electron diffraction pattern from the crystal in figure 6, the zone axis is (001) (Fitzgerald et al[13]).

enables a rapid comparison to be made between possible fits to an electron diffraction pattern from each of the carbon allotropes, some of which have extremely complex crystal structures. The crystal shown in figure 6 has been identified from the electron diffraction pattern (figure 7) as either a crystal of α-carbyne or a crystal of Ries crater carbon. These two hexagonal crystal structures appear to be almost identical, differing principally in the length of the c-axis of the unit cell.

Figure 8 shows a crystallite in a carbon film prepared by means of a triode sputtering gun from a graphite target with argon as the sputtering gas. Figure 9 is the electron diffraction pattern from the crystal in figure 8 that contains forbidden diamond reflections and has been identified as either the α-carbine or Ries crater forms of carbon with (013) zone axis in each case. If the fact that (110) type diamond reflections are forbidden is ignored, this electron diffraction pattern can be indexed as a (111) diamond

Figure 8 Electron micrograph of a crystallite in a film produced by sputtering from graphite using a d.c. triode source (Fitzgerald et al[4]). Scale mark 0.3μm.

orientation. Similar electron diffraction patterns have been obtained from crystallites in thin diamondlike films by Has, St. Mitura and Wendler [17]. Spencer, Schmidt, Joy and Sasalone [18] have also identified single crystal electron diffraction patterns composed of forbidden diamond reflections from ion beam deposited films. In the latter paper the appearance of these forbidden reflections was attributed to the possible formation of a diamond superstructure formed by the ordering of inclusions of noble gas atoms incorporated in the the crystallites during growth. This hypothesis is not unreasonable since Evans and Phaal [19] have obtained similar forbidden reflections in diffraction patterns from natural diamond crystals which they believed to result from the ordered incorporation of nitrogen.

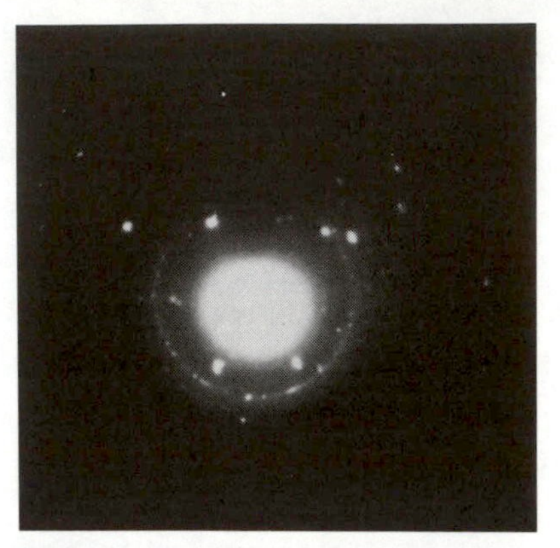

Figure 9 Electron diffraction pattern from the crystal in figure 8, zone axis (013) (Fitzgerald et al[4]).

Some crystallites obtained in carbon films prepared by ion beam techniques have been found to be sensitive to ion and electron beam irradiation. Figure 10 shows electron micrographs obtained from a crystallite of α-carbyne or Ries crater carbon as it is exposed to a prolonged period of electron irradiation.

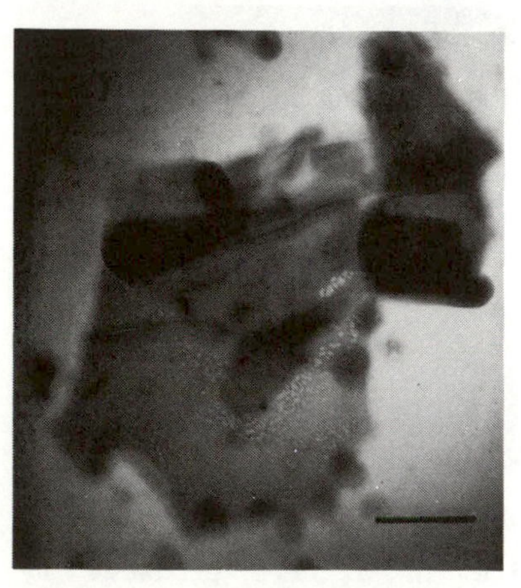

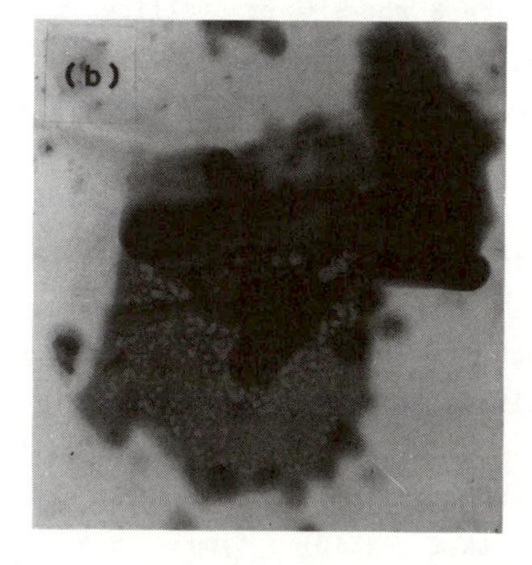

Figure 10 Electron micrograph (a) taken after a short exposure to the electron beam in the STEM, (b) after a longer period of electron beam irradiation, showing enlarged areas that appear to have decomposed. Scale mark 0.3μm.

Areas of the crystal show enhanced electron transmission with increased electron irradiation time and it appears that some of the crystal material has evaporated. The material was produced using a FAB source.

Another unusual property of crystallites dispersed over the surface of a-C:H is the occurrence, in some films, of crystallites with crystal basal plane lattice parameters that are slightly larger than the accepted lattice parameters for α-carbyne and the Ries crater form of carbon. Figure 11 shows a crystal and corresponding electron diffraction pattern from a film with a basal plane lattice parameter in this hexagonal structure of 9.417Å which should be compared with the accepted value of 8.92Å for α-carbyne. Crystallites exhibiting these variations in lattice parameter appear to be confined to films deposited using ion beam sources where argon ion beam irradiation occurs during film deposition [20]. High temperature experiments have been carried out in a heating stage in the STEM. These high temperature anneals result in amorphization of the crystalline material. It is probable that this amorphization is accompanied by the emission of included gas.

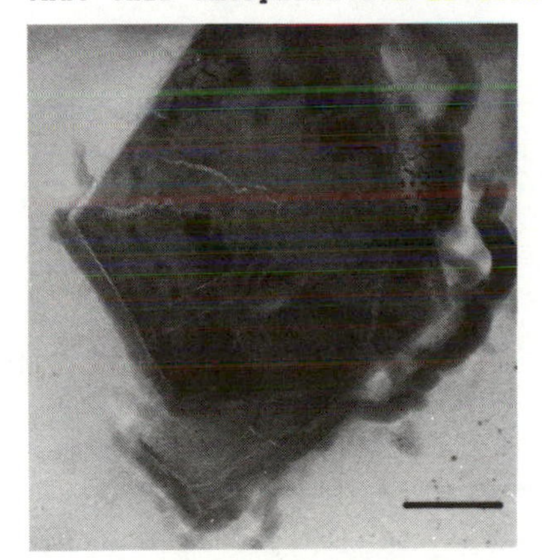

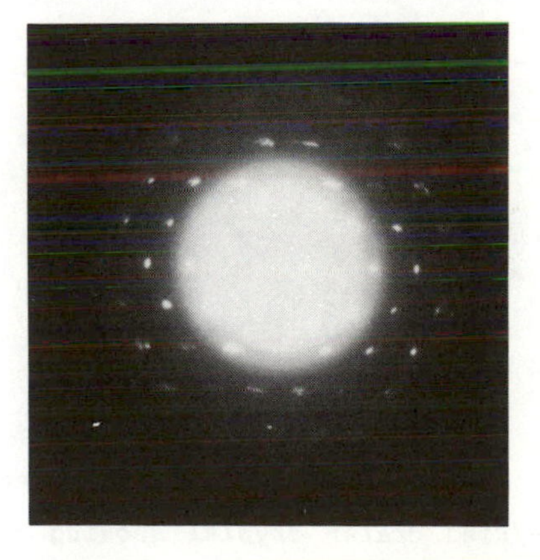

Figure 11 (a) Electron micrograph from a crystal with larger lattice parameters than either α-carbine or Ries crater carbon, (b) electron diffraction pattern from the crystal in (a). The material was produced using a dual ion beam system with methane gas. Scale mark 0.2μm.

ELECTRON ENERGY LOSS SPECTROSCOPY

Electron energy loss spectroscopy is an important technique for characterization of a-C:H films since the complete thickness of the film is analysed. The two main regions of interest in the electron energy loss spectrum from a carbon film are the carbon K-edge and the low loss region of the spectrum. Fink et al [21] have demonstrated the range of near-edge structures that are obtained from a range of a-C:H films subjected to different annealing temperatures, and from amorphous carbon, diamond and graphite. In carbon compounds having only σ bonds (sp^3) the K-edge starts at about 290eV and in diamond this edge is at 288.9eV. Carbon compounds with π electron bonding are characterized by an absorption edge near 284eV as a result of the lower lying antibonding π^* states. Fink et al [21] have shown that a-C:H films contain a large peak at about 285eV indicating a significant proportion of carbon atoms with the sp^2 configuration. This proportion increases as the films are annealed at higher temperatures. Fine structure consisting of a shoulder at 287.5eV is believed to be due to carbon-hydrogen clusters which result in absorption maxima in the gap between π^* and σ^* absorption edges.

Information on the sp^3 to sp^2 electron configuration ratio can be obtained by analysing the valence excitations by evaluating the loss function $Im(1/\varepsilon)$ for a film [9].

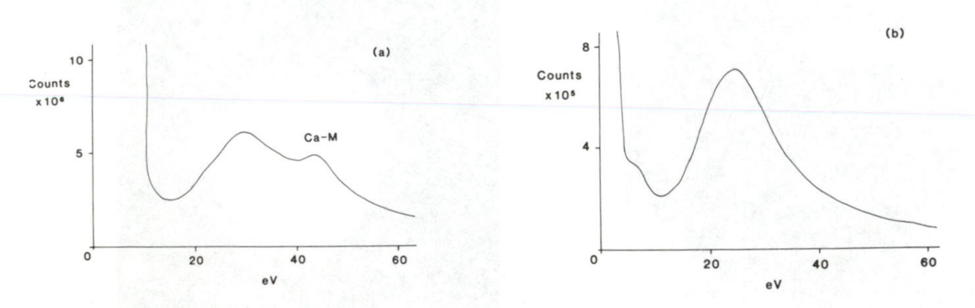

Figure 12 The low loss region in the electron energy-loss spectrum from (a) an α-carbine/Ries crater crystal showing the Ca M-edge and (b) the surrounding amorphous film. Material prepared by hybrid PVD using an exposed tungsten filament with high purity acetone and graphite.

The electron energy loss spectrometer used in the present study did not have an energy resolution capable of enabling a detailed investigation of fine structure in the low loss region of the electron energy loss spectrum. Since the EELS measurements were made in a STEM efforts were concentrated on making full use of the high resolution electron imaging capability of the instrument to enable the electron energy loss spectra from film and crystallites to be compared. Spectra over the low loss region obtained from an α-carbyne crystallite and from the surrounding amorphous film are shown in figure 12.

The spectra from a number of these crystallites have been recorded in this energy region and in each case a calcium M absorption-edge was detected. This edge did not appear in spectra from the amorphous areas of the film.

Some other features were also observed in the low loss region. A broad plasmon peak at 30eV accompanied the Ca M-edge in the energy-loss spectra from the α-carbyne/Ries crater crystals. The electron energy-loss spectrum from the amorphous film surrounding these crystals contained plasmon peaks at 7eV and 24eV. The position of these plasmon peaks should be compared with the position of plasmon peaks found in graphite by Liang and Cundy [22] at 7.2eV and 26.8eV, in diamond by Egerton and Whelan [23] at 33.3eV and in a-C:H by Fink et al [21] at 6.5eV and 22.9eV. The peak in these energy-loss spectra in the 6-7eV region is associated with π oscillations and the higher energy-loss peak is due to the oscillations of all the valence electrons ($\sigma + \pi$). The electron energy-loss spectra presented here indicate that the amorphous regions of the films exhibit peaks characteristic of a-C:H. However, the α-carbyne/Ries crater crystals exhibit broad plasmon peaks centred at an energy-loss of 30eV.

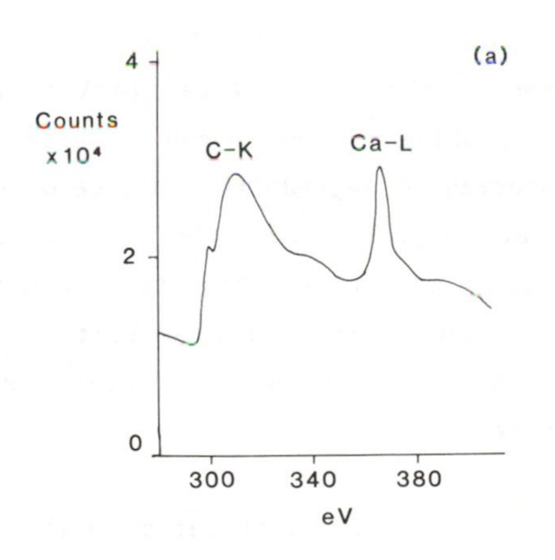

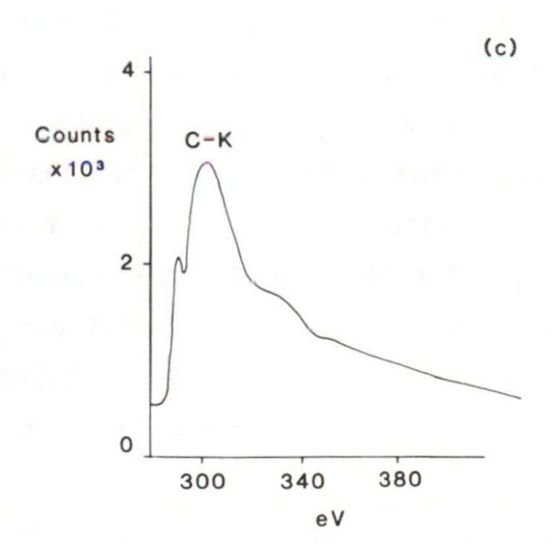

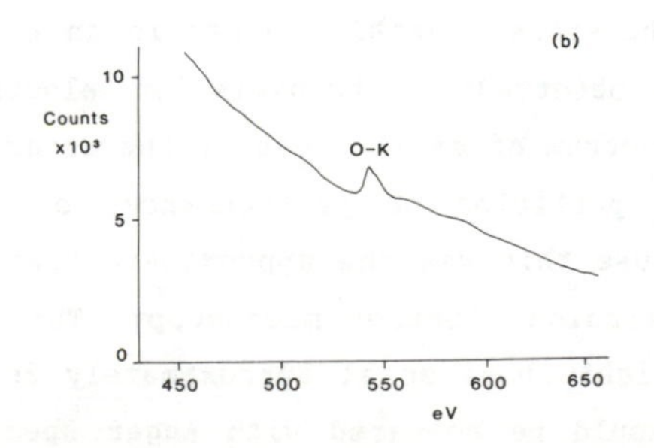

Figure 13 Electron energy-loss spectra from (a) an α-carbine/Ries crater crystal showing the C K-edge and the Ca L-edge, (b) this crystal showing the O K-edge and (c) the surrounding amorphous film. Material prepared by hybrid PVD using an exposed tungsten filament with high purity acetone and graphite.

Spectra have also been obtained from the energy range close to the carbon K-edge from α-carbyne crystallites and from amorphous areas of the film (figure 13). A calcium L-edge and oxygen K-edge were detected in electron energy loss spectra from the α-carbyne crystallites. The form of the carbon K-edge from both the α-carbyne crystallites and the surrounding film resembled the structure of the carbon K-edge obtained from amorphous carbon films by Colliex [24]. Comparison of the carbon K-edge structure with the spectra obtained by Fink et al [21] was difficult because the near-edge structure only was studied in their work, and in more detail, with a much higher resolution electron energy loss spectrometer.

AUGER ELECTRON SPECTROSCOPY

The main difference between Auger electron spectra from the surfaces of graphite, diamond and hydrogenated carbon films lies in the form of the carbon KLL Auger line shape . This difference in line shape reflects the different electron energy band structures of the three forms.

The great difficulty in analysing Auger spectra from these different forms of carbon is in determining the contribution to the carbon Auger peak from atmospheric hydrocarbon contamination. The problem arises because surface cleaning using an argon ion beam has been observed to degrade the surface of a diamond crystal leading to the formation of graphite [25]. The preferred technique for cleaning diamond and other forms of carbon is a high temperature anneal under high vacuum conditions with an intentional hydrogen leak. For hydrogenated carbon films it is likely that this type of surface preparation would modify the composition and microstructure.

The scanning Auger electron microscope used in this investigation enabled areas of the surface potentially as small as 10nm in diameter to be analysed by Auger electron spectroscopy. It was therefore feasible to obtain an Auger analysis of the α-carbyne crystallites observed in transmission electron microscope studies. The Auger electron spectrum of small areas of the order of 2μm in diameter that appeared as small particles or protuberances on the surface have therefore been examined because this was the approximate size of α-carbyne crystallites observed by transmission electron microscopy. The KLL carbon Auger peak was found to have a slight shoulder at approximately 260eV in spectra from these particles. This should be compared with Auger spectra obtained using a larger probe size from the surrounding area. In this case no shoulder was observed, see figure 14 which shows the structure of the KVV carbon Auger peak obtained from a carbon film. Because of the reported ion beam sensitivity of the surface layers of diamond crystals [25] the effects of a light argon ion beam etch on the hydrogenated carbon films have been investigated. The Auger spectrum from areas of film that have been argon ion beam etched also contain a slight shoulder at approximately 260eV in the region of the carbon KVV peak (figure 14). These observations suggest that this shoulder appears as a result of the effects of intense electron or ion beam irradiation rather than from any structural differences between crystalline and amorphous regions of the film. Electron irradiation into the 2μm diameter area occupied by a crystallite results in an electron current density that is many times greater than generated over a 200μm diameter area.

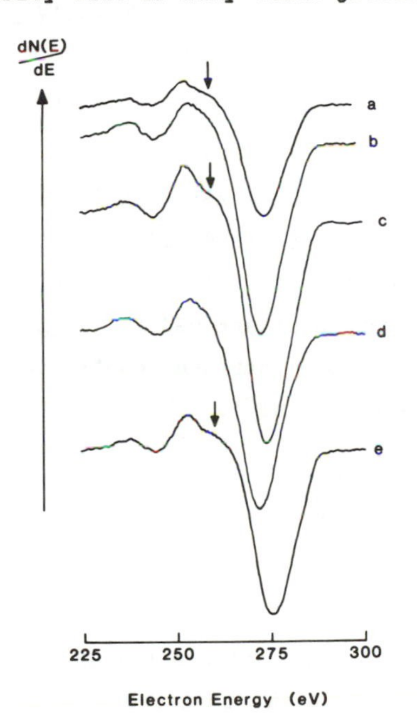

Figure 14 Structure of the KVV carbon Auger peak obtained from a carbon film prepared with a FAB source from (a) a 5 μm diameter crystallite (butane gas), (b) a 100 μm diameter area (butane gas), (c) a 2 μm diameter area (butane plus argon), (d) a 200μm diameter area (butane plus argon) and (e) a 100μm diameter area (butane plus argon) after a 60s argon ion etch. An arrow indicates the position of the shoulder that develops on the carbon KVV Auger peak after intense electron irradiation or an ion etch.

X-RAY PHOTOELECTRON SPECTROSCOPY

The main features of interest in x-ray photoelectron spectroscopy studies of hydrogenated carbon films are the structure of the C 1s core photoelectron peak, the structure of the carbon KVV Auger peak and the form of the carbon valence band spectrum.

The films examined by XPS in this study were found to have the C 1s photoelectron peak located at a binding energy of between 284.5eV and 284.7eV. This is close to the binding energy range reported in previous x-ray photoelectron spectroscopy studies of hydrogenated carbon films [26]. Figure 15 shows the C 1s photoelectron peak from a film prepared using a d.c. triode sputtering gun with a graphite sputtering cathode. A tail (arrowed in figure 15 (a))is observed on this peak on the higher binding energy side after the removal of the shoulder associated with surface hydrocarbon contamination. This tail could result from hydrocarbon bonding within the film or it could also be due to small electron energy losses [27].

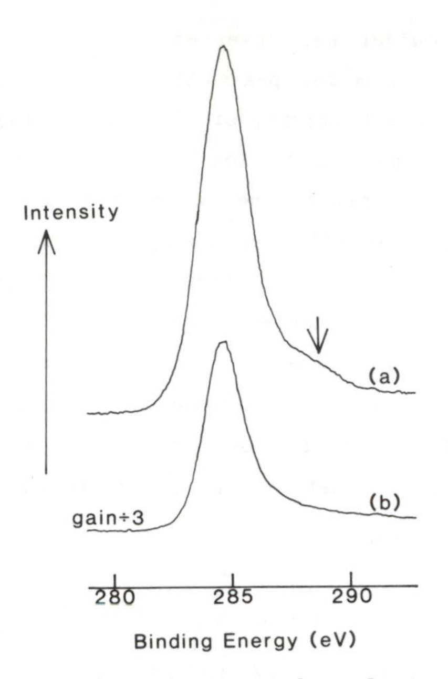

Figure 15 The C 1s x-ray photoelectron peak (a) as deposited and (b) after a 30s argon ion etch. The higher binding energy component arrowed in (a) is due to hydrocarbon surface contamination (Fitzgerald et al[4]).

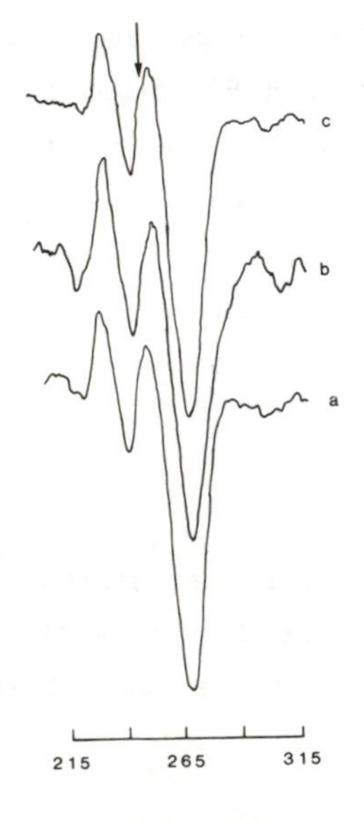

Figure 16 The differentiated carbon KLL Auger peak obtained by x-ray excited Auger electron spectroscopy for a carbon film prepared by (a) physical vapour deposition from a graphite cathode, (b) sputtering using a d.c. triode sputtering gun with a graphite cathode and (c)a FAB source with butane gas. The arrow indicates the position of the weak shoulder associated with sp^3 tetrahedral bonding.

The use of the electron beam excited C KLL Auger peak as a fingerprint to identify carbon bonding on the surface of carbon films has been thwarted in this study and in previous work [26] by the sensitivity of these films to electron beam irradiation. To overcome this problem the first derivative of the C KLL peak in x-ray excited Auger electron spectra (XAES) has also been examined (figure 16). A weak shoulder at approximately 248eV was observed in some of these spectra. This fine structure has been observed by XAES in diamond crystals [28]. This weak shoulder suggests that some tetrahedral sp^3 carbon bonding occurs in these films.

The peak positions in the valence band of diamond, graphite and diamondlike carbon have been tabulated by Mizokawa et al [28]. They have found that peaks occur at photoelectron binding energies of 5eV, 13eV and 20eV for diamond, at 7eV, 10eV and 19ev for graphite and at 5ev, 14ev and 20eV for diamondlike carbon films in the valence band spectra of these materials. Valence band spectra have been recorded for carbon films prepared by the three of the techniques discussed here (figure 17). Broad peaks were observed. The position of peaks in each of the valence band spectra suggest that each film contains a mixture of tetrahedral sp^3 bonds characteristic of diamond and the sp^2 bonds characteristic of graphite.

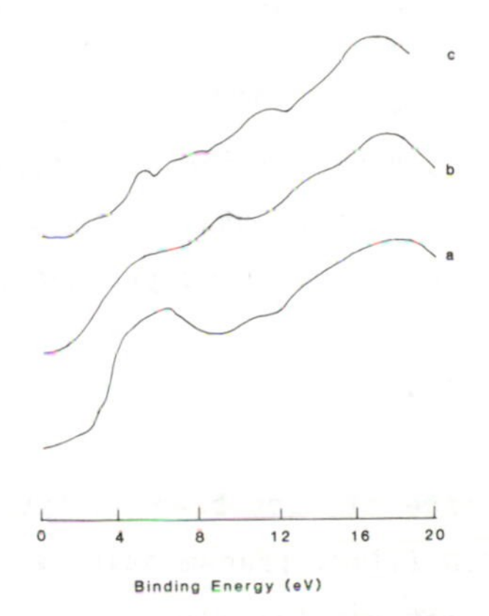

Figure 17 Valence band spectra from a carbon film prepared by (a) a FAB source using propane gas, (b) a d.c. triode sputtering gun with a graphite cathode and (c) hybrid PVD in a low voltage butane plasma from graphite and cast basalt.

SECONDARY_ION_MASS_SPECTROMETRY

Negative ion mass spectra obtained from carbon films prepared by a FAB source and using a d.c. triode source are shown in figure 18. The spectra were acquired using the a fast atom beam source (FAB SIMS). A hydrogen (H_2) peak was detected in all spectra from the films investigated. A carbon peak and a range of carbon-hydrogen peaks (C_xH_y) from mass fragments have also been detected. Traces of contaminants of fluorine, oxygen, hydroxide ions, sulphur, chlorine, calcium, potassium, iron and chromium were also detected.

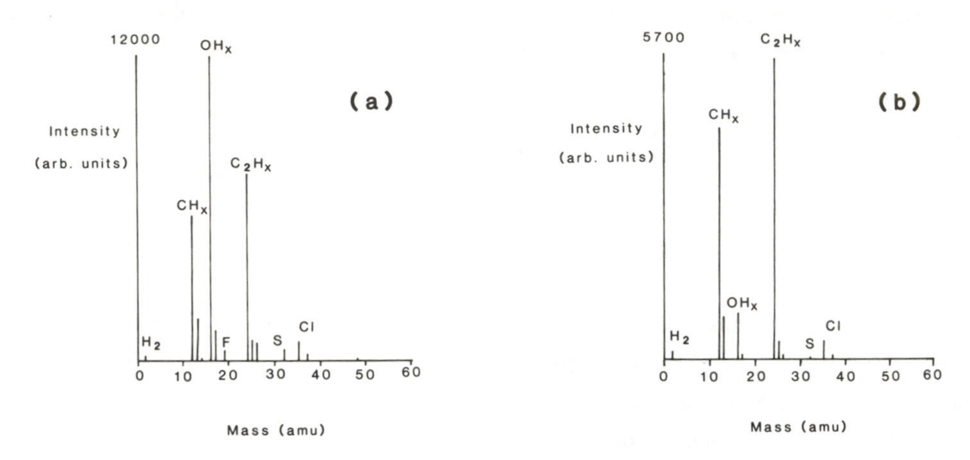

Figure 18 Negative ion secondary ion mass spectrum from a carbon film produced (a) by sputtering using a d.c. triode sputtering gun with a graphite cathode, (b) a FAB source using propane gas. (x = 0,1,2 for hydrocarbon fragments).

CONCLUSIONS

A range of microbeam analytical techniques have been applied to the study of a-C:H films. Transmission electron microscopy and electron diffraction have shown that crystallites with the α-carbyne or Ries crater forms of carbon are formed in most a-C:H films. These crystallites have been shown to have lattice parameters that vary with the preparation conditions. Under certain preparation conditions these crystallites are electron beam sensitive and TEM observations suggest that gas intercalated in the α-carbyne/Ries crater carbon layer structure or chemically bonded hydrocarbon material evolves from these crystals. This observation supports the suggestion that C≡C groups are present in a-C:H films. Since α-carbyne is believed to be composed of polymerized alternately triple bonded and single bonded carbon atoms it is possible that some hydrogen could be incorporated in this chain structure by disruption of the triple bonds to form hydrocarbon chains. The enlarged lattice parameters observed in some crystals could also result from the inclusion or bonding of hydrogen.

Preliminary scanning electron diffraction studies of a-C:H films prepared by rf glow discharge have shown that the bonding present in these films is a combination of both trigonal sp^2 and tetrahedral sp^3 bonds. These results are in good agreement with a continuous random network model for amorphous carbon films.

Electron energy loss spectra from the carbon films studied in this investigation show that calcium and oxygen are associated with the crystallites dispersed over the surface of these films. It is possible that contaminant calcium oxide particles act as nucleation centres for α-carbyne/Ries crater carbon crystals. The peaks arising from plasmon resonances located at low loss from amorphous regions occurred at the positions expected for a-C:H films. The peak at 7eV is associated with plasmon oscillations from π electrons. This peak is strongest in graphite where there is one π electron per atom. The peak in the low loss region observed for α-carbyne/Ries crater crystals was located at a position close to the position of the peak arising from plasmon resonance in diamond. A peak was not observed in the low loss region below 10eV in spectra from these crystals. This suggests that π electrons are not involved in the bonding of carbon atoms in this material.

Auger electron spectroscopy studies have confirmed TEM observations that the a-C:H films investigated in this work are sensitive to high intensity electron beam and ion beam irradiation. The shoulder that develops in the carbon KVV Auger peak as a result of electron or ion irradiation probably results from the decomposition of surface layers under high intensity irradiation leading to a modification of the valence band structure in this material.

The position of the C 1s x-ray photoelectron peak from the films investigated here lies within the range of binding energies associated with hydrogenated carbon films. Fine structure in x-ray excited Auger electron spectra suggest that some tetrahedral sp^3 bonding exists in these films. Valence band spectra also confirm the presence of some sp^3 tetrahedral bonding in these films. Peaks are also detected in the valence band spectrum that indicate sp^2 graphite-like bonding.

SIMS spectra confirm the hydrogenated nature of these films, hydrogen has been detected in films prepared by each of the preparation techniques considered here. Carbon-hydrogen fragments are also detected in the mass spectrum. Traces of calcium and oxygen have also been detected confirming the detection of calcium and oxygen in electron energy loss spectra from crystallites dispersed throughout the film.

The results of microbeam analytical studies of these a-C:H films have demonstrated that the hydrogen contained in these films is readily evolved either as hydrogen or hydrocarbon gas when subjected to ion or electron beam irradiation. Static SIMS studies confirm this effect. The observed lattice expansion observed with some of the crystallites in the a-C:H films probably occurs as a result of the the formation of hydrocarbon bonding within the films and also possibly due to the inclusion of hydrogen or hydrocarbon gas probably by intercalation. The effects of ion beam irradiation suggest that it is likely that the unusual effects observed with crystallites are characteristic of the film in general, and this is confirmed by SIMS observations, that is, hydrogen or hydrocarbon gas is present in the amorphous areas of the films. A probable site for this included material is the large number of voids present in amorphous material.

An interesting observation is that films prepared without hydrocarbon gas as a component appear to contain some hydrogen or hydrocarbon material, for example, hydrogen and hydrocarbon fragments have also been detected in films produced by sputtering in argon with a d.c. triode sputtering gun from a graphite cathode.

These microbeam analytical studies of a-C:H films have shown that the carbon atoms in the films prepared by a range of methods are bonded in a combination of tetrahedral sp^3 and graphite type sp^2 bonding. The presence of the diamond-like sp^3 bonding is confirmed in results from valence band spectra, XAES spectra and radial distribution analysis of the amorphous electron diffraction patterns from the films.

ACKNOWLEDGEMENTS

The authors wish to thank Dr M. Simpson, Dr A. Matthews, Mr D. Tither, Dr B.E. Storey , Mr P.A. Moir and Mr G.A. Dederski for their permission to publish the results of some joint work carried out under the auspices of the Universities Carbon Films and Materials Group, IOP Publishing for permission to reproduce figure 1 from reference 29, and Pergamon Press PLC for permission to publish figures from references 4 and 11. We also wish to thank Dr C.P. McHardy and Dr. W.A.P. Nicholson of the Department of Physics and Astronomy, University of Glasgow for their assistance in obtaining electron energy loss spectra and

Mr.W. Foster of Ion Tech Inc. for his help in the production of carbon films prepared by ion-beam deposition. A.E. Henderson wishes to thank the University of Dundee for financial support through a University of Dundee Centenary Scholarship.

REFERENCES

1) Craig, S. and Harding, G.L.: Thin Solid Films, 1982, 97, 345

2) Mirtich, M.J., Swec, D.M., and Angus, J.C.: Thin Solid Films, 1985, 131, 245

3) Bubenzer, A., Dischler B., Brandt, G. and Koidl, P.: J. Appl. Phys., 1983, 54, 4590

4) Fitzgerald, A.G., Simpson, M., Dederski, G.A., Moir, P.A., Matthews, A and Tither, D.: Carbon, 1988, 26, 229

5) Jansen, F., Kuhman, D.:J. Vac. Sci. Technol. A, 1988, 6(1), 13

6) Robertson, J.: Adv. Phys., 1986, 35, 317

7) Vora, H., Moravec, T.J.: J. Appl. Phys., 1981, 52, 6151

8) Zorin, E.I., Sukhorukov, V.V. and Tetel'baum, D.I.: Sov. Phys. Tech. Phys., 1980, 25, 103

9) Egerton, R.F.: Electron Energy Loss Spectroscopy in the Electron Microscope, Plenum Press, New York 1986

10) McKenzie, D.R., Botten, L.C., and McPhedran, R.C.: Phys. Rev. Lett., 1983, 51, 280

11) Sproul, A., McKenzie, D.R., and Cockayne, D.J.H.: Phil. Mag., 1986, B54, 113

12) Beeman, D., Silverman, J., Lynds, R. and Anderson, M.R.: Phys. Rev. B, 1984, 30, 870

13) Fitzgerald, A.G., Simpson, M, Dederski, G.A., Storey, B.E., Moir, P.A. and Tither, D.: Carbon, 1988, 26, 547

14) Kasatochkin, V.I., Shterenberg, L.E., Kazakov, M.E., Slesarev, V.N. and Belousova, L.V.: Dokl. Akad. Nauk USSR 1973, 209, 388

15) Kasatochkin, V.F., Egorova, O.I., and Aseev, Yu.G.: Dokl. Akad. Nauk USSR, 1963, 151, 570

16) Rhoades, B.L.: Micron, 1975, 6, 123

17) Has, Z., St.Mitura and Wendler, B.: Proc. International Ion Engineering Congress, ISAT'83 and IPAT'83, Kyoto, Japan, 1983, 1143.

18) Spencer, E.G., Schmidt, P.H., Joy, D.C., and Sansalone, F.J.: Appl. Phys. Lett., 1976, 29, 118

19) Evans, T., and Phaal, C.: Proc. Roy. Soc., 1962, A270, 538

20) Guseva, M.B., Savchenko, N.F. and Babaev, V.G.: Sov. Phys. Dokl. 1985, 30, 686

21) Fink, J., Muller-Heinzerling, T., Pfluger, J., Bubenzer, A., Koidl, P., and Crecelius, G.: Solid State Communications, 1983, 47, 687

22) Liang, W.Y., and Cundy, S.L.: Phil. Mag., 1969, 19, 1031

23) Egerton, R.F., and Whelan, M.J.: Phil. Mag., 1974, 30, 739

24) Colliex, C.:'Electron Energy Loss Spectroscopy in the Electron Microscope' in Advances in Optical and Electron Microscopy, eds Barer, R.,and Cosslett, V.E., Academic Press New York, 1984, 9 113

25) Green, A.K., and Rehn, V.: J. Vac. Sci. Technol., 1983, A1(4), 1877

26) Moravec, T.J. and Orent, T.W.: J. Vac. Sci. Technol., 1981, 18(2), 226

27) Baer, Y., Heden, P.F., Heden, J., Klassen, M., Nordling, C. and Siegbahn, K.: Phys. Scripta, 1970, 1, 55

28) Mizokawa, Y., Miyasato, T., Nakamura, S., Gieb, K.M., and Wilmsen, C.W.: J. Vac. Sci. Technol.,1987, A5(5), 2809

29) Fitzgerald, A.G., Henderson, A.E., Storey, B.E.: Inst. Phys. Conf. Ser. No.93, 1988, 1, 97

Materials Science Forum Vols. 52 & 53 (1989) pp. 515-542

CHARACTERIZATION OF HYDROGENATED, AMORPHOUS CARBON THIN FILMS

D.G. Thompson
Kobe Steel Research Laboratories, USA
Electronic Materials Center, Research Triangle Park, NC 27709, USA

ABSTRACT

Hydrogenated, amorphous carbon (a-C:H) thin films, produced by glow discharge decomposition of methane/hydrogen mixtures, were characterized by several analytical techniques. Films were made using 100 KHz and 13.56 MHz plasma frequencies at various powers and pressures. FTIR, Raman, and visible-infrared spectroscopies, ESCA analysis, and van der Pauw measurements yielded information on the chemical structure, bonded hydrogen content, optical bandgaps, and electrical resistivities of the materials. Correlations are drawn between the processing parameters and the film properties. The C-C bonds in the films are mostly sp2, but sp3 bonds also occur. Hydrogen occurs as C-H bonds, 85 to 95% of which are sp3. The electrical resistivities and optical bandgaps are dependent on the amounts of sp2 and sp3 bonds in the films. Low frequency plasmas produce more sp2 bonding than the rf plasmas, hence lower resistivities and bandgaps are associated with the 100 KHz films. The conductivity in the films is shown to occur by the mechanism of electron hopping.

I. INTRODUCTION

Hard, amorphous carbon coatings have been studied since 1971 when Aisenburg and Chabot [1] reported using a vapor deposition technique, ion beam deposition, to produce amorphous carbon thin films that were extremely hard, electrically insulating, and resistant to chemical attack (including acids, bases, and organic solvents). The term "diamond-like carbon (DLC)" was applied. Prior to that time, carbon coatings were produced by evaporation of graphite source materials yielding soft, conductive coatings much like graphite. The method of Aisenburg and Chabot differs from the evaporation techniques in that the carbon source materials is ionized and accelerated toward the substrate at energies much higher than those of evaporated neutral atoms. The energy of an evaporated neutral atom, kT, is of the order of 10^{-1} eV. The ion beam method produces energies ranging from about 10 to 1500 eV. Ion energies of modern ion beams can be controlled to about ±1 eV.

much like graphite. The method of Aisenburg and Chabot differs from the evaporation techniques in that the carbon source materials is ionized and accelerated toward the substrate at energies much higher than those of evaporated neutral atoms. The energy of an evaporated neutral atom, kT, is of the order of 10^{-1} eV. The ion beam method produces energies ranging from about 10 to 1500 eV. Ion energies of modern ion beams can be controlled to about ±1 eV.

Since this initial report many vapor deposition methods have been used to produce amorphous carbon and hydrogenated amorphous carbon coatings. The accepted terminology is a-C and a-C:H, respectively. Some of the techniques include RF glow discharge of methane, ethane, benzene, ethanol, and other hydrocarbon gases; RF and DC sputtering from graphite targets; single and dual ion beam sputtering from graphite targets; ion beam direct impingement using hydrocarbon gases; and microwave glow discharge, electron or laser-assisted CVD and thermal CVD of hydrocarbon gases. In common with the ion beam process of Aisenburg and Chabot, all of these methods have some mechanism for imparting high energy to the carbon atoms as they impinge upon the substrate. Some excellent reviews of this technology have been published [2,3,4,5].

Currently of great interest is the development of methods to produce true diamond thin films. High temperature chemical vapor deposition (CVD), electron-assisted CVD, and microwave plasma deposition are techniques which can produce diamond coatings [6,7,8]. Deposition of diamond thin films by CVD methods is a high temperature process. Thermal CVD requires temperatures greater than 800C, while electron-assisted CVD operates at temperatures as low as 400C. The microwave deposition process also involves temperatures which can exceed 400C.

Though the capabilities to deposit carbon coatings with true diamond crystal structures now exist, there are many reasons for further study of amorphous carbon coatings. The high temperatures associated with the production of diamond thin films limit their applications on low melting point substrate materials such as plastics. The microwave coating technique is also limited by the area over which diamond films can be applied. Generally, a-C films are easier to produce in that they do not require extremely high temperatures or microwave frequencies as do the diamond coating methods. On the other hand, a-C coatings are amorphous with a mixture of bond-types. The properties of the films are highly dependent upon the carbon bonding which is, in turn, dependent on the processing conditions. Thus, obtaining a thorough understanding of the material properties and their relations to processing parameters can be somewhat more challenging for amorphous carbon coatings than for the diamond coatings.

II. PROPERTIES

2.0 INFLUENCE OF THE GROWTH ENVIRONMENT

The properties of a-C and a-C:H coatings are controlled by the conditions in the deposition process. Plasma frequency, self-bias voltage, current, and power; cathode and anode geometry; and gas pressures can all effect the characteristics of the films [2,3,9,10].

Cauderc and Catherine [11] describe the reduced parameters of $J/P^{1/2}$ (for low frequencies) and $V/P^{1/2}$ (for rf) as representative of the ion bombardment energy. P is the pressure during the coating trials, V is the negative self-bias voltage across the dark-space at high frequencies, and J is the ion current density for low frequencies. The sheath thickness of the plasma (also called the dark-space) inversely follows the square root of the pressure. The mean free path between ion collisions is approximated by the sheath thickness [12], so the ion bombardment energy also inversely follows $P^{1/2}$. The plasma frequency for CH_4 ions is shown by Cauderc and Catherine [11] to be 5.4 MHz, while that for electrons is 895 MHz. Thus, at 13.56 MHz the electrons follow the frequency, but to the ions, the plasma frequency appears as a dc potential. For this reason the voltage controls the ion bombardment energy. The resulting reduced parameter is $V/P^{1/2}$.

The situation is different at the lower plasma frequency of 100 KHz. The influence of pressure is the same as for high frequency, but now the ions can follow the frequency. The self-bias voltage, however, is nearly independent of the total pressure and the plasma power. Ionization becomes strongly dependent on the plasma power, so the ion bombardment energy is characterized by the ion current density, J, of the CH_4 ions impinging the substrate, as well as the pressure and we have the reduced parameter $J/P^{1/2}$.

The interrelationship between the cathode and the anode in the deposition system can also effect the properties of the films [13]. For example, when the cathode and the anode are of the same size and shape, and are parallel and closely spaced the electric field lines which run between them are parallel and of equipotential. The result is that the atoms which form the coating are subjected to homogeneous growth conditions. However, when the cathode/anode geometry are not symmetrical, as is the case in the experiments described in this study, the growth conditions are inhomogeneous and properties of the films may vary during deposition.

2.1 CHEMICAL BONDING AND STRUCTURE

In a-C:H films, the favored chemical bonds for the carbon atoms are sp2 (trigonal prismatic) as in graphite and sp3 (tetrahedral) as in diamond. The ratio of sp3 to sp2 bonds is sometimes used to gauge the similarity of the films to diamond or graphite. For 100% sp3 bonding, assuming no hydrogen bonds, the film is true diamond. Similarly, for 100% sp2 bonds the material is graphite. Thus, as the number of tetrahedral bonds increases relative to the number of trigonal prismatic bonds, the more closely the properties of the coating will approach those of natural diamond.

For a carbon atom to condense at the substrate and link to another carbon atom in a tetrahedral coordination it must possess a characteristic energy such that sp3 bonding is favored over sp2 bonding (and sp1, which is rare in these materials). The energy spread at which carbon ions and neutral atoms exist at the moment of impact with the substrate is sufficiently wide that the coatings grow with a mixture of diamond and graphite bonds. Although a carbon atom is bonded tetrahedrally, it is not necessarily bound to four other carbon atoms. In fact, according to Cauderc and Catherine [10], nearly all of the sp3 bonded carbon atoms are bound to one or more hydrogen atoms. The structure is modeled by Weissmantel [14] as distorted benzene rings which are cross-linked by sp3 bonded carbon atoms. Angus [15] suggests a slightly different model, that of a skeletal structure of sp3 bonded carbon with short-range order caused by C=C and C-H bonds. Cauderc and Catherine [10] model the structure as clusters of C=C bonds which contain some bond angle distortion that is locked in place by sp3 bonded carbon atoms. The Weissmantel model and the Angus model are virtually opposites of each other. Weissmantel implies a layered structure. Angus, however, describes a continuous network of diamond bonds with branches of trigonal prismatic bonds. The Cauderc and Catherine model differs from both in that the network is not necessarily continuous.

Analysis of chemical bonds can be performed using FTIR. In the spectral region 2750 cm^{-1} to 3100 cm^{-1} wave numbers, FTIR will show C-H bonds. 3100 cm^{-1} to ~2970 cm^{-1} represents sp2 bonded carbon and 2970 cm^{-1} to 2800 cm^{-1} arises from sp3 carbon bonds. This information alone is insufficient to thoroughly describe the chemical structure of the materials since all of these bonds are carbon which is involved with hydrogen and does not account for carbon-carbon combinations. However, when examined in concert with data from Raman analysis, and electrical conductivity measurements, one can gain a picture of the probable structure of the films.

Raman spectroscopy can give information regarding the disorder of the a-C:H film structures. Tuinstra and Koenig [16] presented Raman spectra for single crystal graphite, stress annealed pyrolite graphite, commercial-grade graphite, and activated charcoal. The single crystal graphite exhibits a first-order Raman shift at 1575 cm^{-1}, called the G-line, which is very sharp. As the amount of "unorganized" carbon (e.g., at the grain boundaries) increases, the G-line shifts downward and broadens, and a first-order line, the D-line, appears at 1355 cm^{-1}. The intensity of this line increases with increasing disorganization of the atomic order. The ratio of the intensities, I_{1355}/I_{1575}, gives a measure of the disorder [17]. Broadening of the bands occurs in a-C:H materials because of local variations in ground states and virtual states of each carbon atom caused by bond angle disorder and resulting in a band of energies emitted by Raman photons.

The first-order Raman line for diamond as given at 1332.5 cm^{-1} by Solin and Ramdas [18]. The Raman diffusion cross-section (a measure of Raman efficiency) is about 55 times lower for diamond than for graphite. Thus, even for films which are predominantly sp3 bonded, the diamond line may be unresolvable due to low intensity or to masking from the broadened and offset 1355 cm^{-1} sp2 disorder line.

2.2 INCORPORATED HYDROGEN

Entrapped H_2 has been mentioned by several authors [9,15,19] as contributing to effects on certain properties of a-C:H such as density and intrinsic stress. For example, increasing concentrations of H_2 cause decreasing densities and increasing internal stress. Entrapped H_2 is difficult to measure and the reports are inconclusive.

Hydrogen, bonded as C-H, can represent as much as 50 at% of the films depending on the method and specific parameters of deposition. Several analytical techniques, such as NMR, microcombustion [15], and Fourier Transform Infrared spectroscopy (FTIR) can give useful information on bonded hydrogen. FTIR is the most commonly reported technique. Using the spectra in the band from 2750 cm^{-1} to 3100 cm^{-1}, which is an absorption region for C-H stretch vibrations, the area of the absorbance per unit thickness is directly proportional to the amount of bound hydrogen in that coating, according to

$$A = \alpha t C \tag{1}$$

where A is area of the absorbance envelope, α is the absorption coefficient, t is the sample thickness, and C is the concentration of C-H bonds. By taking the ratio of each corrected area, A_n/t_n, with n as the sample number, to the area of the sample which exhibits the least area, A_m/t_m, we have a relative measure of the concentration of hydrogen which occurs as C-H in the films. (Note that α drops out for this method).

The effect of bound hydrogen on the properties of these films has been widely discussed. Hydrogen tends to stabilize the random network structure. Cauderc and Catherine [10], and Nir [20} correlated film stress with hydrogen content showing that films with lower compressive stresses have higher hydrogen concentrations. High compressive stresses are due to the differing bond lengths of sp2 bonds (1.34 Å) and sp3 bonds (1.54 Å) and from distortion of the C=C bond angles [2,10]. Cauderc and Catherine [10] suggest that the hydrogen tends to relieve the internal stress of the coatings by minimizing the volume expansion associated with the transformation of sp2 and sp3 bonds. Their annealing experiments have shown that as hydrogen is driven from the coatings, which occurs at about 400C and again at about 650C [15] the sp3 bonds revert to sp2 bonds and films become optically absorbing and electrically conductive similar to graphite. Hydrogen stabilizes sp3 bonds by tying up dangling bonds in the structure. This effect can result in a direct dependence of resistivity and optical bandgap on the concentration of hydrogen in the films. Table 1 lists the range of reported values for the concentration of hydrogen in a-C:H films, the deposition method, and the reference.

Table 1. Reported values of hydrogen concentrations, electrical resistivities, and optical bandgaps for a-C:H films.

Atomic % H		Resistivity (ohm–cm)		Optical Bandgap (eV)		Deposition Method
	Ref		Ref		Ref	
≤50	21					Various
6 to 33	10	10^6 to 10^{13}	22	0.8 to 1.8	10	RF glow discharge
		10^6 to 10^{13}	28	0.5 to 1.9	27	RF glow discharge
				1.8 to 2.3	29	RF glow discharge
≤2	25	10^1 to 10^4	25	0.40 to 0.74	26	DC sputtering
30 to 50	27					Plasma-activated CVD
8 to 19	9	10^{-4} to 10^6	9	0.9 to 1.1	9	Ion beam (methane ions)
		$\geq 10^{11}$	28			Ion beam (methane ions)
		$\geq 10^9$	29	0.7 to 0.9	9	Ion beam (sputtered graphite)

2.3 ELECTRICAL CHARACTERISTICS

Potential exploitations of the electrical properties of amorphous carbon coatings may include integrated circuit heat-sinks and substrates, insulating layers, and possibly, semiconducting devices. a-C:H materials are electrically insulating and can exhibit resistivities from as low as $10^3\Omega$–cm up to $10^{12}\Omega$–cm depending on the film composition and chemical structure. Table 1 shows the reported values, deposition method, and references.

One of the most important factors affecting the film resistivity is the growth environment. Has, *et al* [23], and Staryga, *et al* [22], report that the resistivity decreases as the self-bias voltage increases for glow discharge deposited coatings. The former find that the decrease is rapid at first, then the resistivity becomes constant as the voltage is further increased. They attribute this effect to a polymer component in the structure at low voltages with an increasing contribution from sp2 bonds as the voltage is increased. Savvides, *et al* [26] report a similar effect for sputtered a-C films. Cauderc and Catherine [10] find from annealing studies that the resistivity decreases as the hydrogen concentration decreases. They attribute this to the transition of sp3 C-H bonds to conductive sp2 C=C bonds as hydrogen is lost from the films. They also report that films made from methane gas by glow discharge deposition exhibit resistivities which are dependent on the plasma frequency. Low frequency (50 KHz) plasmas typically result in coatings with resistivities two orders of magnitude less than those from high frequency plasmas (13.56 MHz) [19]. They suggest that the frequency effect is due to a greater number of sp3 bonds in high frequency films as compared to low frequency coatings. Kerwin, *et al* [9], find that the resistivity increases as the concentration of hydrogen in the plasma increases. Summarizing the above, the growth parameters affect the amount of sp2 and sp3 bonds in the films; and the resistivity is directly proportional to the sp2 component of the chemical structure.

2.3.1 CONDUCTION MECHANISM

Conduction in amorphous semiconductors can occur by a combination of mechanisms. Metallic conduction, conduction across an energy gap (called band conduction), and electron hopping can all contribute to conductivity. In amorphous metals, conduction occurs by carriers which are in the conduction band as in crystalline metals, but the disordered lattice increases scattering. The result is a higher intrinsic resistivity for the amorphous phase compared to the crystalline phase. For this mode of conduction to predominate in a-C:H materials, the films must be predominantly sp2 bonded with the carriers coming from pi electrons in the graphite bonds.

For crystalline semiconductors, the band structure consists of extended electronic states in the valence and conduction bands which can be illustrated reasonably well by a nearly-free electron model for conduction. On the other hand, the conduction mechanism in amorphous materials arises from electronic states which are localized within the band structure. These localized electronic states occur because the disordered lattice discourages the delocalization of electrons seen in crystalline materials. These localized states are essentially Bloch functions that are solutions to the Shroedinger equation with non-periodic energy distribution boundary conditions [30]. The length of the local wells is of the order of 10Å [31]. Although the electrons are localized within potential wells, they are relatively weakly bound. The hopping mechanism of conduction results from the interaction between phonons and localized electronic states near the bandgap. An electron is confined (trapped) in a localized potential well until acted upon by a phonon, thus the electron hops from trap to trap.

At higher temperatures, band conduction may dominate. As the temperature is decreased the contribution from electrons elevated to the conduction band decreases and a contribution due to the mechanism of hopping begins to show an effect. At very low temperatures, the conductivity is due entirely to hopping. The relation of log conductivity vs. temperature for amorphous semiconductors typically exhibits four regions. As described by Mott and Davis, Figure 1 [32] shows these regions. Figure 2 [32], also from

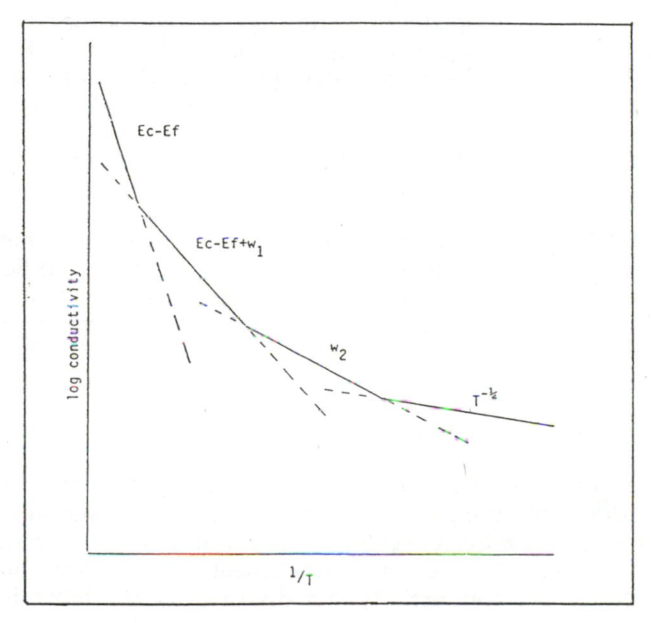

Figure 1. The four regions of log σ versus $1/T$ corresponding to Equations 2 through 5.

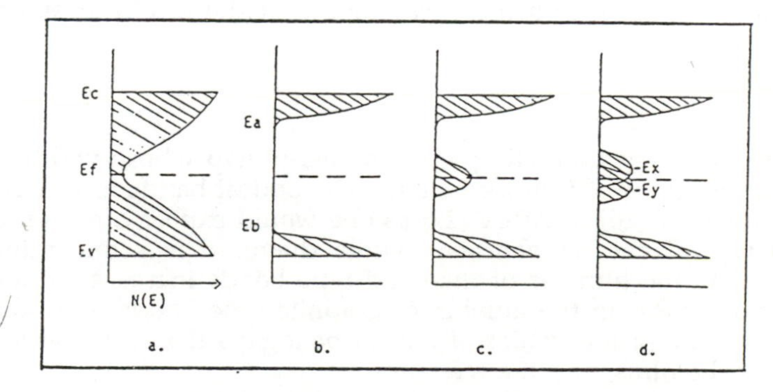

Figure 2. Possible densities of state in amorphous semiconductors.

Mott and Davis, shows four possibilities for the density of states around the bandgap. The conductivity in Region a of Figure 1 is dominated by (b) and conduction and follows

$$\sigma = \sigma_{min}\ exp(-Ec-Ef/kT) \quad (2)$$

where σ_{min} is the minimum metallic conductivity. This mechanism occurs when carriers are elevated from the localized states near Ec to extended states in the conduction band as expected at high temperatures where the electrons no longer remain in their localized states.

As the temperature is decreased, the contribution from the localized states increases. Region b of Figure 1 contains a contribution from carriers in localized states at the band edges (corresponding to Figure 2, Region a) with hopping energies near Ea or Eb (which are equivalent to donor levels, but arise from defect or localized states). Conductivity then follows

$$\sigma = \sigma_1\ exp(-Ea-Ef-w1/kT) \quad (3)$$

where $w1$ is the activation energy for hopping and σ_1 is much less than σ_{min} (due to a lower mobility and a lower effective density of states at Ea compared to Ec). If Ef exhibits a finite density of states (equivalent to Figure 2c), then there is a contribution from carriers with energies near Ef similar to impurity conduction in highly doped crystalline semiconductors. With w2 as the hopping energy, ~1/2 the width of the band of states which are near Ef, and $\sigma_2 << \sigma_1$, then

$$\sigma = \sigma_2\ exp(-w2/kT) \quad (4)$$

At low temperatures where kT is less than the band width, or when the density of localized states that extend into the gap overlap between the conduction and valence bands (Figure 2a), or if, in the case of a defect center, deep donor and accepter states lie near mid-gap (Figure 2d), then hopping occurs not between nearest neighbors, but rather variable range hopping takes place. The conductivity then displays a $T^{-1/4}$ dependence as

$$\sigma = \sigma_2'\ exp\ (-B/T^{1/4}) \quad (5)$$

where B is a constant dependent on the decay length of the localized states and on the density of states at the Fermi level.

2.4 OPTICAL BANDGAP

a-C:H thin films are semiconducting with an energy gap whose width is determined by the method and conditions of film formation. The optical bandgap can exhibit a range of values from about 0.4 eV [26] to 2.8 eV [27] as one would expect, since graphite exhibits metallic conductivity and has no bandgap, while diamond is semiconducting with a bandgap of 5.45 eV. As the number of diamond-type bonds increases in a-C:H coatings with corresponding decreases in the number of graphite type bonds, a bandgap begins to appear. Table 1 lists some of the values of optical bandgaps that have been reported, the method of depositing the films, and the reference.

A well-known method for determining optical bandgap, Eg, in amorphous materials is that presented by Mott and Davis [33] which gives the absorption coefficient as a function of incident energy and provides Eg at the intercept of the ordinant axis. The relation is

$$(\alpha h\nu)^{1/2} = m(Eg + h\nu) \qquad (6)$$

where α is the absorption coefficient, $h\nu$ is the incident energy, and m is a constant. In amorphous materials, the localized electronic states around the bandgap can extend into the gap. This is known as Urbach tailing. Using Eq. 6 one can depict the Urbach tail as the portion of the curve where the slope deviates from linearity and shows an exponential change. The electron mobility in this tail is about 10^3 less than the electron mobility of nearby extended states [3]; hence, a mobility gap exists. The low mobility within this gap, μ, is the direct result of the disorder of the atomic structure, since μ is proportional to the mean free time between collisions, τ, which in turn depends on the length of the localized states.

Several authors [9,10,34] have applied Mott's method to a-C:H thin films. Natarajan, *et al*, studied the dependence of the bandgap on the bias voltage finding an inverse relation, i.e., as the bias is increased the bandgap tended to decrease. A range for the bandgap of as much as 1 eV was observed.

Cauderc and Catherine plot Eg against their reduced parameters, $V/P^{1/2}$ for rf frequency, and $J/P^{1/2}$ for low frequency. They find Eg inversely dependent on both of these parameters. These same authors also examine Eg against the concentration of bound hydrogen and find a linear and increasing trend. Kerwin *et al*, expressing [H] as the ratio H/C, also report an increasing relation for Eg.

III. EXPERIMENTAL PROCEDURE

3.1 PREPARATION OF SAMPLES

Samples in this study were prepared by rf and low frequency plasma decomposition of methane vapor. A gas mixture consisting of 95% H_2 and 5% methane was used for all coating trials since hydrogen appears important in the growth of films with diamond-like properties. Pressures from a few Pascal to a few hundred Pascal were utilized. Pure argon was used at a pressure of 2.7 Pascal for the sputter etch step of each run. Plasma power was supplied at two different frequencies for various experiments. The frequencies used were 13.56 MHz radio frequency and 100 KHz, designated rf and low, respectively. The power output (in watts) of the power supplies and the voltage drop across the plasma dark space were monitored. Deposition parameters that were varied are the plasma frequency, cathode potential (called the negative self-bias voltage or U_b) and the total pressure (P_T). A coating run was allowed to continue for one to four hours resulting in film thicknesses from a few hundred to several thousand angstroms. Substrate materials consisted of 1"× 2" glass microscope slides and single crystal silicon wafers of size approximately 1/4"× 2".

Before starting an actual coating trial, the substrate surfaces were subjected to a sputter-etch step. The frequency for the sputter etching was the same as that of the deposition that was to follow. The power applied to the argon plasma was about 200 watts at U_b = 700-800 volts for low frequency and about 30 watts with U_b = 200-250 volts for rf. Sputter-etch times were 10 to 20 minutes. Typical run conditions are shown in Table 2.

3.2 POST-DEPOSITION ANALYSIS

Coated samples were subjected to analytical techniques which would yield information pertinent to this study. These tests included profilometer measurements for thickness; Fourier Transform Infrared Spectroscopy which observes C-H stretch bands and

Table 2. Typical coating run conditions.

Sputter Etch				Deposition			
Freq.	Argon partial pressure (Pa)	Power (watts)	Time (min)	Power (watts)	DC (volts)	CH_2/H_2 partial pressure (Pa)	Time (min)
100 KHz	2.7 to 27.0	200	10 to 20	100 to 300	500 to 700	2.67×10^3 to 4.67×10^4	107 to 172
13.56 MHz	2.7 to 27.0	30	10 to 20	4 to 45	100 to 450	2.67×10^3 to 2.67×10^4	76 to 242

reveals information on the concentration of strongly bound hydrogen and the amount of sp3 and sp2 bonds; Raman spectroscopy which examines polarizable bonds and gives an indication of the degree of disorder of the atomic structure; Electron Spectroscopy for Chemical Analysis which reveals impurities; Visible and Infrared Absorption which yields the optical bandgap of the material; and van der Pauw analysis which gives electrical resistivity. The first five of these are standard analytical techniques and will not be described further. The resistivity measurements present some difficulties; however, and the method used is presented in detail in the following section.

3.2.1 RESISTIVITY MEASUREMENTS

Materials which exhibit high resistivities can be difficult to measure. Cable capacitance, long time constants, and current leakage can all contribute to inaccuracies in the measurements. A useful method for the measurement of high resistivity thin films is the van der Pauw [35] technique with the adaption of Hemminger [36].
van der Pauw analysis requires only that the sample be of uniform thickness and continuous, i.e., free of pinholes. Four contacts are made anywhere on the periphery of the sample. The measurement procedure corrects for geometry. Figure 3 illustrates the sample configurations. In Figure 3a the current is applied to contacts 1 and 2, and the voltage drop between 3 and 4 is measured. The resistance, Ra, is then V_{34}/I_{12}. In Figure 3b, one obtains Rb as V_{41}/I_{23}. The resistivity is given by

$$\rho = (\pi t/\ln 2)(Ra+Rb/2)f \qquad (7)$$

where t is the sample thickness in centimeters, and f is a correction factor dependent on the symmetry of the contacts and satisfies

$$(Ra-Rb/Ra+Rb)=f(arccosh1/2 \quad exp(ln2/f) \tag{8}$$

Figure 4 shows the function f against Ra/Rb.

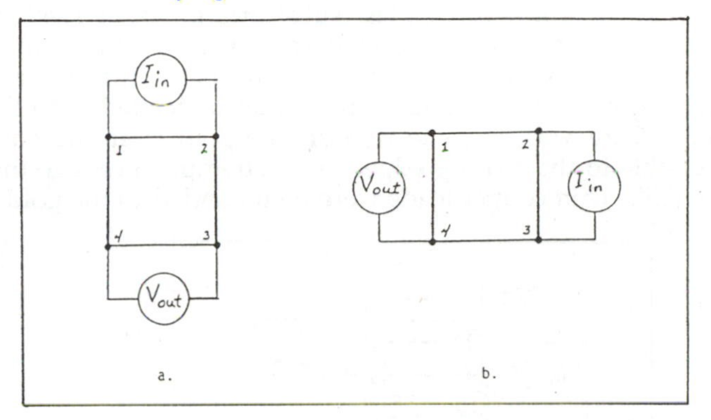

Figure 3. Sample geometries for van der Pauw resistivity measurements.

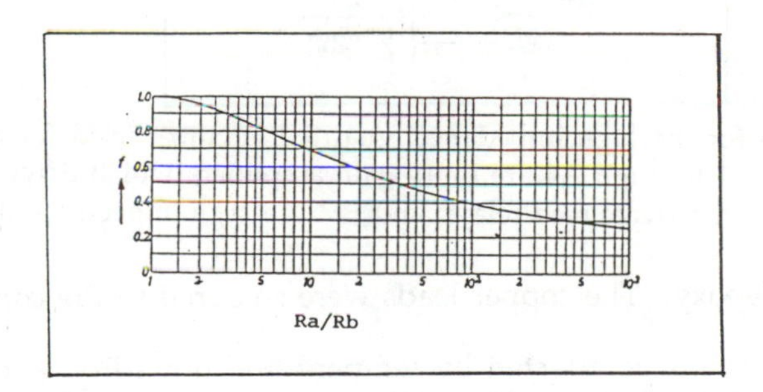

Figure 4. The function f used as a correction factor in the van der Pauw resistivity measurements.

Although only these two measurements are required, an average of the resistivities of all four of the possible configurations of voltage and current is recommended (ASTM). For perfect symmetry, where Ra=Rb, f is unity, and the resistivity calculation becomes

$$\rho = (\pi t/ln2) \ R \tag{9}$$

Hemminger adapts the van der Pauw technique to high resistivity ($>10^7$ Ω-cm) materials through the use of a completely guarded circuit, shown in Figure 5. The high impedance leads are guarded by driving the shields through unity gain amplifiers.

Switching is done by a rotary switch on the low impedance side of the amplifiers. The high impedance reed relays are also guarded. This circuitry reduces cable capacitance, current leakage paths, and the system time constant.

van der Pauw resistivity apparatus was built, including the guarded circuitry described by Hemminger. The sample fixture is shown schematically in Figure 6a. The sample is encased in an aluminum can connected to earth-ground to shield it from electromagnetic interference. Samples were cut from coated glass substrates to about 1/4" to 1/2" square. The samples were taken from areas near the centers of the substrates in order to have thicknesses as uniform as possible. Figure 6b is a cross-section of the substrate/coating/contact system. Gold spots were sputter deposited onto the edges of the coatings through masks. Care was taken to deposit the gold over the smallest possible areas and as near as possible to the coating edges in an attempt to reduce the error caused by contacts of finite size [35]. Thin copper leads were then bonded to the gold spots using

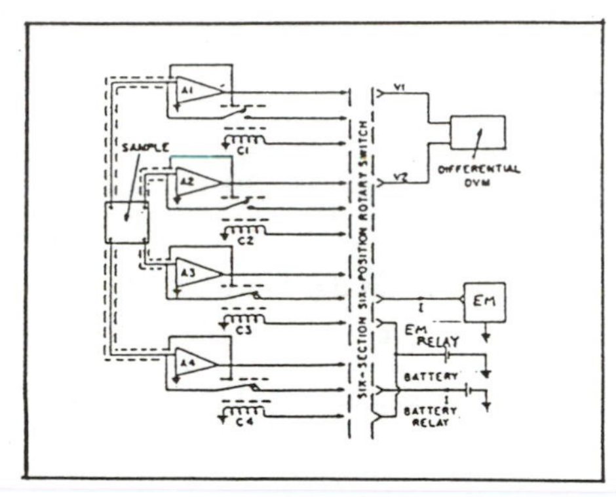

Figure 5. Circuit diagram for the high impedance, guarded circuitry used for the van der Pauw resistivity measurements. A1 through A4 are unity gain amplifiers which drive the guard shield. C1 through C4 are coils which activate reed relays, and EM is a high impedance electrometer.

conductive, silver-filled epoxy. The copper leads were soldered to the contact posts in the sample fixture.

The measurements could be affected by temperature, humidity, and vibrations. Of these, temperature was controlled by placing the sample fixture between large stainless steel blocks which would not be affected by small changes in the ambient temperature. A thermocouple was placed inside the sample fixture to monitor the sample temperature. Humidity and vibrations could not be controlled, but by placing the sample fixture within the stainless steel blocks as described above, the effects were held constant from sample to sample.

Current was provided to the sample by a bank of batteries connected in series having a total input voltage of 7.49 volts. The current varied according to the resistance between the contacts; typically, it fell between 0.05 to 5.0 picoamps. The output voltages ranged between 0.1 and 7.49 volts. The output voltage was measured on a strip chart recorder as was the temperature. The resistance measurements were taken after the voltage output became steady. The measurement was repeated for each of the lead combinations, and the average resistivity of the sample was calculated according to Eq. 7.

Determination of the error experimentally gives ±1/2 an order of magnitude as a reasonable estimate. This was checked in several ways. The apparatus that was built for

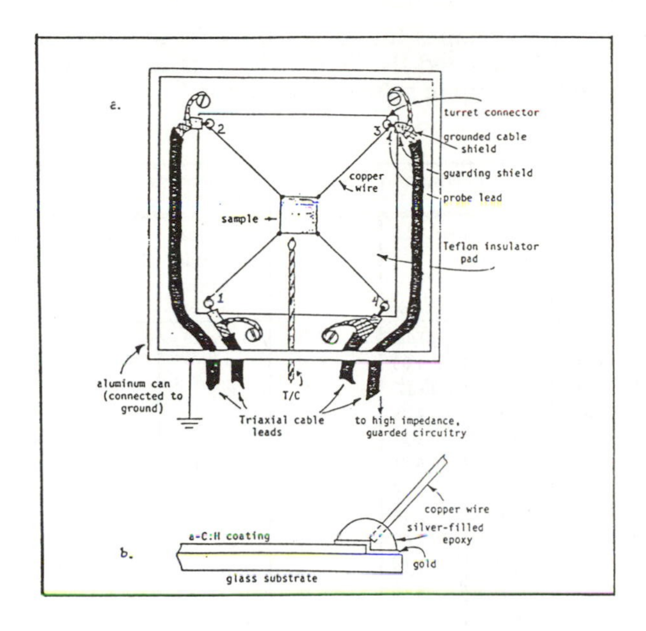

Figure 6. Resistivity sample fixture (a), and electrical contact system (b).

the measurements was calibrated against precision resistors with values ranging from 10^{10} to 10^{12} ohms. The apparatus was determined to be accurate to better than one percent. This is the same method that was used by Keithley Instruments on their commercially available van der Pauw measurement device, Model 7065. Several samples that were measured on the apparatus used in this study were also measured on the Keithley device. The results were always in agreement within half an order of magnitude. An additional check was performed at periodic intervals by removing a sample from the measurement fixture, rotating the connections by 90° and retaking the measurements. Again the values agreed within half an order of magnitude.

Several experiments were performed by varying the temperature of the sample and measuring the resistivity response. For these experiments, Eq. 9 is assumed to provide a reasonable representation of the sample resistivity, so only $Ra=V_{12}/I_{34}$ was monitored as a function of temperature.

IV. RESULTS

In this section the data from the various analytical techniques are presented. The deposition parameters and the experimental data for this study are given in Table 3.

4.1 FOURIER TRANSFORM INFRARED SPECTROSCOPY

Figure 7 shows a typical FTIR spectrum of a hydrogenated, amorphous carbon thin film. By the method previously described [10], the spectra of C-H bonds can be separated into sp2 and sp3 bonds at about 2970 cm^{-1}. The ratio of the areas, A_{sp2}/A_{sp3}, represents the fraction of bond types for the carbon which is involved with hydrogen. The ratios of the areas of the C-H absorption envelopes for each sample to the area of sample No. 17 are shown in Table 3. Sample 17 exhibited the least area. It is seen that about 85% to 95% of the incorporated hydrogen is tied up in sp3-bonded carbon.

Table 3. Experimental results and run parameters.

Run No.	Frequency rf-13.5MHz If-100KHz	Pressure (Pascal)	DC Voltage (volts)	Rel. [H]	A_{sp2}/A_{sp3}	Electrical Resistivity (ohm-cm)	Optical Bandgap (eV)	I_D/I_G (glass)	I_D/I_G (silicon)
16	rf	2.67×10^3	243	10.48	9.38	1.67×10^7	1.50	0.58	0.45
17	rf	2.67×10^3	192	1.00	5.22		1.63	0.64	0.25
18	rf	2.67×10^3	290	13.17	9.45		1.81		0.50
19	rf	2.67×10^3	360	4.08	8.29		1.34	0.41	0.40
20	rf	2.67×10^3	407	3.33	7.11		1.36	0.46	0.44
21	rf	2.67×10^3	450	5.71	7.80	2.47×10^7	1.32	0.62	0.41
22	rf	2.67×10^3	262	3.19	9.09		1.26	0.54	0.40
26	rf	2.67×10^3	100	2.41	6.96		1.27		
27	lf	2.67×10^3	700	2.69	7.22		0.16	0.43	
29	rf	1.07×10^4	250	2.14	10.00		0.85		0.48
30	lf	5.33×10^3	700	3.16	18.40		0.42		0.39
31	lf	1.00×10^4	630			2.51×10^8			
36	lf	1.00×10^4	500	4.87	6.58	5.79×10^7	0.91	0.46	0.44
51	lf	1.00×10^4	315	6.64	8.24	4.62×10^7	0.56		0.41
52	lf	1.00×10^4	415	5.44	7.39	1.79×10^7	0.89	0.32	0.40
53	lf	3.33×10^3	630	6.94	7.22	4.31×10^7	0.68		0.35
54	lf	2.00×10^4	600	3.19	8.70	1.06×10^8	0.98	0.52	0.42
55	lf	3.33×10^4	517	1.00	10.50	2.83×10^8	1.11	0.46	0.46
56	lf	4.67×10^4	500	1.13	7.75	3.45×10^7	0.23		0.43
58	rf	1.33×10^3	222	2.99	9.32		1.33	0.23	0.42
59	rf	6.67×10^3	195	2.68		2.85×10^7	1.32	0.21	0.36
60	rf	2.67×10^4	171	4.06	8.87	3.18×10^7	1.36	0.26	0.37
61	rf	2.67×10^3	100	2.82	8.79	1.77×10^8	1.14		0.37

4.2 VAN DER PAUW MEASUREMENTS

The electrical response of the samples to the application of current typically displayed a non-equilibrium period. Within this period the magnitude of the voltage typically increased rapidly to many times the steady-state value, then slowly decreased to a constant output. This initial voltage response is attributed to the filling of traps in the localized energy states. The sign could be either positive or negative for this initial voltage surge, but the sign of the final output voltage was always the same. The response time taken for the output voltage to reach steady-state was of the order of 5 minutes. Ten minutes was allowed to elapse before taking the data. By arranging the input current and output voltage leads according to the standard method used by Keithley Instruments Company [37] (shown in Figure 3), the sign of the output voltage was made negative.

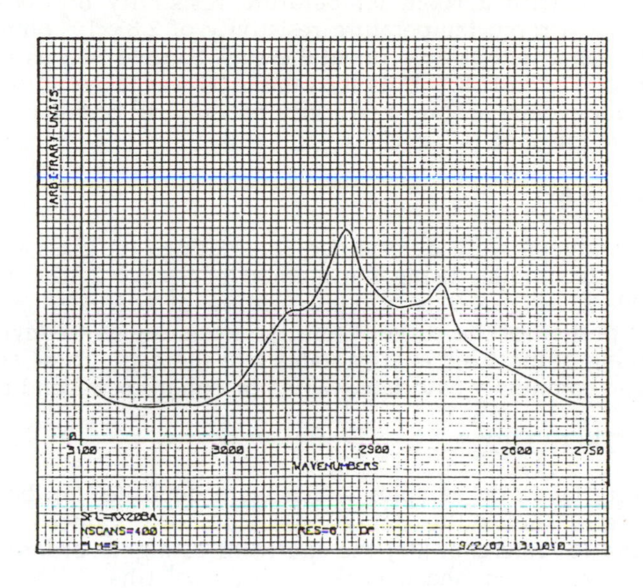

Figure 7. Typical FTIR spectrum.

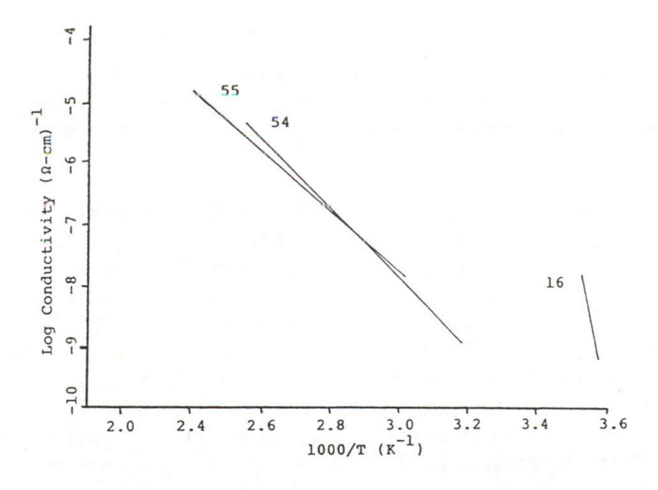

Figure 8. Log conductivity against inverse temperature for samples 16, 54, and 55.

Table 3 lists the resistivities of the samples which were measured by this technique. Electrical resistivity data was taken as a function of temperature. The temperature was varied from 14C to 161C while the output voltage was tracked with a strip-chart recorder and the input current was noted on the same scan. The resistivity was calculated by Eq. 9

and the conductivity plotted against 1/T. The results of this treatment are shown in Figure 8. Sample 54, with a room temperature resistivity of 1.06×10^8 ohm-cm, and sample 55, which has a room temperature resistivity of 2.83×10^8 ohm-cm, exhibit slopes that are nearly the same within experimental error. Sample 54 was measured within the temperature range of 17C to 134C. Data for sample 55 was taken in the range 20C to 161C. Sample 16 had a room temperature resistivity of 1.67×10^7 ohm-cm, and was measured over the temperature range of 14C to 20C

4.3 OPTICAL BANDGAP

The optical bandgaps were determined for coatings on glass substrates by visible-infrared absorption techniques using the method of Section 2.3.2. The energy gaps are contained in Table 3. Figure 9 is a typical curve of $(\alpha h\nu)^{1/2}$ against hv for these experiments. The portion of the curve where the slope begins to deviate from linearity is due to Urbach tailing of localized energy states into the gap. These states are responsible for the mobility gap commonly associated with amorphous semiconductors.

4.4 RAMAN SPECTROSCOPY

Raman spectra were taken on samples which were deposited on both silicon and glass substrates. Figure 10 shows the Raman spectrum of an uncoated silicon substrate, uncoated glass, and for reference [38] the spectrum of highly ordered pyrolytic graphite (HOPG) which shows a very sharp peak at 1575 cm^{-1} (the G line). Peak ratioing was performed on the spectra using 1355 cm^{-1} for the D line and the actual peak height around 1575 cm^{-1} of each sample for the G line.

4.5 ESCA

ESCA was used on Samples 16 and 17 to detect impurities. The spectra for these samples are shown in Figure 11. Oxygen would appear at a binding energy of about 524 eV. Within the resolution of the instrumentation, which is about 4%, none appears.

V. DISCUSSION

Hydrogen is an important component of the coatings produced in this study. It effects chemical structure, electrical resistivity and optical bandgap. The hydrogen occurs predominantly in C-H bonds. The sp2/sp3 ratios from Table 3 indicate that from about 85% to about 95% of the bonded hydrogen is attached to tetrahedrally coordinated carbon atoms. This structural similarity to the methane precursor probably occurs because the four hydrogen atoms are not completely stripped from the methane molecule in the film formation process. The methane is ionized in the plasma forming primarily [11] CH_4^+ and CH_3^+ with other less frequently occurring radicals. Some of the hydrogen atoms are stripped from the carbon by mechanical and chemical energy exchanges at the growing surface, but much of the hydrogen in the coatings probably comes from the original C-H bonds of the methane precursor.

The relative concentration of hydrogen, (H), in the a-C:H films is affected by the plasma frequency, and at high frequency, varies with the ion bombardment reduced parameters. Figure 12 shows the relative concentration of hydrogen, [H], against $V/P^{1/2}$ for

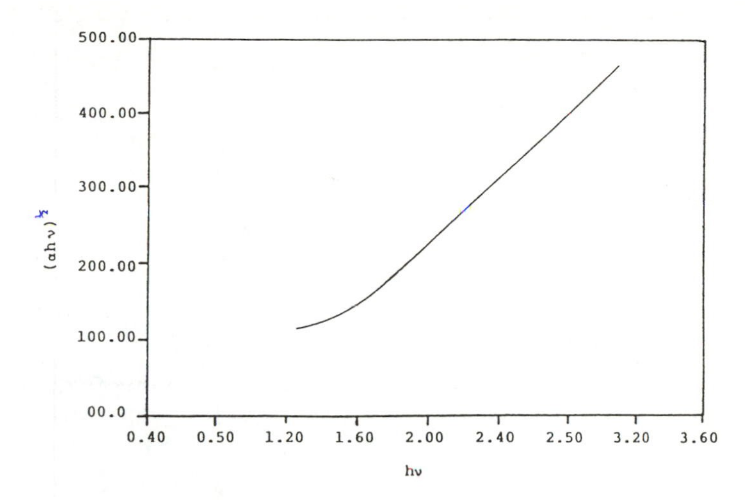

Figure 9. Typical curve of $(ah\nu)^{1/2}$ versus hν used to determine the optical bandgap of the a-C:H films.

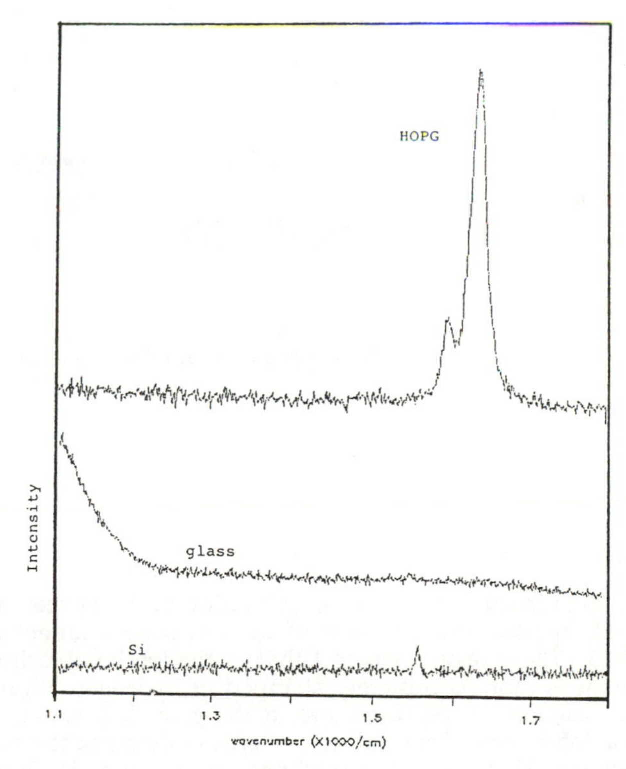

Figure 10. Raman spectrum of the (a) silicon substrate, (b) glass substrate, and (c) highly ordered pyrolytic graphite.

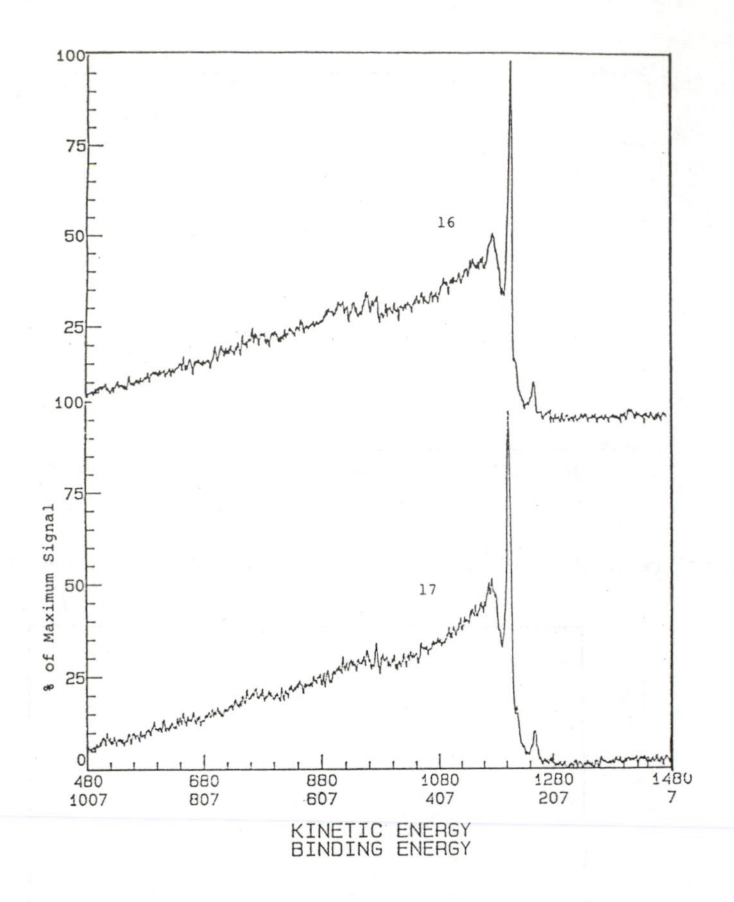

Figure 11. ESCA spectra for samples 16 and 17.

coatings made in high frequency plasmas. A minimum [H] is seen at about $4V/Pa^{1/2}$. At low frequency, [H] appears independent of the ion bombardment reduced parameter as shown in Figure 13. These results suggest that as the ion bombardment increases in the rf plasma the methane is more completely stripped of hydrogen allowing a greater amount of carbon-carbon reactions to predominate in the growth process. However, in the low frequency plasma, ion bombardment does not efficiently crack the methane molecule.

A plot of the relative [H] against the thickness of the films is shown in Figure 14. The thinner coatings contain more hydrogen. This dependence of [H] upon thickness is the result of the general rise in temperature of the substrate from ion bombardment during the course of a run. The longer coating trials reach higher temperatures. The effect is a self-anneal with the result that hydrogen diffuses out of the coatings. From Figure 14, thehydrogen concentration drops rapidly until a thickness of about 2000 Å. For thicker coatings the concentration of hydrogen becomes constant.

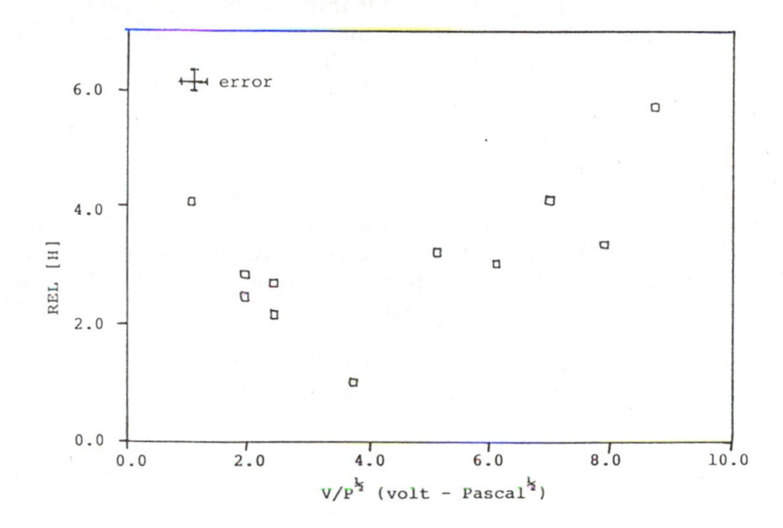

Figure 12. Relative concentration of hydrogen, *[H]*, against the high frequency reduced parameter $V/P^{1/2}$ which represents the ion bombardment energy.

The disorder in the lattice of a-C-H films produced by plasma decomposition of methane/hydrogen mixtures can be examined by Raman data. It should be noted that the D-line displays no measurable shift toward lower wavenumbers as required by Beeman [39] for the occurrence of sp3 clusters. This leads to the conclusion that most of the carbon is sp2 bonded even though the hydrogen is mostly bound to sp3 carbon.

An effect from the substrates is seen by comparing the disorder of films grown on silicon to that of films grown on glass. This can be done by taking the ratio of the intensities of the D and G lines and comparing the range of these ratios for the two substrate types and at the two plasma frequencies. Since the exact position of the D line is difficult to determine, it is taken at 1355 cm^{-1} for all of the spectra. The position of the G line is found at the actual peak of each spectrum. Table 3 lists these ratios. Figure 15 shows the range of these ratios for low frequency (LF) glass and silicon and high frequency (RF) glass and silicon substrates. The ratios are not found to be dependent on the ion bombardment energies, however, the spread of ratios for glass substrates is much greater that for silicon substrates. This implies that the crystalline lattice of the substrate has an effect on the disorder in the film.

The resistivities reported in Table 3 for the hydrogenated amorphous carbon films examined in this study are generally at the low end of values reported in the literature as shown in Table 1. One deduces that these coatings contain a large amount of sp2 bonding which contributes to conductivity by providing carriers. Raman analysis confirms that the majority of carbon bonding is sp2.

Resistivity decreases with increasing ion bombardment energy as seen in Figures 16 and 17. Resistivity also decreases as the hydrogen concentration decreases. This latter result at first appears in conflict with the observations of Cauderc and Catherine [10]. However, their data are for annealed coatings that initially exhibited high resistivities. As hydrogen was lost, resistivities decreased due to the transition of sp3 and sp2 bonds. The data presented here are for as-coated samples. The decrease in resistivity with an increase in [H] may be understood by considering the equation for conductivity,

$$\sigma = ne\,\mu \tag{10}$$

with $\mu = e\tau/\mu^1$, where τ is the mean free time between collisions as the electrons move through the lattice. As previously described, τ is directly related to the amount of disorder in the lattice. It appears then, that hydrogen promotes some ordering in the lattice, thereby increasing τ and decreasing the mobility gap which exists at the band edges, and therefore, increasing the conductivity.

The conduction mechanism in these films is determined from the following argument. The resistivities of the films show no dependence on the bandgaps, as seen in

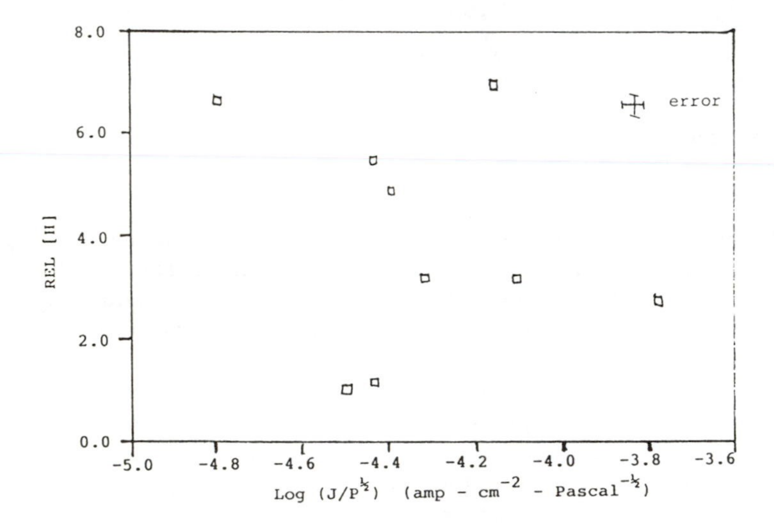

Figure 13. Relative concentration of hydrogen, *[H]*, against the low frequency ion bombardment reduced parameter $J/P^{1/2}$.

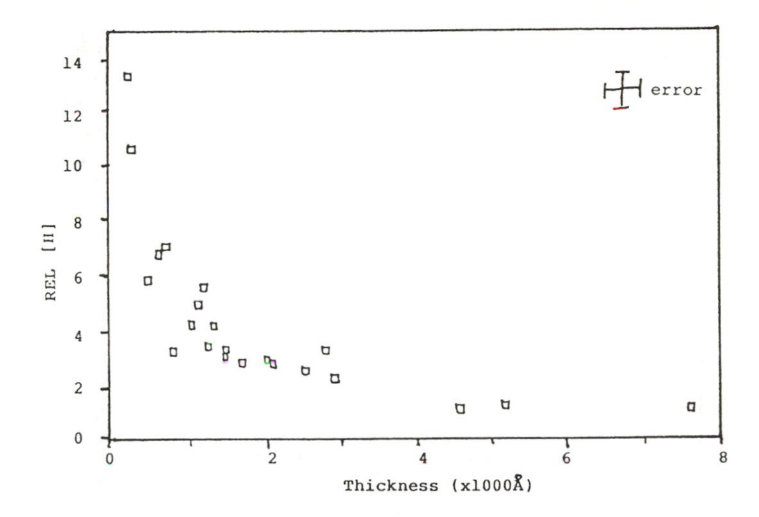

Figure 14. Relative concentration of hydrogen versus the sample thickness.

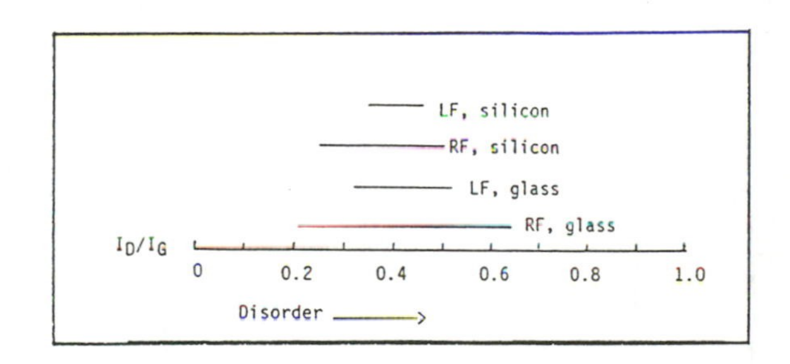

Figure 15. The range of the ratios I_D/I_G representing the amount of disorder in the amorphous lattices. Coatings deposited on silicon and glass substrates in rf and low frequency plasmas are shown.

Figure 18, which effectively rules out band conduction as a conduction mechanism. Metallic conduction can be eliminated because the conductivity is seen to increase with temperature. Since the data was not taken at low enough temperatures for variable range hopping (Eq. 5) to dominate (<180K according to Kerwin, *et al* [19]), and since the log σ data in Figure 8 are all linear against $1/T$, the conduction mechanism in these films must result from a hopping mechanism as described by either Eq. 3 or Eq. 4. The slopes of samples 54 and 55 shown in Figure 8 are essentially the same within experimental error, while that of sample 16 is steeper. It appears, since sample 16 exhibits a full order of magnitude lower resistivity than the other two, that this sample more closely follows Eq. 3, while samples 54 and 55 follow Eq. 4. The pi electrons associated with sp2 bonds apparently contribute more strongly to the conductivity of sample 16 than to the other two samples, and one can conclude that there is a greater amount of sp2 bonds in sample 16. Conversely, samples 54 and 55 contain a greater amount of sp3 bonding.

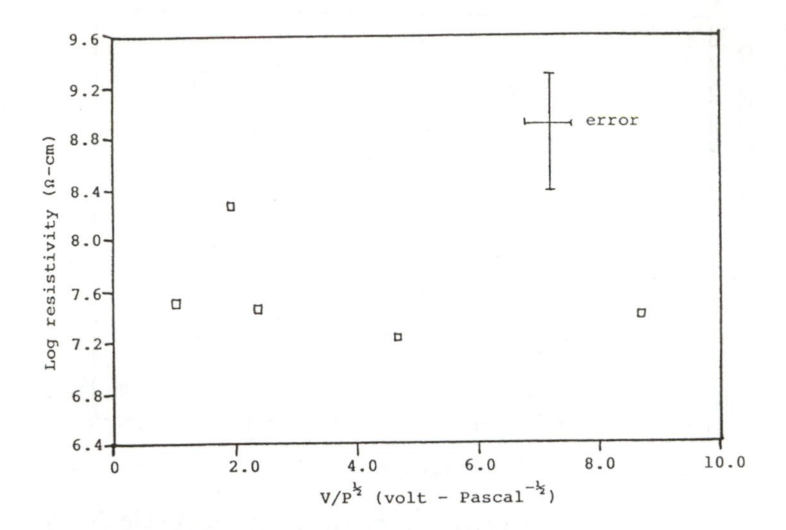

Figure 16. Log resistivity versus the high frequency ion bombardment reduced parameter $V/P^{1/2}$.

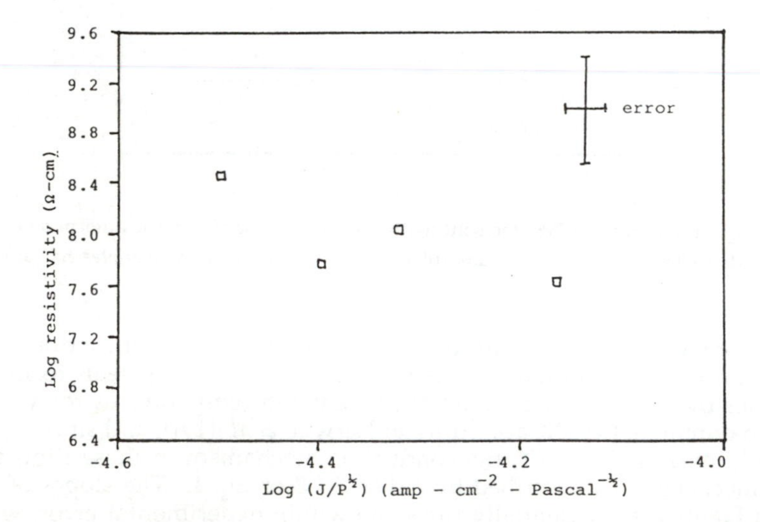

Figure 17. Log resistivity versus the log of the low frequency ion bombardment reduced parameter $J/P^{1/2}$.

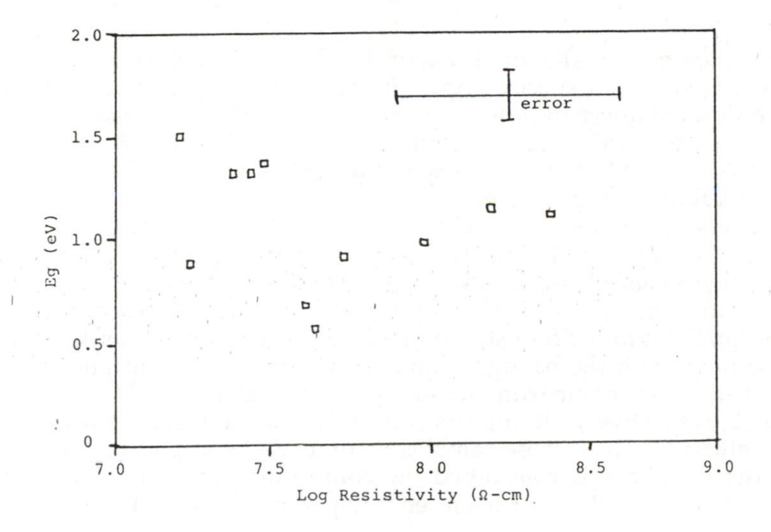

Figure 18. Optical bandgap, Eg, versus log resistivity for all of the samples.

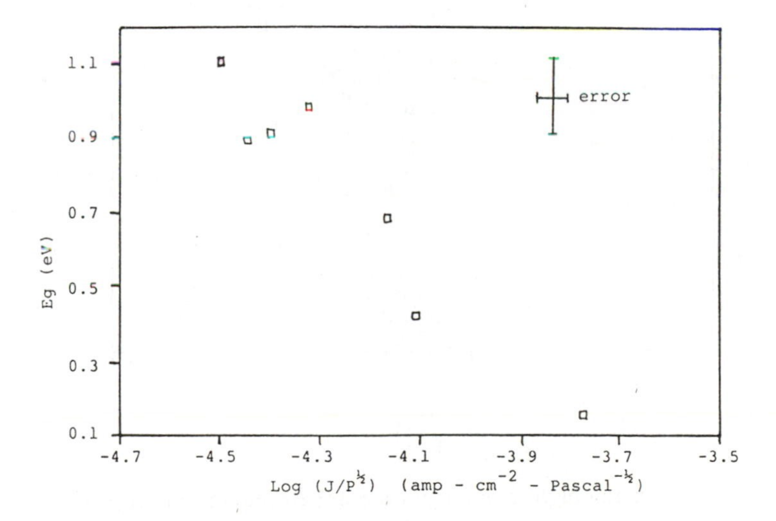

Figure 19. Eg against the log of the low frequency ion bombardment reduced parameter.

An optical bandgap is shown to occur in all of the a-C:H samples which were measured by visable-infrared spectroscopy. It can be seen by examination of the data in Table 3 that the Eg's are lower for low frequency coatings than for coatings deposited in an rf plasmas. The range of values for the former are 0.16 eV to 1.11 eV and for the latter are 0.85 eV to 1.81 eV. This is because more sp3 bonds are formed at the higher frequency as described by Cauderc and Catherine [10].

A strong inverse dependence of the bandgaps against the parameter of $J/P^{1/2}$ is shown at low frequency as seen in Figure 19. When compared to the dependence of the concentration of hydrogen versus this same parameter (Figure 13), one can conclude that more sp2 bonds are formed with increasing ion bombardment energy. The result is less hydrogen incorporated into the film structure and, therefore, lower bandgaps are formed.

In the case of rf, both the bandgaps and the hydrogen concentrations are essentially independent of the ion bombardment energy parameter, $V/P^{1/2}$. This is shown in Figures 20 and 12, respectively. It appears from this that the bandgap reaches a maximum value for the method in which these samples were made by rf plasma. This result can be understood if the frequency is considered the controlling parameter for the formation of sp3 bonds instead of the ion bombardment energy parameter in rf plasmas.

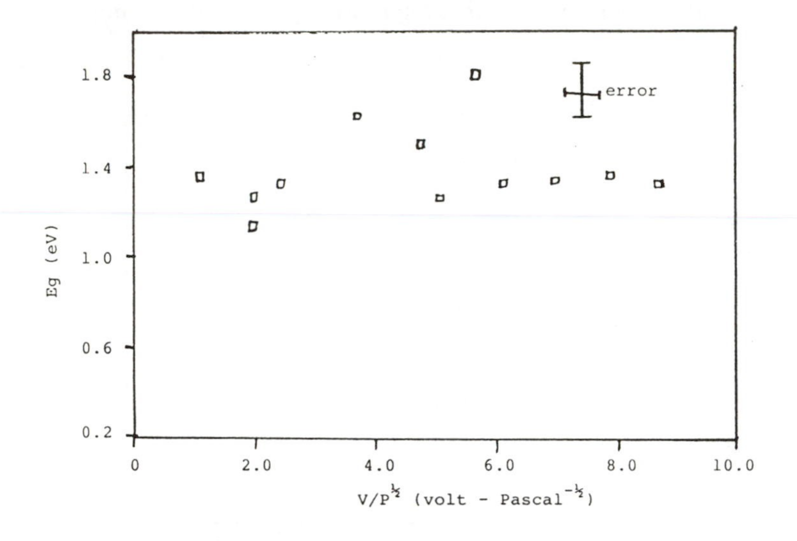

Figure 20. Eg versus the high frequency ion bombardment reduced parameter.

[. CONCLUSIONS

The coatings produced in this study using a 5% CH_4/95% H_2 gas mixture which is decomposed in either a 100 KHz or a 13.56 MHz plasma are hydrogenated amorphous carbon. This is evidenced by FTIR and Raman analysis.

The majority of the carbon-carbon bonding of films produced by rf or low frequency decomposition of the methane/hydrogen mixture is sp2. This is seen from the Raman analysis and the electrical resistivity data. Some of the bonding is sp3 which is required for the occurrence of the optical bandgaps that are shown to exist by visable-infrared spectroscopy.

The hydrogenated amorphous carbon films contain C-H bonds, 85 to 95% of which are sp3, as confirmed by FTIR.

The concentration of bonded hydrogen in the a-C:H coatings is dependent on the thickness of the films. This implies that the coatings undergo a self-anneal during the deposition process, whereby C-H bonds are broken, and hydrogen can diffuse out of the coatings in the longer runs.

The electrical conductivity in the hydrogenated amorphous carbon films produced by plasma decomposition of methane occurs by the mechanism of electron hopping as demonstrated by the temperature dependence of conductivity. Metallic type conduction is ruled out because the conductivity tends to increase with increasing temperature; and band conduction is eliminated as a possible mechanism since the conductivity is independent of the bandgap size.

In the a-C:H coatings of this work, the electrical conductivity is a function of the amount of sp2 and sp3 bonds in the films. A higher ratio of sp2 and sp3 bonds is associated with lower resistivities due to a greater contribution of carriers from the sp2 bonds. This is seen from the slopes of the log σ vs. 1/T curves, where the lower resistivity film exhibits a slope that appears to follow Eq. 3 as compared to the higher resistivity coatings which appear to follow Eq. 4.

The electrical conductivity of films produced by plasma decomposition of 5% methane/95% hydrogen gas mixtures is directly proportional to the concentration of hydrogen which is incorporated into the a-C:H structure. The hydrogen is postulated to affect the local energy states near the bandgap in a way which increases the mobility.

Raman data shows that the silicon substrate effects the amount of disorder in the a-C:H films more so than does the glass substrate. The broadening of the D line, as well as the lack of a measurable downward shift of this same line, indicate that there is no sp3 clustering occuring in the films.

Optical bandgaps are lower for a-C:H films produced in low frequency plasmas than for coatings made under similar conditions using high frequency plasmas. Possibly this occurs because the rf plasma promotes the formation of sp3 bonds more efficiently than the low frequency plasma.

10. The resistivities for samples produced in low frequency plasmas are inversely dependent on the concentration of hydrogen, and the concentration of hydrogen is inversely dependent on the ion bombardment energy. This implies that, at low frequency, higher ion bombardment energies drive the bonding toward more stable sp2 configurations.

11. Coatings made in rf plasmas show no dependence of bandgap on hydrogen or of hydrogen on the ion bombardment energy. It appears from this result that the rf frequency controls the formation of sp3 bonds, rather than the ion bombardment energy as is the case for low frequency plasmas.

VII. ACKNOWLEDGMENTS

The author wishes to acknowledge Battelle Memorial Institute for having provided the financial support for this work. Appreciation is due to Dr. William B. Johnston, formerly of the Ohio State University, for his invaluable service as my thesis advisor. Thanks also go to R. Bowling who gathered the Raman data and to Keithley Instruments who aided in the calibrations of the van der Pauw apparatus. And lastly, I extend my deepest gratitude to my wife, Laurie, for her unwavering faith and support and for her understanding of the many long hours required in completing this work.

VIII. BIBLIOGRAPHY

1. Aisenberg, S., Chabot, R., J. Appl. Phys., 42, (1971) 2953.

2. Weissmantel, C., Bewilogua, K., Bruer, K., Diethich, D., Ubersbach, U., Erler, H., Rau, B., and Reisse, G., Thin Solid Films, 96 (1982) 31–44.

3. Tsai, H., and Bogy, D. B., J. Vac. Sci. Technol., A5, 6 (1987) 3287–3312.

4. Wollam, J. A., Proc. El. Chem. Soc., 86 (1986) 289–303.

5. Thompson, D. G., Proceedings of the 1987 Soc. of Vac. Coaters Conference.

6. Messier, R., Spear, K. E., Badzian, A. R., and Roy, R., J. Metals, (1987) 8–11.

7. Matsumoto, S., Sato, Y., Tsutsumi, M., and Setaka, N., J. Mat. Sci., 17 (1982) 3106–3112.

8. Derjaguin, B. V., and Fedoseev, D. V., Izd. Nauka, Moscow, (1977) (in Russian).

9. Kerwin, D. B., Spain, I. L., and Robinson, R. S., Thin Solid Films, 148 (1987) 311–321.

10. Couderc, P. and Catherine, Y., Thin Solid Films, 146 (1987) 93–107.

11. Catherine, Y. and Couderc, P., Thin Solid Films, 144 (1986) 265–280.

12. Maissel, L. I., and Glang, R., ed. *Handbook of Thin Film Technology*, McGraw-Hill Publ., New York (1970) 1–21.

13. Wendler, B., Bacuum, 36, No. 1–3, (1986) 107–109.

14. Weismantel, C., *Thin Films from Free Atoms and Particles*, Chapt. 4, ed. K. Kablunde, Academic Press, New York (1985).

15. Angus, J. C. *et al*, Thin Solid Films, 118 (1984) 311-320.

16. Tuinstra, F., and Koenig, J. L., J. Chem. Phys., 53, No. 3 (1970) 1126–1130.

17. Dillon, R. O., Woollam, J. A., and Katkanant, V., Phys. Review B, 29, No. 6 (1984) 3482–3489.

18. Solin, S. A. and Ramdas, A. K., Phys. Review B, 1, No. 4 (1970) 1687–1698.

19. Matsumoto, O. and Katagiri, T., Thin Solid Films, 146 (1987) 283–289.

20. Nyaiesh, A. R., and Nowak, W. B., J. Vac. Sci. Technol. A, 1 (1983) 433.

21. Nir, D., Thin Solid Films, 146 (1987) 27–43.

22. Staryga, E. and Lipinski, A., Thin Solid Films, 145 (1986) 17–22.

23. Has, Z., Mitura, S., and Clapa, M., Thin Solid Films, 136 (1986) 161–166.

24. Alterovitz, S. A., Warner, J. D., Liu, D. C., and Pouch, J. J., NASA Technical Memorandum 87135 (1985).

25. Savvides, N., J. Appl. Phys., 59 (12) (1986) 4133–4145.

26. Savvides, N. and Window, B., J. Vac. Sci. Technol. A, 3 (6) (1985) 2386–2390.

27. Memming, R., Philips G.m.b.H Hamburg, Thin Solid Films, 143 (1986) 279–289.

28. Banks, B. A. and Rutledge, S. F., NASA Tech. Memo. 82873 (1982).

29. Khan, A. A., Woollam, J. A., Chung, Y., and Banks, B., IEEE Electron Device Letters 4 (5) (1983) 146–149.

30. Owen, A. E., Contemp. Phys., II, No. 3 (1970) 227–255.

31. Omar, M. A., *Elementary Solid State Physics*, Addison-Wesley Publ., Reading, MA (1975), 582.

32. Mott, N. F. and Davis, E. A., *Electronic Processes in Non-Crystalline Materials*, Second Edition, Clarendon Press, Oxford (1979) 209–222.

33. Mott, N. F. and Davis, E. A., *Electronic Processes in Non-Crystalline Materials*, Clarendon Press, Oxford (1971) 197.

34. Natarajan, V., Lamb, J. D., and Woollam, J. A., Liu, D. C., and Gulino, D. A., J. Vac. Sci. Technol. A, 3 (3) (1985), 681–685.

35. van der Pauw, L. J. Philips Research Reports, 13 (L) (1958) 1–9.

36. Hemenger, P. M., Rev. Sch. Instrum., 44 (6) (1973) 698–700.

37. Hall and van der Pauw Measurements of Semiconductors, Application Note #510, Keithley Instruments, Inc. (1984)

38. Bowling, R., The Ohio State University, Department of Chemistry, from unpublished doctoral dissertation.

39. Beeman, D., Silverman, J., Lynds, R., and Anderson, M. R., Physical Review B, 30 (2) (1984) 870–875.

Materials Science Forum Vols. 52 & 53 (1989) pp. 543-558
Copyright Trans Tech Publications, Switzerland

CHARACTERIZATION OF AS-PREPARED AND ANNEALED HYDROGENATED CARBON FILMS

J. Gonzalez-Hernandez (a), B.S. Chao and D.A. Pawlik (b)
(a) Centro de Investigacion del Instituto Politecnico National, Ap. Postal 14-740
Mexico, D.F., Mexico
(b) Energy Conversion Devices, Inc., Troy, MI 48084, USA

ABSTRACT

Hydrogenated amorphous carbon films were obtained from the decomposition of methane using a rf-couple glow-discharged system at power densities ranging from 0.4 to 4.8 W/cm^2. The structure of as-prepared and annealed films were characterized by Raman spectroscopy (RS), infrared absorption (IR), photoluminescence (PL), energy-loss spectroscopy (ELS), and x-ray diffraction (XRD). The results indicate that the incorporated hydrogen concentration is an importment parameter in determining the structure and properties of the films. Carbon films deposited at a lowest rf power density contain a large amount of hydrogen with most of the C-H bonds in CH_3 configurations, whereas films produced at higher rf powers reveal dominant CH_2 bonding structures. According to Raman scattering measurements the sp^2 domains in as-prepared samples are disordered due perhaps to bond angle distortions. Upon annealing, hydrogen leaves the film at a temperature that depends on the initial hydrogen concentration. Once most of the hydrogen has been driven out, crystallization into the graphite phase takes place. The microcrystallite size was measured by Raman and x-ray diffraction.

I. INTRODUCTION

Recently, amorphous diamondlike carbon films have attracted much attention because of their potential in various technological applications [1,2]. They have been mainly produced by glow-discharge decomposition of hydrocarbon gases and by ion beam sputtering techniques. The properties of these films not only depend on the deposition methods but also on the specific deposition parameters.

Although the chemical composition can be easily determined, much less is known about the atomic bonding and the microstructure of amorphous carbon films. Unlike other tetrahedrally bonded semiconductors where the electronic configuration of individual atoms is the same as carbon, however, the latter can exist with tetrahedral sp^3 as well as trigonal sp^2 bondings [3,4]. The physical properties of these films strongly depend on the type of chemical bonding. If the sp^3 bonding is dominant they are called diamondlike films, whereas those with a predominant sp^2 bonding are called graphitelike films.

In general, carbon films produced by the ion beam sputtering method contain small amounts of hydrogen. Since hydrocarbons are the source of carbon in glow-discharge produced films, significant amounts of hydrogen can be incorporated. In these films hydrogen may play a crucial role in the bonding configuration by stabilizing the tetrahedral coordination and consequently controlling the film properties.

In this paper a systematic study of the structure in as-deposited and annealed glow-discharge produced carbon films has been carried out using infrared absorption (IR), Raman spectroscopy (RS), photoluminescence (PL), energy-loss spectroscopy (ELS) and x-ray diffraction (XRD). We have found that the structure of as-prepared films strongly depends on the rf power density used during deposition. Films prepared at low powers show a polymeric structure. At high rf powers, the rupture of C-H bonds and sputtering of hydrogen become increasingly enchanced and therefore a more developed graphitelike structure is observed [2]. However, our results indicate that the formation of microcrystalline graphite does not exist even at the highest rf power deposited film. At intermediate rf powers the structure is formed by a mixture of both the diamond- and graphite-like phases. An increase in the volume fraction of the latter phase is achieved by increasing the rf power. During the anneal, hydrogen effuses out of the films at temperatures that depend on the rf power density; the higher the deposition rf power density, the higher the annealing temperature required to effuse hydrogen out. Once most of the hydrogen is driven out only trigonally bonded carbon remains and the film transforms into a microcrystalline structure. The dimensions of the crystallites were estimated along the basal graphitic planes (L_a) from the RS measurements and perpendicular to these planes (L_c) from XRD. Typical dimensions were L_a = 40 Å and L_c = 10 Å. We have also observed that the annealing of samples prepared at low rf powers produce larger particles than those samples deposited at higher rf powers annealed similarly.

II. EXPERIMENTAL

The amorphous hydrogenated carbon films (a-C:H) were simultaneously deposited on unheated (100) Si wafer and SiO_2 quartz substrates by rf glow-discharge decomposition of methane (CH_4) gas. The deposition parameters were as follows: (i) a gas flow rate of 100 sccm, (ii) a gas pressure of 0.2 Torr, and (iii) an input rf power density ranging from 0.4 to 4.8 W/cm^2. The deposition time was set to give films of nominal thickness 1.5 μm. Actual thicknesses were measured using an interferometric technique.

The RS and PL measurements were carried out using a Jobin Ivon double spectrometer and an argon-ion laser for excition. The IR absorption spectra in the range 300-4000 cm^{-1} were recorded using a Perkin-Elmer model 580B double-beam recording spectrophotometer interfaced with a Perkin-Elmer model 3600 data acquisition station. Spectral resolution was approximately 3 cm^{-1}. The x-ray induced ELS spectra were carried out in a PHI 550 surface analysis system having a base pressure of $< 10^{-10}$ Torr. The system was operated in a retarding mode with pass energy set at 25 eV giving a binding-energy (BE) uncertainty of ± 0.2 eV. Al Kα radiation (KE = 1486.6 eV) was used as the x-ray source to generate the C 1s core state photoelectron spectrum. The ELS spectra were monitored on films annealed in-situ up to 600 ^{o}C. The XRD traces were recorded using a Philips Θ-2Θ diffractometer with Cu Kα radiation.

III. RESULTS AND DISCUSSIONS

A) As-Prepared Films:

Figure 1 shows the IR transmission spectra of films prepared from methane gas at various rf power densities. In a-C:H films, two of the important regions containing IR vibrational motions are the stretching vibrations near 3000 cm^{-1} and deformation modes near 1400 cm^{-1} [5]. In the former region, the absorption bands are identified as the CH_3 and CH_2 asymmetric stretching modes at 2960 and 2930 cm^{-1}, respectively, and their corresponding symmetric stretching modes at 2870 and 2860 cm^{-1}. Interesting to notice is the fact that vibrations related to isolated monohydride groups (CH) are not observed in any of the spectra.

The CH_3 asymmetric and the CH_2 deformation modes appear at 1460 cm^{-1} and the CH_3 symmetric deformation modes display at 1375 cm^{-1}. Since the thickness and illuminated area are nearly the same for all samples, the integrated absorption area can be used to estimate the amount of bonded hydrogen. As seen in Fig. 1, at the lowest rf power density (0.4 W/cm^2) the absorption bands are narrower, well-defined, and dominant with CH_3 groups. The latter suggests a less polymeric structure. With the increase of rf power density, the absorption bands become broader and weaker in intensity. At those higher rf powers, the CH_2 modes dominant.

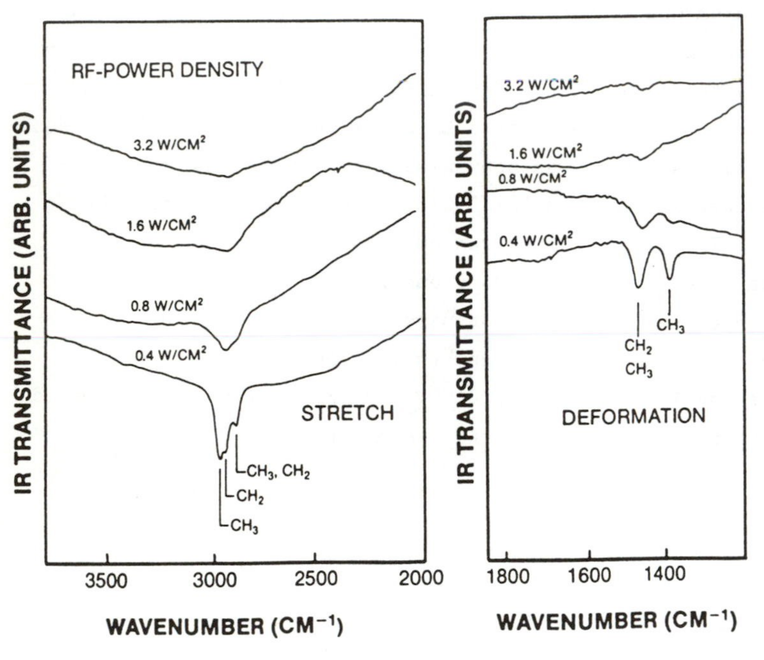

FIG. 1. IR transmission spectra of as-prepared films deposited at rf power densities of 0.4, 0.8, 1.6, and 3.2 W/cm^2.

Figure 2 shows the PL spectra for the as-prepared samples. All the spectra were taken at room temperature using the 4880 Å line from an Ar-ion laser as the excitation. The intensity of the various curves is not normalized, therefore comparison of integrated intensity is shown in Fig. 3. In Fig. 2 the sample

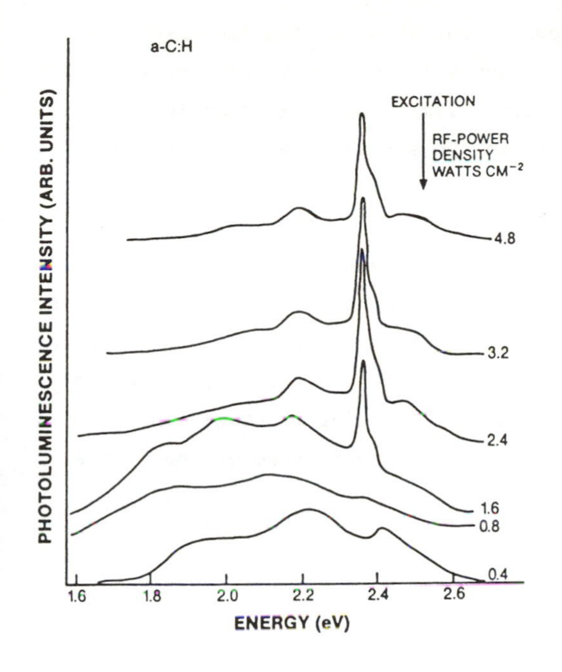

FIG. 2. PL spectra of as-prepared films with various rf power densities.

prepared at an rf power density of 0.4 W/cm^2 exhibits a broad luminescence band composed of three major features centered at ~ 1.9, 2.2, and 2.4 eV. When rf power density increases a red-shift in all three features is observed. However, it is well known that the structure of the PL band depends not only on the type of material but also on the energy of the excitation, provided this energy is lower than the energy gap of the diamondlike phase where the PL emission originates [6,7]. For rf power densities over 1.6 W/cm^2, there are sharper structures located at 2.16, 2.35, and 2.38 eV in addition to the PL band. These peaks are associated with the first-order (2.35 and 2.38 eV) and second-order (2.16 eV) Raman active modes in a graphitic structure. A more detailed analysis of the Raman spectra is presented latter in this section.

Figure 3 presents the integrated PL intensity per absorbed photon of the as-prepared samples plotted as a function of the input rf power density. In order to take into account the increase in optical absorption for increasing powers, the integrated PL intensity was normalized to the number of absorbed photons from the excitation beam. The change in the optical density at that energy as a function of the input rf power density is shown in the insert. The decrease of almost two orders of magnitude in the PL efficiency is related to the decreased hydrogen content for higher rf power deposited films resulting in an increase in both the nonradiative recombination centers in the sp^3 phase and the volume fraction of sp^2 nonradiative phase.

The details of the first-order Raman spectra of samples prepared at different rf power densities are illustrated in Fig. 4. The horizontal scale provides the Raman frequency shift (in wave numbers) from the Rayleigh line. The spectra show two features: i) a relatively sharp line, denoted by G, with frequency increasing from 1530 to 1570 cm^{-1} as a function of increasing rf power density and ii) a broader shoulder (D) located at the low-frequency side of the G line.

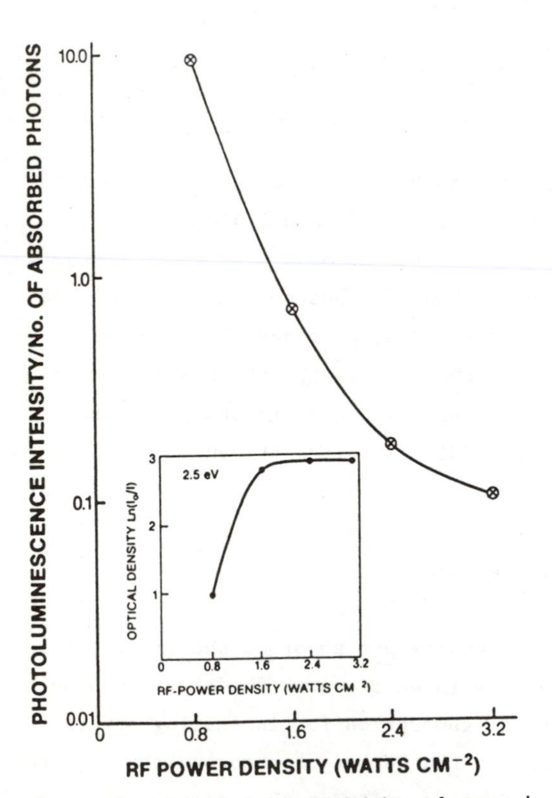

FIG. 3. The integrated PL intensity per absorbed photon of as-prepared samples as a function of the input rf power density. The insert is the change in the optical density at the excitation energy as a function of the rf power density.

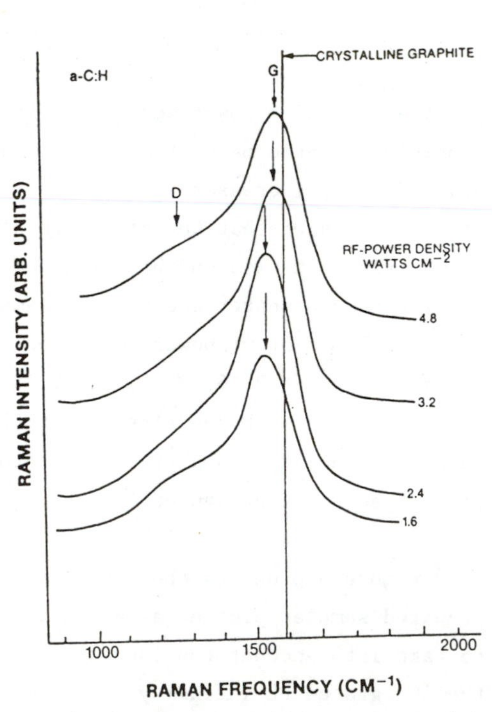

FIG. 4. The first-order Raman spectra of as-prepared films deposited at rf power densities of 1.6, 2.4, 3.2, and 4.8 W/cm^2.

The first-order Raman spectrum of a single crystal graphite reveals only one line (G line) [8,9]. Its position measured in this study is shown by a vertical line in the figure. In graphitic materials where the crystallites are reduced to domains of finite size, the Raman spectrum exhibits an extra line at ~1350 cm^{-1} [8,9]. The origin of this D line has been ascribed to the uncertainty of the wave vector of vibrational modes confined to the crystalline regions and its intensity has been found to be inversely proportional to the particle size [8,9]. For carbon films the long-range translational symmetry may also be lost by introducing bond-angle distortions. Recent calculations by Beeman et al. [10] showed that disorder in an sp^2 coordinated carbon network produced by bond-angle distortions can account for the observed changes in the Raman spectra of samples in Fig. 4. In their model the upward shift in frequency indicates a reduction in the mean value of the bond-angle distortion. Their calculations also show that similar changes in the Raman spectra can be produced by the addition of fourfold coordinated carbon atoms. Further support for a diphasic structural model, in which the sp^2 component has not reached a degree of ordering comparable to that of microcrystalline graphite, is provided by the annealing studies presented in the following sections.

B) Heat Treated Films:

Figure 5 shows the annealing behavior of the IR spectra in the stretching modes region for three samples deposited at power densities of 0.4, 0.8, and 1.6 W/cm^2. Data of those samples prepared at higher power densities are not provided because of their much weaker absorption bands. Similar to Fig. 1, the transmission spectra provide a qualitative comparison of the hydrogen content in the films. The film prepared at 0.4 W/cm^2 displays no appreciable changes in either the hydrogen bonding configuration or the hydrogen content for annealing temperatures (T_a) up to 400 °C (curv e). At T_a = 500 °C (curve f) a sudden reduction in the hydrogen concentration occurs, leaving only traces of bonded hydrogen in the film which finally disappear at T_a = 600 °C (curve g). Fink et al. have reported that the release of hydrogen occurs between 550 and 650 °C on the rf glow-discharge deposited a-C:H films [3]. Films deposited at 0.8 and 1.6 W/cm^2 exhibit similar trends but higher annealing temperatures are required to observe hydrogen effusion. At T_a = 600 °C both spectra show a reduction in hydrogen concentration and the appearence of a new IR absorption band (the band labled "D") at 3040 cm^{-1} ascribed to CH stretching vibrations of aromatic

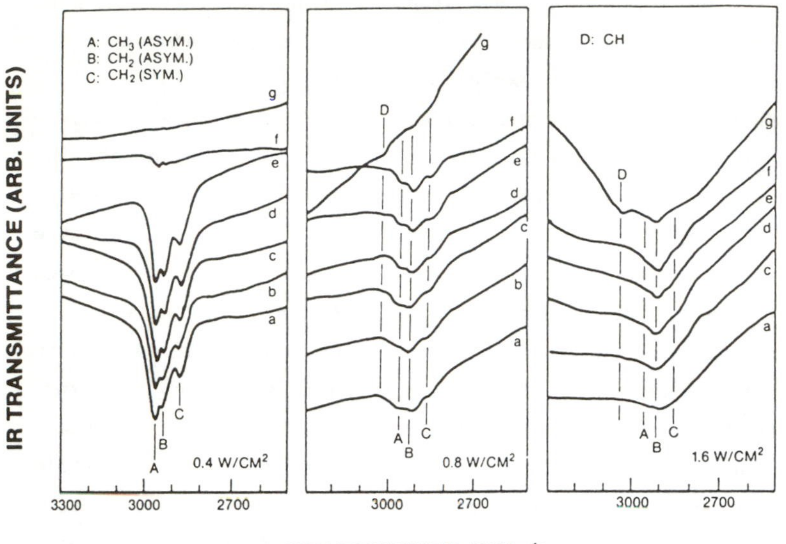

FIG. 5. IR transmission spectra of C–H stretching vibrations for samples prepared at rf power densities of 0.4, 0.8 and 1.6 W/cm^2: (a) As-prepared and after (b) 150 (c) 210 (d) 300 (e) 400 (f) 500, and (g) 600 °C annealing for 1 h.

ring groups. It is interesting to notice that as-prepared films deposited at higher rf powers did not reveal the D band even though the majority of the carbon atoms was in threefold sp^2 coordination. This suggests that the formation of condensed aromatic ring structures during deposition was somehow prohibited by the incorporation of a small amount of hydrogen.

Further evidence for the change in hydrogen bonding configuration due to high-temperature annealing of carbon films arises from different IR vibrational regions. Three regions of interest are shown in Fig. 6 for a sample prepared at 1.6 W/cm^2 before (curve a) and after (curve b) annealing at 600 ^oC for 1 h: the CH stretching, the C==C stretching and the CH wagging modes in the ranges 2800-3100, 1300-1700, and 700-1000 cm^{-1} respectively. As indicated, the as-prepared samples only show absorption bands corresponding to the vibrations of methyl and methylene groups. After 600 ^oC annealing not only the D band, due to aromatic CH stretching, but other vibrations involving benzene rings appear in the spectrum. Both the E and the F bands at 1600 and 1450 cm^{-1} are associated with the carbon-carbon ring stretching vibrations. Bands H, I, and J in the region of 700-900 cm^{-1} are related to the CH out-of-plane bending modes for hydrogen substituted aromatic rings.

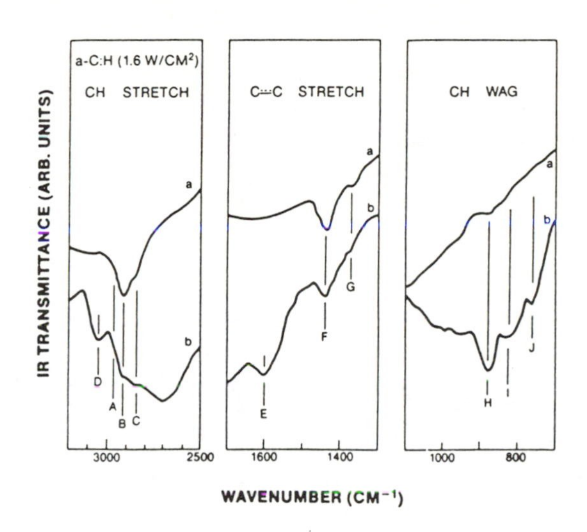

FIG. 6. IR transmission spectra of a film prepared at a rf power density of 1.6 W/cm^2: (a) As-prepared and (b) after 600 °C annealed for 1 h.

Figure 7 shows the PL spectra for the samples prepared at 0.8 W/cm^2 before and after annealing at temperatures up to 600 ^oC. Several changes occurred in the shape of the PL band after the various annealing steps. At T_a = 600 ^oC, a sharp Raman structure at ~ 2.4 eV appears on top of a weak PL background. This indicates that partial graphitization of the amorphous diamondlike phase has taken place. The integrated PL intensity for the same sample is shown in Fig. 8. Similar to Fig. 3, the intensity has been normalized to the number of absorbed photons during excitation. The insert shows the change in the optical density, at the energy of the excitation, as a function of T_a. Although there is no significant change in the optical density for T_a up to 300 ^oC, the intrgrated PL intensity increases by approximately one order of magnitude in the same annealing range. At higher temperatures the integrated PL intensity remains approximately constant and suddenly drops to values just above the detection limit of our apparatus at T_a = 600 ^oC. Notice that at this annealing temperature most of the hydrogen effuses out of the film (Fig. 5) and the sharp Raman structure, characteristic of sp^2 bonding configurations of the microcrystalline graphite, appears (Fig. 7).

The integrated PL intensity of samples at rf power densities higher than 0.8 W/cm^2 has a similar annealing behavior as shown in Fig. 8, the only differences being that in samples prepared at higher rf power densities the initial increase

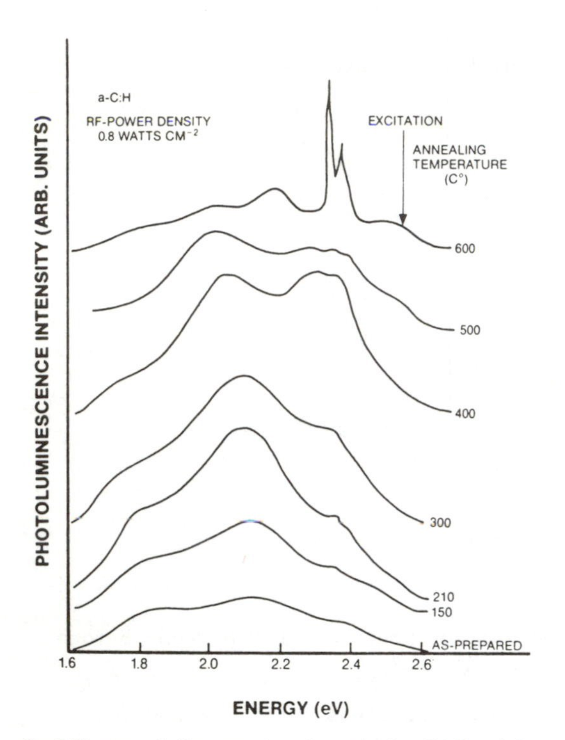

FIG. 7. PL spectra of a film prepared at a rf power density of 0.8 W/cm^2 after heat treatment at various temperatures for 1 h.

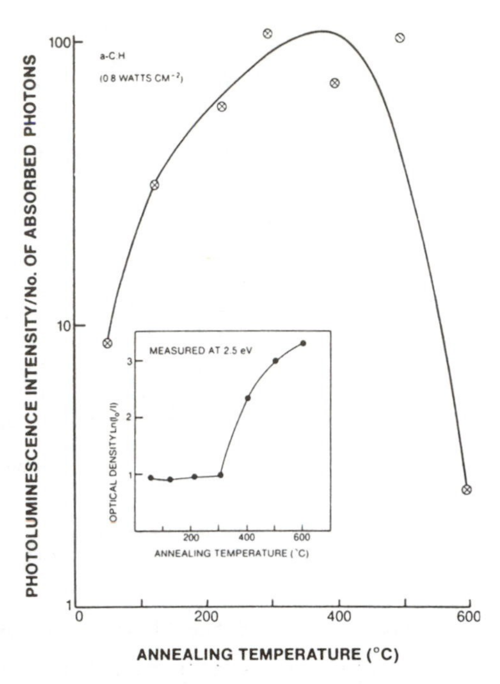

FIG. 8. The integrated PL intensity per absorbed photon of a sample (0.8 W/cm^2) as a function of the annealed temperature. The insert is the change in the optical density at the energy of excitation as a function of the annealed temperature.

in the integrated PL intensity is gradually reduced and its maximum value is achieved at slightly higher annealing temperatures.

ELS has been demonstrated to provide usefully structural informations of various carbon materials [3,11-13]. Two prominent features have been found in the ELS of graphite at 7.2 eV and 26.8 eV [11]. The higher energy peak corresponds to the bulk valence-band plasmon oscillation. The lower energy peak arises from the π-electron plasmon loss. In the case of diamond, only the bulk plasmon peak presents at 32.8 eV owing to the higher electron density [14]. X-ray, not electron beam, was used as an excitation source in this study to examine the ELS feature of the C 1s line. The ELS spectra focusing on the lower energy peak region obtained from a film deposited at 0.8 W/cm^2 are shown in Fig. 9. On air exposed as-prepared surface, it is found that the bulk plasmon peak is ~23 eV from the C 1s line (not shown) with complete absence of the π-plasmon peak [14,15]. The ELS feature remains unchanged on sample after prograssively annealing up to 500 °C, due to the absence of an appreciable amount graphite clusters. After T_a = 600 °C for 2 h, a major change of the ELS spectrum is a development of the π-plasmon loss peak located 6.6 eV from the C 1s line [Fig. 9(b)] originated from the π electrons of aromatic hydrocarbon rings and graphitic constituents in the film [13,15].

The annealing of samples that show the Raman signal in the as-prepared state leads to structural changes of the sp^2 phase similar to those in Fig. 4. That is, if T_a is lower than the temperature at which hydrogen effuses out, there is a narrowing and an upward frequency shift in the G Raman line as T_a increases. For T_a equal or higher than the dehydrogenation temperature, the Raman spectra experiences more drastic changes as is shown in the next figure. Figure 10 displays the Raman spectra for two samples of 0.4 (top) and 1.6 (bottom) W/cm^2

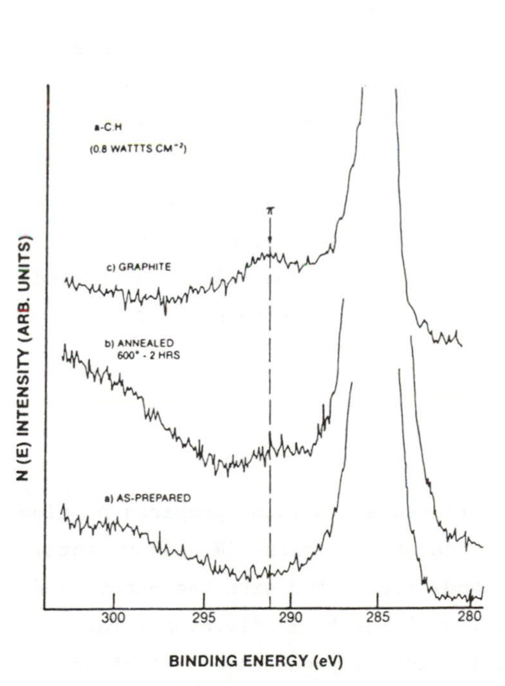

FIG. 9. ELS spectra of a film prepared at a rf power density of 0.8 W/cm^2 for (a) as-prepared and (b) T_a = 600 °C for 2 h. (c) Reference trace obtained from a crystalline graphite.

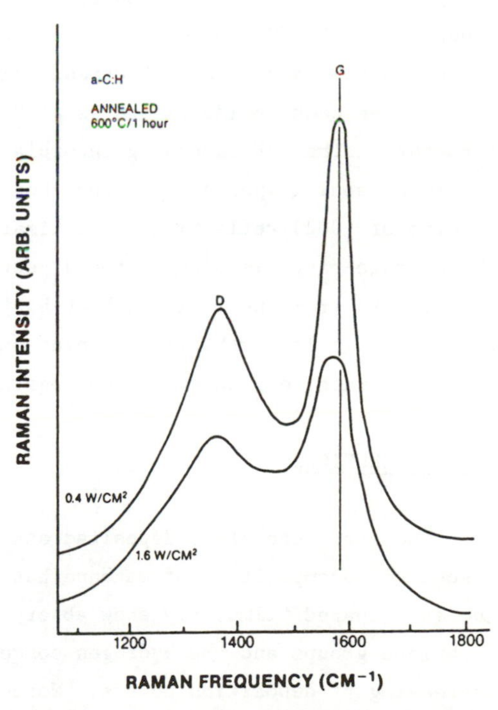

FIG. 10. Raman spectra of films prepared at rf power densities of 0.4 (top) and 1.6 W/cm^2 (bottom) after heat treatment at 600 °C for 1 h.

after annealing at T_a = 600 ^{o}C for 1 h. Comparing these data with those in Fig. 4, the changes due to annealing are clearly demonstrated: both the G and D lines become narrower and the frequency of the former shifts up; in addition, the intensity ratio of the D to the G lines also increases. Furthermore, these changes are more pronounced for samples prepared at lower rf power densities. According to earlier publications the annealed induced changes in the RS can be related to structural changes in the sp^2 phase [16,17]. In the as-prepared samples the graphitelike domains are disordered mainly due to bond-angle distortions; annealing at temperatures below the dehydrogenation temperature, the disorder is gradually reduced. For T_a above this temperature nucleation and growth take place and a characteristic Raman spectrum of microcrystalline graphite, with a well-defined D line, is observed. Following Tuinstra et al. [8], an effective crystallite size L_a in the basal plane of graphite was estimated to be 50 and 30 Å for samples of 0.4 and 1.6 W/cm^2, respectively.

XRD measurements provide further information about the structure of the as-prepared and annealed carbon films. The as-prepared samples exhibit a typical amorphous XRD trace with no detectable diffraction lines. However, those samples annealed at 600 ^{o}C show a single weak peak centered at about $2\Theta = 22.5^{o}$ whose position and linewidth depends slightly on the rf power density. This crystalline peak is identified as a (002) reflection of graphite in a hexagonal structure. The corresponding interplanar distance is 3.36 Å for a large single-crystal graphite. For the above 600 ^{o}C annealed films (Fig. 10), the d spacing of (002) reflection was estimated as 3.97 (top) and 4.00 Å (bottom). The average particle size in the direction normal to the basal planes (c axis) was derived from the full width at half-maximum (FWMH) of the peak to be 12 and 7 Å [18]. It is worthwhile to point out that 7 Å correlates to no more than two graphite planes stacked together.

IV. CONCLUSIONS

The structure of as-deposited and heat treated a-C:H films prepared by glow-discharge decomposition of methane has been studied. From the IR measurements, the as-prepared films only show absorption bands associated with the methyl and methylene groups and the hydrogen concentration in the films diminishes for increasing rf deposition powers. Moreover, in samples produced at low rf power the CH_3 groups are dominant resulting in soft and transparent films. Higher rf powers lead to samples with more opaque and harder structure which the CH_2 groups are now dominant. Large amounts of hydrogen in as-prepared films result in a strong PL emision from the sp^3 phase. The existance of the threefold coordinated carbon was indicated by the presence of the Raman signal at about 1300-1600 cm^{-1}. In samples with high hydrogen concentration the Raman signal may be obscured by the intense PL band. On the other hand, in samples with less hydrogen content, prepared at higher rf power densities, the PL intensity diminishes and the Raman signal appears indicating a larger volume fraction of the graphite-like phase. In addition, from the shape and position of the Raman spectrum, the existence of disorder in the threefold coordinated network is suggested. As rf power density increases, the frequency of the main Raman line (G line) shifts toward the position of observed crystalline graphite, indicating the improvement of local order.

Upon annealing, hydrogen effuses out of the films at temperatures that depend on the hydrogen concentrations, i.e., higher annealing temperatures are required for hydrogen effusion in samples containing lower hydrogen concentrations. In contrast to as-prepared samples, annealed films show IR absorptions of aromatic rings stretching and deformation modes due to the formation of a condensed ring structure with a long-range order. For annealing temperatures above dehydrogenation, the PL quenches and the shape of the Raman signal becomes similar to that of a microcrystalline graphite. The dimensions of the microcrystallites depend on the final annealing temperature; typically for T_a = 600 ^oC they are 30-50 Å along the graphite planes as measured by Raman and < 10 Å in the direction normal to the basal planes according to XRD measurements.

ACKNOWLEDGEMENTS

We would like to thank Professor G. Lucovsky of NC State University, Dr. A. Reyes-Mena and Dr. R. Asomoza of CINVESTAV-Mexico, and G. DeMaggio for their valuable discussions and suggestions. We would also like to acknowledge the encouragement of S.R. Ovshinsky of ECD. Thanks also go to A. Register of OSMC for his assistance in the sample preparation and T. Ammar of ECD for her art work.

REFERENCES

[1] Robertson, J.: Adv. in Phys., 1982, 35, 317.

[2] Tsai, H. and Bogy, D.B.: J. Vac. Sci. Technol., 1987, A 5, 3287.

[3] Fink, J., Muller-Heinzerling, T., Pfluger, J., Bubenzer, A., Koidl, P. and Crecelius, G.: Solid State Commun., 1983, 47, 687.

[4] Nemanich, R.J., Glass, J.T., Lucovsky, G. and Shroder, R.E.: J. Vac. Sci. Technol., 1988, A 6, 1784.

[5] "Introduction to Infrared and Raman Spectroscopy", edited by Colthup, N.B., Daly, L.H. and Wiberley, S.E., (Academic, New York, 1975).

[6] Reyes-Mena, A., Gonzalez-Hernandez, J. and Asomoza, R.: "Amorphous Hydrogenated Carbon Films", in Proceedings of the European MRS Meeting, 1987, edited by Koidl, P. and Oelhafen, P. (Les Evitions de Thysique, France, 1987) Vol. 17, p.229.

[7] Watanabe, I. and Inove, M.: Jpn. J. Appl. Phys., 1983, 22, L176.

[8] Tuinstra, F. and Koeing, J.L.: J. of Chem. Phys., 1970, 53, 1126.

[9] Lespade, P., Al-Jishi, R. and Dresselhans, M.S.: Carbon, 1982, 20, 427.

[10] Beeman, D., Silverman, J., Lynds, R. and Anderson, M.R.: Phys. Rev., 1984, B 30, 870.

[11] Liang, W.Y. and Cundy, S.L.: Philos. Mag., 1969, 19, 1031.

[12] Weissmantal, C., Bewilogua, K., Schurer, C., Breuer, K. and Zscheile, H.: Thin Solid Films, 1979, 61, L1.

[13] Weissmantal, C., Bewilogua, K., Breuer, K., Dietrich, D., Ebersbach, U., Erier, H.J., Rau, B. and Reisse, G.: Thin Solid Films, 1982, 96, 31.

[14] Zaluzec, N.J.: Ultramicroscopy, 1982, 9, 319.

[15] Oelhafen, P., Freeouf, J.L., Harper, J.M.E. and Cuomo, J.J.: Thin Solid Films, 1984, 120, 231.

[16] Nemanich, R.J. and Solin, S.H.: Phys. Rev., 1979, B 20, 392.

[17] Dillon, R.O., Woollam, J.A. and Katkanant, V.: Phys. Rev., 1984, B 29, 3482.

[18] "X-Ray Diffraction", edited by Warren, B.E. (Addison-Wesley, Reading, MA, 1969) p.253.

APPLICATIONS

Materials Science Forum Vols. 52 & 53 (1989) pp. 559-576

AMORPHOUS HYDROGENATED CARBON (a-C:H) FOR OPTICAL, ELECTRICAL AND MECHANICAL APPLICATIONS

K. Enke

Technisch-wissenschaftliche Unternehmensberatung
Am Wingert 19, D-8752 Johannesberg, FRG

ABSTRACT

In this paper the stress is put on a detailed description of the production process of amorphous hydrogenated carbon (a-C:H). Examples are given showing the effect of using different hydrocarbons such as ethylene (C_2H_4), acetylene (C_2H_2), and hexane (C_6H_{14}), and mixtures of hydrocarbons with Ar, H_2, CO_2, O_2, PH_3, B_2H_6, and N_2 on properties of the resultant layers. Dependences of deposition rates, refractive index, optical absorption, hydrogen content, density, microhardness, mechanical stress, friction behavior, and electrical conductivity on the deposition parameters such as gas pressure, self-bias voltage, and choice of gas and gas mixtures are shown.

INTRODUCTION

The great fascination arising from amorphous hydrogenated carbon (a-C:H) has its origin in a remarkable similarity to diamond with respect to many properties such as optical absorption, electrical resistivity, and hardness. As the synthesis of pure diamond crystals normally requires temperatures of thousands of degrees celsius and at the same time a pressure in the kilobar range, many people know that the process of producing artificial diamond is a great technological challenge. And moreover, artificial diamond is not very much cheaper than the natural one.

So it seems easy to understand why at the first euphory, when a-C:H was discovered (for detailed literature surveys see reference [1, 2]) this special metastable modification of carbon very soon got the attribute "diamond-like". But one has to be very careful to use this adjective because "like" means "equal" and "similar" at the same time. This confusion has lead to the fact that usually technicians and scientists mean "similar to diamond" which is the very truth, and merchants mean "equal to diamond" which sells better.

For years, producing a-C:H has been an issue of a scientific art being worked on only by use of small laboratory equipment. The most interesting properties of this modification of hydrogen containing carbon have been investigated using sample holders of only 50 to 200 cm^2. Although many technical applications for a-C:H were predicted from the beginning of the investigations, it is only since about six years that the first successful applications in the field of infrared optics came into a widespread industrial use. Apparently, this was only feasible with the prerequisite of an excellent adhesion of the carbon layers not only to silicon - an easy exercise to begin with - but also to such crucial substrate materials like germanium and steel.

Taking advantage of properties such as microhardness like sapphire, low frictional coefficient, inertness against any aggressive chemistry, no pin holes nor grain boundaries, and the ability to coat geometrically complex bodies, a-C:H and its metal containing derivatives now reluctantly begin being accepted in the market.

DEPOSITION PROCESS

Coating a substrate with a-C:H, requires a vacuum chamber with its pumping unit (see figure 1), a gas manifold, and a power source,

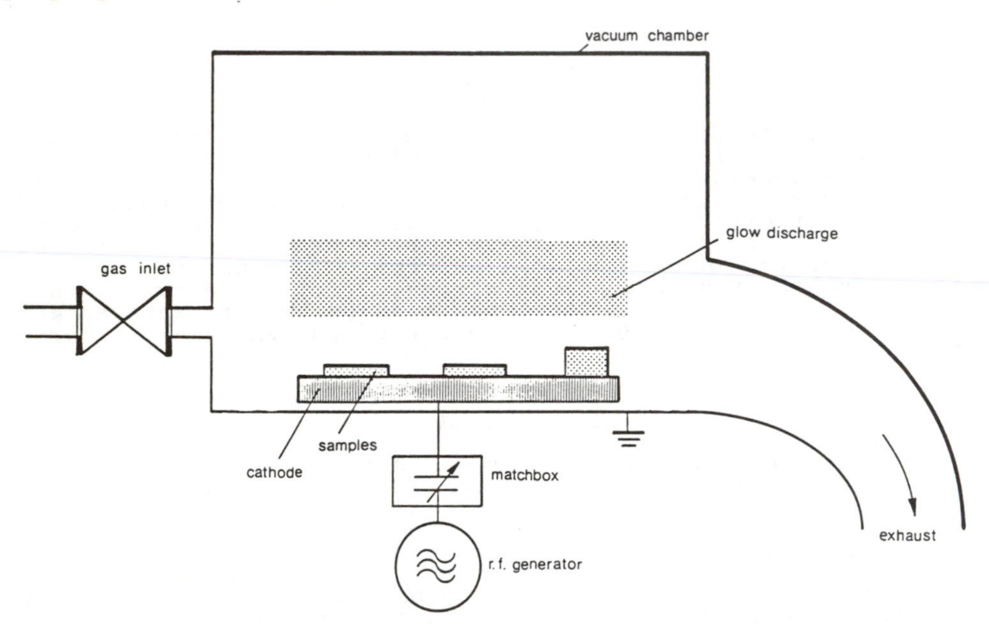

Figure 1: Deposition plant for a-C:H

preferably a radio frequency generator providing a sample holder with a minimum power density of round about half a watt per square centimeter. For a gas pressure between 10^{-3} to 10 mbar and a d. c. or r. f. voltage of a hundred to some thousand volts, a self-sustained glow discharge is ignited which means that a few atoms or molecules are ionised (preferably positively), some of them are activated, the molecules are cracked most probably, and most of them remain unimpressed. The major part of the gas is as cold as it is when it enters the chamber. Only the electrons and those ions which find slowly varying electric fields can accumulate energies which represent temperatures in the order of 10^4 degrees

C. Hence this kind of plasma is said to be unbalanced. The electrodes surround themselves with a poorly conductive dark space through which the ions are accelerated towards the electrode surfaces. For the d. c. case normally the smaller electrode is connected to the negative terminal. Using r. f. the smaller electrode charges negatively because of different mobilities of ions and electrons and different current densities on different electrode areas. The relatively sluggish positive ions "see" only the time averaged negative charge at the electrodes by which they are extracted from the boundary between dark space and luminous zone of the plasma.

Using noble gases, this is the case of sputtering. The cathode is the target of the material which is to coat samples being held opposite to it.

For reactive gases containing halogens, this means reactive ion etching, known as a very powerful tool in microelectronics.

For gases containing carbon, hydrogen and less halogens, this means plasma activated (or enhanced) chemical vapor deposition (PACVD or PECVD or PCVD) by which solid films are deposited at the surfaces of the electrodes. As those films are bombarded with ions while they grow, they undergo a process much like heavy quenching. This most probably is responsible for metastable, mostly amorphous structures of PCVD-layers. On the other hand, ion bombardment also means resputtering of part of the layer while it is being deposited. As this resputtering prefers weak atomic bonds, the stronger ones survive thus imparting high density and microhardness to the resulting layers.

One of the most interesting PCVD layers is undoubtedly the amorphous hydrogenated carbon, abbreviated a-C:H (or sometimes i-C) which can be deposited using hydrocarbon gases. It has turned out that basically the microstructure of hard carbon layers is not affected by the choice of the hydrocarbon gas. The films "forget" completely their origin, i. e. though the mass spectral pattern of a hydrocarbon plasma is a fingerprint of the respective gas, the resulting layer shows no evidence of the molecular structure of the precursor [3]. So attempts of producing pure diamond layers simply by using precursors having diamond structure already in the molecule (like methane) failed up to now. Even the use of adamantane ($C_{10}H_{16}$) yields only a-C:H with no equality to diamond. So the patent [4] describing the production of equal-to-diamond layers by means of using neopentane (dimethyl propane) apparently must be taken as a hoax.

PROCESS PARAMETERS

The most important process parameters for PCVD are

1. electrical power (self-bias, cathode voltage)
2. gas pressure
3. choice of gases
4. mixing ratio of different gases
5. sample temperature
6. sample position with respect to gas flow and plasma
7. electrode distance
8. frequency of the electrical field
9. gas flow (pumping efficiency)
10. power distribution (triode)
11. magnetical field intensity
12. pretreatment of substrates

Of those, parameters no. five to twelve can be taken as weak. But even using only two different gases and combining the third and fourth parameter, the dimension of the parameter space is four. As this can be hardly imagined, most investigators take one- or two-dimensional cuts through that space, i. e. they keep two or three parameters constant and vary the remaining ones step by step. Each step means a single deposition experiment; so total knowledge of the dependences of properties requires many deposition runs.

Working on a two-dimensional parameter space containing the self-bias voltage U_B and the gas pressure p, it can be shown that the influence of the choice of hydrocarbon gas is of minor importance. As has been shown in reference [5], changing the hydrocarbon gas results only in a slight shift and turn of the parameter landscape. Variation of self-bias, gas pressure, and mixing ratio, however, influences the nature of a-C:H layers more drastically [5 to 11]. Generally the deposition rate decreases with decreasing bias-voltage and pressure and the admixture of gases like noble gases, oxygen, nitrogen, and hydrogen. The addition of hydrogen to acetylene for instance [11] leaves the deposition rate nearly unaffected up to about 90 % of hydrogen. Only a further increase of hydrogen concentration changes deposition into etching.

CRYSTALLOGRAPHIC STRUCTURE

Electron diffraction:

a-C:H has been evidenced to be almost completely amorphous sometimes exhibiting microcrystalline inclusions [12, 13] of cubic face centered symmetry. This has also been observed by the author. Moreover, the appearance of cubic microcrystals is closely related to the mechanical properties of a-C:H. Below a critical self-bias voltage of 200 to 300 volts, soft polymeric hydrogen-rich layers are produced. Those layers are free from any crystalline inclusions and electron diffraction shows only a diffuse zone around the center of diffraction. For a self-bias of more than 300 volts the layers are hard and electron diffraction shows sharp rings related to cubic structure besides the strong diffuse amorphous central region.

The identification of the sharp electron diffraction rings as to cubic face centered microstructure has been questioned by Dischler and Brandt [14] who point out that the diffraction patterns of diamond and graphite, respectively, are almost indistinguishable so that there is a great probability that the microcrystals found could also be graphite instead of diamond. But this is not a contradiction to the high microhardness of a-C:H, as in graphite the carbon atoms within a lattice plane are bound more tightly than in diamond.

An interesting question as to amorphicity of hard carbon is, how are tetrahedral (diamond-like) bonds related to trigonal (graphite-like) ones. Using electron energy loss experiments, Fink et al. [15] found a ratio of 2:1 for a statistical distribution of tetrahedral to trigonal bonds which does not change very much even when polymer-like a-C:H layers are made [16]. This ratio obviously explains most of the properties of a-C:H, especially its greater similarity to diamond than to graphite.

Density:

Closely related to microstructure is the mass density of a-C:H [17 to 37]. Using gravimetry, the authors of references [17, 19 to 21, 24 to 27] get the relatively low values between 1.5 and 1.8 g/cm^3. Bubenzer at al. [21] showed a linear relationship between density and the normalized parameter $U_B/p^{0.5}$ where p is the gas pressure of a benzene glow discharge. The decrease of density is associated with an increase of the polymeric nature of a-C:H indicated by an increase of the strength of the infrared absorption band at 2900 cm^{-1} which also means an increase of hydrogen content [33]. Using the floating technique, the maximum density values of a-C:H layers made of acetylene have been found near 2.0 to 2.1. This discrepancy can be explained by an etching effect of the substrate at the very beginning of the deposition [38, 39] which pretends a lighter weight of the coated substrate at the end of deposition than expected without etching.

Extremely high density of carbon layers exceeding even the density of pure diamond (3.5 g/cm^3) have been reported by Bakai and Strel´nitskii [35] who used a carbon plasma beam for deposition and by Matyushenko et al. [36] and Vakula et al. [37] who produced "metallic" carbon by using an arc discharge and a graphite cathode.

OPTICAL PROPERTIES

1 µm thick a-C:H layers deposited on glass appear dark brown when viewed in transmission. This color is due to a very high optical absorption in the violet and ultraviolet part of the light spectrum and a low absorption in the red and infrared part [5]. Above a wavelength of about 1.5 µm a-C:H is nearly completely free from optical absorption up to about 12 µm. Due to vibrations of CH_x groups against the amorphous "crystal lattice", there is some localized absorption at 3.5 and 7 µm which has been studied in great detail in references [16, 17, 40].

The refractive index of a-C:H shows nearly no dispersion [41 to 46] and can easily be adjusted between 1.6 (soft, polymeric) to 2.2 for very hard diamond-like layers [8 to 10, 19 to 21]. The dependence of the refractive index on gas pressure and bias is very different for hexane [8, 9, 10] and benzene [19 to 21], respectively. Whereas the refractive index is nearly constant for half a decade of pressure variation for layers made of hexane, iso-lines are almost 45 degree lines in the U_B-p-plane for layers made of benzene. I. e. refractivity of benzene-made layers shows a strong dependence on both parameters. This behavior again shows the influence of the choice of hydrocarbon gas on a property of a-C:H. It also means that changing the deposition gas requires to change the main parameters U_B and p in order to get the same value of a property as before. This is quickly found out by two or three deposition runs. As the property "landscapes" in the parameter space are different for different properties, in some cases an optimization of property sets is possible. One can e. g. choose a high pressure for a high deposition rate. In order to get a fixed refractive index, one needs only to choose the appropriate selfbias. Unfortunately, this does not work for the two properties refractivity and mechanical stress using hexane. As the iso-lines of refractivity and stress both are completely parallel for a refractive index of 2.0, no optimization with respect to stress is

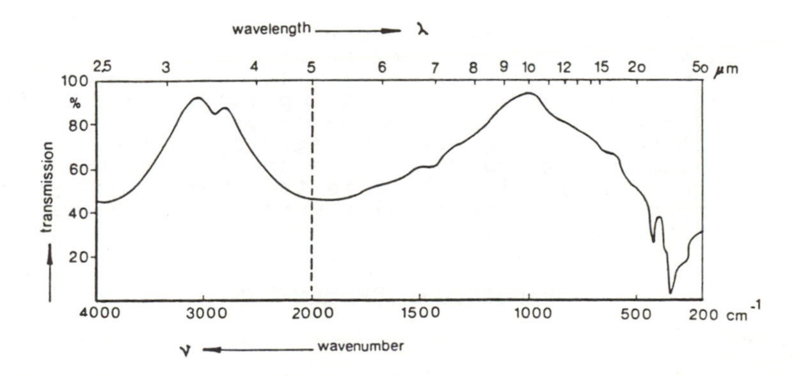

Figure 2: IR transmission spectrum of a both sides carbon coated germanium disk.

possible. Nevertheless, manufacturing anti-reflection layers on germanium for a wavelength of 10 µm is possible, for the necessary thickness is 1.25 µm in this case, well below a critical thickness of about 2 µm where the stress exceeds the adhesion strength of a-C:H to germanium.

This favorable circumstance lead to one of the first industrial applications for a-C:H, namely a scratch resistant, non-corrosive anti-reflection layer for infrared optical windows and lenses made of germanium and silicon. Figure 2 shows an infrared spectrum of a 1 mm thick germanium disk coated on both sides with a 1.25 µm thick a-C:H layer. The transmission minima at wavelengths of 2.5 and 5 µm show the low transmission of 42 % an uncoated disk would have for nearly the whole spectral range. In the thermovision range between 8 and 12 µm the transmission can be raised to a maximum of 92 %. The missing 8 % are due to residual absorption of a-C:H because of weakening of the selection rules of optical transitions in that range due to the amorphicity of the carbon.

ELECTRICAL PROPERTIES

a-C:H is an insulator having a specific resistivity between 10^8 and 10^{13} Ωcm, again proving its dissimilarity to graphite. Analogous to amorphous hydrogenated silicon (a-Si:H), saturation of dangling bonds by hydrogen is said to be responsible for good insulating properties of a-C:H. The incorporation of hydrogen is strongly dependent on the cathode voltage U_B, and so is the specific resistivity [47 to 53]. For voltages of less than 100 volts a-C:H is a good insulator with a resistivity of 10^{13} to 10^{10} Ωcm. Above 100 volts, the resistivity drops over several orders of magnitude and remains at values of about 10^7 Ωcm nearly independent of cathode voltage. Berg and Andersson [47] report values of only 100 and 1 Ωcm for 1200 and 1800 volts, respectively. Those high values of conductivity can be due to high temperatures associated with the intensive ion bombardment and no information on temperature is given in that reference. Indeed, deposition temperatures of up to 1000 degrees C result in resistivities of down to 10^3 Ωcm [54 to 57]. Interestingly, in mixtures of acetylene and argon the resistivity of the carbon layers can be lowered from 10^8 to 10^3 Ωcm by varying the acetylene partial pressure from 8 to 2 mTorr [57]. The argon partial pressure was kept at 45 mTorr.

Like other semiconductors, a-C:H can be doped with elements of group III and V of the periodic system. Preferred elements are boron using diborane (B_2H_6) as doping gas and phosphorous using phosphine (PH_3) [58 to 60] but also nitrogen [60] and tantalum and ruthenium are reported [61, 62].

Another possibility to increase electrical conductivity of a-C:H layers is annealing to temperatures of at least 300 degrees C [16, 17, 43, 44, 63, 64]. Dischler et al. [16, 17] show that upon annealing a-C:H continuously reduces its optical band gap from 1.2 eV to nearly zero and changes the ratio of sp^3 to sp^2 hybridisation from 2 : 1 to 0 : 1. Above 600 degrees C almost no further changes occur indicating that a-C:H has completely changed its metastable nature into graphite which is the energetically preferred state of carbon. The electrical conductivity reaches values of 10^{-1} to 10 Ωcm then.

MECHANICAL PROPERTIES

Microhardness:

From the viewpoint of industrial applications, the two most interesting features of hard carbon are its microhardness of at least 1700 kg/cm² [6, 14] which is much like that of sapphire, and its low frictional coefficient [6, 65, 66], especially in vacuum and dry nitrogen.

In order to eliminate the influence of the microhardness of the substrate, it is indispensible to use layers having a minimum thickness of 2 μm [14, 64]. Therefore hardness values of more than 2500 kg/cm² should be taken with care. Complete elimination of the thickness problem is possible by depositing very thick layers of about 50 μm on aluminum. Coad et al. [67] made microhardness measurements on a polished cross section of such a thick layer and got a value of 1700 kg/cm² for an indentation load of 50 g.

To obtain the elastic properties of a-C:H, two approaches have been found in the literature [66, 68]. Gille and Rau [68] evaluated the geometry of wrinkles of delaminated carbon layers and obtained a Young's modulus Y = 120 to 290 GPa. Memming et al. [66] made use of a microindentation method as commonly used for microhardness measurements but with the possibility to measure the penetration depth in situ during indenting. They found Y = 120 GPa, both values lying close to that of steel which is 250 GPa [66].

Friction:

Presumably the first to report on low friction of carbon layers were Aisenberg and Chabot [69] who used an ion plating process which is characterized by the absence of hydrogen. So, by this means a different modification of carbon is deposited which is more of a hard version of amorphous carbon (a-C) than a-C:H.

A systematic investigation of sliding friction of steel rubbing against a-C:H coated silicon has been undertaken by Enke et al. [6, 65]. They discovered a strong dependence of the friction coefficient on the relative humidity of a nitrogen atmosphere in which the friction experiments took place. In vacuum or very dry nitrogen the frictional coefficient has values of only 0.02. Raising the relative humidity from 10 to 99 % the frictional coefficient increases from 0.05 to 0.3 (see figure 3). The explanation of an

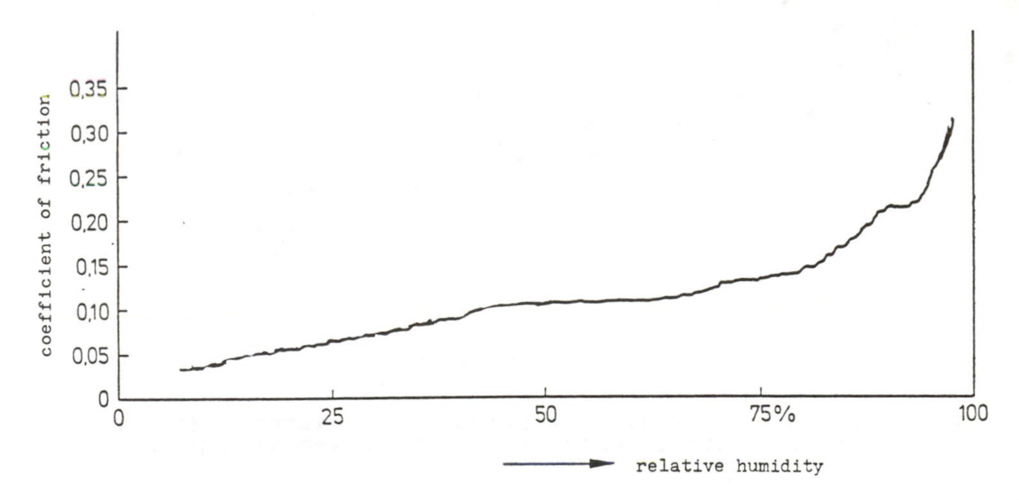

Figure 3: Coefficient of sliding friction of steel against a-C:H versus relative humidity of a nitrogen atmosphere.

easy mobility of lattice planes as for graphite or molybdenum disulfide does not hold for amorphous carbon. As a-C:H evolves hydrogen and hydrocarbons when heated to more than 300 degrees C [70], there is a great probability of gas effusion during sliding of steel against a-C:H. This gas effusion can provide a gas "cushion" on which the sliding steel ball can skim. At least the adhesion between steel and carbon is reduced which in turn reduces the frictional forces a great deal. Surfaces sliding against each other are heated at microscopic spots to temperatures of more than 1,000 degrees C [71 to 73], so effusion of gases takes place in any case.

For soft, low friction materials like molybdenum disulfide, graphite, and PTFE, low wear can only be achieved when a conservation of wear debris within the tribosystem can be provided. For hard carbon this is not necessary as wear of a-C:H against many different frictional partners is low [69, 74, 75].

Tribology is a multiparameter system much like plasma activated CVD. Moreover up to now no closed theory exists which could be able to predict frictional behavior of tribosystems. So the user of friction reducing (as well as enhancing) coatings is dependent on his experence and ingenuity and on trial and error experiments, in order to choose the convenient combinations of layer materials to solve his respective tribological problem. Taking into account, that the system diamond against a-C:H yields very low frictional coefficients over a large range of relative humidity (see figure 4), it can be intriguing to investigate more systems like that and see what happens.

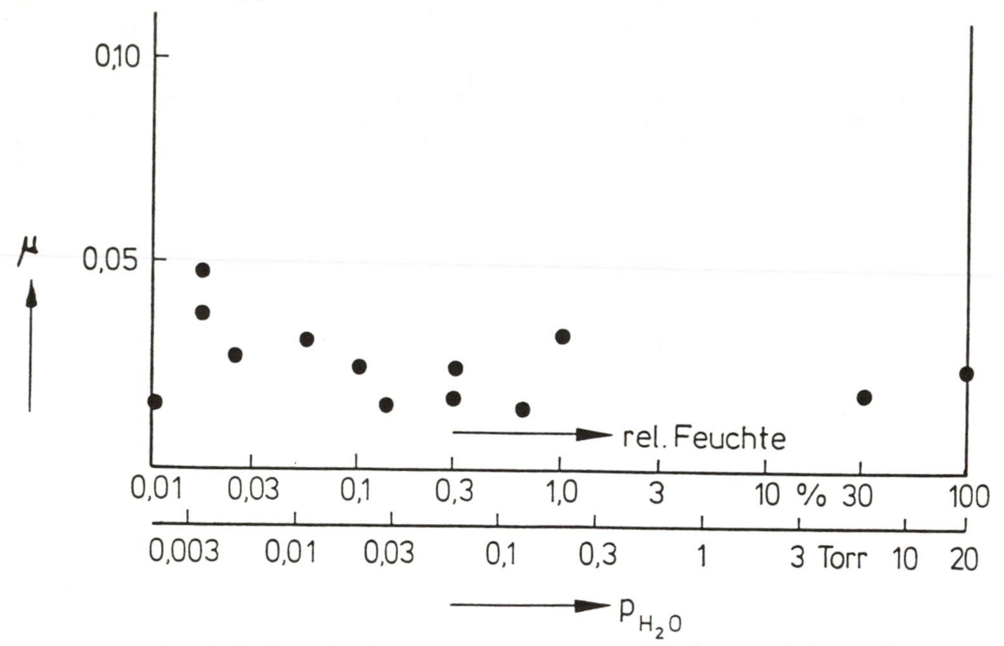

Figure 4: Coefficient of sliding friction of diamond against a-C:H versus relative humidity of a nitrogen atmosphere.

Stress:

Since the early successful attempts to produce a-C:H, it was discovered that the layers exhibit compressive stress. Detailed informations on dependences of stress on deposition parameters are described in references [5 to 7, 11, 76 to 78]. Sometimes the heavy ion bombardment during deposition is made responsible for the stress but structural and compositional effects obviously are superposed. An indication for that can be the fact that increasing the ion energy by increasing U_B lets the stress pass through a maximum [5, 6, 7]. Weissmantel et al. [22, 23, 79] report on a maximum of microhardness associated with ion accelerating voltages of about 900 volts. So one can speculate on a correlation between stress and hardness.

A drastical reduction and even a sign reversal from compressive to tensile stress has been reported by Zelez [80, 81]. As the infrared transmission spectra, the high resistivity of 1.5 x 10^{14} Ωcm, and the high optical gap of 3 eV indicate, his a-C:H layers are polymeric in nature. Repetitions of Zelez's experiments by the author using U_B = 1000 V at the r. f. powered upper electrode and a d. c. voltage of 200 volts at the lower substrate bearing electrode also resulted in low stress highly transparent layers which could easily be scratched off the silicon substrate and again showed polymer-like infrared spectra (see references [16, 82, 83]).

SPECIAL GAS MIXTURES

Mixing reactive or non-reactive gases to hydrocarbons can result in carbon layers with properties which are more or less different from "pure" a-C:H [11, 38, 57 to 60, 84, 85].

Argon:

The physical dilution effect of argon results in a decrease of deposition rate up to a transition from deposition to etching which is near an argon-to-hydrocarbon ratio of 4 : 1 [11] to 10 : 1 [38]. Even more argon results in etching.

Addition of argon to hydrocarbon also makes a-C:H electrically more conducting [57], possibly due to a change of the carbon microstructure towards graphite.

A very welcome effect of argon admixture is its influence on the mechanical stress of hard carbon [11]. Unfortunately, reference [11] gives only little information and it can be worth while to investigate the concentration range of zero to 25 % argon in acetylene in more detail.

Phosphine, diborane, nitrogen:

The former two gases are typical doping means for tetravalent semiconductors. 1 % of them diluted in hydrocarbon but also 5 % nitrogen increase electrical conductivity of a-C:H 10-fold for deposition temperatures between 150 and 300 degrees C [60].

Oxygen:

An oxygen plasma is an excellent means to clean sample holders and the deposition chamber walls from a-C:H by reactively "burning" carbon into CO and CO_2 and hydrogen into water vapor. So only small amounts (25 %) of oxygen in hydrocarbon are sufficient to stop deposition of a-C:H [38].

Hydrogen:

Nishikawa et al. [85] have shown that a-C:H and a-C:H:F can be etched in a hydrogen plasma. Using a mixture between acetylene and hydrogen, deposition and etching occur at the same time. Which one is dominant depends most on the hydrogen-to-acetylene ratio of course. For a fixed partial pressure of 1 mTorr for C_2H_2 the deposition rate astonishingly increases slowly but steadily from 0.8 to 1.4 µm/h when hydrogen is added up to a hydrogen concentration of 90 % [11]. Further addition of hydrogen results in a steep decrease of deposition rate until the transition from deposition to etching is reached at a hydrogen concentration of about 99 %. Though the plasma chemistry is quite different, a similar behavior with respect to hydrogen admixture shows C_2F_6. This gas normally is used as an etch gas for microelectronics because of its high fluorine content. For a fixed flow of 19 sccm for C_2F_6, the deposition rate rises when hydrogen is added in small amounts [85]. Addition of hydrogen neutralizes the etching effect of fluorine by creating HF and/or deactivating the CF_3 radical thus enabling the gas mixture to yield solid a-C:H:F layers. This effect ends for a hydrogen concentration of about 50 %. Further increase of hydrogen flow lets the deposition rate go down until the transition to etching is reached at a flow of about 40 sccm.

In a broad range of hydrogen concentration (5 to 70 %) of an acetylene-hydrogen mixture, the Knoop microhardness of a-C:H remains at a constant value of about 1,800 kg/cm² [11]. For further hydrogen addition the microhardness drops to only 500 kg/cm² at 90 % hydrogen and increases again to 1,400 kg/cm² near the transition from deposition to etching at 99 % hydrogen. The hardness drop at 90 % hydrogen can give space for speculation and one possible explanation is an increased incorporation of hydrogen thus shifting the nature of a-C:H more to the polymer side.

The intrinsic mechanical compressive stress of a-C:H layers made from C_2H_2 can be reduced by a factor of 3 by adding 23 % hydrogen without any expense of deposition rate or hardness [11]. Taking into account that other deposition parameters are e. g. the addition of argon, the self-bias voltage at the cathode, and the overall pressure of the working gas mixture, one can imagine that there is the possibility of optimizing hard carbon layers with respect to high hardness and low stress at the same time.

Adding hydrogen to acetylene also influences the frictional coefficient of a-C:H by a factor of about 3.7 [11]. No conclusive explanations can be given for that now and it will be reserved to future investigations to shed more light on correlations between microstructure, chemistry, hardness, and friction.

Carbon dioxide:

Nir et al. [84] used an ion gun having a graphite cathode as a carbon source and ran it with a mixture of C_2H_2, CO_2, and Ar. The resulting layers are reported to be essentially free from the typical CH_x-vibration to be seen in the infrared transmission spectrum at 2900 cm^{-1}. What Nir et al. do not report but what is revealed by their spectra, is an additional absorption in the short wave region near 2.5 µm probably due to OH-vibrations arising from a plasma-chemical reaction of oxygen from CO_2 and hydrogen from C_2H_2. A slight increase of absorption at a wavelength of 6 µm (H_2O) and at 9 µm (CO) supports this assumption. The findings of Nir et al. could be confirmed by own experiments using the simple diode configuration of figure 1 together with a mixture of hexane (C_6H_{14}) and carbon dioxide. For U_B = 800 volts, a hexane flow of 180 sccm, and a carbon dioxide flow of 10 sccm, the deposition rate was 3.4 µm/h which increased only slightly to 3.8 µm/h by increasing the carbon dioxide flow to 100 sccm. The difference of the IR spectra of a 1.25 µm thick a-C:H layer on germanium, however, was more pronounced. For the higher CO_2 flow, the optical absorption at 7.5 µm was about 3 % (single side coated germanium) where it uses to be nearly zero for no CO_2. Near the water absorption region of 2.9 µm the additional absorption even reaches 12 %. The strong weakening of the CH_x vibration at 2,900 cm^{-1} could be confirmed supporting the assumption that the oxygen in the carbon dioxide burns away plasmachemically much of the hydrogen of the C_6H_{14} molecule. Compare also the results of the authors of references [40, 86] who stepwise substituted the hydrogen in benzene by fluorine. A quite interesting feature is the occurence of a sharp absorption line at 4.3 µm which is assigned to gaseous CO_2. No carbon monoxide peak (to be expected at 4.6 µm) could be found in the spectra.

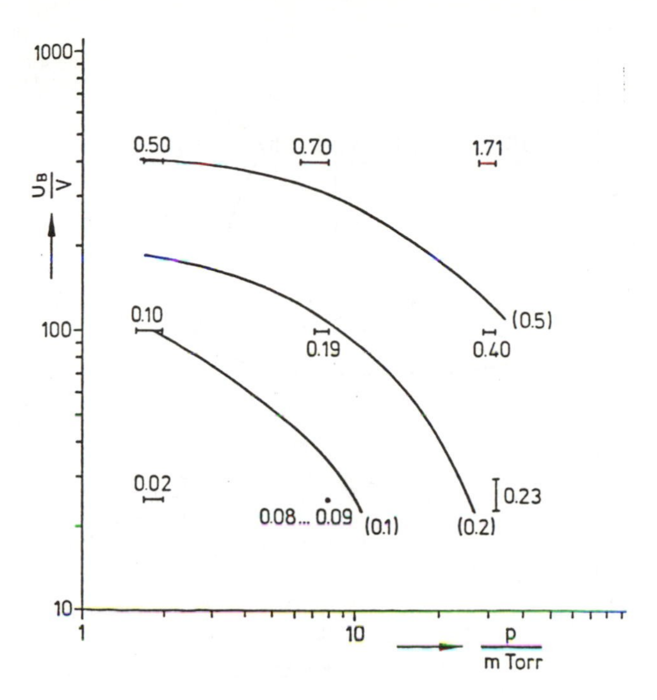

Figure 5: a-C:H from acetone. Iso-lines of deposition rates (in µm/h) in the U_B-p-plane.

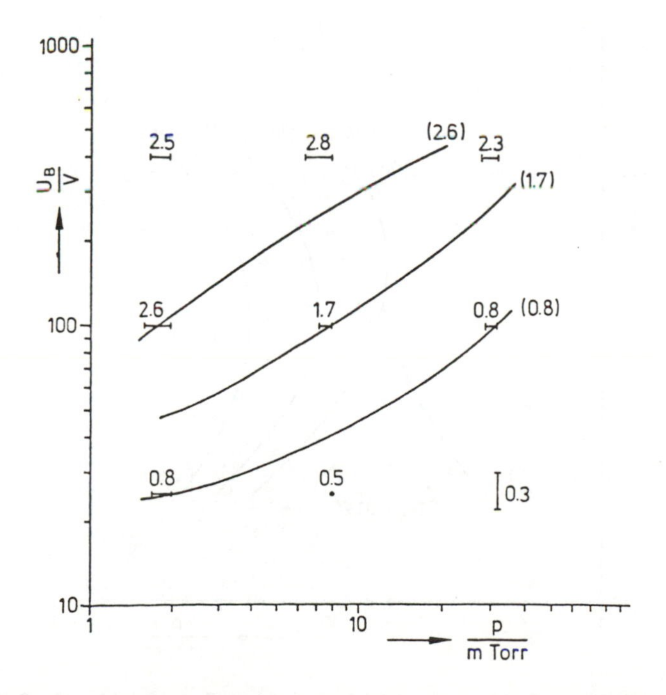

Figure 6: a-C:H from acetone. Iso-lines of compressive stress (in GPa).

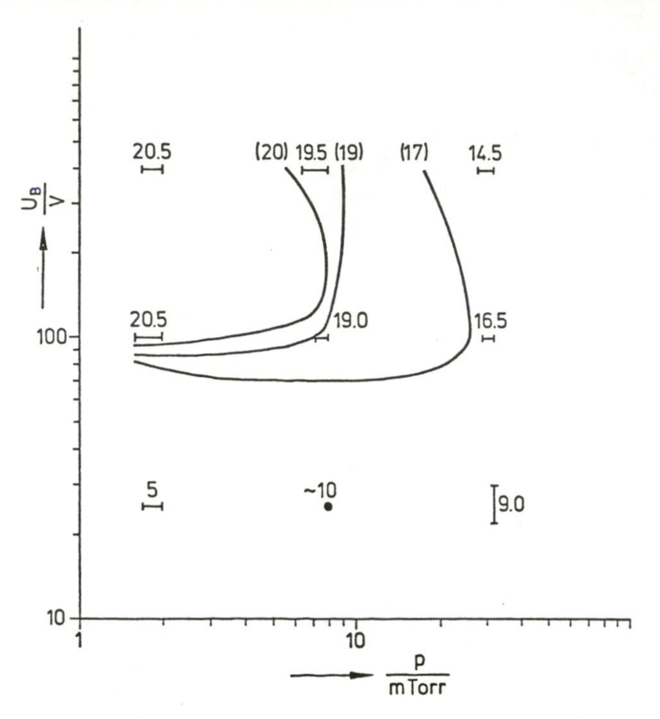

Figure 7: a-C:H from acetone. Iso-lines of Knoop microhardness. Numbers are in GPa.

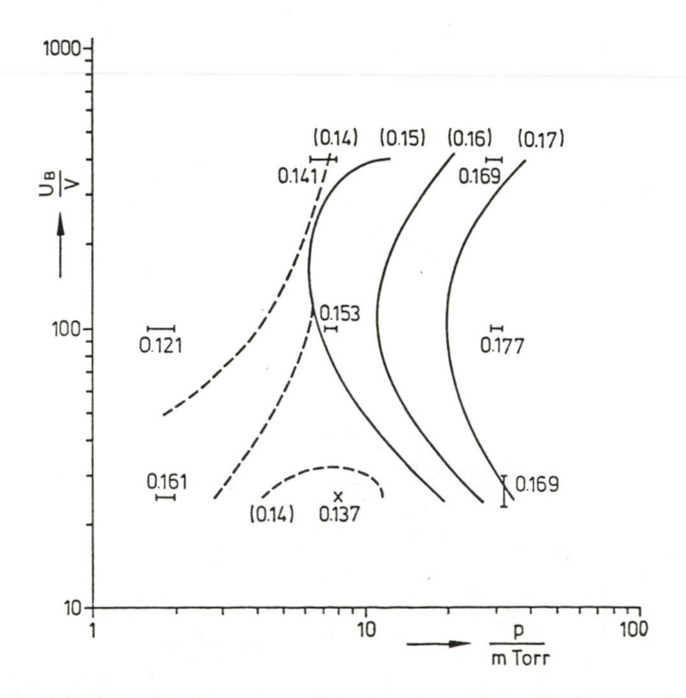

Figure 8: Iso-lines of the coefficient of sliding friction of steel against a-C:H made of acetone. Atmosphere: Nitrogen, 40 % relative humidity.

Acetone:

Acetone (C_3H_6O) is volatile enough and contains little enough oxygen to be appropriate for deposition of a-C:H. As figure 5 shows, the deposition rates are not very high, possibly because of the plasmachemical etching effect of the oxygen.

As figure 6 shows, the intrinsic mechanical stress (numbers are values in GPa) which is compressive is in the order of other a-C:H layers. Each error bar means a single deposition experiment. The number of depositions is small, so it can be speculated if the stress runs through a maximum in the upper left corner of the figure. It is the same region where there is a maximum of Knoop microhardness (see figure 7).

The frictional coefficient of a degreased steel ball (100Cr6) sliding against a-C:H layers made of acetone has been measured using a nitrogen atmosphere having a relative humidity of 40 %. Figure 8 shows iso-lines of the coefficient of sliding friction. As there are only few datum points available, the dotted lines are speculative, indicating that there can be a region in the upper left part of the parameter plane where the frictional coefficient falls below 0.1.

According to the molecular structure, the oxygen content of the acetone molecule is 10 atomic %. As figure 9 shows, however, the

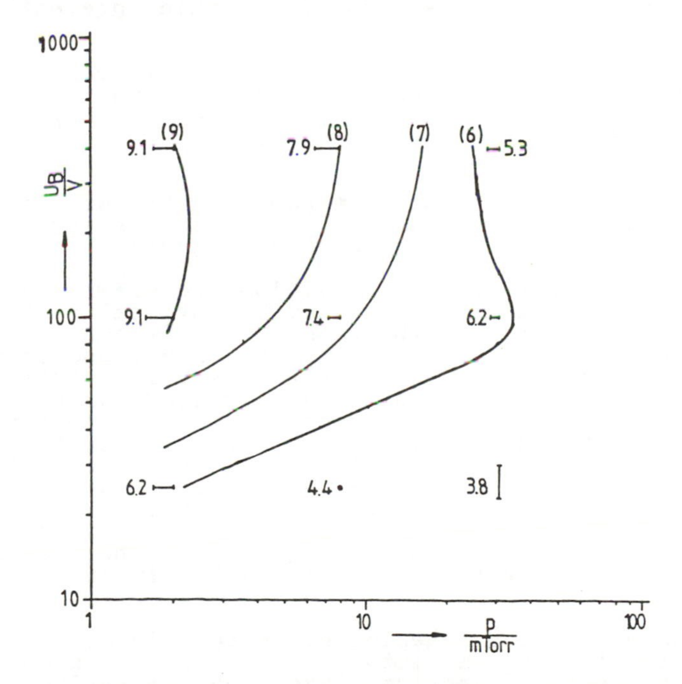

Figure 9: a-C:H from acetone. Iso-lines of atomic % of oxygen.

plasma CVD process removes 9 to 62 % of the oxygen, with respect to the parameter range investigated.

CHEMICAL COMPOSITION

As the term "amorphous hydrogenated carbon" suggests, the main "impurity" in a-C:H is hydrogen. The hydrogen content in a hypothetical carbon compound of the general chemical formula CH_x is determined to be between x = 0.1 and 1.0 [15, 17, 26, 29, 32 to 34, 87 to 94]. Values of under 0.2 have to be classed with graphite-like forms, the hard diamond-like modifications have x-values between 0.3 and 0.5 while CH_x with $x > 0.5$ belongs to hydrogen-rich soft polymeric carbon. The strongest dependence of x is on the self-bias U_B, the deposition temperature, and the annealing temperature in the sense that x is reduced by large U_B and temperatures. Other properties like electrical conductivity and optical properties point into the same respective directions.

INDUSTRIAL APPLICATIONS

The outstanding properties of a-C:H have led to many industrial applications. Its high hardness, wear and corrosion resistance, and its low friction against many materials inspired many people to a lot of practical applications. But the first enthusiasm vanished when the applicants realized that some precautions must be taken for a good adhesion of the hard carbon layers to the respective substrates. To clean and especially to degrease the substrates prior to deposition is self-evident. To elucidate this point, a simple experiment can be made. Make a finger print on a polished and clean silicon substrate. Then deposit hard carbon onto the silicon and look at the finger print. Between the papillary lines the adhesion of the carbon layer is excellent but it is not on the very lines where the grease of the skin prevents adhesion of carbon.

And besides, the adhesion of a-C:H to silicon is so good that layers exceeding a critical thickness of about 6 µm take part of the substrate with them. For this thickness the carbon gathers such a large stress (see section "mechanical properties") that its value is larger than the binding forces between the silicon atoms.

These forces are much less for aluminum than for silicon. But as aluminum can flow plastically, it can relax the stress forces at the interface even for 50 µm thick carbon layers.

Critical substrates for carbon deposition are glass and steel. The best results for adhesion are achieved by physically etching (presputtering) the substrates with argon ions. Precaution has to be taken at the instant when presputtering is changed to deposition by shutting the argon valve and at the same time opening the hydrocarbon valve. The plasma must remain ignited during this transition. An interruption of only seconds is sufficient to deteriorate adhesion to an intolerable extent. The reason for this obviously is the chemical activation of the substrate surface during sputter etching. Once the residual gas or the hydrocarbon touches the activated surface, molecules can condense without the action of ion bombardment. The nature of this condensate is generally undefined but presumably it is soft so that even thin carbon layers chip off easily, leaving the substrate partially or completely uncoated.

Once the problems of bad adhesion had been solved, the first industrial application to come into use was the scratch resistant anti-corrosive, anti-reflection layer on germanium windows and front lenses [5, 7 to 10].

Other applications which will soon appear on the market are:
- biocompatible layers on biological implants
- wear protection of magnetic layers
- wear and corrosion protection of hydraulic and pneumatic elements
- electrical insulators

OUTLOOK

The conference on "Amorphous Hydrogenated Carbon Films" in Strasbourg in 1988 showed a continuing interest in this fascinating carbon modification, though the first publications date back already into the early fifties [95 to 97]. After the question whether or not a-C:H is diamond-like, diamond or similar to diamond seems to have been settled, two offshoots begin to develop. One is the class of metal-carbons (see e. g. [61 to 63]) and the other is the real, genuine diamond. The latter needs deposition conditions which are far away from those necessary for the production of a-C:H [98], so that the issue "diamond" has been neglected completely in this article. Nevertheless, despite the high production temperatures of at least 600 degrees C up to now, diamond will become the successor of a-C:H in importance and scientific and technical interest within the next few years.

Literature

1) Woollam, J.A., Chang, H., Natarajan, V.: Appl. Phys. Commun., 1985-1986, 5, 263.
2) Robertson, J.: Advances in Physics, 1986, 35, 317.
3) Wagner, J., Wild, Ch., Bubenzer, A., Koidl, P.: Proc. Mat. Res. Soc. Symp., 1986, 68, 205.
4) Ovshinsky, S.R., Flasck, J.: European Patent Application 0 175 980, 1986.
5) Enke, K.: Appl. Optics, 1985, 24, 508.
6) Enke, K.: Thin Solid Films, 1981, 80, 227.
7) Enke, K.: Proc. 28th Ann. Tech. Conf. Soc. Vac. Coaters, Philadelphia, 1985, p. 78.
8) Enke, K.: Proc. 5th Intern. Conf. Ion & Plasma Assisted Techniques (IPAT), Munich, 1985, p. 319.
9) Enke, K., Lorenz, G., Stoll, H.: Proc. Intern. Conf. Ion & Plasma Assisted Techniques (IPAT), Brighton, 1987, p. 56.
10) Enke, K., Geisler, M., Kieser, J., Münz, W.-D.: Mat. Res. Soc. Symp. Proc., 1987, 93, 310.
11) Enke, K.: Proc. E-MRS Conf. on Amorphous Hydrogenated Carbon Films, Strasbourg, 1987, XVII, 117.
12) Vora, H., Moravec, T.J.: J. Appl. Phys., 1981, 52, 6151.
13) Moravec, T.J.: SPIE, Vol. 325, Optical Thin Films, 1982, 117.
14) Dischler, B., Brandt, G.: Industrie Diamanten Rundschau, 1984, 18, 249.
15) Fink, J., Müller-Heinzerling, T., Pflüger, J., Bubenzer, A., Koidl, P., Crecelius, G.: Solid State Commun., 1983, 47, 687.
16) Dischler, B., Sah, R.E., Koidl, P., Fluhr, W., Wokaun, A.: Proc. 7th Intern. Symp. on Plasma Chemistry, Eindhoven, 1985, editor: C. J. Timmermans, IUPAC Subcommittee of Plasma Chemistry, Eindhoven, 1985, p. 45.
17) Dischler, B., Bubenzer, A., Koidl, P.: Solid State Commun., 1983, 48, 105.
18) Smith, F.W.: J. Appl. Phys., 1984, 55, 764.
19) Dischler, B., Bubenzer, A., Koidl, P., Brandt, G.: Proc. SPIE, 1983, 400, 122.
20) Dischler, B., Bubenzer, A., Koidl, P., Brandt, G., Schirmer, O. F.: Laser Induced Damage in Optical Materials: 1982, NBS Spec. Publ., 1984, 669, 249.
21) Bubenzer, A., Dischler, B., Brandt, G., Koidl, P.:J. Appl. Phys., 1983, 54, 4590.
22) Weissmantel, C.: Le Vide les Couches Minces, 1986, 41, 45.
23) Weissmantel, C., Bewilogua, K., Breuer, K., Dietrich, D., Ebersbach, U., Erler, H.-J., Rau, B., Reisse, G.: Thin Solid Films, 1982, 96, 31.
24) Bubenzer, A., Dischler, B., Brandt, G., Koidl, P.: Opt. Eng., 1984, 23, 153.
25) Bubenzer, A., Dischler, B., Brandt, G., Koidl, P.: Proc. SPIE, 1983, 401, 321.
26) Dischler, B., Bubenzer, A., Koidl, P.: Appl. Phys. Lett., 1983, 42, 636.
27) Koidl, P., Bubenzer, A., Dischler, B.: Proc. SPIE, 1983, 381, 177.
28) Anttila, A., Koskinen, J., Bister, M., Hirvonen, J.: Thin Solid Films, 1986, 136, 129.
29) Ojha, S.M., Norström, H., McCulluch, D.: Thin Solid Films, 1979, 60, 213.
30) Pompe, W., Scheibe, H.-J., Richter, A., Bauer, H.-D., Bewilogua, K., Weissmantel, C.: Thin Solid Films, 1986, 144, 77.
31) Catherine, Y., Couderc, P.: Thin Solid Films, 1986, 144, 265.
32) Couderc, P., Catherine, Y.: Thin Solid Films, 1987, 146, 93.
33) Angus, J.C.: Proc. E-MRS Conf. on Amorphous Hydrogenated Carbon Films, Strasbourg, 1987, XVII, 179.
34) Ingram, D.C., Woollam, J.A., Bu-Abbud, G.: Thin Solid Films, 1986, 137, 225.
35) Bakai, A.S., Strel'nitskii, V.E.: Sov. Phys. Tech. Phys., 1981, 26, 1425.
36) Matyushenko, N.N., Strel'nitskii, V.E., Gusev, V.A.: JETP Lett., 1979, 30, 199.

37) Vakula, S.I., Padalka, V.G., Strel'nitskii, V.E., Usoskin, A.I.: Sov. Tech. Phys. Lett., 1979, 5, 573.
38) Norström, H., Olaison, R., Andersson, L.P., Berg, S.: Le Vide les Couches Minces, Suppl., 1979, 196, 11.
39) Andersson, L.P., Berg, S.: Vacuum, 1987, 28, 449.
40) Sah, R.E., Dischler, B., Bubenzer, A., Koidl, P.: Appl. Phys. Lett., 1985, 46, 739.
41) Khan, A.A., Mathine, D., Woollam, J.A., Chung, Y.: Phys. Rev. B, 1983, 28, 7229.
42) Lin, S., Chen, S.: J. Mater. Res., 1987, 2, 645.
43) McKenzie, D.R., Briggs, L.M.: Solar Engergy Materials, 1981, 6, 97.
44) McKenzie, D.R., McPhedran, R.C., Savvides, N., Botten, L.C.: Phil. Mag. B, 1983, 48, 341.
45) Moravec, T.J., Lee, J.C.: J. Vac. Sci. Technol., 1982, 20, 338.
46) Pellicori, S.F., Peterson, C.M., Henson, T.P.: J. Vac. Sci. Technol. A, 1986, 4, 2350.
47) Berg, S., Andersson, L.P.: Thin Solid Films, 1979, 58, 117.
48) Has, Z., Mitura, S., Clapa, M., Szmidt, J.: Thin Solid Films, 1986, 136, 161.
49) Reisse, G., Schürer, C., Ebersbach, U., Bewilogua, K., Breuer K., Erler, H.-J., Weissmantel, C.: Wiss. Z. Techn. Hochschule Karl-Marx-Stadt, 1980, 22, 653.
50) Weissmantel, C., Ackermann, E., Bewilogua, K., Hecht, G., Kupfer, H., Rau, B.: J. Vac. Sci. Technol. A, 1986, 4, 2892.
51) Weissmantel, C., Bewilogua, K., Dietrich, D., Erler, H.-J., Hinneberg, H.-J., Klose, S., Nowick, W., Reisse, G.: Thin Solid Films, 1980, 72, 19.
52) Staryga, E., Lipinski, A., Mitura, S., Has, Z.: Thin Solid Films, 1986, 145, 17.
53) Mori, T., Namba, Y.: J. Vac. Sci. Technol. A, 1983, 1, 23.
54) Anderson, D.A.: Phil. Mag., 1977, 35, 17.
55) Meyerson, B., Smith, F.W.: J. Non-Cryst. Solids, 1980, 35 & 36, 435.
56) Onuma, Y., Kato, Y., Nakao, M., Matsushima, H.: Jpn. J. Appl. Phys., 1983, 22, 888.
57) Pirker, K., Schallauer, R., Fallmann, W., Olcaytug, F., Urban, G., Jachimowicz, A., Kohl, F., Prohaska, O.: Thin Solid Films, 1986, 138, 121.
58) Meyerson, B., Smith, F.W.: Solid State Commun. 1982, 41, 23.
59) Meyerson, B., Smith, F.W.: Solid State Commun., 1980, 34, 531.
60) Jones, D.I., Stewart, A.D.: Phil. Mag., 1982, 46, 423.
61) Benndorf, C., Grischke, M., Köberle, H., Memming, R., Thieme, F.: Suppl. à la Revue, Le Vide, les Couches Minces, 1987, 235, 485.
62) Köberle, H., Memming, R.: Proc. E-MRS Conf. on Amorphous Hydrogenated Carbon Films, Strasbourg, 1987, XVII, 485.
63) Memming, R.: Thin Solid Films, 1986, 143, 279.
64) Mackowski, J.M., Berton, M., Ganau, P., Touze, Y.: Le Vide, 1982, 212, 99.
65) Enke, K., Dimigen, H., Hübsch, H.: Appl. Phys. Lett., 1980, 36, 291.
66) Memming, R., Tolle, H.J., Wierenga, P.E.: Thin Solid Films, 1986, 143, 31.
67) Coad, J.P., Dugdale, R.A., Martindale, L.P.: AERE Harwell report no. R 10256 (1981).
68) Gille, G., Rau, B.: Thin Solid Films, 1984, 120, 109.
69) Aisenberg, S., Chabot, R.W.: J. Vac. Sci. Technol., 1973, 10, 104.
70) Wild, C., Koidl, P.: Proc. E-MRS Conf. on Amorphous Hydrogenated Carbon Films, Strasbourg, 1987, XVII, 207.
71) Knappwost, A., Arnason, V., Sohn, J.: Schmiertechnik + Tribologie, 1987, 2, 53.
72) Thiessen, P.A., Heinicke, G., Meyer, K.: Festkörperchemie, Verlag Grundstoffind. Leipzig, 1973, p. 497.
73) Thiessen, P.A., Meyer K.: Naturwissenschaften, 1970, 57, 423.
74) Nir, D.: Thin Solid Films, 1986, 144, 201.
75) Schürer, C., Solondz, D.: Schmierungstechnik, 1983, 14, 78.

76) Nir, D.: Thin Solid Films, 1987, 146, 27.
77) Nir, D.: J. Vac. Sci. Technol. A, 1986, 4, 2954.
78) Nir, D.: Thin Solid Films, 1984, 112, 41.
79) Weissmantel, C.: Proc. IX IVC-V ICSS, Madrid, 1983, 299.
80) Zelez, J.: J. Vac. Sci. Technol. A, 1983, 1, 305.
81) Zelez, J.: RCA Rev., 1982, 43, 665.
82) Wild, C., Koidl, P., Wagner, J.: Proc. E-MRS Conf. on Amorphous Hydrogenated Carbon Films, Strasbourg, 1987, XVII, 137.
83) Dischler, B.: Proc. E-MRS Conf. on Amorphous Hydrogenated Carbon Films, Strasbourg, 1987, XVII, 189.
84) Nir, D., Kalish, R., Lewin, G.: Thin Solid Films, 1984, 117, 125.
85) Nishikawa, S., Kakinuma, H., Fukuda, H., Watanabe, T., Nihei, K.: Jpn. J. Appl. Phys., 1986, 25, 511.
86) Fink, J., Nücker, N., Sah, R.E., Koidl, P., Baumann, H., Bethge, K.: Proc. E-MRS Conf. on Amorphous Hydrogenated Carbon Films, Strasbourg, 1987, XVII, 475.
87) Angus, J.C.: Thin Solid Films, 1986, 142, 145.
88) Angus, J.C., Stultz, J.E., Shiller, P.J., MacDonald, J.R., Mirtich, M.J., Domitz, S.: Thin Solid Films, 1984, 118, 311.
89) Kaplan, S., Jansen, F., Machonkin, M.: Appl. Phys. Lett., 1985, 47, 750.
90) Reimer, J.A., Vaughan, R.W., Knights, J.C., Lujan, R.A.: J. Vac. Sci. Technol., 1981, 19, 53.
91) Sakamoto, Y., Amemiya, H., Ishibe, Y., Okazaki, K., Oyama, H., Yano, K., Akaishi, K., Noda, N., Masuda, T., Tsurita, Y., Amemiya, S., Minagawa, H., Hino, T., Yamashina, T., Matsumoto, S.: J. Vac, Sci. Technol. A, 1987, 5, 2297.
92) Sofield, C.J., Woods, C.J., Cowern, N.E.B., Bridwell, L.B., Butcher, J.M., Freeman, J.M.: Nucl. Instrum. Meth., 1982, 203, 509.
93) Tait, N.R.S.: Nucl. Instrum. Meth., 1981, 184, 203.
94) Tait, N.R.S., Tolfree, D.W.L., John, P., Odeh, I.M., Thomas, M.J.K., Tricker, M.J., Wilson, J.I.B., England, J.B.A., Newton, D.: Nucl. Instrum. Meth., 1980, 176, 433.
95) König, H., Helwig, G.: Z. Phys., 1951, 129, 491.
96) Schmellenmeier, H.: Exp. Tech. Phys.: 1953, 1, 49.
97) Schmellenmeier, H.: Z. Phys. Chem.: 1955-1956, 205, 349.
98) Badzian, A.R., Bachmann, P.K., Hartnett, T., Badzian, T., Messier, R.: E-MRS Conf. on Amorphous Hydrogenated Carbon Films, Strasbourg, 1987, XVII, 63.

Materials Science Forum Vols. 52 & 53 (1989) pp. 577-608

DIAMONDLIKE CARBON APPLICATIONS IN INFRARED OPTICS AND MICROELECTRONICS

J.A. Woollam, B.N. De, S. Orzeszko*, N.J. Ianno and P.G. Snyder (a)
S.A. Alterovitz and J.J. Pouch (b) and
R.L.C. Wu and D.C. Ingram** (c)

(a) Center for Microelectronic and Optical Materials Research, and
Dept. of Electrical Engineering, University of Nebraska
Lincoln, NE 68588-0511, USA
(b) NASA Lewis Research Center, Cleveland, OH 44135, USA
(c) Universal Energy Systems, Dayton, OH 45432, USA

I. Introduction

Numerous authors worldwide have prepared diamondlike carbon (DLC) for various potential applications [1]. The present chapter reviews work done, by ourselves and others on an important aspect of the use of DLC as a protective coating in harsh environments. This article contains new material not previously published, as well as material re-presented, but with a new perspective. The motivation for the original studies was the potential application of DLC on infrared transmitting optics [2], and as protective coatings in microelectronics [3].

There are three sub-topics in this chapter. The first is a description of the preparation of DLC on seven different infrared transmitting materials, and the possibility of using DLC as an anti-reflecting coating at commonly used wavelengths [4]. DLC doesn't bond easily to all materials, and special techniques for bonding have been found both by ourselves and others.

The second topic deals with how well DLC will protect a substrate from moisture penetration. This is an important aspect in numerous uses of DLC, including both infrared optics and integrated circuits [5].

The third sub-topic also involves an environmental aspect, namely the effect of particulate impact on film performance and integrity. For example, an infrared coating may be exposed to a space environment or to conditions of blowing sand or water [2]. It is important to know how well DLC maintains its integrity.

* Now at Nicholas Copernicus University, Torun, Poland.

**Now at Whickham Ion Beam Systems, Ltd., Newcastle-Upon-Tyne, Great Britain.

II. DLC on Infrared Transmitting Materials

Several different infrared substrates were used, including lexan, silicon, fused silica, KG-3 glass, BK-7 glass, ZnS, ZnSe, heavy metal fluoride glass, GaAs, and Ge.

Extensive depositions were carried out using four different systems, including direct ion beam, and three parallel plate capacitively coupled plasma chamber designs.

A. Ion-Beam Deposition

The schematic diagram of the system used to ion beam deposit DLC films is shown in Figure II-1. It consisted of four sections: the ion source, the gas inlet system, the vacuum system, and the target fixture. The ion source was a 2.5 cm Kaufman source made by Ion-Tech Inc. Figure II-2 shows a schematic circuit diagram of the source. The ion beam was produced by a plasma discharge and a typical ion current was 10mA. The ion kinetic energy used in the deposition could be varied from 100 to 1500 eV. The beam profiles of the ion source were extensively characterized under various conditions such as ion energy, and external magnetic and

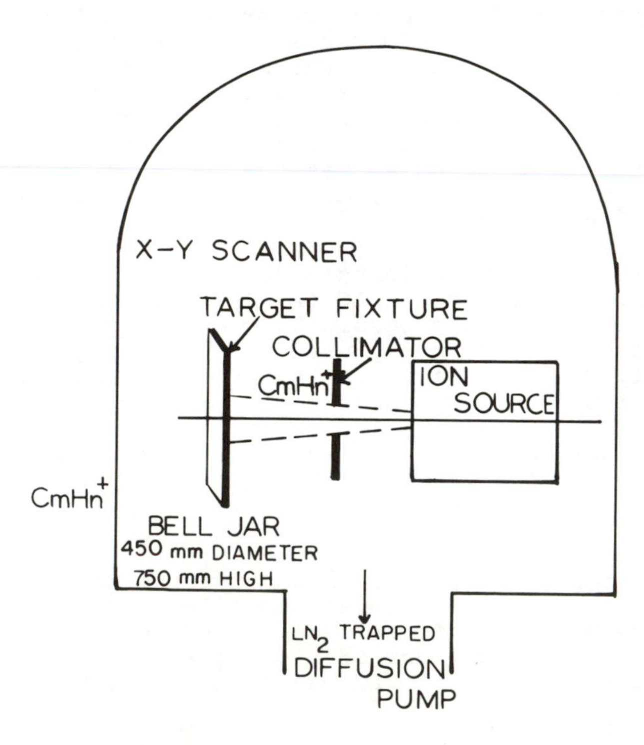

FIGURE II-1. Schematic drawing showing the relationship of the ion source to the target fixture inside the bell jar.

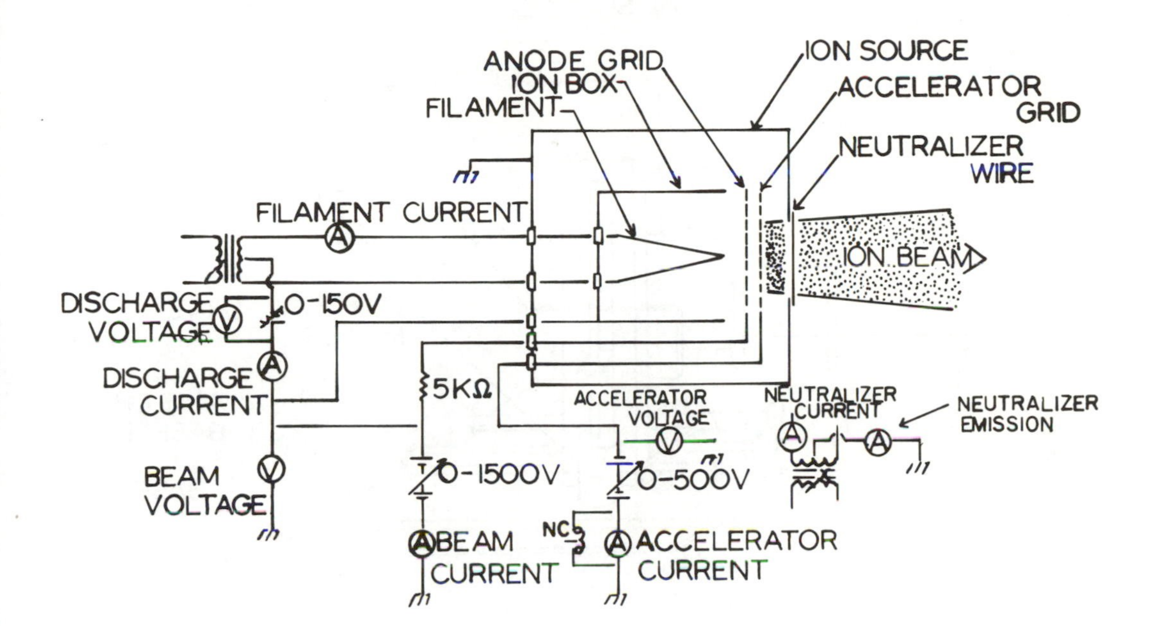

Figure II-2 Circuit diagram of the ion source.

electrostatic fields. In general the beam profile was narrow and peaked.

The sample inlet system permitted introduction of two pre-mixed gases into the ion source. The flow rate of each gas was controlled by an MKS flow controller, and the gases used in this work were methane (99.99%) and hydrogen (99.99%). The background pressure was typically 10^{-6} torr, and the operating pressure was on the order of 10^{-5} ~ 10^{-4} torr.

The target fixture was mounted about 8 cm from the ion source. Using the present ion source, the directly deposited films were found to be nonuniform. In order to obtain a uniform and large area film, an X-Y scanner was constructed, as shown in Figure II-3. The target plate was able to move approximately ± 17.5 cm in two orthogonal directions in a plane perpendicular to the beam. This was accomplished using stepping motors controlled by SLO-SYN indexers (430-PI, Superior Electric Co.). The

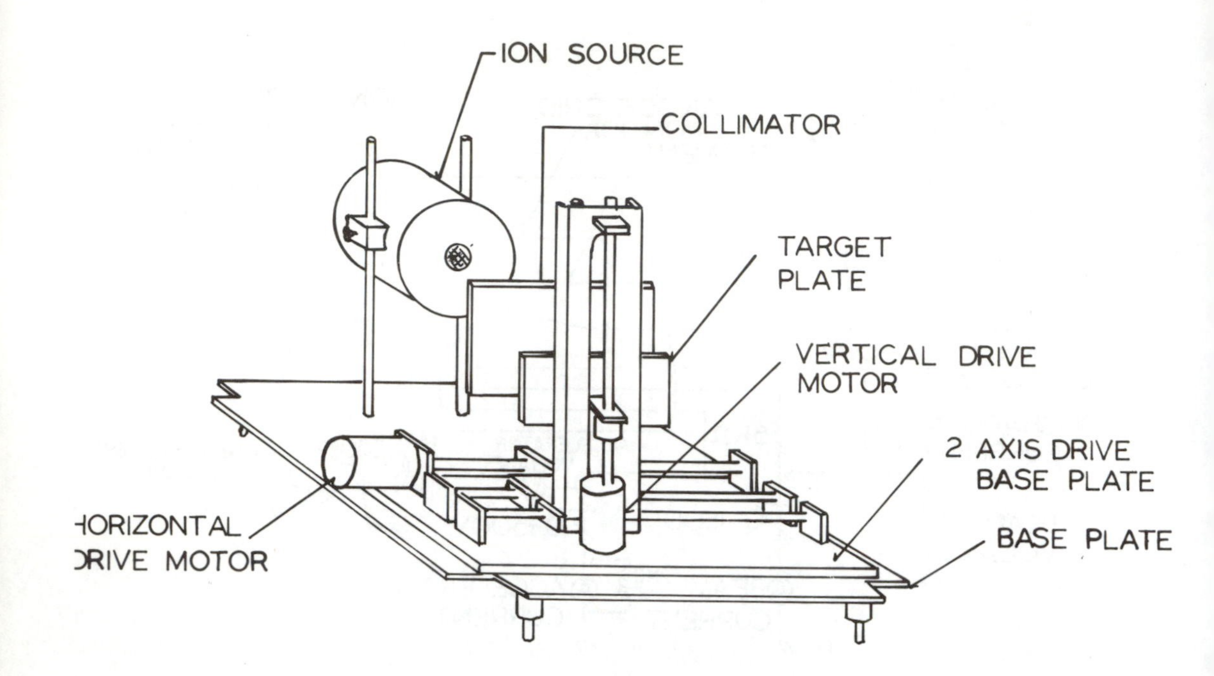

Figure II-3 Schematic layout of the target scanner.

indexers were interfaced to an IBM XT-compatible computer via an RS-232 port, which allowed the user to change parameters, i.e., feed-rate and travel distance, in the indexers. Extensive tests were performed to determine the optimum scan rates. The carbon content of the deposited film was then analyzed for uniformity using Rutherford Backscattering across a 5.5 x 5.5 cm^2 area. The optimum conditions were determined to be 0.04 cm/sec for the X-direction and 1.6 cm/sec for the Y-direction. This set of scan rates was used on all subsequent depositions.

B. <u>RF-Plasma Configuration I</u>

An rf sputtering system was purchased from Cooke Vacuum Corporation, consisting of a cryopumped stainless steel bell jar chamber. Inside the chamber were two parallel plate electrodes which were driven by a 13.56 MHz, 0 to 500 watt rf generator, and a load matching network.

In Configuration I, one electrode was grounded, and the other driven by the rf generator. The driven electrode was much smaller in area than the grounded electrode, as sketched in Figure II-4. This geometry created an intense plasma above the driven electrode which was accompanied by a large self-induced DC bias. Also, a rather diffuse plasma existed at the ground plane with virtually no DC bias. The substrate was placed on the driven electrode to take advantage of the intense plasma and the high deposition rate. During depositions at low power, the self-induced dc bias was moderate and did not appear to alter the film characteristics. However, significant high energy ion bombardment of the depositing film occurred. We feel the main effect of the ion bombardment was to heat the depositing film and drive out the hydrogen, leaving a low band gap material. This conclusion was based on the results obtained when a deposit was made at 500 watts rf power, 140 microns torr total pressure for 30 minutes. Another deposit was made under the same conditions except the discharge was run for 2 minutes and turned off for 3 minutes until a total time of 30 minutes deposition was achieved. The former deposit exhibited a band gap of approximately 0.2eV. the band gap of the latter film was greater than 1 eV, comparable to that observed in films deposited at low powers. In view of this, the Cooke system was modified such that the areas of the driven electrode and ground plane were approximately equal (Configuration II). This eliminated the self-induced bias and the accompanying heating effect. This modified system allowed a more accurate exploration of the effect of various plasma parameters on the film properties.

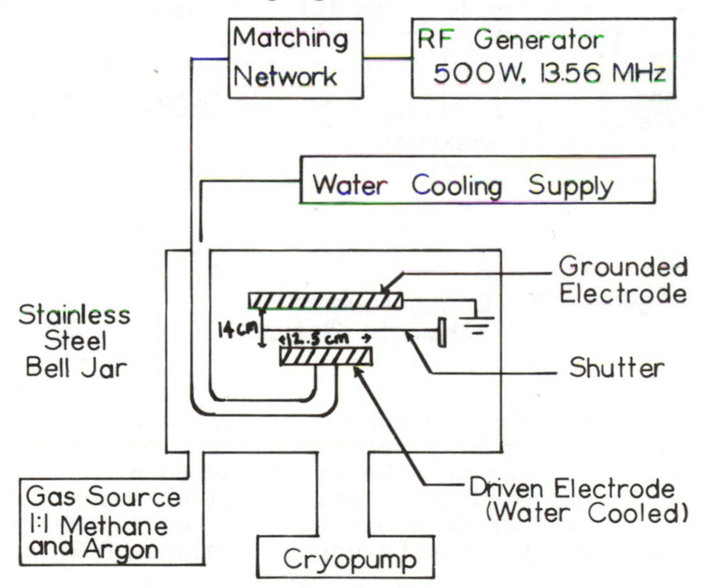

Figure II-4 Configuration I schematic design

C. RF-Plasma Configurations II and III

The reason behind the design for this configuration as discussed above was to eliminate high energy ion impact and loss of hydrogen during deposition. This configuration had a 30 cm diameter stainless cathode on the bottom (Figure II-5), and a 28 cm diameter stainless steel upper electrode with an appropriate ground shield.

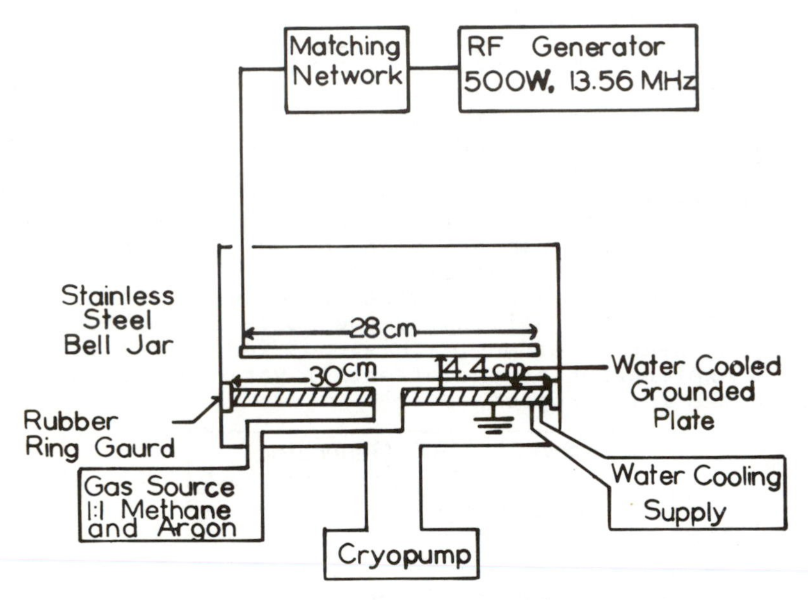

Figure II-5 Configuration II schematic design

The center of the lower grounded electrode was connected to the gas inlet pipe by means of a small plastic tube. A small metal screen was placed over the center of the hole in the lower plate to prevent the gas discharge from igniting in the gas outlet opening. This electrode was also water cooled by a 28 cm diameter stainless steel plate with copper cooling coils soldered to it.

The Configuration II design reduced the self-induced DC bias, by forcing the area of the driven electrode to be approximately equal to the area of the ground plate. This reduced the high energy ion bombardment of the depositing film and the subsequent heating effects. This design provided a uniform gas flow over the lower electrode for a wide range of input gas flow rates and pumping speeds, creating a large area of uniform deposition for samples placed on this electrode. A rubber ring guard was placed around the outside of the lower electrode to prevent the discharge from igniting around the edges. The frequency was 13.56 MHz.

In Configuration III (used at NASA Lewis) the chamber was similar to II but the plates were of equal area (20 cm diameter), and 2 cm separation. The plasma was pure methane at a pressure of 20 microns. The frequency was 30 kHz.

D. Optimization of Experimental Parameters

The physical properties of the DLC films on various substrates were found to be dependent upon the deposition parameters. Optimum deposition parameters, for each technique, were established and are described below.

1. Ion-Beam Deposition

An attempt was made to optimize the hydrogen content of DLC films by varying the hydrogen-to-methane ratio inside the ion source. A series of experiments were performed using silicon substrates. The hydrogen concentration was varied from 0% to 98%. The deposition conditions were as follows: beam voltage - 1000 V; accelerator voltage - 100 V. The discharge voltage increased from 58 V to 98 V with increasing hydrogen content in order to maintain the source discharge. The filament current was typically 6 A; neutralizer current typically 6 A with 3 mA emission, and the gas flow increased from 1.5 to 10 SCCM with increasing hydrogen content.

The deposited film composition was determined by Rutherford backscattering (RBS) analysis for carbon content and proton recoil detection (PRD) for the hydrogen content. The results are shown in Table II-1. Hydrogen content in the deposited DLC film has been found to be dependent on the hydrogen-to-methane ratio inside the ion source. In particular, increased hydrogen concentration in the ion source has been found to result in an increased hydrogen content in the deposited film. Using pure methane in the process of the present study, the final hydrogen concentration in the DLC film is approximately 30%. With 20% hydrogen in the ion source, the hydrogen concentration in the DLC film is approximately 35% while with hydrogen concentration in the ion source of between 50% and 80%, the final hydrogen concentration in the DLC film is approximately 40% to 41%. Accordingly, variation of the hydrogen concentration in the ion source can be utilized to vary the hydrogen concentration of the deposited DLC film. In general, lower hydrogen concentrations in the DLC film render the coating denser and harder than films having higher hydrogen concentrations. Accordingly, in most applications, the use of pure methane is preferred.

The effect of ion-impact energy on the film quality (uniformity, pinhole, adhesion, and optical properties), on the deposition rate, and on the damage to the substrate was investigated. Two impact ion-energies of 500 and 1000 eV were used to deposit diamondlike carbon on silicon substrates. A pure CH_4 gas was used in these experiments. The gas flow rate was 3 SCCM and the chamber pressure was 9.0 x 10^{-5} torr. Both films appeared to be uniform and golden in color. The film growth rate, hydrogen content and carbon content were analyzed. The results are shown in Table II-2. Within the uncertainties of the RBS, PRD and Dektak film thickness analyses, the deposition rate and the hydrogen concentration in the film were found to be the same. The cross-sectional transmission electron micrograph of the DLC film at 1000 eV showed no damage on the surface of the silicon substrate. At high energy ion impact, the ion current seemed more intense and the films stuck well to the substrate.

The increase of methane molecules resulted in ion-molecule reactions inside the ion source; higher molecular weights of hydrocarbon ions were produced. The present experiments were carried out at three different pressures: 2.6 x 10^{-4} torr, 9 x 10^{-5} torr and 6 x 10^{-5} torr brought about by controlling the CH_4 flow rate 7.32, 3.00 and 1.32 SCCM. An ion-impact energy of 1000 eV and silicon substrates were used. Uniform golden color films were observed for deposition pressures of 2.6 x 10^{-4} torr and 9 x 10^{-5} torr. However, darker films were obtained using a pressure of 6 x 10^{-5} torr. The carbon and hydrogen content of the films were analyzed, and results are shown in Table II-3.

It can be seen that at the lowest pressure, the hydrogen content in the film was slightly increased. Therefore, if harder films have less hydrogen, it is better to use higher methane pressures.

The effect of substrate material (silicon, fused silica, lexan, KG-3, BK-7 glass, ZnS, ZnSe and HMF) on DLC film growth rate and film quality was investigated using the same experimental conditions: pure CH_4 at flow rates of 3.0 SCCM, 1 KeV ion energy, source pressure of 9 x 10^{-5} torr, deposition times from 65 to 450 minutes. All films on the above listed substrates were found to be uniform. The film thicknesses were measured using a Dektak instrument. Table II-4 gives the average growth rate of DLC on various substrates over an area of 4.9 cm^2. These were direct ion beam depositions, with the present 2.5 cm diameter ion source.

The effect of cleaning the substrate surface prior to ion deposition of the DLC films was studied extensively. Bonding of the DLC film on various substrates was found to be strongly dependent upon the surface cleaning procedures. The initial cleaning procedures included: (1) washing with 1,1,1 Trichloroethane, (2) washing with acetone, (3) washing with methanol and finally (4) blow drying by dry nitrogen. All samples were cleaned by these four procedures, except the lexan substrate

Table II-1. Effect of Hydrogen in the Source Gas on the Carbon, Hydrogen Contents of Direct Ion Beam Deposited Diamondlike Carbon Films

Sample No.	% Hydrogen in the Source	% Hydrogen in the Film (±5%)	% Carbon in the Film (±5%)
DLC - C	0	30	70
DLC 140	2	40	60
DLC 141	5	36	64
DLC 142	10	38	62
DLC 143	20	35	65
DLC 144	50	40	60
DLC 145	80	41	59
DLC 146	90	NO FILM	NO FILM
DLC 147	90	NO FILM	NO FILM
DLC 149	95	39	61
DLC 148	98	NO FILM	NO FILM

TABLE II-2. Effect of Ion Impact Energy on the Direct Ion Beam Deposited Diamondlike Carbon Films.

Sample No.	Ion Impact Energy (eV)	% Hydrogen in the Film (±5%)	% Carbon in the Film (±5%)	Film Growth Rate A/min±5 A/min
871-365	1000	33.0	67.0	23
871-366	500	37.5	63.5	25

TABLE II-3. Effect of Methane Pressure on the Direct Ion Beam Deposited Diamondlike Carbon Films.

Sample No.	CH4 Source Pressure (torr)	Flow Rate (SCCM)	% Hydrogen in the Film (±5%)	% Carbon in the Film (±5%)
871-367	$6x10^{-5}$	1.32	38.4	61.6
871-365	$9x10^{-5}$	3.0	33	67
871-368	$2.6x10^{-4}$	7.32	33.4	66.7

TABLE II-4. Diamondlike Carbon Film Growth Rate on Various Substrates.

Substrate	Direct ion beam Deposition Rate (Å/sec)
Lexan	11.5
BK-7	8.3
KG-3	8.3
Silicon	6.3
Fused Silica, Glass	5.7
ZnS, ZnSe	6.8
HMF	6.8

which used only procedures (3) and (4). It was found that DLC films adhered to BK-7, KG-3, ZnS, silicon and lexan substrates quite well, and passed the initial "Scotch" tape tests. However, the DLC film on fused silica and heavy metal fluoride glass failed the "Scotch" tape tests. Thus an attempt was made to investigate the surface cleaning procedures for HMF glass and fused silica. The substrates of HMF and fused silica were cleaned by procedures (1), (2), (3), and (4), and cleaned again by 1000 eV Ar^+ ion beam for 20 minutes prior to DLC deposition. The DLC films on both substrates again failed the "Scotch" tape test. Another cleaning procedure was tried to clean the surface by washing with methanol and drying using a heat gun or dry nitrogen. The DLC films on both substrates were found to stick well to these substrates. The new cleaning procedure was thus adapted for cleaning of HMF and fused silica substrates.

In the present ion-beam deposition technique, the temperature of the substrate was constantly monitored by a temperature tape and was found to be less than 60°C. Since optical materials can be temperature sensitive, no attempt was made to heat the substrate.

2. RF Plasma Discharge: Configuration I

Plasma DLC deposition was extremely successful on glass slides (Thickness $\leq$ 1 micron) and Si-wafers. Occasionally there were problems with pinholes when high power was used. At high power (especially 250 and 500 watts), the film quality became significantly degraded, in terms of the uniformity of the film thickness and the ability to adhere to glass. The thickest films spalled off the edges of the glass slides if the power was too high. As measured from UV-VIS absorption measurements on samples deposited on glass slides, the optical energy gap was about 0.2 eV for the 500-watt sample. Substrate heating was suspected to occur at high rf powers. To test for this postulate, sample K3 was prepared on glass, at 500-watt power, 140 microns pressure, and deposited for 10 minutes (1 minute times 10 with 5 minutes cooling interrupt periods in between); the resulting sample had an optical gap of 1.1 eV, the same value that occurred when low powers were used. This supported the hypothesis that heating caused the drop in optical gap. All subsequent depositions were made with the plasma on for only a few minutes, then the plasma off for 3 to 5 minutes to permit cooling.

All glass samples were cleaned by washing with the following sequence: 1) 1,1,1,-Trichloroethane, 2) acetone, 3) methanol, 4) deionized water, and 5) dry nitrogen blow. All Lexan samples were cleaned by washing with methanol, followed by deionized water, and finally by a dry nitrogen blow.

We noticed that pinholes in films were caused by segregated granular carbon deposits. The pinholes were exposed after the film was blown with dry nitrogen. These granular carbon deposits were probably formed before reaching the substrate, and were likely caused by the excessive amount of carbon atoms in the gas phase (in the plasma). Thus, it was logical to reduce the carbon atom density within the plasma. That could be achieved in two ways: 1) By reducing the pressure, and 2) by reducing the power.

We chose the second alternative. Thus, all samples were made at 25 watts power and 140 microns pressure. The results were highly successful.

3. Plasma Discharge: Configuration II

A 1:1 mixture of methane and argon and a 13.6 MHz RF power source, capable of delivering up to 500 watts, were used for the generation of the plasma. The plate areas were made almost equal in order to deliver the power with a minimum DC bias voltage between the plates. The maximum DC vias voltage observed was 550 volts. The lower plate was grounded and the upper plate was driven by an RF power supply. A cryopump was used on the chamber.

When the system shown in Figure II-5 was first designed and operated, the rubber guard ring was not present, and the plasma was very unsteady and sometimes passed beyond the lower ground plate to the bottom of the chamber. Also, sometimes it became difficult to start the plasma; even after using the tesla coil and adjusting the matching network of the RF power supply we could not start the plasma, (at the 80 microns base pressure and the flow rate of 12.5 sccm for both methane and argon). Figure II-6 shows more details of the Figure II-5 design which permitted easier ignition of the plasma due to a better gas flow geometry. We made 16 small holes in the plate and closed the gap between the lower ground plate and the chamber using a vacuum compatible rubber strip as shown in Figure II-6. The gas thus entered the plasma region through the center and flowed radially outward. We planned to use the external DC INPUT to start the plasma (instead of the tesla coil), but found that after modifying the system, the gas plasma was generated rather easily by increasing the gas pressure to 100 microns. Sometimes, use of a tesla coil in combination with some adjustment of the matching network of the RF power supply was helpful in starting the gas discharge. The plasma was found to be confined within the volume above the lower ground plate.

Except for Si, Lexan, HMF glass, and ZnS, all the substrates were first ultrasonically cleaned using 1,1,1-Trichloroethane; then washed with acetone, methanol, and deionized water successively and finally dried by blowing dry nitrogen. Lexan was ultrasonically cleaned using methanol, washed with deionized water and finally dried using dry nitrogen. HMF surface was found to be deteriorated by the use of any of the organic solvents mentioned previously. Thus, we visually looked for the cleanest surface of HMF, and dried it in flowing dry nitrogen. The Si surfaces were clean as received; therefore we used only

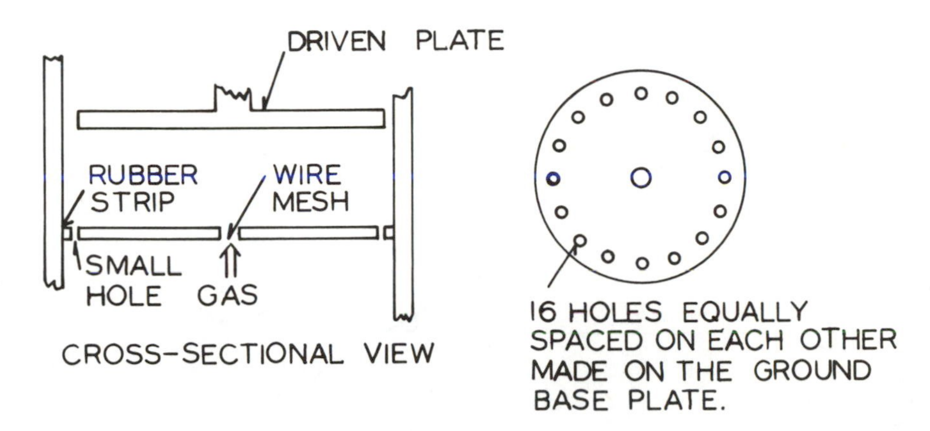

Figure II-6 Cross-sectional view of the plasma discharge

Configuration II.

dry nitrogen to clean them. For ZnS, we followed the same procedure, in order to maximize adhesion.

The best deposition conditions were found to be: 1) flow rate for methane and argon: 13 sccm each, 2) base pressure: 100 microns, 3) power: 200 watts, and 4) DC bias between the two plates: 300 volts.

From the depositions using Configuration I, we found that DLC films on ZnS substrates tended to spall very easily if the film was thicker than a few hundred angstrom units. Because of that, we deposited for 2 minutes (only) on the ZnS substrates, resulting in an estimated film thickness of about 300 Å. We should note that DLC will adhere to ZnS and ZnSe if a thin (300 Å) Ge film is deposited between the semiconductor and the DLC layer [6].

4. 30 kHz Plasma Deposition System: Configuration III

Figures II-7 and II-8 show the strong dependence of deposition rate on substrate temperature, and on power [7]. The configuration for these depositions was similar to Plasma Deposition Configuration II, described above. The plasma was of pure methane gas at a pressure of 20 microns. 100°C substrate temperature and a power level of 200 watts was typical for producing good films.

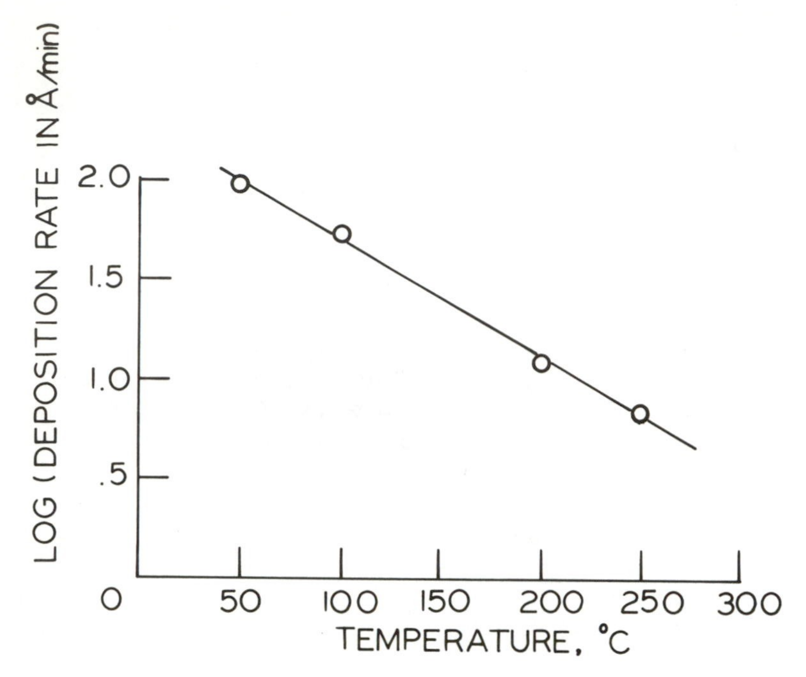

Fig II-7. Logrithmic plot of deposition rate vs temperature

E. Antireflection Conditions Using DLC

As mentioned above, substrates of infrared interest on which we deposited DLC included: lexan, silicon, fused silica, KG-3 glass, BK-7 glass, ZnS, GaAs, Ge, and heavy metal fluoride glass. It was desired to know if DLC could be deposited to the proper thicknesses and with the correct indices of refraction for use as an antireflecting coating on each of these substrates.

Antireflecting coatings provide an important method of enhancing transmission through optical window materials [8].

The reflectivity has a minimum when

$$n_1 d_1 = \lambda_o / 4 \qquad \text{II-1}$$

where n_1 is the index of refraction of the coating, and d_1 its thickness. The reflectivity minimum is at its lowest value (zero) when

$$n_1^2 = n_o n_2 \qquad \text{II-2}$$

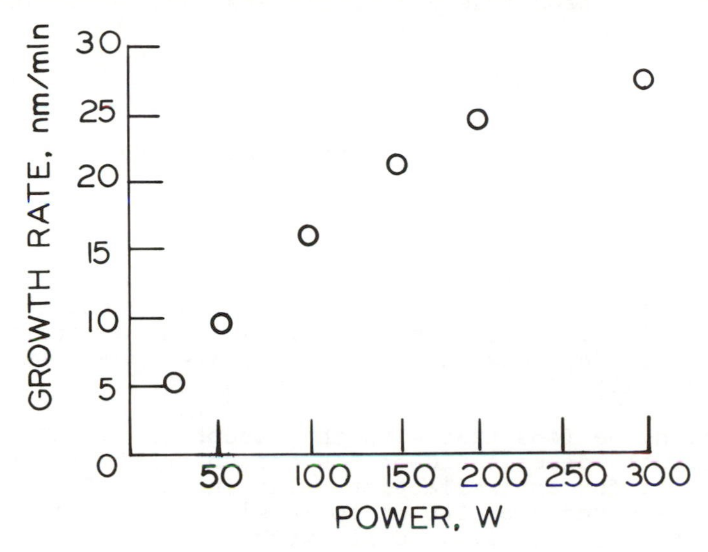

Fig II-8 Growth rate of hydrogenated carbon on n-InP as a function of deposition power.

where n_o is the index for the ambient, which is normally air, so

$$n_1 = \sqrt{n_2}$$

is required, where n_2 is the index of the substrate at the wavelength of interest.

Table II-I lists infrared transmitting substrates, their indexes of refraction, the operating wavelengths of interest, and the required DLC thickness and optical index of refraction. The proper index came from use of Eq. II-3, and the proper thickness from Eq.II-1.

What is noticed immediately from Table II-I is that the required film index of refraction ranges from 1.2 to 2.85 for the examples listed.

The index of refraction of DLC can be controlled by choosing the proper deposition technique and parameters [1]. The range typically found is for

$$1.6 \leq n_1 \leq 2.4 \qquad \text{II-4}$$

which makes the optimum matching substrates have indices from

$$2.5 \leq n_2 \leq 5.8 \qquad \text{II-5}$$

These values are much higher than the indexes for the glasses under consideration (Table II-I), but result in a decent match for ZnS, diamond, TiO_2, As_2S_3-glass, Se-glass, and results in good matches for Si, Ge, GaAs, and InSb. All are common infrared transmitting materials [9].

In conclusion, we find that with the exception of ZnS we were able to directly deposit DLC on the chosen substrates to the desired thicknesses for antireflection. The indexes of refraction were measured from 300 nm to 10 microns, and found to be in the range from 1.6 to 2.0. Other workers have prepared DLC samples with indices up to 2.4. Thus, we have established a range of conditions for use of DLC as an antireflecting coating. Zero reflectance can be achieved on substrates of Si, Ge, GaAs, and InSb. Low reflectance can be achieved on ZnS, diamond, TiO_2, As_2S_3-glass, Se-glass; but DLC will not provide total antireflecting conditions when deposited on the common glasses with index near 1.5.

III. Moisture Protection with DLC

A. Introduction

DLC is amorphous and thus has no grain boundaries through which water might otherwise diffuse. One of the most common uses of thin films is for coatings for moisture protection. Thus DLC seemed to be an ideal candidate material for use as a hermetic seal [5]. Applications, for example, might be to passivate integrated circuits, or to keep water from sensitive infrared transmitting optical windows or lenses.

It is very difficult to measure penetration of liquids into thin films. Common surface analysis techniques such as AUGER, ESCA, and SIMS require ultra high vacuum, and therefore cannot be used.

B. Diagnostics Technique

We have shown that variable angle spectroscopic ellipsometry (VASE) can be used to determine the thickness of ultrasmall amounts of water on, and in a thin film [5]. This spectroscopy is not commonly known, so a brief description will be given [10]. The interested reader can find more details in references listed.

Ellipsometry determines the ratio of complex reflection coefficient

$$\tilde{\rho} \equiv \tilde{R}_p/\tilde{R}_s = \tan\psi \exp j\Delta \qquad \text{III-1}$$

where $\tilde{R}p$ and $\tilde{R}s$ are the complex Fresnel reflection coefficients for components of light parallel (P) and perpendicular (S) to the plane of incidence of the incident and reflected light.[10] In our VASE technique, data are taken from 300 to 800 nm with light incident at an angle ϕ to the normal to the sample. The reflected light polarization state is analyzed with a rotating polarizer. Light intensity is measured with a photomultiplier tube, and the signal is digitized and Fourier analyzed to determine the ψ and Δ parameters of equation III-1.

The measured complex ratio $\tilde{\rho}$ is related to the optical index of refraction, n, and extinction coefficient, k, of the material under study. If complex materials structures are involved, then n and k can be determined for individual layers, and layer thicknesses determined.

Microstructural analysis is performed assuming the nature of the sample under study. For the present samples the model is shown in Fig. III-1. The t_i are layer thicknesses, and f_2 is the fraction of DLC in a DLC plus H_2O Bruggeman effective medium approximation (EMA) mixture layer. The procedure is to calculate using the Fresnel reflection coefficients for a multilayer parallel stack (and EMA mixed layers), for a given initial set of values for thicknesses and fractions. Next, a regression analysis is performed to minimize the mean square error function (MSE) defined by

$$MSE = \frac{1}{N} \sum_{i=1}^{N} (\psi_i^{exp} - \psi_i^{calc})^2 + (\Delta_i^{exp} - \Delta_i^{calc})^2 \qquad \text{III-2}$$

where "exp" means experimentally measured, and "calc" means calculated. The psi and delta are functions of wavelength λ and angle of incidence ϕ. A large range of both λ and ϕ are chosen so that an "overdetermination" of measurements with respect to the number of unknown parameters is made, and correlation problems avoided. In our analysis programs we can use Eq. (III-2) as formulated, or we can use psi alone or delta alone, or the minimization can be done with respect to tan ψ and cos Δ. The final outcome is a set of values for thicknesses, EMA fractions, and optical constants for any of the layers. The optical constants can take on several forms: 1) index of refraction n and extinction coefficient k, 2) real, E1 and imaginary, E2, parts of the optical dielectric function, or 3) the amplitude, position, and width of Lorentz oscillators.

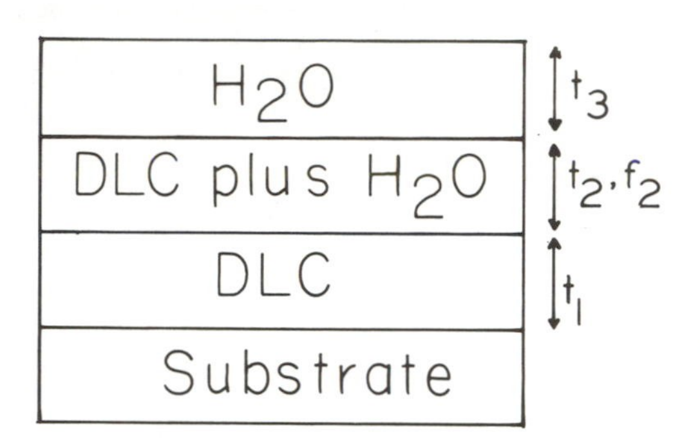

FIG. III-1 Structural model assumed for analysis of moisture. t_i are thicknesses and f_2 is the fraction of H_2O in layer 2.

Since we used the Lorentz oscillator model extensively it will be described further. When the wavelength-dependent n, k (or E1, E2) values are solved for, the number of unknown parameters equals the number of wavelengths times two, plus the number of layers, plus the number of unknown fractions in an EMA. Thus if 10 wavelengths are used and there are two unknown thicknesses there will be (2 X 10) + 2 = 22 unknowns. In the Lorentz oscillator model there are fewer unknowns, since the following equation represents the spectral dependence of optical constants:

$$E = 1 + \sum_{i=1}^{M} Ai \left(\frac{1}{\lambda + P_i + jW_i} - \frac{1}{\lambda - P_i + jW_i} \right), \qquad \text{III-3}$$

where E is the complex dielectric function, and the sum is over the total number of oscillators M. For the presently reported work, the maximum number of oscillators used was one. (In section IV the results of a two oscillators analysis are presented). In Eq. (3), A_i, P_i, and W_i are the amplitude, position, and width of the ith oscillator, respectively, and λ is the photon wavelength. In the oscillator analysis A_i, P_i, and W_i thicknesses and EMA fractions are solved for in the regression analysis. Thus a typical one oscillator DLC analysis has five unknowns: three oscillator parameters and two layer thicknesses; or seven unknowns: three oscillator parameters, three layer thicknesses, and an EMA fraction (see the structural model shown in Fig. III.1).

C. Samples

The films of DLC used for moisture penetration studies were prepared using the 30kHz parallel-plate plasma deposition system (Configuration III) described above [7]. Pure methane and a chamber pressure of 20 microns was used. Power levels of 100, 200, and 300 watts were used, but results for 200 watts are reported here. Substrate temperatures ranged from 23°C to 250°C.

Moisture was introduced to the films in two ways; from immersion in 23°C water, and from a steam jet at 100°C.

D. Results

1. As-deposited films

Fig. III-2 shows the dependence of average (over the 300 to 800 nm range) index of refraction, n on substrate temperature for films made with a plasma power level of 300 watts. In general the index rises from about 1.7 at low temperature to above 1.9 at high temperature. A measurement of hydrogen content in the film by proton recoil analysis showed that lower hydrogen was present in films with a higher index of refraction [11]. Measurements of the optical absorption coefficient showed that the bandgap increased with increasing hydrogen concentration. Similar trends have been seen by other workers as well [1].

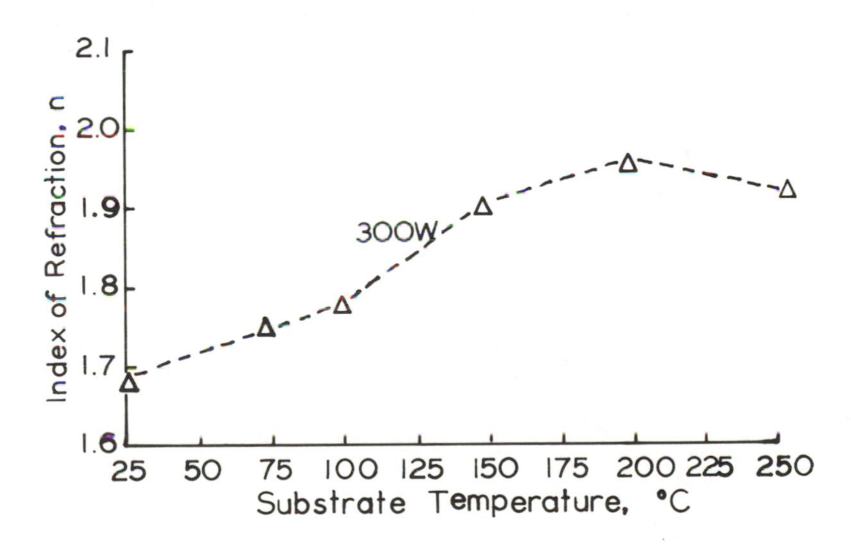

FIG. III-2 Optical index of refraction vs substrate temperature for a power level of 300 Watts.

The environmental stability of DLC films was studied on a large number of samples deposited at various power and temperature values onto polished silicon single-crystal wafers and subjected to immersions in trichlorethane, acetone, ethyl alcohol, sulfuric acid, nitric acid, hydrochloric acid, and hydrofluoric acid. After each immersion samples were subjected to rubber eraser abrasion tests and "scotch tape" pull tests.

These tests served as a comparative measurement. They had no effect on samples deposited with 100 W. For 200 W and 300 W depositions there was partial removal for samples deposited at room temperature, but there was no effect for substrate temperatures of 74 up to 250°C. Thus DLC samples prepared under all but a few conditions survived very stressing environmental tests.

2. Moisture Introduced

A large number of samples were prepared and investigated for water penetration. The final result was that DLC films were not pentrated by water. The DLC surface had a small amount of roughness, and moisture was found to penetrate the valleys of this roughness, but no further.

Example ellipsometric data are shown in Figure III-3. The data at 0 hours after H_2O indicate that water was introduced, then the bulk of it allowed to run off a vertical surface. At this time the maximum amount of water remained, and the Δ parameter was lowest. Twenty four hours later some water had evaporated, and Δ increased. After exposure to a heat lamp much of the water on the surface was evaporated (but not all!). After exposure to the laboratory 23°C atmosphere a small amount of moisture from the ambient air deposited, with an associated decrease in Δ.

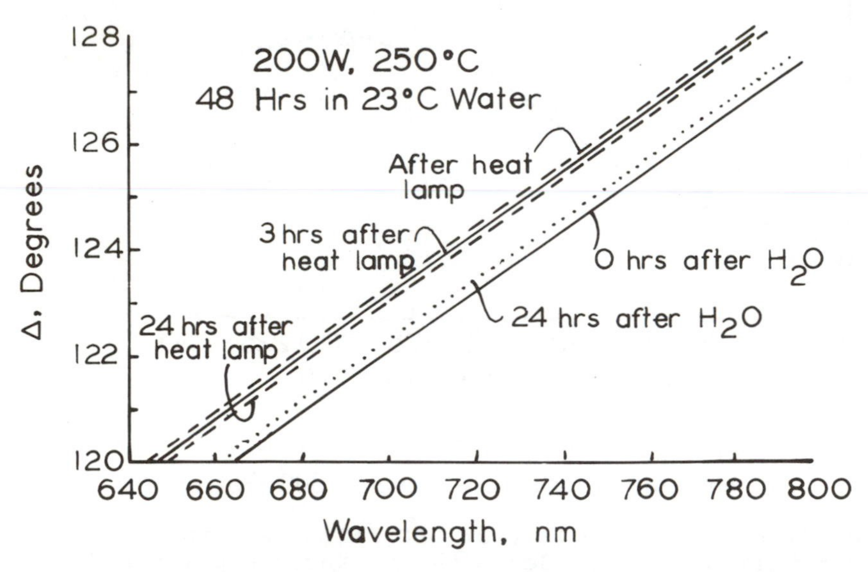

FIG. III-3 Ellipsometric delta parameter vs wavelength at various times after removing the sample from 23°C water. Sample was deposited at 200W, and 250°C, and soaked in room temperature water for 48h. After removing from water, it was kept in a room atmosphere.

Using the "oscillator" regression analysis for this sample, and a two-layer (t_2 = 0 in Fig. III-1) model in our ellipsometric analysis, we determined that the water layer was 66-Å thick on top of a 344-Å thick DLC film. The three-layer (nonzero t1, t2, t3, in Fig. III-2) analysis was consistent with this result: 330 Å of DLC, 29 Å of 50%-50% mixture of DLC and water, and 48 Å of pure water on top.

An interesting result was that the heat lamp removed only 12 Å of water and that 54 Å of water still remained! We propose that this thin water layer was being held to the surface by an unusually high surface tension associated with DLC surface micropores.

A detailed analysis of VASE data for all samples, assuming the model of III.2, was carried out. Tables III.1-4 summarize some of the results.

A summary of the results of DLC moisture studies are that: a) Moisture resided mainly on the surface of DLC independent of water temperature. b) In cases where there was an apparent penetration (≤50 Å) it is likely that water was merely filling the void regions of a "rough" surface. The substrates were highly polished oriented single-crystal (semiconductor grade) wafers. 50Å of roughness layer was reasonable for these materials. c) A heat lamp removed water, but not all of it. This suggests that surface micropores provide a strong surface tension for these very thin water layers. d) The amount of water on the surface after wetting (but without further treatment) did not depend on the parameters of DLC deposition such as power or substrate temperature, nor did it depend on the DLC film thickness in any systematic manner. Thus, DLC protects effectively against moisture penetration [5].

IV. Ion-Beam Modified, Ion-Beam Deposited DLC

A. Introduction

In this section we present results of variable angle spectroscopic ellipsometric (VASE) studies of ion beam deposited DLC films [11]. These films have been modified by directing 1 MeV gold ions, as well as 6.4 MeV fluorine ions through the DLC and into the underlying silicon substrates. The percentage of hydrogen in the film was measured vs. fluence using proton recoil analysis. Optical analysis was performed assuming the Lorentz oscillator model, using two oscillators with spectral position, width, and amplitude all variable. This model fit the VASE data extremely well. With ion modification the oscillators shifted to lower photon energy, consistent with reduction in hydrogen concentration and possible increased graphitization.

Ion beam modification of DLC films has not been extensively investigated. The purpose of the present work was to study the effects of high energy ion beam irradiation when the ion beam passed <u>through</u> the DLC films. Two ions (gold and fluorine) and a

Table III-1. 200 Watt, 250°C deposited DLC (3 layer analysis). 23°C water.

DLC thickness	Mixture (50-50) layer thickness	Water layer thickness
330 Å	29 Å	48 Å

Table III-2. 200 Watt, 250°C deposited DLC, 23°C water (3 layer analysis).

Condition	DLC Thickness	Mixture (50-50) layer thickness	Water layer thickness
No water	326 Å	...	...
2 h in water	326 Å	24 Å	37 Å
4 h in water	328 Å	28 Å	51 Å

Table III-3. 200 W, 250°C deposited DLC sample in 100°C water (3 layer analysis).

Condition	DLC thickness	Mixture (50-50) layer thickness	Water layer thickness
No water	338 Å	...	...
2 h in water	337 Å	46 Å	36 Å
4 h in water	339 Å	54 Å	56 Å

TABLE III-4. Best fit ellipsometric solutions for indicated samples (steam jet was used to introduce water).

75°C	100°C	150°C	200°C	250°C
200-W DLC samples without H_2O				
t_1 = 915 Å	t_1 = 810 Å	t_1 = 840 Å	t_1 = 448 Å	t_1 = 252 Å
oscillator parameters	oscillator parameters	oscillator parameters	oscillator parameters	oscillator parameters
11.8;7.7;3.8	13.2;7.9;3.8	14.4;7.9;3.8	14.8;7.9;3.8	12.5;7.4;4.05
MSE = 10	MSE = 11.5	MSE = 7	MSE = 2.5	MSE = 0.18
200-W DLC samples with H2O, oscillator parameters (as given above) fixed				
t1 = 943 Å f2 = 97% MSE = 5.1	t1 = 847 Å f2 = 100% MSE = 8.4			
t3 = 4.5 Å t1 = 936 Å MSE = 2.8	t3 = 3 Å t1 = 847 Å MSE = 8.3	t3 = 74 Å t1 = 904 Å MSE = 13	t3 = 139 Å t1 = 473 Å MSE = 15.1	t3 = 123 Å t1 = 271 Å MSE = 0.21
t2 = 31 Å f2 = 99% t1 = 906 Å MSE = 2.6	t2 = 12 Å f2 = 58% t1 = 841 Å MSE = 8.2			

range of fluences were used. Diagnostics techniques included Rutherford Backscattering, proton recoil, hydrogen analysis, and variable angle of incidence spectroscopic ellipsometry (VASE).

B. Experimental

Samples were prepared in the ion-beam system described in section I above. Methane gas in the 2.5 cm diameter Kaufman type neutralized ion beam gun system (IonTech Corporation) was used, and the chamber had a base pressure of ~10^{-7} Torr. During deposition the chamber pressure rose to the range 10^{-4} to 10^{-5} Torr, and the gas flow rate was 20 sccm. In these experiments all substrates were silicon.

Rutherford Backscattering (RBS) and proton-recoil experiments were performed at Universal Energy Systems using a tandem (1.6 MeV) accelerator (Tandetron Corp.) which was used to provide 1 MeV gold ions, or 6.4 MeV fluorine ions for the ion beam modifications. Fluences ranged from 3×10^{14} cm^{-2} to 1×10^{16} cm^{-2}.

In the present VASE optical diagnostics, spectral data were taken at several angles of incidence, and the data analyzed with respect to a two oscillator model with all six parameters variable. In addition, the thickness of the DLC film was a variable.

RBS was used to determine the number of carbon atoms per cm^2 in each DLC film. Film thicknesses on these same samples were then determined by ellipsometry. From these combined measurements the film density was determined. The limits of error in RBS measurements were approximately ± 10 percent, and in ellipsometry were ± 5 percent for DLC films. The combined error of ± 15 percent covers the range of values measured on all eight samples. Thus we conclude that the density was 1.5 ± 0.2 gms/cm^3 independent of thickness for this set of samples. The density could easily depend on deposition technique and parameter setting, however.

C. Fluorine (6.4 MeV) Irradiated Samples

Table IV-I summarizes results of ellipsometric analysis of "fluorine-beam" processed DLC on silicon. The shorthand notation has the following meanings: A, P, and W are the one oscillator amplitude, position (in eV), and width (in eV), respectively. Subscripts I and II refer to first oscillator, and second oscillator in the two oscillator analyses, respectively. "Th" signifies thickness, and the MSE defined by Equation (2) in the last section.

Notice from Table IV-I that irradiation shifted the position of the oscillators to lower photon energy. Another universal trend was for the amplitude in the lower energy oscillator to increase with fluence.

Figure IV.1 shows the effect of fluorine irradiation on the imaginary part of the dielectric function, E2, analyzed allowing all six oscillator parameters as well as the film thickness to be variables in the regression analyses. The trends are obvious: a downward shift of the E2 maxima in energy, and an increase in the E2 amplitude. At the same time, the higher energy oscillator position decreases (Table IV-1).

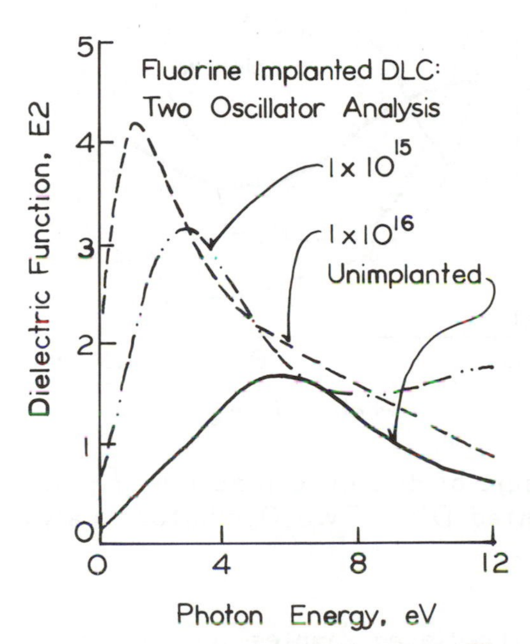

FIG. IV-1 Imaginary part of dielectric function, E_2, for Fluorine Implanted DLC: Two Oscillator Analysis.

In Figure IV.2 the real part of the dielectric function E1 is shown. Again there are shifts to lower energies, and an increase in the amplitude of E1.

The extinction coefficient, k, vs. wavelength for the various fluences are shown in Fig. IV.3. Note the general rise in magnitude of k with increasing fluence. This general increase in k with fluence gives rise to a decrease in optical bandgap, and a decrease in hydrogen content, as determined by proton recoil. The main effects of fluence of both the fluorine and gold species on hydrogen content are shown in Figure IV.4. A nearly linear relationship between energy gap and hydrogen content was found.

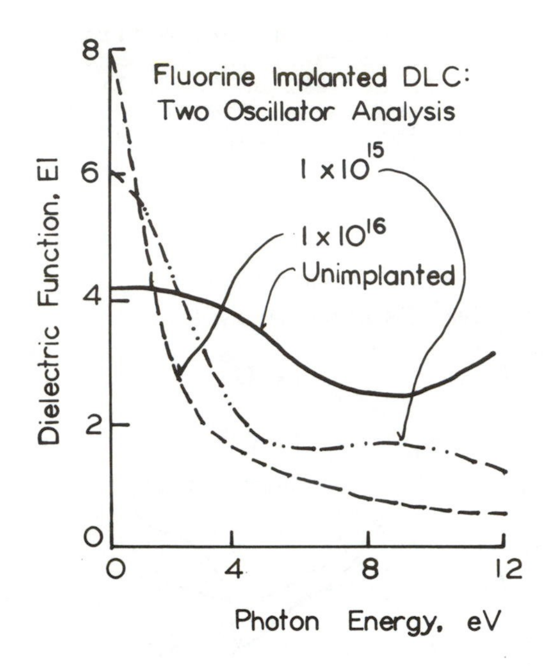

FIG. IV-2 Real part of dielectric function, E_1, for Fluorine Implanted DLC: Two Oscillator Analysis.

D. Gold (1MeV) Irradiated Samples

Results of ellipsometric analysis of samples irradiated with gold ions are shown in Table IV.2. A two oscillator model was used, with all six oscillator parameters variable.

Again there are downward shifts in the oscillator positions with increasing fluence. Other trends were less obvious. By fitting data over our spectral range of 300 to 800 nm, the oscillator analysis allowed us to extend our knowledge of the optical constants over a wider range (with an admitted danger of errors far from the measured range). The results show that the two oscillator model is appropriate for both F and Au irradiation. In both cases the downward shifts are consistent with there being a loss of hydrogen, a decrease in the optical gap, and a tendency towards graphitization.

Table IV-1. Comparison results of 2-osc. model analysis for F-implanted (4.6 MeV) DLC ion beam deposited samples (-fit only; u = unimplanted, I = implanted). Except for thicknesses, all units are electron volts. fluence, cm^{-2}sec.

2-oscillator model - Position of 2nd oscillator variable

	1st oscillator			2nd oscillator			Tk (Å)	MSE
	AI	PI	WI	AII	PII	WII		
0	7.5	5.58	4.0	12.2	17.4	0.44	1221	1.22
3	10.4	3.4	3.6	8.37	14.5	1.64	1307	0.13
10	11.1	2.56	2.92	10.22	12.8	6.0	790	0.015
30	12.8	2.49	2.92	8.6	11.0	7.8	757	0.1
100	22.6	0.68	2.28	18.7	4.57	7.8	2624	0.47

Table IV-2. Results of 2-oscillator models for Au (1 MeV) -implanted DLC ion beam deposited samples (u = unimplanted, I = implanted). Except for thicknesses, all units are electron volts. = fluence, cm^{-2}sec.

2-oscillator model

	1st oscillator			2nd oscillator			Thickness	MSE
	AI	Position	WI	AII	Position	WII		
0	3.34	4.33	1.45	17.0	16.7	0.55	1832 Å	1.18
3	1.40	3.24	0.78	21.4	16.5	0.39	2096 Å	2.60
10	3.60	3.53	1.38	9.9	11.6	1.70	1173 Å	0.34
30	3.50	3.44	1.53	9.7	9.2	2.30	913 Å	0.26
100	3.29	3.33	1.18	12.5	10.8	1.21	1314 Å	0.77

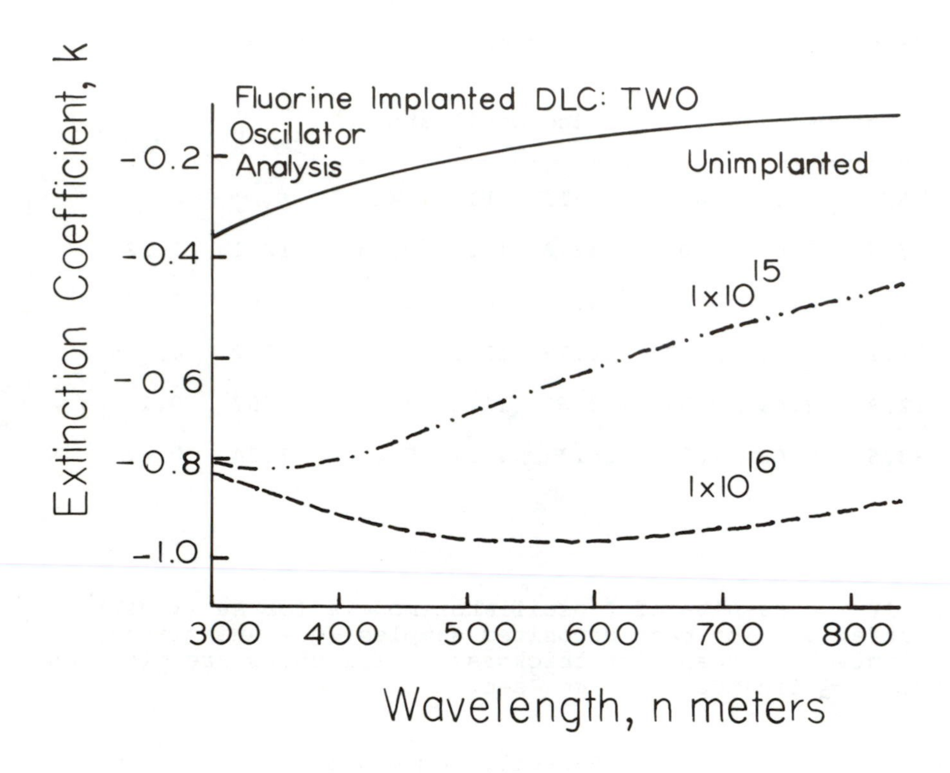

FIG. IV-3 Extinction Coefficient, for Fluorine Implanted DLC: Two Oscillator Analysis.

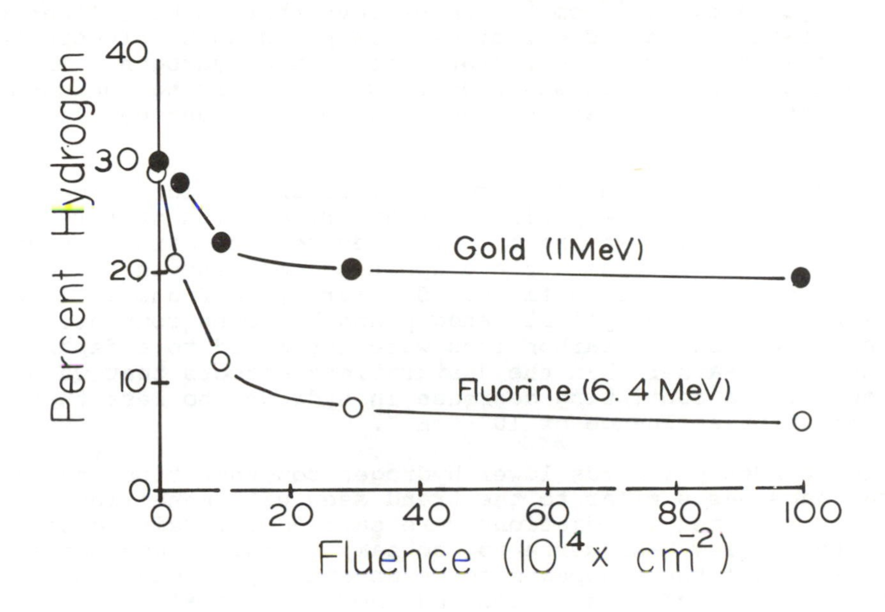

FIG. IV-4 Hydrogen concentration for Au and F implanted DLC.

For gold implants, an optical gap slightly above 1 eV is seen to decrease to about 1 eV with increasing fluence. The effect of irradiation on bandgap isn't as great for Au as it is with F irradiation. However, the loss of hydrogen with gold fluence was not as great as it was for fluorine.

E. <u>Discussion of Results</u>

We found that irradiation with a lighter ion (F) at 6.4 MeV through a DLC film (composed of light elements) had a greater effect on the hydrogen content and optical properties of DLC than did Au irradiation at 1 MeV.

In general a two oscillator model yielded an excellent fit to the optical data, and the two oscillators were centered near 5 eV and 17 eV in the unirradiated samples. These positions shift to lower photon energies with fluence, and proton recoil experiments show that the hydrogen content decreased from near 30% (in unirradiated materials) to near 20% (Au), or near 6% (F) with fluence. The loss of hydrogen was not linear with fluence;

rather there was a near "saturation" of the irradiation effects for fluences above 10^{15} cm^{-2}. We believe that at high fluence the ion beams introduced defect centers produced by electronic energy loss which acted to retrap part of the hydrogen. The magnitude of this effect was reduced with increasing ion mass as the proportion of nuclear to electronic energy loss was increased.

It has been argued [13] from optical absorption, Raman, conductivity, and ESR experiments that irradiation at first had only the effect of lowering the hydrogen content, and then at high fluences it increased the graphitic component in their glow discharge produced carbon films. Similarly, we found a linear relationship between optical bandgap and hydrogen content. However, their 50 keV carbon ions were implanted to a depth of only 100 nm. Rather than the "saturation" effects that we saw, they observed a continuing decrease in hydrogen to less than a few percent at a fluence of 10^{17} cm^{-2}.

The tendency towards lower hydrogen concentration and lower optical gaps was similar to the trend seen with annealing. (The presently reported irradiations were carried out using a cool substrate and low beam currents, so sample heating was not a problem). Thus there appears to be an analogy between the physical effects of irradiation and those of annealing.

References

1. J.C. Angus, P. Koidl, and S. Domitz, in Plasma Deposited Thin Films, edited by J. Mort and F. Jansen (CRC, Boca Raton, FL, 1986). Chapter 4.

2. R.L.C. Wu, D.C. Ingram, and J.A. Woollam, Final Report: "Diamondlike Carbon Coatings for Optical Systems", U.S. Army Materials Technology Laboratory, Contract No. DAAL04-86-C-0030, October 1988.

3. J.A. Woollam, S. Orzeszko, B.N. De, N.J. Ianno, A.R. Massengale, J.J. Pouch, and S.A. Alterovitz, "Diamondlike Carbon Dielectrics for Integrated Circuits", Final Report, Control Data Corporation, December 1987.

4. J.A. Woollam, B.N. De, L.Y. Chen, J.J. Pouch, and S.A. Alterovitz, MRS Proceedings Volume: Optical Materials: Processing and Science, 1989.

5. S. Orzeszko, B.N. De, J.A. Woollam, J.J. Pouch, S.A. Alterovitz, and D.C. Ingram, J. Appl. Phys. 64, 4175 (1988).

6. M.J. Mirtich, D. Nir, D. Swec, and B. Banks, J. Vac. Sci. Technol. A4, 2680 (1986).

7. J.J. Pouch, S.A. Alterovitz, J.D. Warner, D.C. Liu, and W.A. Lanford, Mat. Res. Soc. Symp. 47, 201 (1985).

8. H.J. Hovel, "Solar Cells", in Semiconductors and Semimetals, 11, edited by R.K. Willardson, and A.C. Beer, Academic Press, New York, 1975, p. 203.

9. The Infrared Handbook, edited by W.L. Wolfe, and G.J. Zissis, The Infrared Information and Analysis Center, Environmental Research Institute of Michigan, Ann Arbor, MI, 1978.

10. J.A. Woollam, P.G. Snyder, and M.C. Rost, Thin Solid Films 166, 317 (1988).

11. S. Orzeszko, J.A. Woollam, D.C. Ingram, and A.W. McCormick, J. Appl. Phys. 64, 2611 (1988).

12. S. Prawer, R. Kalish, M. Adel, and V. Richter, J. Appl. Phys. 61, 4492 (1987).

13. M.E. Adel, R. Kalish, and S. Prawer, J. Appl. Phys. 62, 4096 (1987).

Materials Science Forum Vols. 52 & 53 (1989) pp. 609-644

MICROSTRUCTURE AND PHYSICAL PROPERTIES OF METAL-CONTAINING HYDROGENATED CARBON FILMS

C.P. Klages (a) and R. Memming (b)
(a) Philips GmbH, Forschungslaboratorium Hamburg, D-2000 Hamburg 54, FRG
(b) Institut für Solarenergieforschung GmbH, Sokelantstr. 5, D-3000 Hannover 1, FRG

ABSTRACT

The properties of various types of metal - containing hydrogenated carbon films are reviewed with particular emphasis on the effect of structure and composition on electrical conductivity, mechanical and tribological behavior. It is shown that the electrical conductivity can be varied over many orders of magnitude. The conduction mechanism is interpreted by a hopping process below the insulator-metal transition. The material exhibits unusual tribological properties, such as low friction at small metal concentrations and a maximum wear resistance at metal concentrations around 10-20 at%. The latter properties are discussed in terms of hardness and elasticity of the films.

I. INTRODUCTION

Carbon thin films, especially the dense hydrogen-free or hydrogenated films which can be obtained from solid carbon sources by ion-beam assisted sputtering or by evaporation and arc techniques or from gaseous carbon sources (hydrocarbons) in RF or DC glow discharges, have already been subject of several recent review articles [1,2,3]. Carbides, on the other hand, especially those of transition-metals, have already found widespread applications as hard coatings. Their properties have also been reviewed [4]. The present review article deals mainly with metal containing hydrocarbon films, i. e. a class of thin film material taking - at least as far as their composition is concerned - an intermediate position between amorphous carbon and metals or metal carbides. Properties of pure hydrogenated carbon films will only be described in so far as they are of importance for the understanding of metal containing films. Different abbreviations for metal-containing hydrogenated carbon films have been used in the literature, which, besides characterizing the composition qualitatively, point to the close relationship to amorphous hydrogenated carbon films: Me-C:H,

Me-a-C:H, a-C:H(Me). In this article, the first of these acronyms will exclusively be used, as it is the simplest.

Although films containing a metal besides carbon and hydrogen can be prepared with compositions ranging from very small carbon contents to nearly pure hydrogenated carbon, in the present article the main emphasis will be laid on carbon-rich layers with metal to carbon atomic ratios Me/C below 1. Several techniques have been used for the deposition of corresponding films such as co-evaporation of polymer and metal, plasma polymerization of volatile organo-metal compounds or simultaneous plasma polymerization and evaporation of metals. The thin film materials considered in this paper are prevailingly prepared by reactive sputtering or arc-discharge evaporation of a metal in a hydrocarbon environment. This narrowed choice is made because films prepared by these methods have turned out to possess outstanding tribological and interesting electrical properties. These materials have found great interest since some tribological properties were reported by Dimigen and Hübsch in 1983 [5,6].

II. DEPOSITION METHODS FOR ME-C:H FILMS

As already mentioned in the introduction, several methods can in principle be used for preparing metal-containing carbon-hydrogen films.

By co-evaporation of a polymer (polyethylene [7] or polypropylene [8]) and copper, for example, metal-polymer composites with finely dispersed copper grains were prepared and characterized with respect to their electrical properties.

In a large number of papers, plasma polymerization of a metal containing monomer, pure or together with another organic compound, is described. Preparation conditions are generally characterized by relatively high working pressure, up to 10 Pa or above, and low powers frequently coupled "inductively" into the glow discharge used for decomposition of the starting monomers.
This method is naturally restricted to accessible volatile metal-organic compounds as for instance vinylferrocene, plasma polymerized at 26 Pa and 10 W [9], tetraalkylyltin (6.5-9.0 Pa, 37 W, inductive coupling [10]; 20 Pa, 0.1-1 W/cm^2, RF-diode [11]) or allylcyclopentadienyl-palladium (20 Pa, 0.1-1 W/cm^2, RF-diode [11]). Also the preparation of tin-containing films investigated by Shuttleworth [12] should be considered as plasma polymerization, in spite of the authors terminology "sputtering": These films were produced from a tin foil in an Ar/ethylene atmosphere at 26 Pa using an inductively coupled glow discharge process. In view of the experimental setup, the high pressure used, and the experimental results, a kind of a transport reaction between metallic tin and reactive species generated in the glow discharge seems to be a more probable mechanism.

Metal/polymer cosputtering was used by Roy and colleagues to prepare composites containing platinum, gold, and copper [13,14]. Oxygen- and fluorine-containing polymers only were used in that work. Cosputtering of a metal and a C-H polymer has seemingly not yet been applied to prepare dense metal-containing carbon-hydrogen films. Simultaneous plasma deposition from an organic monomer and evaporation of a metal has, for example, been used by Asano [15] and Yoneda et al. [16] to prepare metal-containing polymer films. An inductively coupled glow discharge and a remote, inductively coupled discharge, resp., were applied in these experiments.

In most of the papers mentioned above, the films were characterized with regard to their microstructure; in some cases electrical and optical properties were reported. Mechanical or tribological properties, on the other hand, were in general not investigated and can, in fact, be assumed to be generally poor due to the expected "polymeric" nature of the layers prepared. A considerably better tribological performance can be attained with films consisting of a dense network of highly crosslinked carbon atoms (a-C:H films). The most important factor determining the extent of densification and crosslinking of the carbon-containing film is the flux density and energy of energetic species impacting on the surface of the growing film. Depending on the type of deposition apparatus used, these species may be either ions, extracted from a plasma towards a biased or self-biased electrode, or neutrals, originating from unloaded and reflected ions bombarding a sputter target. As typical examples of sputter deposition of Me-C:H films from metallic or carbidic targets in a hydrocarbon atmosphere, the preparation of Ti-C:H and Ta-C:H layers by DC-magnetron and RF-diode reactive sputtering will now be described in more detail [17,18]. Films with excellent tribological properties, which will be discussed below, have been prepared by these methods.

The configurations used for RF-diode and DC-magnetron sputtering have been reported in many papers. RF-diode sputtering of TaC in an Ar/acetylene atmosphere was used in a study of the influence of acetylene percentage, substrate bias and substrate inclination on film composition, morphology and properties. Experimental parameters and their ranges are collected in Table 1.

Table 1.: RF-diode sputtering of Ta-C:H, experimental parameters

Target:	TaC, hot pressed
Target diameter:	15.2 cm
Target-substrate distance:	7.8 cm
Target bias:	-1.75 kV
Substrate bias:	+100....-120 V
Base pressure:	$5*10^{-4}$ Pa
Working pressure:	2.0 Pa
Acetylene flow percentage:	0...2.7 %
Substrate inclination:	0, 45, 75, 90°

Before starting the deposition, a substrate cleaning and target presputtering process was run in pure Ar (2.0 Pa) at -1.75 kV and -800 V target and substrate bias, resp., with a grounded shutter being placed between both electrodes. In order to improve the film adherence, frequently an adhesion layer is then deposited after removing the shutter and changing the substrate bias to +20 V. This procedure is of special importance for tribological measurements or applications. The thickness of this layer is usually about 10% of the total film thickness. Parameters (acetylene flow, substrate bias) are then gradually (within about 5 to 10 min) changed to grow the Ta-C:H film. By using this procedure, films showing good adherence in a Rockwell indentation test as well as in a scratch test (20 to 35 N) were obtained.

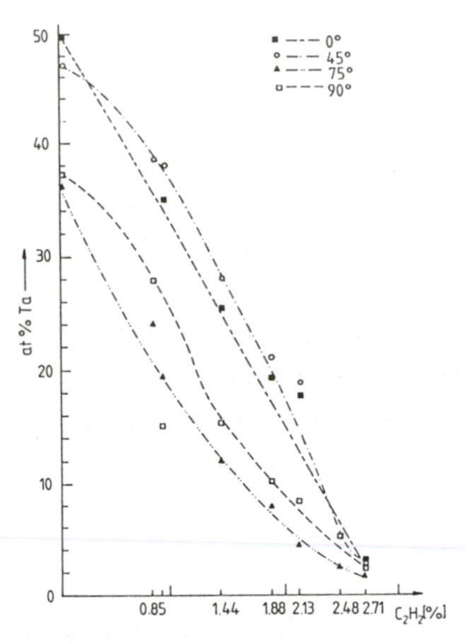

Fig. 1: Metal content of Ta-C:H films (at%) grown by RF-sputter deposition on differently inclined substrates as a function of the gas phase composition. Elemental analysis of tilted films performed at the center of circular substrates located 5.5 cm from the target.

Growth rates are typically in the order of 10 nm/min. The composition of films depends mainly on the flow of acetylene: Ta-contents from about 50 down to a few at% can be obtained by increasing the acetylene flow from 0 to 2.7 % of the total gas flow. It is important to note, that all concentrations in "at%" used in this article have been determined, (usually by EPMA analysis of the metal, carbon, argon and oxygen) neglecting the hydrogen content. 1 at% Ta, to take an example, indicates an atomic number ratio of Ta/(Ta+C+Ar+O) = 0.01!

The effect of an inclination of the substrate against the target plane on Ta incorporation can be seen in Fig. 1 (substrate bias: +20 V). The composition of films grown on inclined substrates was determined at the center of circular substrates, which were located 5.5 cm from the target. The appearance of curves in Fig. 1 and the consequences of the substrate inclination for tribological behavior will be discussed in section VIII b.

Much higher growth rates, compared with reactive RF-diode sputtering, can be attained in DC-magnetron processes. Typical experimental conditions, as they were used for the deposition of Ti- and Ta-C:H films in a Balzers BAS 450 sputtering facility, are given in Table 2:

Table 2.: DC-magnetron sputtering of Ti- or Ta-C:H, experimental parameters

Targets:	Ti, Ta
Target dimensions:	21.0x11.0 cm^2
Target-substrate distance:	6.3 cm
Target power:	2 kW
Substrate bias:	+50....-50 V
Base pressure:	< 10^{-3} Pa
Working pressure:	0.4 Pa
Acetylene flow percentage:	0...57 % (Ta)
	0...35 % (Ti)

A similar procedure as described for RF-diode sputtering is used to clean target and substrate as well as to deposit an adhesive and a graded transition layer. Film compositions ranging from pure metal to Me-C:H films containing only few at% of Ta or Ti are accessible by the proper choice of hydrocarbon/Ar flow ratio. The qualitative appearance of metal content vs. acetylene percentage curves is similar to that of Fig.1. Me-C:H film growth rates are typically around 100 nm/min, roughly an order of magnitude higher than with RF-diode sputtering. The price, which has to be paid for this advantage, is an instability of the process at high acetylene flows (low metal content in the film grown), related to the coverage of the target with carbon-rich electrically insulating deposits growing from the edges into the erosion zone which, eventually, lead to a closure for the electric current flowing through the target surface. Similar phenomena are also typical for other reactive DC-sputtering processes [19]. Usually, they are connected with a hysteresis in the sputter rate vs. reactive gas flow rate relation. Electrically, these processes manifest themselves characteristically as sudden increases of the voltage necessary to maintain a preselected power, if a critical hydrocarbon flow is exceeded. Simultaneously, sparking is observed on the target. Due to the hysteresis effect, the deposition process can not be continued by just decreasing the hydrocarbon flow below its critical value. It was verified experimentally, that deposition processes near the critical point can be run more safely, if gas flows are controlled instead of pressure in order to minimize flow fluctuations. Films with Ta-contents of only 3 at% are thus accessible.

Instead of a sputter target, an arc-discharge can be used as a metal source, generating a highly ionized (ion content up to more than 90 %) beam of particles of low energy (a few tenth of an eV) On negatively biased substrates, Ti-C:H films of good tribological properties have been grown in an acetylene atmosphere [20].

III. MICROSTRUCTURE AND COMPOSITION

Hydrogenated carbon films produced by a plasma enhanced CVD process are amorphous. Their composition with respect to carbon and hydrogen depends on plasma power and deposition temperature. The highest hydrogen concentrations, typically around 50 at %, are obtained with layers at substrate temperatures of about 20 °C at very low plasma powers (low substrate self bias) and decrease considerably with increasing plasma power. In the same direction (increasing power) the cross-linking and the chemical stability increases.

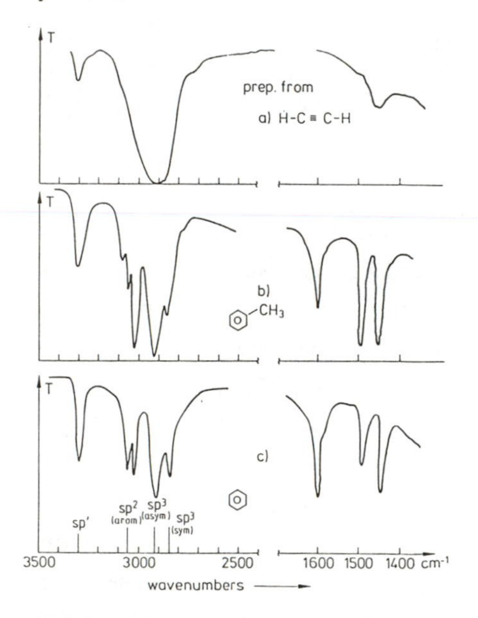

Fig. 2: IR absorption spectra (ATR technique) of layers prepared from different monomers at low plasma power (low self bias) [21]

Various methods have been applied to get detailed structural informations [1]. The functional groups were mainly investigated by IR-spectroscopy. With respect to the C-H stretching vibrations in the range of 2800 - 3400 cm^{-1} great differences have been found depending on the monomer gas and plasma powers being used. One example is shown in Fig. 2, in which the IR-absorption spectra are given for a-C:H layers produced from aliphatic (acetylene) and aromatic (toluene, benzene) monomers at very low plasma powers (negative self bias - 25 V) [21]. The absorption peaks can be correlated to various hybridizations sp^3, sp^2 and sp^1. In the case of layers produced from an aromatic monomer gas a large portion of aromatic sp^2-bonds was found whereas for acetylene the sp^3 hybridization dominates. At higher plasma powers (e. g. substrate DC self bias of - 400 V) the fine structure disappears and all spectra have a shape similar to that of layers made from acetylene (Fig. 2). Accordingly sp^3 hybridization is then dominating. The structure also changes upon annealing above around 400 °C. According to a detailed analysis performed by Dischler et al. the amount of carbon-carbon sp^2 bonding increases at the expense of tetrahedral sp^3 bonding [22]. This structural change is accompanied by effusion of hydrogen and some hydrocarbon fragments. Corresponding effusion measurements have further shown that hydrogen effusion does not occur in a small temperature range but was observed even at temperatures above 1000 °C [23].

The microstructure of metal-containing hydrocarbon films (Me-C:H) is considerably different. According to X-ray investigations small crystalline metal carbide clusters are formed which are embedded in the a-C:H matrix provided that a carbide forming metal, such as e. g. Ta, is used in the deposition process. A corresponding X-ray diagram is given in Fig. 3 [25, 24].

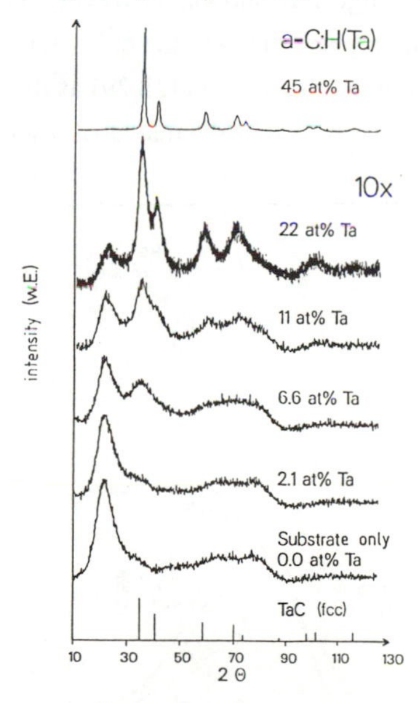

Fig. 3: X-ray diffraction pattern of Ta-C:H films with different Ta-contents (powder diffractometer, Cu-K_α-radiation, quartz substrates) [24]

Similar results have been obtained with other carbide forming metals such as Ti, V, W, Mo, Nb. On the other hand, using metals such as Au and Ag which are not capable of forming a carbide, pure metal clusters are produced distributed within the a-C:H matrix. As shown in Fig. 3 the X-ray diagram corresponds to that of cubic TaC. The X-ray signals are visible down to a concentration of about 2 at% Ta. The width of the spectral lines increases with decreasing Ta-concentrations indicating a smaller cluster size. The latter can be estimated by using the Scherrer equation [47]

$$D = \frac{\lambda}{\Gamma \cos\Theta} \tag{1}$$

in which λ is the wavelength of the X-ray (1.54 Å), Γ the full width at half maximum of the spectral lines, on the 2 Θ scale expressed in radians, and Θ the diffraction angle. Particle sizes between 10 and several 100 Å have been determined, depending on the type of metal and its concentration. In the case of Ta-C:H with 22 at% Ta for instance a cluster size of about 20 Å has been obtained.

The structure of Me-C:H films (mainly Ta-C:H and Ti-C:H) has been further studied by X-ray photoelectron-spectroscopy (XPS). Investigations of the carbon-1s signal clearly indicate the existance of a carbidic phase due to the formation of a metal carbide and a second one corresponding to the a-C:H matrix [26,27]. These results confirm the X-ray data. A detailed analysis of the spectra over a large metal concentration range has shown, however, that the spectra could only be deconvoluted keeping the carbon binding energies and the half width of each line within each phase constant, if a third phase ('pseudocarbidic') is assumed [28,29] as illustrated in Fig. 4. The chemical nature of the 'pseudocarbidic' phase, however, is still a matter of discussion.

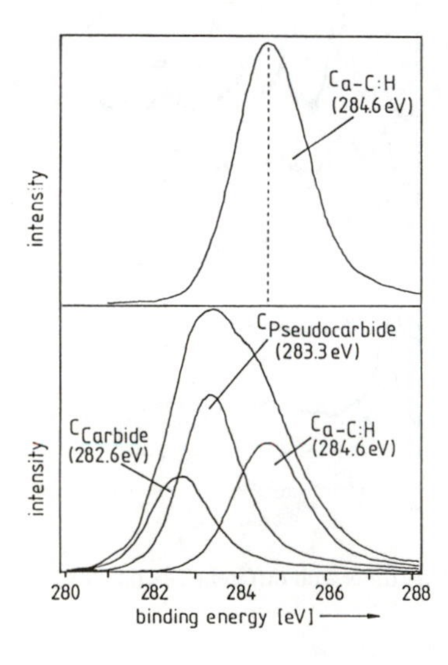

Fig. 4: XPS-spectra of a-C:H and Ta-C:H films. Top: C1S signal of a metal-free a-C:H film. Bottom: Deconvolution of the C1S signal of a Ta-C:H film (25 at% Ta) into three components [29]

The quantitative distribution of carbon in the three phases as a function of the Ta-concentration is given in Fig. 5 [29]. Qualitatively similar results have been obtained with Ti-C:H [29].

The microstructure of the a-C:H matrix could be analyzed by IR-technique only at very low metal concentrations (up to about 4 at% Ta). The absorption peaks clearly indicate that sp^3-bonding

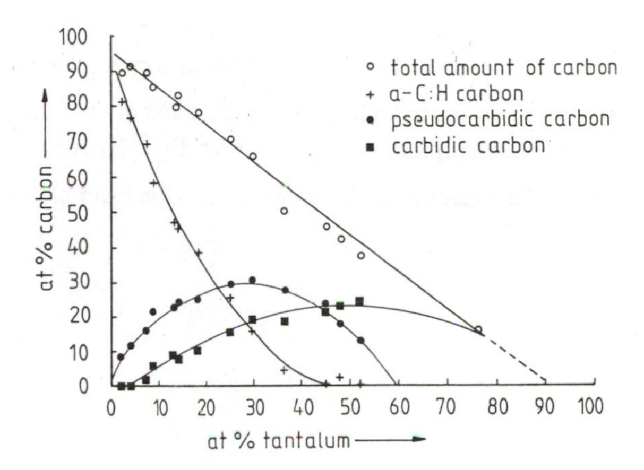

Fig. 5: Quantitative distribution of carbon states in Ta-C:H films prepared from methane, according to XPS analysis [29]

dominates. This result is reasonable because the production of Me-C:H-films by reactive sputtering is only possible at relatively high plasma powers. Metal-free a-C:H films deposited under these conditions exhibit a hydrogen concentration of 35 at%. The H/C-ratio decreases with increasing metal concentrations as shown in Fig. 6, [29]. Effusion studies have further shown that - similar as for pure a-C:H films - hydrogen effuses over a large temperature range. The onset temperature for effusion is shifted towards lower temperatures with increasing metal content [23].

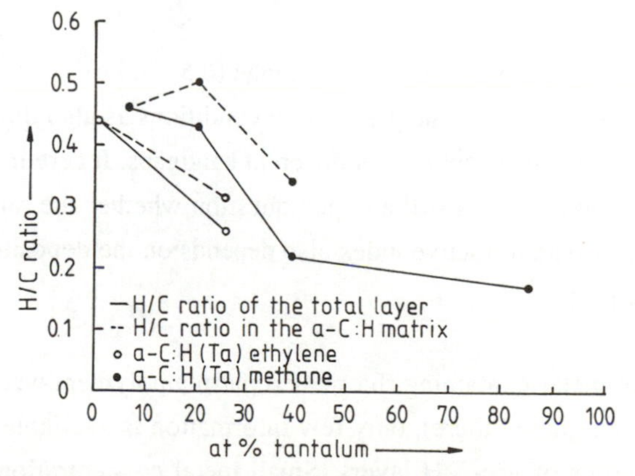

Fig. 6: H/C ratios of Ta-C:H films according to RBS analysis [29]

IV. OPTICAL PROPERTIES

The electronic absorption of a-C:H films has been measured by many authors. A collection of data is given by J. C. Angus et al [1]. In most cases the absorption follows the Tauc relation

$$(E_{h\nu}\,\alpha)^{1/2} = G\,(E_{h\nu} - E_G) \tag{2}$$

in which $E_{h\nu}$ is the photon energy, E_G the optical energy gap, a the absorption coefficient and G is a constant. Although this equation is derived for crystalline compounds and not for amorphous material mostly straight lines have been obtained by plotting $(E_{h\nu}\,\alpha)^{1/2}$ vs. $E_{h\nu}$. Nevertheless this method provides a simple parameterization of the electronic absorption edge. One example is given in Fig. 7, [21].

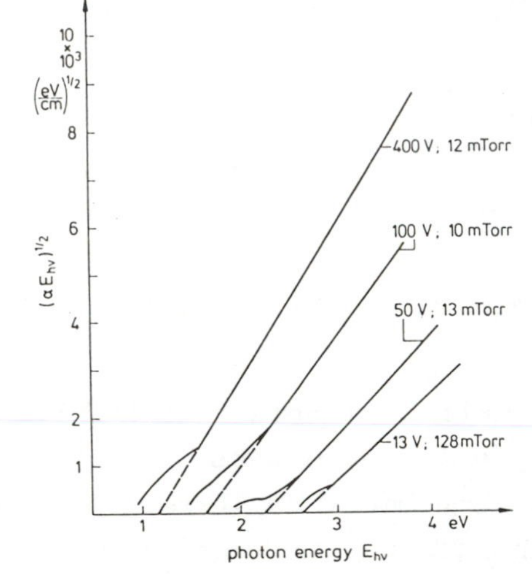

Fig. 7: Tauc plot of absorption vs. photon energy $E_{h\nu}$ for a-C:H films prepared under various conditions [21]

The band gaps reported in the literature vary over a large range (0.5 - 3.3 eV). This is not surprising because the energy gaps depend strongly on the preparation conditions as also illustrated by Fig. 7. Angus et al. discussed the origin for the occurance of different bandgaps. It certainly depends on the content of bonded hydrogen. However, it is still an open question whether the ratio of sp^3 and sp^2 bonding determines the bandgap. The refractive index also depends on the deposition parameters, it covers the range $1.8 \leq n \leq 2.3$ [30].

While the optical properties of metal-containing fluorinated plasma polymers were studied in great detail see ([31,32] and literature quoted there), only few information is available from the present literature about optical properties of Me-C:H layers: Small metal concentrations have already a detectable effect on the absorption spectra [25]. Systematic investigations of the optical properties were performed with Fe-C:H-layers of a metal content above 10 at% [33]. Typically the reflectance is low from the visible range up to about 1.5 μm whereas it increases strongly above this wavelength. Craig et al. have determined the optical constants n and k from absorption and reflectance spectra for a large metal concentration range (10 - 100 at %) as given in Fig 8. Both, refractive index and extinction coefficient, rise continuously with the metal content in the long wavelength region, reaching finally values being typical for metals. In films containing Au or Ag, the optical absorption was found to decrease in the near UV region and to increase at longer wavelengths. Simultaneously an absorption maximum at 550 and 400 nm, resp., is developed, which is characteristic for small metallic grains in a dielectric matrix ("dielectric anomaly") [31,34,35].

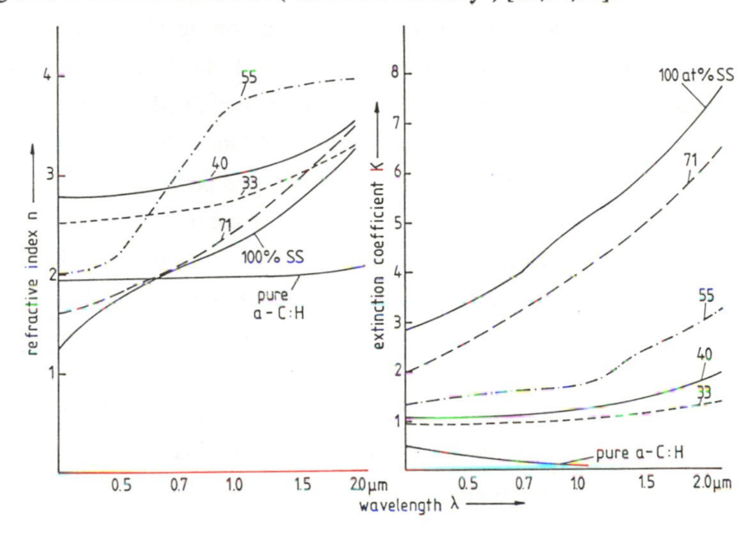

Fig. 8: Refractive indices and extinction coefficients vs. wavelength for Me-C:H (Me: stainless steel (ss)), Data selected from [33]

V. ELECTRICAL CONDUCTIVITY

As already discussed in the previous chapter the electronic structure of pure a-C:H films can be described by energy bands. The conductivity is, therefore, a thermally activated process. Values reported in the literature (see [1] and literature cited there) range from $\sigma = 10^{-15} - 10^{-8}$ $(\Omega\ cm)^{-1}$at room temperature, the greater values being obtained with materials deposited at higher substrate temperatures. Activation energies have been found between 0.2 and 1 eV. The conductivity of a-C:H films was considerably enhanced by doping. N-type conductivity was achieved by doping with phosphorus and p-type with boron [36.], i. e. transport of charge carriers via energy bands was obtained similarly as in a-Si:H films. However, a very high doping was required. This result may be partly due to a low mobility of electrons and holes in the amorphous material but is mainly caused by a high density of energy states located near the middle of the energy gap which are occupied by unpaired electrons. According to EPR measurements the density of unpaired electrons is ranging between 10^{17} and 10^{20} cm^{-3} depending on the preparation conditions [37].

The conductivity of a-C:H films can also be considerably increased by 'metal-doping'. However, the conduction mechanism is completely different (see below). On the one hand, metal incorporation can bring about a steady increase of room temperature conductivity, starting at one atomic percent of metal, as it was observed by Dimigen et al. [6] with Ru- and W-containing films. On the other hand, a percolation type of behavior was found in Co-C:H and Au-C:H films, sputtered reactively in Ar/propane at 2.6 Pa, with room temperature conductivities increasing steeply by at least six orders of magnitude at 22 and 35 vol% of precipitated metal [38]. Different locations of the insulator/conductor transition in the Co and Au system were attributed to differences in the characteristic metal cluster shapes at equivalent volume fractions: At about 40 vol% Au incorporated, TEM studies revealed a change from roughly spherical, well isolated particles (8.5 nm diameter at 30 vol%) to a more wormlike composite structure. In the case of Co, the metal was clustered in much (factor of about five) smaller particles, changing from spherical to wormlike beyond 15 vol% already. Interestingly, conductivity-vs.-metal volume fraction curves for Au- and Co-C:H resemble very closely to those of Au-containing plasma polymerized perfluoropropane films [39] and cosputtered Co/Al_2O_3 films [40], resp.. These Me-C:H films can thus be considered as granular metal/dielectric composites. A comparison of annealing effects on conductivities of Au-C:F and Au-C:H revealed irreversible decreases of sheet resistances in the former material starting already at 80 °C, while irreversible changes occured in Au-C:H only above 220 °C [34,31].

As an example of a carbide-forming metal, Ta-C:H films with metal contents between 0 and 100 % have been studied in detail [41,42,25,43]. Conductiviteis of Ta-C:H films deposited at about 270 ° C with concentrations between 0 and 94 at% Ta are shown in Fig. 9. In contrast to Au containing films, the conductivity rises already at very low Ta-concentrations [25]. The temperature dependence of the conductivity (Fig. 10) investigated in the range of 4-300 K, strongly depends on the Ta-content [43,25].

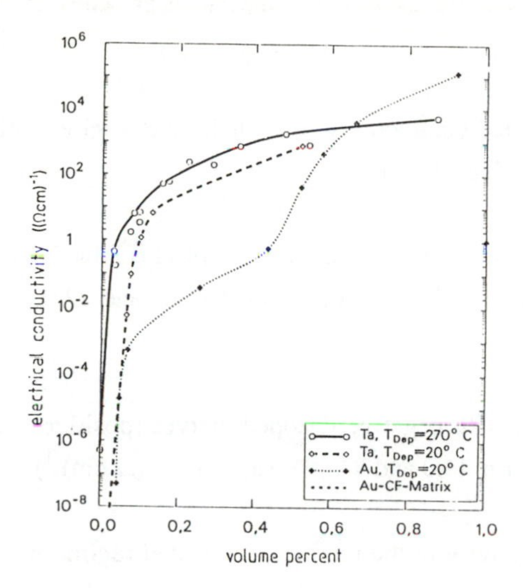

Fig. 9: Dependence of room temperature conductivities of Ta-C:H films on volume fraction of segregated TaC at room temperature and 5 K [25]

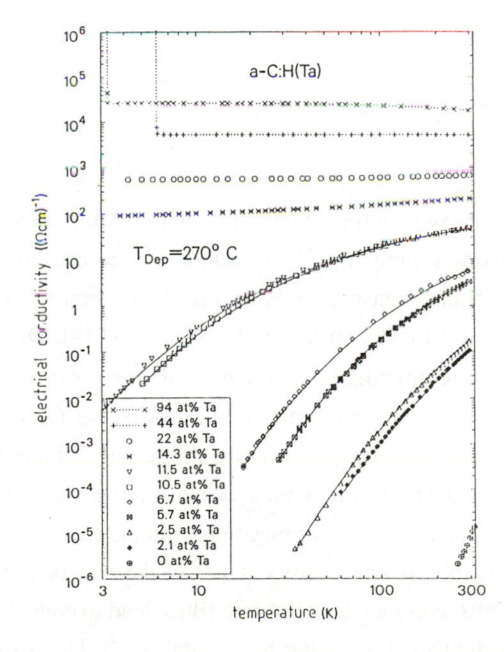

Fig. 10: Electrical conductivity of Ta-C:H films with different Ta-contents as a function of temperature [25]

A similar picture emerges from measurements with samples deposited at 20 °C (room temperature conductivities then cover a range of 15 orders of magnitude, starting at 10^{-11} $(\Omega\ \mathrm{cm})^{-1}$ for a metal-free film). As far as the T-dependence is concerned, the results presented in this figure can be divided into three categories:

* Thermally activated conductivity in a "dielectric" region with less than 11.5 at% Ta (<13 at% in films deposited at 20 °C),

* a transition region with metallic behavior (residual conductivity at zero temperature), but still a positive or zero slope of log (s) vs. T (14.3 and 22.0 at% Ta, s > 10^2 $(\Omega\ \mathrm{cm})^{-1}$), and

* a metallic region with negatively sloped curves (positive temperature coefficient of resistivity, TCR, for 44.0 and 94 at% Ta, s > 10^3 $(\Omega\ \mathrm{cm})^{-1}$).

In an earlier paper, the conductivity in the thermally activated regime was attributed to a thermal excitation of electrons from localized states into the conduction band beyond the mobility edge [41]. The explanations were modified after the granular structure of the films was realized [26] and more conductivity data were available [25]. An evaluation of conductivity data in terms of equation 3 yields values of

$$\sigma = \sigma_0 * \exp [-(T_0/T)^q] \qquad (3)$$

q between 0.4 and 0.55 for layers with > 5 at% Ta, about 0.3 for smaller metal contents. In Fig.11 conductivity data are plotted against $T^{-1/2}$. A behavior according to equation 3 with $q \approx 1/2$ is commonly found in electrical measurements on classical cermets, consisting of granular metal precipitates within an oxidic matrix (often Al_2O_3 or SiO_2) [44]. Different theories have been developed to explain the temperature dependence of the conductivity ($T^{-1/2}$-dependence) of granular conductors, see for example the paper of Mostefa and Olivier [45] as one of the recent models, and the literature quoted in that paper. While the question of electrical properties of a material consisting of metal spheres in an ideal homogeneous insulating matrix is still a matter of discussion among theoreticians, the situation is even more complex in the case of metal-containing a-C:H films. This is due to the presumably heterogeneous nature of the matrix containing extended π-bonded carbon clusters which themselves can, in metal-free films, lead to considerable electric conductivities if films are annealed or deposited at higher temperatures [3]. Deposition of Ta-C:H at increased substrate temperature leads to larger conductivities and a smaller temperature dependence (Fig.11). As already mentioned above, q becomes smaller for very low Ta-concentrations (≤ 3 at %). In fact the conductivity follows even the relation ln $\sigma \sim (T/T_0)^{1/4}$ as proved by plotting corresponding conductivity data vs. $T^{1/4}$ [25]. A theory, the so-called variable range hopping model, as derived by

Mott [46] actually leads to q = 1/4. Here, hopping of charge carriers between localized energy states is assumed. A quantitative relation exists between T_0 and the density of these states and an evalution

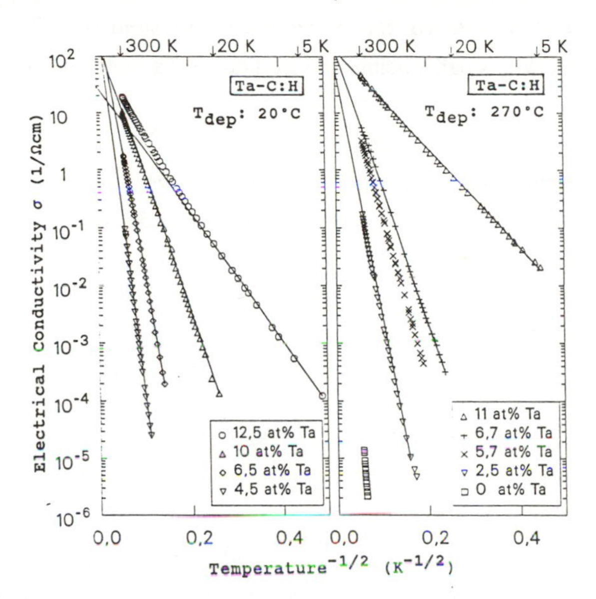

Fig. 11: Electrical conductivities of Ta-C:H films (logarithmic scale) vs. $T^{1/2}$ [25]

of experimentally determined T_0-values yields reasonable density values. However, even such a result is not a proof for this model because q = 1/4 can also be derived for certain granular metal distributions [45].

Contrary to Cu-C:H and Au-C:H films (see above), no sudden increase of room temperatur conductivity occurs at any metal concentraion in the case of Ta-C:H, i.e. no insulator/metal transition is visible. On the other hand, using the low temperature conductivity values, an insulator/metal transition occurs around a carbide volume fraction of f_c = 0.2 (Fig. 9). This value is near to that determined by Laurent and Kay for Co-C:H (f_c = 0.22) [38, 47]. At present, it is not clear, whether this relatively low value (compared, e.g., to Au-C:H or classical cermets [44]) can also be attributed to a specific geometry of the precipitates, like in Co-C:H, or whether the matrix structure, possibly modified during growth by the presence of reactive Ta, is responsible for the transition to metallic-like behavior at a relatively low volume fraction.

Concerning the electrical behavior another material property has to be taken into account when comparing different precipitates within a given matrix, namely the work function Φ of the respective segregation: Φ is 3.14 eV for TaC_{1-x} with x=0.04 [48], while values of 5.0 and >5.0 eV have been measured on cobalt and gold, respectively [49]. The value of Φ determines the position of the Fermi level in the precipitates relative to the electronic states of the matrix. Smaller tunneling barriers and hence larger tunneling rates are expected for smaller Φ's. In a matrix with unoccupied localized states, a higher electron transfer rate is expected from a precipitate with smaller Φ. The pronounced room temperature conductivity rise with small amounts of incorporated Ta might at least partly be due to the low ø of segregated TaC.

Finally it should be mentioned that also Hall measurements have been performed with Ta-C:H films of different metal concentrations in order to get data on carrier density and mobility. Using the standard equations for Hall constant R_H and conductivity

$$R_H = \frac{A}{ne} \tag{4b}$$

and

$$\sigma = en\mu \tag{4a}$$

reasonable carrier densities in the order of n ~10^{23} cm^{-3} and mobilities of $\mu = 0.5$ cm^2/Vs have been obtained with layers of high Ta-concentrations assuming A = 1 [25]. According to the sign of the Hall voltage, Ta-C:H is an electron conductor (pure Ta-films show hole conduction). In the case of much lower metal-concentrations, however, unreasonably high carrier densities were obtained. For instance, $n = 3 \cdot 10^{22}$ cm^{-3} and $\mu = 0{,}02$ cm^2/Vs were obtained with 14 at% Ta, again assuming A = 1. This assumption, however, is only valid for $k_F l > 1$ (k_F = Fermi velocity and l = mean free path length) [50], a condition being not fulfilled here [25] (with A = 1 the mean free path length is 0.14 A, $k_F l \approx 0.14$). The same kind of experience has been made with a-Si:H films where drift mobilities were found to be by an order of magnitude larger than the corresponding Hall mobilities [51]. Accordingly, A « 1 for disordered systems. Since no theoretical values of A are available Hall measurements are not very useful at presence.

VI. MAGNETIC BEHAVIOR OF Fe-C:H AND Co-C:H

Both, Co-C:H and Fe-C:H exhibit magnetical ordering, due to the presence of metal or metal carbide precipitates. In Co-C:H the magnetic behavior below electrical percolation threshold (22 vol%, see above) can be explained by the presence of non-interacting superparamagnetic particles without significant shape anisotropy and a small dispersion in size. Above the electrical percolation threshold a ferromagnetic ordering among individual Co particles begins to manifest itself in the magnetization curves starting to show hysteresis [52].

In a series of Fe-C:H films magnetic ordering was observed in all samples prepared (down to 10.2 at% Fe). Apart from samples with more than 60 at% Fe, no hysteresis could be detected in room temperature magnetization curves [53].

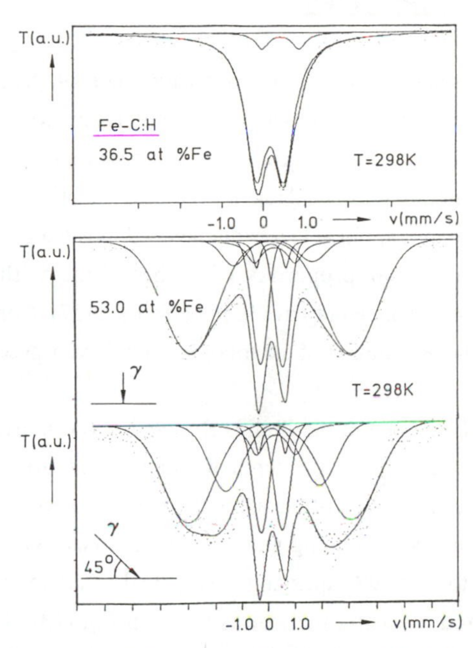

Fig. 12: Room temperature Mößbauer spectra of Fe-C:H with 36.5 (top) and 53.0 (bottom) at% Fe. The orientation dependence of the lower two spectra hints towards a perpendicular orientation of the magnetization within the film [43]

Two samples containing 36.5 and 53 at% Fe were studied in more detail using Mößbauer spectroscopy and Kerr rotation Θ_K experiments in addition to vibrating sample magnetization measurements [43]. From the vanishing of Θ_K at 520+/4 K, the presence of Fe_5C_2 ("Hägg carbide") precipitates was inferred. An evaluation of the low metal concentration (36.5 at% Fe corresponding to 41 vol% Fe_5C_2) curve in terms of superparamagnetism yields an average diameter of 44 Å , if the saturation magnetization value of bulk Fe_5C_2 is used. The magnetization curve of the higher concentration sample (66 vol%) could not be interpretated in terms of non-interacting particles. In the Mößbauer spectra, measured with Fe-C:H films conaining only 36.5 at% Fe, any magnetic hyperfine splitting is absent, because the superparamagnetic relaxation time becomes smaller than the characteristic time scale of the Mößbauer experiment (Fig. 12). The upper value of anisotropy energy, which can be estimated from this observation, suggest the absence of any appreciable shape anisotropy of the particles. The Mößbauer spectra obtained with the film of higher Fe-concentration (53 at%) are magnetically split, their appearance being dependent on the angle between photon wave-vector and film normal. A preferred orientation of the individual particle magnetization vectors parallel to the film normal is a possible explanation of this effect.

VII. MECHANICAL PROPERTIES

Me-C:H films exhibit outstanding wear properties which will be presented in section VIII b. The mechanisms of abrasive wear processes proposed in the literature are generally based on mechanical material properties which are in principle accessible in simple bulk deformation experiments [54,55, 56]. Although standard techniques are available for the determination of elastic and plastic properties of materials, they are mostly not applicable for thin layers deposited on a substrate. In order to obtain informations on the mechanical properties of a thin films without substrate interference, measurements with a nanometer-indenter have been performed [57]. Corresponding investigations are of importance for a better understanding of the tribological behavior described in section VIII.

The apparatus used was developed at the Philips Research Laboratories in Eindhoven/Netherlands [58]. It allows continuous recording of the depth of indentation (0 to 5 μm, resolution 1.5 nm) of an indentor into the material as a function of the applied force (20 to 50,000 μN, resolution 5 μN). In the "quasi-static" measurement mode loading and unloading curves were recorded using a triangular pyramidal diamond indentor with 97° apex angle. A typical result concerning the deformation of the material upon applying and releasing the load is given in Fig [13]. This result shows clearly that elastic and plastic deformations occur. Recently Doerner and Nix [59] have described a method which can be used to determine film hardness from the final indentation depth (intercept of upper curve with the ordinate in Fig. 13). It has been checked with bulk materials that this method yields the same hardness as with standard techniques [57]. In addition, it was found that the hardness values obtained by this method remain independent of the load even at very small forces. This rather unexpected result makes it possible to determine hardnesses of thin layers. The influence of the substrate is negligible for a maximum indentation depth of about 10% of the film thickness in the case of hard films on soft substrates (hardness ratio 10). According to Doerner et al. the Young modulus can be obtained from the same set of measurements, i.e. from the slope of the unloading curve in Fig.13. The slope is related to the Young modulus by the equation [59]

$$\frac{d\varepsilon}{dF} = \frac{1}{2\varepsilon_p} \left(\frac{\pi}{A}\right)^{1/2} \frac{1}{E_r} \qquad (5)$$

where the reduced Young modulus E_r is given by

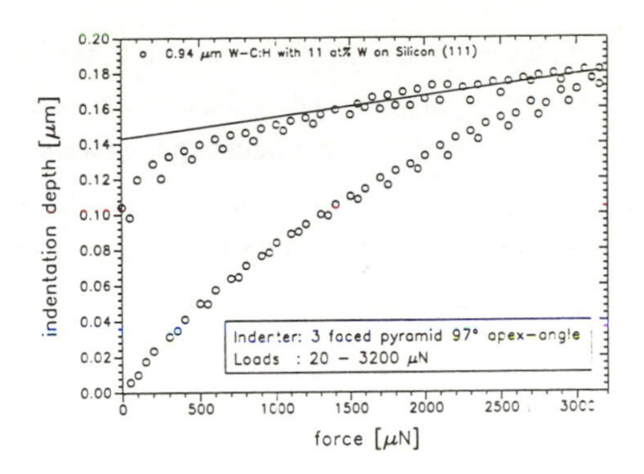

Fig. 13: Depth of indentation during loading and unloading for W-C:H (11.0 at% W) using the Philips nanometer-indenter [73]

$$\frac{1}{E_r} = \frac{1-\mu^2}{E} + \frac{1-\mu_i^2}{E_I} \tag{5a}$$

with E and μ being Young modulus and Poisson ratio, respectively, E_i and μ_i, the same parameters for the indenter tip, A the ratio of the sectional pyramidal area to the square of the indentation depth ε. F the force and ε_p the indentation depth obtained by extrapolating the unloading curve to F = 0. Hardnesses (expressed in GPa, 1 GPa corresponding to about 100 kg/mm[2]) of W- and Ta-C:H films are plotted in Fig.14 as a function of the volume fraction of segregated carbide. In order to calculate volume fractions complete segregation of the total metal content as stoichiometric cubic monocarbide

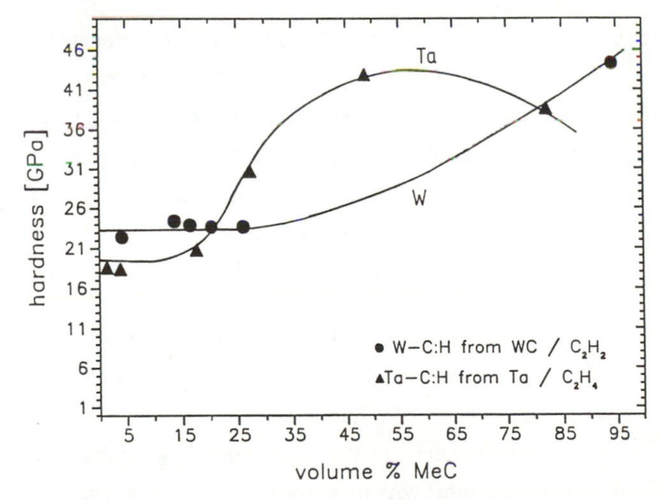

Fig. 14: Hardnesses of W-C:H and Ta-C:H films as a function of the volume fraction of segregated carbides MeC [57]

MeC was assumed, the density of metal-free a-C:H films was used as that of the matrix. 30 vol% carbide corresponds then roughly to about 20 at% metal. Qualitatively similar results with hardness values, i. e. 15 to 18 GPa, were obtained with Ti-C:H films containing up to 30 vol% TiC. The differences in hardness as found with layers containing different metals are most probably due to the hydrocarbon used for deposition: Acetylene, ethylene and methane were used for preparing the films containing W, Ta, and Ti, resp.. The pressure-distance products pd_{TS} (see section VIII) used in the experiments were 120, 180, and 30 Pa · mm, resp.. Most probably even larger differences between hardnesses of W- and Ti-C:H would have resulted, if the films had been prepared at the same pd_{TS}.

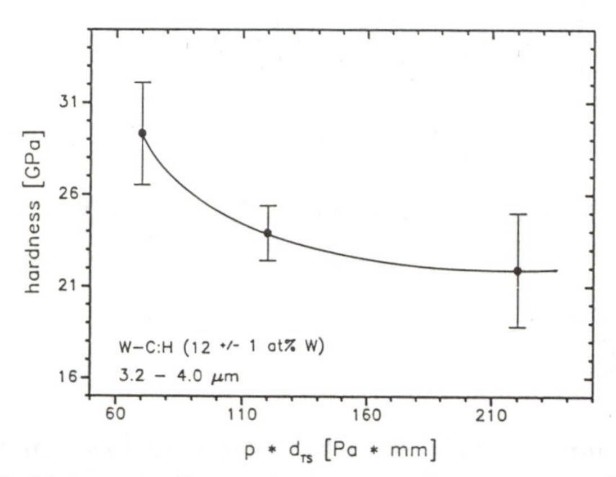

Fig. 15: Dependence of the hardness of W-C:H films (approximately 12 at% W) on the pd_{TS} product [57]

This can be concluded from Fig. 15, where hardnesses of W-C:H films with nearly identical metal content (12 at%), prepared at different pd_{TS} are compared. The influence of this quantity on tribological behavior will be discussed in section VIII.

The constancy of hardness values below a critical volume fraction of segregated hard phase suggests, that in this compositional regime the yield behavior of the composite system is mainly determined by the yield behavior of the matrix. For larger volume fractions, hardness values increase distinctly, probably due to the formation of large extended clusters of particles coming in close contact with each other, similar to a percolation phenomenon. Although the influence of the type of hydrocarbon used in the preparation of the layers, has not yet been studied in detail, it seems to be justified to assert, that an increase of the H/C ratios in the carbon source gas leads to decreasing film hardnesses.

Incorporation of Al, Cu, Ag, Ni, and Au, i. e. metals being unable to form hard carbidic precipitates, by evaporation into a-C:H films grown on the RF-powered electrode in a plasma-CVD process from benzene or butane was found to decrease film hardness. Au, Ag, and Cu are dispersed as nearly spherical metal clusters (3-10 nm diameter) in the amorphous hydrogenated carbon matrix [31]. Decreasing hardness and Young's modulus with increasing metal content was also found for Au-C:H films deposited by reactive sputtering in an ethylene atmosphere [57] and for Cr-containing films prepared by Weissmantel and coworkers by an ion plating process [60].

Contrary to the plastic behavior of the material, Young's moduli evaluated by using equation (4) increase with carbide volume fraction even for small metal contents (Fig.16). This behavior is qualitatively in accordance with theories of the elastic behavior of multiphase systems [61]. Like hardness, also Young's modulus increases, for a given layer composition, with decreasing pd_{TS}: Young's moduli of about 140 to 190 GPa were measured with W-C:H (12 at%W) layers which also the hardness was determined (compare with Fig.15). More detailed investigations are necessary, however, in order to get more information on the influence of preparation conditions, like pd_{TS} and the hydrocarbon monomers, on hardness and Young's modulus of Me-C:H films.

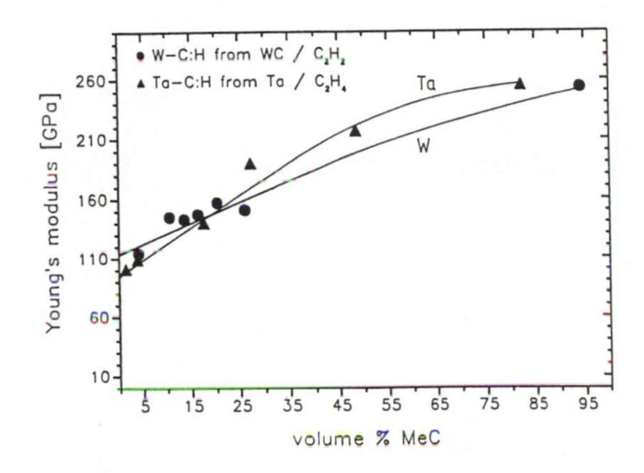

Fig. 16: Young's moduli of W-C:H and Ta-C:H films as a function of the volume fraction of segrgated carbides MeC [57]

VIII. TRIBOLOGICAL PROPERTIES

Because of the outstanding tribological properties of a-C:H and Me-C:H films, they have been subject of many investigations. However, only few data have been published. The results presented here are largely based on unpublished results obtained by Dimigen and coworkers [62].

a) FRICTION

A typical friction result as obtained with an a-C:H film by using a ball-on-disc apparatus (insert of Fig. 17) at various partial water pressures (or relative humidities) is given in Fig. 17 [63,64]. Especially at low pressures the friction becomes extremely low ($\mu \sim 0.015$). This behavior is rather unusual in so far as materials such as graphite, diamond and glassy carbon reach very high friction values ($\mu > 0.5$) at low pressures although at humidities above 10 % similar values were found as those obtained with a-C:H.The friction behavior of a-C:H films deposited at cathodic bias was found to be independent of the preparation conditions. (Gas pressure, plasma power).

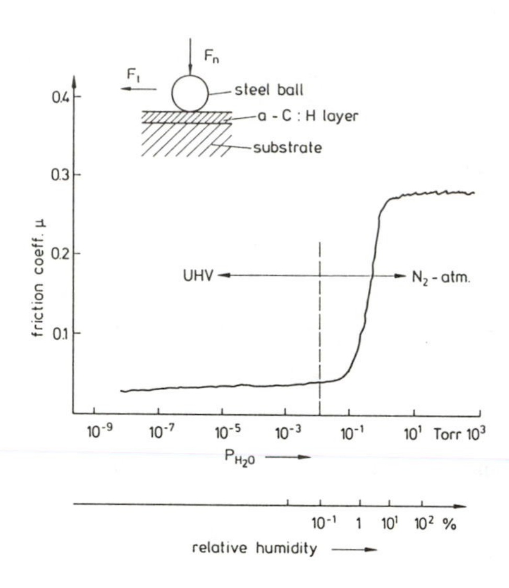

Fig. 17: Friction coefficient vs. H_2O pressure for an a-C:H film produced from acetylene [63]

Small coefficients of sliding friction against steel or cemented carbide (WC/Co 6%) have also been measured with Me-C:H films provided that the contents of a carbide-forming metal is not too large. Friction coefficients between cemented carbide and Ta-C:H films deposited in a reactive RF sputtering process from a TaC target in an Ar/acetylene ambient are plotted in Fig. 18 as a function of tantalum content for different loads. The experiments were performed in the same arrangement as above with an uncoated WC/Co ball (5 mm ø) and a disc coated with Ta-C:H at a constant speed of 167 cm/s in ambient atmosphere (18-41% relative humidity). No significant metal concentration dependence can be detected with very small loads below 0.1 N. With larger loads, increasingly smaller friction coefficients are obtained in measurements with layers of a metal atomic percentage

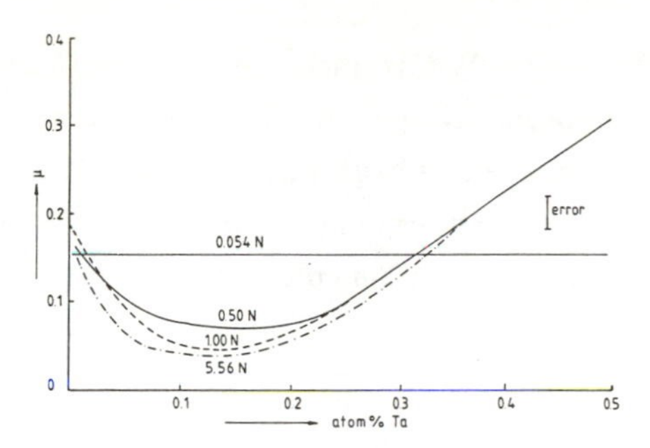

Fig. 18: Friction coefficients of Ta-C:H films sliding against an uncoated cemented carbide ball (5 mm ø) at 167 cm/s speed in ambient atmosphere (18-41% rel. humidity) [62]

smaller than 30 (Ta/C< 0.42). Films with about 13 at% Ta measured under a load of 5.36 N have friction coefficients of only 0.04. No significant load dependence of μ is to be found for metal contents larger than 30 at%, if at least 0.5 N are applied.

Qualitatively similar friction results have been obtained for a number of other carbide-forming metals as far as the dependence on metal concentration and load is concerned. In Fig. 19 the load dependence of the friction coefficient of a W-C:H film with 12.0 at% W is compared with that of TiN and TiAlN coatings, the latter being deposited by reactive sputtering.Again, besides the result that the absolute values of μ are considerably smaller at high loads, the distinct decrease by a factor of roughly two is the most prominent feature of W-C:H in this plot.

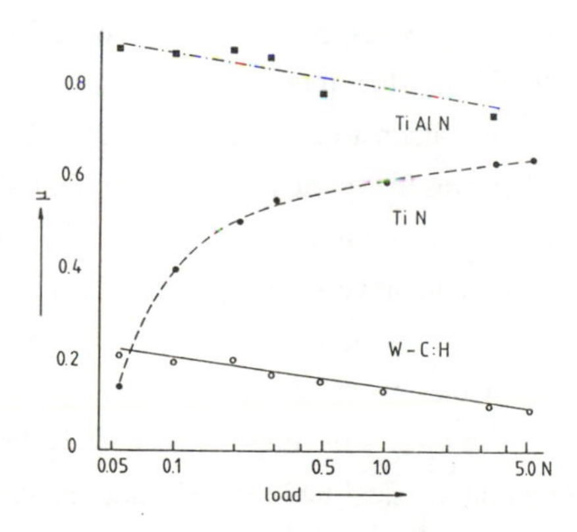

Fig. 19: Load dependence of friction coefficients of W-C:H (12.0 at% W), compared to TiAlN and TiN [18]

Interestingly, not only the load dependence but also the humidity dependence of the friction coefficient changes its sign at a critical metal content. Coefficients of sliding friction between a stainless steel ball (steel code 1.4034, 4.8 mm ø, load 3.1 N) and W-C:H films, measured in dry and humid nitrogen, are plotted vs. the metal-to-carbon atomic ratio in Fig. 20. In dry nitrogen with relative humidity smaller than 0.1%, very small values of about 0.02 have been obtained (similarly as in pure a-C:H films, Fig. 17) as long as W/C is smaller than 0.15 (<13 at% W). Considerably higher values are obtained in humid nitrogen, as long as the W/C ratio is smaller than 0.4. A reversal of the humidity influence is observed, however, for larger metal contents. At a W/C ratio of 0.4, layers with humidity-independent friction coefficients can thus be prepared. Qualitatively similar results have been obtained for Fe-C:H films, the reversal taking place at an Fe/C ratio of about 0.04 [17].

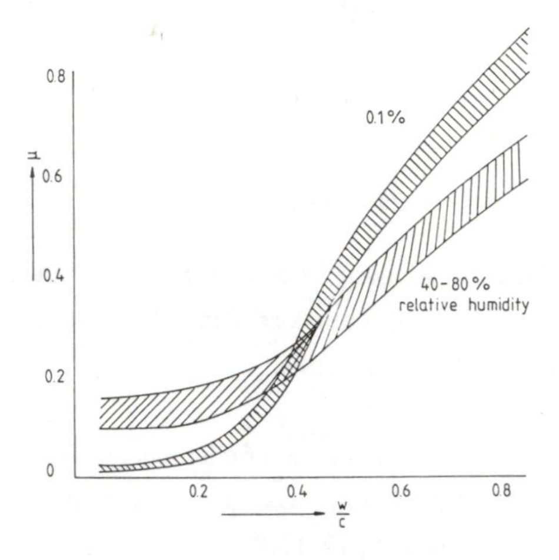

Fig. 20: Compositional dependence of the sliding friction coefficient of a stainless steel ball on W-C:H in low and high humidity nitrogen atmosphere

It is an interesting question which factors are responsible for the low friction coefficient in UHV as found with pure a-C:H and Me-C:H films of low metal concentration. As already mentioned before most other (inorganic) materials sliding against each other exhibit in UHV a considerably higher friction coefficient than in air [65]. This was interpreted by oxide formation on the metal surface or by adsorption of H_2O in air leading to a lower friction coefficient. In addition material transfer during sliding may be involved. Various attempts have been made to study a possible material transfer between a sliding ball and disc, one of them being coated with a hydrocarbon film. In one case material transfer between a steel ball and an a-C:H-coated substrate, both sliding against each other under UHV-conditions, was investigated by ex situ Auger analysis [63]. In fact a significant increase of the carbon Auger peak was found on the steel ball after sliding, indicating some carbon transfer from an a-C:H film to the bare steel ball. This result was supported by the observation that a low friction of $\mu = 0.02$ was only observed after about 20 -30 sliding cycles of the steel ball over the track. The low value occured immediately, however, if both, ball and disc, were coated.

Dimigen and Hübsch [66] also investigated this problem and came to another conclusion. They used an arrangement in which a steel ball coated with W-C:H (8.0 at% W) was sliding on a bare steel disc in a spiral like movement in an ambient atmosphere. Accordingly, the coated ball was always passing virgin steel which had not come into contact with the film material before. In this experiment they obtained a constant friction coefficient of $\mu = 0.12$ over a total track length of 21 m. In a control experiment in which the ball was cycling repeatedly over a circular track the same friction coefficient was measured. In addition it was found that the amount of material worn from the film was far too little for producing a monolayer during sliding on the spiral like track. Accordingly, any influence of carbon transfer on the friction coefficient was excluded.

It is impossible to judge whether these controversal results are due to the fact, that one experiment was conducted in UHV leading to extremely low friction and the other at higher humidity. In an other case, however, it has been clearly shown that material transfer is involved. For instance, if a steel ball is sliding on an a-C:H layer in dry N_2-atmosphere the friction coefficient is very low (0.02) which increases to high values above $\mu = 0.6$ upon switching from N_2-to O_2-atmosphere. In this case it could be proved by Auger analysis that Fe-transfer from the steel ball to the a-C:H film is responsible for the increase in friction [63].

b) WEAR

b1. Abrasive Wear

The abrasive wear resistance of a coating characterizes its ability to resist mechanical damage on sliding contact with small hard particles. The frictional work associated with this process will be dissipated within a surface zone, the thickness of which is in the order of the contact length [54]. In order to avoid problems in abrasive wear tests, (e. g. decohesion at the substrate/film interface), the dimensions of test particles should not be too large compared to the film thickness. Good results are being obtained using suspensions of 5 µm Al-oxide or 1 µm diamond particles in water for films of a few micron thickness. The mechanical load on the substrate/film interface can be reduced further by using a ball as a counterbody which is rotating stationarily around a horizontal axis instead of sliding a fixed ball on a disc.

The abrasive wear data quoted in the following have been determined using an apparatus, originally designed for measuring layer thicknesses, available commercially under the name "Calo-Test". Its principle of operation becomes evident from Fig. 21. The wear is quantified as the volume of material worn from the film per minute. The dependence of abrasive wear rates, expressed in mm^3/min, on metal concentration is given in Fig. 22 for several refractory metals and for the "half-metals" boron and silicon. For comparison, also the wear rate of a-C:H, prepared on the RF-powered electrode in a plasma-CVD process, is represented as a horizontal line. The most prominent characteristic of nearly all curves is the steep increase of wear rate towards vanishing metal content and the occurence of a more or less pronounced minimal wear around 10 ± 3 for films containing Ti, Mo, or W and about

20 at% for Nb. Interestingly, boron and silicon behave qualitatively similar as the refractory metals, although the usual Si- and B-carbides have structures completely different from those of refractory metal carbides (The compositions of Si-C:H films investigated do not extend to below 20 at%, but a minimum has to be expected here, too).

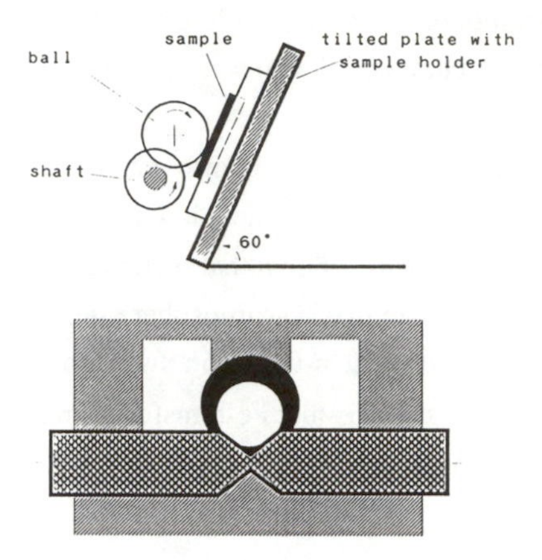

Fig. 21: "Calo-Test" apparatus used to determine abrasive wear rates. A fixed amount of the abrasive medium is applied to the steel ball

All refractory metals included in Fig. 22 are able to form cubic carbides MeC_{1-x} with wide ranges of composition, in the case of Mo and W at least as high temperature modifications. According to X-ray measurements cubic carbides are actually formed in Me-C:H layers containing Ti, Ta, Mo, Nb and W. Attempts to prepare under comparable process conditions Fe-C:H and Cr-C:H layers with wear resistances similar to other films, have hitherto failed: The minimal wear rates (not contained in the figure) were $7.0 \cdot 10^{-4}$ (Fe-C:H) and $9.5 \cdot 10^{-4}$ mm^3/min. (Cr-C:H), i. e. values being at least an order of magnitude greater than those of the other Me-C:H films (Fig. 22). For both, iron and chromium, no cubic carbide MeC_{1-x} is known. Monoclinic "Hägg carbide", Fe_5C_2, is present in Fe-C:H as proven by magnetic measurements [43]. Seemingly only the formation of hard fcc carbides can lead to the very small wear rates given in Fig. 22. It must be pointed out, however, that the microstructure of B-C:H and Si-C:H films studied tribologically has not been investigated yet and it is not known until now, if there are really carbidic precipitates in these films. From electron-diffraction [67] and XPS studies [68] it is known, that a-$Si_{1-x}C_x$:H films prepared by glow discharge deposition from silane and methane are more or less chemically ordered, i.e. heteronuclear bonds (Si-C) dominate over homonuclear (Si-Si and C-C) bonds. Whether this tendency does actually lead to a precipitation of SiC in the hydrogenated carbon film has still to be investigated in more detail, but preliminary X-ray measurements indicate, that a segregation of particles with dimensions beyond 5 Å does not take place.

Abrasive wear data of Me-C:H films containing metals unable to form carbides are quite scarce. Preliminary results of abrasive wear measurements on Ru-C:H indicate, that the wear resistance is improved by incorporation of ruthenium, which is unable to form stabel carbides. The location of the minimum of wear is similar to that found in typical refractory metals. The actual microstructure of the Ru-C:H is still unknown.

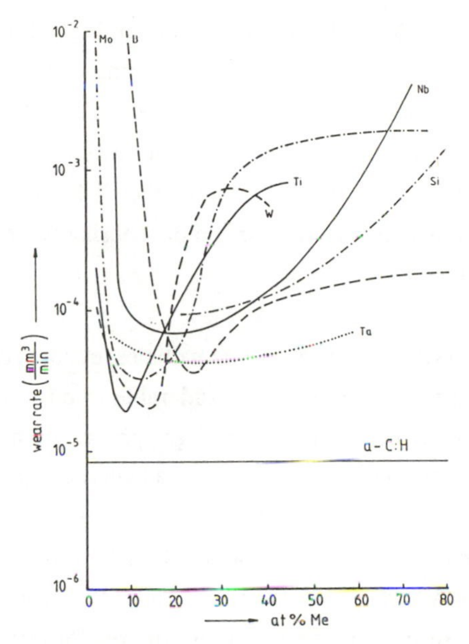

Fig. 22: Compositional dependence of abrasive wear rates of Me-C:H films, determined using the "Calo-Test" apparatus with aquous suspensions of 5 μm Al_2O_3 particles. Elemental targets were used for the film preparation except for Ta-C:H and B-C:H, where TaC and B_4C were used. Only Ti-C:H films were DC-magnetron sputtered (0.8 Pa total pressure, substrate-to-target distance d_{TS} 63 mm, substrate electrically floating), RF-diode sputtering was used otherwise (2 Pa, d_{TS} = 55 mm, +20 V substrate bias). The horizontal line refers to an a-C:H film deposited on the powered electrode cathodic bias in a parallel plate RF-diode reactor from acetylene [18]

The width of wear rate curves in Fig. 22 increases from Ti (subgroup IVa of the periodic system) over Mo and W (Va) to Nb and Ta (VIa). Within one subgroup the minimum wear rate decreases with increasing atomic mass (compare Mo and W, V with a minimal wear rate $1*10^{-4}$ mm^3/min. which is not plotted, Nb and Ta). A comparison with the horizontal line for "cathodic" a-C:H shows, that Me-C:H films can be prepared with wear resistances being only a factor of two smaller. On the other hand the adhesion of Me-C:H to a steel substrate is much better than for a-C:H because a graded junction can be prepared which is very advantageous for technical applications.

Fig. 22 shows that the metal-free films deposited on a non-powered electrode have wear rates at least four orders of magnitude beyond that of a-C:H grown under strong ion-bombardment. Why the formation of carbidic precipitates leads to the strong wear reduction observed, can not be satisfactorily explained.

A sequence of minimal wear rates which differs from that in Fig. 22, was reported by Bergmann and Vogel [69] for Me-C:H with Me = Cr, Ti, and W. According to their abrasive wear measurements, performed with unrotating spheres sliding on coated disks in a paste consisting of grinding oil and SiC powder, Ti-C:H has a significantly smaller wear resistance than W-C:H, whereas with Cr-C:H a value between the two others was found. Possibly the differences in the wear tests used are responsible for the deviating results, as abrasive and decohesive wear processes enter with different weights into the results of either test.

The bombardment of the growing film by energetic particles, either neutral ones or positively charged ions, during the deposition process can play a considerable role for the wear resistance of Me-C:H coatings. A suitable measure is the product of total gas pressure p and target-to-substrate distance d_{TS}. This product ($p{\cdot}d_{TS}$) determines the energy loss of an energetic particle travelling from the target to the substrate and being slowed down by elastic collisions with the sputter gas atoms. According to the results of Monte Carlo calculations carried out by Somekh [70], for instance a sputtered W atom with 17 eV initial energy arrives at the substrate with only 25 % and 3 % of this energy, if $p{\cdot}d_{TS}$ is equal to 70 and 190 Pamm, resp.. The corresponding figures for reflected Ar neutrals with 340 eV (1100 eV) initial energy are 75 % (85 %) and 38 % (65 %), resp.. These numbers illustrate the decrease of collision cross sections with increasing projectile energy.

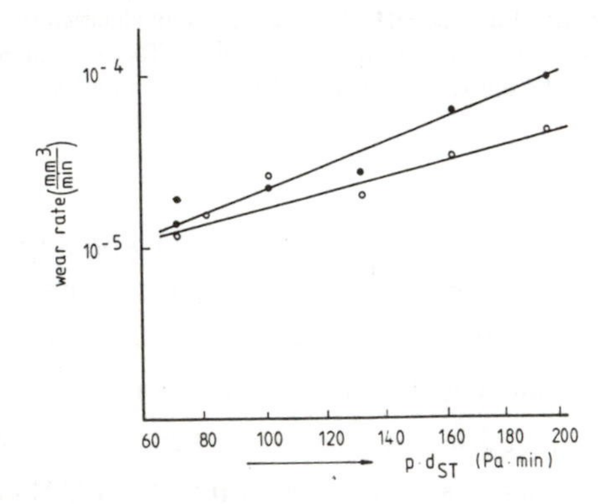

Fig. 23: Dependence of abrasive wear rates of Ta-C:H films with 12 to 13 at% Ta (closed symbols) and 13 to 15 at% Ta (open symbol) on the pressure-distance product $p \cdot d_{TS}$ used for the preparation [18]

The influence of particle bombardment on wear rate has been investigated in two series of experiments with Ta-C:H systems [62]. According to Fig. 23 the wear resistance of reactively sputtered Ta-C:H depends indeed significantly on $p{\cdot}d_{TS}$; i. e. smaller pressure-distance products allow preparation of films with better abrasive wear properties. Evidently a "strengthening" against abrasive attack by impact of energetic particles plays a very important role in this material. As discussed in section VII a pronounced $p{\cdot}d_{TS}$ dependence was also detected - taking W-C:H as an example - in hardness and Young's modulus; i. e. a decrease of pd_{TS} yields harder and stiffer films, presumably due to enhanced crosslinking of the films as a consequence of increased bombardment during growth.

The effect of energetic particle bombardment is considerably reduced, if the substrate is no longer parallel to the target surface. In Fig. 24 data of abrasive wear rates are presented of Ta-C:H films prepared at different substrate voltages and different angles between substrate and target plane. With positive bias voltage, at which bombardment of the growing film with positive ions is excluded, the flux of energetic neutrals per unit area of the substrate is decreased considerably by a tilt of 75° or 90°. A tilt of 90° does not completely exclude the impingement of energetic particles, because of an angular distribution of reflected projectiles. For the same reason, the energy flux onto the 90° substrate might even be larger than that onto the 75° substrate, depending on the details of the geometric arrangement of the substrates relative to the target. The reduced flux of energetic particles onto 75° and 90° substrates manifests itself in a marked increase of the wear rate of the layers. Simultaneously, the amount of Ta incorporated is strongly decreased compared to films on 0° and 45° substrates (cp.

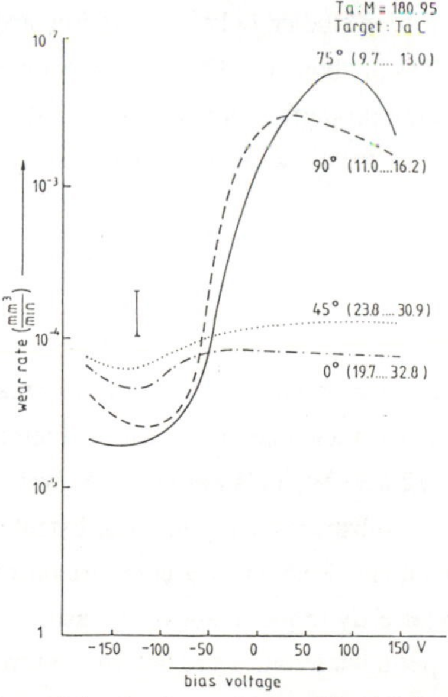

Fig. 24: Substrate bias voltage dependence of the abrasive wear rate of Ta-C:H films deposited by reactive RF-sputtering from a TaC target on inclined substrates. Ranges of Ta-contents (at%) are given in parentheses. [18]

ranges of at% quoted in brackets and Fig.1), presumably due to the fact, that trajectories of the heavy Ta atoms are hardly disturbed by collisions with light Ar atoms, in contrast to carbon containing species, which are more strongly scattered in the Ar atmosphere. The difference in Ta content alone, however, is not sufficient to explain the marked differences in wear resistance (cp. Fig. 22).

The situation changes considerably, if a negative DC voltage is applied to the substrate. Positive ions are now attracted from the plasma to the growing film, leading to a decrease in wear rate, which is especially pronounced for the 75° and 90° substrates, thus compensating for the negative effects of a reduced neutral flux. This experiment demonstrates nicely the necessity of controlling ion bombardment via the substrate bias if films are deposited on nonplanar substrates.

The abrasive wear rate of standard W-C:H films was compared to that of TiN and TiAlN PVD layers using a ball-on-disc arrangement with an SiC abrasive suspension. In these experiments W-C:H proved to be worn a factor of two slower than TiA1N and eight times slower than TiN (Dimigen, Hübsch, unpublished).

Finally the question arises whether the wear behavior of Me-C:H films and especially its dependence on metal concentration (Fig. 22) can be linked to the mechanical properties such as hardness and Young modulus. A comparison of the influence of increasing metal content on wear rate (Fig. 22) and hardness or Young's modulus (Figs. 14 and 16) indicates, that elastic and yield behavior of Me-C:H alone is not sufficient to explain the tribological properties of these films. Until now, fracture properties of Me–C:H layers have been studied only very preliminarily (fracture strain measurements on Fe-C:H yielded maximum values of about 1 % [71]), a closer investigation of this topic appears to be indispensible. A study of fracture properties should also include the question of internal stresses, which might play a considerable role for the wear resistance. ('compressive strengthening') [72].

b2. Adhesive Wear

Adhesive wear processes between two bodies involve the dissipation of frictional work in thin regions near the interface. Transfer wear and chemical wear are their main constituents [54]. Adhesive wear data of a W-C:H coating (12 at% W) on steel (1.3505) were compared to those of TiN and TiAlN PVD layers and of the bare substrates using uncoated steel and cemented carbide (WC/Co) spheres in a ball-on-disc arrangement. Results of a comparison of these coatings with regard to sliding friction coefficients were already reported above. Measurements were performed in ambient atmosphere. Volume wear rates per track length were determined from the dimensions of the calotte formed on the ball and the track profiles formed on the disc. Film and substrate wear rates are plotted in Fig. 25 on a logarithmic scale. Compared with the steel balls sliding on a bare steel substrate, a reduction of the wear by nearly two orders of magnitude can be brought about by a W-C:H coating with an optimum W-content. Wear reduction by TiN is less than an an order of magnitude, TiAlN coating is of no advantage in this situation. Sliding against cemented carbide balls, W-C:H reduces the wear rates by a factor of about 3, whereas the application of TiN and TiAlN is even disadvantageous.

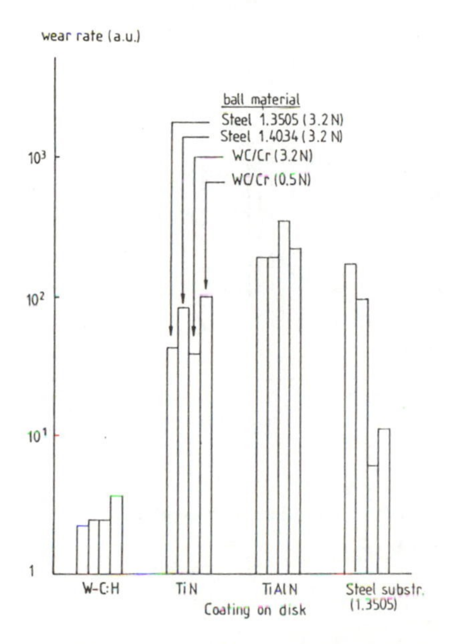

Fig. 25: Adhesive wear rates (a.u.) of W-C:H, TiN, and TiAlN films and bare steel substrates sliding against uncoated steel and cemented carbide balls [18]

The excellent tribological behavior of W-C:H is even more pronounced, compared to the nitrides, if the wear of the uncoated ball is considered (cp. Fig. 26). Cemented carbide spheres were not measurably attacked during sliding against W-C:H, in contrast to steel, TiN or TiAlN; the adhesive wear of steel balls is reduced by three to four orders of magnitudes, compared with the bare substrate or Ti-based nitride coatings. The strong adhesive attack of nitride coatings on uncoated sliding counterparts is a consequence of extensive material transfer onto the coating. In sliding friction measurements this transfer manifests itself in high friction coefficients. The neglegible transfer to Ta-C:H in contrast to TiN is nicely documented in the Fig. 26, showing a micrograph of a disk coated partly with TiN and partly with Ta-C:H which was taking after sliding against a steel ball. A track of transferred iron is clearly to be seen on TiN (right side of Fig. 27) but not on Ta-C:H [74].

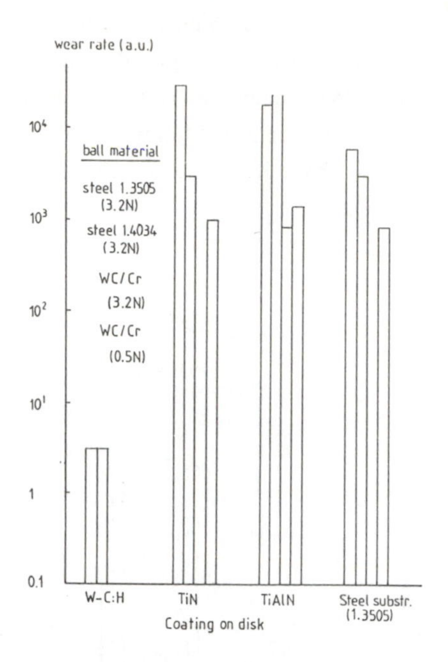

Fig. 26: Adhesive wear rates (a.u.) of uncoated steel and cemented carbide balls sliding against W-C:H, TiN, and TiAlN films and bare steel substrates [18]

Fig. 27: Micrograph of a disc wated partly with TiN (right side) and Ta-C:H (13 at% Ta) (left side); after sliding against a steel ball (100 Cr 6, 5 mm diameter, load 100 g)

The advantages of W-C:H over TiN and TiALN with regard to adhesive wear are clearly demonstrated by Figs. 25-27. In situations where considerable frictional work is dissipated in the sliding contact, however, the applicability of Me-C:H films will be limited, due to its thermal sensitivity, to temperatures below 300 °C in ambient atmosphere and 450 °C in a non-oxidizing gas.

CONCLUSION

It has been shown that the electrical properties can be interpreted in terms of a hopping mechanism similar as in granular metals. The tribological properties such as friction and wear resistance were compared with the mechanical properties. It was not possible, however, to find a relation between abrasive wear resistance and hardness of the films. These properties are not yet understood and further investigations are required.

ACKNOWLEDGEMENTS

The authors are indebted to Dr. Dimigen and coworkers for making many experimental results available prior to publication. This work was in part sponsored by the German Federal Minister for Research and Technology (BMFT) under grant No. 13 N 5374/7, 13 N 5609/6 and 03T0004D5.

VI. REFERENCES

[1] Angus, J.C., Koidl, P. and Domitz, S., Plasma Deposited Thin Films, in: (J. Mort and Jansen, F., Eds), CRC Press, Inc., Boca Raton, Florida 1986

[2] Tsai, H. and Bogy, D.B., J. Vac. Sci. Technol., 1987, A5, 3287

[3] Robertson, J., Adv. Phys., 1986, 35, 317

[4] Sundgren, J.E., Hentzell, H.T.G., J. Vac. Sci. Technol. 1986, A4, 2259

[5] Dimigen, H. and Hübsch, H., Philips Techn. Rev., 1983/4, 41, 186

[6] Dimigen, H., Hübsch, H. and Memming, R., Appl. Phys. Lett., 1987, 50, 1056

[7] Boonthanom, N. and White, M., Thin Solid Films, 1974, 24, 295

[8] Al-Ismail, S.A.Y. and Hogarth, C.A., J. Mater. Sci., 1987, 22, 415

[9] Nowak, R.J., Schultz, F.A., Umana, M., Lam, R. and Murray, R.W., Anal. Chem., 1980, 52, 315

[10] Kny, E., Levenson, L.L., James, W.J. and Auerbach, R.A., Thin Solid Films, 1981, 85, 23

[11] Suhr, H., Etspü´ler, A., Feurer, E. and Oehr, C., Plasma Chem. Plasma Proc., 1988, 8, 9

[12] Shuttleworth, D., J. Phys. Chem., 1980, 84, 1629

[13] Roy, R.A., Messier, R. and Cowley, J.M., Thin Solid Films, 1981, 79, 207

[14] Roy, R.A., Messier, R. and Krishnaswamy, S.V., This solid Films, 1983, 109, 27

[15] Asano, Y., Thin Solid Films, 1983, 105, 1

[16] Yoneda, T., Hori, M., Yamada, H., Morita, S. and Hattori, S., Proc. of Int. Symp. Plasma Chem., ISPC-7, Eindhoven 1985, Paper number P-6-4, p. 1278

[17] Dimigen, H., Hübsch, H. Schaal, U. and Schön, D., Proc. of the 4th Int. Congr. on Surface Technology, Berlin 1987, p. 373

[18] Dimigen, H., Hübsch, H., Schaal, U., Schön, D., Spolert, M., Tode, T. and Werner, W. unpublished results

[19] Berg, S., Larsson, T., Nender, C. and Blom, H.-O., J. Appl. Phys., 1988, 63, 887

[20] Dimigen, H. and Schaal, U., unpublished results

[21] Memming, R., Thin Solid Films, 1986, 143, 279

[22] Dischler, B., Bubenzer, A. and Koidl, P., Sol. State Comm. 1983, 48, 105

[23] Gerstenberg, K.W. and Grischke, M., to be published

[24] Köberle, H., Grischke, M., Thieme, F., Benndorf, C. and Memming, R., Proc. Int. Conf. on Metallurgical Coatings, San Diego, USA, 1989

[25] Köberle, H., PhD Thesis, Hamburg 1989

[26] Grischke, M., Brauer, A., Benndorf, C., Memming, R. and Thieme, F., Proc. of E-MRS, Les Editions de Physique 1987, 17, 491

[27] Benndorf, C., Grischke, M., Köberle, H., Memming, R., Brauer, A. and Thieme, F., Surface and Coatings Tech. 36, 1988, 171

[28] Grischke, M., Harnack, J.T., Benndorf, C. and Thieme, F., PSE, Proc. First Int. Conf. on Plasma Surface Eng., Garmisch Partenkirchen 1988, in press

[29] Grischke, M. PhD Thesis, University of Hamburg 1989, to be published as VDI-Fortschrittsberichte, Ser. 5

[30] Bubenzer, A., Dischler, B., Brandt, G. and Koidl, P., J. Appl. Phys. 1983, 54, 4590

[31] Biedermann, H., Martino, L., Slavínská, D. and Chudaced, I., Pure & Appl. Chem., 1988, 60, 607

[32] Perrin, J., Despax, B. and Kay, E., Phys. Rev., 1985, B 32, 719

[33l Craig, S. and Harding, G.L., Thin Solid Films, 1983, 101, 97

[34] Biedermann, H., Chjok, Hong jon, Martino, L., David, J., Kadlec, S. and Lukác, P., Proc. of E-MRS, Les Editions de Physique, 1987, 17, 499

[35] Biedermann, H., Howson, R.P. and McCabe, I., Proc. of IPAT ´87 (6th Int. Conf. on Ion & Plasma Assisted Techniques), Brighton 1987, p. 152

[36] Meyerson, B. and Smith, F.W., Sol. State Comm., 1980, 34, 531

[37] Winterfeld, H.J., Diploma Thesis, FH Lübeck, 1986

[38] Laurent, C. and Kay, E., J. Appl. Phys., 1989, 65, 1717

[39] Perrin, J., Despax, B., Hanchett, V. and Kay, E., J. Vac. Sci. Technol., 1986, A2, 46

[40] Niklasson, G.A. and Buhrman, R.A., J. Appl. Phys., 1984, 55, 3382

[41] Köberle, H. and Memming R., Proc. of E-MRS, Les Editions de Physique, 1987 , 17, 485

[42] Köberle, H., Grischke, M., Thieme, F., Benndorf, C. and Memming, R., Proc. of Metallurgical Coatings 1989, in press

[43] Klages, C.-P., Köberle, H., Bauer, M and Memming, R., Proc. of 1st Int. Symp. on Diamond and Diamond-Like Films, (Los Angeles) 1989,Vol. 89-12, 225

[44] Abeles, B., in Wolfe, R. (Ed.), Applied Solid State Science, Vol. 6, Academic Press, New York - San Francisco - London, 1976, p.1

[45] Mostefa, M. and Olivier, G., Physica , 1986, 142 B, 80

[46] Mott, N.F., J. Non-Cryst. Sol. 1987, 1, 1

[47] Klug, H.P. and L.E. Alexander, X-Ray Diffraction Procedures, John Wiley & Sons, New York - London 1974

[48] Kosolapova, T.Ya., Carbides, Plenum Press, New York - London 1971, p. 36

[49] Weast (Ed), R.C., CRC Handbook of Chemistry and Physics, CRC Press, Inc., Boca Raton, Florida 1981, p. E-79

[50] Joffe, A.F., Regel, A.R., Prog. Semicond. 1960, 4, 237

[51] Nagels, P., in Amorphous Semiconductors (Brodsky, M.H. ed), Springer Verlag, Berlin, 1970

[52] Laurent, C., Mauri, D., Kay, E. and Parkin, S.S.P., J. Appl. Phys., 1989, 65, 2017

[53] Köberle, H. unpublished results

[54] Briscoe, B., Tribology International August 1981, 231

[55] Zum Gahr, K.-H., Microstructure and wear of materials, Elsevier, Amsterdam - Oxford - New York - Tokyo 1987

[56] Hornbogen, E. Metall 1980, 34, 1079

[57] Fryda, M., Taube, K. and Klages, C.-P., Proc. of the 11th Int. Vac. Congress (IVC11), Köln 1989, to be published in: Vacuum (GB)

[58] Wierenga, P. and Franken, A.J.J., J. Appl. Phys., 1984, 55, 4244

[59] Doerner, M.F. and Nix, W.D., J. Mater. Res., 1986, 1, 601

[60] Weissmantel, C., Breuer, K. and Winde, B., Thin Solid Films, 1983, 100, 383

[61] Kerner, E.H., Proc. Phys. Soc., 1956, B 69, 808

[62] Dimigen, H., Hübsch, H., Schaal, U. and Schön, D., unpublished results

[63] Memming, R., Tolle, H.J. and Wierenga, P.E., Thin Solid Films, 1986, 143, 31

[64] Enke, K., Dimigen, H. and Hübtsch, H., Appl. Phys. Lett., 1980, 36, 291

[65] Buckley, D.H., Surface Effects in Adhesion, Friction, Wear and Lubrication, Elsevier, Amsterdam 1981

[66] Dimigen, H., private communication

[67] McKenzie, D.R., Smith, G.B. and Liu, Z.Q., Phys. Rev., 1988, B 37, 8875

[68] Smith, G.B., McKenzie, D.R. and Martin, P.J., phys. stat. sol.(b), 1989, 152, 475

[69] Bergmann, E. and Vogel, J., J. Vac. Sci. Technol., 1986, A4, 2867

[70] Somekh, R.E., J. Vac. Sci. Technol., 1984, A2, 1285

[71] Taube, K., Fryda, M. and Klages, C.-P., unpublished results

[72] De Beurs, H. and De Hosson, J.Th.M., Appl. Phys. Lett., 1988, 53, 663

[73] Taube, K. and Klages, C.-P., VDI-Berichte, 1988, 702, 183

[74] Dimigen, H. and Hübsch, H., unpublished results

Materials Science Forum Vols. 52 & 53 (1989) pp. 645-656
Copyright Trans Tech Publications, Switzerland

PLASMA-DEPOSITED AMORPHOUS HYDROGENATED CARBON FILMS AND THEIR TRIBOLOGICAL PROPERTIES

K. Miyoshi, J.J. Pouch and S.A. Alterovitz
National Aeronautics and Space Administration
Lewis Research Center, Cleveland, OH 44135, USA

ABSTRACT

Recent work on the properties of "diamondlike" carbon films and their dependence on preparation conditions are reviewed. The results of the study indicate that plasma deposition enables one to deposit a variety of amorphous hydrogenated carbon (a-C:H) films exhibiting more diamondlike behavior to more graphitic behavior. The plasma-deposited a-C:H can be effectively used as hard, wear-resistant, and protective lubricating films on ceramic materials such as Si_3N_4 under a variety of environmental conditions such as moist air, dry nitrogen, and vacuum.

INTRODUCTION

Carbon films exhibiting unique properties can be formed on different substrates by ion-beam deposition, ion-beam sputtering, and plasma deposition of gaseous hydrocarbons [1 to 8]. The properties are sensitive to the deposition conditions. These resulting films can exhibit high electrical resistivity, semitransparency, mechanical hardness, and chemical inertness. The carbon films show promise as wear-resistant, hard solid lubricating coatings for mechanical systems such as bearings and optical components. In addition, carbon films are useful as gate dielectrics and passivating layers in semiconductor device processing, insulators for metal-insulator-metal fabrication, and masks for nanometer lithography [9 to 11].

This chapter is principally concerned with the chemical, physical, and tribological characteristics of amorphous hydrogenated carbon (a-C:H) films grown on different substrates (Si_3N_4, GaAs, InP, Si, and fused silica) by means of plasma chemical vapor deposition at 30 kHz. The influence of growth conditions on the chemical and physical properties of these films was studied by Auger electron spectroscopy (AES), secondary ion mass spectroscopy (SIMS), x-ray photoelectron spectroscopy (XPS), ellipsometry, and N^{15} nuclear reaction techniques. The nuclear reaction techniques provide the hydrogen concentration information. These analysis techniques and procedures are described in references 12 to 18. Tribological studies have also been conducted with the a-C:H films to better understand those chemical and physical properties of the films that will affect their tribological behavior when in contact with a ceramic material. The friction, wear, and lubricating behavior of the a-C:H films were examined with flat specimens (composed of an a-C:H film and Si_3N_4 substrate) in contact with Si_3N_4 riders in two processes. The first was done in dry nitrogen gas in moist air to determine the environmental effects on friction and resistance to wear of the a-C:H films. The second was done in an ultrahigh vacuum system to determine the effect of temperature on adhesion and friction of a-C:H films.

AMORPHOUS HYDROGENATED CARBON (a-C:H) FILMS

Plasma Deposition

Amorphous hydrogenated carbon films were formed on the different substrates from the 30 kHz ac glow discharge by using a planar plasma reactor [12 to 18]. All substrate materials were first cleaned in acetone and ethanol baths and then rinsed in deionized water. The substrates were placed on the ground anode of the parallel plate reactor in the chamber; the upper electrode was capacitively coupled to the 30 kHz power source. The background pressure was typically 2.7 Pa (20 mtorr). The gas sources were CH_4 and C_4H_{10} (methane and butane, 99.97 percent pure). The deposition gas (CH_4 or C_4H_{10}) was used to flush the system three times prior to each run. The chamber pressure was controlled by the input gas flow rate and pumping speed. The power density and flow rate settings covered the ranges 0.4 to 5 kW m^{-2} (25 to 300 W) and 3 to 9×10^{-5} m^3 min^{-1} (30 to 90 SCCM), respectively. The initial substrate temperature was 25 °C, and it increased a few degrees during each deposition.

The film growth rate varied monotonically with deposition power. Figure 1 illustrates the typical dependence of growth rate on deposition power. The specimens are a-C:H films grown on InP substrates using a CH_4 flow rate of 70 SCCM. This growth rate increases from 5 to 27 nm min^{-1} as the power increases from 25 to 300 W.

Film Characteristics

The AES and XPS measurements indicated that the a-C:H films contained only carbon; no other element was observed to the detection limits (0.1 at. %) of the instrument [12 and 13]. Figure 2 shows a typical AES profile of a-C:H films on the InP and GaAs. Oxygen was not present in the films, but there was a small percentage of oxygen at the carbon-InP interface. This suggests that the CH_4 (methane) plasma removes all of the native oxides from the GaAs surfaces and most of it from the InP surfaces.

Relative counts of hydrocarbon ions sputtered from a-C:H films deposited on the InP substrate were determined by means of SIMS depth-profiling studies performed with 3 keV Ar^+ ions [12]. In figure 3(a) the distribution of ion counts is plotted as a function of mass-to-charge ratio for various deposition conditions using a C_4H_{10} plasma. The predominant ion is CH^+; it is interesting that a higher CH^+ level is obtained from films produced at the higher power densities. Additional ions are presented in figure 3(a): CH_2^+, CH_3^+, C_2H^+, $C_2H_2^+$, and $C_2H_3^+$.

The ion distributions extracted from a-C:H films prepared by a CH_4 (methane) discharge are shown in figure 3(b). Evidently, CH^+ has a higher probability of being sputtered from each film. At 50 SCCM (32.7 Pa), more CH^+ is generated from the a-C:H deposit made at 2.45 kW m^{-2}. In addition, some of the films obtained from the C_4H_{10} discharge (fig. 3(a)) have higher amounts of incorporated C_2H_3 relative to the CH_4-derived films (fig. 3(b)). Figure 3 indicates that the lowest populations are associated with CH_3^+. The ion distributions thus reflect some of the bonding arrangements that result from the interaction of the plasma radicals with the growing film [19 to 21].

A SIMS depth profile (3 keV Ar^+ ions) of carbon deposited onto GaAs using C_4H_{10} is presented in figure 4. The CH_x^+ ($x = 0,1,2,3$) distributions are uniform in the bulk of the film, and they drop to lower levels in the vicinity

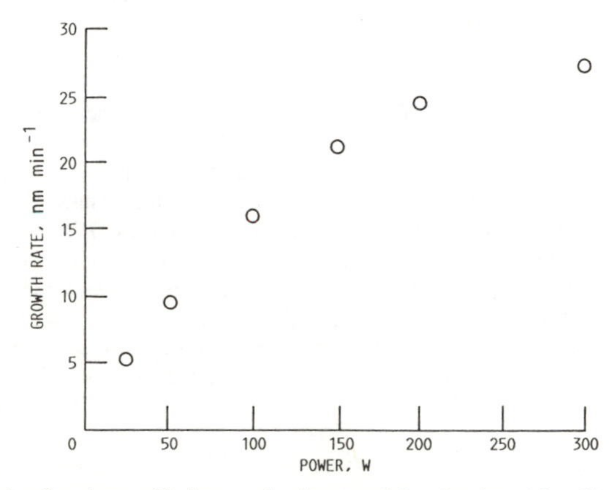

Figure 1.—Growth rate of hydrogenated carbon on *n*-InP as function of deposition power.

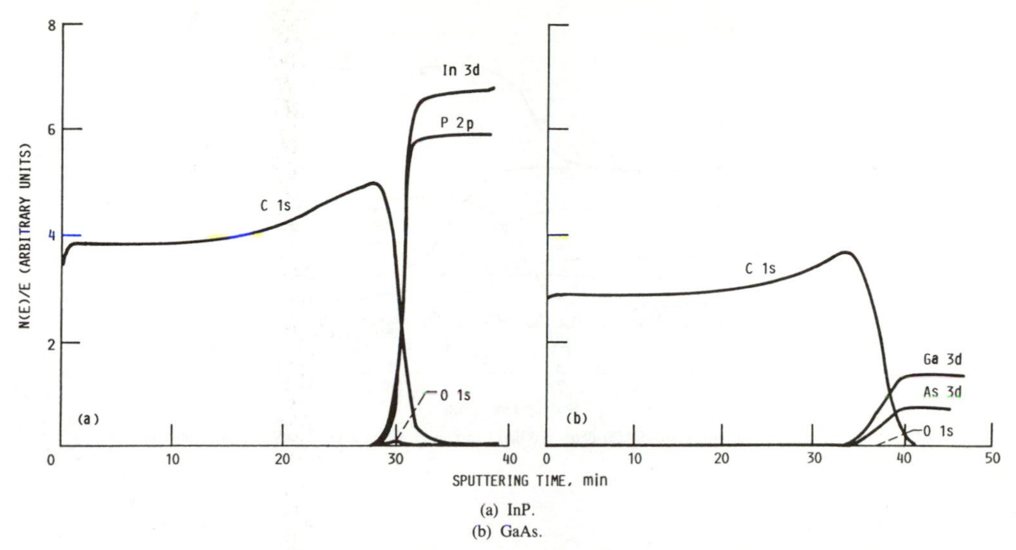

(a) InP.
(b) GaAs.

Figure 2.—Auger electron spectroscopy (AES) profiles of carbon films on InP and GaAs, using 25 mA 3 keV Ar^+ ions.

of the carbon-GaAs interface. Moreover, figure 4 shows oxygen to be present throughout the film. This determination cannot be made with the less-sensitive AES technique. It is apparent that Ga^+ and As^+ are readily detected as the carbon film is sputtered away.

The argon ion etching rate of the a-C:H films is shown as a function of deposition power in figure 5(a). An inverse relationship between argon ion etching rate and deposition power is observed. The etching rate drops from 80 to 50 nm min^{-1} when the deposition power is increased from 25 to 300 W. This suggests that films grown at higher powers are denser than those grown at lower powers. Figure 5(b) shows the nuclear reaction analyses data. The hydrogen concentration in the carbon film decreases slightly with increasing power. The hydrogen concentrations are in the 7.2 to 7.7×10^{22} cm^{-3} range, which gives an approximate value of 0.8 for x in the formula CH_x.

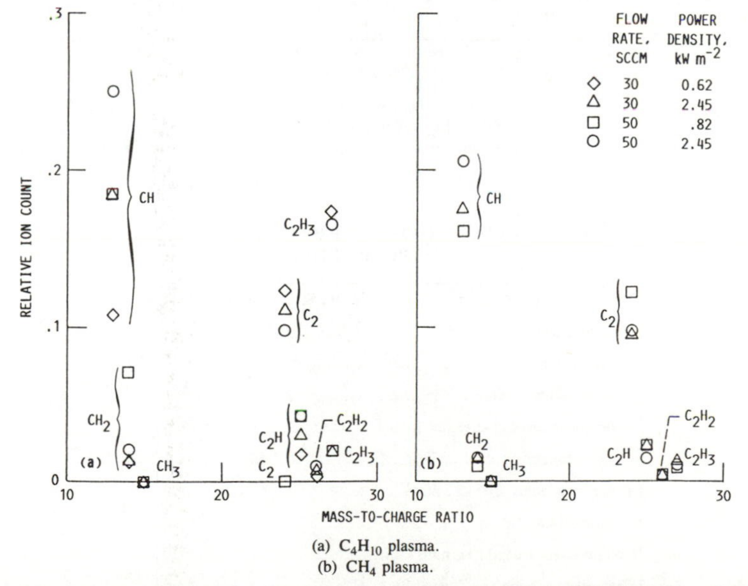

(a) C_4H_{10} plasma.
(b) CH_4 plasma.

Figure 3.—Relative ion count as function of mass-to-charge ratio for carbon deposited onto InP using C_4H_{10} and CH_4 plasmas.

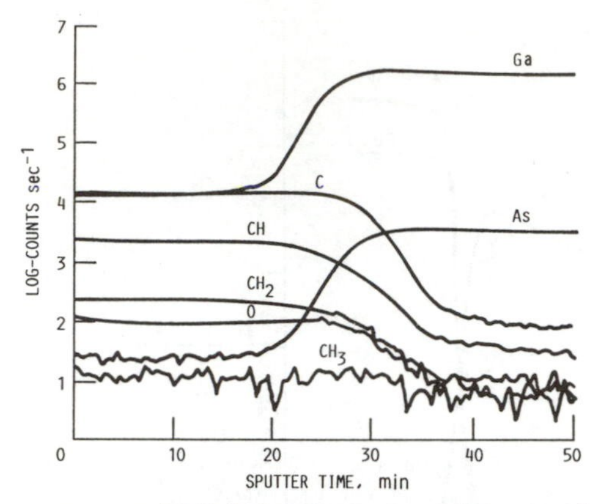

Figure 4.—Secondary ion mass spectroscopy (SIMS) depth profile of carbon on GaAs using 3 keV Ar^+ ion bombardment.

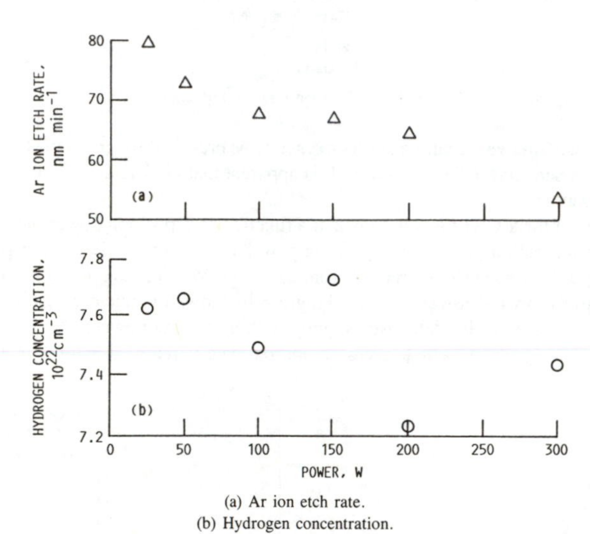

(a) Ar ion etch rate.
(b) Hydrogen concentration.
Figure 5.—Ar ion etching rate and hydrogen concentration as function of deposition power for carbon film grown on *n*-InP.

TABLE I.—COMPOSITION AND PROPERTIES OF HOT-PRESSED SILICON NITRIDE

Nominal composition, wt %	92 Si_3N_4-4MgO-4Y_2O_3
Structural phase	β-Phase
Density, g cm^{-3}	3.27
Vickers hardness, GPa	16.1
Three-point bending strength, MPa	980
Fracture toughness, MN $m^{-3/2}$	9.4
Young's modulus, 10^4 kg mm^{-2}	2.9
Poisson's ratio	0.27
Compressive strength, kg mm^{-2}	330
Thermal expansion coefficient, 10^{-6} °C^{-1}	3.6

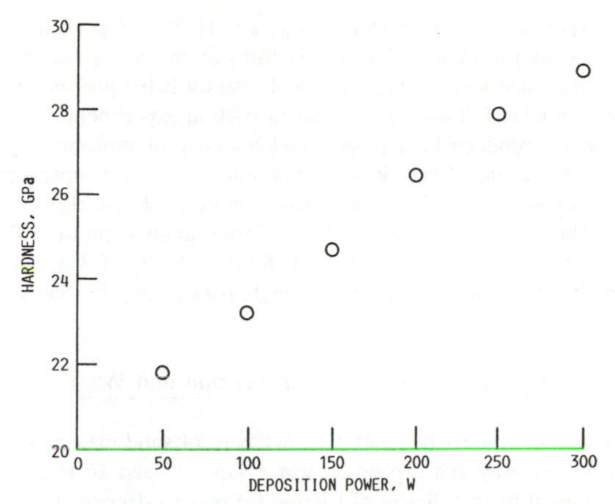

Figure 6.—Vickers hardness as function of deposition power for a-C:H film deposited on Si_3N_4 (hardness measuring load, 0.25 N; hardness of Si_3N_4 substrate, 17.1 GPa).

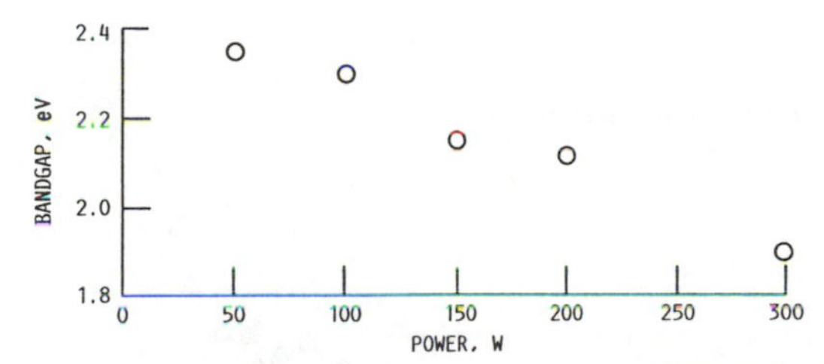

Figure 7.—Optical bandgap as function of deposition power for carbon film grown on *n*-InP.

The microhardness data measured for the carbon films on Si_3N_4 substrate (table I) at various deposition powers are presented in figure 6. The microhardness increases as the power increases. Thus it appears that a decrease in hydrogen concentration is accompanied by an increase in film density and/or c-c bondings, and in hardness.

The optical energy gap is shown as a function of the deposition power in figure 7. A decrease in the optical energy gap is clearly observed with increasing power.

At this stage, we will compare our results with conjecture forwarded by S. Kaplan et al. [22]. They claim that since double-bond hydrogenation is an exothermic process, ''graphitic'' behavior is favored over tetrahedral bonding in higher energy growth environments. They show evidence of this assumption by comparing a-C:H films made by five different experimental configurations. As a-C:H properties are dependent quite strongly on the many variables encountered in different preparation conditions, it seems that a better test of this assumption is in order. In addition, their results show a rather striking feature: a-C:H films exhibiting more ''diamondlike'' behavior (i.e., larger bandgap and more tetrahedral bonding) show a steep decrease in their hardness as compared with the more ''graphitic'' films. Our results confirm this model, including the hardness measurements. The higher the plasma deposition power, the more sp^2 versus sp^3 bonds are made, giving a more ''graphitic'' film, with smaller bandgap (fig. 7) and higher density and hardness (figs. 5 and 6, respectively).

TRIBOLOGICAL PROPERTIES

In the preceding section there have been indications that a-C:H films have diamondlike behavior in lower energy growth environments as compared with the more graphitic behavior in higher energy growth environments. Therefore, the objective of this section is to compare the tribological properties of a-C:H films made by different deposition powers.

Sliding friction and wear experiments were conducted with a-C:H films deposited on Si_3N_4 flats in contact with hemispherical Si_3N_4 riders (1.6 mm in radius). The a-C:H films on the Si_3N_4 flat substrates were approximately 0.06 μm. The Si_3N_4 used for flat substrates and hemispherical riders was hot pressed, and its composition and some of its properties are presented in table I. Two types of sliding friction experiments were conducted with the a-C:H films [23 to 25]. The first type was conducted in nitrogen and laboratory air atmospheres with a load of 1 N (Hertzian contact pressure, 910 MPa) and at a sliding velocity of 8 mm min^{-1} at room temperature. The specimen rider was made to traverse on the surface of a-C:H film. The motion was reciprocal. The a-C:H films were subjected to multipass sliding by the Si_3N_4 riders. The second type was conducted in ultrahigh vacuum (10^{-8} to 10^{-9} torr) with loads up to 1.7 N (Hertzian contact pressure, 1.5 GPa) and at a sliding velocity of 3 mm min^{-1} at temperatures up to 700 °C. In this case, the a-C:H films were subjected to single-pass sliding by the Si_3N_4 riders.

Environmental Effects on Friction and Wear

Environment significantly changes the friction and wear behavior of solid materials. Friction and wear of a-C:H films are consistent with this generality and depend on water vapor [1 and 26 to 28].

Figure 8 presents typical plots of the coefficient of friction for plasma-deposited a-C:H films at low (50 W) and high (250 W) deposition power as a function of the number of repeated passes in dry nitrogen and humid air environments. The values of coefficient of friction given are typical, but the trends with number of passes are quite reproducible. With the 50-W plasma-deposited a-C:H films, the coefficient of friction was generally found to increase,

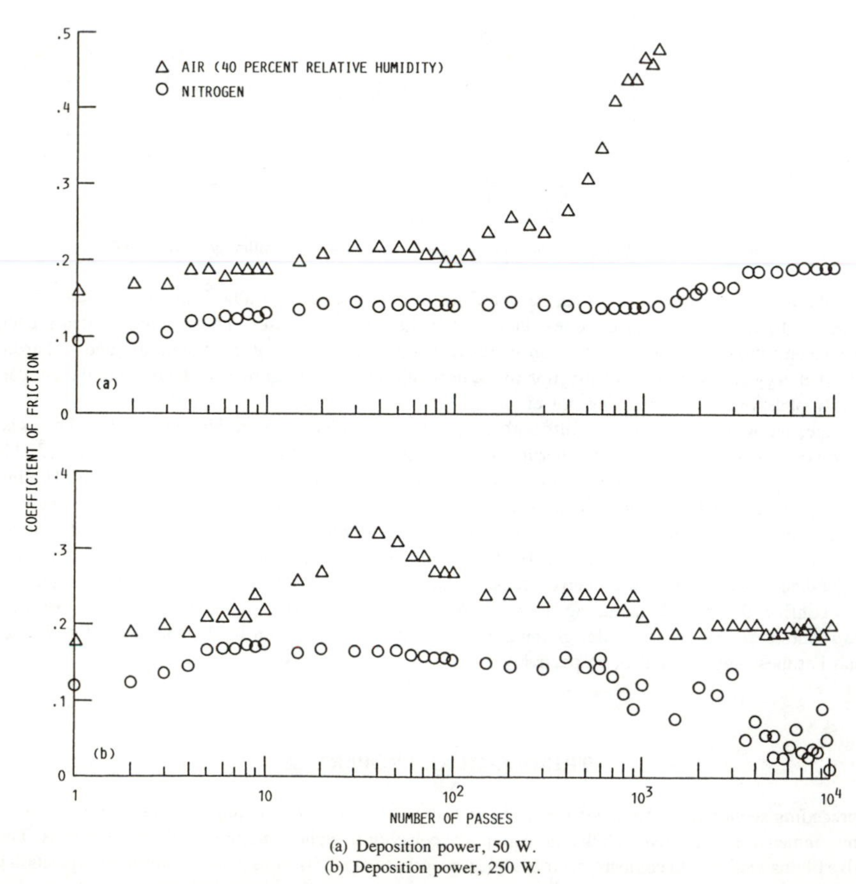

(a) Deposition power, 50 W.
(b) Deposition power, 250 W.

Figure 8.—Average coefficient of friction as function of number of passes of Si_3N_4 rider across a-C:H film surface in laboratory air and in dry nitrogen at deposition powers of 50 and 250 W.

as shown in figure 8(a). This increase, however, was small in a dry nitrogen environment even after it had been in contact with the rider for about 10 000 passes. On the other hand, in humid laboratory air a significant increase in the coefficient of friction occurred at about 500 passes and increased sharply to 1000 passes. The sliding action caused breakthrough of the film and removed it from the sliding contact area at about 1000 passes. Note that among the a-C:H films deposited at various deposition powers (50 to 300 W), the film deposited at 50 W has the lowest initial coefficient of friction (0.08 to 0.09) in the dry nitrogen environment. The value of the coefficient of friction was similar to that of a hemispherical diamond pin (radius, 0.2 mm) in sliding contact with a Si_3N_4 flat. The friction of the diamond was low (0.05 to 0.1) in dry nitrogen. It is well known that diamond has a low coefficient of friction in contact with various types of materials [29].

With the 250-W plasma-deposited a-C:H films (fig. 8(b)), although the coefficient of friction increased with increasing number of passes for about 10 passes in the dry nitrogen environment, it generally decreased in the range 10 to 10 000 passes.

At 600 to 700 passes, the coefficients of friction became very erratic and variable, as presented in figure 8(b). Optical microscopic examination indicated that some wear debris particles formed in the front region of the rider and on the wear track of the a-C:H film. Thus the wear particles so produced were caught up in the sliding mechanism and affected the coefficient of friction.

At 1000 passes and above, the coefficient of friction became low, but still variable (0.01 to 0.1). At this range the coefficients of friction for the 250-W plasma-deposited film were lower than those for the film deposited at 50 W.

In a humid air environment, the coefficients of friction for the 250-W plasma-deposited a-C:H film were higher as compared with those in dry nitrogen by a factor of 1.5 to 3 up to 10 000 passes. The film, however, did not wear off from the substrate even in the humid air environment.

Thus, water vapor greatly increases friction and reduces the wear life of plasma-deposited a-C:H film at low deposition powers. In general, a-C:H films deposited with lower deposition powers were more susceptible to water vapor, when compared with the films deposited with higher deposition power. Particularly, the deposition power greatly affects the wear life of the films in a humid air environment. The greater the deposition power (the more graphitic the film), the greater the wear life in humid air.

Annealing Effects on Friction and Wear

Thermal annealing significantly affects the properties of a-C:H films. For example, an abrupt decrease of the optical bandgap has been observed for the thermal annealing process [14].

Figure 9 presents the optical bandgap of the a-C:H films deposited on the quartz substrates by using 150-W, 70-SCCM-flow-rate CH_4 plasma as a function of annealing time at 400 and 600 °C. The thermal processing of the films was accomplished in nitrogen gas with tungsten halogen light. The main part of the reduction in the optical bandgap is obtained at short annealing time. This fact can also be deduced by the result obtained by laser annealing [30],

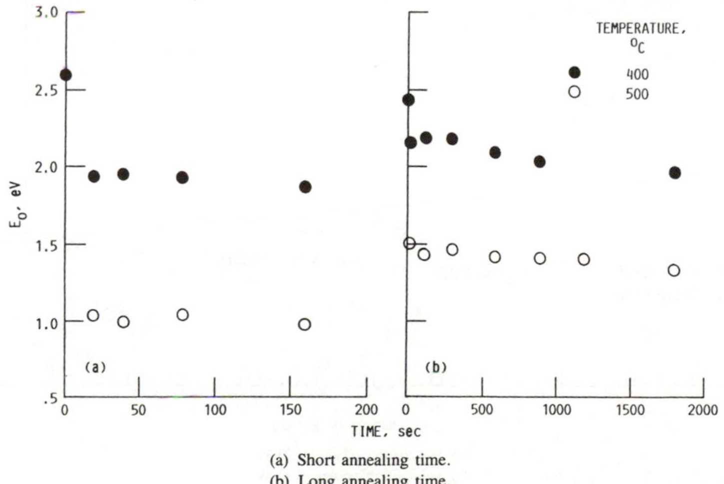

(a) Short annealing time.
(b) Long annealing time.

Figure 9.—Optical energy gap E_o as function of annealing time for a-C:H films on quartz annealed at two temperatures.

when processing time is much shorter than that reported here. The mechanism involved should be a two-step process. There is known to be a two-stage pyrolysis of organic material into graphite [31] for temperatures in this range, namely carbonization and polymerization. The carbonization stage includes loss of volatile matter, which we identify with hydrogen loss in this case [1]. This stage occurs in the temperature range 400 to 600 °C in a-C:H. The polymerization stage includes the formation of graphitic crystallites or sheets. If we assume that the polymerization is a diffusion-dependent process with a relatively long time constant (on the order of 10^3 sec), then we can deduce that the two processes of carbonization and polymerization occur simultaneously in our a-C:H films. The abrupt decrease of the bandgap versus time at very short processing time is due to the hydrogen loss, while the subsequent decrease in optical bandgap is due to an increase in cluster size [31].

Further, absorption in the UV-visible range was measured with a-C:H films on quartz substrates [14]. The absorbance-versus-wavelength plot at 600 °C shows a decrease in peak height and a shift in the peak position. The shift indicates changes in the carbon bonding. The decrease in peak height is attributed mostly to loss of material in this case.

Thermal annealing also changes the friction and wear characteristics of a-C:H films. Figure 10 presents the friction data for annealed a-C:H films in sliding contact with Si_3N_4 riders in dry nitrogen and humid air environments. The

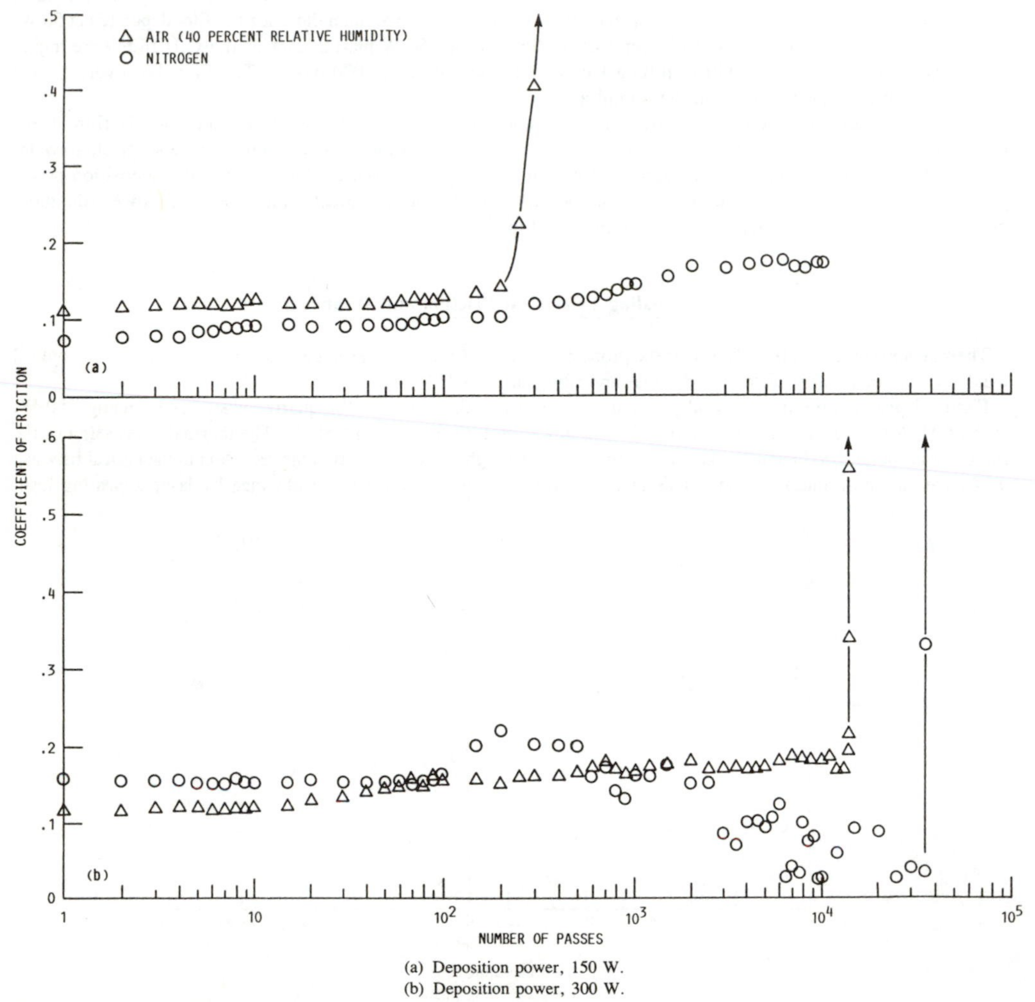

(a) Deposition power, 150 W.
(b) Deposition power, 300 W.

Figure 10.—Average coefficient of friction as function of number of passes of Si_3N_4 rider across a-C:H film surface in laboratory air and in dry nitrogen, after annealing a-C:H in vacuum at 700 °C, at deposition powers of 150 and 300 W.

annealing of the a-C:H films deposited on the Si_3N_4 substrates was accomplished at 700 °C in ultrahigh vacuum (10^{-8} to 10^{-9} torr).

Both in dry nitrogen and in humid air, the initial coefficients of friction for the annealed film deposited at 150 W (fig. 10(a)) were reduced by about a factor of 2 as compared with those for as-deposited film. The annealed film did not wear off from the substrate in dry nitrogen even after it had been in contact with the rider for about 10 000 passes, while in humid nitrogen it wore off at 300 passes and had shorter wear life.

With the a-C:H deposited at 300 W (fig. 10(b)), the results show an interesting feature: the annealed film exhibited more graphitic behavior; that is, at up to 100 passes the initial coefficients of friction in humid air were lower than those obtained in dry nitrogen. This is contrary to the results obtained from the as-deposited a-C:H films (e.g., fig. 10) and the annealed film at 150 W power (fig. 10(a)). Further, in the humid air environment the coefficients of friction were reduced by about a factor of 2 as compared with the as-deposited film. The generally accepted theory (that graphite lubricates because of adsorbed water or gaseous films) seems capable of explaining these results [32 and 33]; namely, the a-C:H film deposited at 300 W is believed to be more graphitic than the films deposited at lower power. Moreover, the annealing of the film gives a more graphitic film. Effective lubrication is possible with the very graphitic film provided both by the high-power plasma deposition and by the annealing process when an adsorbed water vapor film is present. Thus, the annealed a-C:H film deposited at 300 W has very graphitic friction behavior.

Temperature Effects on Adhesion and Friction in Vacuum

An increase in the surface temperature of a-C:H films tends to cause chemical changes, as discussed in the preceding subsection. These chemical changes can alter their friction and wear behavior. For simplicity of discussion, the effect of temperature on tribological properties of concern is investigated in a nonoxidizing environment (i.e., in an ultrahigh vacuum). The in situ friction experiments were conducted in a vacuum with the as-received plasma-deposited a-C:H films in contact with the ion-sputter-cleaned, hemispherical monolithic Si_3N_4 rider specimens.

Typical plots of the coefficient of friction for a-C:H films plasma-deposited at 150 and 300 W as a function of surface temperature are presented in figures 11(a) and (b), respectively. Comparative data for an uncoated Si_3N_4 flat in contact with a hemispherical Si_3N_4 rider are presented in figure 12.

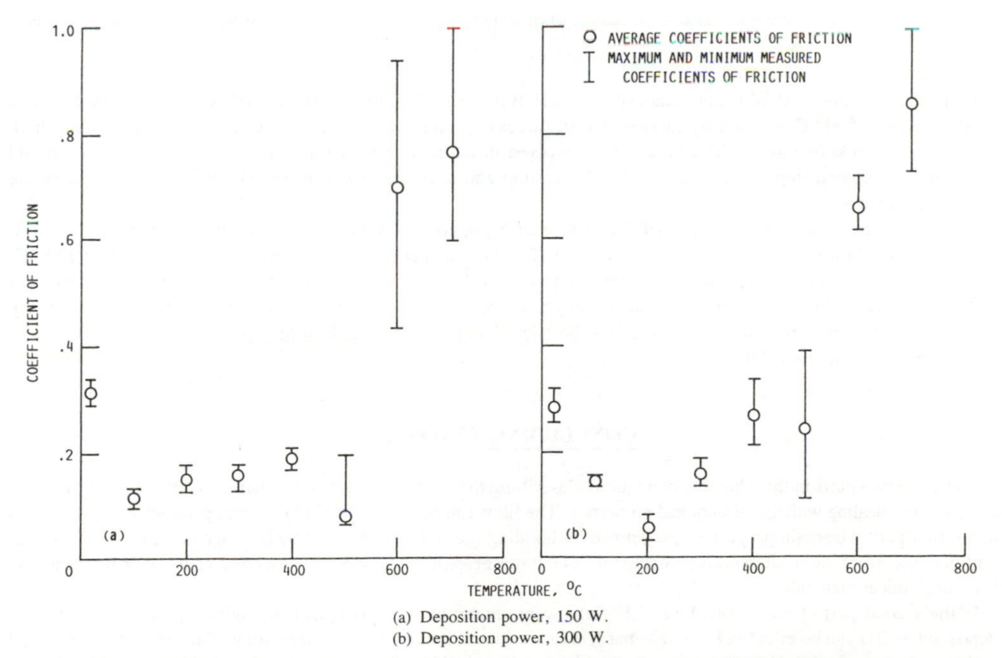

(a) Deposition power, 150 W.
(b) Deposition power, 300 W.

Figure 11.—Coefficients of dynamic friction as function of temperature for a-C:H films in contact with Si_3N_4 rider specimens in vacuum.

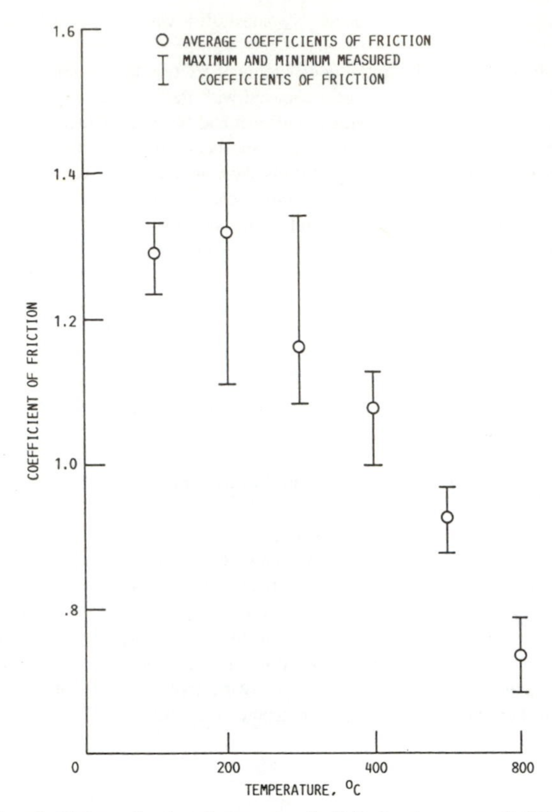

Figure 12.—Coefficients of dynamic friction as function of temperature for Si_3N_4 flats in contact with Si_3N_4 rider specimens in vacuum.

With the 150- and 300-W plasma-deposited a-C:H films (fig. 11), the coefficient of friction remained low at temperatures to 500 °C and rapidly increased with increasing temperatures at 600 °C and above, remaining high in the range of 600 to 700 °C. The mechanism involved in the rapid increase in friction at 600 to 700 °C should be related to the two-step process, namely carbonization and polymerization of a-C:H films, as discussed in the preceding subsection.

When compared with the coefficient of friction for Si_3N_4 in contact with Si_3N_4 itself (fig. 12), the coefficient of friction for a-C:H films in contact with a Si_3N_4 rider (fig. 11) was generally much lower at temperatures to 500 °C. It is also interesting to note that the coefficient of friction for the film deposited at 150 W had a very low coefficient of friction (about 0.08 at 500 °C) even in an ultrahigh vacuum environment (fig. 11(a)), and that the film effectively lubricated Si_3N_4 surfaces. Note that in vacuum the friction behavior of a-C:H film deposited at 50 W was similar to that shown in figure 11(a).

CONCLUDING REMARKS

There are two parts in this chapter, one mainly describing the growth and physical characterization of a-C:H films and the other dealing with the tribological properties. The films can be characterized by several parameters, as described in the first part. The main properties are defined by bonding ratios (sp^2/sp^3) and by hydrogen concentration, which will give the graphitic or diamondlike behavior. The more graphitic behavior is associated with lower bandgap and low mechanical etch rate.

In the second part of the work, the a-C:H films were shown to be capable of tribological applications. Plasma-deposited a-C:H can be effectively used as hard, wear-resistant, and protective lubricating films on ceramic material under a variety of environmental conditions such as moist air, dry nitrogen, and vacuum. More specifically, we found a very good correlation of the tribological behavior with the physical properties, as described in the first part of the work. For a-C:H films deposited at low power, which are more diamondlike, we found friction behavior similar to that of bulk diamond. The present experiments show that among the a-C:H films deposited at various deposition powers (50 to 300 W), the film deposited at 50 W had the lowest initial coefficient of friction in dry nitrogen. The value of the coefficient of friction (0.08 to 0.09) was similar to that of hemispherical diamond (radius, 0.2 mm) in sliding contact with a Si_3N_4 flat. Conversely, for the a-C:H films deposited at higher power, a graphitic tribological behavior was found. Effective lubrication is possible with the graphitic films like bulk graphite when adsorbed water vapor is present.

Lastly, a simple physical characterization of the films can partially predict the tribological properties.

REFERENCES

1. Angus, J.J., Koidl, P., and Domitz, S.: "Carbon Thin Films," in Plasma Deposited Thin Films, ed. by Mort, J. and Jansen, F. CRC Press Inc., 1986, pp. 89–127.
2. Aisenberg, S., and Chabot, R.: J. Appl. Phys., 1971, **42**, 2953.
3. Holland, L., and Ojha, S.M.: Thin Solid Films, 1978, **48**, L21.
4. Berg, S., and Andersson, L.P.: Thin Solid Films, 1979, **58**, 117.
5. Meyerson, B., and Smith, F.W.: Solid State Commun., 1980, **34**, 531.
6. Moravec, T.J., and Orent, T.W.: J. Vac. Sci. Technol., 1981, **18**, 226.
7. Vora, H., and Moravec, T.J.: J. Appl. Phys., 1981, **52**, 6151.
8. Banks, B.A., and Rutledge, S.K.: J. Vac. Sci. Technol., 1982, **21**, 807.
9. Khan, A.A., Woollam, J.A., Chung, Y., and Banks, B.A.: IEEE Electron Devices Lett., 1983, **4**, 146.
10. Khan, A.A., Woollam, J.A., and Chung, Y.: Solid-State Electron., 1984, **27**, 385.
11. Kikuchi, M., Hikita, M., and Tamamura, T.: Appl. Phys. Lett., 1986, **48**, 835.
12. Pouch, J.J., Warner, J.D., Liu, D.C., and Alterovitz, S.A.: Thin Solid Films, 1988, **157**, 97.
13. Warner, J.D., Pouch, J.J., Alterovitz, S.A., Liu, D.C., and Lanford, W.A.: J. Vac. Sci. Technol. A, 1985, **3**, 900.
14. Alterovitz, S.A., Pouch, J.J., and Warner, J.D.: "Rapid Thermal Annealing of Amorphous Hydrogenated Carbon (a-C-H) Films," in Rapid Thermal Processing of Electronic Materials, MRS Symp. Proc. Vol. 92, ed. by Wilson, S.R., Powell, R., and Davies, D.E., Materials Research Society, 1987, pp. 311–318.
15. Pouch, J.J., Warner, J.D., and Liu, D.C.: Carbon Films Grown from Plasma on III–V Semiconductors. NASA TM-87140, 1985.
16. Pouch, J.J., Alterovitz, S.A., Warner, J.D., Liu, D.C., and Lanford, W.A.: "Optical Properties of Hydrogenated Amorphous Carbon Films Grown from Methane Plasma," in Thin Films: The Relationship of Structure to Properties, MRS Symp. Proc. Vol. 47, ed. by Aita, C.R. and Sreeharsha, K.S., Materials Research Society, 1985, pp. 201–204 (Also, NASA TM-86995).
17. Alterovitz, S.A., Warner, J.D., Liu, D.C., and Pouch, J.J.: "Ellipsometric and Optical Study of Some Uncommon Insulator Films on III–V Semiconductors," in Dielectric Films on Compound Semiconductors, Electrochemical Society Symp. Proc. Vol. 86-3, ed. by Kapoor, V.J., Connolly, D.J., and Wong, Y.H., The Electrochemical Society, Pennington, NJ, 1986, pp. 59–77 (Also, NASA TM-87135).
18. Pouch, J.J., Alterovitz, S.A., and Warner, J.D.: "Optical and Compositional Properties of a-C:H and BN Films," in Plasma Processing, MRS Symp. Proc. Vol. 68, ed. by Coburn, J.W., Gottscho, R.A., and Hess, D.W., Materials Research Society, 1986, pp. 211–216 (Also, NASA TM-87258).
19. Benninghoven, A.: Surf. Sci., 1975, **53**, 596.
20. Wagner, J., Wild, Ch., Pohl, F., and Koidl, P.: Appl. Phys. Lett., 1986, **48**, 106.
21. Kobayashi, K., Mutsukura, N., and Machi, Y.: J. Appl. Phys., 1986, **59**, 910.
22. Kaplan, S., Jansen, F., and Machonkin, M.: Appl. Phys. Lett. 1985, **47**, 750.
23. Miyoshi, K., and Rengstorff, G.W.P.: Corrosion, 1989, **45**, 266.
24. Miyoshi, K., and Buckley, D.H.: Wear, 1986, **110**, 295.
25. Miyoshi, K., Pouch, J.J., Alterovitz, S.A., Pantic, D.M., and Johnson, G.A.: in Wear of Materials, Vol. 2, ed. by Ludema, K.C., ASME, 1989, p. 585.
26. Enke, K., Dimigen, H., and Hubsch, J.: Appl. Phys. Lett., 1980, **36**, 291.
27. Memming, R.: Thin Solid Films, 1986, **143**, 279.
28. Okada, K., and Namba, Y.: J. Vac. Sci. Technol. A, 1989, **7**, 132.
29. Bowden, F.P., and Tabor, D.: "The Friction and Lubrication of Solids—Part II," Clarendon Press, Oxford, 1964, pp. 158–185.
30. Prawer, S., Kalish, R., and Adel, M.: Appl. Phys. Lett., 1986, **48**, 1585.
31. Robertson, J.: Adv. Phys., 1986, **35**, 317.
32. Bisson, E.E.: "Nonconventional Lubricants," in Advanced Bearing Technology, ed. by Bisson, E.E. and Anderson, W.J., NASA SP-38, 1965, pp. 203–258.
33. Bisson, E.E., Johnson, R.L., and Anderson, W.J.: "Friction and Lubrication with Solid Lubricants at Temperatures to 1000 °F with Particular Reference to Graphite," in Proceedings of the Conference on Lubrication and Wear, Institution of Mechanical Engineers, London, England, 1957, pp. 348–354.

Materials Science Forum Vols. 52 & 53 (1989) pp. 657-670

INVESTIGATION OF THE MECHANICAL PROPERTIES OF 25 nm AND 90 nm, HARD CARBON FILMS RESIDING ON Al

W.R. LaFontaine (a), T.W. Wu (b), P.S. Alexopoulos (b) and D. Stone (c)

(a) Department of Materials Science and Engineering, Bard Hall
Cornell University, Ithaca, NY 14853, USA
(b) IBM Research Division, Almaden Research Center
San Jose, CA 95120-6099, USA
(c) Formerly at the Dept. of Materials Science and Engineering
Cornell University, Ithaca, NY 14853, USA
Presently at the Dept. of Materials Science and Engineering
1509 University Ave., U. Wisconsin, Madison, 53706, USA

ABSTRACT

A continuous indentation technique is utilized to investigate mechanical properties of 25 nm radio frequency-sputtered and 90 nm plasma enhanced chemical vapor-deposited films on 1 μm Al. The carbon films introduce strengthening effects up to depths of 0.6-0.8 μm. At shallow depths (0.0 to 0.3 μm) the extra load, ΔL, due to the 25 nm carbon film is proportional to indent depth, x. A model is presented to explain this effect. Beginning at about 0.3 μm, the loading rate (dΔL/dx) increases, then suddenly becomes negative at about 0.6 μm. The increase in dΔL/dx at 0.3 μm appears to be associated with the development of pileup of Al around the indent, whereas the subsequent drop in ΔL at 0.6 μm may be caused by fracture of carbon film that covers the pileup. Based on analysis of the loading data and a model for the strengthening effect of the film, stresses developed in the 25 nm film due to stretching and bending around the indent are estimated at 10-20 GPa. Stresses in the 90 nm film may be substantially lower than these.

INTRODUCTION

Recently, researchers have undertaken to develop hard carbon coatings desired for their optical and mechanical properties. Extremely high hardness can be achieved in such materials, but their structure and properties are sensitive to process parameters [1]. Because some applications can be foreseen to require overcoats of extreme thinness, novel techniques are needed to characterize the mechanical properties of these films.

This article reports use of an indentation technique to determine the role of thin (25 nm and 90 nm) carbon overcoats in improving mechanical integrity of surfaces. Because these films are so thin, it is difficult to measure their properties without receiving "interference" from the substrate. One option in evaluating the film is to remove it from the substrate prior to mechanical testing. This option is often undesirable, because near-surface mechanical properties reflect interaction between substrate and surface layer.

In the work reported below, a continuous indentation test is utilized to probe the properties of Al films on which thin (25 nm and 90 nm) carbon films are deposited. Indentation experiments are performed to depths up to 1 μm. Despite these depths, the films still influence the loading data; removal of the film from the substrate prior to testing to obtain meaningful mechanical properties data may be unnecessary for such hard films.

EXPERIMENTAL WORK

In the experimental work, Al films were sputtered to 1 μm thickness on wafers of <100> semiconductor grade silicon. Process parameters for deposition of the Al are given in reference [2]. Some aluminum film samples were used as controls; others were sputter-etched, then deposited with 25 nm-thick carbon films using a radio-frequency (RF) sputtering technique. Other samples were deposited with 90 nm carbon through the use of plasma enhanced chemical vapor deposition (PECVD).

Raman spectroscopy revealed a higher degree of diamond-like bonding in the RF-sputtered films than the PECVD films, as indicated by spectra in figure 1. According to Knight and White [3] the graphitic peak falls near 1580 wavenumber, and the peak for tetrahedrally coordinated carbon falls near 1330 wavenumber.

Indentation tests were performed using a displacement controlled indentation apparatus [2,4,5]. A commercial Vickers indentor (4-sided pyramid) with a 143° angle between opposite faces and a 1.2 μm^2 flat area at the tip was used. This relatively blunt tip is less likely to puncture the carbon film than a 3-sided, or Berkovitch indentor. The Berkovitch indentor is suitable for making shallow indents due to better tip definition at such depths.

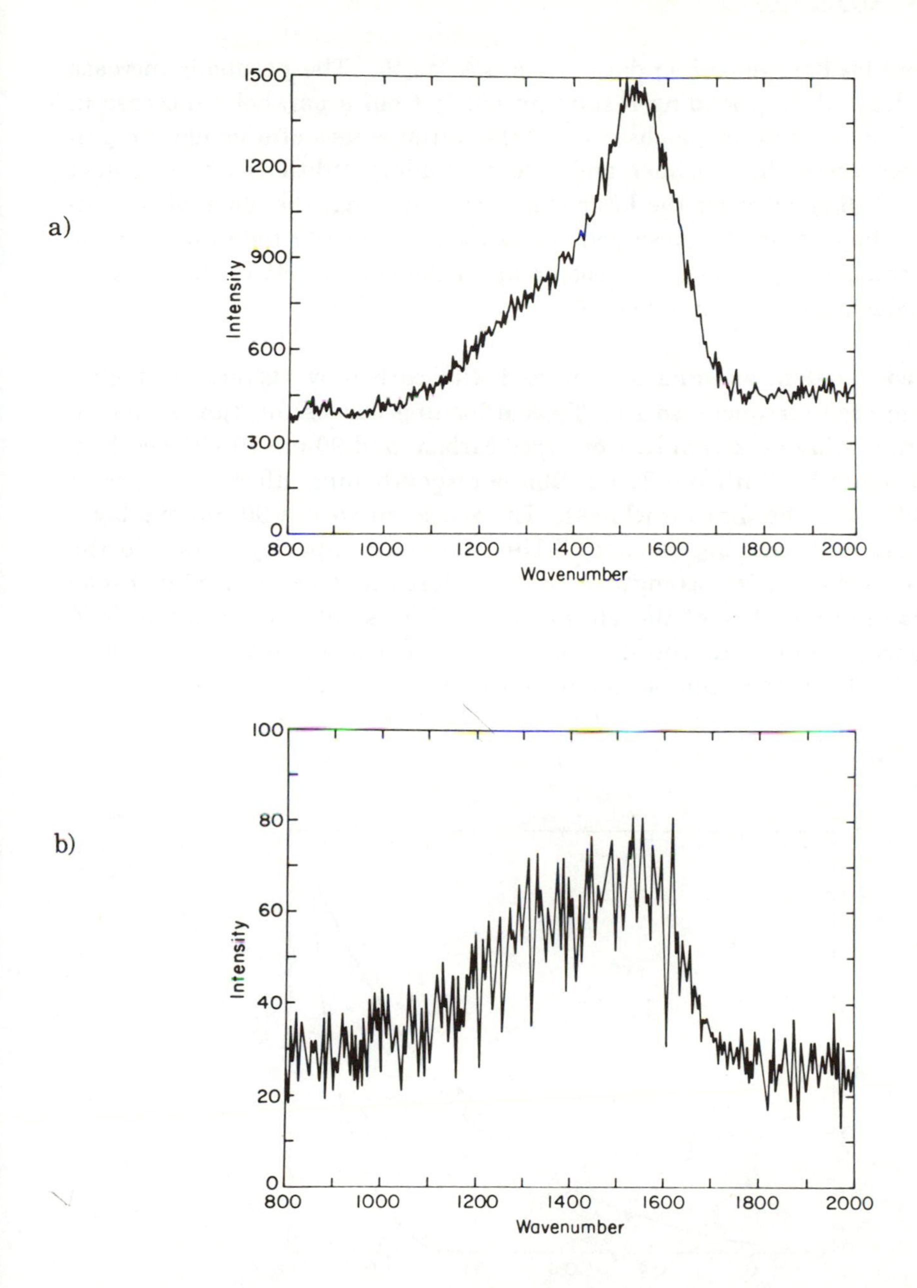

Figure 1. Raman spectra of (a) the 90 nm-thick PECVD film and (b) the 25 nm-thick RF-sputtered film.

EXPERIMENTAL RESULTS

Figure 2 shows loading-unloading data of 1 μm Al on Si. The parabolic increase in load with depth during loading results primarily from a parabolic increase in indent area. The constant slope observed in the initial stages of unloading represent elastic recovery of the specimen and indentor under conditions where contact area is constant [6,7]. During the later stages of unloading, the slope of the unloading curve changes as the specimen "peels away" from the indentor and the contact area changes [6]. The hysteresis areaS in the curves reflect the work required to nonelastically deform the specimen.

Loads required to indent specimens reinforced with carbon overlayers are higher than those required for uncoated Al. Typical loading curves for 1μm Al on Si, alone and with overlaying 25 nm RF-sputtered carbon and 90 nm PECVD carbon, are shown in figure 3. With the 25 nm film, a strengthening effect continues at depths over 30 times the film's thickness. The specimen with a 90 nm overlayer has the highest load-carrying capacity. This improved capacity is due to the film's greater thickness; its strength (elastic modulus and yield strength) may actually be lower than that of the 25 nm film. The sudden decrease in load carrying capacity evident in the data of 25 nm carbon at ≈0.6 μm depth is associated with a fracture in the carbon film outside the area of the indent.

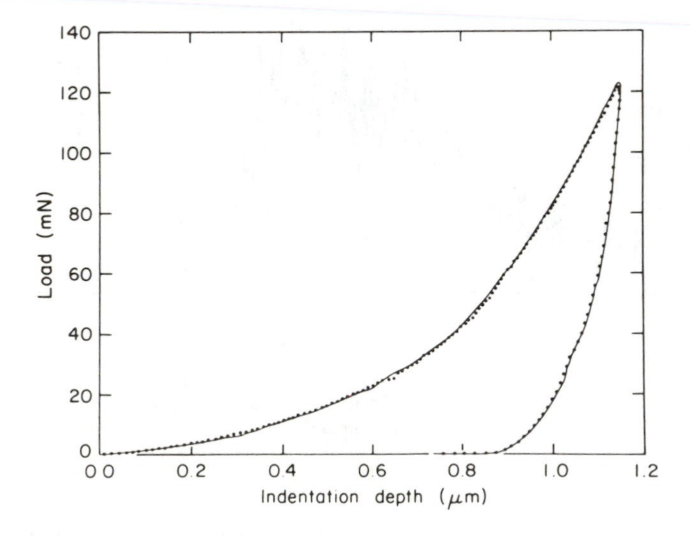

Figure 2. Loading-unloading data of 1 μm Al on Si. Two sets of data are illustrated, one as a line and the other as individual points.

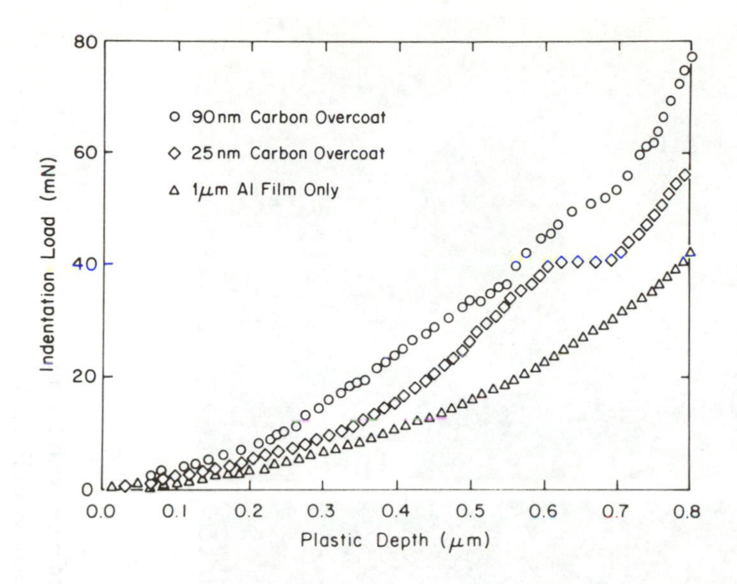

Figure 3. Loading curves of indents in specimens with: 25 nm RF-sputtered carbon, 90 nm PECVD carbon, and no overcoat.

Scanning electron micrographs of remnant impressions made in the 25 nm RF carbon coated sample are shown in figures 4a-c. The last micrograph was taken at a 30° angle from the surface to display more clearly the pileup of aluminum surrounding the indent.

For shallow indents the carbon film stretched to conform to the surface of the plastically deformed Al. Fractures in the film were rarely observed for indents less than 0.2 μm deep (figure 4a). In slightly deeper indents (0.3 μm) fractures developed along the indent diagonals (as indicated in figures 4b and 4c). These fractures do not appear to be associated with any sudden decreases in the load (figure 3).

Fractures were regularly observed in the pileup around indents deeper than 0.4 μm, as shown in figure 4c. The propagation of large-scale fractures coincides with the reduction in strengthening effect, figure 3.

The extra load, ΔL, carried by the specimen with a 25 nm RF-sputtered carbon film is plotted in figure 5. Several sets of data from different runs are illustrated. These curves were obtained by subtracting the load values of the lowest curve in

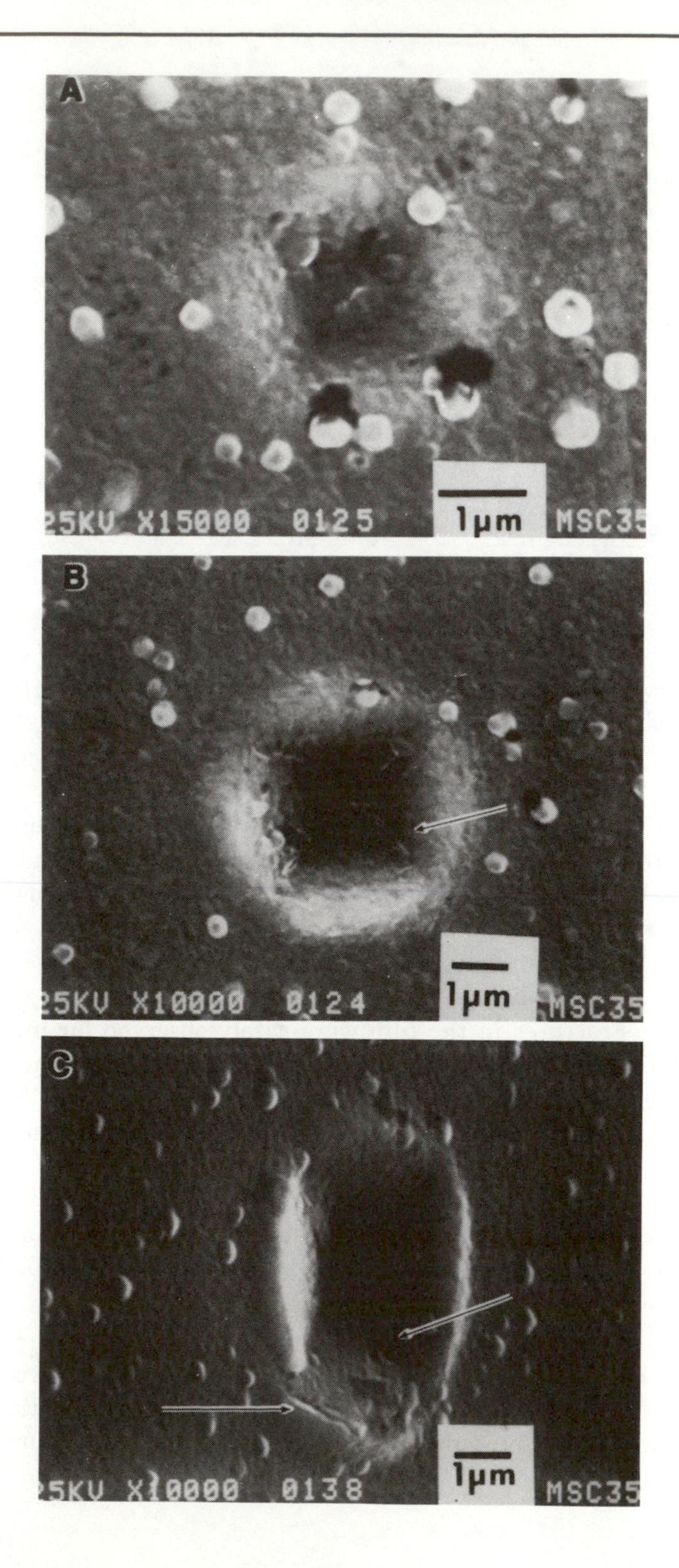

Figure 4. Remnant impressions of indents in the specimen with 25 nm carbon. Indent depths (μm): (A) 0.16, (B) 0.32, (C) 0.62. The arrows indicate fractures.

figure 3 from those of the intermediate curve and similar data curves taken from the same specimen. Because these data represent the difference between measurements whose values are close together, there is considerable uncertainty (perhaps as much as 50%) in the load values shown in Fig. 5. In part, this uncertainty arises from difficulty in choosing the zero of indentation depth.

Independent of how the zero of indentation depth is chosen, three trends observed in figure 5 persist. These are: (1) Initially, ΔL increases linearly with depth of penetration, x. (2) Beginning at 0.3 μm depth (corresponding to an indent diagonal of about 4 μm) dΔL/dx increases. (3) Finally, at 0.6-0.7 μm depth dΔL/dx becomes negative. Note that the strengthening effect of the carbon film extends to indent depths many times its thickness. This type of effect from hard surface layers has also been reported by Ross and coworkers [8].

An indent in the specimen coated with 90 nm PECVD film is shown in figure 6. In most respects indents in the 90 nm film appear similar to those in the 25 film. The indent in figure 6 is deeper (0.78 μm) than those in figures 4a-c (0.16-0.62 μm). For indents this deep, large fractures can be found in the carbon film covering the pileup. The fractures along the diagonals are substantially wider than those in figure 4b. Micrographs of deep indents in the 25 nm film look similar to figure 6.

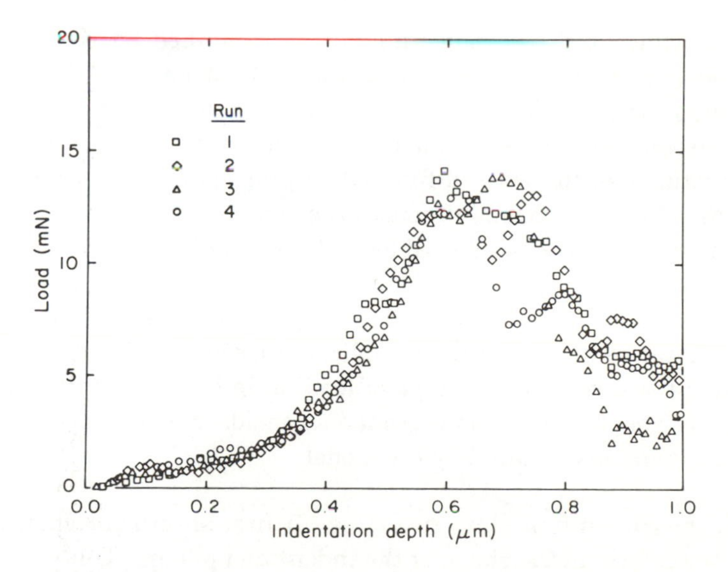

Figure 5. Increase (ΔL(x)) in load carried by the specimen with a 25 nm-thick surface layer of carbon, compared to that of the uncoated Al. The data from several runs at different sites are shown.

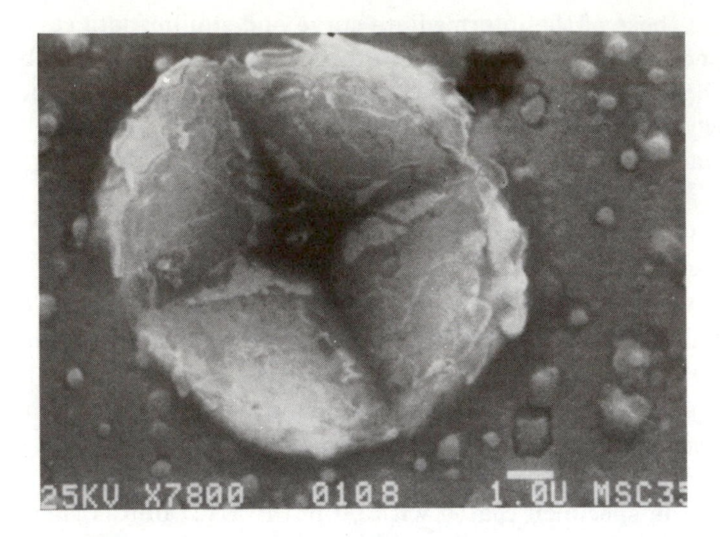

Figure 6. Remnant impression of an indent in the specimen with a 90 nm-thick overlayer of PECVD carbon. The indent is approximately 0.78 μm deep. The white spots are hillocks in the Al film

DISCUSSION

The reinforcing effect of the carbon film can be explained based largely on its rigidity derived from high elastic modulus and yield strength. Clearly, the tendency to undergo large strains without fracturing is important. Good adhesion between the carbon film and Al promotes strengthening because it reduces the tendency of fractures as observed in figure 4 to propagate along the interface between C and Al. Indentor shape influences the observed strengthening effect; the carbon film is less likely to be punctured by a blunt indentor.

A model is presented below to account for strengthening effect. This model must be considered tentative in view of the complicated mechanics involved. Nevertheless, the slope of the initial, linear stage of loading in figure 5 can be predicted to within a factor of about 2, which is reasonable considering the uncertainty in the data and the assumptions required by the model.

In this model, the carbon film plays two roles. It first absorbs mechanical energy by stretching to conform to the shape of the indent and pileup. This contributes an amount ΔL_a to the load in figure 5. Second, the carbon increases the effective area of contact between indentor and Al. This contributes an amount ΔL_b. The sum of these two contributions is the total increase in load, ΔL.

The contribution to the additional load from stretching is expressed, as a function of indentation depth, x, as:

$$\Delta L_a(x) = dU/dx \quad (1)$$

where U is the work required to stretch the carbon film. This energy is the total volume, $t \times A$, where A is the stretched area and t is the thickness, times the work per unit volume required to stretch the film. Beneath the indentor the strain in the film is about 5% due to the 143° apical angle (from geometry; 1-cos((180-143)/2) = 0.05). We may assume that, due to the relatively high hardness/modulus ratio for these types of films [9,10], strains of 5% in the film can remain elastic. When only the carbon film touching the indentor is considered, it contributes approximately 0.5 (σ^2/E) t dA/dx to ΔL_a where from indentor geometry dA/dx ≈ 70x.

In addition, carbon overlaying the pileup stretches. The pileup covers an area roughly equal to that of the indent, and prior to the development of a sharp edge to pileup the average strain in the carbon covering the pileup is estimated to be 1/2 that in the carbon directly beneath the indentor. Therefore, we estimate that, at shallow depths (< 0.3 μm in figure 3), the strain energy contribution of the pileup is about 1/4 that of the film directly beneath the indentor.

The elastic modulus of the film, E, is unknown because it is sensitive to process variables and was not measured in our experiments. (A technique developed recently to measure elastic moduli is mentioned briefly at the end of the discussion). Based on experiments by Tsukamoto and coworkers [11], we can estimate that the elastic modulus is about 300-400 GPa. Therefore, the stress in the film is 15-20 GPa beneath the indentor. If 17.5 GPa is taken as a representative stress, then stretching of the carbon contributes an additional load carrying capacity to the Al given by

$$\Delta L_a(x) \approx 44\ (\sigma^2/E)\, t \times x \approx 1\ \text{mN}/\mu\text{m} \times x \quad (2)$$

at shallow depths. The contribution of ΔL_a to the initial loading slope in figure 5 is approximately 1/5-1/3 that observed.

A second contribution, ΔL_b, is now estimated. Two methods are used to construct the estimate; their results agree closely so it is difficult to decide which best describes the experimental data.

In the first method we assume that the carbon film, simply by contacting the surface of the indentor, increases its effective diameter. The load-carrying area is larger than the diameter of the indentor. The same effect would exist had the carbon film been deposited on the diamond indentor instead of the Al. The distance by which the indent radius increases is given by $\Delta D = \arctan(143°/2) \times t \approx 3\ t$ where again t is the film thickness.

For the second method the carbon film is assumed to act as a membrane to transfer load away from the edge of the indent. At the edge of the indent, the film bends in a radius R when transferring from beneath indentor to the adjacent free surface. The Al accomodates this bending by plastically deforming beneath the curved carbon film but resists with a pressure equal to the hardness, H, of the Al. Throughout the region where the carbon film is bent the average membrane stress is roughly $\varepsilon E/2$, where ε (≈ 0.05) is the amount by which carbon film beneath the indentor is stretched and E is the elastic modulus of the carbon. A force balance between the carbon film, which pushes into the surface, and the Al, which pushes away from the surface, provides the distance, ΔD, that the film increases the diameter of the indent: $\Delta D = \sin(\theta) \times (\varepsilon E/2H) \times t$, where θ is the angle between the plane of the indentor face and surface of the Al outside the indent. (This expression for ΔD predicts that ΔD goes to zero in the limits that the Al is very hard or the carbon modulus is very low - a reasonable prediction).

In the absence of pileup and elastic deflection of the specimen surface, $\theta = 18.5°$; both effects can be expected to decrease θ. Knowing that the Al has a hardness of 0.75 GPa [2] and E of the carbon is about 350 GPa, then we can estimate ΔD to range between 2t and 4t, depending on the value of θ (9-18.5°).

For either of the above methods a value $\Delta D \approx 3t$ can be taken to be representative. The increase in load encountered by the indentor due to the spreading of the indentor load by the carbon film is:

$$\Delta L_b(x) = \Delta D\, P(x)\, H \tag{3}$$

where H is the hardness of the Aluminum and P(x) is the perimeter of the indent. For this indentor, $P(x) \approx 24\ (x+.2)$ where x is in μm. The increased load is therefore given by:

$$\Delta L(x) \approx (2.5x+0.3)\ \text{mN}. \tag{4}$$

This prediction may be compared with the loading data in figure 5, where the initial portion is approximately linear with a slope of 3-5 mN/μm.

Beginning at a depth of about 0.3 μm, the loading rate (dΔL/dx) begins to increase dramatically (figure 5). The change in loading rate appears to be associated with the formation of an abrupt edge in the pileup, that requires the carbon film to bend sharply over the surface. Below depths of 0.3 μm this sharp, outer edge is not observed; its formation may be inhibited by the carbon overcoat. The minimum radius of curvature in the carbon film covering the pileup is somewhere between 0.2 and 0.5 μm, and the maximum bending stress in the film is E (t/2R), where R is the radius of curvature. This formula translates to a maximum bending stress of between 10 and 20 GPa. Such stresses tend to peel the carbon film away from the Al when the carbon fractures.

The PECVD film, while being 3-4 times as thick as the RF-sputtered film, only introduces about twice the strengthening effect (figure 3). The above models suggest, therefore, that stresses supported by the CVD film are only about 1/2 of the RF-sputtered film. The PECVD film may have a lower elastic modulus as well as a lower ratio of hardness to elastic modulus. The extra thickness of the 90 nm-thick film may induce it to fracture more easily than the 25 nm film, explaining the more frequent (but less abrupt) fracture events in its loading curve, figure 3.

Vital to understanding the strengthening effect of hard carbon film is identification of its yield strength and elastic modulus. Since the time the data presented in figures 3 and 5 were collected, it has now become possible to utilize the continuous indentation technique to investigate thin film elastic moduli [7,12,13]. In figure 7, unloading data of a 100 nm-thick hard carbon film are shown based on which the elastic modulus of the film is estimated at 350±50 GPa. An elasticity analysis for indentation of a bilayer is required for the effort [13]. A Berkovitch indentor was used to obtain the data [4]. (The film from which the data in figure 7 were collected was manufactured differently from either of the films investigated here; it would be unwise to suggest that these data present strong evidence to corroborate the value of elastic modulus used in the above calculations).

CONCLUSIONS

When deposited on a relatively soft substrate, a hard carbon film can increase the strength of the surface to depths many times the film's thickness. Two mechanisms have been proposed to account for the early increase in load: stretching the film requires extra energy, which translates into an increased load necessary to create an indent of any given area, and spreading of the load by the rigid film effectively increases the indent perimeter, requiring a larger volume of Al to deform. A detailed model has not been provided to account for the abrupt increase in dΔL/dx in figure 5 beginning at 0.3 μm. This increase is clearly an important effect, but uncertainties in models prevent reliable predictions. The effect appears to be associated with development of a sharp edge in the pileup and the need of

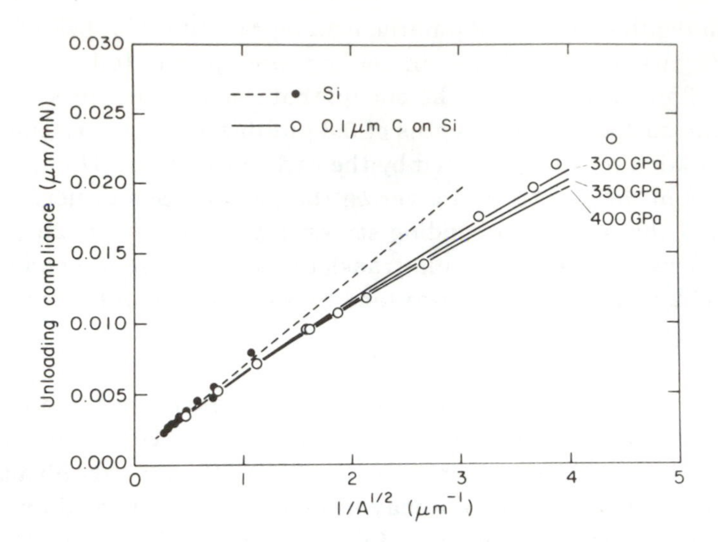

Figure 7. Unloading compliance vs $1/A^{1/2}$ for Si and 100 nm carbon on 100 Si. Here, A is the area of the indent projected in the plane of the surface. An elasticity analysis to take into account the effects of the substrate indicates the Young's modulus of the carbon to be 350±50 GPa. The symbols indicate data, and the lines indicate simulations based on an elasticity analysis [13].

the carbon film (due to its adhesion to the surface) to conform to the shape of the pileup. Fractures in the carbon film along the perimeter of the pileup greatly reduce its strengthening effect .

ACKNOWLEDGEMENTS

A portion of this work was supported by the National Science Foundation through the Materials Science Center and the National Nanofabrication Facility (Grant ECS-8619049) at Cornell. The constructive comments of Che-Yu Li and Mary F. Doerner were greatly appreciated. B.D. Stone is acknowledged for her help in editing this manuscript, J. Jorgensen is appreciated for having provided the technical drawings, and the authors are grateful to D. Whealon for photographing the figures.

REFERENCES

[1] Meissier, R., Badzian, A.R., Badzian, T. Spear, K.E., Bachmann, P., and Roy, R.: *Thin Solid Films,* 153, 1987, 1.

[2] Stone, D., LaFontaine, W.R., Alexopoulos, P.-S., Wu, T.W., and Li, C.-Y.: *J. Mater. Res.,* 3(1), 1988, 141.

[3] Knight, D.S., and White, W.B.: *J. Mater. Res.,* 4(2), 1989, 385

[4] Hannula, S.-P., Stone, D., and Li., C.-Y.: Mater. Res. Soc. Symp. Proc. 40, 1985, 217.

[5] Wu, T.W., Hwang, C., Lo, J., and Alexopoulos, P.S.: *Thin Solid Films*, 166, 299.

[6] Loubet, J.L., Georges, J.M., and Meille, G.: in ASTM STP 889, Blau and Lawn, Eds., American Society for Testing and Materials, Philadelphia, PA, 1986, 72.

[7] Doerner, M.F., and Nix, W.D.: *J. Mat. Res.,* 1(4), 1986, 601.

[8] Ross, J.D.J., Pollock, H.M., Pivin, J.C., and Tzakadoum, J.: *Thin Solid Films,* 148, 1987, 171.

[9] Pethica, J.B., Koidl, P., Gobrecht, J, and Schuler, C.: *J. Vac. Sci. Technol. A,* 3(6), 1985, 2391.

[10] Antilla, A., Koskinen, J., Lappalainen, R., Hirvonen, J.-P., Stone, D., and Paszkiet, C.: *Appl. Phys. Lett.,* 50(3), 1987, 132.

[11] Tsukamoto, Y., Yagamuchi, H., and Yanagisawa, M.: *Thin Solid Films,* 154, 1987, 171.

[12] Oliver, W.C., McHargue, C.J., and Zinkle, S.J.: *Thin Solid Films*, 153, 1987, 185.

[13] Stone, D.S., Wu, T.W., Alexopoulos, P.S., and LaFontaine, W.R.: Mater. Res. Soc. Symp. Proc. 130,J.C. Bravman, D.M. Barnett, W.D. Nix, and D.A. Smith Eds. Materials Research Society, 1989. pp. 105-110.

Materials Science Forum Vols. 52 & 53 (1989) pp. 671-688

AMORPHOUS CARBON FILMS FOR SENSOR APPLICATIONS

F. Olcaytug, K. Pirker, R. Schallauer, F. Kohl, G. Urban,
A. Jachimowicz, O. Prohaska (a), W. Fallmann and K. Riedling
Institut für Allgemeine Elektrotechnik und Elektronik
Technische Universität Wien, Austria
and
Ludwig Boltzmann Inst. für biomedizinische Mikrotechnik
Gusshausstrasse 27-29, A-1040 Wien, Austria
(a) Case Western Reserve University, Cleveland, OH 44106, USA

ABSTRACT

Low temperature glow discharge decomposition of acetylene in argon and nitrogen has been employed for the formation of low resistivity carbon films which should serve as electrodes and temperature sensors in microprobes exposed to biological tissue or tissue fluid. The films were grown in an r.f. activated diode system operated at 7 or 13.56 MHz, under conditions optimized for the formation of low resistivity films, typically 100 Ωcm, whose properties are otherwise very similar to those of conventional a-C:H films. Deposition modes with and without self bias of the substrate electrode were compared. Results obtained with the non-self bias mode are similar to the values published for layers deposited with bias mode. The chemical stability of the layers was verified electrochemically. Contact and temperature sensor structures based upon the carbon films were successfully realized and tested.

1. INTRODUCTION

In the last two decades, amorphous hydrogenated carbon (a-C:H) films have roused considerable interest because of their attractive characteristics: They are chemically inert, hard, wear resistant, transparent in infrared and partially in the visible range, have a diamond-like microstructure, and a high electrical resistivity and dielectric strength.

In general, there are several methods for the production of amorphous carbon layers. Most of them utilize ion assisted techniques, based on low pressure plasmas [1,2]. Usual starting materials are hydrocarbons, e.g., acetylene. In addition, fluorinated hydrocarbons have been tested as starting materials [3]. A comparison between the results obtained with two different ion assisted methods, namely, ion beam sputtering, and r.f. plasma enhanced chemical vapor deposition (PECVD), is given in [4]. One of the less common deposition methods is photo induced chemical vapor deposition [5]. Recently, PECVD with electron cyclotron resonance (ECR) has been reported [6]. The film properties may be varied over a relatively wide range, depending on the method of the layer formation. In fact, a-C:H films cover a very distinct range in the relatively wide scale of the characteristics of carbon coating layers. Several definitions for a-C:H films have been given in the literature; a recent review of these topics is presented in [1].

One of the film formation techniques commonly used in modern semiconductor and thin film technology is plasma enhanced chemical vapor deposition [1,7]. This method is also well suited for the production of a-C:H films [8]. Many authors have already investigated the properties of a-C:H films deposited under the various conditions which the plasma enhanced deposition techniques offer. Only few papers deal with the details of the formation process itself [9-11]. Interface formation kinetics are mentioned in [12]. One of the most important features of a-C:H films, the free hydrogen content in the layer, has been studied in [13] and [14].

Research on carbon films concentrates on finding correlations between the deposition conditions and the resulting film characteristics [15], aiming to understand the deposition process and to be able to tailor a process for a requested set of film properties.

In all the above activities, interest is focused on hard "diamond-like" films with good electrical insulator characteristics. A discussion of the structure and the definition of a-C:H films can be found in [1] and [16]. Protective coatings and antireflection layers for optical components for the infrared range [17,18] and laser devices [19], wear protection [20], dielectric materials [15,21-23], and coating experiments on fusion reactor walls [6] are typical areas for their application. Even superlattices of a-C:H and a-Si:H have been realized [24]; compound layers of a mixture of carbon and silicon (a-Si:C:H) have been reported in [25]. Process specific studies can be found in [10,15], and [26]. Some studies on the structural changes leading to graphitic or polymeric films obtained by modified processing during or after the deposition were also carried out [27,28]. Some of these films exhibited a very low resistivity [29]. All these investigations indicate the importance of carbon coating films, beyond those which comply exactly with the definition of a-C:H layers.

The carbon films described in this contribution were developed for integrated thin film electrode probes for biological and medical applications; they should serve as chemically inert conductive coatings on electrodes immersed into biological material, and as active components of thin film thermistors which are inert enough to permit direct contact with biological tissue or tissue fluid. Carbon is known to be bio-compatible; carbon compounds are used as prosthesis material in the human body. The compatibility with blood, particularly of PECVD carbon layers, has been treated in [30] from the point of view of a material's energy band structure. For our sensors, we required thin conducting layers which ideally should have the conductivity and chemical inertness of graphite and, as an additional feature, the mechanical properties of a-C:H films. Evidently, we had to find a compromise between these opposing demands. Our intent was therefore to reduce the layer resistance as much as possible, without losing too much of the characteristics of a-C:H films.

A low temperature deposition method was required for compatibility with other process steps which permit the production of an integrated multiple probe system by thin film technology. Preferably, the deposition temperature should not exceed room temperature, particularly because of photoresist and micromechanical structures which are already present on the substrates to be coated.

In the following sections, a description of the film deposition, a general characterization of the resulting layers, and some application results will be presented. In a first step, we were able to essentially reproduce the results on a-C:H films published in the literature, although our films were grown with short-circuited self-bias. This is particularly interesting because many authors consider the bias as one of the most important parameters for the formation of amorphous carbon films, and deposition with short circuited self bias has not yet been described in the literature.

2. FILM DEPOSITION

2.1. FILM DEPOSITION SYSTEMS

Two separate film deposition systems were constructed and tested. One of these systems, and some of the results obtained with it, are described in detail in [31]. The two systems differed essentially by their reactor designs. In addition, the second system features improved gas supply and vacuum components, but the changes in these areas do not affect the operation of the system principally. The deposition reactor of the first system is depicted in figure 1. The chamber consists of pyrex glass, with an inner diameter of 10 cm. The substrates are placed on top of the r.f. activated lower electrode whose diameter is 7 cm, and which can be directly cooled and indirectly heated. The grounded counter electrode with a diameter of 10 cm and the guarding rings provide an area ratio of greater than three, in relation to the substrate electrode, which is necessary for self biasing according to [32]. The distance between the electrodes is 7 cm.

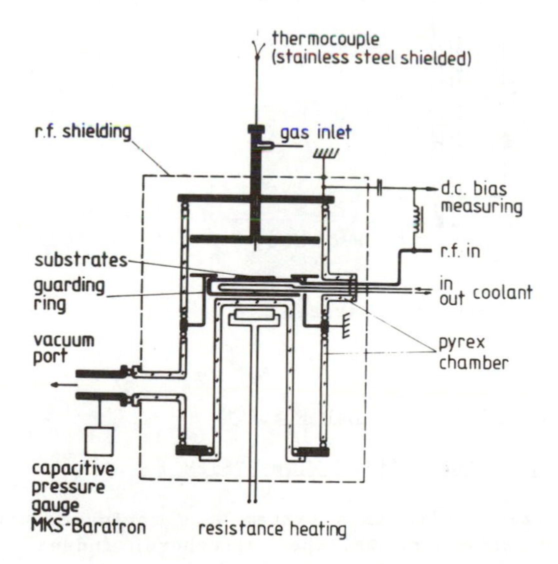

Figure 1: Reactor of deposition system 1.

The second deposition system features a significantly simplified reactor design. It does not permit to cool or heat the substrate electrode, but it allows to monitor the substrate temperature continuously with a radiation thermometer. A mass spectrometer and a pressure transducer are connected to a port in the reaction chamber wall which is exposed to the plasma. This configuration permits more accurate monitoring of the process parameters, compared to a measurement taken at some distance from the reactor. The ratio of the electrode areas is identical to the first system. The gas inlet shower is integrated into the grounded upper electrode; it could be designed for an improved film thickness homogeneity. Mass flow controllers regulate the gas flow rates, and a booster pump with variable speed permits process pressure control. The design of the second reactor is shown in figure 2.

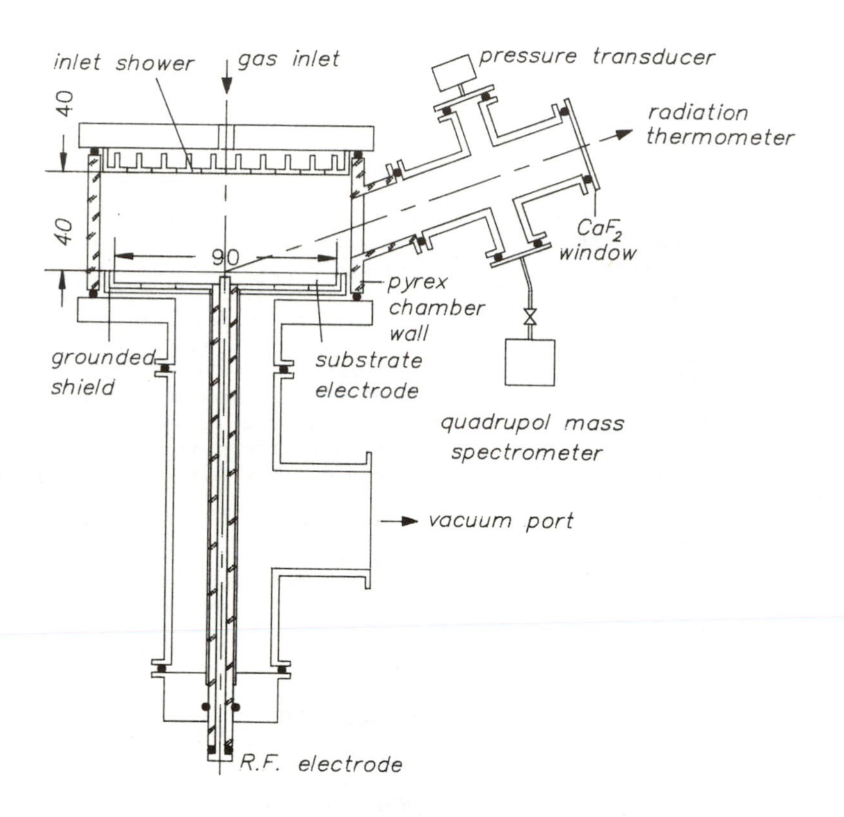

Figure 2: Reactor of deposition system 2.

2.2. FILM DEPOSITION PARAMETERS AND RESULTS

2.2.1. FILM DEPOSITION AND RESULTS WITH SYSTEM 1

Most of the experiments with this system were performed without self bias, in contrast to the experiments published elsewhere; indeed, almost no data obtained without bias are available in the literature. The operation parameters were selected in a range where self biasing would normally occur. However, self biasing was suppressed by a d.c. short circuit between the electrodes introduced with an inductor coil immediately outside the reaction chamber. The frequency was 7 MHz, power densities were about 1 - 2 Wcm^{-2}, and the total discharge pressure was between 10 and 100 mTorr. The effective pumping speed was 3 m^3h^{-1} at the pumping flange of the reactor.

Argon was used as a carrier gas; the carbon containing gas component was mostly acetylene. In addition, methane, dichlormethane, and octofluorocyclobutane were used in rare cases. The results given below were exclusively obtained with acetylene. With the other precursors, roughly similar trends were observed, which seems to be in accordance with later studies in [10] which claim that the formation of CH-groups in the immediate vicinity of the substrate surface is the process which determines the layer properties; therefore, the film characteristics can be expected to be independent of the original gas.

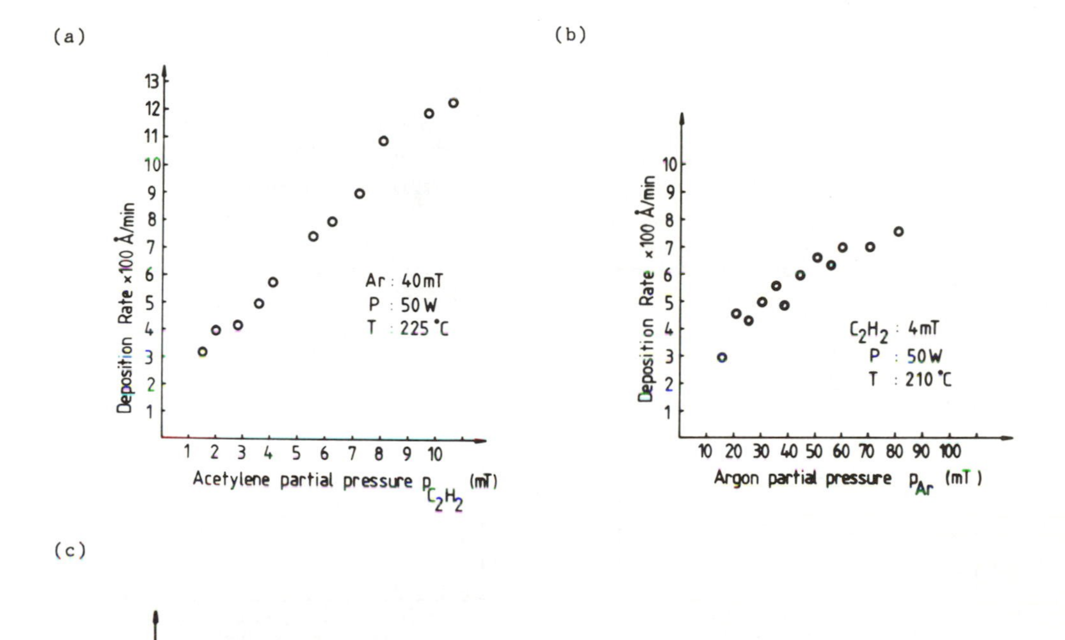

Figure 3: Deposition rate with system 1 as a function of acetylene (a) and argon (b) partial pressures, and of the r.f. power (c).

Figure 3 shows the dependence of the deposition rate on some process parameters. In figure 4, the influence of operation parameters on the layer resistivity is depicted.

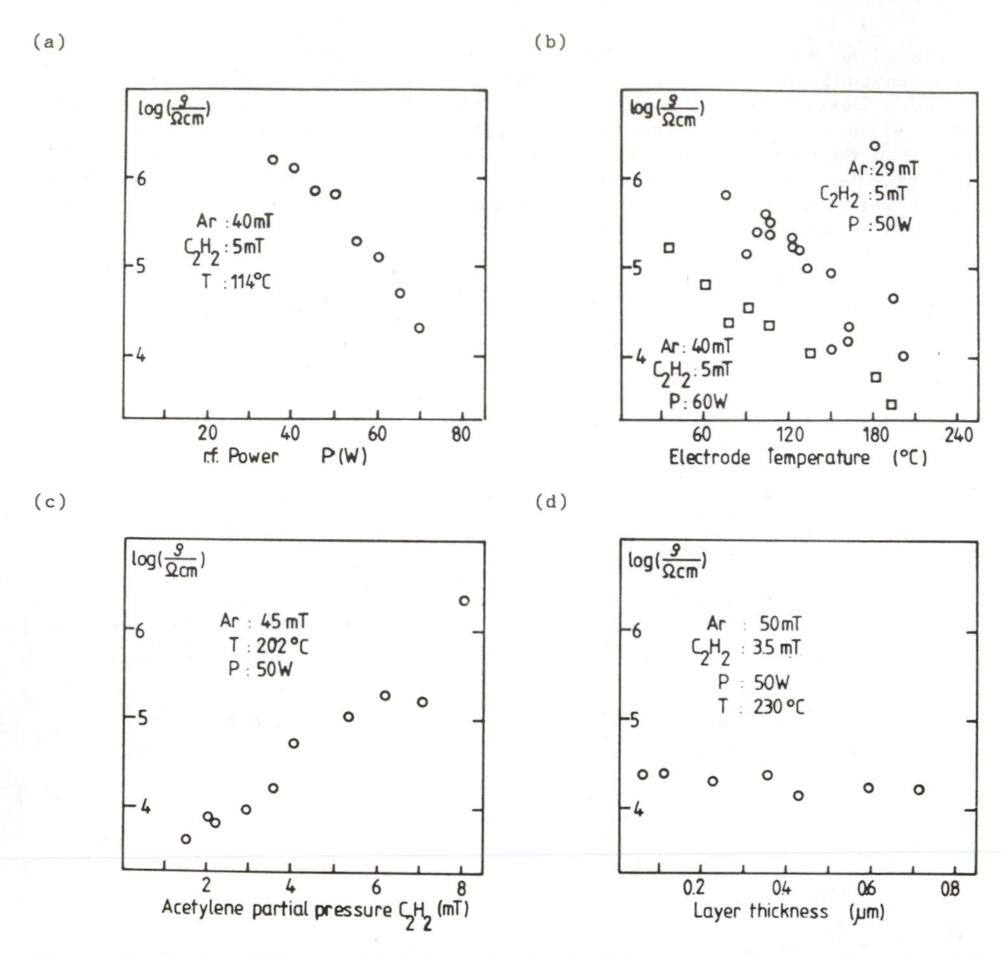

Figure 4: Carbon film resistivity obtained with system 1 as a function of r.f. power (a), electrode temperature (b), acetylene partial pressure (c), and layer thickness (d).

Optical and IR absorption studies indicate similarities to a-C:H layers described by other authors. However, a very low C-H absorption peak in the IR spectrum seems to be characteristic for our deposition process.

Figure 5 shows the distribution of the layer components and of some contaminants as a function of the distance from the sample surface. The graphs display the raw SIMS measurement data and ought to be scaled with the different sensitivities for the materials presented in order to give an actual mass percentage. The main contaminations are sodium (from the environment) and aluminum which is redeposited from the electrodes.

The Vicker's hardness of the films ranged from 2400 to 2800.

Our experiments with the first system indicated that a reduced resistance of "quasi a-C:H" films is feasible. For resistance values as low as 5.10^3 Ωcm, the hardness and stability of conventional a-C:H layers are still maintained. However, substrate heating was necessary for achieving low resistance values, and even the lowest resistance values obtained are much higher than required. For technological reasons, these high substrate temperatures, around 250°C, are rather prohibitive.

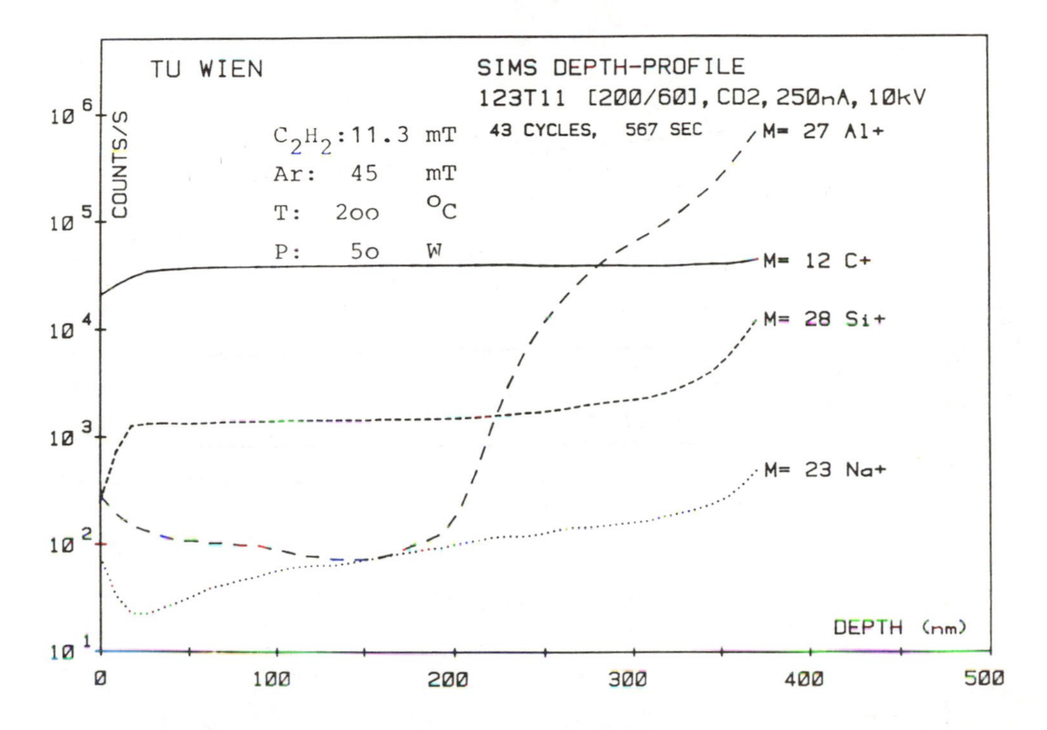

Figure 5: SIMS depth profile of a carbon layer grown with system 1.

2.2.2. FILM DEPOSITION AND RESULTS WITH SYSTEM 2

The two major changes in the course of the experiments with system 2 were the use of nitrogen instead of argon as a carrier gas, and the re-introduction of self biasing of the substrate electrode. For technical reasons, the excitation frequency was changed to 13.56 MHz. A set of suitable deposition parameter combinations was determined experimentally. The total pressure measured before the glow discharge was turned on was kept at 120 mTorr in the reactor for all experiments. The atmosphere generally consisted of acetylene and nitrogen. The C_2H_2 flow rate was varied between 1 and 6 sccm, and a complementary amount of nitrogen was added to maintain a total flow rate of 17 sccm, which resulted in the above total pressure of 120 mTorr.

Figure 6 shows that the resistivity hardly depends on the acetylene content of the reactant gas mixture within a technologically relevant range. However, a strong dependence on the deposition power and the closely correlated self bias can be observed.

The results of figure 6 indicate that resistance values are feasible which are acceptable for the applications in mind. Furthermore, these values were obtained under relatively "soft" deposition conditions, close to room temperature and with low r.f. power levels. Other desirable layer properties, like scratch resistance, hardness, and adherence on glass and silicon, could be kept close to the ideal characteristics of a-C:H films.

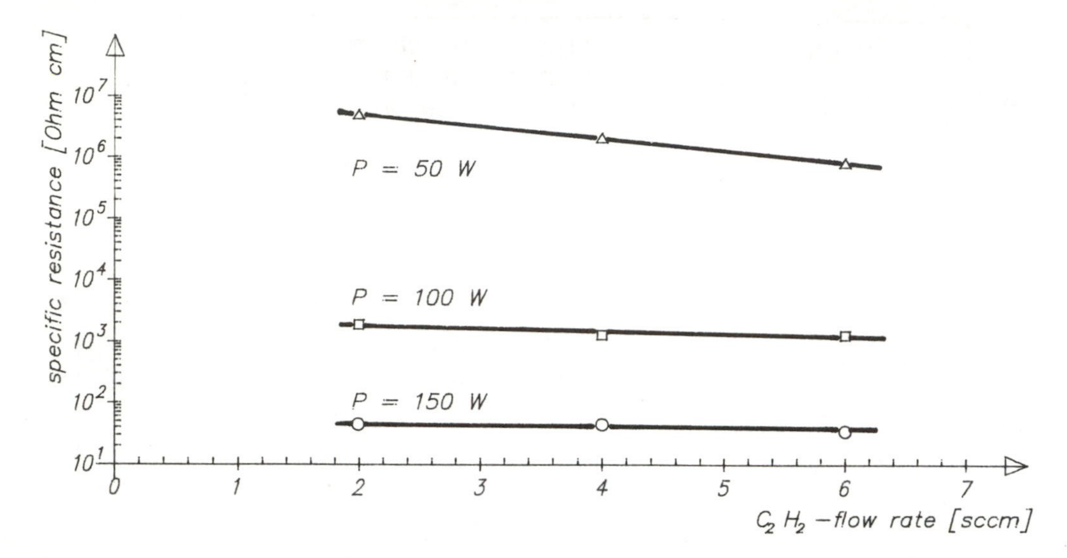

Figure 6: Dependence of the film resistance on the acetylene flow rate and on the r.f. power for system 2.

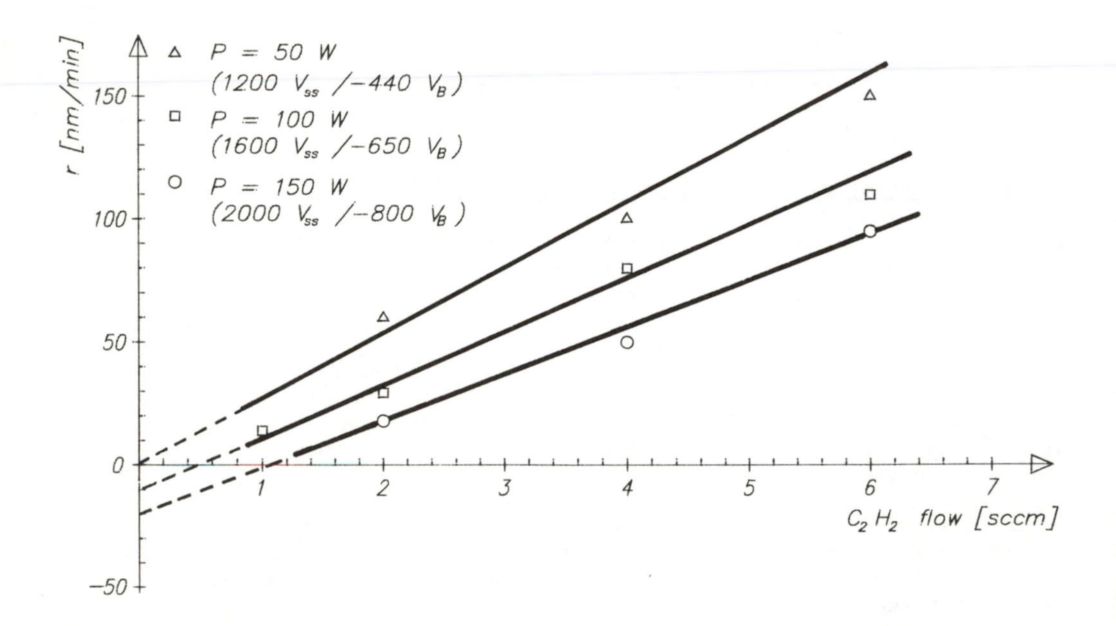

Figure 7: Dependence of the deposition rate on the acetylene flow rate and the r.f. power for system 2.

Figure 7 shows the deposition rate versus the acetylene content in the reactor, with the excitation power as a parameter. The shift at higher power values towards lower deposition rates is a very clear indication of two competing mechanisms, namely, layer formation and simultaneous ablation due to sputtering [22,33].

Possibly, this ion bombardment during the layer formation is responsible for a modification of the microscopic structure of the films, and hence of their electrical characteristics. A process has been reported in the literature which transforms existing insulating carbon layers into a graphitic structure [27]. In our case, the deposition and the transformation may happen simultaneously while the film is growing, although the ion energy of 600 eV in our system is by orders of magnitude lower than the energy used in [27].

2.3. DATA MEASURED DURING THE DEPOSITION PROCESS

In the first reactor, measurements of the sample temperature were only possible by touching the sample surface with a moveable thermocouple immediately before and after the deposition. During these measurements, the r.f energy had to be switched off. The pressure was monitored next to the pumping port, which gave indications of discharge pressure changes but did not allow an accurate pressure measurement.

In the second reactor, a port next to the discharge area (see figure 2) permitted to monitor the total and partial pressures of the glow discharge atmosphere as a function of time, and to measure the surface temperature of the samples *in situ* by means of a radiation thermometer. Furthermore, we registered the partial pressures in the reaction chamber and the mass spectra with a quadrupole mass spectrometer (BALZERS QMG 420).

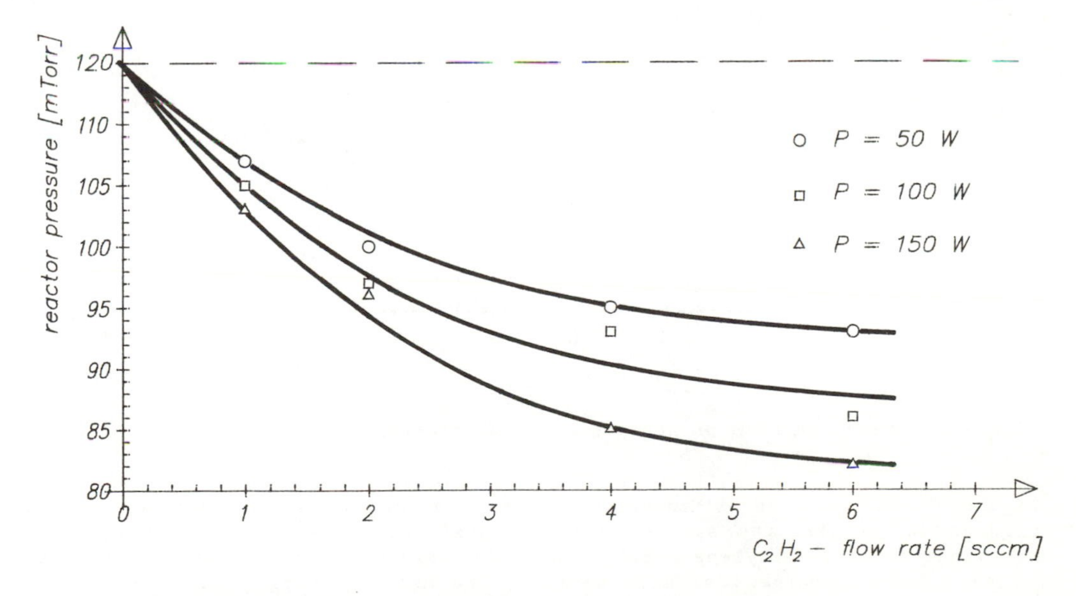

Figure 8: Reactor pressure during the deposition as a function of the acetylene flow rate and the r.f. power.

One of the characteristic events in PECVD is the pressure decrease when the discharge is turned on, because gas is consumed and deposited as a solid on the surfaces of the substrate and the reactor. This effect has already been reported in [33] in connection with plasma polymerization. The PECVD process is governed by similar mechanisms. Figure 8 shows the pressure decrease under equilibrium conditions under the influence of the glow discharge, depending on the acetylene ratio and the power level.

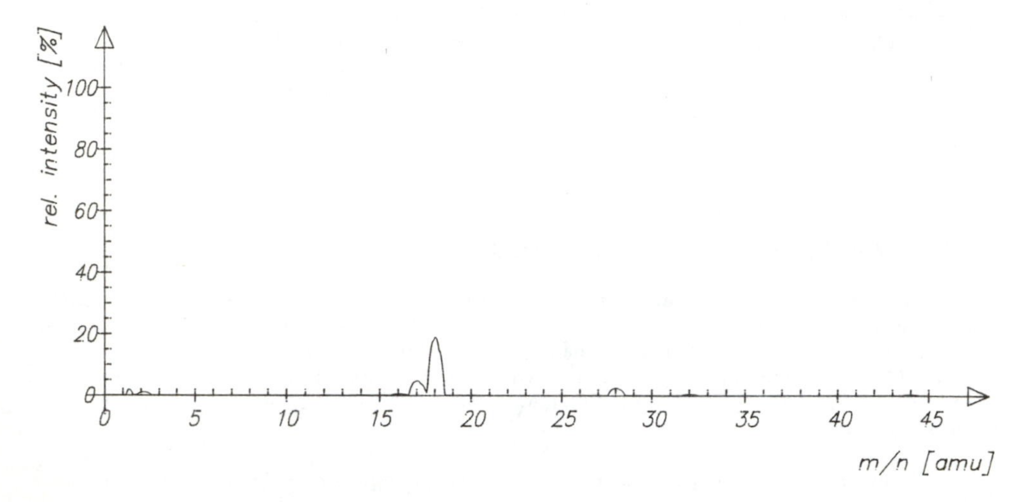

Figure 9: Mass spectrum of the reactor background atmosphere.

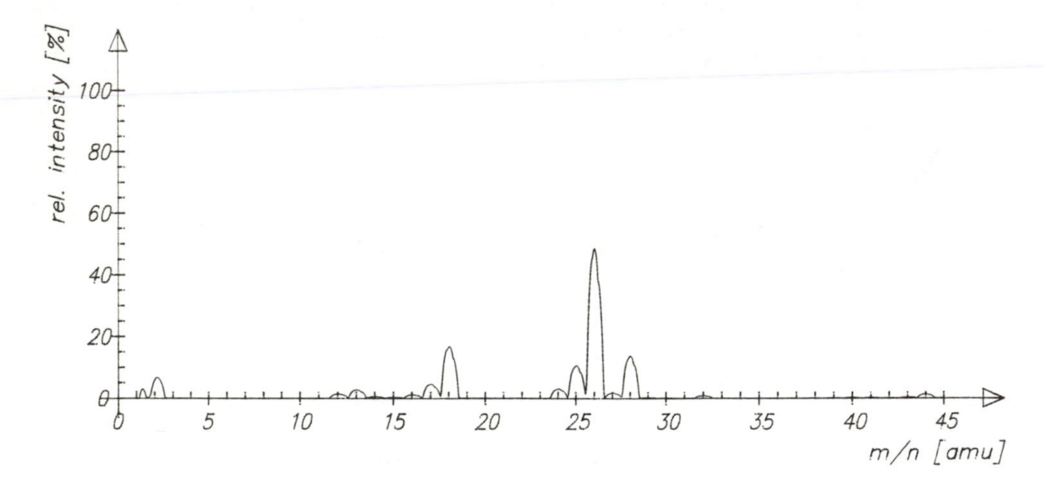

Figure 10: Mass spectrum of acetylene in the reactor.

Figure 9 shows the mass spectrum of the reactor background atmosphere. In figures 10 and 11, the spectra are presented which result after introducing acetylene and the acetylene-nitrogen mixture, respectively, but without turning on the glow discharge. The main peeks at the mass-to-charge ratios 26 and 28 correspond to C_2H_2 and N_2, respectively. Figure 12 depicts the spectrum during deposition. Mass number 26 vanishes completely here, indicating a mass limited deposition mode, whereas hydrogen is generated. The time dependence of the intensities of the typically involved components is illustrated in figure 13 for no discharge, and for various discharge power levels. Evidently, acetylene can no more be consumed completely at power levels below 25 W. The flow conditions referred to in figure 13 were 6 sccm C_2H_2, and 11 sccm N_2, resulting in a total pressure of 120 mTorr without discharge.

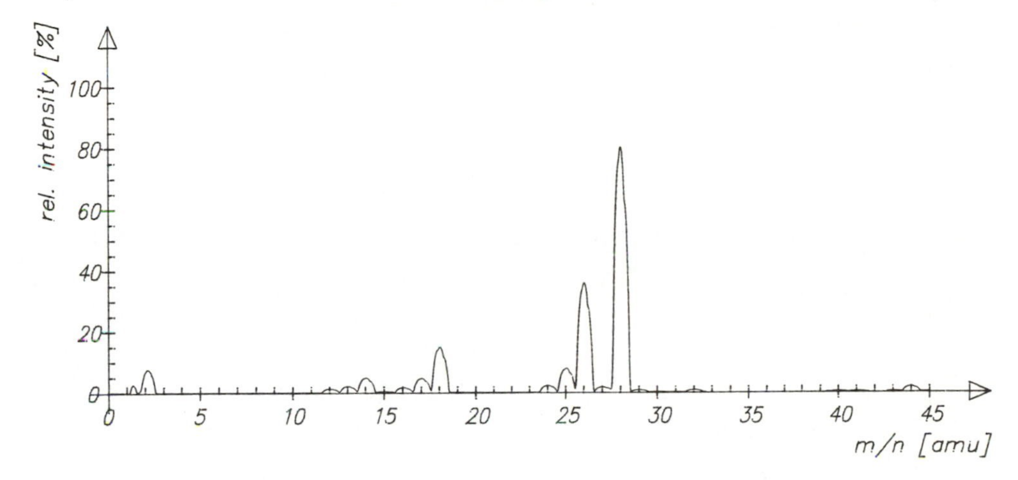

Figure 11: Mass spectrum of the acetylene-nitrogen mixture in the reactor without a glow discharge.

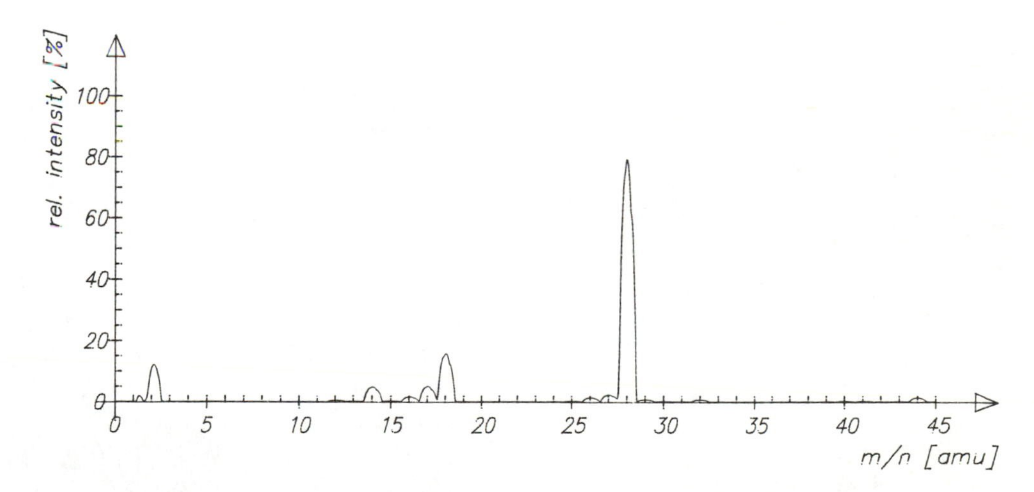

Figure 12: Mass spectrum of the reactor atmosphere during film deposition.

The substrate surface temperature was measured during the deposition with a radiation thermometer (HEIMANN KT 79, sensitivity window 7.5 - 8.2 μm). A remarkable temperature increase was observed which can be interpreted as the surface temperature because it responds very quickly to the discharge power levels and, incidentally, to an impedance mismatch.

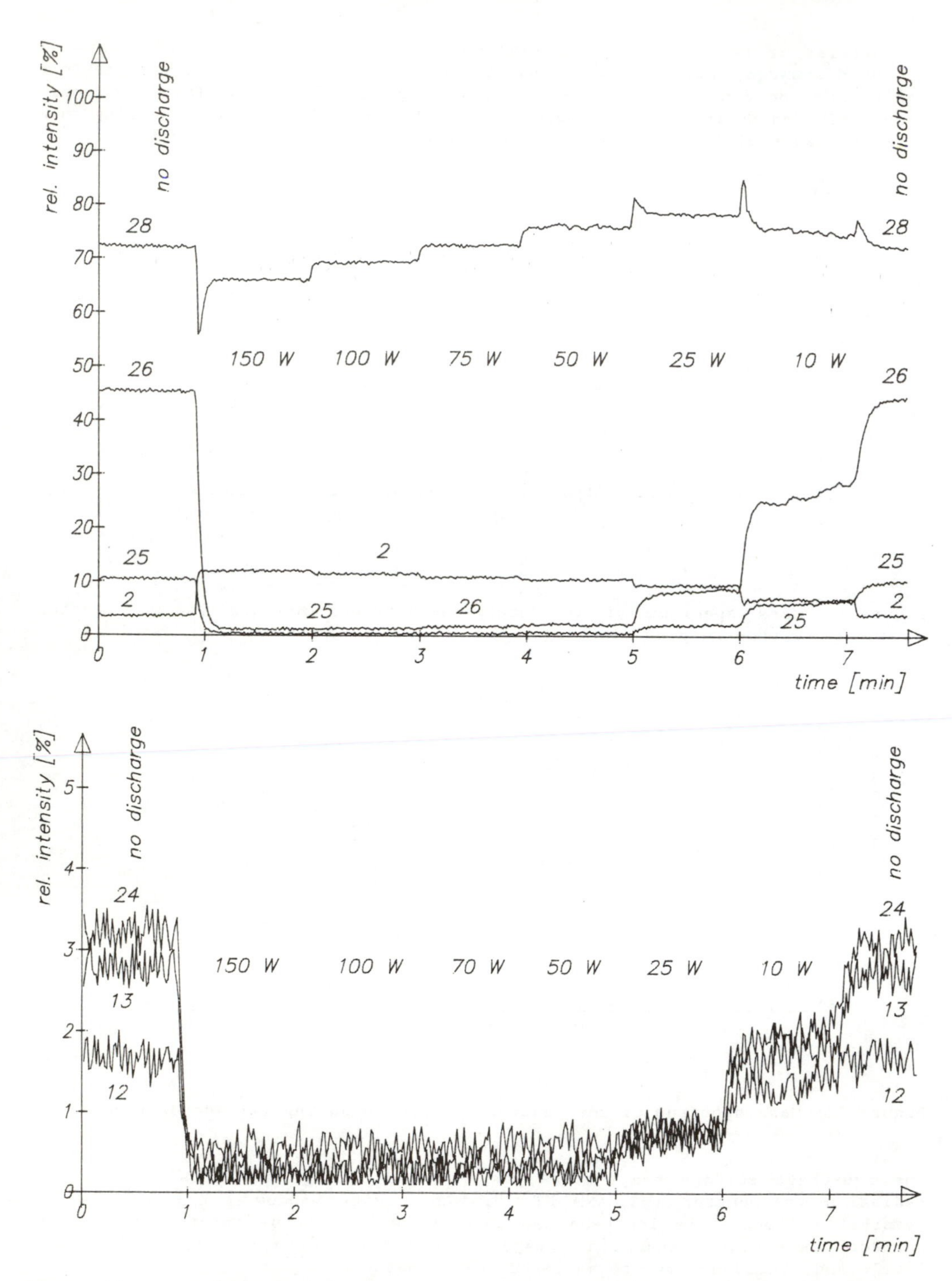

Figure 13: Partial pressures of various gaseous components (identified by their mass numbers) as functions of the r.f. power.

Figure 14 shows the measured temperatures, depending on the discharge power. The significant temperature reduction caused by a mismatch introduced on purpose (i.e., a SWR of 2.8 - 20 W reflected at 100 W incident power) indicates the importance of exact matching and automatic tuning correction during deposition. The increase of the surface temperature without discharge after processing at higher r.f. values like 150 W can be interpreted as the real temperature rise of the substrate. The emission coefficient of glass substrates coated with the carbon layers was found to be equal to 0.95, very similar to the emission coefficient of uncoated substrates. This can be interpreted as an indication of the good infrared transparency of our carbon films. The emission coefficient was determined by comparing covered and uncovered substrates heated externally to temperatures typical for the discharge process.

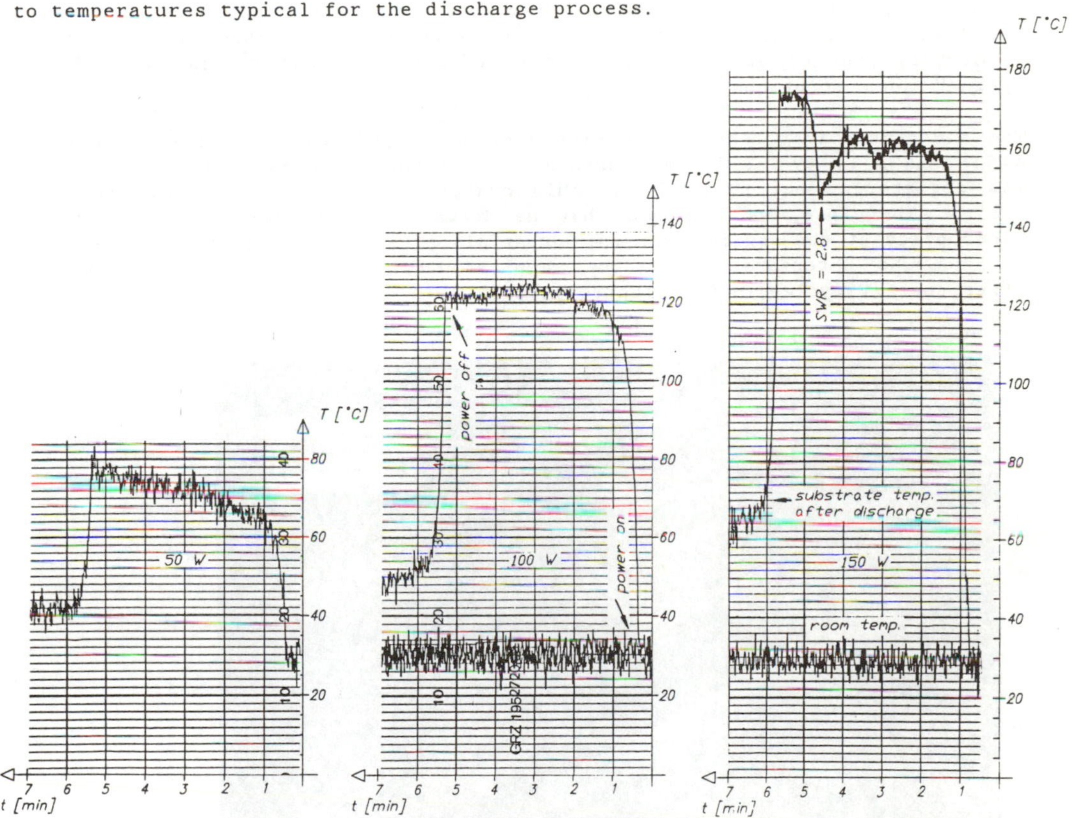

Figure 14: Substrate surface temperature as a function of time at various r.f. power levels.

3. TECHNOLOGICAL APPLICATION OF THE FILMS

3.1. CONDUCTIVE COATINGS

The lowest resistivities obtained with layers which fulfilled the various boundary requirements like sufficient hardness, adherence to glass, and corrosion resistance in sodium chloride solutions were of the order of 100 Ωcm. Resistivity was measured with a four point probe. Generally, films used for device applications were grown with 4 sccm acetylene and 13 sccm nitrogen at 120 mTorr, excited by 100 W deposition power without external substrate heating but also without cooling. Photoresist exhibited to these conditions is still removable in organic solvents, and lift off techniques are applicable. Micromechanical structures are preserved. Insulating a-C:H layers may be deposited *in situ* on top of the conductive films by a slight variation of the deposition parameters, i.e., by reducing the power and the acetylene content (see figure 6). The films can be processed with conventional photolithography, and structured by RIE in an oxygen plasma. The etch rate is about 40 $nm.min^{-1}$ at 35 mTorr and 70 W.

These conductive carbon layers were either used to form laterally conducting interconnections, or protective conductive coatings on top of metallic conductors, mainly on Ti:Au:Ti sandwich structures. The latter approach permits to build electrodes with a lower impedance; however, the carbon layers must be absolutely free of pinholes in order to effectively seal the metal conductor against ionic solutions.

The electrochemical behavior of single carbon layers was tested in a solution of 1N H_2SO_4 by cyclic I/V measurements. A slight hysteresis of the I/V curve was registered, but no shift of this hysteresis could be observed after 15 cycles. Therefore, we conclude that no dramatic modification of the surface takes place at the applied maximum potentials of ±3.5 V and maximum currents of ±0.5 μA. No deterioration of the samples could be observed microscopically after the I/V cycles in the electrolyte.

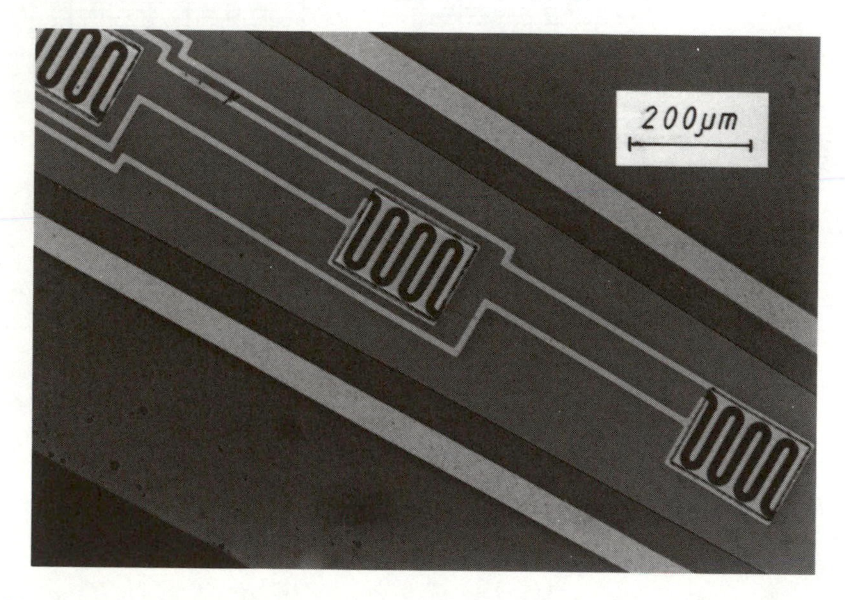

Figure 15: Integrated sensor probe for biological applications.

3.2. TEMPERATURE SENSITIVE LAYERS

Films produced under the above conditions exhibit a negative temperature coefficient which permits their application as temperature sensors. Figure 15 shows the sensor probe arrangement. Each of the three meander patterns constitutes one temperature sensor element. The sensor surface is passivated with plasma deposited silicon nitride in order to insulate the conducting paths against the ionic test solutions. For process compatibility reasons, silicon nitride was preferred to insulating a-C:H films in this case. Temperature tests were carried out in a glycerol bath. Figures 16 and 17 show the dependence of the resistance on the temperature, and the I/V characteristics of the sensor device. The resistance hysteresis over six cycles between 10 °C and 60 °C is less than 0.2 percent of the resistance value. The temperature coefficient is 2.1 percent per °C at 30 °C, and 1.9 percent per °C at 60 °C. Resistance drift due to ageing is 0.062 percent per hour at 33 °C, corresponding to a temperature error of 31 mK/h. Resistor noise corresponds to 10 mK.

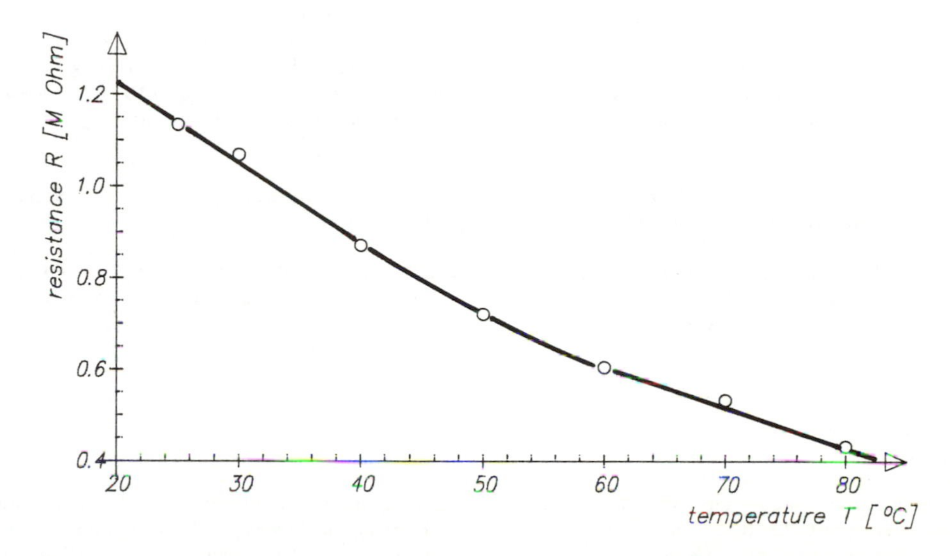

Figure 16: Temperature sensor resistance as a function of temperature.

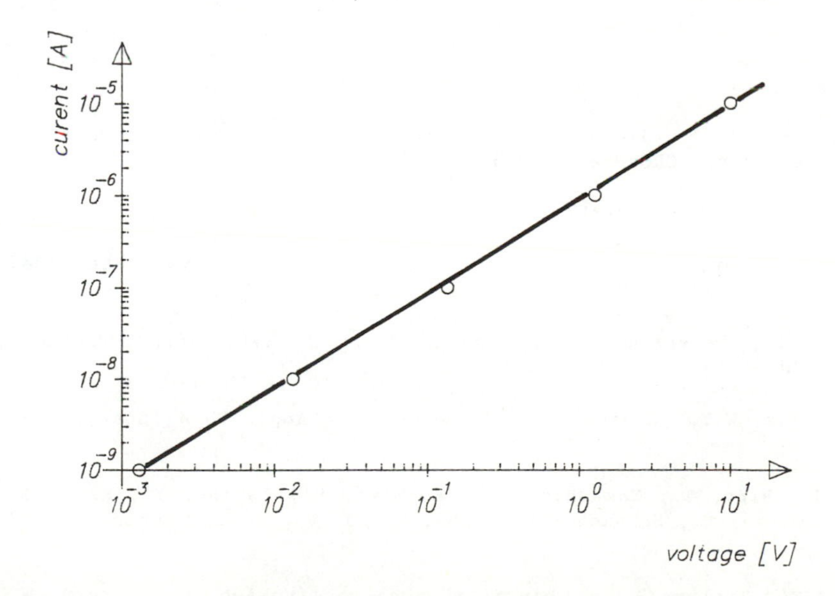

Figure 17: I/V characteristic of a temperature sensor.

4. SUMMARY AND CONCLUSIONS

Amorphous carbon coatings were produced in a parallel plate system with an r.f. glow discharge. Acetylene was decomposed in an argon or nitrogen carrier gas. Operation modes with self bias and with short circuited self bias were used. In contrast to the predictions given in the literature, the non-self bias mode produced results very similar to those published for self bias which is commonly considered to be a very important factor of the film formation process.

Parameters for the deposition of layers with low resistivity could be successfully determined. No external substrate heating was required; during the deposition, the substrate surface temperature increased to up to 200 °C, depending on the discharge power. For the deposition parameters used for technologically applied films, the substrate surface temperature did not exceed 130 °C. Quadrupole mass spectra indicate a mass limited process in the parameter range chosen.

The films were mechanically stable. Electrochemical investigations documented the inertness and the chemical stability of the layers.

Miniaturized thermistor structures on glass substrates with a-C:H film sensors could be realized with thin film technology. The determined process parameters are compatible with the current production steps of integrated multiple sensor probe systems [34].

ACKNOWLEDGMENT

This work was supported by the Austrian National Bank, the Fonds zur Förderung der wissenschaftlichen Forschung, and the Ludwig Boltzmann-Gesellschaft. We highly appreciate the permission of Elsevier Sequoia for the reproduction of figures 1 and 3 - 5 which have been previously published in Thin Solid Films.

REFERENCES

1) Angus, J.C., Koidl, P., Domitz, S.: Plasma Deposited Thin Films (Mort, J., Jansen, F., editors), CRC Press, Florida, 1986

2) Holland, L.: J. Vac. Sci. Technol., 1977, 14, 5

3) Sah, R.E., Dischler, B., Bubenzer, A., Koidl, P.: Appl. Phys. Lett. 1985, 46, 739

4) Pellicori, S.F., Peterson, C.M., Henson, T.P.: J. Vac. Sci. Technol., 1986, A4, 2350

5) Tanaka, T., Kim, W.Y., Konaga, M., Takahashi, K.: Appl. Phys. Lett., 1984, 45, 865

6) Minagawa, H., Hino, T., Yamashina, T., Amemiya, H., Ishibe, Y., Kato, S., Okazak, K., Oyama, H., Sakamoto, Y., Yano, K.: J. Vac. Sci. Technol., 1987, A5, 2293

7) Hollahan, J.R., Rosler, R.S.: Thin Film Processes (Vossen, J.L., Kern, W., editors), Academic Press, New York, 1978

8) Holland, L., Ojha, S.M.: Thin Solid Films, 1976, 38, L17

9) Wagner, J., Wild, Ch., Pohl, F., Koidl, P.: Appl. Phys. Lett., 1986, 48, 106

10) Wild, Ch., Wagner, J., Koidl, P.: J. Vac. Sci. Technol., 1987, A5, 2227

11) Catherine, Y., Couderc, P.: Thin Solid Films, 1986, 144, 265

12) Berg, S., Gelin, B., Suardstrom, A., Babulanam, S.M.: Vacuum, 1984, 34, 969

13) Wild, Ch., Koidl, P., Appl. Phys. Lett. 1987, 51, 1506

14) Nyaiesh, A.R., Nowak, W.B.: Vacuum, 1985, 35, 203

15) Natarajan, V., Lamb, J.D., Woollam, J.A., Liu, D.C., Gulino, D.A.: J. Vac. Sci. Technol., 1985, A3, 681

16) Angus, J.C., Jansen, F.: J. Vac. Sci. Technol., 1988, A6, 1778

17) Moravec, T.J., Lee, J.C.: J. Vac. Sci. Technol., 1982, 20, 338

18) Bubenzer, A., Dischler, B., Brandt, G., Koidl, P.: J. Appl. Phys., 1983, 54, 4590

19) Furuse, T., Suzuki, T., Matsumoto, S., Nishida, K., Nannichi, Y.: Appl. Phys. Lett., 1978, 33, 317

20) Pethica, J.B., Koidl, P., Gobrecht, J., Schueler, C.: J. Vac. Sci. Technol, 1985, A3, 2391

21) Warner, J.D., Puoch, J.J., Alterovitz, S.A., Liu, D.C., Lanford, W.A.: J. Vac. Sci. Technol., 1985, A3, 900

22) Has, Z., Mitura, S., Clapa, M., Szmidt, J.: Thin Solid Films, 1986, 136, 161

23) Pouch, J.J., Warner, J.D., Liu, D.C., Alterovitz, S.A. Thin Solid Films, 1988, 157, 97

24) Chen, Z.M., Wang, J.N., Mei, X.Y., Kong, G.L.: Solid State Communications, 1986, 58, 379

25) Tafto, J., Kampas, F.J.: Appl. Phys. Lett., 1985, 46, 949

26) Kobayashi, K., Mutsukura, N., Machi, Y.: Thin Solid Films, 1988, 158, 233

27) Gonzales-Hernandez, J., Asomoza, R., Reyes-Mena, A., Rickards, J., Chao, S.S., Pawlik, D.: J. Vac. Sci. Technol., 1988, A6, 1798

28) Webb, A.P., El-Hossary, F.M., Fabian, D.J., Maydell-Ondrusz, E.A.: Thin Solid Films, 1985, 129, 281

29) Berg, S., Andersson, L.P.: Thin Solid Films, 1979, 58, 117

30) Bolz, A., Schaldach, M.: Proceedings of BME-Austria 88 (Schuy, S., Leitgeb, M., editors), Graz/Austria, 1988

31) Pirker, K., Schallauer, R., Fallmann, W., Olcaytug, F., Urban, G., Jachimowicz, A., Kohl, F., Prohaska, O.: Thin Solid Films, 1986, 138, 121

32) Chapman, B.: Glow Discharge Processes, Wiley & Sons, Inc., 1980

33) Yasuda, H.K.: Plasma Polymerization, Academic Press, 1985

34) Prohaska, O.J., Kohl, F., Goiser, P., Olcaytug, F., Urban, G., Jachimowicz, A., Pirker, K., Chu, W., Patil, M., LaManna, J., Vollmer, R., Proc. 4th Conf. on Solid State Sensors and Actuators, Jap. Inst. Elect. Eng., Tokyo, 1987

Materials Science Forum Vols. 52 & 53 (1989) pp. 689-700

THE PHOTOCONDUCTION OF THE a-Si:H ON CARBON GRADED a-SiC:H STRUCTURE

T. Itoh and M. Nakatsugawa
Department of Electronics and Communication
School of Science and Engineering, Waseda University
Shinjuku-ku, Tokyo 169, Japan

Carbon graded a-SiC:H films came to be fabricated using PC controlled mass flow controller. The properties of a-Si:H/a-SiC:H systems, the carbon contents of which were graded, have been evaluated by current-voltage (I-V) characteristics measurements with Xe lamp irradiation and monochromatic light through the iterference filter.

The photocurrent (I_{ph}) induced by low energy photons was dominant at the low applied voltage, and I_{ph} was influenced by high energy photons at the middle range. When high voltage was applied to the films, I_{ph} was again influenced by low energy photons. When applied voltage was low, I_{ph} was dominated by the electrons generated in a-Si:H layer. At the middle voltage, I_{ph} induced by high energy photons was due to the electrons generated in the surface region of a-SiC:H(/a-C:H). When high voltage was applied to the film, I_{ph} was influenced by the cha racteristics of a-Si:H. Because of the high resistivity of a-SiC:H or a-C:H, the electric field in the a-SiC:H (or a-C:H) layer was stronger than that in the a-Si:H. The barrier for the holes generated in a-Si:H layer became low with increase of the applied voltage, and the holes passed through the wide gap materials.

INTRODUCTION

The devices used amorphous sem iconductors films utilized in the various fields such as solar cells, pho todetectors, pho toreceptors, image sensor and TFT panels because of its advantage of high absorption efficiency to visible light and of suitability for forming large area devices.

To improve their per formances, the investigation on the fabrication techniques of films and on the kinds of het ero structural films has been carried out. The former is the tudy to progress the amorphous materials in themselves, such as photo assisted CVD[1,2], ECR plasma CVD[3], separate reaction chambers[4] and ultrahigh vacuum reaction cha mbers[5,6] to obtain low gap state densities. The latter is aimed at research on the structures for high quality devices e.g. silicon carbide window layers[7], superlattice a-Si window lay ers[8],and graded layers at the interface of p/i or i/n[9,10] junction.

Since the optical energy gap of a-SiC:H can vary from 1.7eV to 2.8eV according to the ratio of Si to C, it is possible to obtain the favorable structure of the energy gap. We deposit the graded carbon content a-SiC:H films. The built-in field effect was expected, also the reduction of mismatch between a-Si:H and a-C:H could be expected by the use of this structure.

EXPERIMENTAL

The a-Si:H, a-SiC:H, and a-C:H were deposited on PYREX glass coated with indium-tin-oxide (ITO). Figure 1 shows the schematics of the apparatus. The deposition system consisted of a reaction chamber and a plasma column connected to a turbo molecular pump and the gas control equipment. Source gases used to deposit films were SiH_4, SiH_4/C_2H_2 , and C_2H_2 for a-Si:H, a-SiC:H, and a-C:H, res pectively. After the deposition system was evacuated below $5x10^{-7}$ Torr, the source gases were introduced into the chamber. The flow rates of SiH_4 and C_2H_2 were controlled by mass flow systems and a computer. The film was deposited in the reaction chamber with the rf power of 5W. The substrate was kept at 250 ^{o}C by the halogen lamp irradiation through the VYCOR window. During the film deposition, the pressure of the chamber was maintained at 5×10^{-2} Torr, which was monitored by the capacitance manometer (MKS Baratron BHS-1). The substrate holder and mesh-type electrode were biased positively 50V to prevent the exposure to the energetic ions, and the rf plasma was formed between the mesh-type electrode and rf electrode.

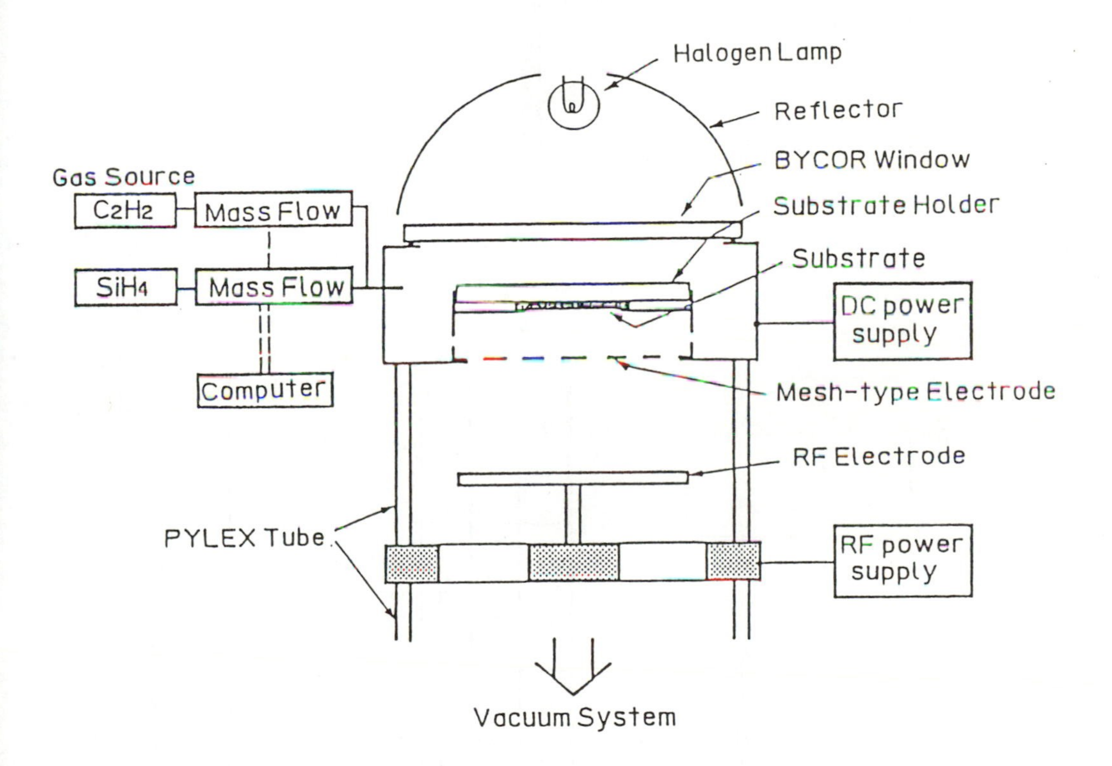

Fig. 1. The schematics of the apparatus.

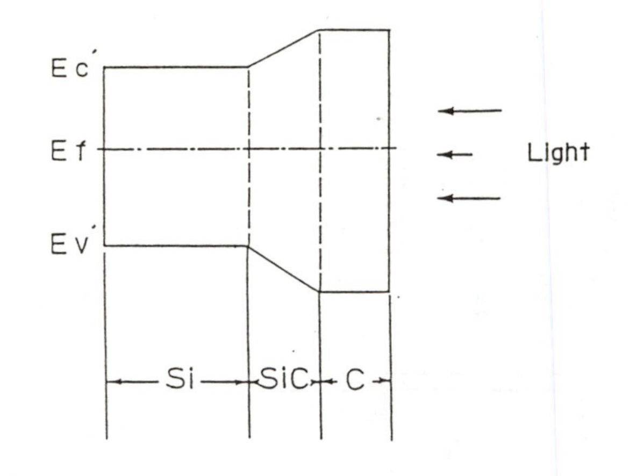

	Si (nm)	SiC (nm)	C (nm)
A	0	320	0
A'	320	320	0
B	0	160	160
B'	320	160	160
C	0	80	240
C'	320	80	240

Fig. 2. The structure of the specimens.

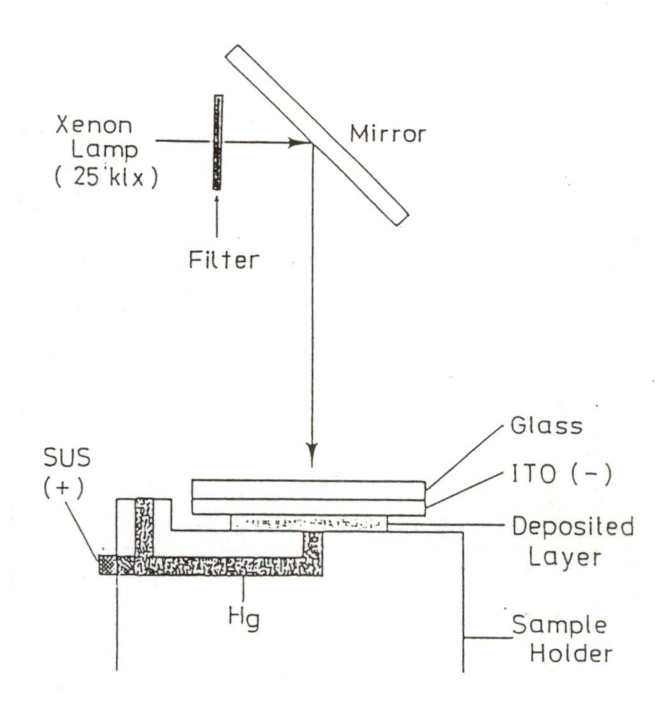

Fig. 3. The method of I–V measurement.

we deposited a-SiC:H(/a-C:H) films to evaluate the graded layer itself and deposited a-Si/SiC(/a-C:H) films to compare both structural films as shown in Fig. 2.

Figure 3 shows the method of the I-V measurement. The ITO was used for the cathode and the 0.8 mm^2 Hg contact was used for the positive electrode. The photocurrent measurement was carried out under 25 klx Xe lamp irradiation. Monochromatic light at with wavelength of 370,430, 500, 570 and 720nm, was used.

RESULTS

The thickness of obtained films was uniform and the surface was very smooth. The optical energy gaps of a-Si:H and a-C:H were 1.7eV and 2.8eV, respectively, and the gap of a-Si_nC_{1-n}:H varied between these values. In this experiment a-SiC:H is representing the component of a-Si_nC_{1-n}:H and n is varied $0<n<1$ in the film.

Figures 4, 5 and 6 show the I-V characteristics of the A structure with a-SiC(320nm):H, B structure with a-SiC(160nm) /C(160nm):H and C structure with a-SiC(80nm)/C(240nm):H, respectively. The dark currents of these films were saturated at the low voltage region and increased with an increase of the applied voltage. As shown in Fig. 4, the characteristics of the I_{ph} could be classified into two regions. At lower than 36V, the I_{ph} increased according to the increase of the energy of irradiated photons, and these tendencies could be observed in the case of a-C:H. When higher than 36V was applied to the film, the I_{ph} for the wavelength of 370nm decreased compared to the case of the other wavelength. The currents shown in Fig. 5 and Fig. 6 indicated almost the same trend, and the I_{ph} increased with increase of the photon energy. As shown in Fig. 5, the I_{ph} for the wavelength of 370nm was decreased only at the voltage of 128V.

Figure 7 shows the I-V characteristics of the A' structure with a-Si(320nm)/SiC(320nm):H. The dark current of the A' structure was similar to that of the A structure. The value of the dark current was considered mainly defined by the conductivity of a-SiC:H. The I_{ph} has three ranges. When a low voltage of <16V was applied, the I_{ph} decreased in the order of

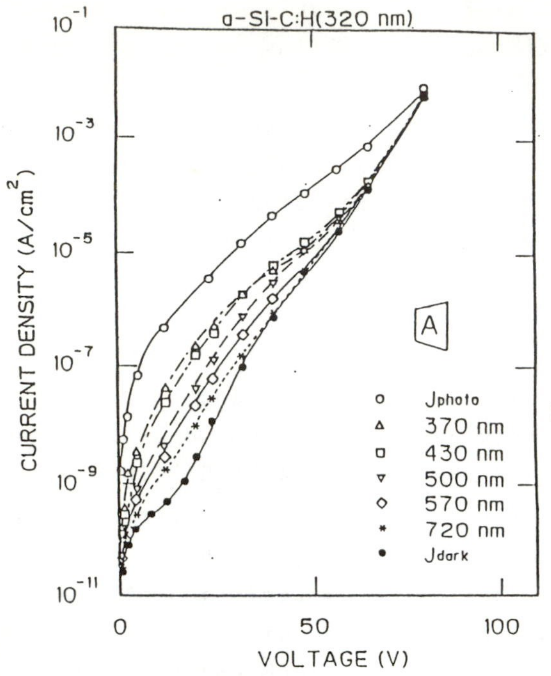

Fig. 4. I-V characteristics of the a-SiC(320nm):H structure.

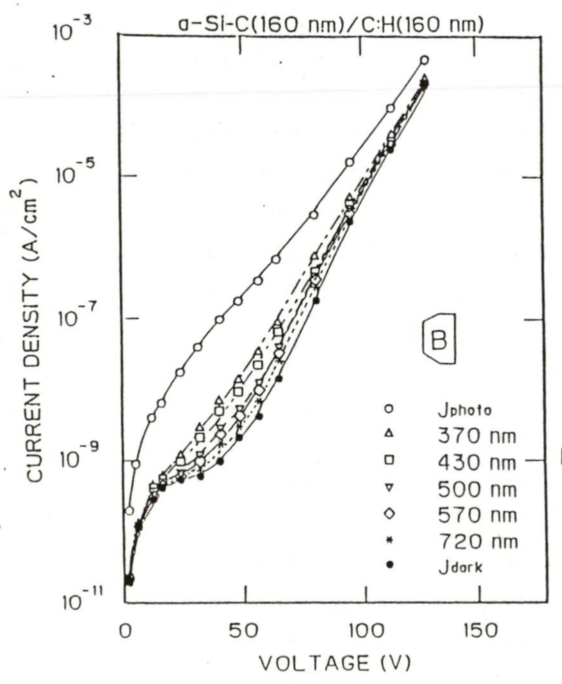

Fig. 5. I-V characteristics of the a-SiC(160nm)/C(160nm):H structure.

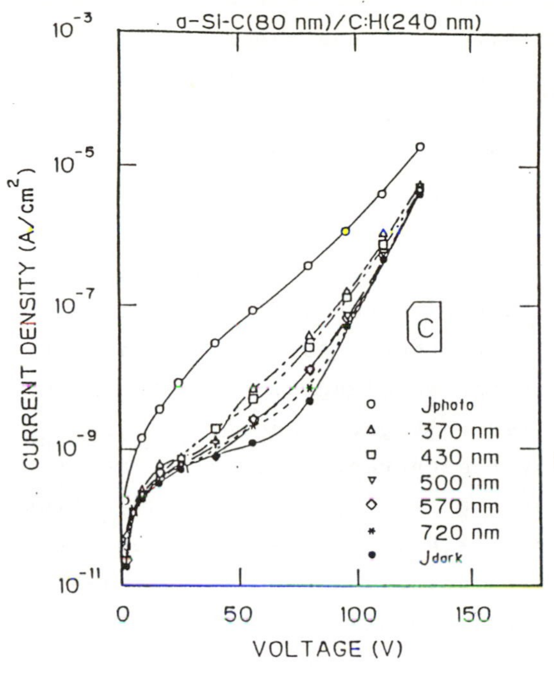

Fig. 6. I-V characteristics of the a-SiC(80nm)/C(240nm):H structure.

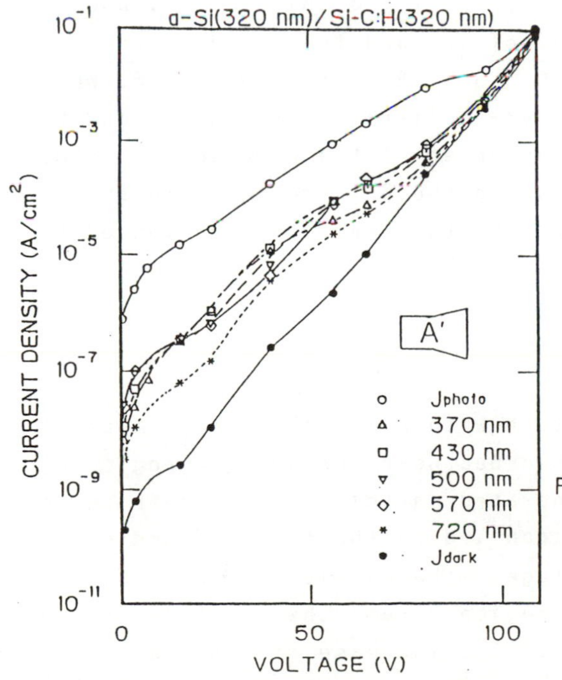

Fig. 7. I-V characteristics of the a-Si(320nm)/SiC(320nm):H

the wavelength; 570, 500, 430, 370 and 720nm. This order could be observed in the measurement of a-Si:H. At the range where middle voltage was applied, the I_{ph} decreased in the order of the wavelength; 370, 430, 500, 570 and 720nm. When the high voltage of >72V was applied, the I_{ph} decreased in the order of the wavelength; 570, 500, 430, 370 and 720nm again.

Figures 8 and 9 show the photo and dark currents of the B' structure with a-Si(320nm)/SiC(160nm)/C(160nm):H and the C' structure with a-Si(320nm)/SiC(80nm)/C(240nm):H, respectively. These curves were similar with each other. As shown in Fig. 8, the I_{ph} decreased in the order of the wavelength; 570, 500, 430, 370 and 720nm under the low applied voltage. Therefore, the I_{ph} was considered to be affected by the characteristics of a-Si:H. For middle voltages, the I_{ph} of high energy photon illumination was larger than that of low. When high voltage was applied, the characteristics of the I_{ph}s were influenced by that of a-Si:H again. The I_{ph} decreased in the order of the wavelength; 570, 430, 500, 370 and 720nm in Fig. 9, and it was considered to be affected by the characteristics of both a-Si:H and a-C:H in the C' structure. For middle voltages, the I_{ph} of high energy photon illumination was larger than that of low and the I_{ph} decreased in the order of the wavelength; 430, 370, 500, 570 and 720nm. When high voltage was applied, the characteristics were influenced by a-Si:H. The I_{ph} decreased in the order of the wavelength; 570, 430, 500, 370 and 720nm. Comparing the B' structure to the C' structure, the I_{ph} of the C' structure was more influenced by a-C:H at the low voltage.

DISCUSSION

In the case of the A structure with a-SiC(320nm):H, most of the high energy photons were absorbed near the surface region of the film, and the carriers generated in this region contributed to the I_{ph} when low voltage was applied to the film. According to the increase of the applied voltage, the carriers induced by low energy photons in the deep region from the surface could contribute to the I_{ph} because of the increase of the electric

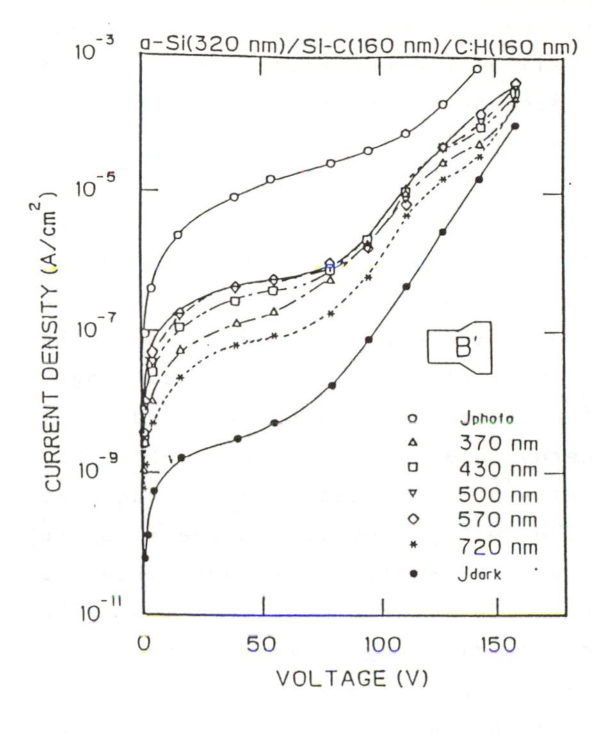

Fig. 8. I-V characteristics of the a-Si(320nm)/SiC(160nm)/C(160nm):H structure.

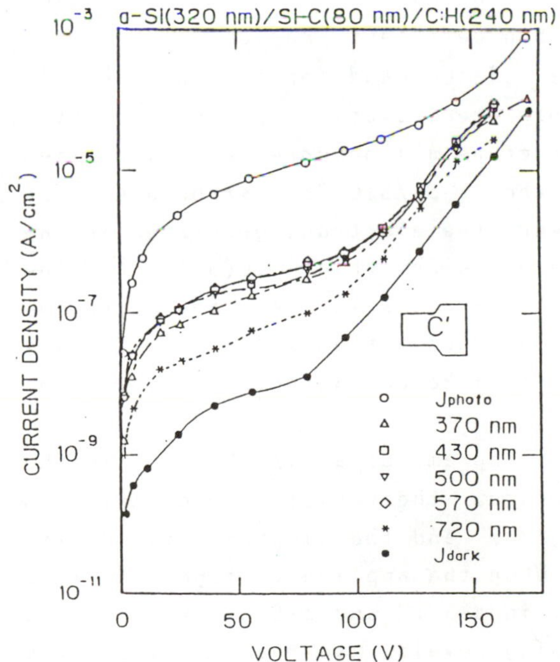

Fig. 9. I-V characteristics of the a-Si(320nm)/SiC(80nm)/C(240nm):H structure.

field and the reduction of the barrier height for the holes. Comparing Fig. 5 to Fig. 6, all of the currents in Fig. 6 were smaller than those in Fig. 5. Because the highly resistive a-C:H layer was thicker in a-SiC(80nm)/C(240nm):H of the C structure than in a-SiC(160nm)/C(160nm):H of B, the currents of the C structure were suppressed. In the case of the a-SiC/C:H structure, high energy photons were especially absorbed in the a-C:H layer and the characteristics of a-C:H dominanted. At the applied voltage of 128V (Fig. 5), the carriers generated in the deep region contributed to I_{ph}, so that the I_{ph} of the wavelength of 370nm decreased and I_{ph}s of another wavelength increased.

Figure 10 shows the energy diagrams of the a-Si/SiC:H structure. Since the resistivity of a-C:H was greater than that of a-Si:H, the proportion of the voltage applied to the film to form an electric field in carbon rich a-SiC:H was considered to be greater than that in carbon deficient a-SiC:H. Thus the electric field formed in the carbon rich region was stronger than that of carbon deficient region. In the case of carbon content graded a-SiC:H, the potential of the carbon rich a-SiC:H region rose more than that of the carbon deficient region. Therefore, the hock-shaped barrier in the valence band for the holes hardly appeared, and the changes in characteristics of a-Si:H, to a-C:H, to a-Si:H in the a-Si/SiC:H structure all occurred at low applied voltage compared to those in the a-Si/SiC/C:H structure. The carriers contributing to I_{ph} were the electrons generated in a-Si:H layer at the low voltage of <16V (Fig. 10(a)). As the applied voltage increased, the carriers in the a-C:H region contributed to I_{ph}s (Fig. 10(b)). When a high voltage (>72V) was applied, the I_{ph}s indicated a-Si:H-like characteristics, as shown in Fig. 10(c).

Figure 11 shows the energy diagrams of a-Si/SiC/C:H. In the case of the a-Si/SiC/C:H structure, the resistivity of the film was greatest in the a-C:H region, and the electric field was concentrated in this region. When the applied voltage was low, the gradient of the potential in a-Si:H and a-C:H regions was slight and a-C:H had some trapping levels, so that the I_{ph}s was due to the electrons generated in a-Si:H region, as shown in

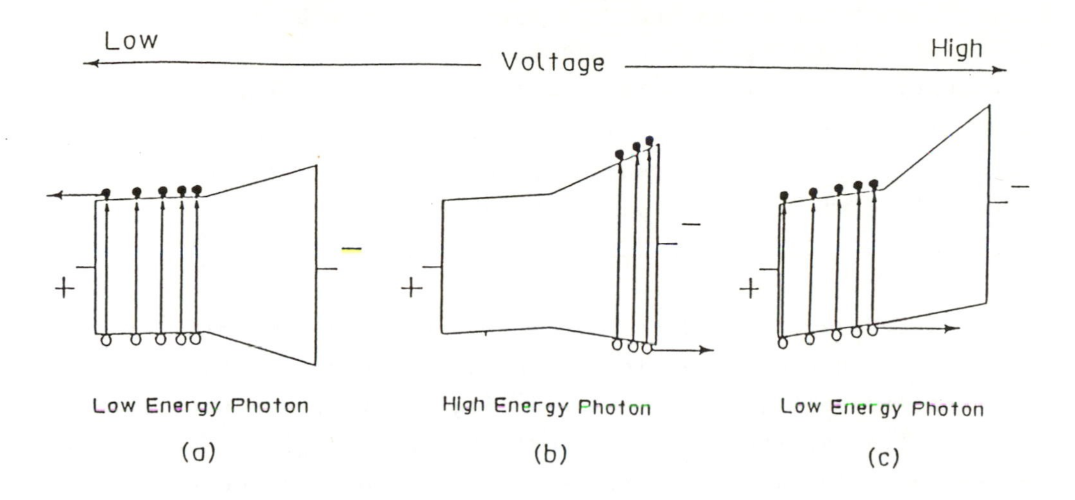

Fig. 10. Energy diagrams of the a-SI/Si-C:H structure.

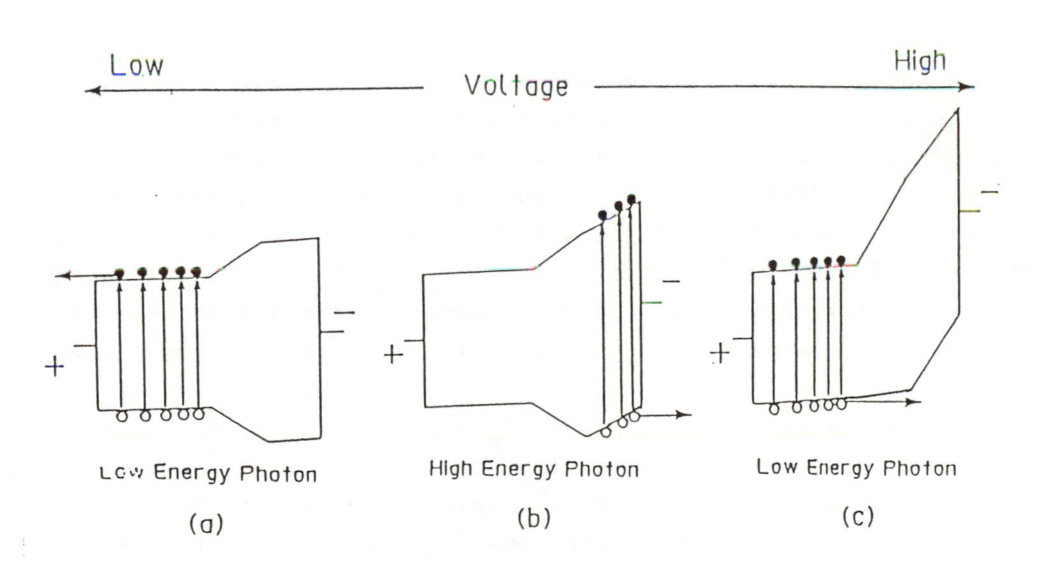

Fig. 11. Energy diagrams of the a-SI/Si-C/C:H structure.

Fig. 11(a). Comparing the C' structure to the B' structure, the I_{ph} of the former was more influenced by a-C:H characteristics than that of the latter. This is caused by the difference of the a-C:H thickness. When the a-C:H layer was thick, the electric field was especially concentrated in this region and the potential of this region was much graded. Thus, the carriers generated in a-C:H region were able to pass through here. As the applied voltage increased, the potential of the film mainly rose in the a-C:H layer, and the barrier height for the holes induced in the a-Si:H region remained unchanged despite the increase of the applied voltage, as shown in Fig. 11(b). Therefore, for middle voltages, the carriers contributing to I_{ph} were the holes generated in the a-C:H layer. When higher voltages were applied, the barrier height for the holes became low with an increase of the applied voltage, and the holes induced in a-Si:H injected into a-C:H region and contributed to the I_{ph}, as shown in Fig.11(c).

CONCLUSION

In the case of the films without the a-Si:H layer, the I_{ph}s the high energy photon illumination was larger than that for low. This tendency was more noticeable in the a-C:H films. As the applied voltage increased, I_{ph} (370nm illumination) decreased.

In the case of the films with a-Si:H layers, the I_{ph}s was affected by the characteristics of a-Si:H at the low voltages and the carriers contributing to I_{ph} were electrons generated in the a-Si:H. However, I_{ph} also was influenced by the holes generated in the a-C:H region when the a-C:H layer was thick. When middle voltages were applied, I_{ph} was influenced by that of a-C:H and the carriers for the I_{ph} were the holes generated in the a-C:H region. At the high voltages, I_{ph} was affected by the characteristics of a-Si:H and the I_{ph} was due to the holes generated in the a-Si:H region. In the a-Si/SiC:H structure, the characteristics of I_{ph} changed at lower applied voltages than in the a-Si/SiC/C:H structure.

ACKNOWLEDGMENT

The authors would like to thank Kohjundo Chemical Laboratory for the preparation of ITO/PYREX substrates.

REFERENCES

(1) T. Tanaka, W. Y. Kim, K. Konagai, and K. Takahasi, Appl. Phys. Lett., 45, 865 (1984).

(2) H. Takei, T. Tanaka, W. Y. Kim, M. Konagai, and K. Takahasi, J. Appl. Phys., 58, 3664 (1985).

(3) S. Kato, and T. Aoki, J. Non-Cryst. Solids, 77&78 813 (1985).

(4) Y. Kuwano, M. Onishi, S. Tsuda, Y. Nakashima, and N. Nakamura, Jpn. J. Appl. Phys., 21, 413 (1982).

(5) C. C. Tsai, J. C. Knights, R. Lujan, B. Wacker, B. L. Stafford, and M. J. Thompson, J. Non-Cryst. Solids, 59&60, 731 (1983).

(6) S. Tsuda, T. Takahama, M. Isomura, H. Tarui, Y. Nakashima, Y. Hishikawa, N. Nakamura, T. Matsuoka, H. Nishiwaki, S. Nakano, M. Ohnishi, and Y. Kuwano, Jpn. J. Appl. Phys., 26, 33 (1987).

(7) Y. Tawada, T. Yamaguchi, S. Nonomura, S. Hotta, H. Okamoto, and Y. Hamakawa, Jpn. J. Appl. Phys., 21, Suppl, 21-1, 297 (1982).

(8) S. Tsuda, H. Tarui, T. Matsuyama, T. Takahama, S. Nakayama, Y. Nishikawa, N. Nakamura, T. Fukatsu, M. Ohnishi, S. Nakano, and Y. Kuwano, Jpn. J. Appl. Phys., 26, 28 (1987).

(9) K. S. Lim, M. Konagai, and K. Takahashi, J. Appl. Phys., 56, 538 (1984).

(10) R. R. Arya, A. Catalano, and R. S. Oswald, Appl. Phys. Lett., 49, 1089 (1986).

AUTHOR INDEX